PRECISION ASTEROSEISMOLOGY

IAU SYMPOSIUM No. 301

COVER ILLUSTRATION:

Photograph of Wojtek Dziembowski taken in August 2010 by science historian and journalist Dr. Andrzej M. Kobos. Dr. Kobos conducted an interview with Wojtek as one of a series of interviews of members of the Polish Academy of Arts and Sciences (in Polish: Polska Akademia Umiejętności, PAU, http://pau.krakow.pl/index.php/en/) published by Kobos.

IAU SYMPOSIUM PROCEEDINGS SERIES

Chief Editor

THIERRY MONTMERLE, IAU General Secretary
Institut d'Astrophysique de Paris,
98bis, Bd Arago, 75014 Paris, France
montmerle@iap.fr

Editor

PIERO BENVENUTI, IAU Assistant General Secretary
University of Padua, Dept of Physics and Astronomy,
Vicolo dell'Osservatorio, 3, 35122 Padova, Italy
piero.benvenuti@unipd.it

INTERNATIONAL ASTRONOMICAL UNION

UNION ASTRONOMIQUE INTERNATIONALE

International Astronomical Union

PRECISION ASTEROSEISMOLOGY

PROCEEDINGS OF THE 301st SYMPOSIUM OF THE INTERNATIONAL ASTRONOMICAL UNION HELD IN WROCLAW, POLAND AUGUST 19–23, 2013

Edited by

JOYCE A. GUZIK
Los Alamos National Laboratory, Los Alamos, NM USA

WILLIAM J. CHAPLIN
School of Physics and Astronomy, University of Birmingham, UK
Stellar Astrophysics Centre (SAC), Department of Physics and Astronomy,
Aarhus University, Denmark

GERALD HANDLER
Nicolaus Copernicus Astronomical Center, Poland

and

ANDRZEJ PIGULSKI
Instytut Astronomiczny Uniwersytetu Wrocławskiego, Poland

CAMBRIDGE
UNIVERSITY PRESS

Shaftesbury Road, Cambridge CB2 8EA, United Kingdom

One Liberty Plaza, 20th Floor, New York, NY 10006, USA

477 Williamstown Road, Port Melbourne, VIC 3207, Australia

314–321, 3rd Floor, Plot 3, Splendor Forum, Jasola District Centre, New Delhi – 110025, India

103 Penang Road, #05–06/07, Visioncrest Commercial, Singapore 238467

Cambridge University Press is part of Cambridge University Press & Assessment, a department of the University of Cambridge.

We share the University's mission to contribute to society through the pursuit of education, learning and research at the highest international levels of excellence.

www.cambridge.org
Information on this title: www.cambridge.org/9781107045170

First published 2014

A catalogue record for this publication is available from the British Library

ISBN   978-1-107-04517-0   Hardback

Table of Contents

Part 1. TALKS

Session 1. Introduction
Chairs: Jadwiga Daszyńska-Daszkiewicz, Jørgen Christensen-Dalsgaard

Session 2. Observations: from ground to space
Chairs: Jørgen Christensen-Dalsgaard, Werner Weiss, William Chaplin

Session 3. Resolving the rich oscillation spectra
Chairs: Annie Baglin, Juan Carlos Suárez

A special session of Wojtek Dziembowski
Chair: Hiromoto Shibahashi

Session 4. Applications of pulsating stars in astrophysics
Chairs: Marcella Marconi, Gilles Fontaine, Karen Pollard, Joyce Guzik

Splinter session. What is the role of asteroseismology in the era of precision cosmology?

Chair: Hilding R. Neilson

Session 5. New solutions to old problems and new challenges

Chairs: Gilles Fontaine, Karen Pollard, Jørgen Christensen-Dalsgaard

Session 6. From the Sun to the stars: the helio-asteroseismology connection
Chairs: Hiromoto Shibahashi, Jadwiga Daszyńska-Daszkiewicz

Photos from the conference (taken by Alosha Pamyatnykh)

Part 2. POSTERS

Preface

The IAU Symposium 301 on *Precision Asteroseismology* was held in Wrocław, Poland, from the 19th to the 23rd of August 2013. This was also the 21st pulsation meeting, and the first to receive the status of an IAU Symposium. These meetings started in Los Alamos in 1971, and very quickly became a regular tradition for the stellar pulsation community. The Astronomical Institute of Wrocław University took great pleasure in hosting this conference, in particular because it was also an opportunity to celebrate the scientific opus of Wojtek Dziembowski.

The science motivation of this conference centred around seismic studies of pulsating stars which, in the era of high-precision data, is one of the most rapidly developing branches of astrophysics. Owing to photometric observations made by space missions such as *MOST*, *CoRoT*, and *Kepler*, and to high-resolution, high-signal-to-noise spectroscopic observations from the ground, the numbers of known oscillation frequencies of distant stars have increased by orders of magnitude, and new classes of pulsators have become available for seismic studies.

The program of the Symposium was divided into seven sessions: the five main sessions, a special session and a splinter session. Each session was accompanied by interesting and fruitful discussions. The program consisted of 35 invited talks, 28 contributed talks and 75 posters. The Symposium gathered 145 participants from 26 countries all over the world. There were 56 women, i.e., almost 40 % of participants.

This conference discussed what physics is missing from stellar structure and evolution theory, and how analysis of stellar oscillation data provides insights to improve our models of stars. In particular the following questions/problems were addressed: How can we make best use of rich, but irregular, oscillation spectra? Where do limits to the application of asteroseismology lie, and what inferences can be made on the underlying physics? How will these data help to solve the problems and uncertainties in stellar physics? How far are we from unravelling the mode-selection mechanism? Are we still missing something from predictions of astrophysical opacities (an announcement of the new opacity bump near $\log T = 5.06$ was made at the meeting)?

The other vital subject raised during the Wrocław Symposium was the efficiency of convection and its interaction with pulsation. Moreover, the most recent understanding of the role of rotation, magnetic fields and element mixing in pulsation excitation was presented. During the last day, the synergies between helio- and asteroseismology were discussed as well as the prospects for seismic studies of planet-hosting stars.

From the observational side, we heard about the results on pulsating variables from the massive ground-based surveys, OGLE, ASAS and Araucaria, as well as recent seismic results from three space missions, *MOST*, *CoRoT* and *Kepler*. At the time of the meeting, some preliminary results from two new asteroseismic projects were also presented. The first is the *BRITE* mission, a Canadian-Austrian-Polish project of six nanosatellites aiming to obtain for the first time two-colour space photometry of bright pulsating stars. The second project is SONG, a network of telescopes that will collect Doppler velocity data for asteroseismology (and also detect planets by gravitational lensing).

The symposium honoured Wojtek Dziembowski, one of the world's leaders in the study of solar and stellar pulsations. He is a pioneer in the research of non-radial oscillations of stars and in carrying out non-adiabatic calculations of such oscillations. Twenty years ago he succeeded in explaining pulsations in main-sequence B-type stars, which resolved a long-standing problem for the field. On the second day of the Symposium, we held a special session devoted to Wojtek. There were talks by colleagues recalling previous collaborations and highlights: *"Some memories of collaboration with Wojtek"* by Phil Goode and Alexander Kosovichev and on *"Ups and downs in understanding stellar variability"* by Wojtek Dziembowski. After that, a very interesting discussion took place between Wojtek and the audience, titled *"Questions, Answers & Advice"*. It was a great pleasure and privilege to have such a special opportunity to celebrate the great scientific opus of Wojtek.

On the third day of the Symposium, we held a splinter session on *"Asteroseismology, standard candles and the Hubble Constant: what is the role of asteroseismology in the era of precision cosmology?"* This session was proposed and chaired by Hilding Neilson. There were several presentations highlighting progress in the field and setting the stage for open discussion on how asteroseismology can advance our understanding of the distance scale and help to measure the Hubble constant.

The Symposium was also a forum for two public lectures to disseminate the knowledge of asteroseismology and to communicate its role for understanding our place in the Universe. These public talks were given by two of the best speakers within the pulsation community: Joyce Guzik on *"The origin and fate of the Sun and helioseismology"*, and Don Kurtz on *"Planets and pulsations: The new Keplerian revolution"*.

In closing, we would like to thank once again all the participants for coming and making this Symposium possible. Many thanks to all invited speakers for giving impressive and engrossing talks as well as to the participants who delivered contributed talks or presented posters that helped to broaden the scientific exchange at the meeting. The members of the Scientific Organizing Committee are acknowledged for their work on setting the conference program and important comments on the proposal for the IAU Symposium. Special thanks are also due to Gerald Handler for his significant contribution to the preparation of the IAU Symposium proposal. Finally, we are very grateful to the members of the Local Organizing Committee for their huge efforts on organizing the Wrocław Symposium.

Post-symposium thanks go to all authors for their contributions and to the editors of the proceedings of the IAU Symposium 301 for their hard work. We believe that for many of us this volume will be an important memento for many years.

Jadwiga Daszyńska-Daszkiewicz & Hiromoto Shibahashi, SOC chairs

Wrocław, Tokyo, November 4, 2013

THE ORGANIZING COMMITTEES

Scientific

A. Baglin (France)
W.J. Chaplin (UK)
J. Christensen-Dalsgaard (Denmark)
M. Cunha (Portugal)
J. Daszyńska-Daszkiewicz (Poland, co-chair)
G. Fontaine (Canada)

J.A. Guzik (USA)
M. Marconi (Italy)
K. Pollard (New Zealand)
H. Shibahashi (Japan, co-chair)
J.C. Suárez (Spain)
W.W. Weiss (Austria)

Local

U. Bąk-Stęślicka
B. Cader-Sroka
J. Daszyńska-Daszkiewicz (co-chair)
Z. Kołaczkowski

G. Kopacki
A. Pigulski (co-chair)
M. Stęślicki
P. Walczak

Acknowledgements

The symposium is sponsored and supported by the IAU Division G (Stars and Stellar Physics); and by the IAU Commissions No. 27 (Variable Stars) and No. 35 (Stellar Constitution).

The Local Organizing Committee operated under the auspices of the Instytut Astronomiczny, Uniwersytet Wrocławski.

Funding by the
International Astronomical Union,
Ministerstwo Nauki i Szkolnictwa Wyższego
and
Fundacja Astronomii Polskiej im. M. Kopernika,
is gratefully acknowledged.

CONFERENCE PHOTOGRAPH

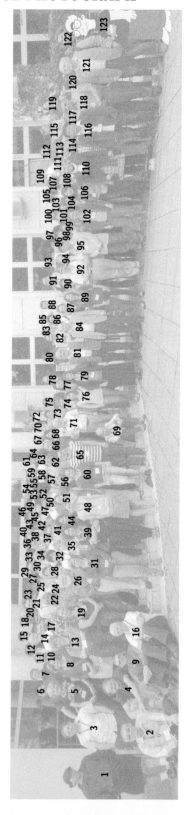

1. Mike Casey
2. Young-Beom Jeon
3. Chris Geroux
4. Gerald Handler
5. Ceren Ulusoy
6. Alexandre Gallenne
7. Daniel Reese
8. Magdalena Polińska
9. Paul Bradley
10. Elżbieta Zocłońska
11. Varvara Butkovskaya
12. Hideyuki Saio
13. Siobahn Morgan
14. Slavek Rucinski
15. Andrew Tkachenko
16. Gilles Fontaine
17. Krzysztof Kamiński
18. Jørgen Christensen-Dalsgaard
19. Olivera Latković
20. Steven Kawaler
21. Warrick Ball
22. Marcella Marconi
23. Wojciech Szewczuk
24. Alexey A. Pamyatnykh
25. Gisella Clementini
26. Douglas Gough
27. Vichi Antoci
28. Katrien Uytterhoeven
29. Frank Grundahl
30. Coralie Neiner
31. Valentina Schmid
32. Takafumi Sonoi
33. Stéphane Mathis
34. Annie Baglin
35. Konstanze Zwintz
36. Richard Townsend
37. Luis Balona
38. Michel Breger
39. Maria Pia Di Mauro
40. Barry Smalley
41. Mathieu Grosjean
42. Mariejo Goupil
43. Jesper Schou
44. Mélanie Godart
45. Masao Takata
46. Hilding Neilson
47. Hiromoto Shibahashi
48. Gemma Whittaker
49. Dieter Nickeler
50. Charles Kuehn
51. Sophie Saesen
52. Maëlle Le Pennec
53. Travis Metcalfe
54. Daniel Holdsworth
55. Michaela Kraus
56. Emese Plachy
57. Wiebke Herzberg
58. Giovanni Mirouh
59. Tunç Şenyüz
60. Evelin Bányai
61. Chris Engelbrecht
62. Georges Alecian
63. Nancy Remage Evans
64. Cyrus Zalian
65. László Molnár
66. Robert Szabó
67. Paweł Moskalik
68. Filiz Kahraman Aliçavuş
69. Wojtek Dziembowski
70. Zbigniew Kołaczkowski
71. Gérard Vauclair
72. Patricia Lampens
73. Thomas Constantino
74. Monika Fagas
75. Elisabeth Guggenberger
76. Joanne Breitfelder
77. Ádám Sódor
78. Irina Kitiashvili
79. Joyce A. Guzik
80. Nathalie Themessl
81. Marek Dróżdż
82. Sanja Tomić
83. Jadwiga Daszyńska-Daszkiewicz
84. Friedrich Kupka
85. Rafael Garrido Haba
86. Javier Pascual Granado
87. Katrien Kolenberg
88. Jurek Krzesiński
89. Aneta Wisniewska
90. Nicolas Nardetto
91. Chow-Choong Ngeow
92. Nami Mowlavi
93. Simon Jeffery
94. Andrzej Pigulski
95. Jianning Fu
96. Sébastien Deheuvels
97. Donald W. Kurtz
98. Agnieszka Słowikowska
99. Emily Brunsden
100. Paulina Sowicka
101. Géza Kovács
102. Aliz Derekas
103. Lucie Alvan
104. Gabriela Michalska
105. Urszula Kaszubkiewicz
106. Erika Pakštienė
107. Monika Jurković
108. Radosław Smolec
109. Jakub Ostrowski
110. Joanna Molenda-Żakowicz
111. Márcio Catelan
112. Markus Hareter
113. Dominik Drobek
114. Sylvie Vauclair
115. Karolina Kubiak
116. Lester Fox-Machado
117. Karen Pollard
118. Alexander Kosovichev
119. Ewa Niemczura
120. Noemi Giammichele
121. Henryk Cugier
122. Stéphane Charpinet
123. Jaymie M. Matthews (simpsonised)

Missing: Kévin Belkacem, József Benkő, Marek Biesiada, Daniel Bramich, Przemysław Bruś, Barbara Castanheira, Bill Chaplin, Yvonne Elsworth, Phil Goode, Martin Groenewegen, Saskia Hekker, Mikołaj Jerzykiewicz, Noé Kains, Grzegorz Kopacki, Ernst Paunzen, Grzegorz Pietrzyński, Suzanna Randall, Marek Skarka, Juan Carlos Suárez, Valerie Van Grootel, Przemysław Walczak, Werner Weiss, Ewa Zahajkiewicz.

Participants

Georges **Alecian**, Observatoire de Paris, LUTH, France — georges.alecian@obspm.fr
Lucie **Alvan**, CEA Saclay, Gif-sur-Yvette, France — lucie.alvan@cea.fr
Victoria **Antoci**, University of Aarhus, Denmark — antoci@phys.au.dk
Annie **Baglin**, Observatoire de Paris-Meudon, France — annie.baglin@obspm.fr
Warrick **Ball**, Institut für Astrophysik, Göttingen, Germany — wball@astro.physik.uni-goettingen.de
Luis **Balona**, South African Astronomical Observatory, Cape Town, South Africa — lab@saao.ac.za
Evelin **Bányai**, Konkoly Observatory, Budapest, Hungary — ebanyai@konkoly.hu
Kévin **Belkacem**, Observatoire de Paris-Meudon, LESIA, France — kevin.belkacem@obspm.fr
József **Benkő**, Konkoly Observatory, Budapest, Hungary — benko@konkoly.hu
Marek **Biesiada**, University of Silesia, Katowice, Poland — biesiada@us.edu.pl
Paul **Bradley**, Los Alamos National Laboratory, Los Alamos, USA — pbradley@lanl.gov
Daniel **Bramich**, European Southern Observatory, Garching, Germany — dbramich@eso.org
Michel **Breger**, University of Vienna, Austria — michel.breger@univie.ac.at
Joanne **Breitfelder**, European Southern Observatory, Chile — jbreitfe@eso.org
Emily **Brunsden**, University of Canterbury, Christchurch, New Zealand — stellarwings@gmail.com
Przemysław **Bruś**, Uniwersytet Wrocławski, Poland — brus@astro.uni.wroc.pl
Varvara **Butkovskaya**, Crimean Astrophysical Observatory, Ukraine — itbiz@mail.ru
Mike **Casey**, Saint Mary's University, Halifax, Canada — mcasey@ap.smu.ca
Barbara **Castanheira**, University of Texas at Austin, USA — barbara@astro.as.utexas.edu
Márcio **Catelan**, Pontificia Universidad Católica de Chile, Santiago, Chile — mcatelan@astro.puc.cl
William J. **Chaplin**, University of Birmingham, United Kingdom — w.j.chaplin@bham.ac.uk
Stéphane **Charpinet**, IRAP/CNRS, Toulouse, France — stephane.charpinet@irap.omp.eu
Jørgen **Christensen-Dalsgaard**, University of Aarhus, Denmark — jcd@phys.au.dk
Gisella **Clementini**, INAF - Osservatorio Astronomico Bologna, Italy — gisella.clementini@oabo.inaf.it
Thomas **Constantino**, Monash University, Melbourne, Australia — thomas.constantino@monash.edu
Henryk **Cugier**, Uniwersytet Wrocławski, Poland — cugier@astro.uni.wroc.pl
Jadwiga **Daszyńska-Daszkiewicz**, Uniwersytet Wrocławski, Poland — daszynska@astro.uni.wroc.pl
Sébastien **Deheuvels**, Observatoire Midi-Pyrénées, Toulouse, France — sebastien.deheuvels@irap.omp.eu
Aliz **Derekas**, Konkoly Observatory, Budapest, Hungary — derekas@konkoly.hu
Maria Pia **Di Mauro**, INAF/IAPS, Rome, Italy — maria.dimauro@inaf.it
Dominik **Drobek**, Uniwersytet Wrocławski, Poland — drobek@astro.uni.wroc.pl
Marek **Dróżdż**, Mt. Suhora Astronomical Observatory, Cracow, Poland — sfdrozdz@cyf-kr.edu.pl
Wojtek **Dziembowski**, Warsaw University Observatory, Poland — wd@astrouw.edu.pl
Yvonne **Elsworth**, University of Birmingham, United Kingdom — y.p.elsworth@bham.ac.uk
Chris **Engelbrecht**, University of Johannesburg, South Africa — engelbrecht.chris@gmail.com
Nancy Remage **Evans**, SAO, Cambridge, USA — nevans@cfa.harvard.edu
Monika **Fagas**, A. Mickiewicz University, Poznań, Poland — astrobserver@gmail.com
Gilles **Fontaine**, Université de Montréal, Canada — fontaine@astro.umontreal.ca
Lester **Fox-Machado**, Universidad Nacional Autónoma de México, Mexico — lfox@astrosen.unam.mx
Jian-Ning **Fu**, Beijing Normal University, China — jnfu@bnu.edu.cn
Alexandre **Galenne**, Universidad de Concepción, Chile — agallenne@astro-udec.cl
Rafael **Garrido Haba**, IAA - CSIC, Granada, Spain — garrido@iaa.es
Chris **Geroux**, University of Exeter, United Kingdom — geroux@astro.ex.ac.uk
Noemi **Giammichele**, Université de Montréal, Canada — noemi@astro.umontreal.ca
Mélanie **Godart**, University of Tokyo, Japan — melanie.godart@gmail.com
Phil **Goode**, BBSO, New Jersey Institute of Technology, Newark, USA — pgoode@bbso.njit.edu
Douglas **Gough**, University of Cambridge, United Kingdom — douglas@ast.cam.ac.uk
Marie-Jo **Goupil**, Observatoire de Paris, LESIA, France — mariejo.goupil@obspm.fr
Martin **Groenewegen**, Royal Observatory of Belgium, Brussels, Belgium — martin.groenewegen@oma.be
Mathieu **Grosjean**, Université de Liège, Belgium — grosjean@astro.ulg.ac.be
Frank **Grundahl**, University of Aarhus, Denmark — fgj@phys.au.dk
Elisabeth **Guggenberger**, University of Vienna, Austria — elisabeth.guggenberger@univie.ac.at
Joyce A. **Guzik**, Los Alamos National Laboratory, Los Alamos, USA — joy@lanl.gov
Gerald **Handler**, N. Copernicus Astronomical Center, Warsaw, Poland — gerald@camk.edu.pl
Markus **Hareter**, Konkoly Observatory, Budapest, Hungary — hareter@konkoly.hu
Saskia **Hekker**, MPISSR, Kaltenburg-Lindau, Germany — hekker@mps.mpg.de
Wiebke **Herzberg**, KIS, Freiburg, Germany — wiebke@kis.uni-freiburg.de
Daniel **Holdsworth**, Keele University, United Kingdom — d.l.holdsworth@keele.ac.uk
Simon **Jeffery**, Armagh Observatory, Northern Ireland, United Kingdom — csj@arm.ac.uk
Young-Beom **Jeon**, KASSI, Daejeon, South Korea — ybjeon@kasi.re.kr
Mikołaj **Jerzykiewicz**, Uniwersytet Wrocławski, Poland — mjerz@astro.uni.wroc.pl
Monika **Jurković**, Astronomical Observatory, Belgrade, Serbia — mojur@aob.rs
Filiz **Kahraman Aliçavuş**, Çanakkale Onsekiz Mart University, Turkey — filizkahraman01@gmail.com
Noé **Kains**, European Southern Observatory, Garching, Germany — nkains@eso.org
Krzysztof **Kamiński**, A. Mickiewicz University, Poznań, Poland — chrisk@amu.edu.pl
Steven **Kawaler**, Iowa State University, Ames, USA — sdk@iastate.edu
Irina **Kitiashvili**, Stanford University, USA — irinasun@stanford.edu
Zbigniew **Kołaczkowski**, Uniwersytet Wrocławski, Poland — kolaczkowski@astro.uni.wroc.pl
Katrien **Kolenberg**, Harvard-Smithsonian CfA, Cambridge, USA — kkolenbe@cfa.harvard.edu
Grzegorz **Kopacki**, Uniwersytet Wrocławski, Poland — kopacki@astro.uni.wroc.pl
Alexander **Kosovichev**, Stanford University, USA — sasha@sun.stanford.edu
Géza **Kovács**, Konkoly Observatory, Budapest, Hungary — kovacs@konkoly.hu
Michaela **Kraus**, Astron. Institute, AV ČR, Ondřejov, Czech Republic — kraus@sunstel.asu.cas.cz
Jurek **Krzesiński**, Pedagogical University, Cracow, Poland — jk@astro.as.up.krakow.pl
Karolina **Kubiak**, University of Vienna, Austria — karolina.kubiak@gmail.com
Charles **Kuehn**, University of Sydney, Australia — kuehn@physics.usyd.edu.au
Friedrich **Kupka**, University of Vienna, Austria — friedrich.kupka@univie.ac.at
Donald W. **Kurtz**, University of Central Lancashire, Preston, United Kingdom — kurtzdw@gmail.com
Patricia **Lampens**, Koninklijke Sterrenwacht van België, Brussels, Belgium — patricia.lampens@oma.be
Olivera **Latković**, Astronomical Observatory, Belgrade, Serbia — olivia@aob.rs
Maëlle **Le Pennec**, CEA Saclay, Gif-sur-Yvette, France — maelle.le-pennec@cea.fr
Marcella **Marconi**, INAF - OAC, Napoli, Italy — marcella.marconi@oacn.inaf.it
Stéphane **Mathis**, Laboratoire AIM Paris-Saclay, France — stephane.mathis@cea.fr
Jaymie M. **Matthews**, University of British Columbia, Vancouver, Canada — matthews@astro.ubc.ca
Travis **Metcalfe**, Space Science Institute, Boulder, USA — travis@spacescience.org
Gabriela **Michalska**, Uniwersytet Wrocławski, Poland — michalska@astro.uni.wroc.pl
Giovanni **Mirouh**, IRAP/CNRS, Toulouse, France — gmirouh@irap.omp.eu
Joanna **Molenda-Żakowicz**, Uniwersytet Wrocławski, Poland — molenda@astro.uni.wroc.pl
László **Molnár**, Konkoly Observatory, MTA CSFK, Budapest, Hungary — molnar.laszlo@csfk.mta.hu

Siobahn **Morgan**, University of Northern Iowa, Cedar Falls, USA — siobahn.morgan@uni.edu
Paweł **Moskalik**, N. Copernicus Astronomical Center, Warsaw, Poland — pam@camk.edu.pl
Nami **Mowlavi**, Geneva Observatory, Switzerland — nami.mowlavi@unige.ch
Nicolas **Nardetto**, Observatoire de la Côte d'Azur, Nice, France — nicolas.nardetto@oca.eu
Hilding **Neilson**, East Tennessee State University, Johnson City, USA — neilsonh@etsu.edu
Coralie **Neiner**, Observatoire de Paris-Meudon, LESIA, France — coralie.neiner@obspm.fr
Chow-Choong **Ngeow**, National Central University, Zhongli, Taiwan — cngeow@astro.ncu.edu.tw
Dieter **Nickeler**, Astron. Institute, AV ČR, Ondřejov, Czech Republic — dieter.nickeler@asu.cas.cz
Ewa **Niemczura**, Uniwersytet Wrocławski, Poland — eniem@astro.uni.wroc.pl
Jakub **Ostrowski**, Uniwersytet Wrocławski, Poland — ostrowski@astro.uni.wroc.pl
Erika **Pakštienė**, Vilnius University, Lithuania — erika.pakstiene@tfai.vu.lt
Alexey **Pamyatnykh**, N. Copernicus Astronomical Center, Warsaw, Poland — alosza@camk.edu.pl
Javier **Pascual Granado**, IAA - CSIC, Granada, Spain — javier@iaa.es
Ernst **Paunzen**, Masaryk University, Brno, Czech Republic — epaunzen@physics.muni.cz
Grzegorz **Pietrzyński**, Warsaw University Observatory, Poland — pietrzyn@astrouw.edu.pl
Andrzej **Pigulski**, Uniwersytet Wrocławski, Poland — pigulski@astro.uni.wroc.pl
Emese **Plachy**, Konkoly Observatory, Budapest, Hungary — eplachy@astro.elte.hu
Magdalena **Polińska**, A. Mickiewicz University, Poznań, Poland — polilena@gmail.com
Karen **Pollard**, University of Canterbury, Christchurch, New Zealand — karen.pollard@canterbury.ac.nz
Suzanna **Randall**, European Southern Observatory, Garching, Germany — srandall@eso.org
Daniel **Reese**, Université de Liège, Belgium — daniel.reese@ulg.ac.be
Slavek **Rucinski**, University of Toronto, Canada — rucinski@astro.utoronto.ca
Sopie **Saesen**, Geneva Observatory, Switzerland — sophie.saesen@unige.ch
Hideyuki **Saio**, Tohoku University, Sendai, Japan — saio@astr.tohoku.ac.jp
Valentina **Schmid**, Katholieke Universiteit Leuven, Belgium — valentina.schmid@ster.kuleuven.be
Jesper **Schou**, MPISSR, Katlenburg-Lindau, Germany — schou@mps.mpg.de
Tunç **Şenyüz**, Çanakkale Onsekiz Mart University, Turkey — tuncsenyuz@gmail.com
Hiromoto **Shibahashi**, University of Tokyo, Japan — shibahashi@astron.s.u-tokyo.ac.jp
Marek **Skarka**, Masaryk University, Brno, Czech Republic — maska@physics.muni.cz
Agnieszka **Słowikowska**, University of Zielona Góra, Poland — aga@astro.ia.uz.zgora.pl
Barry **Smalley**, Keele University, United Kingdom — b.smalley@keele.ac.uk
Radosław **Smolec**, N. Copernicus Astronomical Center, Warsaw, Poland — smolec@camk.edu.pl
Ádám **Sódor**, Royal Observatory of Belgium, Brussels, Belgium — sodor@konkoly.hu
Takafumi **Sonoi**, University of Tokyo, Japan — sonoi@astron.s.u-tokyo.ac.jp
Paulina **Sowicka**, Jagiellonian University, Cracow, Poland — paula@byk.oa.uj.edu.pl
Juan Carlos **Suárez**, IAA - CSIC, Granada, Spain — jcsuarez@iaa.es
Robert **Szabó**, Konkoly Observatory, MTA CSFK, Budapest, Hungary — rszabo@konkoly.hu
Wojciech **Szewczuk**, Uniwersytet Wrocławski, Poland — szewczuk@astro.uni.wroc.pl
Masao **Takata**, University of Tokyo, Japan — takata@astron.s.u-tokyo.ac.jp
Nathalie **Themessl**, University of Vienna, Austria — nathalie.themessl@univie.ac.at
Andrew **Tkachenko**, Katholieke Universiteit Leuven, Belgium — andrew@ster.kuleuven.be
Sanja **Tomić**, University of Belgrade, Serbia — sun.freckle@gmail.com
Richard **Townsend**, University of Wisconsin-Madison, USA — townsend@astro.wisc.edu
Ceren **Ulusoy**, University of South Africa, Pretoria, South Africa — cerenulusoy@gmail.com
Katrien **Uytterhoeven**, Instituto de Astrofísica de Canarias, La Laguna, Tenerife, Spain — katrien@iac.es
Valerie **Van Grootel**, University of Liège, Belgium — valerie.vangrootel@ulg.ac.be
Gérard **Vauclair**, IRAP/CNRS, Toulouse, France — gerard.vauclair@irap.omp.eu
Sylvie **Vauclair**, IRAP/CNRS, Toulouse, France — sylvie.vauclair@irap.omp.eu
Przemysław **Walczak**, Uniwersytet Wrocławski, Poland — walczak@astro.uni.wroc.pl
Werner **Weiss**, University of Vienna, Austria — werner.weiss@univie.ac.at
Gemma **Whittaker**, University of Toronto, Canada — whittaker@astro.utoronto.ca
Aneta **Wisniewska**, KIS, Freiburg, Germany — aneta.wisniewska@kis.uni-freiburg.de
Ewa **Zahajkiewicz**, Uniwersytet Wrocławski, Poland — zahajkiewicz@astro.uni.wroc.pl
Cyrus **Zalian**, Université Nice Sophia-Antipolis, France — cyruszalian@gmail.com
Elżbieta **Zocłońska**, N. Copernicus Astronomical Center, Warsaw, Poland — ela@camk.edu.pl
Konstanze **Zwintz**, Katholieke Universiteit Leuven, Belgium — konstanze.zwintz@ster.kuleuven.be

TALKS

Precision Asteroseismology
Proceedings IAU Symposium No. 301, 2013
J. A. Guzik, W. J. Chaplin, G. Handler & A. Pigulski, eds.
© International Astronomical Union 2014
doi:10.1017/S1743921313013999

A personal view of the scientific career of Wojtek Dziembowski (perceived by an admirer from abroad)

Douglas Gough

Institute of Astronomy & Department of Applied Mathematics and Theoretical Physics,
Madingley Road, CB3 0HA, UK
email: douglas@ast.cam.ac.uk

Abstract. I present a personal view of Wojtek Dziembowski's scientific career, derived mainly from my direct interactions with Wojtek. Necessarily this presentation is biased towards the earlier days, partly because we interacted more then, and partly because the presentation after mine is by a local admirer who has been much more involved than I with Wojtek's later work.

Keywords. instabilities, Sun: abundances, Sun: evolution, Sun: helioseismology, Sun: interior, Sun: magnetic fields, Sun: oscillations, Sun: rotation, stars: binaries, stars: chemically peculiar, stars: interiors, stars: oscillations (including pulsations), stars: rotation, stars: giants, stars: variables: Cepheids, stars: variables: δ Scuti, stars: white dwarfs

---

1. Introduction

Wojtek Dziembowski has many friends; it could hardly be otherwise of such a kind and gentle person. It is demonstrated at this conference by the large gathering who have come to wish him well. Indeed, it is always a great pleasure to be in Wojtek's company. He is amusing – see the twinkle in his eye – and he is very knowledgeable of many subjects: art, literature, music, philosophy, politics; and, of course, he is a superb scientist, whose achievements we are here to celebrate. Wojtek always thinks deeply about all that he studies, taking much more into consideration than is often evident from his published work. That becomes apparent as one witnesses his questions and responses at conferences, as I am sure that we all shall experience this week.

It is a great pleasure and honour to have been invited to whet your anticipation with a personal view of a small sample of his opera, and to try to convey some of the gratitude that I and many others owe to his teaching and his friendship. In so doing, I shall be able to refer explicitly to but a few of Wojtek's 250 or so publications.

2. The pulsations of B stars

I first met Wojtek in 1967 at Columbia University, New York. He had just arrived as a postdoctoral research associate to work with Norm Baker on theoretical aspects of pulsating stars. I was also working with Norm, in my spare time, on the role played by convection in the excitation and damping of the pulsations of RR Lyrae stars, Cepheids and Miras, so it was inevitable that we should soon meet. At that time Wojtek was interested mainly in hotter stuff than I, and he continued to be so for long after, as I shall allude later when describing his teaching in Aci Trezza. In particular, he was trying to understand how β Cephei pulsations are driven. I recall his having a great deal of trouble (as did I, but for a different reason), and he didn't publish on the subject until long afterwards (and I likewise). We had each chosen difficult problems that took a long time to resolve, and in those days, fortunately, one was not pressured as one is today to publish come what may. Wojtek was concerned also with other stars, on which he did publish soon afterwards, such as planetary nebulae (1977a), low-mass stars (with Maciek Kozłowski, 1974), giants (1977b) and also White Dwarfs (1977c, also with Ryszard Sienkiewicz, 1977), having been attracted to the last, one might surmise, by their initials.

Notwithstanding his early inability to obtain pulsationally overstable models, Wojtek did publish, with Marcin Kubiak in 1981, the outcome of an investigation of the efficacy of the κ mechanism in exciting low-degree nonradial oscillations in β Cephei stellar models. Their procedure was to compute linearized oscillations, nonadiabatic in only the outer layers and forced by an artificially imposed inner boundary condition to be neutrally stable, a technique that Wojtek had learned from Norm, who earlier had investigated RR Lyrae instability that way with Rudi Kippenhahn. The work expended by the motion of the inner boundary determines the stability of the star, whose actual sources of driving and damping, because they are all weak, can be estimated from the neutrally stable 'eigenfunctions'. The procedure affords a substantial improvement on the quasi-adiabatic approximation, which Wojtek (1971) had used previously. Even though Wojtek and Marcin were unable to find overstability in their models, they nonetheless concluded that it is the κ mechanism operating in the He II ionization zone that is likely to be the cause of the oscillations in the actual stars, suggesting that future, appropriately higher, calculated opacities would do the trick for theory. Subsequently, in response to helioseismologists' pressure arising from a different matter, Carlos Iglesias and Forrest Rogers at Livermore provided by moonlight what was required – revised opacity having a local hump at a temperature of about 2×10^5 K – which enabled Paweł Moskalik and Wojtek (1992) to solve that principal β Cephei problem, obtaining an instability strip that was more-or-less in agreement with observation. Wojtek had waited a quarter of a century for that result. I'm sure that it made him very happy.

The flood gates were now opened to admit an outpouring of further fruitful investigation. First, an extension of the new results, in collaboration with Alosha Pamyatnykh (1993) and also Paweł (1993) to a wider range of models, including SPB stars. Then asteroseismology of β Cephei stars with Mike Jerzykiewicz (1998, 1999) and Alosha (2008), and also the slow g-mode oscillations of rotating B stars, joining Jagoda Daszyńska-Daszkiewicz (2007) for assessing their visibility. Work of this genre has been applied to other kinds of star, such as δ Scuti, but I shall leave discussion of such matters to Alosha in the next presentation.

3. The excitation and damping of solar gravity modes

In 1972 I proposed with Fisher Dilke that a low-order low-degree g mode in the Sun has at times been excited by the ϵ mechanism, and that it grew to such an amplitude as to have triggered a nonlinear direct mode in a manner analogous to a similar triggering of direct overturning motion in thermohaline convection. The outcome would have been an episodic redistribution of heat and the chemical products of the nuclear reactions in the core, which would have caused the solar neutrino flux temporarily to have declined. The Sun's luminosity would have declined too, with important consequences for the terrestrial climate. The proposal attracted considerable attention from climatologists, and also from two giants in astrophysics: Martin Schwarzschild and Paul Ledoux, Martin partly because it overturned an important assumption in solar-evolution theory, Paul because of his interest in the intrinsic dynamics. However, a decade later, a new young giant emerging in the field demonstrated that the idea is very likely to be wrong. That giant was Wojtek.

A digression on the ϵ mechanism is not out of place here. The programme of Geophysical Fluid Dynamics held annually at the Woods Hole Oceanographic Institution, Massachussetts, USA, is housed in an old wooden 'cottage' in which the participants work. 'Geophysical', in this context, is interpreted quite broadly, often encompassing the astrophysical and even what one might call the syllastrophysical. Each morning and afternoon the participants assemble for coffee or tea, and, of course, scientific intercourse. The water for the beverages was boiled on an electric stove in a somewhat battered aluminium pan which had been overheated too many times and had consequently developed a rounded bottom. The pan could rock on the flat hotplate, and, if the quantity of water it contained were right for a surface gravity mode to resonate with the rocking, an oscillation would be sustained. With each rock, alternate sides of the bottom of the pan would receive a thermal impulse as it momentarily came into contact with the hotplate, thereby driving the motion. Here was the ϵ mechanism visibly operating on a human scale. Then, once the water was boiling, turbulent viscosity quenched the oscillation. If it could happen in the pan, then surely it could happen even more easily in a star, thought I, because in a star the diffusion coefficients (appropriately scaled to Reynolds and Rayleigh numbers) are much smaller, and the dissipation correspondingly lesser. I was wrong. In 1982 Wojtek published a paper in *Acta Astronomica* in which he demonstrated that grave low-degree g modes in a main-sequence star are likely to couple to resonating high-degree daughter pairs who drain energy from their parent so efficiently that the parent cannot grow to a respectable stature. There was a follow-up discussion the following year applying the result explicitly to the Sun. I illustrate in Fig. 1 part of the analysis, extracted, perhaps a little unfairly, from the first paper; it is typical of Wojtek's early style. After reading it for the first time I was left wondering how it could be that the lower the damping rates γ_2 and γ_3 of the daughters, the lower the limiting amplitude Q_1 of the parent. Surely greater amplitude limitation requires the daughters to dissipate energy faster. The explanation, of course, is clearly evident from a more careful reading of Wojtek's lucid mathematics, which speaks louder than words, as Fig. 1 illustrates. I have to confess, however, that it took a second reading for me to realize that. One further conclusion that I can therefore draw, from generalizing this anecdote, is that no paper by Wojtek should be read only once. I should perhaps point out also that in his later years (e.g. 2012a,b) Wojtek has become rather more expansive.

4. Wojtek's longest collaborations

In 1981, following a meeting at the Crimean Astrophysical Observatory on stellar and solar oscillations, I stopped off in Warsaw on the way home – my first visit to Poland – to work with Wojtek, starting one of the longest, yet, I am quite sure, the most unsuccessful of his collaborations. On my arrival I was immediately told that there was hardly any food in Warsaw; the canteen at the Copernicus Center, where I was housed, was permanently closed for lack of food. However, I hasten to add that that does not explain the paucity of our overt scientific achievements; in fact, it might even have mitigated our apparent failure, for it probably induced us to spend more time working. Wojtek was well prepared for averting famine: he had purchased 250 g of butter for me before leaving the Soviet Union, and then each day in Warsaw he gave me a hunk of dry bread, a hard-boiled egg and an apple or pear to stave of hunger until evening. I was worried that I was depriving him and his family of essential sustenance, but of course, as is characteristic of Wojtek, he denied it vehemently. (It is only on occasions such as that that one cannot trust

4. Parametric resonance instability

This effect, also called a "decay instability", consists of an instability of a linearly driven mode ($\gamma_1 > 0$) to growth of two modes that have frequencies σ_2 and σ_3 such that $\sigma_1 \approx \sigma_2 + \sigma_3$ and which are linearly damped.

The equations governing the initial growth of modes 2 and 3 may be obtained from Eq. (2 26) for $k = 2$ (say) and the complex conjugate of this equation for $k = 3$. Setting $Q_k = S_k \exp(i\Delta\sigma\tau/2)$ to get rid of the explicit time-dependence, we obtain

$$\frac{dS_2}{d\tau} = \gamma_2 S_2 + i\frac{H}{2\sigma_2 I_2}Q_1 S_3^*,$$ (4.1)

$$\frac{dS_3^*}{d\tau} = \gamma_3 S_3^* - i\frac{H}{2\sigma_3 I_3}Q_1^* S_2.$$ (4.2)

We assume here that Q_1 is constant which is justified in the initial phase of the development of the instability if γ_1 is sufficiently small.

Assuming now $S_k \sim \exp(\nu\tau)$ we get a characteristic equation that yields

$$\nu = \frac{1}{2}[\gamma_2 + \gamma_3 \pm \sqrt{(\gamma_3 - \gamma_2 + i\Delta\sigma)^2 + \nu_{pr}^2|Q_1|^2}},$$ (4.3)

where

$$\nu_{pr} = \frac{H}{\sqrt{I_2 I_3 \sigma_2 \sigma_3}},$$ (4.4)

which is equivalent to Vandakurov's (1981) Eq. 7. The instability takes place ($Re(\nu) > 0$) if

$$|Q_1|^2 > \frac{1}{\nu_{pr}^2}\left\{\Delta\sigma^2\left[1 - \frac{(\gamma_2 - \gamma_3)^2}{(\gamma_2 + \gamma_3)^2}\right] + 4\gamma_2\gamma_3\right\}.$$ (4.5)

It is interesting to notice that except in the case $\gamma_2 = \gamma_3$ the criterion in the limit $\gamma_{2,3} \to 0$ differs from the strictly adiabatic one that is obtained from (4.3) setting there $\gamma_2 = \gamma_3 = 0$. This paradox is easily resolved by the realization that the instability occurring at γ_2 or γ_3 passing through zero is a vibrational, not a dynamical, instability. It remains important, however, that with grossly unequal damping rates the dissipation may promote instability.

Substituting Eqs. (3.21) and (3.23) in Eq. (4.4), we get

$$\nu_{pr} = Z_{123}\left\langle \frac{C}{A}[(H_1 - H_3' - H_3)\cos(\psi_2 - \psi_3) + \varkappa_{g,2}(H_2 - H_3)\sin(\psi_2 - \psi_3)]\right\rangle_g$$

where $\langle\ \rangle_g = \frac{1}{\psi_2}\int d\psi_2$ (4.6)

It is easy to see that the expected minimum frequency mismatch, $\Delta\sigma$, decreases like l_2^{-2} with increasing l_2.

On the other hand, γ_2 increases with increasing l_2. If the quasiadiabatic approximation is valid there is a simple estimate of γ_2 (see e.g. Dziembowski 1971).

$$\gamma_2 = \frac{\Lambda_2}{\sigma_2^2 \tau_{th}} \sim l_2^2,$$

where

$$\tau_{th} = 18\frac{GM^2}{RL}(4\pi G\langle\varrho\rangle)^{-1/2}\left\langle \frac{L_r}{L}V V_{ad}\left(1 - \frac{V_{ad}}{V}\right)\left(\frac{R}{r}\right)^5\frac{\langle\varrho\rangle}{\varrho}\right\rangle^{-1}$$ (4.8)

is the thermal time-scale of the g-mode propagation zone in our time-units. This quantity is of the order of $10^{10} - 10^{12}$ in the interior of main sequence stars.

The final amplitude of mode 1 may be written in the form

$$|Q_1| = \frac{2}{\nu_{pr}}\sqrt{(1 + q^2)\gamma_2\gamma_3}.$$

Figure 1. Part of Wojtek's (1982) lucid derivation of the expected amplitude of a low-order solar g mode (with permission from *Acta Astronomica*).

Wojtek to tell the truth.) In the evenings I went into town to seek an open restaurant. I was finding the Polish language totally inscrutable, but it didn't matter: I wanted to eat, a restaurant would be open only if it had food to serve, and Wojtek assured me that no restaurant would have more than one dish, so I would encounter no difficulty with communication; all I needed to do was to smile and nod. It worked well. Only on my last day, when I chose a grander-looking establishment in celebration of a most stimulating visit, did I nearly fail, for the waitress tried to engage me in conversation. I could make neither head nor tail of what she was trying to tell me, and eventually she gave up and disappeared, apparently back into the kitchen. Was I going to be fed? Should I stay, or should I leave? I waited. And I was rewarded: after quite a while the waitress returned, armed with pictures of a chicken and a pig. Wojtek was wrong! I had a choice. Subsequent visits to Poland have been more salubrious, not least this wonderful stay in Wrocław.

Weekends in Warsaw were very enjoyable. Many of the members of the Copernicus Center would go to the Observatory, just outside the city. There we foraged and subsequently feasted on wild mushrooms, and had much discussion over and after dinner on a wide variety of subjects, except politics: that subject was reserved for walks in the forest.

But I digress. The intellectual outcome of this first visit to Warsaw was the start of a collaboration to seek an approximate second-order differential equation to describe nonradial adiabatic stellar oscillations that takes some account of the perturbation to the gravitational potential, at least at high order. The purpose was for using it as a basis for obtaining simple formulae that could be applied straightforwardly to diagnosing aspects of the structure of the underlying star from its oscillation eigenfrequencies. We found a relatively simple equation, which we named the first post-Cowling approximation. Then we embarked on improving it. We didn't complete our task, and a year or two later Wojtek made his first visit to Cambridge, where, I believe, we almost achieved the second post-Cowling approximation. Later I made a second visit to Warsaw, well after the lifting of Martial Law; Wojtek was so happy with the new freedom that he insisted on exchanging currency with me in public in the middle of the main square in the old town. But the scientific adventure was not as successful as we had hoped, because what we thought was a brilliant final manoeuvre subsequently became irretrievably lost (actually stolen), and neither of us have been able to recover the derivation since. Wojtek then said that he washed his hands of the matter, for it was so easy to compute frequencies numerically. I held on to the surviving remnants, and, with Wojtek's permission, even published the formula of the first approximation in some lecture notes for a Les Houches summer school. I doubt that anyone has noticed it because it is unobtrusively buried in the text. I still think that it would be useful for this first result to see the light of day, and I intend to make that happen. Therefore, after 32 years, I continue to consider this dormant collaboration not to have died, despite our failure to have published. Incidentally, when Wojtek made another visit to Cambridge we wisely decided to work on something else: the pulsations of red-giant stars such as α UMa, in collaboration with Günter Houdek and Ryszard Sienkiewicz (2001).

Not all of Wojtek's long collaborations have been so fraught. On the contrary, a still-ongoing and highly successful collaboration began with Phil Goode when Wojtek was visiting Henry Hill in Tucson, helping him organize one of the most memorable conferences in helioseismology. Wojtek and Phil have worked much together since, mainly on some of the more interesting aspects of helioseismology, often away from the main stream. Their first paper, published in 1983, concerned imposing seismologically inferred limits to the strength of the magnetic field in the core of the Sun. Since that time, they have contributed significantly to almost every facet of the subject, publishing to date some 50 papers, some of which I shall mention later.

Wojtek has also collaborated very fruitfully with Alosha Pamyatnykh; I leave discussion of that work to Alosha's presentation. His collaboration with Jagoda Daszyńska-Daszkiewicz is notable because, like his association with Luis Balona, it has drawn him closer to the techniques of observation. Strictly speaking, I should not have included it in a discussion of the longest collaborations, for it has taken place only during the current millennium; however, we all hope that it will continue for many years to come.

5. Scientific leadership

In 1983 Wojtek gave an extensive introductory talk at a conference in Catania on the combined effect of rotation and a magnetic field on degeneracy lifting of stellar acoustic oscillations, work that was in progress with Phil Goode. Application to the Sun was principally in mind. The necessarily aspherical perturbations were regarded as being globally, if not locally, small, so perturbation theory about the oscillations of a spherically symmetrical star could be employed. The speaker who followed was stunned; the talk he had intended to give appeared to address the identical subject, and suddenly he found himself in front of the audience wondering what there was left to say. Pointing out to the audience the embarrassing situation in which he found himself, he stalled for time by asking Wojtek if he could borrow one of his transparencies ('overhead' projection of transparencies was the technology of the day), a diagram illustrating the manner in which the eigenfrequencies are split, in the hope that Wojtek would not be too hasty. Since Wojtek had established his main point in his usual mathematical fashion – recall Fig. 1 – the second speaker offered to explain it in physical terms, which is why he had asked for that particular transparency. But to his bewilderment, when the transparency had been placed on the projector plate, and he had inspected it more carefully than he had had time to do during Wojtek's presentation, it was apparent that Wojtek's conclusion was very different from his own: Wojtek's frequency splitting had the opposite sign, and the magnitude was 40 times greater! What was the second speaker to do now? Evidently the truth had to be established. So he chose the usual scientific procedure: democracy. He called for a vote from the audience. Wojtek won by precisely 2:1. Wojtek was delighted, not for having won, but for witnessing a procedure which resonated with his political ideals. Nevertheless, there remained the task of finding the root of the discrepancy. It turned out to be a matter of principle, rather than of algebra; and it was resolved some weeks afterwards, in time for the two published versions of the presentations (1984) to be made consistent. The event was just one of many which demonstrates the faith the scientific community has in Wojtek's work.

One afternoon during a gap between lectures, some of the participants of the Catania conference climbed Mount Etna – it was not forbidden in those days – some of us intent on peering into the throat. We didn't see much other than the thick choking sulphurous smoke. I recall my feet nearly burning despite my thick-soled shoes, and later Wojtek and I being taken into a small hut and being plied with strong brandy to help recover from the fumes. I recall also remarking that I didn't know that such brandy was made in Sicily, and being told that what I was drinking could not be found in any shop. Somewhat revived, and certainly undeterred, Wojtek and I then decided to venture with our colleagues even closer to the action. Not presuming to emulate Empedocles, we descended into an ostensibly dormant fumarole, although by now the adventurous wing of the party had dwindled to just us two. Later, as we emerged, choking yet elated, it was Wojtek, of course, who was leading.

Most of the conferees were housed in a hotel in Catania. The invited lecturers, however, were honoured with a luxurious resort in a nearby village, Aci Trezza, overlooking the

Cyclops. Each day, after the formal activities, we were driven to the resort and left for an hour to swim and relax. Then, those other participants who were not yet too exhausted to continue talking science were driven to meet us by the pool. I noticed that most of us each had accumulated similar small groups of enthusiastic, mainly male, young people. But Wojtek's was different: he was surrounded by a bevy of beautiful ladies. Evidently, Wojtek is no ordinary scientific leader.

There was another interesting occurrence concerning Wojtek at this conference. On the traditional free Wednesday afternoon there was an outing to Syracusa. Just before we boarded the coaches that were to take us, six police arrived on motorcycles. They watched us from afar. Lucio Paternò, the organizer of the conference, went over to ask them why they had come, but they would not say. They escorted us to Syracusa, some riding ahead, others behind, and some to the side when there was no oncoming traffic. We thought, somewhat arrogantly, that they had come to protect us from possible Mafia intervention. It later transpired that we were quite wrong: they had been there to protect the Sicilian people from us, for they had heard that somewhere on board was a delegate from an Eastern-Bloc country.

It turned out that the police escort actually became beneficial to us. On the way home we encountered a serious traffic jam – several kilometres of stationary traffic on a narrow road. The reason was a road-widening operation, causing the traffic to wait while a gap in an outcrop of rock through which the road passed was enlarged with dynamite. One of the police rode ahead. When he returned, we were then escorted on the wrong side of the road past the waiting vehicles and the scene of the road work, where explosions had been halted temporarily, affording us, thanks evidently to Wojtek, a timely arrival home.

6. Rapidly oscillating Ap stars

The analysis reported at the Catania meeting can also be applied to the rapid oscillations of Ap stars. As in the paper that follows in the proceedings of that meeting, the point of view is that of an Earthbound observer, as is the case also of a companion paper with Phil Goode in 1985, resolving the oscillations into normal modes as seen from an inertial frame. The stars generally have two antipodal spots, produced, we believe, by a large-scale, predominantly dipolar, magnetic field. The magnetic axis, and that of the spots, is usually inclined from the axis of rotation. The oscillations also appear to be mainly dipolar, with their axes of symmetry more-or-less aligned with the spots. This is a natural consequence of degeneracy lifting if the spot-induced asphericity of the star dominates over that due to rotation. Wojtek subsequently went into some detail with the problem, mainly with Phil Goode (1996) and with Lionel Bigot (2002, 2003) and other collaborators (2000). With Lionel (2002), he also noted the possibility that the effect on the oscillations of the second-order, centrifugally induced asphericity could, in some circumstances, exceed that of both the first-order Coriolis force and, more pertinently, the asphericity due to the spots, and then the strong connexion between the oscillations and the spots is broken. But always the discussion was just about normal modes.

There are circumstances in which some may consider it to be prudent to describe the oscillations as (precessing) standing waves, superpositions of suitable pairs of modes propagating in opposite senses around the star. The two descriptions, are, of course, equivalent. Yet, as witness to Wojtek's well deserved influence even on scientists outside our field, there have been times when referees of putative journal articles inadequately familiar with wave dynamics, yet conversant with some of what has been written at least in some of Wojtek's abstracts, have failed to recognize the equivalence, and have recommended rejection on the mistaken ground that Wojtek has proved otherwise, so confident are they in their incomplete interpretation of Wojtek's words. That is surely another measure of the extent to which confidence in Wojtek has spread.

7. Rotation of the Sun and other stars

Wojtek was amongst the first to analyse rotational splitting data from the Sun and obtain a (nonuniformly weighted, spherically averaged) estimate of the angular velocity almost to the energy-generating core (Duvall *et al.* 1984; see also Brown *et al.* 1989). Subsequently, with Phil Goode and Ken Libbrecht (1989), he obtained one of the two-dimensional views, describing the (near) discontinuity across the tachocline in terms of the continuity of the (uniformly weighted) spherical average of the angular-momentum density (1993), which had earlier been discussed as a consequence of a putative local latitudinally uniform stress-strain relation. Wojtek pursued further the rotational perturbation to (mainly acoustic) stellar oscillations from a theoretical point of view (with Sasha Kosovichev 1987a,b, Maciek Kozłowski 1987, Phil Goode 1992, and Fatma Soufi and Marie-Jo Goupil 1998), taking higher-order terms into account. As I have mentioned already, there are circumstances in which the centrifugal force can dominate over the inertial (Coriolis) effects, a result which is pertinent particularly to work on roAp stars.

Rotational effects on the oscillations of γ Doradus and SPB stars have also attracted Wojtek's attention, hardly surprising in view of his early love in New York. Here, gravity modes are the centre of interest, and they can have frequencies comparable with or less than the angular velocity of the star. That adds richness to the investigation, for it is no longer the case that rotation imparts only a small perturbation to the dynamics. The oscillation problem is no longer straightforwardly nearly separable in radial and angular coordinates, and more extensive argument is called for. Together with Alosha and Jagoda, Wojtek (2007, 2009, 2010) adopted what geophysicists call the traditional approximation, which ignores the contribution of the horizontal component of the angular velocity of the star to the Coriolis force acting on the pulsations, restoring the differential system to one that admits separable solutions, and thereby making the problem much more tractable.

8. WET

Although Wojtek is primarily a theorist, he has been involved directly with observation (e.g. Breger *et al.* 1995; Handler *et al.* 1997). I am not sure what precisely was Wojtek's role.

9. Structural helioseismology

Wojtek seized on helioseismology very soon after its power had been established. I have already mentioned his visit to Tucson with Henry Hill, and his consequent meeting with Phil Goode, resulting in a long and fruitful collaboration. Most of that joint work concerned the Sun. And that spawned other investigations in collaboration with Alosha Pamyatnykh and with Ryszard Sienkiewicz (1990, 1992). From these investigations emerged one of the early seismic solar models (1990, 1994, 1995), an early estimate of the protosolar helium abundance (1991, 1998), and the realization (1999) that, subject to the usual assumptions of solar evolution theory, a helium-abundance estimate (Kosovichev *et al.* 1992, Richard *et al.* 1998) appeared to be consistent with a main-sequence age that itself is more-or-less consistent with the ages of the oldest meteorites (Dziembowski *et al.* 1999). There were also examinations of solar models (Richard *et al.* 1996, degl'Innocenti *et al.* 1997), the solar core (1996), the theoretical neutrino luminosity (1996, 2000), a seismic radius (Schou *et al.* 1997) and the opacity. Moreover, Wojtek published a suite of papers with Phil Goode (1991, 2002, 2004, 2005) on the seismic response to magnetic activity, and diagnosis of the solar cycle, sometimes in collaboration with Jesper Schou (2001), and also Steve Tomczyk (1997) and Sasha Kosovichev (2000). The work was not undertaken in isolation; there were others elsewhere with similar pursuits. Wojtek's contributions certainly added credence to the other work, and invariably stimulated further thought.

10. Nonlinear matters

The matters to which I refer all concern interactions between distinct modes of stellar oscillation. They have pervaded Wojtek's entire career. I discussed at some length the nonlinear suppression of grave solar g modes by their daughters, but that work is but a tiny fraction of Wojtek's contribution to the subject. My emphasis was chosen because it related directly to my own interests, and my presentation here is supposed to illustrate my personal exposure to Wojtek's work. But most of Wojtek's nonlinear studies (some with Małgosia Królikowska 1985) arose from his attempts to understand δ Scuti pulsations, and the effect of rotation (with Sasha Kosovichev, 1987, 1988), not to mention amplitude limitation of double-mode Cepheids (with Géza Kovács 1984 and with Radek Smolec 2009), and amplitude modulation in RR Lyrae stars (with Rafał Nowakowski 2001, 2003, and with Tomek Mizerski 2004) and the Blazhko effect. Indeed, I would dare suggest that Wojtek regards that work as his most important; and I'm sure that there are many others who think likewise. However, I shall not dwell on these matters here because Alosha, who has been much closer than I to Wojtek when these studies were being carried out, will address them more authoritatively in the following presentation.

11. Giants and supergiants

Much of Wojtek's attention in this arena (e.g. 1977, 2001b, 2012a, also Van Hoolst *et al.* 1998) has been devoted to addressing issues concerning mixed modes: frequency pulling, as it is sometimes called, to describe how the frequency of a mixed mode – g-mode-like near the centre and p-mode-like in the outer envelope – is an appropriately weighted average of the frequencies of a putative pure g mode and a corresponding pure p mode that are considered to be isolated from one another by the evanescent zone that separates their corresponding cavities. The work on a theoretical model of α UMa, which I mentioned earlier, is such an example. Understanding this subject is crucial to the asteroseismology of these stars, for which it is necessary to identify the modes responsible for the oscillation frequencies observed. It is a burgeoning issue for those trying to analyse the enormous number of data newly acquired from the *Kepler* space mission, a mission with which Wojtek is involved directly. Moreover, in my opinion, the pertinent theoretical issues are yet inadequately understood. The (linear) "interactions", as some call them, between the acoustic and gravity components of a mixed mode of oscillation lead to oscillation spectra with complicated nonuniformities which demand to be sorted out. That can be a difficult, tedious task, which many would shun. Wojtek, despite his penchant for the clean mathematical argument, has had considerable experience with similar issues related to δ Scuti pulsations, and, by virtue of his extremely strong theoretical background, has been an invaluable contributor to that endeavour, and will be so in this case too if he decides to put his mind to it. Life is limited, however, and one cannot do everything. So it is possible that Wojtek will decide to devote his energies elsewhere.

12. An astrophysical miscellany

Wojtek is concerned not just with theory, but also with the interpretation of astronomical observations. It is that, I believe, that has led him to study so many different phenomena. He has contributed, albeit not as thoroughly as the other subjects that I have mentioned, to the stability of accreting white dwarfs (Sienkiewicz & Dziembowski 1977), to explaining the acoustic flux of Sirius B as the outcome of resonant interactions between short-period acoustic modes and travelling waves (Dziembowski & Gesicki 1983), to the study, with Paul Bradley, of PG1159 stars, and to the theory of tidal friction in close binary systems (1967), to name but a few. He has been brought much closer to observation by collaborating particularly with Luis Balona and Jagoda Daszyńska-Daszkiewicz, and he has also joined much larger groups to explain observations of red giants (Soszyński *et al.* 2011a,b). His very latest publication to date (2012b) concerns the puzzling periods of first-overtone Cepheids: they are typically 0.6 of the fundamental period, yet it seems to be difficult to explain the driving mechanism in models that can reproduce that period ratio. Wojtek's is an excellent discussion of the pertinent unresolved issues, setting up the groundwork and paving the way for future research, on which I am sure Wojtek is ready to embark.

13. Postscript

Wojtek seems to have covered almost the entire arena of stellar pulsation. The possible pulsations of pulsars is a notable exception; they rotate and harbour large-scale magnetic fields, just as do roAp stars, which is right up Wojtek's street, although the physical conditions are very different. Moreover, Wojtek has paid only scant attention to the driving of very red variables, such as Miras. That is because in these stars convection is so important, and that is an area of fluid dynamics that Wojtek has chosen to avoid. He has made many studies of the nonlinear interactions between distinct modes of oscillation, but, so far as I am aware, he has not paid dynamical attention to the nonlinear interaction of a mode with itself. These matters he has kindly left to others, like me, who can therefore feel that they are not living completely in the shadow of the giant. However, this giant is always very happy to discuss the issues with whomever so wishes, and his insight is always very illuminating.

Recently, Wojtek told my wife, Rosanne, how wonderful it is to become old. I'm not sure that he really explained why, although, being not very much younger than he, I no doubt appreciate some of the reasons. One thing that it implies is that we have been able to know each other for a long while: 46 years, to be precise. I have truly valued that time, not just the time we have spent in each other's company, but also time apart with the knowledge that Wojtek is a true friend for whom I have enormous respect. Wojtek is a source of energy and inspiration to us all, much of which has been made possible by the continual encouragement and support from his dear wife Anna.

References

Balona, L. A. & Dziembowski, W. A. 1999, *MNRAS*, 309, 221
Balona, L. A., Dziembowski, W. A., & Pamyatnykh, A. A. 1997, *MNRAS*, 289, 25
Bigot, L. & Dziembowski, W. A. 2002, *A&A*, 391, 235

Bigot, L. & Dziembowski, W. A. 2003, *Ap&SS*, 284, 217

Bigot, L., Provost, J., Berthomieu, G., Dziembowski, W. A., & Goode, P. R. 2000, *A&A*, 356, 218

Bradley, P. A. & Dziembowski, W. A. 1996, *ApJ*, 462, 376

Breger, M., Handler, G., Nather, R. E. *et al.* 1995, *A&A*, 297, 473

Brown, T. M., Christensen-Dalsgaard, J., Dziembowski, W. A., Goode, P., Gough, D. O., & Morrow, C. A. 1989, *ApJ*, 343, 526

Daszyńska-Daszkiewicz, J., Dziembowski, W. A., Pamyatnykh, A. A., & Goupil, M.-J. 2002, *A&A*, 392,151

Daszyńska-Daszkiewicz, J., Dziembowski, W. A., & Pamyatnykh, A. A. 2003a, *Ap&SS*, 284, 133

Daszyńska-Daszkiewicz, J., Dziembowski, W. A., & Pamyatnykh, A. A. 2003b, *A&A*, 407, 999

Daszyńska-Daszkiewicz, J., Dziembowski, W. A., & Pamyatnykh, A. A. 2007, *AcA*, 57, 11

degl'Innocenti, S., Dziembowski, W. A., Fiorentini, G., & Ricci, B. 1997, *Astroparticle Physics*, 7, 77

Duvall Jr, T. L., Dziembowski, W. A., Goode, P. R., Gough, D. O., Harvey, J. W., & Leibacher, J. W. 1984, *Nature*, 310, 22

Dziembowski, W. A. 1963, *AcA*, 13, 157

Dziembowski, W. A. 1967, in: J. Dommanget (ed.), *On the Evolution of Double Stars*, Communications Serie B, No. 17 Computes Rondus, p. 105

Dziembowski, W. A. 1971, *AcA*, 21, 289

Dziembowski, W. A. 1977a, in: R. Kippenhahn, J. Rahe, & W. Strohmeier (eds.), *The Interaction of Variable Stars with their Environment*, Proc. IAU Colloquium No. 42 (Bamberg: Remeis-Sternwarte), p. 342

Dziembowski, W. A. 1977b, *AcA*, 27, 95

Dziembowski, W. A. 1977c, *AcA*, 27, 1

Dziembowski, W. A. 1977d, *AcA*, 27, 203

Dziembowski, W. A. 1982, *AcA*, 32, 147

Dziembowski, W. A. 1983, *Solar Phys.*, 82, 259

Dziembowski, W. A. 1984, *Adv. Space Res.*, 4, 143

Dziembowski, W. A. 1996, *Bull. Ast. Soc. India*, 24, 133

Dziembowski, W. A. 2000, *Acta Phys. Pol.*, Ser. B, 31, 1389

Dziembowski, W. A. 2010, *Highlights of Astronomy*, 15, 360

Dziembowski, W. A. 2012a, *A&A*, 539, A83

Dziembowski, W. A. 2012b, *AcA*, 62, 323

Dziembowski, W. A. & Cassisi, S. 1999, *AcA*, 49, 371

Dziembowski, W. A. & Gesicki, K. 1983, *AcA*, 33, 183

Dziembowski, W. A. & Goode, P. R. 1983, *Nature*, 305, 39

Dziembowski, W. A. & Goode, P. R. 1984, *MemSAIt*, 55, 185

Dziembowski, W. A. & Goode, P. R. 1985, *ApJ*, 296, L27

Dziembowski, W. A. & Goode, P. R. 1989, *ApJ*, 347, 540

Dziembowski, W. A. & Goode, P. R. 1991, *ApJ*, 376, 782

Dziembowski, W. A. & Goode, P. R. 1992, *ApJ*, 394, 670

Dziembowski, W. A. & Goode, P. R. 1993, *PASP*, 42, 225

Dziembowski, W. A. & Goode, P. R. 1996, *ApJ*, 458, 338

Dziembowski, W. A. & Goode, P. R. 2004, *ApJ*, 600, 464

Dziembowski, W. A. & Goode, P. R. 2005, *ApJ*, 625, 548

Dziembowski, W. A. & Jerzykiewicz, M. 1996, *A&A*, 306, 436

Dziembowski, W. A. & Jerzykiewicz, M. 1999, *A&A*, 341, 480

Dziembowski, W. A. & Kosovichev, A. G. 1987a, *AcA*, 37, 313

Dziembowski, W. A. & Kosovichev, A. G. 1987b, *AcA*, 37, 341

Dziembowski, W. A. & Kovács, G. 1984, *MNRAS*, 206, 497

Dziembowski, W. A. & Kozłowski, M. 1974, *AcA*, 24, 245

Dziembowski, W. A. & Królikowska, M. 1985, *AcA*, 35, 5

Dziembowski, W. A. & Kubiak, M. 1981, *AcA*, 31, 153

Dziembowski, W. A. & Mizerski, T. 2004, *AcA*, 54, 363

Dziembowski, W. A. & Pamyatnykh, A. A. 1993, *MNRAS*, 262, 204

Dziembowski, W. A. & Pamyatnykh, A. A. 2008, *MNRAS*, 385, 2061

Dziembowski, W. A. & Smolec, R. 2009, *AcA*, 59, 19

Dziembowski, W. A. & Soszyński, I. 2010, *A&A*, 524, A88

Dziembowski, W. A., Kosovichev, A. G., & Kozłowski, M. 1987, *AcA*, 37, 331

Dziembowski, W. A., Królikowska, M., & Kosovichev, A. G. 1988a, *AcA*, 38, 61

Dziembowski, W. A., Paternò, L., & Ventura, R. 1988b, *AcA*, 38, 61

Dziembowski, W. A., Goode, P. R., & Libbrecht, K. G. 1989, *ApJ*, 337, L53

Dziembowski, W. A., Pamyatnykh, A. A., & Sienkiewicz, R. 1990, *MNRAS*, 244, 542

Dziembowski, W. A., Pamyatnykh, A. A., & Sienkiewicz, R. 1991, *MNRAS*, 249, 602

Dziembowski, W. A., Pamyatnykh, A. A., & Sienkiewicz, R. 1992, *AcA*, 42, 5

Dziembowski, W. A., Moskalik, P., & Pamyatnykh, A. A. 1993, *MNRAS*, 265, 588

Dziembowski, W. A., Goode, P. R., Pamyatnykh, A. A., & Sienkiewicz, R. 1994, *ApJ*, 432, 417

Dziembowski, W. A., Goode, P. R., Pamyatnykh, A. A., & Sienkiewicz, R. 1995, *ApJ*, 445, 509

Dziembowski, W. A., Goode, P. R., Schou, J., & Tomczyk, S. 1997, *ApJ*, 323, 231

Dziembowski, W. A., Fiorentini, G., Ricci, B., & Sienkiewicz, R. 1999, *A&A*, 343, 990

Dziembowski, W. A., Goode, P. R., Kosovichev, A. G., & Schou, J. 2000, *ApJ*, 537, 1026

Dziembowski, W. A., Goode, P. R., & Schou, J. 2001a, *ApJ*, 553, 897

Dziembowski, W. A., Gough, D. O., Houdek, G., & Sienkiewicz, R. 2001b, *MNRAS*, 328, 601

Dziembowski, W. A., Daszyńska-Daszkiewicz, J., & Pamyatnykh, A. A. 2007, *MNRAS*, 374, 248

Goode, P. R. & Dziembowski, W. A. 2002, in: A. Wilson (ed.), *From Solar Min to Max: Half a Solar Cycle with SOHO*, ESA SP-508, p. 15

Goode, P. R., Dziembowski, W. A., Korzennik, S. G., & Rhodes, E. J., Jr. 1991, *ApJ*, 367, 649

Handler, G., Pikall, H., Dziembowski, W. A. *et al.* 1997, *MNRAS*, 286, 303

Hill, H. A. & Dziembowski, W. A., (eds.), 1980, *Lecture Notes in Physics*, Vol. 125 (Springer: Berlin)

Kosovichev, A. G., Christensen-Dalsgaard, J., Däppen, W., Dziembowski, W. A., Gough, D. O., & Thompson, M. J. 1992, *MNRAS*, 259, 536

Moskalik, P. & Dziembowski, W. A. 1992, *A&A*, 256, L5

Moskalik, P. & Dziembowski, W. A. 2005, *A&A*, 434, 1077

Nowakowski, R. M. & Dziembowski, W. A. 2001, *AcA*, 51, 5

Nowakowski, R. M. & Dziembowski, W. A. 2003, *Ap&SS*, 284, 273

Pamyatnykh, A. A., Handler, G., & Dziembowski, W. A. 2004, *MNRAS*, 350, 1022

Richard, O., Vauclair, S., Charbonnel, C., & Dziembowski, W. A. 1996, *A&A*, 312, 1000

Richard, O., Dziembowski, W. A., Sienkiewicz, R., & Goode, P. R. 1998, *A&A*, 338, 756

Schou, J., Kosovichev, A. G., Goode, P. R., & Dziembowski, W. A. 1996, *ApJ*, 489, L197

Sienkiewicz, R. & Dziembowski, W. A. 1977, in: R. Kippenhahn, J. Rahe, & W. Strohmeier (eds.), *The Interaction of Variable Stars with their Environment*, Proc. IAU Colloquium No. 42 (Bamberg: Remeis-Sternwarte), p. 327

Soszyński, I., Dziembowski, W. A., Udalski, A., *et al.* 2011a, *VizieR Online Data Catalog*, 1206, 10001

Soszyński, I., Dziembowski, W. A., Udalski, A., *et al.* 2011b, *AcA*, 61, 1

Soufi, F., Goupil, M.-J., & Dziembowski, W. A. 1998, *A&A*, 334, 911

Van Hoolst, T., Dziembowski, W. A., & Kawaler, S. D. 1998, *MNRAS*, 297, 536

Precision Asteroseismology
Proceedings IAU Symposium No. 301, 2013
J. A. Guzik, W. J. Chaplin, G. Handler & A. Pigulski, eds.

© International Astronomical Union 2014
doi:10.1017/S1743921313014002

Wojtek Dziembowski: selected photos and events of his scientific biography

Alexey A. Pamyatnykh

Nicolaus Copernicus Astronomical Center,
Bartycka 18, 00-716, Warsaw, Poland
email: `alosza@camk.edu.pl`

Abstract. Selected events of the scientific biography of Wojtek Dziembowski are briefly described, and several related photos are presented. The full version of the presentation is available at the IAU Symposium 301 webpage.

Keywords. biographies, Sun: helioseismology, stars: oscillations

Wojtek Dziembowski was born in Warsaw in 1940 into the family of a lawyer. In 1943 the family moved to the village Zawoja in Southern Poland, and later, after the end of World War II, to the small town Żnin about 250 km to the north-west of Warsaw. Already in primary school, being motivated by popular books of Sir James H. Jeans, Wojtek decided to be an astronomer. In 1962 he obtained an MSc degree in astronomy from the Jagiellonian University in Kraków. On the basis of his MSc Thesis, Wojtek published (just 50 years ago!) his first paper, "On the equations of internal constitution of components in close binaries" (Dziembowski 1963).

The next part in Wojtek's career was played by Prof. Paul Ledoux (!) who visited Poland in 1963 and whom Wojtek told about his scientific work. Paul Ledoux wrote about Wojtek to Prof. Stefan Piotrowski who invited Wojtek for graduate studies at Warsaw University. In 1967 Wojtek defended his PhD thesis on tides in binary systems and, a few months later, thanks to the recommendation of Prof. Józef Smak, obtained a postdoctoral fellowship at Columbia University, New York, where he worked until 1969 with Prof. Norman Baker. The main goal of their studies was an attempt to find an excitation mechanism for the β Cephei pulsations. Wojtek was the first to develop a computer code for linear nonadiabatic nonradial oscillations. He obtained instability for realistic models of some classical variables like δ Scuti stars but not for the β Cephei-type variables. The results were never published, and much later, at IAU Symposium 162 in France in 1993, Wojtek wrote in retrospect: *"For many years, explaining the cause of β Cephei variability has been a major challenge to stellar pulsation theory. The problem is now solved, but we owe the solution to progress in opacity calculation and not to new astrophysical ideas. In fact, if the OPAL opacities were available, the problem would have been solved many years ago (Baker & Dziembowski 1969, unpublished)"* (Dziembowski 1994). It is absolutely fair that just Wojtek was among the first who really solved the puzzle of the β Cephei pulsations (Moskalik & Dziembowski 1992)!

From 1969 to the present, Wojtek has been working at Nicolaus Copernicus Astronomical Center (CAMK) in Warsaw, and, since 1997, also at the Astronomical Observatory of the Warsaw University. In 1977 he defended his Habilitation thesis on theoretical aspects of nonradial stellar oscillations, and in 1978 − 1979 he was working as Visiting Professor at the University of Arizona. In 1988 Wojtek obtained the title of Professor; since 1989 he has been a Corresponding Member, and since 2007 a Member of the Polish Academy of Sciences. Since 1997 he is also the Corresponding member of the Polish Academy of

A. A. Pamyatnykh

Figure 1. Wojtek and his older sister Teresa (Wojtek has three sisters) in Zawoja, a village in Southern Poland, 1943. Photo from Archive of WAD.

Figure 2. Wojtek and Bohdan Paczyński at the summer astronomical practice, Observatory Piwnice near Toruń, July 1960. Photo: M. Górski.

Figure 3. Onboard the motor yacht "Podhalanin" during a voyage of Polish astronomers on their way to the 14th IAU General Assembly in Brighton, August 1970. From the left to the right: Wojtek Dziembowski, Janusz Zieliński, Barbara Kołaczek, Janusz Ziółkowski, 2nd officer Konstanty Pelak, Mike Jerzykiewicz, Jan Bieniewski, and Adam Spodenkiewicz. Photo from Archive of WAD.

Arts and Sciences, an elite scientific corporation of the Polish scientists. In 1987 − 1992 Wojtek was the Director of the CAMK, and in 2003 − 2006 he was the President of IAU Commission 35 "Stellar Constitution".

According to the SAO/NASA Astrophysics Data System (ADS, as of August 2013), Wojtek has 244 scientific publications with total number of citations equal to about 6800. Sixteen publications have each more than 100 citations, and Wojtek's Hirsch index is equal to 44. He is sole author of 52 publications; most of them are invited talks at many astrophysical meetings - due to obvious restrictions for conference proceedings, they are relatively short but written extremely carefully, clearly and precisely. As an example, I would like to refer again to Dziembowski (1994), where a very short and clear description of the κ-mechanism is given with special attention to B-star pulsations. At that time Wojtek and collaborators published results of detailed studies of β Cephei and SPB pulsations (Dziembowski & Pamyatnykh 1993; Dziembowski *et al.* 1993).

Note that two of Wojtek's single-author papers have been published only recently − on dipolar modes in red giants (Dziembowski 2012a) and on puzzling frequencies in first-overtone Cepheids (Dziembowski 2012b).

A few of Wojtek's papers have become classics, as they gave rise to the creation of new directions for theoretical studies of stellar pulsations and for interpretations of observations of multiperiodic variables. Wojtek was one of the first who developed detailed algorithms and codes for computation of linear nonadiababic nonradial stellar oscillations (Dziembowski 1971, 1977a), a method of nonradial mode identification from observations of light and radial velocity variations (Dziembowski 1977b), the theory of nonlinear mode coupling in oscillating stars (Dziembowski 1982), a method of taking into account effects of differential rotation on stellar oscillations (Dziembowski & Goode 1992), an algorithm for numerical solution of the inverse problem in helioseismology (Dziembowski *et al.* 1990;

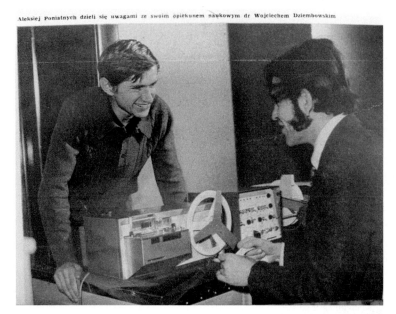

Aleksiej Poniatnych dzieli się uwagami ze swoim opiekunem naukowym dr Wojciechem Dziembowskim

Figure 4. "Alexey Pamyatnykh shares his considerations with his scientific adviser Dr Wojciech Dziembowski", Warsaw, Autumn 1972. (Computer ODRA 1204, read-punch unit for paper tape with an Algol program.) Photo and caption from the magazine "Przyjaźń" ("Friendship"), January 1973.

see the contribution by Douglas Gough in this volume for Wojtek's many other achievements in helioseismology).

A distinctive property of Wojtek's main papers is that they became most popular and cited more than 20 years after publication – this means that these papers were ahead of their time and they predicted or even defined the direction of stellar pulsation studies! For example, Wojtek's most cited paper on light and radial velocity variations in a nonradially oscillating star (Dziembowski 1977b) with a total number of 221 citations has been cited 42 times in 1978–1987 and 82 times in 2003–2012. Indeed, owing to recent cosmic missions and high-precision ground-based observations, a lot of multiperiodic variables have been detected, and elaboration of the methods of nonradial mode identifications is of highest importance for asteroseismology. Jagoda Daszyńska-Daszkiewicz and Wojtek, in collaboration with other colleagues, continue to refine his method, taking into account effects of rotation (see Daszyńska-Daszkiewicz *et al.* 2002 and later papers).

Many of Wojtek's investigations have been performed in collaboration with colleagues both from Poland and from other countries. In my opinion, all of Wojtek's collaborators are happy to work with him; they get creative scientific impulses from his comprehensive knowledge and from his fantastic intuition.

Wojtek was the scientific adviser of 7 students who successfully defended their PhD theses (Ryszard Sienkiewicz, Zbigniew Loska, Paweł Moskalik, Małgorzata Królikowska, Paweł Artymowicz, Jan Zalewski, Rafał Nowakowski).

When one considers 225 of Wojtek's publications with less than 6 authors, his closest collaborators appear to be Philip Goode (as coauthor of 48 publications), Alexey Pamyatnykh (40), Jagoda Daszyńska-Daszkiewicz (16), Ryszard Sienkiewicz (15), Marie-Jo Goupil (11) and Aleksander Kosovichev (11). It can be undoubtedly stated that he is the creator of the Polish school of asteroseismology, and now his informal group includes

Figure 5. With Arthur Cox, Douglas Gough, Philip Goode and other colleagues, IAU Coll. 121 "Inside the Sun", Versaille, May 1989. Wojtek and Philip Goode delivered a talk "Magnetic field in the Sun's interior from oscillation data". From Archive of WAD.

Figure 6. With Paweł Moskalik after the end of IAU Coll. 155, "Astrophysical Applications of Stellar Pulsation", February 1995, Cape Town. At the conference, Wojtek with coauthors had four presentations, and Paweł presented the up-to-date results of our group on pulsating OB-stars. From Archive of P.M.

Figure 7. On the way to the Los Alamos Meeting "A Half Century of Stellar Pulsation Interpretations. A Tribute to Arthur N. Cox", Wojtek with Jurek Krzesiński and Andrzej Pigulski. New Mexico, June 1997 (near VLA of NRAO). At the conference, Wojtek gave an introductory talk on helio- and asteroseismology. Photo: A. Pamyatnykh.

Figure 8. Conference "Stellar Pulsation and Evolution", Monte Porzio Catone, Italy, June 2005. Jørgen Christensen-Dalsgaard, Arthur Cox and Wojtek. In his introductory talk on asteroseismology, Wojtek gave examples of interpretations of rich oscillation spectra of different stars, including solar-like, sdB and β Cep-type pulsators. Photo: J.A. Guzik

Figure 9. With Jagoda Daszyńska-Daszkiewicz and Mike Breger. The HELAS Workshop "Interpretation of Asteroseismic Data", Wrocław, June 2008. Photo: A. Pamyatnykh.

Figure 10. From right to left: Paweł Moskalik, Wojtek Dziembowski, Radek Smolec, Andrzej Pigulski, Alexey Pamyatnykh. CAMK, Warsaw, September 2009, Defense of PhD dissertation by Radek Smolec. Paweł was the scientific adviser of Radek, and Wojtek was many years ago the scientific adviser of Paweł. This is a part of our group headed by Wojtek and now also by Gerald Handler.

researchers from Warsaw and Wrocław, and collaborates with many other scientists around the world (I would like to note that one of the most active researchers of pulsating stars, Gerald Handler, works now at the Copernicus Center in Warsaw.)

Figure 11. Conference "Impact of new instrumentation and new insights in stellar pulsations", Granada, September 2011. With Arthur Cox at the conference dinner. Photo: A. Pamyatnykh.

Acknowledgements

I acknowledge partial financial support from the Scientific Organizing Committee and from the Polish NCN grant 2011/01/B/ST9/05448. I am grateful to Wojtek Dziembowski, Joyce Guzik, Mike Jerzykiewicz and Paweł Moskalik for the photos which have been used in this contribution and presented at the Symposium.

The extended version of the presentation at the Symposium is available at http://www.astro.uni.wroc.pl/IAUS301_Talks/Day1-0950-Pamyatnykh.pdf.

References

Daszyńska-Daszkiewicz, J., Dziembowski, W. A., Pamyatnykh, A. A., & Goupil, M.-J. 2002, *A&A*, 392, 151
Dziembowski, W. A. 1963, *AcA*, 13, 157
Dziembowski, W. A. 1971, *AcA*, 21, 289
Dziembowski, W. A. 1977a, *AcA*, 27, 95
Dziembowski, W. A. 1977b, *AcA*, 27, 203
Dziembowski, W. A. 1982, *AcA*, 32, 147
Dziembowski, W. A. 1994, in: L. A. Balona, H. F. Henrichs, & J. M. Le Contel (eds.), *Pulsation, rotation and mass loss in early-type stars*, Proc. IAU Symposium No. 162 (Kluwer Academic Publishers) p. 55
Dziembowski, W. A. 2012a, *A&A*, 539, A83
Dziembowski, W. A. 2012b, *AcA*, 62, 323
Dziembowski, W. A. & Goode, P. R. 1992, *ApJ*, 394, 670
Dziembowski, W. A. & Pamyatnykh, A. A. 1993, *MNRAS*, 262, 204
Dziembowski, W. A., Pamyatnykh, A. A., & Sienkiewicz, R. 1990, *MNRAS*, 244, 542
Dziembowski, W. A., Moskalik, P., & Pamyatnykh, A. A. 1993, *MNRAS*, 265, 588
Moskalik, P. & Dziembowski, W. A. 1992, *A&A*, 256, L5

Precision Asteroseismology
Proceedings IAU Symposium No. 301, 2013
J. A. Guzik, W. J. Chaplin, G. Handler & A. Pigulski, eds.

© International Astronomical Union 2014
doi:10.1017/S1743921313014014

What can we expect from precision asteroseismology?

G. Handler

Nicolaus Copernicus Astronomical Center, Bartycka 18, 00-716 Warsaw, Poland
email: `gerald@camk.edu.pl`

Abstract. Precision asteroseismology is the determination of accurate stellar parameters from oscillation data. At first successful for pulsating white dwarf stars, it is now applied to more and more types of stars. We give a number of selected examples where precision asteroseismology, but also asteroseismology based on few observables may lead to considerable improvement of stellar astrophysics in the near future.

Keywords. stars: abundances, stars: atmospheres, stars: early-type, stars: fundamental parameters, stars: interiors, stars: late-type, stars: oscillations, convection, diffusion, atomic data

1. Introduction

The term "precision asteroseismology" was coined by Steve Kawaler and Paul Bradley in the early 1990s, when they found considerable precision in their model fits to the oscillation spectra of pulsating DO and DB white dwarf stars (e.g., Kawaler & Bradley 1994). This work may be the first that yielded global stellar properties (for, obviously, a star other than the Sun) from the analysis of oscillation data only. Note that precise asteroseismology does not necessarily mean accurate asteroseismology, but when being able to match large numbers of observed oscillation frequencies unambiguously, there must be something fundamentally correct about the models used.

The range of topics to be covered in a contribution with the present title is wide. Hence, a rigorous selection of topics needs to be made. I have decided to restrict myself to a personally biased selection of problems related to stellar physics that may be solved in the near future thanks to asteroseismology. Other highly interesting topics, such as the study of pulsating stars as an end in itself, or what can be learned about companions to pulsating stars, or stellar aggregates that harbour them, remain ignored here.

2. Precision asteroseismology: selected applications

2.1. Internal rotation

A recent major breakthrough in the seismic study of stochastically excited pulsators concerned the determination of the interior rotation rates of some 300 red-giant stars (Mosser *et al.* 2012) in combination with the identification of their basic interior structure (inert or nuclear burning He-core, cf. Bedding *et al.* 2011). It turned out that the transition between these two phases coincides with rapid braking of core rotation, larger than the expansion of the core only would cause, and consistent with white dwarf star rotation rates. Theoretical investigations in this direction (Marques *et al.* 2013) suggest that rotationally induced angular momentum transfer is not efficient enough to explain this rotational braking. Efforts to obtain better resolved rotation profiles and to extend the sample of investigated stars are underway (e.g., Deheuvels *et al.*, in preparation). The analysis of year-long observations of solar-like oscillators is also expected to facilitate the study of differential interior rotation of main sequence stars (Metcalfe *et al.* 2012).

However, it is not only the stochastically driven oscillators that can be examined this way. There is such a potential for any type of pulsator that has modes sampling the deep interior and excited to observable amplitude. For instance, the existence of differential interior rotation has been proven for some β Cephei stars (e.g., Aerts *et al.* 2003, Pamyatnykh *et al.* 2004). It is also present in pulsating white dwarf stars (Kawaler *et al.* 1999), and model computations imply the possibility of rapid differential interior rotation in sdB pulsators (Kawaler & Hostler 2005). Whereas it is clear that it will not be possible to investigate a comparably large number of κ-driven pulsators for differential rotation, it is hoped that a consistent and accurate picture of angular momentum evolution over a large part of the stellar life cycle will be obtained in the near future.

2.2. *Convective cores*

For stars more massive than the Sun, the size of their convective cores is the most important parameter that determines their main sequence lifetimes. There have been several measurements of convective core sizes for β Cephei stars, expressed through the overshooting parameter α_{ov}. However, these do not yet result in a consistent picture (cf., e.g., Pamyatnykh *et al.* (2004): $\alpha_{\mathrm{ov}} < 0.12$ for ν Eri and Briquet *et al.* (2007): $\alpha_{\mathrm{ov}} = 0.44 \pm 0.07$ for θ Oph). Whereas it is not expected that α_{ov} is constant all over the H-R diagram, it should at least be similar for stars of similar mass and evolutionary state. Perhaps we deal with precise as opposed to accurate asteroseismology here.

A first direct detection of a convective core in a stochastically excited main sequence pulsator has recently been made (Silva Aguirre *et al.* 2013). It was found to extend beyond the Schwarzschild convective boundary, which could be a hint towards the presence of overshooting. However, the imprints of a convective core present in the main sequence stage of evolution can still be found in the oscillation spectra of subgiant stars (Deheuvels & Michel 2011), yielding another possibility to determine convective core sizes.

2.3. *Opacities*

An example of the differences between precise and accurate asteroseismology can be found in Daszyńska-Daszkiewicz & Walczak (2010): fitting β Cep star pulsation spectra with different opacity tables will yield different global parameters. More importantly, these authors showed (Daszyńska-Daszkiewicz & Walczak 2009) that the observed ratio of the bolometric flux to radius variation in hot pulsating stars can be used to constrain which opacities better reproduce the observations — and the results for the stellar interior may be different than for the surface regions. This topic is discussed in detail by Walczak (these proceedings).

3. Precise asteroseismology from few observables

It does not always require to fit large numbers of pulsation modes to obtain important asteroseismic results. This does of course not mean that this type of asteroseismology is not precise, this separation is rather a question of concept.

3.1. *The ϵ mechanism*

Pulsational driving via variations in nuclear energy generation has historically often been suspected to be a feasible mechanism (e.g., see Kawaler 1988), but observational confirmation has never been attained. Recently, this topic has reappeared in several different contexts. For instance, the long pulsation periods of a He-rich subdwarf B star could only be reproduced with an ϵ mechanism operating in unstable He-burning shells. It has also been suggested that a small fraction of the remnants of double He-white dwarf mergers should show such oscillations (Miller Bertolami *et al.* 2013).

In an attempt to explain the variability of Rigel, Moravveji *et al.* (2012) found that the ϵ mechanism in the H-burning shell would be able to excite high-order g-mode pulsations in blue supergiants. If so, asteroseismology of immediate core-collapse supernova progenitors would become possible.

The ϵ mechanism may also be responsible for low-order g-mode excitation in some GW Vir stars (see Córsico *et al.* 2009), but an observational detection has not yet been convincingly been made. Other recent theoretical investigations showing the possibility of pulsations excited by the ϵ mechanism comprise Sonoi & Shibahashi (2009) for models of very low metallicity in the domain of the γ Dor stars, and Rodríguez-López *et al.* (2012) for models of M dwarfs.

3.2. *Further sources of excitation*

The rapidly oscillating Ap stars are believed to be excited by the κ mechanism in the H/He I ionization zone. However, theoretical calculations of their instability strip (Cunha 2002) do not match the observed locations of the stars well. In addition, a few stars show pulsations with frequencies higher than the acoustic cut-off frequency.

Similarly, some δ Scuti stars have very large ranges of oscillation frequencies excited that are too wide to be driven only by the κ mechanism in the He II ionization zone (e.g., Antoci *et al.* 2011). These problems may be resolvable if the turbulent pressure in the near-surface convection zone acted as an additional driving agent (e.g., see Antoci, these proceedings).

Atomic diffusion may also add to pulsational driving (e.g. in main sequence B stars, sdB pulsators or γ Dor stars) by producing suitable accumulation of elements to enhance the κ effect (e.g., see Théado *et al.* 2009). A new opacity bump has recently been identified (Cugier 2012) from improved model atmosphere data and its effects on stellar pulsation spectra are being investigated (Cugier, these proceedings).

3.3. *Surface convection*

Three-dimensional simulations (Trampedach & Stein 2011) allow to predict the mixing-length parameter of surface convection in the cooler part of the H-R diagram. These results can be tested by model calculations based on the large frequency separation and frequency of maximum power observed in stochastically excited pulsators (Bonaca *et al.* 2012). From the latter study it is evident that there are still problems. First, the rather strong dependence of the mixing-length parameter on temperature predicted by the models could not yet be confirmed. Additionally, using the solar value for the mixing length instead, a problem with the resulting helium abundance arises: it is in many cases below the primordial one. Whatever the solution to this dilemma, a better understanding of surface convection or other physics of solar-like stars is to be expected.

Acknowledgements

I am grateful to the scientific organizers of this meeting for entrusting me with this presentation. My attendance of the conference has been supported by the Polish NCN grant 2011/01/B/ST9/05448.

References

Aerts, C., Thoul, A., Daszyńska, J., *et al.* 2003, *Science*, 300, 1926

Antoci, V., Handler, G., Campante, T., *et al.* 2011, *Nature*, 477, 570

Bedding, T. R., Mosser, B., Huber, D., *et al.* 2011, *Nature*, 471, 608

Bonaca, A., Tanner, J. D., Basu, S., *et al.* 2012, *ApJ*, 755, L12

Briquet, M., Morel, T., Thoul, A., *et al.* 2007, *MNRAS*, 381, 1482

Córsico, A. H., Althaus, L. G., Miller Bertolami, M. M., González Pérez, J. M., & Kepler, S. O. 2009, *ApJ*, 701, 1008

Cugier, H. 2012, *A&A*, 547, A42

Cunha, M. S. 2002, *MNRAS*, 333, 47

Daszyńska-Daszkiewicz, J. & Walczak, P. 2009, *MNRAS*, 398, 1961

Daszyńska-Daszkiewicz, J. & Walczak, P. 2010, *MNRAS*, 403, 496

Deheuvels, S. & Michel, E. 2011, *A&A*, 535, A91

Kawaler, S. D. 1988, *ApJ*, 334, 220

Kawaler, S. D. & Bradley, P. A. 1994, *ApJ*, 427, 415

Kawaler, S. D. & Hostler, S. R. 2005, *ApJ*, 621, 432

Kawaler, S. D., Sekii, T., & Gough, D. O. 1999, *ApJ*, 516, 349

Marques, J. P., Goupil, M. J., Lebreton, Y., *et al.* 2013, *A&A*, 549, A74

Metcalfe, T. S., Chaplin, W. J., Appourchaux, T., *et al.* 2012, *ApJ*, 748, L10

Miller Bertolami, M. M., Córsico, A. H., Zhang, X., Althaus, L. G., & Jeffery, C. S. 2013, in: J. Montalbán, A. Noels, & V. Van Grootel (eds.) *Ageing Low Mass Stars: From Red Giants to White Dwarfs*, EPJ Web of Conferences, Vol. 43, id. 04004

Moravveji, E., Moya, A., & Guinan, E. F. 2012, *ApJ*, 749, 74

Mosser, B., Goupil, M. J., Belkacem, K., *et al.* 2012, *A&A*, 548, A10

Pamyatnykh, A. A., Handler, G., & Dziembowski, W. A. 2004, *MNRAS*, 350, 1022

Rodríguez-López, C. & MacDonald, J., Moya A. 2012, *MNRAS*, 419, L44

Silva Aguirre, V., Basu, S., Brandão, I. M., *et al.* 2013, *ApJ*, 769, 141

Sonoi, T. & Shibahashi, H. 2012, *MNRAS*, 422, 2642

Théado, S., Vauclair, S., Alecian, G., & LeBlanc, F. 2009, *ApJ*, 704, 1262

Trampedach, R. & Stein, R. F. 2011, *ApJ*, 731, 78

Precision Asteroseismology
Proceedings IAU Symposium No. 301, 2013　　　　© International Astronomical Union 2014
J. A. Guzik, W. J. Chaplin, G. Handler & A. Pigulski, eds.　　　　doi:10.1017/S1743921313014026

Pulsating variables from the OGLE and Araucaria projects

G. Pietrzyński[1,2]

[1]Warsaw University Observatory, Al. Ujazdowskie 4, 00-478 Warszawa, Poland
email: pietrzyn@astrouw.edu.pl

[2]Departamento de Astronomía, Universidad de Concepción, Casilla 160-C, Concepción, Chile

Abstract. We present some results of long-term studies of pulsating stars conducted in the course of the OGLE and Araucaria projects. In particular, very scarce eclipsing binaries containing pulsating stars are discussed. Such systems provide a unique opportunity to improve calibration of the cosmic distance scale and to better calibrate stellar evolutionary models.

1. OGLE and Araucaria projects

Optical Gravitational Lensing Experiment (OGLE) is one of the biggest astronomical surveys. This project has been in operation over the last 21 years. With the current instrumental setup, the OGLE team is capable to observe about one billion stars every night. The observations are mainly conducted in the Milky Way (bulge and disc) and Magellanic Clouds. Based on the collected data, an enormous amount of precise light curves for basically all kinds of pulsating stars were already published: Cepheids (Soszyński *et al.* 2008, 2010a), RR Lyrae (Soszyński *et al.* 2010b, 2011a), long-period variables (Soszyński *et al.* 2011b, 2013), etc., and several new catalogs are in preparation. Thanks to an exquisite statistics and high quality of the data, many extremely scarce and very interesting objects have been also discovered, including very good candidates for pulsating stars in eclipsing binary systems (Soszyński *et al.* 2008, 2011a) and a unique sample of eclipsing binary systems composed of clump giants (Graczyk *et al.* 2011).

Araucaria project. The main goal of this project is to significantly improve the calibration of the cosmic distance scale based on observations of several distance indicators in nearby galaxies. As a first step, we performed an optical survey of nine nearby galaxies discovering about 700 Cepheids (Pietrzyński *et al.* 2002, 2004, 2006, 2007, 2010a). We also performed infrared observations of discovered Cepheids (Gieren *et al.* 2005b, 2006, 2009, 2013; Pietrzyński *et al.* 2006), and RR Lyrae stars selected from the OGLE catalogs in the LMC, SMC and Sculptor galaxies (Pietrzyński *et al.* 2008, Szewczyk *et al.* 2008, 2009). Recently, a very important part of the Araucaria project is related to detailed studies of eclipsing binary systems discovered by the OGLE project, which have very large potential for precise and accurate distance determination and also for improving our knowledge of basic physics of pulsating stars. During my talk, I will focus on some results obtained for some of these systems.

2. Setting the zero point for P-L relation of pulsating stars

The distance to the LMC is widely adopted as the zero point in the calibration of the cosmic distance scale. Therefore, a precise and accurate distance determination to this galaxy is paramount for astrophysics in general.

Detached eclipsing double-lined spectroscopic binaries offer a unique opportunity to measure distances directly (e.g. Kruszewski & Semeniuk 1999). Recently, we measured distance to the LMC with 2% precision, based on the analysis of eclipsing binary systems composed of red clump giants (Pietrzyński *et al.* 2013). For eight such systems linear sizes of both components were measured with a 1% accuracy based on modeling of high-quality photometric light curves obtained by the OGLE project and radial velocity curves constructed by the Araucaria team. Having first ever discovered late-type eclipsing binaries (G type), we used a well calibrated relationship between angular diameter and $V - K$ color (Di Benedetto 2005, Kervella *et al.* 2004) and measured corresponding angular sizes of the components of our systems with an accuracy of 2%. As the result we obtained the most accurate and reliable LMC distance, which provides a strong basis for the determination of the Hubble constant with an accuracy of about 3%. At present, we are working on improving the surface brightness-color calibration and measuring the LMC distance with an accuracy of 1%.

Since the OGLE group constructed outstanding period-luminosity (P-L) relations for several different pulsating stars in the Large Magellanic Cloud (e.g. Soszyński *et al.* 2010a, 2011b, 2013), our distance determination provides also an opportunity to calibrate uniform fiducial P-L relations for basically all P-L relations of pulsating stars used for distance determination.

3. Cepheids in eclipsing binaries

The OGLE project provided also very good candidates for classical Cepheids in eclipsing systems (e.g. Soszyński *et al.* 2008). Such systems provide a unique opportunity to measure precisely and accurately stellar parameters of Cepheids. In consequence, they provide very strong constraints on stellar evolutionary and pulsation models. One can also measure distances to such targets using three independent techniques: P-L relation, Baade-Wesselink method, and eclipsing binaries described above. Comparing the independent distances, the potential systematic errors associated with each of these methods can be precisely traced out. Because of the huge potential of these systems for improving our capability of measuring distances with classical Cepheids and to better understand basic physics of these stars, as a part of the Araucaria project we started a long-term program to characterize them. In 2010 we confirmed that one of the OGLE candidates — OGLE-LMC-CEP-227 — is indeed a physical system containing a Cepheid. Based on high-quality data, we measured the dynamical mass of the Cepheid with an accuracy of 1% (Pietrzyński *et al.* 2010b). Recently, we significantly improved the accuracy of determination of the physical parameters of this system and measured directly the p factor for the Cepheid (Pilecki *et al.* 2013). This analysis complements our previous study on the calibration of the p factor (Gieren *et al.* 2005a, Nardetto *et al.* 2011, Storm *et al.* 2011). Pietrzyński *et al.* (2011) measured the dynamical mass of another Cepheid in an eclipsing system with similar accuracy. These results already triggered several theoretical investigations (Cassisi & Salaris 2011, Neilson *et al.* 2011, Prada Moroni *et al.* 2012, Marconi *et al.* 2013).

Unfortunately, there are no Cepheids in eclipsing binary systems known so far in the Milky Way. However, some of the Cepheids in binary systems are sufficiently close to observe them interferometrically (Gallenne *et al.* 2013). Combining spectroscopic and interferometric data one should be also able to precisely measure distances, masses and other physical parameters for several Cepheids in the Milky Way (Gallenne *et al.* 2013; see also Gallenne *et al.*, these proceedings).

4. Binary Evolution Pulsating (BEP) stars

After many years of intense searching for an RR Lyrae star in an eclipsing binary system, a very good candidate, OGLE-BLG-RRLYR-02792, was discovered by the OGLE team (Soszyński *et al.* 2010b). Together with high-resolution spectra obtained by the Araucaria team, we modeled this system and determined its physical parameters. Surprisingly, the mass of the primary-component pulsating star classified as an RR Lyrae star based on properties of its light curve, turned out to be $0.261 \pm 0.015\,M_{\odot}$ (Pietrzyński *et al.* 2012). This is of course incompatible with the theoretical predictions for a classical horizontal-branch star evolving through the instability strip. However, the relatively short orbital period of this system (15.24 days) suggests that mass exchange between the components should have occurred during its evolution.

Inspired by this possibility, we calculated a series of models for Algol systems and found that a system which initially contained two stars with masses $M_1 = 1.4\,M_{\odot}$ and $M_2 = 0.8\,M_{\odot}$ orbiting each other with an initial period of 2.9 days would, after 5.4 Gyr of evolution, have exchanged mass between the components as classical Algols do, and today would form a system very similar to RRLYR-02792 (e.g. with $M_1 = 0.268\,M_{\odot}$, $M_2 = 1.665\,M_{\odot}$, and $P_{\rm orb} = 15.9$ days). We therefore conclude that the primary component of our observed system is not a classical RR Lyrae star with its well-known internal structure, but rather a star which possesses a partially degenerated helium core and a small hydrogen-rich envelope (shell burning). The primary component has lost most of its envelope to the secondary during the red giant branch phase due to the mass exchange in a binary system, and is now crossing the main instability strip during its evolution towards the hot subdwarf region of the H-R diagram. Pulsational properties of this star were investigated by Smolec *et al.* (2013). As a result, we discovered a new evolutionary channel of producing binary evolution pulsating (BEP) stars — new inhabitants of the classical instability strip. They mimic classical RR Lyrae variables, but have a completely different origin. Since the primary components can have very different masses during the mass exchange, they can be expected in different regions of the instability strip. Very recently, Maxted *et al.* (2013) discovered another BEP star crossing the instability strip of δ Scuti stars. Since close binary systems composed of two intermediate-mass stars orbiting each other with periods of a few days are relatively frequent, the newly discovered evolutionary channel should produce a significant fraction of the white dwarfs in the Universe.

References

Cassisi, S. & Salaris, M. 2011, *ApJ*, 728, L43
Di Benedetto, G. P. 2005, *MNRAS*, 357, 174
Gallenne, A., Monnier, J. D., Mérand, A., *et al.* 2013, *A&A*, 552, A21
Gieren, W. Storm, J., Barnes, T. G., III, Fouqué, P., Pietrzyński, G., & Kienzle, F. 2005a, *ApJ*, 627, 224
Gieren, W., Pietrzyński, G., Soszyński, I., *et al.* 2005b, *ApJ*, 628, 695
Gieren, W., Pietrzyński, G., Nalewajko, K., *et al.* 2006, *ApJ*, 647, 1056
Gieren, W., Pietrzyński, G., Soszyński, I., *et al.* 2009, *ApJ*, 700, 1141
Gieren, W., Górski, M., Pietrzyński, G., *et al.* 2013, *ApJ*, 773, 69
Graczyk, D., Soszyński, I., Poleski, R., *et al.* 2011, *AcA*, 61, 103
Kervella, P., Thévenin, F., Di Folco, E., & Ségransan, D. 2004, *A&A*, 426, 297
Kruszewski, A. & Semeniuk, I. 1999, *AcA*, 49, 561
Marconi, M., Molinaro, R., Bono, G., *et al.* 2013, *ApJ*, 768, L6
Maxted, P. F. L., Serenelli, A. M., Miglio, A., *et al.* 2013, *Nature*, 498, 463
Nardetto, N., Mourard, D., Tallon-Bosc, I., *et al.* 2011, *A&A*, 525, 67

Neilson, H. R., Cantiello, M., & Langer, N. 2011, *A&A*, 529, L9

Pietrzyński, G., Gieren, W., Fouqué, P., & Pont, F. 2002, *AJ*, 123, 789

Pietrzyński, G., Gieren, W., Udalski, A., *et al.* 2004, *AJ*, 128, 2815

Pietrzyński, G., Gieren, W., Soszyński, I., *et al.* 2006, *AJ*, 132, 2556

Pietrzyński, G., Gieren, W., Udalski, A., *et al.* 2007, *AJ*, 134, 594

Pietrzyński, G., Gieren, W., Szewczyk, O., *et al.* 2008, *AJ*, 135, 1993

Pietrzyński, G., Gieren, W., Hamuy, M., *et al.* 2010a, *AJ*, 140, 1475

Pietrzyński, G., Thompson, I. B., Gieren, W., *et al.* 2010b, *Nature*, 468, 542

Pietrzyński, G., Thompson, I. B., Graczyk, D., *et al.* 2011, *ApJ*, 742, L20

Pietrzyński, G., Thompson, I. B., Gieren, W., *et al.* 2012, *Nature*, 484, 75

Pietrzyński, G., Graczyk, D., Gieren, W., *et al.* 2013, *Nature*, 495, 76

Pilecki, B., Graczyk, D., Pietrzyński, G., *et al.* 2013, *MNRAS*, 436, 953

Prada Moroni, P. G., Gennaro, M., Bono, G., *et al.* 2012, *ApJ*, 749, 108

Smolec, R., Pietrzyński, G., Graczyk, D., *et al.* 2013, *MNRAS*, 428, 3034

Soszyński, I., Poleski, R., Udalski, A., *et al.* 2008, *AcA*, 58, 163

Soszyński, I., Poleski, R., Udalski, A., *et al.* 2010a, *AcA*, 60, 17

Soszyński, I., Udalski, A., Szymański, M. K., *et al.* 2010b, *AcA*, 60, 165

Soszyński, I., Dziembowski, W. A., Udalski, A., *et al.* 2011a, *AcA*, 61, 1

Soszyński, I., Udalski, A., Szymański, M. K., *et al.* 2011b, *AcA*, 61, 217

Soszyński, I., Udalski, A., Pietrukowicz, P., *et al.* 2013, *AcA*, 63, 37

Storm, J., Gieren, W., Fouqué, P., *et al.* 2011, *A&A*, 534, A94

Szewczyk, O., Pietrzyński, G., Gieren, W., *et al.* 2008, *AJ*, 136, 272

Szewczyk, O., Pietrzyński, G., Gieren, W., Ciechanowska, A., Bresolin, F., & Kudritzki, R.-P. 2009, *AJ*, 138, 1661

Precision Asteroseismology
Proceedings IAU Symposium No. 301, 2013
J. A. Guzik, W. J. Chaplin, G. Handler & A. Pigulski, eds.

© International Astronomical Union 2014
doi:10.1017/S1743921313014038

A review of pulsating stars
from the ASAS data

Andrzej Pigulski

Instytut Astronomiczny, Uniwersytet Wrocławski, Kopernika 11, 51-622 Wrocław, Poland
email: `pigulski@astro.uni.wroc.pl`

Abstract. The All-Sky Automated Survey (ASAS) appeared to be extremely useful in establishing the census of bright variable stars in the sky. A short review of the characteristics of the ASAS data and discoveries based on these data and related to pulsating stars is presented here by an enthusiastic user of the ASAS data.

Keywords. stars: pulsating, databases

1. Introduction

The last two decades witnessed the onset of a large number of wide-field photometric surveys accomplished with small robotic telescopes or even telephoto lenses. There were several good reasons for that: (i) the invention of large CCD detectors, (ii) the invention of appropriate software enabling robotisation of small telescopes, (iii) the development of wide-band Internet, and finally (iv) a good motivation for this kind of scientific research. As far as the latter reason is concerned, some projects were focused on narrow scientific topics such as a fast (on a timescale of seconds) optical follow-up of γ-ray bursts (e.g. ROTSE; Akerlof *et al.* 2003) or a detection of Kuiper Belt asteroids (TAOS; Chen 2002). There were, however, projects that from the beginning were aimed at monitoring a large part of the sky for variability. The idea of monitoring the whole sky for variability is very old, but it was Prof. Bohdan Paczyński who indicated how many scientific areas can benefit from this kind of observations. For many years he used to encourage observers to set up telescopes that would make such monitoring (Paczyński 1997, 2000). The All-Sky Automated Survey (ASAS, Pojmański 1997, 2009) is probably the best example of a response for this call, though other projects such as the Hungarian Automated Survey (HAT, Bakos *et al.* 2002) can also be regarded as Paczyński's legacy.

Given the characteristics of the ASAS data (see Sect. 3), it is obvious that they can be used mainly for the studies of those pulsating stars that have the largest amplitudes. Out of many classes of pulsating stars, classical pulsators (all types of Cepheids, RR Lyrae stars), high-amplitude δ Scuti (HADS) stars and Miras should be easily searched for in the ASAS data. On the other hand — since the project is ongoing — the growing number of data points leads to a lower and lower detection threshold and enables also studies of pulsating stars with lower amplitudes, like δ Scuti stars with small amplitudes and β Cephei stars. In this paper, we characterise the ASAS data to show their limitations and provide a short review of the up-to-date searches for pulsating stars with these data.

2. The ASAS project and catalogs

The ASAS project started in 1996 with observations of selected fields in the southern sky (Pojmański 1997). The first two phases of the project, ASAS-1 and 2, resulted in the discovery of about 3800 variable stars (Pojmański 1998, 2000). In 2000, the third phase of the project, ASAS-3, was initiated. It lasted ten seasons, from 2000 until 2009. With the change of CCDs in 2010, the project entered the fourth phase, ASAS-4. ASAS-3 and 4 cover about 70 percent of the sky, for declinations $\delta < +28°$ (Pojmański 2001). A detailed description of the equipment located at Las Campanas Observatory, Chile, can be found in the cited papers and the ASAS web page† and will not be repeated here. The observations are conducted in two filters, V and I. Starting in 2006, the project was extended to the northern hemisphere (ASAS-3 North) by placing a telescope at Maui, Hawaii. The telescope covers the northern hemisphere with a large overlap with observations from the southern site. Since 2006, ASAS is therefore a real all-sky survey. Recently, another project related to ASAS has been initiated. This is the All-Sky Automated Survey for Supernovae (ASAS-SN)‡ aimed at detecting bright supernovae and other transient phenomena in the entire sky.

The analysis of the acquired ASAS data resulted in the publication of three catalogs of variable stars. The first one, published by Pojmański (2000), contains nearly 3800 variable stars. It resulted from the analysis of the I-filter ASAS-2 data. Variable stars in this catalog were divided into two broad categories: periodic and miscellaneous. The second catalog is the largest one. It is the ASAS Catalog of Variable Stars (ACVS) containing results of the analysis of the ASAS-3 V-filter observations. It was published in five parts (Pojmański 2002, 2003; Pojmański & Maciejewski 2004, 2005; Pojmański *et al.* 2005). The catalog contains about 50 000 variable stars that were classified automatically with the use of the following information: periods, amplitudes, Fourier coefficients, 2MASS colours and *IRAS* infrared fluxes. Thirteen classes of variability were defined including nine for pulsating stars: classical Cepheids, fundamental ($DCEP_{FU}$), and first-overtone ($DCEP_{FO}$), W Vir stars (CW), anomalous Cepheids (ACEP), RR Lyrae stars of Bailey type ab (RRAB) and c (RRC), Miras (M), δ Scuti (DSCT) and β Cephei (BCEP) stars. Multiple classifications were allowed. It is worth noting, however, that there are also other, independent classifications of the ASAS-1/2 (Eyer & Blake 2005) and the ACVS (Richards *et al.* 2012).

Finally, the third catalog is the catalog of variable stars in the *Kepler* field. It resulted from the analysis of the first 17 months of VI observations by the ASAS-3 North project. The catalog contains only 947 stars (Pigulski *et al.* 2009) classified with a slightly different scheme than that applied in the ACVS.

It is worth noting that both I-filter ASAS-2 data and V-filter ASAS-3 data are available not only for stars from the ASAS catalogs of variable stars, but for all 15×10^9 observed stars, allowing different in-depth variability searches.

3. Characteristics of the ASAS data

The way the observations are carried out in the ASAS project has a large impact on the variability we can expect to discover and is subject to several limitations that are inherent to these data. A potential user of the ASAS data must be aware of those characteristics, so it is reasonable to list the most important ones. They are the following:

DEPTH OF THE SURVEY. For the V-filter observations that were used to produce the ACVS, a typical exposure times lasted 3 minutes. This allowed one to obtain reasonable photometry for stars in the V magnitude range between 7 and 14, albeit some stars brighter than $V = 7$ mag and fainter than $V = 14$ mag can also be found in the catalogs. The best photometry, with a typical scatter of about 0.01 mag, was obtained for stars in

† http://www.astrouw.edu.pl/asas
‡ http://www.astronomy.ohio-state.edu/~assassin/index.shtml

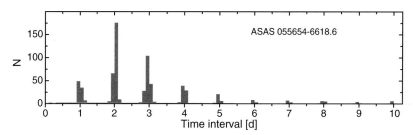

Figure 1. Histogram of time intervals between consecutive observations for one of the ASAS targets.

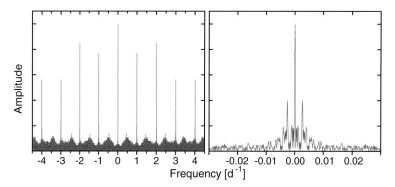

Figure 2. *Left:* Window function in the frequency interval $[-4.5, 4, 5]$ d^{-1} for the ASAS data of the same star as in Fig. 1. *Right:* The same as in the left panel, but in the frequency interval $[-0.03, 0.03]$ d^{-1} to show the yearly aliases at ± 0.0027 d^{-1}. Amplitude is given in arbitrary units.

the magnitude range between 8 and 10. With almost ten years of ASAS-3 observations, the detection limit for periodic signals is, however, much lower than 0.01 mag. As can be judged from Fig. 16 of Pigulski & Pojmański (2008b), signals with amplitudes as low as 4–5 mmag can be detected. This shows that variable stars with low amplitudes can also be studied with the ASAS data (see Sect. 4.4).

SAMPLING AND ALIASING. The all-sky character of the survey makes the sampling of the ASAS data quite sparse. Typically, a single field was observed every one, two or three days. The histogram of time intervals between consecutive observations for one of the ASAS targets is shown in Fig. 1. One may suppose that this type of observing would lead to problems with unambiguous determination of frequencies for stars with periods much shorter than 1 day, e.g. δ Scuti stars (with one-day sampling the Nyquist frequency amounts to 0.5 d^{-1}). Fortunately, this is not the case for the ASAS data: not counting the alias problem (see below) there is no difficulty in identifying a correct frequency in frequency spectra of the ASAS data. This is because the data *are not exactly* evenly spaced.

The sampling shown in Fig. 1 results — as one may expect — in very strong daily aliases in Fourier spectra calculated for the ASAS data. An example of a spectral window, for the same star as in Fig. 1, is shown in Fig. 2. One can easily recognize very strong daily aliases, a consequence of the very short interval during the nights when the star was observed. On the other hand, the yearly aliases (right panel) are quite low; they reach about 40 percent of the height of the central peak. This is the consequence of the observing strategy: a given star was observed whenever possible so that usually longer gaps in data (but still lasting a few months at most) occur only for stars located close to the ecliptic. The alias pattern seen in Fig. 2 has a direct consequence for the study of short-period pulsators with the ASAS data. While short follow-up observations can easily resolve the problem with daily aliases we get from these data, the long interval they cover allows to remove ambiguities related to yearly aliases from other observations. An example of the advantage of using ASAS data for this purpose is the β Cephei-type star V403 Car (NGC 3293-16). The ASAS data helped to identify the correct frequency in this star, which was subject to the yearly aliasing problem (Pigulski & Pojmański 2008a).

FREQUENCY RESOLUTION. As a consequence of over 9 years (3300 days) of observing within the ASAS-3 project, we obtain very good frequency resolution of 0.0003 d^{-1} in the frequency spectra of the ASAS data. This is not the only advantage of the ASAS data in terms of the length of observations. Such long observations can be used to monitor secular changes of amplitudes and pulsation periods. Examples can be found in the papers by Pigulski & Pojmański (2008a) and Wils *et al.* (2007); see also Sect. 4.1.

SPATIAL RESOLUTION. The large fields observed within the ASAS project resulted obviously in a poor spatial resolution. For the equipment used to obtain ASAS-3 *V*-filter data, the scale amounted to about 15.5 arcsec per pixel. In addition, the ASAS magnitudes originate from aperture photometry. The data come in five different apertures with a diameter of 2–6 pixels corresponding to 0.5–1.5 arcmin on the sky. In some conditions, the problem of contamination can be resolved by analysing data in different apertures. However, this is not always possible and only follow-up observations can solve the problem. For example, in their search for pulsating components of eclipsing binaries using the ASAS data, Pigulski & Michalska (2007) found that the massive eclipsing O6 V((f)) + early B-type binary ALS 1135 shows, in addition to eclipses, periodic variability with a frequency of 2.31 d^{-1}. The frequency was tentatively attributed to pulsations of one of the components. Unfortunately, follow-up observations showed (Michalska *et al.* 2013) that it is a nearby eclipsing binary, unresolved in the ASAS frames, that is responsible for this variability.

4. Pulsating stars in the ASAS data

The algorithm used to select variable stars for the ASAS-2 and ACVS catalogs was based on the magnitude-dispersion relation and therefore was biased toward large-amplitude stars. However, many more variables, especially with small amplitudes could be found using the ASAS data (see, e.g., Sect. 4.4 or David *et al.* 2013). Given the limited information used for the automatic classification of variable stars in the ASAS catalogs, no-one can expect that the result will be unambiguous. The variability needs to be verified using other sources of information, e.g. spectral types. The ASAS classification is therefore a good starting point for the study of a given class of variable stars, but needs to be verified.

4.1. Cepheids

The ACVS contains 1669 stars with a first classification assigned to one of the four types of Cepheids (see Sect. 2). There is no complete verification of the classification in this group, but large samples were included in several follow-up programs aimed mainly at the study of the structure of our Galaxy. From multi-colour photometry (Berdnikov *et al.* 2009a, 2009b, 2011, Schmidt *et al.* 2009, Schmidt 2013) and spectroscopy (Schmidt *et al.* 2011) it became clear that genuine Cepheids amount to considerably less than a half of stars classified as such in the ACVS. Nevertheless, the ASAS data increased considerably the number of known bright Galactic Cepheids and triggered several interesting projects. For example, the number of known Galactic double-mode Cepheids was increased almost twofold with the ASAS data (e.g. Wils & Otero 2004, Khruslov 2009) and includes now almost 40 members; for a complete list of Galactic double-mode Cepheids see the references in Sect. 3 of the paper by Smolec & Moskalik (2010).

The ASAS data appeared also to be important in studies of the brightest (i.e. long-period) Cepheids in the Magellanic Clouds. In particular, combined with archival data spanning a century or so, they were used to study secular evolutionary period changes in these Cepheids (Pietrukowicz 2001, 2002, Karczmarek *et al.* 2011). Due to the overlap in brightness with the OGLE survey, these observations allowed also to supplement the long-period tail of the period-luminosity relation for Cepheids in the Large Magellanic Cloud, only recently completed by observations from the OGLE itself (Ulaczyk *et al.* 2013).

4.2. *RR Lyrae stars*

Of over 2200 stars classified as RR Lyrae stars in the ACVS, about 66 percent are RRab stars recognizable by their large amplitudes and characteristic light curves with a steep ascending branch. For this reason, the automatic ACVS classification for these stars worked very well. This was not the case for the other group of RR Lyrae stars, of type RRc. Their light curves are more sinusoidal in shape and for faint stars with a large scatter they are barely distinguished from very common W UMa-type over-contact eclipsing binaries (see, e.g. Kinman & Brown 2010). The most thorough analysis of the ASAS RR Lyrae stars was done by Szczygieł & Fabrycky (2007) and Szczygieł *et al.* (2009). Szczygieł *et al.* (2009) analysed only the ASAS RRab sample of stars, finding a clear manifestation of the Oosterhoff dichotomy among them in the period-amplitude diagram. Szczygieł & Fabrycky (2007) focused on the study of multiperiodicity among these stars, finding over a hundred stars (5.5 percent of the whole sample) showing the Blazhko effect. They also indicated a bi-modal distribution of Blazhko periods in RRc stars (see also Skarka, these proceedings). Finally, they found four new double-mode fundamental/first overtone, (F/1O), i.e., RRd stars. At present, the total number of RRd stars found with the ASAS data amounts to 22, i.e. about 40 percent of all known Galactic stars of this type. Two likely first overtone/second overtone (1O/2O) RR Lyrae stars were also found using the ASAS data (Khruslov 2010, 2012). In order to illustrate the importance of the ASAS project in studying classical pulsators, we show in Fig. 3 the Petersen diagram for three classes of classical pulsators that we discuss here (Sect. 4.1–4.3).

4.3. *High-amplitude δ Scuti (HADS) stars*

The HADS stars form the third group of large-amplitude pulsators located in the classical instability strip. Like Cepheids and RR Lyrae stars, they are recognized by dominating radial modes. SX Phe stars, their Population II counterparts, are strongly related to this group and distinguished only by much lower metallicities. Obviously, in the ACVS the HADS stars could be found mainly among 1275 stars classified as δ Scuti stars. However, we carried out our own search for these stars in the whole ACVS database. We found 148 HADS stars including 18 double-mode HADS (Fig. 3) of which eight are new discoveries. The full results of this study will be published elsewhere (Pigulski, in preparation).

4.4. *Stars with low amplitudes*

Probably the best illustration of the fact that ASAS data can also be used to study low-amplitude pulsating stars are discoveries related to β Cephei stars. Of 40 stars classified as β Cephei stars in the ACVS only 12 were confirmed as such (Pigulski 2005). On the other hand, six other β Cephei stars were classified differently in the ACVS. Even more

A. Pigulski

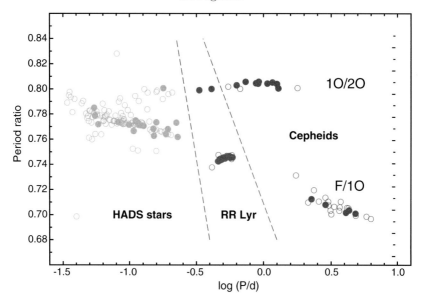

Figure 3. Petersen diagram for known double-mode Galactic classical pulsators: HADS stars (green in the on-line version), RR Lyrae stars (red on-line), and classical Cepheids (blue on-line). Two clear sequences of the F/1O and 1O/2O pulsators can be seen. Filled symbols denote stars that were discovered using the ASAS data.

importantly, Pigulski (2005) recognized that all β Cephei stars included in the ACVS have amplitudes exceeding $\sim$35 mmag whereas a detection threshold of 5–10 mmag can be achieved. In consequence, a more thorough search for β Cephei stars was carried out leading to the discovery of about 300 β Cephei stars in the ASAS-3 data (Pigulski & Pojmański 2008b, 2009, Pigulski & Pojmański, in preparation) which means a fourfold increase in the number of known stars of this type. Five β Cephei stars were also found using the ASAS-2 data (Handler 2005).

There are other types of low-amplitude pulsating stars that were not classified within the ACVS classification scheme. Can we detect them and study with the ASAS data? For at least two types, slowly pulsating B (SPB) stars and γ Doradus stars, the answer is obvious: yes, we can. They have periods ranging from about half a day to several days and amplitudes comparable to β Cephei stars. We may therefore expect to detect them easily. To illustrate this, we show frequency spectra for two γ Dor stars discovered in the ASAS data (Fig. 4). The only complication for SPB and γ Dor stars is strong aliasing which for frequencies below 2–3 d^{-1} is more severe than for δ Sct or β Cep stars due to the presence of additional aliases mirrored from negative frequencies.

Unfortunately, there is only a small chance to detect pulsating white dwarfs, hot sub-dwarfs, and rapidly oscillating Ap stars in the ASAS data, for the following reasons. First, the detection threshold of a few mmag for the ASAS data may be still much higher than the amplitudes observed in most of these pulsators. This is also the case for stars showing solar-like oscillations. Second, the short periods they exhibit are comparable to the exposure time for the ASAS-3 data (3 minutes). Due to averaging, this leads to a considerable reduction of the observed amplitudes. However, in favourable conditions (long periods, high amplitudes) even stars of the types mentioned above can be detected with the ASAS data.

ASAS variable stars

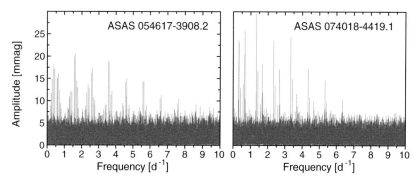

Figure 4. Frequency spectra of two γ Doradus-type stars discovered in the ASAS data: ASAS 054617−3908.3 (CD −39°2175, left) and ASAS 074018−4419.1 (HD 62110, right). Four modes, with frequencies equal to 1.62557, 1.57170, 1.71452, and 1.76774 d^{-1} were detected in the former star and three (frequencies of 1.33265, 1.42350, and 1.70049 d^{-1}) in the latter.

5. Conclusions

We are entering the golden age of wide-field surveys of different depth and extent that will provide an enormous amount of photometric data suitable for different kinds of studies. The ASAS project we have briefly summarized here already showed how the subject of pulsating stars can benefit from this type of project. With the ASAS and other all-sky surveys, different types of statistical studies become possible. Such surveys are also ideal for selecting best targets for follow-up studies and targets that may help to understand different aspects of pulsations. The ASAS survey already triggered a lot of follow-up studies. Some examples were presented even during this conference (Drobek & Pigulski, Jurković & Szabados, these proceedings). Next, the almost uninterrupted data provided by ASAS over a long time base allow for different studies of secular changes of amplitudes and periods including evolutionary changes or dynamical evolution and multiplicity of stars via the light-time effect (Pilecki *et al.* 2007).

The ASAS data (also ACVS) are still unexploited and await new discoveries. We believe that digging into these data will bring as much fun as was experienced by the author of this note.

Acknowledgements

I would like to thank Grzegorz Pojmański for his cooperation in working on the ASAS data, of which I am an enthusiastic user. This work was supported by the NCN grant 2011/03/B/ST9/02667.

References

Akerlof, C. W., Kehoe, R. L., McKay, T. A., *et al.* 2003, *PASP*, 115, 132

Bakos, G.Á., Lázár, J., Papp, I., Sári, P., & Green, E. M. 2002, *PASP*, 114, 974

Berdnikov, L. N., Kniazev, A. Y., Kravtsov, V. V., Pastukhova, E. N., & Turner, D. G. 2009a, *Astron. Lett.*, 35, 39

Berdnikov, L. N., Kniazev, A. Y., Kravtsov, V. V., Pastukhova, E. N., & Turner, D. G. 2009b, *Astron. Lett.*, 35, 311

Berdnikov, L. N., Kniazev, A. Y., Sefako, R., Kravtsov, V. V., Pastukhova, E. N., & Zhuiko, S. V. 2011, *Astron. Rep.*, 55, 816

Chen, W. P. 2002, in: H. Rickman (ed.), Proc. XXIVth General Assembly of the IAU, (San Francisco: Astron. Soc. of the Pacific), *Highlights of Astronomy*, 12, 245

David, M., Hensberge, H., & Nitschelm, C. 2013, *A&A*, 557, A47

Eyer, L. & Blake, C. 2005, *MNRAS*, 358, 30

Handler, G. 2005, *IBVS*, 5667

Karczmarek, P., Dziembowski, W. A., Lenz, P., Pietrukowicz, P., & Pojmański, G. 2011, *AcA*, 61, 303

Khruslov, A. V. 2009, *Peremennye Zvezdy Prilozhenie*, 9, No. 17

Khruslov, A. V. 2010, *Peremennye Zvezdy Prilozhenie*, 10, No. 28

Khruslov, A. V. 2012, *Peremennye Zvezdy Prilozhenie*, 12, No. 7

Kinman, T. D. & Brown, W. R. 2010, *AJ*, 139, 2014

Michalska, G., Niemczura, E., Pigulski, A., Steślicki, M., & Williams, A. 2013, *MNRAS*, 429, 1354

Paczyński, B. 1997, in: R. Ferlet, J.-P. Maillard & B. Raban (eds.), Proc. 12th IAP Astrophysics Meeting: *Variables Stars and the Astrophysical Returns of the Microlensing Surveys* (Gif-sur-Yvette: Editions Frontieres), p. 357

Paczyński, B. 2000, *PASP*, 112, 1281

Pietrukowicz, P. 2001, *AcA*, 51, 247

Pietrukowicz, P. 2002, *AcA*, 52, 177

Pilecki, B., Fabrycky, D., & Poleski, R. 2007, *MNRAS*, 378, 757

Pigulski, A. 2005, *AcA*, 55, 219

Pigulski, A. & Michalska, G. 2007, *AcA*, 57, 61

Pigulski, A. & Pojmański, G. 2008a, *A&A*, 477, 907

Pigulski, A. & Pojmański, G. 2008b, *A&A*, 477, 917

Pigulski, A. & Pojmański, G. 2009, *AIP-CS*, 1170, 351

Pigulski, A., Pojmański, G., Pilecki, B., & Szczygieł, D. M. 2009, *AcA*, 59, 33

Pojmański, G. 1997, *AcA*, 47, 467

Pojmański, G. 1998, *AcA*, 48, 35

Pojmański, G. 2000, *AcA*, 50, 177

Pojmański, G. 2001, *ASP-CS*, 246, 53

Pojmański, G. 2002, *AcA*, 52, 397

Pojmański, G. 2003, *AcA*, 53, 341

Pojmański, G. 2009, *ASP-CS*, 403, 52

Pojmański, G. & Maciejewski, G. 2004, *AcA*, 54, 153

Pojmański, G. & Maciejewski, G. 2005, *AcA*, 55, 97

Pojmański, G., Pilecki, B., & Szczygieł, D. 2005, *AcA*, 55, 275

Richards, J. W., Starr, D. L., Miller, A. A., et al. 2012, *ApJS*, 203, 32

Schmidt, E. G. 2013, *AJ*, 146, 61

Schmidt, E. G., Hemen, B., Rogalla, D., & Thacker-Lynn, L. 2009, *AJ*, 137, 4598

Schmidt, E. G., Rogalla, D., & Thacker-Lynn, L. 2011, *AJ*, 141, 53

Smolec, R. & Moskalik, P. 2010, *A&A*, 524, A40

Szczygieł, D. M. & Fabrycky, D. C. 2007, *MNRAS*, 377, 1263

Szczygieł, D. M., Pojmański, G., & Pilecki, B. 2009, *AcA*, 59, 137

Ulaczyk, K., Szymański, M. K., Udalski, A., et al. 2013, *AcA*, 63, 1

Wils, P. & Otero, S. A. 2004, *IBVS*, 5501

Wils, P., Otero, S. A., & Hambsch, F.-J. 2007, *IBVS*, 5765

Precision Asteroseismology
Proceedings IAU Symposium No. 301, 2013
J. A. Guzik, W. J. Chaplin, G. Handler & A. Pigulski, eds.

© International Astronomical Union 2014
doi:10.1017/S174392131301404X

Asteroseismology with SuperWASP

Barry Smalley

Astrophysics Group, Keele University, Staffordshire ST5 5BG, United Kingdom
email: b.smalley@keele.ac.uk

Abstract. The highly successful SuperWASP planetary transit finding programme has surveyed a large fraction of both the northern and southern skies. There now exists in its archive over 420 billion photometric measurements for more than 31 million stars. SuperWASP provides good quality photometry with a precision exceeding 1% per observation in the approximate magnitude range $9 < V < 12$. The archive enables long-baseline, high-cadence studies of stellar variability to be undertaken. An overview of the SuperWASP project is presented, along with results which demonstrate the survey's asteroseismic capabilities.

Keywords. asteroseismology, instrumentation: photometers, stars: chemically peculiar, stars: oscillations, stars: rotation, stars: variables: δ Scuti, stars: variables: roAp

1. Introduction

The Wide Angle Search for Planets (WASP) is one of the world's leading ground-based surveys for transiting exoplanets (Pollacco *et al.* 2006). The project has two robotic telescopes in roll-off roof enclosures, one at the Observatorio del Roque de los Muchachos on the island of La Palma in the Canary Islands, and the other at the Sutherland Station, South African Astronomical Observatory (SAAO). Both instruments consist of an array of eight 200-mm, f/1.8 Canon telephoto lenses and 2048×2048 Andor CCDs, provide a field of view of $7.8° \times 7.8°$ and pixel size of $13.7''$. A broad-band filter gives a 400–700 nm bandpass.

SuperWASP observes a set of pre-determined 'planet fields' each night, subject to their visibility and Moon avoidance. At each pointing, two sequential 30-second exposures are taken. The cameras return to same field with a typical cadence of around 10 minutes, but this is variable. Typically, in a given observing season, some 3000 points per star are taken over a period of 100 to 150 days.

The SuperWASP automated photometry extraction pipeline (Collier Cameron *et al.* 2006) uses the Tycho-2 and USNO-B1.0 catalogues to determine an astrometric solution for each field. Photometry is performed on all objects, using three different sized apertures, with a 3.5-pixel ($48''$) aperture being the default used in lightcurves. Given the size of the apertures, one has to be aware of object blending and dilution. The photometry is transformed to Tycho-2 V using around 100 stars per field, yielding WASP pseudo V magnitudes. Lightcurves are sent to the project archive in Leicester. Further trend removal is performed using the SYSREM algorithm (Tamuz *et al.* 2005), which is effective at removing some of the correlated 'red noise' from the lightcurves (Smith *et al.* 2006). Overall photometric performance is better than 1% for $V < 11.5$ mag and 0.5% for $V < 9.4$ mag.

The SuperWASP archive currently holds over 420 billion data points covering 31 million stars, obtained from 12 million images taken during 2100 nights since 2004. The coverage of the survey is now virtually the entire sky, with the exception of the Galactic plane where the stellar density is too high to permit useful aperture-photometry of objects on

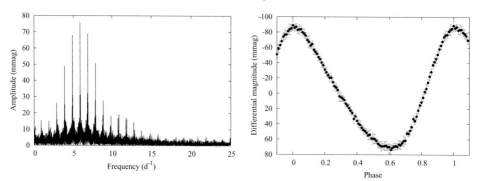

Figure 1. WASP periodogram (left) and lightcurve folded on the 0.17-day period (right) of the δ Sct variable UY Col (HD 40765).

account of the instrument's large pixel size. This multi-season and multi-site photometry is an excellent resource for studying stellar variability.

2. Variability studies with SuperWASP

There are many stars in the WASP archive that exhibit high amplitude pulsations over different timescale from hours to several days, including numerous RR Lyrae, Cepheid and δ Scuti variables (Fig. 1). This is, of course, in addition to many eclipsing and short-period binary systems (e.g. Norton *et al.* 2013). By cross-matching WASP photometry with high-precision lightcurves from *Kepler*, Holdsworth (Ph.D. thesis, in preparation) found a pulsation detection limit of 0.5 mmag is possible for stars brighter than magnitude 10.

2.1. *Pulsations in Am stars*

The large sky coverage of the SuperWASP data allows for statistical studies of the pulsational properties of selected groups of stars. In an extensive study, Smalley *et al.* (2011) took the Renson & Manfroid (2009) catalogue of Am stars and selected stars with greater than 1000 WASP data points. Excluding eclipsing binaries, 1620 Am stars were studied using Lomb periodograms to select candidate pulsating stars. These were subsequently examined in more detail using PERIOD04 (Lenz & Breger 2005). A total of 227 pulsating Am stars were identified, representing 14% of the sample. Pulsations in Am stars are more common than previously thought, but not where expected (Turcotte *et al.* 2000).

While the amplitudes are generally low, the presence of pulsation in Am stars places a strong constraint on atmospheric convection, and may require the pulsation to be laminar. While some pulsating Am stars had been previously found to be δ Sct stars, the vast majority of Am stars known to pulsate have been found by SuperWASP, thus forming the basis of future statistical studies of pulsation in the presence of atomic diffusion.

2.2. *Rotational modulation in Ap stars*

WASP photometry can be used to investigate the rotational modulation of stars, as is routinely performed as part of the planet detection programme (Maxted *et al.* 2011). Rotational modulation is also seen in the magnetic Ap stars. For example, WASP photometry was used, in conjunction with that from ASAS and *Hipparcos*, to improve the rotational ephemeris for HD 96237 (Elkin *et al.* 2011).

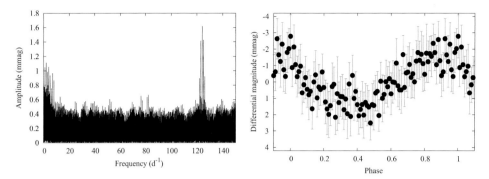

Figure 2. WASP periodogram (left) and lightcurve folded on the 11.6-minute period (right) of the roAp star BN Cet (HD 12932).

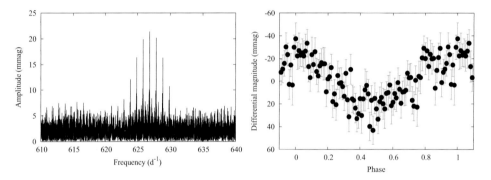

Figure 3. WASP periodogram (left) and lightcurve folded on the 2.3-minute period (right) of the sdB pulsator QQ Vir.

2.3. *Rapidly oscillating Ap stars*

The rapidly oscillating Ap (roAp) stars are a relatively rare subset of the magnetic Ap stars which exhibit short period oscillations. There exist lightcurves of several known roAp stars in the WASP archive, including HD 12932 (BN Cet). This star exhibits $124.1\,\mathrm{d}^{-1}$ (11.6 min) oscillations (Schneider & Weiss 1990) and are clearly detected in the WASP periodogram (Fig. 2). The amplitude in the WASP filter is approximately half that in B, consistent with the expected variation of amplitude with filter wavelength (Medupe & Kurtz 1998).

For stars with sufficient WASP photometry, no new roAp stars have been found among the Ap stars in the Renson & Manfroid (2009) catalogue. Nevertheless, a systematic search of over 1.5 million A- and F-type stars in the WASP archive has yielded over 200 stars with pulsation frequencies higher than $50\,\mathrm{d}^{-1}$. Subsequent spectroscopic follow-up has confirmed that at least ten of these stars are new roAp stars (Holdsworth & Smalley, these proceedings).

2.4. *Ultra-high frequencies*

Pushing to higher frequencies, WASP photometry can be used to search for frequencies up to $\sim 1000\,\mathrm{d}^{-1}$. For example, the sdB star QQ Vir has $626\,\mathrm{d}^{-1}$ (2.3 min) pulsations (Silvotti *et al.* 2002). Figure 3 shows the WASP periodogram where the pulsations are present with an amplitude of 0.02 mag. A preliminary search of the WASP archive for very rapidly pulsating stars has resulted in the identification of a hot sdBV star with $636\,\mathrm{d}^{-1}$ pulsations and an amplitude of only 8 mmag.

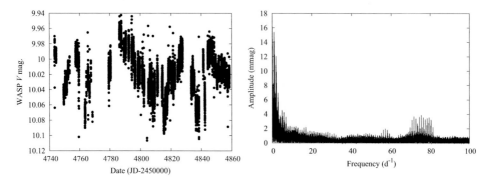

Figure 4. WASP lightcurve (left) and periodogram (right) of the pre-main sequence star V1366 Ori (HD 34282) showing random variations due to dust, but with periodic short period δ Scuti-type pulsations.

2.5. *Signal in noise*

During the WASP project's trawl for planetary transits, lightcurves are encountered which appear very poor and unsuitable for planet hunting. However, closer inspection reveals that some are actually due to real stellar variability. For example, HD 34282 (V1366 Ori) has an apparently random noise lightcurve (Fig. 4) but the periodogram recovers the $79.5\,\mathrm{d}^{-1}$ and $71.3\,\mathrm{d}^{-1}$ δ Scuti-type pulsations known to be hiding in the dusty environment of this pre-main sequence star (Amado *et al.* 2004).

3. Summary

The WASP archive contains broadband photometry for over 31 million stars with a precision of <0.01 mag. The observing strategy yields two consecutive 30-second exposures every 10 minutes. The relatively large $14''$ pixels and $48''$ photometry aperture means that blending and dilution can be an issue for certain stars. Nevertheless, the photometry has proved invaluable in the investigation of the statistical occurrence of variability and allowed for the identification of new members of several classes of variable stars. There are certainly a lot of interesting stars in the WASP archive.

References

Amado, P. J., Moya, A., Suárez, J. C., *et al.* 2004, *MNRAS*, 352, L11
Collier Cameron, A., Pollacco, D., Street, R. A., *et al.* 2006, *MNRAS*, 373, 799
Elkin, V. G., Kurtz, D. W., Worters, H. L., *et al.* 2011, *MNRAS*, 411, 978
Lenz P. & Breger, M. 2005, *CoAst*, 146, 53
Maxted, P. F. L., Anderson, D. R., Collier Cameron, A., *et al.* 2011, *PASP*, 123, 547
Medupe, R. & Kurtz, D. W. 1998, *MNRAS*, 299, 371
Norton, A. J., Payne, S. G., Evans, T., *et al.* 2011, *A&A*, 528, A90
Pollacco, D. L., Skillen, I., Collier Cameron, A., *et al.* 2006, *PASP*, 118, 1407
Renson, P. & Manfroid, J. 2009, *A&A*, 498, 961
Schneider, H. & Weiss, W. W. 1990, *IBVS*, 3520
Silvotti, R., Østensen, R., Heber, U., *et al.* 2002, *A&A*, 383, 239
Smalley, B., Kurtz, D. W., Smith, A. M. S., *et al.* 2011, *A&A*, 535, A3
Smith, A. M. S., Collier Cameron, A., Christian, D. J., *et al.* 2006, *MNRAS*, 373, 1151
Tamuz, O., Mazeh, T., & Zucker, S. 2005, *MNRAS*, 356, 1466
Turcotte, S., Richer, J., Michaud, G., & Christensen-Dalsgaard, J. 2000, *A&A*, 360, 603

Precision Asteroseismology
Proceedings IAU Symposium No. 301, 2013
J. A. Guzik, W. J. Chaplin, G. Handler & A. Pigulski, eds.

© International Astronomical Union 2014
doi:10.1017/S1743921313014051

On the new late B- and early A-type periodic variable stars

Nami Mowlavi, Sophie Saesen, Fabio Barblan and Laurent Eyer

Geneva Observatory, University of Geneva,
51 chemin des Maillettes, 1290 Versoix, Switzerland
email: Nami.Mowlavi@unige.ch

Abstract. We summarize the properties of the new periodic, small-amplitude, variable stars recently discovered in the open cluster NGC 3766. They are located in the region of the Hertzsprung-Russell diagram between δ Sct and slowly pulsating B stars, a region where no sustained pulsation is predicted by standard models. The origin of their periodic variability is currently unknown. We also discuss how the *Gaia* mission, launched at the end of 2013, can contribute to our knowledge of those stars.

Keywords. open clusters: NGC 3766, stars: variables, stars: pulsations

1. Introduction

Stars are known to pulsate if certain conditions are met in their interiors. A pulsation mechanism must be active which, for main-sequence stars, is the κ mechanism acting on H and He (δ Sct stars), the κ mechanism acting on the iron-group elements (β Cep, slowly pulsating B – or SPB – stars, and most probably rapidly oscillating Ap stars), the 'convective blocking' mechanism (γ Dor stars), or turbulence in the outer convection zone (solar-type oscillations). Stringent conditions must be met for those pulsation mechanisms to operate, which translate in the existence of 'instability regions' in the Hertzsprung-Russell (H-R) diagram, specific to each type of pulsating star (Pamyatnykh 1999, Christensen-Dalsgaard 2004). On the main sequence, solar-like pulsators, γ Dor stars, δ Sct stars, SPB stars and β Cep stars form an almost continuous sequence with increasing luminosity, except for luminosities between δ Sct stars and SPB stars where no pulsators are expected to be found, neither from model predictions, nor from observations, at least until recently. This 'gap' is clearly seen in, for example, Fig. 3 of Pamyatnykh (1999) or in Fig. 22 of Christensen-Dalsgaard (2004).

Recently, we have completed the analysis of photometric data gathered on NGC 3766 following a 7-year observation campaign, with the aim of reaching a census on the content of periodic variables in the $11.5' \times 11.5'$ field of view centered on the open cluster (Mowlavi *et al.* 2013, MBSE13 hereafter). The obvious advantages of studying variable stars in a cluster are related to, among other properties, their being coeval and at the same distance from the Sun. As a result, the knowledge of their magnitudes enables a direct inference of their relative positions in the H-R diagram. We found γ Dor, δ Sct and SPB stars at the expected magnitude and frequency ranges. But we also found, quite unexpectedly, a population of periodic variables in the magnitude range between those of δ Sct and SPB stars, at the milli-magnitude level of variability (see MBSE13 for details).

We summarize the properties of these new variables in Sect. 2, and briefly discuss the origin of their variability in Sect. 3. The potential contribution of the *Gaia* mission to our understanding of those stars is given in Sect. 4. Conclusions are drawn in Sect. 5.

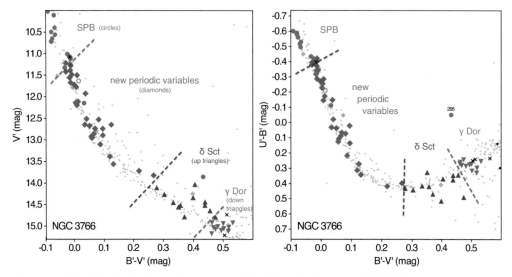

Figure 1. Distribution in the color-magnitude (left) and color-color (right) diagrams of main-sequence periodic variables detected in NGC 3766 in the magnitude range between those of γ Dor and SPB stars. Different symbols represent different groups of periodic variables as indicated in the figures. Black crosses identify eclipsing binaries. Grey dots represent constant or non-periodic variables. The dashed lines separate the regions in the diagrams where the different groups of variables, identified from our data, are found.

2. Properties of the new class of late B- and early A-type variables

The new periodic variables discovered in NGC 3766 have periods between 0.1 and 0.7 days and amplitudes of variability below few mmag (see Fig. 19 in MBSE13; four stars have periods up to 1.1 d, but they may be outliers). The low amplitudes would explain why they are not easily detectable from ground. We refer to Sect. 7.1 of MBSE13 for a summary of all the properties of those variables.

The striking characteristic of these variables is the fact that they lie outside the predicted instability strips in the H-R diagram of both SPB and δ Sct stars (see Fig. 1). In addition, their periods differ from the typical periods expected for those two groups. While SPB stars have periods between 0.5 and 5 d and δ Sct stars between 0.02 and 0.25 d, the majority of our new group of variables have periods in the range $0.25-0.5$ d. Those two properties by themselves make those variables very peculiar, suggesting a new class of periodic variable stars. The facts that their periods are stable over seven years (the duration of the observation campaign) and that up to one third of them are multiperiodic are further clues to understand the origin of their variability.

The fraction of main-sequence stars in NGC 3766 that are detected to be periodically variable is shown in Fig. 2. It is about 50% at magnitudes between 11 and 11.5 mag, i.e. at the lower luminosity end of SPB stars (our data unfortunately cannot tell the number of SPB stars at higher luminosities, due to photometric saturation of the bright stars in the survey). In the region between SPB and δ Sct stars, where the new variables are found, this fraction decreases continuously from $\sim$40% at 11.5 mag to $\sim$10% at 13.5 mag. This is of interest in the exploration of the possible origin of the variability.

3. Possible origins of the periodic variability

Three scenarios can be thought of to explain periodic photometric variability of B- and A-type main-sequence stars: pulsation, stellar rotation (spotted stars), and non-sphericity

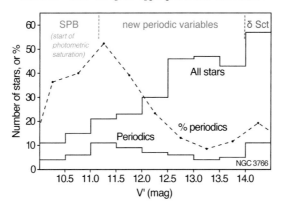

Figure 2. Total number of stars (upper histogram) and number of periodic stars (lower histogram) in NGC 3766 as a function of magnitude. Only stars with $B' - V' < 0.4$ mag are considered to exclude non-members of the cluster. Markers connected with the dashed line give the percentage of periodic stars per magnitude bin. The sharp decrease of the fraction of periodic variables at magnitudes below 11 mag is most probably due to photometric saturation in our data at those high luminosities. The magnitude ranges where the different groups of variable stars are found in our data are indicated on top of the figure, separated by dashed lines.

due to binarity (ellipsoidal variations). They are reviewed in Mowlavi *et al.* (2014) in relation to the new periodic variables found in NGC 3766. None of those scenarios can currently provide a satisfactory explanation for the observations of those stars, though: pulsation is not predicted in this region of the main sequence by *standard* pulsation models; spots at the surface of rotating stars are not predicted in late B- and early A-type stars; and binarity, at the origin of ellipsoidal systems, cannot explain multiperiodicity.

Of relevance to this conference is the question whether a pulsation origin is a viable scenario for those periodic stars, despite the fact that they lie in the pulsation 'gap' on the main sequence. It is interesting to note in this respect that four of those periodic variables for which spectra are available in the literature are rotating faster than half their critical equatorial velocity (see MBSE13). While small number statistics, it suggests that fast rotation may be a key issue in understanding their origin.

No pulsation model prediction exists yet for very rapidly rotating stars. However, models with rotational velocities approaching half the critical velocity do predict pulsation frequencies significantly different than the ones predicted for non-rotating stars (Townsend 2005, Ushomirsky & Bildsten 1998). They also show that rapid rotation can excite modes that are damped in the absence of rotation, opening the possibility to find pulsation outside the classical instability strips. Interestingly, the fraction of stars that are periodic variables reaches a maximum close to the region of SPB stars, and decreases for fainter stars as shown in Sect. 2. This may be a further clue supporting pulsation in rapidly rotating stars.

4. Potential contribution of *Gaia*

The understanding of the new periodic B- and A-type variables would benefit from an all-sky census. The space-based *Gaia* mission†, to be launched towards the end of 2013, is potentially a very good candidate to achieve this, for several reasons:

– It is a multi-epoch mission, with a mean of about 75 measurements per star over the 5-year mission duration.

† See http://www.rssd.esa.int/index.php?project=GAIA

– The photometric uncertainty equals about 1 mmag up to 14 mag in G, the main *Gaia* photometric band (see Fig. 19 of Jordi *et al.* 2010).

– The scanning law leads to a time sampling well suited to detect periodic variables with periods below several days. The spectral window presents only few peaks, that are unlikely to affect the frequency range of interest (below 10 d^{-1}) when convolved with a periodic signal (see Fig. 1 of Eyer *et al.* 2009). As a result, the expected period recovery rate is very high for the periods typical for the new periodic variables, even for low signal-to-noise ratios. Assuming an uncertainty (Gaussian noise) of 1 mmag in the measurement, Eyer & Mignard (2005) predict a period recovery rate above 85% for a sinusoidal signal of amplitude 1.1 mmag and a frequency of 3 d^{-1}, and above 95% for an amplitude of 1.3 mmag.

– *Gaia* will provide parallaxes with unprecedented accuracy, allowing the positioning of the stars in the H-R diagram with high accuracy.

– *Gaia*'s photometric instrument includes spectro-photometric measurements in two color bands. In addition, *Gaia* also includes a spectroscopic instrument. From those data, astrophysical information such as atmospheric parameters and rotational velocities will be derived for stars brighter than about 12 mag† in G_{RVS}, the band of *Gaia*'s spectrometer. This possibility of *Gaia* to characterize the nature of bright stars is a great advantage to explore the nature of periodic stars.

– Finally, *Gaia* is an all-sky survey.

5. Conclusions

The discovery in NGC 3766 of late B- and early A-type stars that show periodic variability in their light curve provides a challenge for stellar physics. Neither pulsation is expected in those stars, because they lie in the pulsation 'gap' on the main sequence, nor spots on their surface, because they have too high effective temperatures.

Not all thirty-six stars in this new class of periodic variables need necessarily to be of the same type of variability, but pulsation in a rapidly rotating star may, we believe, be a good scenario for a fraction of them. Which fraction is a question that must be addressed by further observations, that we have already planned to undertake.

References

Christensen-Dalsgaard, J. 2004, in: D. Danesy (ed.), *SOHO 14 Helio- and Asteroseismology: Towards a Golden Future*, ESA SP-559, p. 1

Eyer, L. & Mignard, F. 2005, *MNRAS*, 361, 1136

Eyer, L., Mowlavi, N., Varadi, M., *et al.* 2009, in: M. Heydari-Malayeri, C. Reyl and R. Samadi, (eds.), *Proceedings of the Annual meeting of the French Society of Astronomy and Astrophysics*, SF2A-2009, p. 45

Jordi, C., Gebran, M., Carrasco, J. M., *et al.* 2010, *A&A*, 523, 48

Mowlavi, N., Barblan, F., Saesen, S., & Eyer, L. 2013, *A&A* 554, A108 (MBSE13)

Mowlavi, N., Saesen, S., Barblan, F., & Eyer, L. 2014, in: *Putting A Stars into Context: Evolution, Environment, and Related Stars* (June 3-7 , 2013, Moscow), Press-Menu, in press (arXiv: 1310.6488)

Pamyatnykh, A. A. 1999, *AcA*, 49, 119

Townsend, R. H. D. 2005, *MNRAS*, 364, 573

Ushomirsky, G. & Bildsten, L. 1998, *ApJ*, 497, L101

† From http://www.rssd.esa.int/index.php?project=GAIA&page=Science_Performance (estimate as of 09/2013).

Precision Asteroseismology
Proceedings IAU Symposium No. 301, 2013 © International Astronomical Union 2014
J. A. Guzik, W. J. Chaplin, G. Handler & A. Pigulski, eds. doi:10.1017/S1743921313014063

Some highlights of the latest *CoRoT* results on stellar physics

S. Deheuvels[1,2] and the *CoRoT* team

[1]Université de Toulouse; UPS-OMP; IRAP; Toulouse, France
email: sebastien.deheuvels@irap.omp.eu

[2]CNRS; IRAP; 14, avenue Edouard Belin, F-31400 Toulouse, France

Abstract. Since its launch in December 2006, the *CoRoT* satellite has provided photometric data precise down to the micro-magnitude level for about 150 bright stars and 150 000 fainter ones. These stars have been observed over runs covering up to 160 days with a 90% duty cycle. Seismic data of such precision had been longed for by the scientific community for decades, and expected as a way of making progress in our understanding of stellar structure and evolution. The analysis and interpretation of *CoRoT* seismic data have indeed made it possible to place observational constraints on several key aspects of stellar structure and evolution, such as the size of mixed convective cores, magnetic activity, mass loss... We here present some highlights of the *CoRoT* results and their implications in terms of internal stellar structure.

1. The *CoRoT* mission

After almost six years of observations, the *CoRoT* satellite has observed 156 bright stars ($5.4 \leqslant m_V \leqslant 10.5$), and more than 160 000 fainter ones ($11 \leqslant m_V \leqslant 16$) that lie at the intersection of the Galactic plane and the equator. In November 2012, the satellite encountered an electric breakdown, which could unfortunately not be repaired despite great efforts. *CoRoT* yielded seismic data of unprecedentedly high quality by observing over long observation runs (ranging from about 20 days to more than 160 days) with duty cycles over 90%. This led to several breakthroughs such as the precise characterization of solar-like oscillation modes in stars other than the Sun (Appourchaux *et al.* 2008, Michel *et al.* 2008), the detection of mixed modes in several thousands of red giants (De Ridder *et al.* 2009, Mosser *et al.* 2011), the evidence for a correlation between oscillations and the occurrence of an outburst in a Be star (Huat *et al.* 2009), the detection of solar-like oscillations in massive stars (Belkacem *et al.* 2010, Degroote *et al.* 2010), the detection of a magnetic activity cycle in the solar-like pulsator HD 49933 (García *et al.* 2010)... Several reviews already presented these results from *CoRoT*, along with many others (Michel & Baglin 2012, Baglin *et al.*, in preparation). We here give a description of some of the latest results that were obtained with *CoRoT* data, focusing on the main scientific objectives of the mission: making progress in our understanding of stellar convection and of the effects of rotation on the structure and evolution of stars.

2. Rotation

Reaching a better understanding of the effects of rotation on stellar structure and evolution was one of the main objectives of the *CoRoT* mission. Progress in this field has been hindered by the lack of observational constraints on the rotation profiles of stars. By providing quasi-uninterrupted photometric observations over periods of several months, the *CoRoT* satellite offered the opportunity to measure rotation periods both

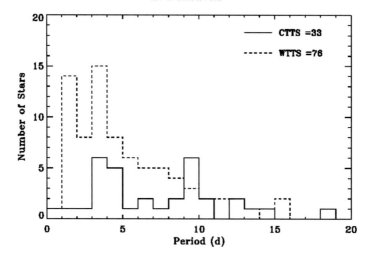

Figure 1. Distribution of rotation periods of weak-line T-Tauri stars and classical T-Tauri stars in the NGC 2264 star forming region (figure from Affer *et al.* 2013).

from stellar activity and from seismology for stars with different masses, from the pre-main sequence to more advanced stages. Also, one of the objectives of *CoRoT* was to help us understand the oscillation spectra of fast rotators whose great complexity have severely limited our seismic diagnostics for these stars so far.

2.1. *Rotation periods of T-Tauri stars in NGC 2264*

The star forming region NGC 2264, which is one of the best known in the solar vicinity, falls in the field of view of *CoRoT*. This gave a unique opportunity to study young stars still in a formation phase, and in particular to estimate their rotation periods. This is particularly relevant to study the role of *disk locking* (magnetic interaction between young stars and their accretion disk) in the braking of stars during the pre-main sequence contraction and thus to better understand the evolution of angular momentum at this time. If disk locking is indeed responsible for the existence of a slow-rotating population in young clusters, there should be a correlation between accretion and stellar rotation. Several previous studies had failed to detect such correlation (e.g. Stassun *et al.* 1999, Cieza & Baliber 2006). Using *CoRoT* observations of the cluster NGC 2264, Affer *et al.* (2013) have measured the rotation periods of both classical T-Tauri stars (CTTS), which are assumed to have an accretion disk, and weak T-Tauri stars (WTTS), which have no disk. They found that the two populations have significantly different rotation distributions with the WTTS rotating faster than the CTTS (Fig. 1). This suggests that the presence of accretion affects the rotational period and is consistent with the disk-locking scenario.

2.2. *Surface rotations of low-mass main-sequence stars*

By observing stars almost continuously over periods of several months, *CoRoT* made it possible to reach the level of precision and the frequency resolution required to measure rotational splittings in the oscillation spectra of solar-like stars. For instance, a mean rotational splitting for p modes has been estimated in the F star HD 181420 (Barban *et al.* 2009). This measure was found consistent with the surface rotation period inferred from the periodic modulation of the lightcurve attributed to starspots.

Another particularly interesting *CoRoT* target is the K-type solar-like pulsator HD 52265. This star was selected as a primary target because it hosts a planetary companion,

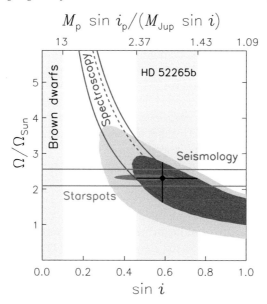

Figure 2. Constraints on the surface rotation of HD 52265 from seismology (black diamond with 1-σ error bars), from spectroscopy (blue lines) and from the modulation of the lightcurve caused by star spots (horizontal green lines). Figure taken from Gizon *et al.* (2013), see this reference for more information.

and it was observed during 117 days by *CoRoT*. A modeling of the star was performed by Escobar *et al.* (2012) and yielded precise estimates of its mass and age. Recently, Gizon *et al.* (2013) obtained an unambiguous measurement of the average surface rotation of the star by confronting three different measurements:

• The signature of starspot modulation, which was found in the *CoRoT* lightcurve of the star (Ballot *et al.* 2011).

• The mean rotational splitting of the acoustic modes, which was extracted from the oscillation spectrum of the star by Gizon *et al.* (2013).

• The combination of the spectroscopic $v \sin i$, the radius of the star from the seismic modeling, and its inclination angle i, which could be determined from seismology (Gizon *et al.* 2013).

The results of these three methods are remarkably consistent, as shown by Fig. 2.

It is well known that low-mass stars are braked during the main sequence because they lose angular momentum through a magnetized wind generated by the convective envelope. Having access to precise and reliable rotation periods for stars whose mass and age can also be constrained by seismology, as was the case for HD 52265, could give precious observational constraints to calibrate theoretical laws of angular momentum loss (e.g. Kawaler 1988), which could be helpful for gyrochronology.

2.3. *Fast rotation in intermediate-mass main-sequence stars*

Intermediate-mass stars that fall in the δ Scuti Instability Strip have extremely rich oscillation spectra. Hundreds of modes were detected in the oscillation spectra of several *CoRoT* δ Scuti stars (e.g. Poretti *et al.* 2009, García Hernández *et al.* 2009). Moya & Rodríguez-López (2010) showed that δ Scuti stars have in principle enough energy to excite such a large number of modes.

The complexity of the observed spectra of δ Scuti stars is probably related to their fast rotational velocities. Indeed, these stars are not braked by a magnetic wind during the main sequence and they can rotate at non-negligible fractions of the break-up velocity. Both theoretical and numerical works have been led to explore the impact of such fast rotation on the mode frequencies, involving the integration of the oscillation equations taking rotation into account in 2D models (Lignières *et al.* 2006, Reese *et al.* 2006). These studies showed that for fast rotators, two frequency subsets should be visible: a regular one, which can be approximated by an asymptotic theory analogous to the slow-rotation case (Pasek *et al.* 2012), and an irregular one, with specific statistic properties. This was proposed as an explanation for the extreme richness of the spectra of δ Scuti stars. Despite the large number of expected modes, recent theoretical work predicts that equidistances corresponding to the large separation and to twice the rotational splitting should still be identifiable in the frequency distribution (Lignières *et al.* 2010), giving hope to interpret the spectra of these stars.

García Hernández *et al.* (2013) analyzed the *CoRoT* lightcurve of the δ Scuti star HD 174966 and extracted 185 significant independent frequencies. They found a significant periodicity of about 64 μHz in the frequency set and built an échelle diagram of the detected frequencies folded with this equidistance. By using spectroscopic measurements, they derived an upper limit of 29 μHz for the rotational splitting and concluded that the observed periodicity is more likely to correspond to the large separation of the star. Paparó *et al.* (2013) also found regular spacings in the mode pattern of the δ Scuti star *CoRoT* 102749568, but it remains uncertain whether they are caused by rotational effects or to the large separation of p modes. These studies constitute the first step toward the understanding of the complex spectra of fast rotators and give the perspective of measuring the mean densities of δ Scuti stars regardless of their rotational velocities.

2.4. *Rotation periods of CoRoT subgiants*

Estimating the surface rotation of post-main sequence stars can place valuable constraints on the mechanisms of angular momentum transport inside stars, which remain poorly understood. Indeed, the rotation periods of subgiants and red giants are the result of a competition between the expansion of the star which tends to spin it down and the deepening of the convective envelope, which dredges up material from the core and tends to spin up the envelope. Redistribution of angular momentum inside the star also influences the surface rotation.

By searching for periodicity in the lightcurves of *CoRoT* subgiants, do Nascimento *et al.* (2013) obtained rotation periods ranging from 30 to 100 days for 30 of these targets. They also found that these observed periods are consistent with the surface rotations predicted by stellar evolution models that include rotationally-induced transport of angular momentum as described by Zahn (1992). However, recent observations from the *Kepler* satellite yielded estimates of the core rotation of subgiants (Deheuvels *et al.* 2012) and red giants (Beck *et al.* 2012, Mosser *et al.* 2012), which showed that the transport of angular momentum à la Zahn (1992) is not efficient enough and that another mechanism must be at work (e.g. Marques *et al.* 2013). The origin of this extra mechanism is not yet known, but a more efficient internal transport of angular momentum should imply faster envelope rotation. On the other hand, van Saders & Pinsonneault (2013) recently showed that the rotation periods found by do Nascimento *et al.* (2013) are smaller than would be expected if solid-body rotation at all times (i.e. instantaneous transport of angular momentum) is assumed, which is consistent with the detection of differential rotation in these stars.

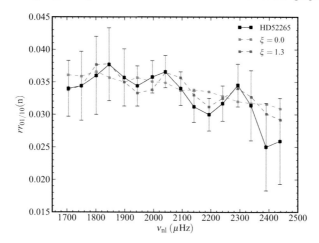

Figure 3. Variations in the frequency ratios rr_{01} and rr_{10} (see Roxburgh & Vorontsov 2003 for a definition) of the solar-like pulsator HD 52265 (black circles). The colored lines indicate the frequency ratios of the best-fit models (purple: no overshooting, red: overshooting over a distance of $0.95\,H_p$). (figure from Lebreton & Goupil 2012)

3. Convection

3.1. *Core and envelope overshooting*

Several processes are expected to extend the size of convective regions beyond the Schwarzschild frontier (overshooting, rotational mixing, ...), but since we lack a realistic description of these mechanisms, the actual size of convective zones remains uncertain. This generates large uncertainties in our determination of stellar ages for stars that have a convective core. Observational constraints on the size of convective regions are therefore needed.

Core overshoot. CoRoT observations have produced indications in favor of an extension of the mixed core for stars of different masses and evolutionary stages, and using different seismic diagnostics. Since these results have already been presented by Michel & Baglin (2012), we only briefly recall them here. For the main-sequence solar-like pulsator HD 49933, Goupil *et al.* (2011) found that an extension of the mixed core over a distance of about $0.2\,H_p$ is required to reproduce the observed frequency separation $\delta\nu_{01}$. The authors showed that rotational mixing as it is currently modeled cannot account in itself for such a large extension, suggesting that core overshoot is needed. On the other hand, Escobar *et al.* (2012) found that no extra mixing beyond the convective core is required for the main-sequence star HD 52265. Since this star has a higher abundance of heavy elements than HD 49933, this result raises the question of a possible dependency of core overshooting on metallicity. By using mixed modes, which are sensitive to the chemical composition of the core, Deheuvels & Michel (2011) found that models with a core overshoot between 0.18 and $0.2\,H_p$ best reproduce the observed frequencies, but they could not exclude the case of a very small overshoot below $0.05\,H_p$. Evidence for extended convective cores was also obtained for more massive stars. Degroote *et al.* (2010) detected a deviation from the regular spacing in period of gravity modes in the B star HD 50230, which they could explain only by assuming a core overshoot above $0.2\,H_p$. Neiner *et al.* (2012) detected groupings of modes in two Be stars at the end of the main sequence and they showed that a core extended over at least $0.3\,H_p$ is needed to explain both the excitation of the modes and the frequencies of the observed mode groupings.

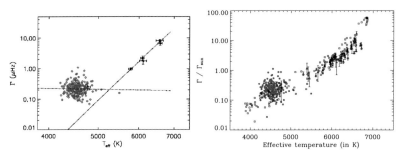

Figure 4. *Left:* Measured mode linewidth versus T_{eff} for *CoRoT* main-sequence (black diamonds) and red-giant (red diamonds) solar-like pulsators (from Baudin *et al.* 2011). *Right:* Mode linewidths predicted from non-adiabatic calculations (open squares) versus T_{eff}. The blue triangles represent main-sequence stars observed with *Kepler*, and red circles correspond to red giants observed with *CoRoT* (from Belkacem *et al.* 2012).

Envelope overshoot. Overshooting from the convective envelope has a less direct impact on stellar evolution than core overshooting, but it can help us understand this theoretically challenging phenomenon. In favorable cases, the depth of the boundary of the convective envelope can be estimated through seismology because the abrupt transition between convective and radiative energy transport induces a glitch in the sound speed profile, to which acoustic modes are sensitive. Lebreton & Goupil (2012) showed that the observed seismic ratios rr_{01} and rr_{10} (as introduced by Roxburgh & Vorontsov 2003) of the *CoRoT* solar-like pulsator HD 52265 are reproduced at closest when assuming an envelope overshooting extending over $0.95\,H_p$ (see Fig. 3). By comparison, Christensen-Dalsgaard *et al.* (2011) recently showed that an overshooting over a distance of $0.37\,H_p$ is needed at the base of the convective envelope of the Sun. The difference between the overshooting distances of the two stars might be linked to the fact that HD 52265 has roughly twice the abundance of heavy elements of the Sun.

3.2. *Convection in the super-adiabatic layer*

The structure of the outer convective envelope remains poorly understood because convective transport is inefficient in this region, which makes the mixing length theory inappropriate. This is currently a major problem for seismology because the frequencies of individual modes depend on the structure of the super-adiabatic layer. Empirical corrections of these so-called *near surface effects* have been proposed (Kjeldsen *et al.* 2008) but a better understanding of the super-adiabatic layer is needed to solve this problem. The seismic data from *CoRoT* and now *Kepler* made it possible to measure the amplitudes and linewidths of solar-like pulsators other than the Sun. This can yield valuable observational constraints on the super-adiabatic layer because the mode amplitudes and widths depend on the excitation and damping of the modes.

Mode amplitudes. Several scaling laws have been proposed for the mode amplitudes of main-sequence solar-like pulsators (e.g. Houdek *et al.* 1999, Samadi *et al.* 2007). They predict mode amplitudes to scale with $(L/M)^s$, where s ranges from 0.7 to 1.5. *CoRoT* observations are in agreement with this scaling and favor values of s in the lower end of the interval $[0.7, 1.5]$ (Baudin *et al.* 2011). Detailed comparisons between 3D simulations and *CoRoT* observations were performed for the solar-like pulsator HD 49933 by Samadi *et al.* (2010). They showed that the effects of metallicity must be taken into account, and while they could reproduce the observed amplitudes for low-frequency modes, differences remain at higher frequency.

The detection of oscillations in thousands of red giants with *CoRoT* and *Kepler* gave the opportunity to test whether or not these scaling relations can be extended to the red-giant branch. Baudin *et al.* (2011) extracted mode amplitudes for several hundreds of *CoRoT* red giants. To compare the mode amplitudes that are predicted from 3D simulations with *CoRoT* observations, a velocity-intensity relation is needed. Samadi *et al.* (2012) showed that for red-giant stars, non-adiabatic effects need to be taken into account in this conversion. Even then, the predicted mode amplitudes for red giants are underestimated by about 40%. Solving this discrepancy will probably require a better knowledge of the depth at which the mode inertia need to be computed and a more realistic treatment of the interaction between convection and pulsations (Samadi *et al.* 2012).

Mode linewidths. Baudin *et al.* (2011) also extracted the mode linewidths for the solar-like pulsators observed by *CoRoT*. They found that the mode linewidths of main-sequence pulsators vary very sharply with the star's temperature ($\Gamma \propto T_{\mathrm{eff}}^m$ with $m = 16.2 \pm 2$), which was later confirmed by Appourchaux *et al.* (2012) with *Kepler* data. This sheds new light on the unexpectedly large width of the modes of F stars such as HD 49933 (Appourchaux *et al.* 2008). According to Baudin *et al.* (2011), this scaling law does not extend to red giants (see Fig. 4, left panel), which led them to suggest that a different damping mechanism might be at work in these stars. However, Belkacem *et al.* (2012) recently performed fully non-adiabatic calculations with the MAD code (Dupret 2001) including time-dependent convection, and they could reproduce the observed mode linewidths of both *Kepler* main-sequence stars and *CoRoT* red giants (see Fig. 4, right panel). *Kepler* observations, which are longer than those of *CoRoT*, should yield mode linewidth for red giants higher in the red-giant branch (T_{eff} lower than 4200 K), which will make it possible to test the predictions of Belkacem *et al.* (2012).

4. Conclusion

The *CoRoT* data have now brought contributions in many different fields of stellar physics. *CoRoT* is providing us with valuable observational constraints on several physical processes that are theoretically challenging, such as the evolution of rotation profiles with time, the transport of angular momentum in stars, core and envelope overshooting, the properties of convection in the super-adiabatic layer... *CoRoT* leaves the scientific community with an extremely rich seismic data set, which has certainly not been used to its full potential yet. The reason for this is of course the very large number of targets that were observed, but also the difficulties that we currently encounter in identifying the observed modes in many cases, or in interpreting the mode frequencies due to near-surface effects. It is exciting to see that we are now making progress in these domains, with for instance spectroscopic ground-based follow-up campaigns that can provide mode identification, interesting developments in our understanding of the oscillation spectra of fast rotators, or the prospect to better describe the structure of the super-adiabatic layers using the constraints given by the amplitudes and lifetimes of solar-like modes. This shows that new scientific breakthroughs can definitely be expected from *CoRoT* data in the years to come.

Acknowledgements

I am very grateful to E. Michel and A. Baglin for their advices and help to write this review.

References

Affer, L., Micela, G., Favata, F., Flaccomio, E., & Bouvier, J. 2013, *MNRAS*, 430, 1433

Appourchaux, T., Michel, E., Auvergne, M., *et al.* 2008, *A&A*, 488, 705

Appourchaux, T., Benomar, O., Gruberbauer, M., *et al.* 2012, *A&A*, 537, A134

Ballot, J., Gizon, L., Samadi, R., *et al.* 2011, *A&A*, 530, A97

Barban, C., Deheuvels, S., Baudin, F., *et al.* 2009, *A&A*, 506, 51

Baudin, F., Barban, C., Belkacem, K., *et al.* 2011, *A&A*, 535, C1

Beck, P. G., Montalbán, J., Kallinger, T., *et al.* 2012, *Nature*, 481, 55

Belkacem, K., Dupret, M. A., & Noels, A. 2010, *A&A*, 510, A6

Belkacem, K., Dupret, M. A., Baudin, F., *et al.* 2012, *A&A*, 540, L7

Christensen-Dalsgaard, J., Monteiro, M. J. P. F. G., Rempel, M., & Thompson, M. J. 2011, *MNRAS*, 414, 1158

Cieza, L. & Baliber, N. 2006, *ApJ*, 649, 862

De Ridder, J., Barban, C., Baudin, F., *et al.* 2009, *Nature*, 459, 398

Degroote, P., Aerts, C., Baglin, A., *et al.* 2010, *Nature*, 464, 259

Deheuvels, S. & Michel, E. 2011, *A&A*, 535, A91

Deheuvels, S., García, R. A., Chaplin, W. J., *et al.* 2012, *ApJ*, 756, 19

do Nascimento, Jr., J.-D., Takeda, Y., Meléndez, J., *et al.* 2013, *ApJ*, 771, L31

Dupret, M. A. 2001, *A&A*, 366, 166

Escobar, M. E., Théado, S., Vauclair, S., *et al.* 2012, *A&A*, 543, A96

García, R. A., Mathur, S., Salabert, D., *et al.* 2010, *Science*, 329, 1032

García Hernández, A., Moya, A., Michel, E., *et al.* 2009, *A&A*, 506, 79

García Hernández, A., Moya, A., Michel, E., *et al.* 2013, *A&A*, 559, A63

Gizon, L., Ballot, J., Michel, E., *et al.* 2013, *Proc. of the Nat. Acad. of Sciences*, 110, 13267

Goupil, M. J., Lebreton, Y., Marques, J. P., *et al.* 2011, *Journal of Physics: Conference Series*, 271, 012032

Houdek, G., Balmforth, N. J., Christensen-Dalsgaard, J., & Gough, D. O. 1999, *A&A*, 351, 582

Huat, A.-L., Hubert, A.-M., Baudin, F., *et al.* 2009, *A&A*, 506, 95

Kawaler, S. D. 1988, *ApJ*, 333, 236

Kjeldsen, H., Bedding, T. R., & Christensen-Dalsgaard, J. 2008, *ApJ*, 683, L175

Lebreton, Y. & Goupil, M. J. 2012, *A&A*, 544, L13

Lignières, F., Rieutord, M., & Reese, D. 2006, *A&A*, 455, 607

Lignières, F., Georgeot, B., & Ballot, J. 2010, *AN*, 331, 1053

Marques, J. P., Goupil, M. J., Lebreton, Y., *et al.* 2013, *A&A*, 549, A74

Michel, E. & Baglin, A. 2012, arXiv: 1202.1422

Michel, E., Baglin, A., Auvergne, M., *et al.* 2008, *Science*, 322, 558

Mosser, B., Barban, C., Montalbán, J., *et al.* 2011, *A&A*, 532, A86

Mosser, B., Goupil, M. J., Belkacem, K., *et al.* 2012, *A&A*, 548, A10

Moya, A. & Rodríguez-López, C. 2010, *ApJ*, 710, L7

Neiner, C., Mathis, S., Saio, H., *et al.* 2012, *A&A*, 539, A90

Paparó, M., Bognár, Z., Benkő, J. M., *et al.* 2013, *A&A*, 557, A27

Pasek, M., Lignières, F., Georgeot, B., & Reese, D. R. 2012, *A&A*, 546, A11

Poretti, E., Michel, E., Garrido, R., *et al.* 2009, *A&A*, 506, 85

Reese, D., Lignières, F., & Rieutord, M. 2006, *A&A*, 455, 621

Roxburgh, I. W. & Vorontsov, S. V. 2003, *A&A*, 411, 215

Samadi, R., Georgobiani, D., Trampedach, R., *et al.* 2007, *A&A*, 463, 297

Samadi, R., Ludwig, H.-G., Belkacem, K., Goupil, M. J., & Dupret, M.-A. 2010, *A&A*, 509, A15

Samadi, R., Belkacem, K., Dupret, M.-A., *et al.* 2012, *A&A*, 543, A120

Stassun, K. G., Mathieu, R. D., Mazeh, T., & Vrba, F. J. 1999, *AJ*, 117, 2941

van Saders, J. L. & Pinsonneault, M. H. 2013, *ApJ*, 776, 67

Zahn, J.-P. 1992, *A&A*, 265, 115

Precision Asteroseismology
Proceedings IAU Symposium No. 301, 2013
J. A. Guzik, W. J. Chaplin, G. Handler & A. Pigulski, eds.

© International Astronomical Union 2014
doi:10.1017/S1743921313014075

Subtle flickering in Cepheids:
Kepler and *MOST*

Nancy Remage Evans[1], Robert Szabó[2], Laszlo Szabados[2],
Aliz Derekas[2], Jaymie M. Matthews[3], Chris Cameron[3],
and the *MOST* Team

[1]SAO, MS 4, 60 Garden St., Cambridge, MA 02138, USA
email: nevans@cfa.harvard.edu

[2]MTA CSFK Konkoly Thege Miklos ut 15-17, H-1121 Budapest Hungary

[3]Dept. of Physics and Astronomy, Univ. of British Columbia, Vancouver, BC, Canada

Abstract. Fundamental mode classical Cepheids have light curves which repeat accurately enough that we can watch them evolve (change period). The new level of accuracy and quantity of data with the *Kepler* and *MOST* satellites probes this further. An intriguing result was found in the long time-series of Kepler data for V1154 Cyg the one classical Cepheid (fundamental mode, P = 4.9 d) in the field, which has short term changes in period ($\simeq$20 minutes), correlated for $\simeq$10 cycles (period jitter). To follow this up, we obtained a month long series of observations of the fundamental mode Cepheid RT Aur and the first overtone pulsator SZ Tau. RT Aur shows the traditional strict repetition of the light curve, with the Fourier amplitude ratio R_1/R_2 remaining nearly constant. The light curve of SZ Tau, on the other hand, fluctuates in amplitude ratio at the level of approximately 50%. Furthermore prewhitening the RT Aur data with 10 frequencies reduces the Fourier spectrum to noise. For SZ Tau, considerable power is left after this prewhitening in a complicated variety of frequencies.

Keywords. stars: variables: Cepheids, pulsation, *Kepler*, *MOST* satellite photometry

1. Introduction

The quality and quantity of satellite data is revealing subtle features in Cepheid pulsation. Specifically, we discuss here recent findings from two long series of observations by the *Kepler* and *MOST* satellites.

2. *Kepler* observations

There is only one classical Cepheid (V1154 Cyg) in the *Kepler* field, and it is a fundamental mode pulsator with a period of 4.9 d. Analysis of the *Kepler* data by Derekas *et al.* (2012) found small cycle to cycle variations of the period. Typically the period excursions might be 20 minutes and last up to 15 cycles before the sign of the period change ($\dot{P}$) reverses.

3. *MOST* observations

To look further for the small fluctuations made visible by the long continuous strings of accurate satellite data, we observed two Cepheids with the *MOST* satellite (Walker *et al.* 2003, Matthews *et al.* 2004). Motivation for the structure of the observation request is provided by known period changes in Cepheids (Fig. 1). The data taken from Szabados (1983) show the well known increase in period fluctuations with period or luminosity of the Cepheid. This is consistent with period changes ($\dot{P}$) determined by evolution through the instability strip, with more luminous stars evolving more quickly. The exception to the

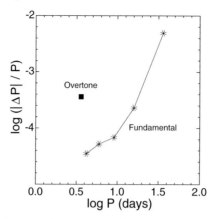

Figure 1. Fractional period change (absolute value) as a function of the log of the pulsation period in days. Fundamental mode pulsators (binned) are denoted by asterisks; overtone pulsators are a filled square.

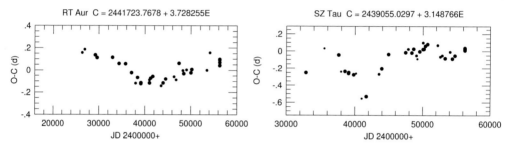

Figure 2. The long-term period variations (O−C diagrams: O−C in days as a function of Julian date − 2 400 000) for RT Aur (left) and SZ Tau (right). As is typical of their pulsation modes, the variation for RT Aur is smooth, where for SZ Tau it is both positive and negative and also constant for a long period. Symbol size denotes the significance of the O−C value.

trend, however, is the group of Cepheids pulsating in an overtone mode which apparently show unusually large period jitter rather than the smooth and continuous evolution of fundamental mode pulsators.

For the *MOST* observations we wanted to contrast the behavior of a fundamental mode pulsator (RT Aur) with a first overtone pulsator (SZ Tau). The background information on $\dot{P}$ is shown in Fig. 2. For RT Aur the period change O−C (observed minus computed) diagram has a decrease followed by an increase, which is typically fitted with a parabola. For SZ Tau, the O−C diagram shows both increasing and decreasing periods, as well as about 10 000 days when it remains constant.

The *MOST* observations (Evans et al., in preparation) phased for pulsation period are shown in Fig. 3. For RT Aur, the observations were interleaved with another target, resulting in gaps in the light curve. However, the light curve repeats very precisely from cycle to cycle. For SZ Tau, the overtone pulsator, only the maximum of the phased light curve is shown. This emphasizes the fact that there are variations in maximum brightness from cycle to cycle (also easily visible at minimum light). A small instrumental signal (differing earthshine through the satellite orbit) is seen in Fig. 3b, which is being removed through additional processing.

The data from the two stars was subjected to a number of comparisons. In Fig. 4, the Fourier spectra are shown after pre-whitening for a Fourier fit of 10 terms. The instrumental signal is seen at the orbital frequency ($\simeq$14 d^{-1}). The Fourier series describes

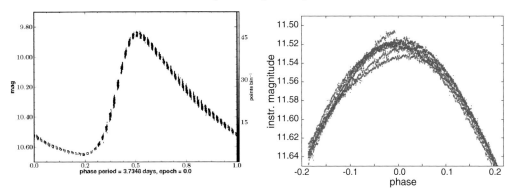

Figure 3. The *MOST* light curves (magnitudes as a function of pulsation phase) for RT Aur (left) and SZ Tau (right). The RT Aur figure also shows points per bin on the right side scale. The RT Aur observations were interleaved with another target, resulting in the gaps in the data. Only the maximum light is shown for SZ Tau, emphasizing the variation in brightness between cycles.

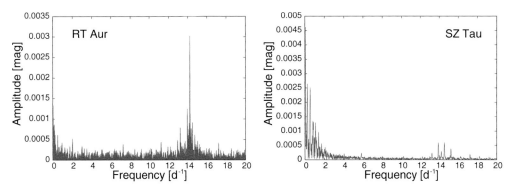

Figure 4. The Fourier spectra of the *MOST* data of RT Aur (left) and SZ Tau (right). The data for both stars have been prewhitened by a Fourier series with 10 terms. In both stars the instrumental signature is seen at about $14\,\mathrm{d}^{-1}$. For RT Aur, little power is left at low frequencies. For SZ Tau, power still remains in a complicated set of low frequencies.

the fundamental mode pulsator (RT Aur) well and little power is left at low frequencies. In SZ Tau, on the other hand, power still remains in a complicated set of low frequencies, indicating that the single periodicity does not fully describe the pulsation.

The Fourier parameters themselves were compared for the two stars. As an example, Fig. 5 shows the amplitude ratio R_{21} changes for the six cycles observed for each of the stars. For RT Aur, the variation is estimated to be 3%. For SZ Tau the variation is much larger (45%). The variation in Fig. 5 (right) is typical of the variation in other Fourier coefficients of the overtone Cepheid.

Thus the *MOST* observations show that the overtone pulsator SZ Tau has a larger variability in the parameters we examined, or a larger instability in the pulsation than the fundamental mode Cepheid.

4. Summary

We present here a brief summary of the period changes or possible period changes in classical Cepheids and the characteristics of pulsation which they suggest. Most probably the $\dot{P}$ which we observe results from a combination of causes.

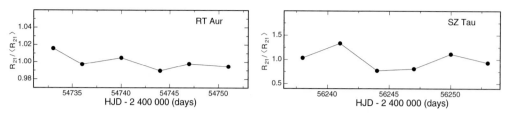

Figure 5. Representative Fourier parameters for the *MOST* data ($R_{21}/\langle R_{21}\rangle$ as a function of HJD - 2 400 000) for each of the six cycles observed, with RT Aur on the left and SZ Tau on the right. We estimate that the $R_{21}/\langle R_{21}\rangle$ parameter varies by 3% in RT Aur and 45% in SZ Tau.

• Evolution through the instability strip: This would have $\dot{P}$ in one direction (at least for a century of observations).

• Light-time effects in binaries: $\dot{P}$ would be cyclic but with a long period. (Known Cepheid binaries have orbital periods longer than 1 year in the Milky Way.)

• Mass loss: $\dot{P}$ would be in one direction,

• Star spots and rotation: $\dot{P}$ would be cyclic and roughly periodic as spots come and go.

The two phenomena discussed here have the following characteristics:

– Flickering $\dot{P}$ (*Kepler*): It is cyclic and reasonably short term. It could be due either to a pulsation phenomenon or star spots.

– Instability in overtones (*MOST*): This seems to be most probably pulsation related.

Acknowledgements

Financial Support was provided by CXC NASA Contract NAS8-03060 (NRE), ESTEC Contract No. 4000106398/12/NL/KML (LS), European Community's Seventh Framework Programme (FP7/2007-2013) under Grant Agreement No. 269194 (IRSES/ASK) (RS, AD), the Hungarian OTKA grant K83790, the KTIA URKUT 10-1-2011-0019 grant, János Bolyai Research Scholarship of the Hungarian Academy of Sciences (RS, AD), the Lendület-2009 Young Researchers Program of the Hungarian Academy of Sciences (AD), and IAU travel grant (RS).

References

Derekas, A., Szabó, Gy. M., Berdnikov, L., *et al.* 2012, *MNRAS*, 425, 1312
Matthews, J. M., Kuschnig, R., Guenther, D. B., *et al.* 2004, *Nature*, 430, 51
Szabados, L. 1983, *Ap&SS*, 96, 185
Walker, G., Matthews, J. M., Kuschnig, R., *et al.* 2003, *PASP*, 115, 1023

Precision Asteroseismology
Proceedings IAU Symposium No. 301, 2013
J. A. Guzik, W. J. Chaplin, G. Handler & A. Pigulski, eds.
© International Astronomical Union 2014
doi:10.1017/S1743921313014087

CoRoT target HD 51844: a δ Scuti star in a binary system with periastron brightening†

Markus Hareter and Margit Paparó

Konkoly Observatory MTA CSFK
Konkoly Thege M. út 15-17, H-1121 Budapest, Hungary
email: hareter@konkoly.hu, paparo@konkoly.hu

Abstract. The star HD 51844 was observed in the *CoRoT* LRa02 as a target in the seismology field, which turned out to be an SB2 system. The 117-day light curve revealed δ Scuti pulsations in the range of 6 to $15\,\mathrm{d}^{-1}$ where four frequencies have amplitudes larger than 1.4 mmag, and a rich frequency spectrum with amplitudes lower than 0.6 mmag. Additionally, the light curve exhibits a 3-mmag brightening event recurring every 33.5 days with a duration of about 5 days. The radial velocities from spectroscopy confirmed that the star is an eccentric binary system with nearly identical masses and physical parameters. The brightening event in the light curve coincides with the maximum radial-velocity separation showing that the brightening is in fact caused by tidal distortion and/or reflected light. One component displays large line-profile variations, while the other does not show significant variation. The frequency analysis revealed a quintuplet structure of the four highest-amplitude frequencies, which is due to the orbital motion of the pulsating star.

Keywords. stars: individual: HD 51844, stars: variables: δ Scuti, binaries: spectroscopic

1. Introduction

Pulsating components of binary systems are of special interest because the masses of the components can be derived directly if the inclination angle i can be constrained sufficiently well. These masses can then be compared to theoretical predictions from evolution theory and pulsation theory.

Thompson *et al.* (2012) propose a class of eccentric binary systems with dynamic tidal distortions found among *Kepler* stars. Typical for this class is a periastron brightening of various forms akin to an electrocardiogram. Hence, they coined the term "heartbeat stars" for this class. Such stars may or may not show eclipses. The light variation allows to constrain orbital elements such as orbital period, inclination, argument of periastron and eccentricity.

† Based on observations made with the Mercator Telescope, operated on the island of La Palma by the Flemish Community, at the Spanish Observatorio del Roque de los Muchachos of the Instituto de Astrofísica de Canarias. Based on observations obtained with the HERMES spectrograph, which is supported by the Fund for Scientific Research of Flanders (FWO), Belgium, the Research Council of K.U. Leuven, Belgium, the Fonds National Recherches Scientific (FNRS), Belgium, the Royal Observatory of Belgium, the Observatoire de Genève, Switzerland and the Thüringer Landessternwarte Tautenburg, Germany. Based on *CoRoT* space-based photometric data; the *CoRoT* space mission was developed and operated by the French space agency CNES, with the participation of ESA's RSSD and Science Programmes, Austria, Belgium, Brazil, Germany, and Spain. Based on observations collected at La Silla Observatory, ESO (Chile) with the HARPS spectrograph at the 3.6-m telescope, under programme LP185.D-0056.

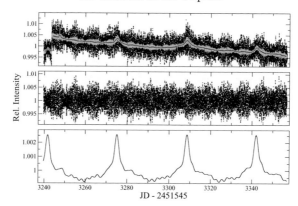

Figure 1. *CoRoT* N2 light curve (top panel, full dots) together with running averages (300 and 3000 points) to highlight the pulsation and the light variation due to the binarity and instrumental effects. To obtain the pure pulsation light curve (middle panel), the binary model (shown in the bottom panel) was subtracted. Note that the y-axes are of different scales.

2. The *CoRoT* light curve

The binary HD 51844 was observed in the *CoRoT* LRa02 for 117 days continuously. The high-quality light curve shows multi-mode δ Sct pulsation and a periodic brightening event. Figure 1 shows the *CoRoT* light curve together with two running averages (300 and 3000 points) to illustrate the p-mode pulsation and the light variation due to the orbital motion of the stars. The outliers and jumps were removed in an iterative approach by involving prewhitening of the δ Sct-type pulsations and correcting the residuals. The removal of outlying data points involved a 3-σ clipping on the residuals rather than the original data. Since the pulsation frequencies are well separated from the binary light variation, we constructed a simple binary light curve by calculating a fit using the first fifteen harmonics of the orbital frequency.

3. Pulsation and frequency modulation

The subsequent frequency analysis was performed on the corrected data where the binary model was subtracted. In this paper, we focus on the effects of the binarity. A more detailed frequency analysis will be published elsewhere, though a brief summary is given below.

The frequency analysis using SigSpec (Reegen 2007) resulted in more than 700 peaks using a significance limit of 6, roughly corresponding to a signal-to-noise ratio (S/N) of 4 in the case of white noise. The pulsation frequencies range from 5 to $15\,d^{-1}$ where four frequencies have amplitudes larger than 1.4 mmag (in the integrated light from both components). These dominant frequencies (f_1 to f_4) are equal to 12.213, 7.054, 6.943 and $8.141\,d^{-1}$, respectively. There are closely spaced frequencies around the two dominant modes (f_1 and f_2) forming distinctive groups. f_4 is a single frequency outside of the groups and reveals an exact 3/2 ratio to the dominant mode. This ratio was also detected among a few frequencies with lower amplitude.

The high-quality of the *CoRoT* data in the seismology field allows one to detect frequency modulation due to the orbital motion. In an eccentric binary system a more complex structure of the side-lobes is expected (Shibahashi & Kurtz 2012). In our case we find side-peaks around the four dominant frequencies spaced by the orbital frequency and sometimes twice the orbital frequency, hereafter referred to as the first- and second-order side-lobes. The amplitudes of the latter side-lobes range from 6.7 to 20 ppm. The average

Table 1. Frequencies, amplitudes and phases used for the determination of theoretical RV curves.

f_c	f_{1+}	f_{1-}	f_{2+}	f_{2-}	a_c	a_{1+}	a_{1-}	a_{2+}	a_{2-}	ϕ_c	ϕ_{1+}	ϕ_{1-}	ϕ_{2+}	ϕ_{2-}
		[d^{-1}]					[ppm]					[rad]		
12.213	12.243	12.183	12.274	12.150	2099	62	67	6.7	12.6	2.28	−2.26	1.94	−1.68	3.09
7.054	7.084	7.026	7.113	6.990	1953	35	48	9.7	8.1	−3.10	−1.81	−0.41	−2.54	0.97
6.943	6.972	6.912	7.007	6.988	1679	33	43	15	8.5	−2.70	−2.90	0.40	0.05	1.91
8.141	8.169	8.110	8.194	8.081	1371	25	50	13	20	−2.12	−1.76	−2.53	1.89	0.72

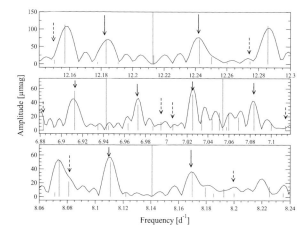

Figure 2. Identified candidate side-lobes at the four dominant frequencies. The solid line shows the Fourier transform after prewhitening with the four dominant frequencies and the vertical solid lines show the frequencies identified by SigSpec. The first-order side-lobes spaced by $0.03\,\mathrm{d}^{-1}$ are marked by solid-line arrows while the second-order side-lobes are marked by dashed–line arrows.

noise of the residuals after prewhitening is 1.8 ppm, which transforms to a S/N of 3.7 for the lowest-amplitude side-lobe. The amplitudes of the lower-than-central-frequency side-lobes (a_{i-}) are systematically higher than those of the higher-than-central-frequency (a_{i+}) side-lobes.

Candidate frequencies for the second-order side-lobes were found around the four dominant frequencies but no side-lobes were detected for the other frequencies. The amplitudes of these second-order side peak candidates range from only 6.7 to 21 ppm. We show that due to their low amplitudes and accordingly large relative errors, they are not sufficient to recover the RV curve purely from photometry.

Figure 2 shows the structure of side-lobes around the four dominant frequencies, with central frequencies subtracted. Table 1 lists the four dominant pulsation frequencies (column f_c), the identified side-lobes ($f_{i\pm}$), the corresponding amplitudes ($a_{i\pm}$) and phases ($\phi_{i\pm}$, zero point: beginning of *CoRoT* observations, JD 2454784.565).

4. The orbit

The binary nature of this star was discovered in the ESO LP185.D-0056 HARPS spectroscopy programme (led by E. Poretti). The light curve clearly shows a recurring brightening event with a period of 33.498 days. Thus, a spectroscopic campaign was organised with good phase coverage that confirmed this period as the binary orbit period. These spectra include 19 HERMES and five McDonald spectra.

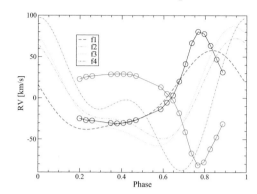

Figure 3. Comparison of the RV curves from spectroscopy to the RV curves from the frequency modulation. The spectroscopic RV measurements are corrected for V_0 and their errors are below $3 \, \mathrm{km \, s^{-1}}$, which is of the order of the symbol size.

The radial velocities (RVs) were determined by fitting synthetic spectra to the observations using the software BinMag3 (by O. Kochukhov, www.astro.uu.ee/~oleg). First, the $v \sin i$ was determined by fitting synthetic spectra (both with $T_{\mathrm{eff}} = 6800 \, \mathrm{K}$ and $\log g = 3.5$) to the observations in which the lines of both components are well separated. In total, 60 measurements were averaged and the mean values of $v \sin i = 41.4 \pm 1.5$ and $41.7 \pm 1.7 \, \mathrm{km \, s^{-1}}$ respectively, were determined. Then, assuming $v \sin i$ for both components equal to $42 \, \mathrm{km \, s^{-1}}$, we determined the RV curves for all spectra with the same software. For each spectrum, 16 frequency ranges were selected to fit the RVs. The standard deviations for the phases where the lines of the components are well separated are $1.5 \, \mathrm{km \, s^{-1}}$ and for the phases where the lines are blended $\approx 3 \, \mathrm{km \, s^{-1}}$. A preliminary fit to these RVs yielded an eccentricity of 0.48 and the projected semi-major axis of $65 \, R_{\odot}$.

The theoretical RV curves calculated from the frequencies (Table 1) are compared to the observed RVs in Fig. 3. For f_1 the agreement is acceptable, while for f_2 to f_4 the agreement is poor. Because the second order side-lobes have extremely low amplitudes, they are prone to significant relative errors. The frequency f_1 is consistent with the slightly lower mass star, which shows the LPV. We estimate that a noise level of less than 1 ppm in the Fourier domain is required to reconstruct the RV curve purely from photometry for such low-amplitude δ Sct stars.

The brightening of the system occurs when the RVs of the components are at maximum separation. Thus, a tidal deformation and reflected light are likely to cause the brightening ("heartbeat effect").

Acknowledgements

M.H. and M.P. acknowledge financial support of the ESA PECS project 4000103541/ 11/NL/KML. M.H. is grateful to J. Benkő and L. Fosatti for valuable discussion. We are grateful to the observers, P. de Cat, P. Lampens, P.M. Arenal, J. Vos, and M. Rainer, for collecting the spectra.

References

Breger, M. 2000, *ASP-CS*, 210, 3
Pamyatnykh, A. A. 2000, *ASP-CS*, 210, 215
Reegen, P. 2007, *A&A*, 467, 1353
Shibahashi, H. & Kurtz, D. W. 2012, *MNRAS*, 422, 738
Thompson, S. E., Everett, M., Mullally, F., *et al.* 2012, *ApJ*, 753, 86

Precision Asteroseismology
Proceedings IAU Symposium No. 301, 2013
J. A. Guzik, W. J. Chaplin, G. Handler & A. Pigulski, eds.

© International Astronomical Union 2014
doi:10.1017/S1743921313014099

The occurrence of non-pulsating stars in the γ Dor and δ Sct instability regions

Joyce A. Guzik[1], Paul A. Bradley[1], Jason Jackiewicz[2],
Katrien Uytterhoeven[3,4], and Karen Kinemuchi[5]

[1]Los Alamos National Laboratory
MS T086, Los Alamos, NM 87545 USA
email: joy@lanl.gov, pbradley@lanl.gov

[2]Dept. of Astronomy, New Mexico State University
P.O. Box 30001, MSC 4500, Las Cruces, NM 88003 USA
email: jasonj@nmsu.edu

[3]Instituto de Astrofísica de Canarias (IAC), 38200, La Laguna, Tenerife, Spain
[4]Departamento de Astrofísica, Universidad de La Laguna, 38200 La Laguna, Tenerife, Spain
email: katrien@iac.es

[5]Apache Point Observatory
P.O. Box 59, 2001 Apache Point Road, Sunspot, NM 88349 USA
email: kinemuchi@apo.nmsu.edu

Abstract. We examine the light curves of over 2700 stars observed in long cadence by the *Kepler* spacecraft as part of the Guest Observer program. Most of these stars are faint (*Kepler* magnitude > 14), and fall near or within the effective temperature and log g range of the γ Dor and δ Sct instability strips. We find that the pulsating stars are obvious from inspection of the light curves and power spectra, even for these faint stars. However, we find that a large number of stars are 'constant', i.e. show no frequencies in the 0.2 to 24 d^{-1} range above the 20 ppm level. We discuss the statistics for the constant stars, and some possible physical reasons for lack of pulsations. On the other hand, γ Dor and δ Sct candidates have been found in the *Kepler* data spread throughout and even outside of the instability regions of both types that were established from pre-*Kepler* ground-based observations. We revisit mechanisms to produce g- or p-mode pulsations in conditions when these modes are not expected to be unstable via the He-ionization κ effect (δ Sct) or convective blocking (γ Dor) pulsation driving mechanisms.

Keywords. stars: variables: δ Scuti, stars: variables: γ Doradus, pulsation driving mechanisms

1. Target selection and data reduction

δ Sct and γ Dor stars in the *Kepler* field have been monitored at an unprecedented level, and the photometric data have generated many questions for asteroseismologists to address. We present analysis of a sample of 633 stars requested for *Kepler* Guest Observer (GO) Cycles 2 and 3, Quarters 6 – 13. These stars were selected from the *Kepler* Input Catalog (KIC) with the criteria: 6200 K < $T_{\rm eff}$ < 8300 K; 3.6 < log g < 4.7; contamination factor < 10^{-2}; *Kepler* Flag 0 (no prior *Kepler* observations).

The stars in our sample are relatively faint, with *Kepler* magnitudes 14 – 15.5. Most of the brighter stars were observed by the *Kepler* Asteroseismic Science Consortium (KASC) (see Uytterhoeven *et al.* 2011). We requested long-cadence data only (29.4 minute integration, Nyquist frequency 24.4 d^{-1}). Since γ Dor stars have frequencies of about 1 d^{-1}, and most δ Sct stars have frequencies of 10 – 20 d^{-1}, long-cadence data are adequate to identify γ Dor and most δ Sct candidates. We used the corrected data from MAST, and a MATLAB script written by J. Jackiewicz that removes outlier points, interpolates the light curves to an equidistant time grid, converts *Kepler* flux to parts per million (ppm), and calculates the power spectrum. We flagged as 'constant' the stars with no peaks in their power spectrum above 20 ppm for frequencies 0.2 to 24.4 d^{-1}. We excluded stars that appear to be eclipsing binaries with stellar components that are either 'constant' or variable, once the binary signal is removed (see Gaulme & Guzik, these proceedings). By these criteria, we find 359 'constant' stars out of the 633 stars in our sample ($\sim$60%).

2. Distribution of 'constant' stars compared to instability regions

We applied a +229 K increase to the effective temperature given by the KIC. This offset roughly accounts for the systematic difference between the KIC T_{eff} and the T_{eff} found using Sloan Digital Sky Survey photometry for the temperature range of our stellar sample, as determined by Pinsonneault *et al.* (2012). We plotted our sample stars that are 'constant' in the $\log g$ - T_{eff} diagram, along with δ Sct and γ Dor instability strip boundaries established from pre-*Kepler* ground-based observations (Rodriguez & Breger 2001; Handler & Shobbrook 2002) (Fig. 1). Only a small number (six) of the 'constant' stars fall within the instability regions. However, the lack of pulsations in these stars requires explanation. Some possibilities are: 1) The stars are pulsating at δ Sct frequencies above the Nyquist frequency 24.4 d^{-1} (no short-cadence data are available); 2) the stars may be pulsating in higher spherical harmonic degree ($l > 3$) modes that aren't easily detectable photometrically. However, photometric variations probably would not completely average out for modes with l up to 10-20 (see Daszyńska-Daszkiewicz 2006; Balona & Dziembowski 1999); 3) The pulsation amplitudes are at or below the noise level of these data, and might be detectable with a longer time series or by reducing noise levels, possibly using the *Kepler* pixel data; 4) A physical mechanism is inhibiting pulsations for some stars. For example, diffusive helium settling might turn off δ Sct pulsations. Diffusion of metals in γ Dor stars may cause the convection zone to become too shallow to enable the convective blocking mechanism; 5) The $\log g$ or T_{eff} in these 'constant' stars may be in error, so that the stars are actually outside the pulsation instability regions. If the observational errors are random, some stars may move out of the instability regions, but it is just as likely that some stars may move into the instability region.

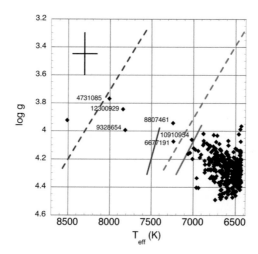

Figure 1. 'Constant' stars from Q6–13 data along with ground-based pulsation instability boundaries. Six 'constant' stars lie within the pulsation instability regions.

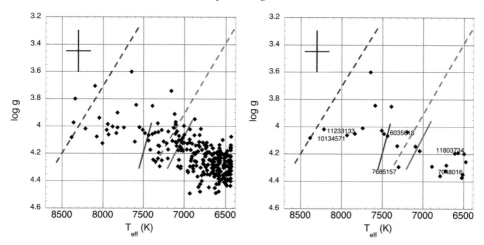

Figure 2. 'Non-constant' stars (left), and twenty-six stars with obvious δ Sct or γ Dor pulsations (right); not all of the pulsating stars lie within their expected instability regions.

3. Distribution of 'non-constant' stars

Figure 2 shows the distribution of the non-constant stars, and the distribution of 26 obviously pulsating γ Dor and δ Sct stars from this sample (see Guzik *et al.* 2013 for further discussion). The stars that fall outside the instability regions, or are within the instability region but pulsating in unexpected frequencies, also require explanation. Additional pulsation driving mechanisms beyond the He-ionization κ effect (δ Sct stars) or convective blocking (γ Dor stars) may be necessary to explain the oscillations of many of the *Kepler* stars. Some ideas that deserve further investigation include: 1) Convective driving at the top of the envelope convection zone, as in DA ZZ Ceti white dwarfs (Wu & Goldreich 1999); 2) κ-effect from Fe concentration due to settling and levitation (Turcotte *et al.* 2000; Théado *et al.* 2009); 3) Interaction and phase offset between Fe and He ionization κ-effect (Löffler 1999); 4) Stochastic excitation (Pereira *et al.* 2011; Antoci *et al.* 2011); 5) Tidally induced pulsations (e.g., KOI-54, Welsh *et al.* 2011).

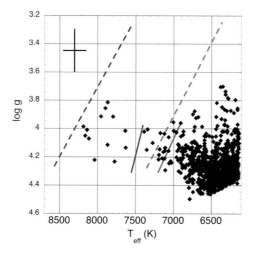

Figure 3. 'Constant' stars from Q14–16 data along with ground-based pulsation instability boundaries. 34 'constant' stars lie within the pulsation instability regions.

4. Q14 – 16 data

We have now received data on an additional 2138 stars proposed for GO Cycle 4. This sample is not unbiased, as it was cross-correlated with stars showing variability in full-frame images. The selection criteria included stars with lower T_{eff}, down to 5800 K. Using the same criteria as above, we find 984 'constant' stars (46%); 34 out of 984 of these stars (3.5%) are within pre-*Kepler* instability region boundaries (Fig. 3).

5. Summary and future work

In our *Kepler* GO Q1 – 13 data for 633 A-F main-sequence stars observed, 359 stars, or about 60%, were found to be constant, defined as showing no frequencies above 20 ppm in their light curves between 0.2 and 24.4 d^{-1}. Only six photometrically non-varying stars are within the γ Dor and δ Sct pulsation instability regions established from pre-*Kepler* ground-based observations. Twenty-six stars, or about 10% of the non-constant stars are obvious γ Dor or δ Sct candidates. However, many of these stars do not lie within their expected instability regions, or are outside of both instability regions. From an additional 2138 stars observed in Q14 – 16, we find 984 (46%) constant stars; of these, 34 stars lie within the instability region. These data have revealed many new γ Dor/δ Sct candidates (see Bradley *et al.*, these proceedings). We plan to use stellar modeling to determine whether diffusive settling can eliminate pulsations in γ Dor or δ Sct stars, and to explore alternate pulsation driving mechanisms to explain stars that show pulsations at unexpected frequencies or lie outside of predicted instability boundaries.

Acknowledgements

The authors gratefully acknowledge support from the NASA *Kepler* Guest Observer program. K.U. acknowledges support from the Spanish National Plan of R&D for 2010. This project greatly benefitted from Project FP7-PEOPLE-IRSES:ASK No. 269194.

References

Antoci, V., Handler, G., Campante, T. L., *et al.* 2011, *Nature*, 477, 570
Balona, L. A. & Dziembowski, W. A. 1999, *MNRAS*, 309, 221
Daszyńska-Daszkiewicz, J., Dziembowski, W. A., & Pamyatnykh, A. A. 2006, *MemSAIt*, 77, 113
Grigahcène, A., Antoci, V., Balona, L., *et al.* 2010, *ApJ*, 713, L192
Guzik, J. A., Bradley, P. A., Jackiewicz, J., Uytterhoeven, K., & Kinemuchi, K. 2013, *Astronomical Review*, published on-line on 3 October 2013
Handler, G. & Shobbrook, R. R. 2002, *MNRAS*, 333, 251
Löffler, W. 2000, *ASP-CS*, 203, 447
Pereira, T. M. D., Suárez, J. C., Lopes, I., *et al.* 2007, *A&A*, 464, 659
Pinsonneault, M. H., An, D., Molenda-Żakowicz, J., Chaplin, W. J., Metcalfe, T. S., & Bruntt, H. 2012, *ApJS*, 199, 30
Rodriguez, E. & Breger, M. 2001, *A&A*, 366, 178
Théado, S., Vauclair, S., Alecian, G., & Le Blanc, F. 2009, *ApJ*, 704, 1262
Turcotte, S. 2002, *ApJ*, 573, L129
Uytterhoeven, K., Moya, A., Grigahcène, A., *et al.* 2011, *A&A*, 534, A125
Welsh, W. F., Orosz, J. A., Aerts, C., *et al.* 2011, *ApJS*, 197, 4
Wu, Y. & Goldreich, P. 2000, *ASP-CS*, 203, 508

Precision Asteroseismology
Proceedings IAU Symposium No. 301, 2013
J. A. Guzik, W. J. Chaplin, G. Handler & A. Pigulski, eds.
© International Astronomical Union 2014
doi:10.1017/S1743921313014105

BRITE-Constellation: Nanosatellites for precision photometry of bright stars

W. W. Weiss[1†], A. F. J. Moffat[2†], A. Schwarzenberg-Czerny[3†],
O. F. Koudelka[4†], C. C. Grant[5], R. E. Zee[5], R. Kuschnig[1†],
St. Mochnacki[6†], S. M. Rucinski[6†], J. M. Matthews[7†], P. Orleański[8†],
A. A. Pamyatnykh[3†], A. Pigulski[9†], J. Alves[1†], M. Guedel[1†],
G. Handler[3†], G. A. Wade[10†], A. L. Scholtz[11],
and the CCD Tiger Team‡

[1]University of Vienna, Institute for Astrophysics, Tuerkenschanzstr. 17, 1180 Vienna, Austria;
email: `werner.weiss@univie.ac.at`
[2]Dept. de physique, Université de Montréal, Canada; [3]Copernicus Astronomical Center,
Warsaw, Poland; [4]Graz University of Technology, Graz, Austria; [5]Space Flight Laboratory,
University of Toronto, Canada; [6]Dept. of Astronomy and Astrophysics, University of Toronto,
Canada; [7]Dept. of Physics and Astronomy, University of British Columbia, Canada; [8]Space
Research Center of the Polish Academy of Sciences, Warsaw, Poland; [9]Astron. Institute,
University of Wrocław, Poland; [10]Dept. of Physics, Royal Military College of Canada, Ontario,
Canada; [11]Institute of Telecommunications, Vienna University of Technology, Austria

Abstract. *BRITE-Constellation* (where *BRITE* stands for *BRIght Target Explorer*) is an international nanosatellite mission to monitor photometrically, in two colours, brightness and temperature variations of stars brighter than $V \approx 4$, with precision and time coverage not possible from the ground. The current mission design consists of three pairs of 7 kg nanosats (hence "Constellation") from Austria, Canada and Poland carrying optical telescopes (3 cm aperture) and CCDs. One instrument in each pair is equipped with a blue filter; the other, a red filter. The first two nanosats (funded by Austria) are *UniBRITE*, designed and built by UTIAS-SFL (University of Toronto Institute for Aerospace Studies-Space Flight Laboratory) and its twin, *BRITE-Austria*, built by the Technical University Graz (TUG) with support of UTIAS-SFL. They were launched on 25 February 2013 by the Indian Space Agency, under contract to the Canadian Space Agency.

Each *BRITE* instrument has a wide field of view (≈ 24 degrees), so up to 15 bright stars can be observed simultaneously in 32×32 sub-rasters. Photometry (with reduced precision but thorough time sampling) of additional fainter targets will be possible through on-board data processing. A critical technical element of the *BRITE* mission is the three-axis attitude control system to stabilize a nanosat with very low inertia. The pointing stability is better than 1.5 arcminutes rms, a significant advance by UTIAS-SFL over any previous nanosatellite.

BRITE-Constellation will primarily measure p- and g-mode pulsations to probe the interiors and ages of stars through asteroseismology. The *BRITE* sample of many of the brightest stars in the night sky is dominated by the most intrinsically luminous stars: massive stars seen at all evolutionary stages, and evolved medium-mass stars at the very end of their nuclear burning phases (cool giants and AGB stars). The Hertzsprung-Russell diagram for stars brighter than mag $V = 4$ from which the *BRITE-Constellation* sample will be selected is shown in Fig. 1. This sample falls into two principal classes of stars:

(1) Hot luminous H-burning stars (O to F stars). Analyses of OB star variability have the potential to help solve two outstanding problems: the sizes of convective (mixed) cores in massive stars and the influence of rapid rotation on their structure and evolution.

(2) Cool luminous stars (AGB stars, cool giants and cool supergiants). Measurements of the

---

† Member of the *BRITE-Constellation* Executive Science Team (BEST).
‡ M. Chaumont, C. Grant, J. Lifshits, A. Popowicz, M. Rataj, P. Romano, M. Unterberger, R. Wawrzaszek, T. Zawistowski & BEST.

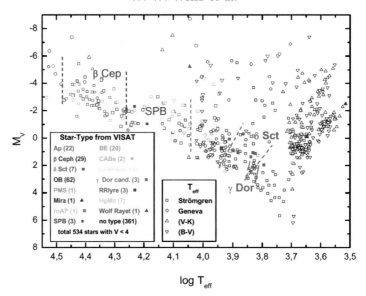

Figure 1. H-R diagram for stars brighter than mag $V = 4$ with instability strips indicated.

time scales involved in surface granulation and differential rotation will constrain turbulent convection models.

Mass loss from these stars (especially the massive supernova progenitors) is a major contributor to the evolution of the interstellar medium, so in a sense, this sample dominates cosmic "ecology" in terms of future generations of star formation. The massive stars are believed to share many characteristics of the lower mass range of the first generation of stars ever formed (although the original examples are of course long gone).

BRITE observations will also be used to detect some Jupiter- and even Neptune-sized planets around bright host stars via transits, as expected on the basis of statistics from the *Kepler* exoplanet mission. Detecting planets around such very bright stars will greatly facilitate their subsequent characterization. *BRITE* will also use surface spots to investigate stellar rotation.

The following Table summarizes launch and orbit parameters of *BRITE-Constellation* components.

Designation	Name	F	Vehicle	Launch date	Orbit km	Descending node	Drift min
Owner: Austria
BRITE-A	*BRITE-Austria*	B	PSLV-21	25 Feb 2013	800	18:00	0
BRITE-U	*UniBRITE*	R	PSLV-21	25 Feb 2013	circular	18:00	0
Owner: Canada
BRITE-C1	*Toronto*		DNEPR	Q3-4/2014	629×577	10:30	40
BRITE-C2	*Montréal*		DNEPR	Q3-4/2014	629×577	10:30	40
Owner: Poland
BRITE-P1	*Lem*	B	DNEPR	21 Nov. 2013	600×900	10:30	100
BRITE-P2	*Heweliusz*	R	China LM-4	2014	SSO/630		

The full version of this paper describing in more detail *BRITE-Constellation* will be published separately in a journal. The symposium presentation is available at http://iaus301.astro.uni.wroc.pl/program.php

Keywords. space vehicles, instrumentation: photometric, stars: general, interiors, oscillations

Precision Asteroseismology
Proceedings IAU Symposium No. 301, 2013
J. A. Guzik, W. J. Chaplin, G. Handler & A. Pigulski, eds.

© International Astronomical Union 2014
doi:10.1017/S1743921313014117

Stellar Observations Network Group: The prototype is nearly ready

**Frank Grundahl[1], Jørgen Christensen-Dalsgaard[1], Pere L. Pallé[2,3],
Mads F. Andersen[1], Søren Frandsen[1], Kennet Harpsøe[4,5],
Uffe Græe Jørgensen[4,5], Hans Kjeldsen[1], Per K. Rasmussen[3],
Jesper Skottfelt[4,5], Anton N. Sørensen[4], and Andrea Triviño Hage[2,3]**

[1]Stellar Astrophysics Centre, Aarhus University, Ny Munkegade 120,
8000 Aarhus C, Denmark, email: `fgj@phys.au.dk`

[2]Instituto de Astrofísica de Canarias (IAC), E–38200 La Laguna, Tenerife, Spain
[3]Dept. Astrofísica, Universidad de La Laguna (ULL), E–38206 Tenerife, Spain
[4]Niels Bohr Institute, University of Copenhagen, Juliane Mariesvej 30,
2100 København Ø, Denmark
[5]Centre for Star and Planet Formation, Natural History Museum,
University of Copenhagen, Østervoldgade 5–7, 1350 København K, Denmark

Abstract. The prototype telescope and instruments for the Stellar Observations Network Group (SONG) are nearing completion at the Observatorio del Teide on Tenerife. In this contribution we describe the current status (autumn 2013) of the telescope and its instrumentation. Preliminary performance characteristics are presented for the high-resolution spectrograph based on daytime observations of the Sun and a 4 hour test series obtained for the sub-giant β Aquilae.

Keywords. instrumentation: spectrographs, techniques: radial velocities, Sun: oscillations, stars: oscillations

1. Introduction

During the past 10 years the field of asteroseismology has undergone a virtual revolution in the amount and quality of the data which have become available. This is primarily due to space-based missions such as *WIRE* (Buzasi 2004), *MOST* (Matthews *et al.* 2004) *CoRoT* (Baglin *et al.* 2012) and *Kepler* (Chaplin & Miglio 2013). In particular *CoRoT* and *Kepler* have provided exciting new results for a wide range of stars and stellar types. In common for all these missions is that the stellar oscillations are observed as intensity variations (photometry). The first ground-based detections of solar-like oscillations used spectroscopy to determine variations in radial velocity or equivalent width of Balmer lines (Kjeldsen *et al.* 2003). The ground-based efforts to detect the oscillations via photometry were largely unsuccessful due to the effects of the Earth's atmosphere. It is well known that for solar-like oscillators radial-velocity measurements have lower background levels (Harvey 1988) and that low-frequency and $l = 3$ modes can be detected more easily than for intensity measurements (Dziembowski 1977, Harvey 1988).

The Stellar Observations Network Group (SONG) is an initiative to develop a ground-based network of 1m-class telescopes which makes it possible to obtain high-quality oscillation measurements for the brightest solar-like oscillators in the sky. The second driver for SONG is to carry out studies of extra-solar planets using the microlensing method by following up on real-time triggers from large surveys such as OGLE† and MOA‡. In

† `http://ogle.astrouw.edu.pl/ogle4/ews/ews.html`
‡ `http://www.phys.canterbury.ac.nz/moa/index.html`

this contribution we describe the component of SONG related to asteroseismology – its design, instruments and status of the network efforts.

2. The SONG node at Tenerife

For the development of the network to proceed as smoothly as possible it was decided to complete the construction of a full prototype network node before requesting funding for more nodes such that the feasibility could be determined.

2.1. *Telescope*

Based on the results by Butler *et al.* (2004) from UVES we found by scaling of their results that a telescope of 1m diameter would be sufficient for the detection of oscillations in bright stars. The telescope is carried on an alt-az mount and driven by magnetic-torque motors which allow fast re-pointing (speeds up to 20° per second). It is a very compact design ensuring a very stiff structure which gives a measured pointing model performance of 3″ with a pointing model based on only 30 stars. The telescope is housed in a 5m diameter dome which has been insulated to limit thermal buildup during daytime – at present the insulation does not provide the desired reduction in dome temperature but a cooling unit will be installed to remedy it.

At the time of writing (September 2013) the control of the active primary mirror, which is only 5 cm thick, is undergoing final tests. The short-term wavefront reconstruction and control is working very well and delivers a nearly diffraction limited performance of the telescope optics based on Shack-Hartmann measurements of bright stars. The Coudé path will undergo final alignment during the autumn once the work on the active optics is completed.

While not fully commissioned yet, our tests with remote observations show that we can acquire and observe objects over the whole sky with fully automatic scripts and operating remotely.

The two instruments for SONG consist of a high-resolution optical spectrograph mounted at the telescope Coudé focus to obtain high precision radial velocities and two EMCCD cameras mounted at one of the two Nasmyth foci. The EMCCD cameras are used for high spatial-resolution photometry and lucky imaging (see for example Harpsøe *et al.* 2012a).

2.2. *Spectrograph*

In order to facilitate precision radial-velocity measurements, a high-resolution spectrograph is needed. The SONG spectrograph is a "classical" cross-dispersed, white pupil, echelle spectrograph with a design inspired by UVES and HARPS. The basic parameters for the spectrograph are provided in Table 1. The spectrograph is located in a standard, insulated shipping container next to the telescope pier. The legs for the spectrograph optical table are secured to a concrete foundation which is not in mechanical contact with the container or telescope foundations. Inside the container are the control computers, power supplies, and calibration lamps for the spectrograph which itself is housed in an insulated, temperature regulated box.

Inspired by the excellent results by Butler *et al.* (2004) we decided to used the iodine method for the determination of the precise radial velocities required for asteroseismic studies. We employ an iodine cell of 10 cm length and 5 cm diameter, which is operated by a ThorLabs TC200 controller and kept at a temperature of 65°C to ensure full evaporation of the iodine. For the calculation of radial velocities a high-resolution template of the

Table 1. Characteristics of the SONG spectrograph

Resolution[1]	35 000 – 112 000
Pixel scale (average)[1]	0.02 Å
Wavelength range	4400 – 6900 Å
Number of spectral orders[2]	51
Collimated beam diameter	75 mm
Collimator focal ratio	F/6 off-axis parabola
Echelle grating	Newport, R4, 31.6 l/mm, 9×33 cm² ruled area
Detector	Andor, IKon-L, 2K×2K
Readout speed/time	3 MHz, 2.3 s full-frame
Number of slits	6
Slit length	10″
Order separation	24 pixels (bluest) and 7 pixels (reddest)

Notes:
[1] Average value over all orders. The resolution and pixel scale varies slightly across the detector area. The reddest orders have resolutions up to 125 000.
[2] Only orders with wavelengths shorter than 5250 Å are fully covered by the detector. The camera has been designed to cover the orders fully with a larger detector.

iodine absorption spectrum is needed. This was obtained using the Fourier Transform Spectrometer at Lund Observatory.

In addition to light from the telescope it is possible to feed the spectrograph with light from a ThAr lamp for wavelength calibration and a halogen lamp for flat fielding. Furthermore it was decided at an early stage to provide also the possibility to observe the Sun by pointing the telescope to the blue daytime sky (looking through a dome window). It was, however, found during laboratory tests that an optical fibre pointing directly to the Sun provided a very high flux – allowing exposure times up to 0.5 s through the iodine cell without saturating the detector. Our current setup comprises a standard, circular, optical fibre from ThorLabs with a core diameter of 400 µm giving a fully illuminated slit. Outside the container the entrance of the fibre is mounted on a solar tracker ensuring pointing to the Sun. In order to provide scrambling of the light before it enters the fibre we have mounted a piece of grey masking tape 1 cm in front of the fibre. The results from daytime solar observations are described below.

2.3. Lucky-imaging camera

The second main science goal for SONG is to study exoplanets via the microlensing method, requiring that precise photometry can be obtained in the crowded fields towards the Galactic bulge where most such events occur. Photometry in crowded fields benefits greatly from high spatial resolution and to allow for this a two-colour lucky imaging instrument is placed at the Nasmyth focus of the telescope. Via a dichroic beamsplitter (split at 6500 Å) light is directed towards two identical science cameras allowing simultaneous imaging in two bands. In addition, each channel has a filter wheel. The cameras are Andor iXon model 897, which can run at full frame speeds of up to 33 Hz – in this way we can carry out lucky imaging.

The field of view for the cameras is $46''$ on a side, providing a $0''\!.09$ pixel scale in order to provide $\sim$2 pixel sampling of the telescope point-spread function at a wavelength of 800 nm. The on–sky performance of the cameras has not yet been tested. However, a nearly identical camera setup (same camera and pixel scale) has been established at the Danish 1.54 m telescope in La Silla and results of test observations can be found in Harpsøe *et al.* (2012b).

3. Spectroscopic measurements

With the prototype telescope nearly operational and a dedicated fibre-feed for solar observations it has been possible to carry out limited tests of the velocity precision that can be obtained with the spectrograph.

3.1. *Solar observations*

During June 11-16, 2012, we carried out a solar-observing experiment. Each morning of the 6 days the input end of the fibre was pointing towards the Sun and the tracker started. We collected approximately 12 500 spectra each day, totalling nearly 70 000 spectra. The data were reduced using a dedicated pipeline written in IDL (see Triviño Hage *et al.* 2012). As mentioned previously, the exposure time was set at 0.5 s to avoid saturation in the wavelength region covered by the iodine absorption lines (5000 – 6200 Å). In conjunction with a 2.3 s readout time this led to a sampling rate of 2.81 s between each successive spectrum. Each day, before starting the solar observations, calibration frames were acquired, consisting of 120 flat fields, 120 bias images and 3 ThAr spectra. In addition, in order to construct a solar template spectrum needed for the velocity extraction, 50 solar spectra without the iodine cell were acquired followed immediately by 50 flat fields through the iodine cell – these were used to determine the spectrograph point-spread function. A detailed account of the observations required for the iodine method was provided by Butler *et al.* (1996). All the spectra were obtained with a resolution of 120 000.

A short segment of the time-series is shown in Fig. 1. The well known 5-minute solar oscillation is easily seen in the data – a clear indication that the measurement precision is very high. We have computed the power spectrum for the entire series – see Fig. 2. An analysis of the high-frequency noise indicated an average measurement precision of $0.8\,\mathrm{m\,s^{-1}}$ per point, with some degradation when the Sun is low in the sky, in late afternoon.

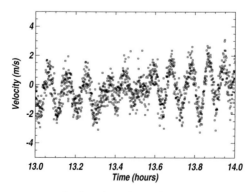

Figure 1. A one hour segment of solar data from June 11, 2012. The solar oscillations are clearly detected and the measured precision per data point (0.5 s exposure + 2.3 s readout) in this segment is close to $65\,\mathrm{cm\,s^{-1}}$.

A more detailed account and comparison to dedicated solar instruments is presented by Pallé *et al.* (2013). This work showed that the measurements from SONG compare extremely well to simultaneous data from the well-characterised Mark-I instrument of the BiSON network (Brookes et al. 1978) and the GOLF (Gabriel *et al.* 1995) instrument on-board the SoHO† spacecraft. This comparison showed that SONG has a very low noise at high frequencies – even lower than found by the Mark-I and GOLF instruments. While this points to excellent measurements, the lower noise is in part likely caused by differences in the details of how the three instruments measure the oscillations.

3.2. *β Aquilae*

Following the initial telescope installation the sub-giant β Aquilae was observed for four hours during July 11-12, 2012. The spectrograph employed a slit with a projected width of $1''03$ – providing a spectral resolution of 90 000. The exposure time was 120 s for the first 100 spectra, and 180 s for the last 20. Template observations of the star were also acquired during the same night. Corsaro *et al.* (2012) have previously provided a detailed study of β Aquilae based on spectra from the SARG spectrograph at the Italian telescope TNG on La Palma (Gratton *et al.* 2001).

In Fig. 3 the radial velocity for the 4 h observing segment is shown. It is clear that β Aquilae has a higher oscillation amplitude and lower frequency than the Sun. A direct comparison to the Corsaro *et al.* data shows that the amplitude and frequency is in excellent agreement with the SARG results (which were analysed using the same software).

4. Towards other nodes

While the prototype node is nearing completion at Observatorio del Teide, it is important to start network expansion – Arentoft *et al.* (2014) have shown the benefits of multiple sites for asteroseismic studies and how stars can be optimally selected in case of a 1, 2 or 3-node network.

† Solar and Heliospheric Observatory

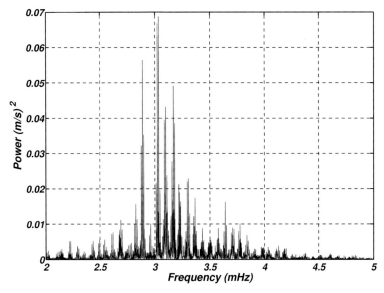

Figure 2. The resulting power spectrum for the 6-day solar observing series obtained June 11-16, 2012 from the Tenerife SONG node.

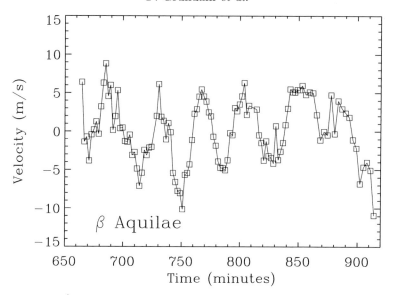

Figure 3. A 4^h test sequence obtained for β Aquilae in July 2012 with SONG.

Work on the installation of the second SONG node is nearly completed at the Delingha Observing Station in the Quinghai province in China‡ with first light expected in late 2013. We anticipate that regular two-node network observations can begin around mid-2014.

SONG is also actively pursuing funding possibilities for other nodes and hope to be successful in this endeavour within 1-2 years. To this end we plan to re-design the spectrograph pre-slit table to allow feeding light from the telescope via an optical fibre. This will allow the use of existing 1 m-class telescopes to be used fully or partly as a SONG node. It is thus our hope that before the end of 2015 the northern half of SONG can be operational.

Acknowledgements

We gratefully acknowledge support from the Villum Foundation, the Carlsberg Foundation and Forskninsgrådet for Natur og Univers (FNU) towards the development of the SONG prototype. Funding for the Stellar Astrophysics Centre and the Centre for Star and Planet Formation is provided by The Danish National Research Foundation (Grant DNRF106 and grant DNRF97, respectively). The research is supported by the ASTERISK project (ASTERoseismic Investigations with SONG and *Kepler*) funded by the European Research Council (Grant agreement no.: 267864). Support from the Spanish Ministry of Economy and Competitiveness (MINECO, grant AYA2010-17803) is gratefully acknowledged.

References

Arentoft, T., Tingley, B., Christensen-Dalsgaard, J., Kjeldsen, H., White, T. R., & Grundahl, F. 2014, *MNRAS*, 437, 1318
Baglin, A. & Michel, E., *CoRoT* team 2012, *ASP-CS*, 462, 492
Brookes, J. R., Isaak, G. R., & van der Raay, H. B. 1978, *MNRAS*, 185, 1

Butler, R. P., Marcy, G. W., Williams, E., McCarthy, C., & Dosanjh, P., Vogt S. S. 1996, *PASP*, 108, 500

Butler, R. P., Bedding, T. R., Kjeldsen, H., *et al.* 2004, *ApJ*, 600, L75

Buzasi, D. L. 2004, in: F. Favata, S. Aigrain, & A. Wilson (eds.), *Stellar Structure and Habitable Planet finding*, ESA Special Publication, 538, 205

Chaplin, W. J. & Miglio, A. 2013, *ARAA*, 51, 353

Corsaro, E., Grundahl, F., Leccia, S., Bonnano, A., Kjeldsen, H., & Paternò, L. 2012, *A&A*, 537, 8

Dziembowski, W. 1977, *AcA*, 27, 203

Gabriel, A. H., Grec, G., Charra, J., *et al.* 1995, *Solar Phys.*, 162, 61

Gratton, R. G., Bonanno, G., Bruno, P., *et al.* 2001, *Experimental Astronomy*, 12, 107

Harpsøe, K. B. W., Andersen, M. I., & Kjærgaard, P. 2012a, *A&A*, 537, A50

Harpsøe, K. B. W., Jørgensen, U. G., Andersen, M. I., & Grundahl, F. 2012b, *A&A*, 542, A23

Harvey, J. W. 1988, in: J. Christensen-Dalsgaard, & S. Frandsen (eds.), *Advances in Helio- and Asteroseismology*, IAU Symposium No. 123 (Dordrecht: D. Reidel Publishing Co.), p. 497

Kjeldsen, H., Bedding, T. R., Baldry, I. K., *et al.* 2003, *AJ*, 126, 1483

Matthews, J. M., Kuschnig, R., Guenther, D. B., *et al.* 2004, *Nature*, 430, 51

Pallé, P. L., Grundahl, F., Triviño Hage, A., *et al.* 2013, *Journal of Physics, Conf. Series*, 440, 012051

Triviño Hage, A., Uytterhoeven, K., Grundahl, F., & Pallé, P. L., and the SONG team 2012, *AN* 333, 1107

Precision Asteroseismology
Proceedings IAU Symposium No. 301, 2013   ⓒ International Astronomical Union 2014
J. A. Guzik, W. J. Chaplin, G. Handler & A. Pigulski, eds.   doi:10.1017/S1743921313014129

Extracting oscillation frequencies from data: various approaches

C. A. Engelbrecht

Department of Physics, University of Johannesburg, PO Box 524, Auckland Park 2006,
South Africa
email: chrise@uj.ac.za

Abstract. Asteroseismology depends absolutely on the detection of authentic pulsation signatures in stars. A variety of mathematical and statistical tools have been developed to extract such signatures from photometric and spectroscopic time series. The earliest tools were developed on the platform of Fourier analysis, and Fourier-based methodology still plays a major part in the detection of pulsation signatures in the present day. Alternative approaches have been gaining ground in recent years. This article offers a brief but broad review of the various methodologies for detecting authentic periodic signals that have been developed over the past few decades, including examples of their pitfalls and successes.

Keywords. methods: data analysis, methods: statistical, stars: oscillations

1. Introduction

Asteroseismology has vastly expanded the scope, range and depth of our understanding of stellar structure and evolution in recent years. As but one prominent example of this, Bedding *et al.* (2011) found a very powerful correlation between the spacing of *g*-modes and the dominant nuclear energy source (i.e. helium-core or hydrogen-shell fusion) in red giants. Such breakthroughs depend critically on the accuracy of the oscillation parameters (frequencies, amplitudes and phases) that are determined from the analysis of observational data. Fourier-based methods of analysis have been used successfully for over a hundred years, starting with the work of Schuster (1897). However, these methods have their pitfalls and other mathematical and statistical points of departure have received increasing attention as asteroseismology has matured to its present strength. This article briefly reviews practically the full range of asteroseismologically relevant methods that have been presented in the general literature. Their basic premises, strengths and weaknesses are briefly summarized and examples of their successful applications are presented.

The scope of published methods may be naturally classified under the two headings of Fourier-based methods and Dispersion/Entropy-based methods, respectively. These two classes will be discussed separately, as sections 2 and 3. The impact of Bayesian approaches on existing methods of period-finding is discussed in section 4. Section 5 contains a summary of ancillary issues of importance and some suggestions for further exploration of this topic.

At the outset, it might be useful to emphasise that the *spectrum* of a time series represents the respective *probability amplitudes* (quantified via one or other measure statistic) of regular variations present in the time series, as a function of variation frequency. This article considers the various methods available for computing such a spectrum.

2. Fourier-based methods

The popularity of Fourier-based methods is due to the mathematical properties of harmonic functions, which make them attractive as the basis of algorithms designed to search for periodicity in the frequency domain. The original Fourier-based method is the Fourier transform, which applies to a continuous function in the time domain. Measurements taken of real signals are, of necessity, not continuous but discrete. Real measurements are also not infinitely long in total extent. The adaptation of the Fourier transform to discrete data (i.e. a discrete sampling of a supposed underlying continuous function) constitutes the Discrete Fourier Transform (DFT), originally applied in cases where the discrete sampling is exactly equally-spaced in the time domain. Both the original DFT and the efficiently-computable Fast Fourier Transform (FFT) can be attributed to Gauss (Heideman *et al.* 1984). Astronomical measurements are almost never exactly equally-spaced. Therefore, a modification to the equally-spaced DFT for application to non-equally-spaced data (which also have a finite total extent), was proposed by Deeming (1975). As pointed out by Ferraz-Mello (1981), amongst others, there is no mathematical foundation supporting the Deeming formula. Although it appears to work well in many cases, it remains, in its essence, a heuristic device. Given the importance of accuracy and precision of period detections in asteroseismology, attention should be given to alternatives to the raw Deeming formula that provide answers with a higher level of confidence. Ferraz-Mello (1981) introduced the so-called Date-Compensated DFT which introduces a non-zero constant mean into the calculated transform. The major advantage of this refinement of the Deeming formula for a non-equally-spaced DFT (called the Deeming DFT in the remainder of this article) is a more accurate estimation of the *amplitudes* of harmonic signals present in the observed data, especially when the number of observations is relatively small. Ferraz-Mello also introduced a weighting algorithm to deal with observations that do not all have exactly the same measurement error. The Deeming DFT and its refinements have been used very extensively in the practice of asteroseismology over many decades, through to the present. It works very well when applied to quasi-equally-spaced data (like the data in the *Kepler* database). However, as stated already, it remains a heuristic device and it should always be used with circumspection.

Assigning a *statistical significance* to suspected oscillation frequencies harvested from a Deeming DFT is problematic, as the statistical conclusions that attach to purely equally-spaced time series do not apply to non-equally-spaced data. This makes it impossible to attach statistical significance *a priori* to peaks in Deeming DFT periodograms. Scargle (1982) attempted to address this particular problem by deriving an alternative algorithm to the Deeming DFT, as discussed below. Breger *et al.* (1993) offered a helpful rule of thumb to estimate the statistical significance of periodogram peaks obtained with a Deeming DFT: if the harmonic wave associated with a frequency peak has an amplitude greater than 4 times the residual noise level in the periodogram (see section 5.4 of Aerts *et al.* (2010) for a detailed discussion), it is statistically significant. It is unfortunate that this commonly used rule of thumb is being called a "four-sigma detection" with growing regularity, since this might create considerable confusion in the future. The terms "four-sigma detection", "five-sigma detection", etc. have a very well-established meaning in the theory of normal distributions and such phrases are commonly used in many areas of modern science. The Breger rule of thumb means something totally different and it should rather be referred to as the "4:1 rule" or something similar. Useful as the rule has proven to be, the brief experiments conducted by Koen (2010) show that the quantitative statistical meaning of the "4:1 rule" can vary wildly from one dataset to the next. The present author has independently found similar results (unpublished). The discussion in

Aerts *et al.* (2010) reminds the reader of the very particular and limited context wherein Breger found the "4:1 rule" to be valid. Considerable caution is required when applying it in general.

Scargle's seminal paper (Scargle 1982) followed on foundations laid by Barning (1963), Vaníček (1969) and Lomb (1976). Scargle claimed that the algorithm that he put forward (commonly called the Lomb-Scargle periodogram or LS periodogram for short) retrieves the statistical properties of the classical (equally-spaced) DFT; therefore, the probability of obtaining a peak of height X in the LS periodogram of a time series consisting of normally-distributed noise is proportional to $\exp(-X)$. However, this claim encounters an insurmountable practical difficulty, as has been pointed out repeatedly (see, for example, Horne & Baliunas (1986), Koen (1990), Frescura *et al.* (2008) and Scargle's own recognition of the problem in his paper): when computing the periodogram for more than one frequency, using non-equally-spaced data, one requires the exact number of statistically independent frequencies in the set of frequency values at which the periodogram is being calculated. There is no way to determine this number (see Horne & Baliunas 1986 and the follow-up work by Frescura *et al.* 2008). Various workers have concluded that Monte Carlo procedures are the only reliable way of determining the statistical significance of peaks in a LS periodogram (for instance, see the papers just quoted). An interesting and thorough discussion of this question appears in Vio *et al.* (2010).

Mathematically, the Lomb-Scargle periodogram calculates the goodness-of-fit of a harmonic (i.e. sinusoidal) function compared with the data, for a selected grid of frequencies. It therefore works particularly well at extracting the frequencies of small-amplitude oscillations in general (since small excursions from an equilibrium state are always well approximated by a harmonic variation). It has been used with considerable success for more than thirty years and remains one of the most widely-used methods of extracting oscillation frequencies from astronomical data. It should be noted that the LS periodogram and the Deeming DFT do not always produce similar results; a good example of discrepant results is discussed in the very thorough study of the LS periodogram by Vio *et al.* (2013). The LS periodogram is generally less susceptible to aliasing than the Deeming DFT is – see e.g. Reegen (2007). Furthermore, Frescura *et al.* (2008) demonstrated that it is vitally important to "oversample" the frequency grid when calculating the LS periodogram.

A substantial overhaul of the classical LS periodogram is contained in the work of Cumming *et al.* (1999) and Zechmeister & Kürster (2009). Cumming *et al.* introduced a floating mean into the LS periodogram calculation, meaning that a constant term is included in the fit for each trial frequency. Their detailed study led them to conclude that "allowing the mean to float is crucial if the number of observations is small, the sampling is uneven, or there is a period comparable to the duration of the observations or longer". Uneven sampling is almost ubiquitous in the data used in asteroseismology, hence their conclusions deserve careful scrutiny. These authors caution that periodic signals could be totally missed, or their amplitudes miscalculated, when the floating mean is omitted. Zechmeister & Kürster added a weighting procedure, to accommodate observations that do not all have exactly the same precision. They chose the name "Generalised Lomb-Scargle periodogram" for their more refined procedure, abbreviated as GLS (although they point out one previous use of the same phrase in a different context). They concluded that, compared to the classical LS periodogram, GLS provides a more accurate frequency determination, is less susceptible to aliasing, and gives a much better determination of the spectral intensity. A recent, comprehensive comparison of various period-finding methods by Graham *et al.* (2013a) provides strong endorsement for the accuracy of the GLS algorithm.

A detailed study of an alternative approach for calculating statistical significances of detected periodicities in unequally-spaced data was presented as the SigSpec algorithm by Reegen (2007), who claimed period-finding accuracy on a par with LS but with less susceptibility to aliasing. The most prominent new feature of SigSpec is its inclusion of *phase* information, while it also incorporates the floating mean and weighting refinements of GLS. SigSpec has been widely used in the literature, often in conjunction with other period-finding methods. Zechmeister & Kürster (2009) demonstrated many equivalences between the GLS and SigSpec approaches to period-finding.

A quite different methodology for period-finding, SparSpec, was presented by Bourguignon *et al.* (2007). The authors explain that the name "SparSpec" derives from the method's character as a "multisine fitting [...] addressed as the *sparse* representation of data in an overcomplete dictionary of frequencies". They model the given data using the sum of an arbitrarily large number of pure frequencies, discretised on a fixed grid, and seek that particular representation that produces the fewest non-zero amplitudes. Following various detailed tests, the authors conclude that SparSpec: i) correctly locates the frequencies embedded in some test data while classical methods fail to do so; ii) outperforms the familiar sequential prewhitening methods in countering sampling artifacts; iii) accurately estimates both frequencies and amplitudes; and iv) is less sensitive to low-frequency perturbations, e.g. to those caused by orbital movements, than familiar methods. Bourguignon *et al.* (2007) actively encourage the community to apply their code and test its efficacy. Bourguignon *et al.*'s (2007) SparSpec paper includes a thorough application of the CLEAN and CLEANest methods. CLEAN refers to the algorithm presented by Roberts *et al.* (1987), with the picturesquely phrased aim to "undo the damage inflicted" by the incomplete sampling of the physical signal. Essentially, the aim of the CLEAN approach is to subtract the spectral window from the so-called "dirty spectrum". Foster (1995) followed this up with CLEANest, specifically tailored to deal with very long time series. Foster (1996) contains a thorough study of wavelet theory applied to period-finding in time series, with the specific aim to treat *variable* periods and/or amplitudes.

In summary, the Fourier-based methods readily available for period-finding at present include the Deeming DFT (often employed in the form of the Period04 package), the classical LS periodogram (available in various packages but also very simple to self-code), GLS, SigSpec and SparSpec. A positive feature of the various modern alternatives to the classical methods is their common presentation to potential users in a very accessible format. The SigSpec and SparSpec codes are readily available on the web, together with substantial user manuals. Refinements to classical methods, like GLS, are simple enough to self-code, with the algorithms readily provided in the source papers.

3. Dispersion/Entropy-based methods

A useful overview of the mathematical basis for these methods appears in Aerts *et al.* (2010). In brief terms, these methods identify the most probable period (or frequency) of a regular variation present in a time series by minimising the scatter in phase bins associated with each respective period in a chosen grid (instead of calculating transforms of time series or fitting functions to them, as the Fourier-based methods do). The original dispersion/entropy-based method ("D/E method" in what follows) is the "String-length" (STR) method introduced by Lafler & Kinman (1965), expressly to deal with light curves that deviate substantially from sinusoidal shapes. The meaning of their test statistic Θ is quite transparent: For a time series consisting of measurements m_j, with mean $\bar{M}$,

$$\Theta = \frac{\sum_i (m_i - m_{i+1})^2}{\sum_i (m_i - \bar{M})^2} \tag{3.1}$$

This simple algorithm for a dispersion-based statistic was followed by a succession of increasingly sophisticated approaches, continuing to the present day (see below). However, the simple STR approach has also received continuing attention. Dworetsky (1983) pointed out the advantages offered by STR when dealing with sparsely sampled signals, while Clarke (2002) refined the original STR statistic, to render it independent of sample size. In summary: STR handles non-sinusoidal signals rather well, but the statistic produces a plot that is rich in false peaks ("aliases" of a kind) and the method is computing-intensive.

Not long after Lafler & Kinman's introduction of STR, Jurkevich (1971) proposed what he called "a statistically more natural" approach, containing a set of statistics with a stronger foundation in formal statistical theory. This procedure was generalised by Stellingwerf (1978) in the format that has been very widely used as "Phase Dispersion Minimisation" or PDM. Similarly to STR, PDM produces very noisy plots of the PDM statistic against frequency (this type of plot is the D/E methods' equivalent of the Fourier-based methods' periodogram).

An important development in D/E methods is the "Analysis of Variance" (or AoV) algorithm proposed by Schwarzenberg-Czerny (1989; SC89). This algorithm is rooted very deeply in fundamental statistical theory and Schwarzenberg-Czerny obtains very impressive results when applying AoV to a variety of test cases. SC89 also makes pertinent statements about STR and PDM. The AoV method appears to be remarkably more powerful at extracting periodicities from time series (at a given significance level) than many other methods. SC89 also points out that Fourier-based methods are superior to D/E methods for the detection of sinusoidal signals; however, the converse is true when dealing with substantially non-sinusoidal signals (e.g. sharp pulses). Schwarzenberg-Czerny (1996; SC96) extends AoV to the so-called "multi-harmonic" case. He exploits a correspondence between Fourier series and series of complex polynomials to set up an algorithm which appears exceptionally powerful at damping aliases (and therefore at detecting weaker physical signals in the data). The comparative study by Graham *et al.* (2013a), mentioned earlier, concluded that the SC96 methodology (labelled AoVMHW) has great merit as an accurate and sensitive period-finding procedure. Baluev (2009) conducted a meticulous statistical study of AoVMHW, focusing on the treatment of non-sinusoidal light curves, and confirmed the merits of AoVMHW as a versatile and powerful period-finding method. Baluev (2008) constitutes a similar treatment of the LS periodogram. The work in both of these papers is carried further in Baluev (2013a), where the Von Mises function is exploited to obtain thought-provoking results. Baluev (2013a) also supplies a weblink to the C++ code for his algorithm with an invitation for its widespread use.

A couple of period-finding algorithms based on various definitions of *entropy* have been introduced recently. Cincotta *et al.* (1995) use the Shannon entropy to select an optimal period. The analytical theory behind SE is presented in Cincotta *et al.* (1999). They test their algorithm with simulated data containing non-sinusoidal variability and find that their method (SE, for Shannon entropy) is more sensitive than the classical periodograms at period detection and that it is very good at resolving closely spaced frequencies. As with the other D/E methods, the actual shape of a periodic variation in the data is not important and the method is well suited to the detection of non-sinusoidal variations. One substantial advantage of the method is its mathematical simplicity and rapid computability. However, it is quite vulnerable to gap aliasing. On the contrary, the "Conditional Entropy" (CE) method introduced by Graham et al. (2013b) is found to be "particularly robust against common aliasing issues". The authors back these claims up with persuasive test results. Essentially, CE is a simple but effective modification of SE. The authors put CE through a robust test run in their paper comparing various period-finding methods (Graham et al. 2013a), where it emerges as a very powerful tool for the accurate detection of periods in large survey databases. Huijse et al. (2012) present a useful summary of some other D/E methods while introducing their CKP (Correntropy Kernelised Periodogram) algorithm.

In summary, the following D/E-methods have been well-tested in the past or recently added to the range of options: STR, PDM, AoV, AoVMHW, SE, CE and CKP.

4. The impact of Bayesian approaches

Asteroseismology relies on the meticulous matching of theoretical calculations of pulsation parameters with the values of those parameters derived from observational data. Besides the actual *values* of frequencies, amplitudes, phases and modal indices we derive from data, a key quantity is how *reliable* the derived values are. Many procedures exist for estimating the statistical significance of data-derived quantities. Bayesian statistics offers a methodology for estimating data-derived quantities as well as modeled quantities. An informative example of estimating modeled quantities is found in Bazot et al. (2012), while Brewer & Stello (2009) present a good example of assigning probabilities to data-derived quantities. Marsh et al. (2008) present an application to solar oscillations. Bourguignon & Cartanfan (2008) demonstrate efficient period extraction with Bayesian methods when traditional methods fail, while White et al. (2010) offer a balanced comparison of Bayesian and Fourier methods and point out the former's advantage in avoiding spurious aliases. Wang et al. (2012) find distinct advantages in applying Bayesian methods to non-sinusoidal light curves. Stoica et al. (2009) (also see He et al. 2009) combine their RIAA method with a Bayesian analysis.

5. Ancillary issues and suggestions

The thorough study by Cumming et al. (1999) makes important conclusions regarding the appropriate *normalization* of the LS periodogram. Koen (2006, 2010b) explores the interpretation of the *Nyquist frequency* for irregularly spaced time series. Eyer & Bartholdi (1999) and Pelt (2009) also consider the topic. Süveges (2012) presents an interesting treatment of the *statistical significance* of periodogram peaks. Further treatments of various aspects of period-finding may be found in Jetsu & Pelt (1999), Palmer (2009), Pelt et al. (2011) and Leroy (2012).

The content of the many published papers on period-finding in astronomical time series is very diverse. Meticulous statistical studies of this issue are rare; however, a few authors have each produced a substantial set of papers that explore important questions regarding the statistical content and meaning of various aspects of observed time series. The reader is referred to the following papers of these authors (in alphabetic order) as an overview: Baluev (2008, 2009, 2012, 2013a, 2013b), Koen (1990, 1999, 2000, 2006, 2009, 2010b), Koen & Lombard (1993), Schwarzenberg-Czerny (1989, 1991, 1996, 1997, 1998, 1999). The work of Stahn & Gizon (2008) on the analysis of time series containing many gaps is also worth consulting.

The aim of this article has been to be as inclusive as possible in pointing the reader to the full scope of methods for extracting oscillation frequencies from time series. There are many important features that attend to the respective methods and many subtle points that thread through them. A detailed study of the papers cited here (and their references) is recommended for further elucidation.

The organisers are thanked for the invitation to compile this review. Financial support from the University of Johannesburg/DHET and the IAU is gratefully acknowledged.

References

Aerts, C., Christensen-Dalsgaard, J., & Kurtz, D. W. 2010, *Asteroseismology* (Springer)
Baluev, R. V. 2008, *MNRAS*, 385, 1279
Baluev, R. V. 2009, *MNRAS*, 395, 1541
Baluev, R. V. 2012, *MNRAS*, 422, 2372
Baluev, R. V. 2013a, *MNRAS*, 431, 1167
Baluev, R. V. 2013b, *MNRAS*, 436, 807
Barning, F. J. M. 1963, *BAN*, 17, 22
Bazot, M., Bourguignon, S., & Christensen-Dalsgaard, J. 2012, *MNRAS*, 427, 1847
Bedding, T. R., Mosser, B., Huber, D., *et al.* 2011, *Nature*, 471, 608
Bourguignon, S., Carfantan, H., & Böhm, T. 2007, *A&A*, 462, 379
Bourguignon, S., Carfantan, H., & Böhm, T. 2008, *Statistical Methodology*, 5, 318
Breger, M., Stich, J., Garrido, R., *et al.* 1993, *A&A*, 271, 482
Brewer, D. J. & Stello, D. 2009, *MNRAS*, 395, 2226
Cincotta, P. M., Mendez, M., & Nunez, J. A. 1995, *ApJ*, 449, 231
Cincotta, P. M., Helmi, A., Mendez, M., Nunez, J. A., & Vucetich, H. 1999, *MNRAS*, 302, 582
Clarke, D. 2002, *A&A*, 386, 763
Cumming, A., Marcy, G. W., & Butler, R. P. 1999, *ApJ*, 526, 890
Deeming, T. J. 1975, *Ap&SS*, 36, 137
Dworetsky, M. M. 1983, *MNRAS*, 203, 917
Eyer, L. & Bartholdi, P. 1999, *A&AS*, 135, 1
Ferraz-Mello, S. 1981, *AJ*, 86, 619
Foster, G. 1995, *AJ*, 109, 1889
Foster, G. 1996, *AJ*, 112, 1709
Frescura, F. A. M., Engelbrecht, C. A., & Frank, B. S. 2008, *MNRAS*, 388, 1693
Graham, M. J., Drake, A. J., Djorgovski, S. G., *et al.* 2013a, *MNRAS*, 434, 3423
Graham, M. J., Drake, A. J., Djorgovski, S. G., Mahabal, A. A., & Donalek, C. 2013b, *MNRAS*, 434, 2629
He, H., Li, J., & Stoica, P. 2009, *Digital Signal Processing Workshop and 5th IEEE Signal Processing Education Workshop*, p. 375
Heideman, M. T., Johnson, D. H., & Burrus, C. S. 1984, *IEEE ASSP Magazine*, 1, 14
Horne, J. H. & Baliunas, S. L. 1986, *ApJ*, 302, 757
Huijse, P., Estévez, P. A., Protopapas, P., Zegers, P., & Príncipe, J. C. 2012, *IEEE Transactions on Signal Processing*, 60, 5135
Jetsu, L. & Pelt, J. 1999, *A&AS*, 139, 629
Jurkevich, I. 1971, *Ap&SS*, 13, 154
Koen, C. 1990, *ApJ*, 348, 700
Koen, C. 1999, *MNRAS*, 309, 769
Koen, C. 2000, *MNRAS*, 316, 613
Koen, C. 2006, *MNRAS*, 371, 1390
Koen, C. 2009, *MNRAS*, 392, 190
Koen, C. 2010a, *Ap&SS*, 329, 267
Koen, C. 2010b, *MNRAS*, 401, 586
Koen, C. & Lombard, F. 1993, *MNRAS*, 263, 287
Lafler, J. & Kinman, T. D. 1965, *ApJS*, 11, 216
Leroy, B. 2012, *A&A*, 545, A50
Lomb, N. R. 1976, *Ap&SS*, 39, 447

Marsh, M. S., Ireland, J., & Kucera, T. 2008, *ApJ*, 681, 672

Palmer, D. M. 2009, *ApJ*, 695, 496

Pelt, J. 2009, *Baltic Astronomy*, 18, 83

Pelt, J., Olspert, N., Mantere, M. J., & Tuominen, I. 2011, *A&A*, 535, A23

Reegen, P. 2007, *A&A*, 467, 1353

Roberts, D. H., Lehar, J., & Dreher, J. W. 1987, *AJ*, 93, 968

Scargle, J. D. 1982, *ApJ*, 263, 835

Schuster, A. 1897, *Terrestrial Magnetism*, 3, 14

Schwarzenberg-Czerny, A. 1989, *MNRAS*, 241, 153

Schwarzenberg-Czerny, A. 1991, *MNRAS*, 253, 198

Schwarzenberg-Czerny, A. 1996, *ApJ*, 460, L107

Schwarzenberg-Czerny, A. 1997, *ApJ*, 489, 941

Schwarzenberg-Czerny, A. 1998, *MNRAS*, 301, 831

Schwarzenberg-Czerny, A. 1999, *ApJ*, 516, 315

Stahn, Th. & Gizon, L. 2008, *Solar Phys.*, 251, 31

Stoica, P., Li, J., & He, H. 2009, *IEEE Transactions on Signal Processing*, 57, 843

Stellingwerf, R. F. 1978, *ApJ*, 224, 953

Süveges, M. 2012, arXiv: 1212.0645

Vaníček, P. 1969, *Ap&SS*, 4, 387

Vio, R., Andreani, P., & Biggs, A. 2010, *A&A*, 519, A85

Vio, R., Diaz-Trigo, M., & Andreani, P. 2013, *Astronomy & Computing*, 1, 5

Wang, Y., Khardon, R., & Protopapas, P. 2012, *ApJ*, 756, 67

White, T. R., Brewer, B. J., Bedding, T. R., Stello, D., & Kjeldsen, H. 2010, *CoAst*, 161, 39

Zechmeister, M. & Kürster, M. 2009, *A&A*, 496, 577

Precision Asteroseismology
Proceedings IAU Symposium No. 301, 2013
J. A. Guzik, W. J. Chaplin, G. Handler & A. Pigulski, eds.

© International Astronomical Union 2014
doi:10.1017/S1743921313014130

On the necessity of a new interpretation of the stellar light curves

J. Pascual-Granado, R. Garrido, and J. C. Suárez

Instituto de Astrofísica de Andalucía (CSIC), Glorieta de la Astronomía s/n 18008, Granada,
Spain. email: `javier@iaa.es`

Abstract. The power of asteroseismology relies on the ability to infer the stellar structure from the unambiguous frequency identification of the corresponding pulsation mode. Hence, the use of a Fourier transform is in the basis of asteroseismic studies. Nevertheless, the difficulties with the interpretation of the frequencies found in many stars lead us to reconsider whether Fourier analysis is the most appropriate technique to identify pulsation modes. We have found that the data, usually analyzed using Fourier techniques, present a non-analyticity originating from the lack of connectivity of the underlying function describing the physical phenomena. Therefore, the conditions for the Fourier series to converge are not fulfilled. In the light of these results, we examine in this talk some stellar light curves from different asteroseismology space missions (*CoRoT*, *Kepler* and *SoHO*) in which the interpretation of the data in terms of Fourier frequencies becomes difficult. We emphasize the necessity of a new interpretation of the stellar light curves in order to identify the correct frequencies of the pulsation modes.

Keywords. methods: data analysis, stars: oscillations

---

1. Overview

Asteroseismology has been very fruitful since the launch of space satellites but some old problems remain unsolved: the non-detection of the solar g-modes in the solar spectrum is a good example. Another persistent problem is the poor modelling of outer layers of the Sun requiring an ad-hoc correction now extended to solar-like stars by Kjeldsen *et al.* (2008). No physical justification for this correction has ever been given. The presence of constant stars within the instability strip of δ Scuti and γ Dor stars (Guzik *et al.*, these proceedings) is another puzzle. More importantly, the range and number of frequencies detected in many pulsating stars observed by satellites are not yet understood. For instance, the observed instability range of the 422 frequencies of the δ Scuti star HD 174936 could not be reproduced by any model (García Hernández *et al.* 2009). For HD 50844 (Poretti *et al.* 2009), prewhitening of more than a thousand frequencies yielded a bushy structure in its residuals, very different from the expected white noise. HD 50870 is another example showing the same phenomenon (Mantegazza *et al.* 2012). Additionally, the analysis of the γ Dor star HD 49434 by Chapellier *et al.* (2011) showed that the amplitudes of new frequency components decrease exponentially as the number of fitted frequencies increases linearly. All the mentioned cases come from the observations made by the *CoRoT* satellite (Auvergne *et al.* 2009), but recently some paradigmatic cases have appeared in *Kepler*'s sample of pulsating stars (Gilliland *et al.* 2010): high frequencies in the oscillation spectrum of the δ Scuti star KIC 4840675 of unknown nature (Balona *et al.* 2012), or KIC 8677585, the only roAp star in which low frequencies are clearly detected (Balona *et al.* 2013).

All the efforts made to explain frequency spectra have been always on the side of the theoretical modelling, never questioning the analysis, which is supposed to be consistent but never has been tested. We put the focus here on this possibility.

2. Methods

Normally we use a DFT to calculate the periodogram (Scargle 1982) and perform a frequency detection based on a Fisher test. The consistency of this operation is due to Parseval's theorem (Kaplan 1992). The DFT and its inverse are approximations to an expansion in Fourier series. When the Fourier series does not converge the function has no expansion and the DFT has no sense. Therefore it is very important to check if the underlying function of a discrete series can be expanded in Fourier series and this is only guaranteed when the function is analytic (van Dijk 2009). Then, before applying any further analysis we should test the analyticity characterized here through the differentiability.

Now, to perform the differentiability analysis of the underlying function our approach is to study how compactly connected the discrete data are through a property called connectivity. We remark here that what we are studying is the differentiability of the continuous function underlying the discrete data and not the discrete data themselves.

The connectivity $\mathcal{C}_n$ of a data point x_n is defined as

$$\mathcal{C}_n = \epsilon_n^f - \epsilon_n^b \qquad \text{with } \epsilon_n^f, \epsilon_n^b \text{ defined as} \tag{2.1}$$
$$\epsilon_n^f = x_n^f - x_n$$
$$\epsilon_n^b = x_n^b - x_n$$

where x_n^f and x_n^b are forward and backward extrapolations respectively made from the data bracketing a datapoint x_n. From these equations the numerical approximation of the point derivative $\mathcal{D}_n$ at x_n can be expressed through the connectivity as:

$$\mathcal{D}_n = \frac{\mathcal{C}_n + x_{n+1} - x_{n-1}}{2\Delta t} \tag{2.2}$$

which reduces to the typical point derivative for discrete data when $\mathcal{C}_n = 0$. When connectivities are not zero but independent randomly distributed values, the derivative is still well-defined. In that case, the connectivities can be considered simply as deviations from the derivative at this point. Otherwise, the derivability condition is not fulfilled.

As it is defined the connectivity resembles the non-differentiability coefficient introduced by Wiener (Wiener 1923) but involving extrapolations in our case.

To calculate those extrapolations we need a model that can approximate arbitrarily well any continuous function. The Stone-Weierstrass theorem (Royden 1988) states that a function uniformly continuous in a closed interval can be approximated arbitrarily well by a polynomial of degree n. Therefore, provided a sufficient amount of data, a spline function (de Boor 1978) should provide a good extrapolation for the datapoint - the residuals form an independent random sequence, i.e. white noise. We also make use of an ARMA (autoregressive moving average) approximation (Box & Jenkins 1976) which is capable of fitting non-analytic functions. Both results are compared.

3. Results

We have calculated the connectivities in the following cases:
• The light curve of a δ Scuti star observed by *CoRoT*, and the best analytic model that can be obtained from the Fourier analysis of its light curve.
• A hybrid star observed by *Kepler*.
• Radial velocities of the Sun observed by the GOLF instrument onboard *SoHO*.

The first case is the δ Scuti HD 174936, one of the enigmatic cases having a range of excited frequencies not predicted by any model. An analytic model for this star has

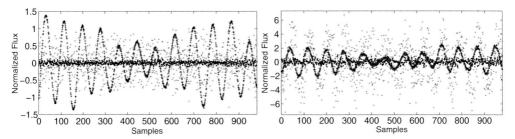

Figure 1. *Right: CoRoT* data for the δ Scuti star HD 174936. *Left:* the corresponding analytic model. Connectivities - ARMA (points), splines (circles), and the original light curve (crosses). Note the different scaling of the panels.

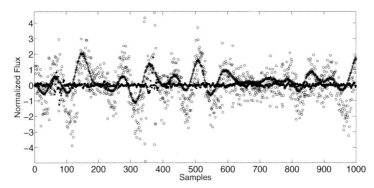

Figure 2. *Kepler* data for the hybrid star KIC 006187665. Connectivities – ARMA (points), spline (circles), and the original light curve (crosses).

been constructed using the 422 frequencies detected using Fourier techniques in García-Hernández *et al.* (2009). It is expected that both time series show the same properties.

In Fig. 1 we show the connectivities calculated with both approaches: splines (analytic fitting) and ARMA (non-analytic fitting). We emphasize here that the points should show an independent random distribution in case of a well-behaved underlying function that could be interpreted in terms of Fourier frequencies. However, when considering only the spline connectivities, while the analytic model shows that kind of distribution, the connectivities of the *CoRoT* data are completely correlated. Two conclusions come to mind from this first analysis:

• *CoRoT* data for the δ Scuti HD 174936 do not originate from an analytic function.

• The analytic model built from Fourier frequencies does not represent the same function as *CoRoT* data.

When we consider ARMA connectivities, in contrast to the analytic approach given by the splines, these connectivities show a distribution typical of a white noise in both panels of Fig. 1, giving more weight to the previous conclusions.

The second light curve studied here is a hybrid pulsating star observed by *Kepler*. The connectivities calculated with splines (Fig. 2) are correlated, whereas those calculated with the non-analytic approach are not. Since we are analyzing data from another instrument, no instrumental effect can be involved in the non-differentiability of the function, but rather an intrinsic effect of the light curves.

Finally, the results corresponding to the radial velocities of the Sun taken by the GOLF instrument (García *et al.* 2005) are shown in Fig. 3. In this case the connectivities calculated using splines present again a higher dispersion than the ARMA ones. Connectivities

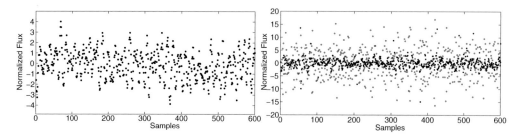

Figure 3. *Left*: original *SoHO*/GOLF data. *Right*: connectivities – ARMA (points) and splines (circles).

are also correlated with the original time series, confirming the non-differentiability of the function in this case too.

In summary, the non-differentiability is ubiquitous in the stellar light curves.

4. Conclusions

The tests carried out here show that the underlying function describing the light variations of some pulsating stars is non-analytic. This implies that the conditions for the Parseval theorem might be not satisfied. Therefore, it is not guaranteed that the signal can be represented by a Fourier series and thereby the periodogram could be not a consistent estimator of the frequency content of the underlying function. We have demonstrated that this phenomenon is neither due to an instrumental effect nor caused by the type of measurement. It does not depend on the intrinsic variability type either. The non-analyticity of the function underlying the light curves is a fine structure that could be the origin of some puzzling features like the flickering observed in classical Cepheids (Evans *et al.*, these proceedings). We conclude that the standard periodogram interpretation of the time series of pulsating stars should be accordingly revised. Work is in progress to investigate the origin of this inconsistency in the harmonic analysis.

References

Auvergne, M., Bodin, P., Boisnard, L., *et al.* 2009, *A&A*, 506, 411
Balona, L. A., Breger, M., Catanzaro, G., *et al.* 2012, *MNRAS*, 424, 1187
Balona, L. A., Catanzaro, G., Crause, L., *et al.* 2013, *MNRAS*, 432, 2808
Box, G. E. P.. & Jenkins, G. M. 1976, *Time Series Analysis, Forecasting and Control* (Holden-Day Series in Time Series Analysis)
Chapellier, E., Rodríguez, E., Auvergne, M., *et al.* 2011, *A&A*, 525, A23
de Boor, C. 1978, *A Practical Guide to Splines*, (1st ed., Series: Applied Mathematical Sciences, 27, New York Berlin Heidelberg: Springer)
Gilliland, R. L., Brown, T. M., Christensen-Dalsgaard, J., *et al.* 2010, *PASP*, 122, 131
García, R. A., Turck-Chièze, S., Boumier, P., *et al.* 2005, *A&A*, 442, 385
García Hernández, A., Moya, A., Michel, E., *et al.* 2009, *A&A*, 506, 79
Kaplan, W. 1992, *Advanced Calculus* (4th ed. Reading, MA: Addison-Wesley)
Kjeldsen, H., Bedding, T. R., & Christensen-Dalsgaard, J. 2008, *ApJ*, 683, L175
Mantegazza, L., Poretti, E., Michel, E., *et al.* 2012, *A&A*, 542, A24
Poretti, E., Michel, E., Garrido, *et al.* 2009, *A&A*, 506, 85
Royden, H. L. 1988, *Real Analysis* (Prentice-Hall)
Scargle, J. D. 1982, *ApJ*, 263, 835
van Dijk, G. 2009, *Introduction to harmonic analysis and generalized Gelfand pairs* (Berlin, New York: Walter De Gruyter)
Wiener, N. 1923, *J. Math. and Phys.*, 2, 131–174

Precision Asteroseismology
Proceedings IAU Symposium No. 301, 2013
J. A. Guzik, W. J. Chaplin, G. Handler & A. Pigulski, eds.

© International Astronomical Union 2014
doi:10.1017/S1743921313014142

Theoretical properties of regularities in the oscillation spectra of A-F main-sequence stars

Juan Carlos Suárez[1], Antonio García Hernández[2], Andrés Moya[3], Carlos Rodrigo[3,4], Enrique Solano[3,4], Rafael Garrido[1], and José R. Rodón[1]

[1]Instituto de Astrofísica de Andalucía (CSIC)
CP3004 , Granada, Spain.
email: jcsuarez@iaa.es

[2]Centro de Astrofísica, Universidade do Porto
Rua das Estrelas 4150-762, Porto, Portugal
email: agh@astro.up.pt

[3]Dept. Astrofísica CAB (INTA-CSIC). ESAC Campus
P.O. Box 78. 28691 Villanueva de la Cañada, Madrid, Spain
email: amoya@cab.inta-csic.es,esm@cab.inta-csic.es, crb@cab.inta-csic.es

[4]Spanish Virtual Observatory

Abstract. We study the theoretical properties of the regular spacings found in the oscillation spectra of δ Scuti stars. A linear relation between the large separation and the mean density is predicted to be found in the low-frequency domain (i.e. radial orders spanning from 1 to 8, approximately) of the main-sequence δ Scuti stars' oscillation spectrum. This implies an independent direct measure of the average density of δ Scuti stars, analogous to that of the Sun, and places tight constraints on the mode identification and hence on the stellar internal structure and dynamics, and allows a determination the radii of planets orbiting around δ Scuti stars with unprecedented precision. This opens the way for studying the evolution of regular patterns in pulsating stars, and its relation to stellar structure and evolution.

Keywords. stars: evolution, oscillations (including pulsations), stars: interiors, stars: variables: δ Scuti.

1. Introduction

Thanks to great efforts made in long ground-based multi-site campaigns of A-F type pulsating stars, it was possible to find regularities in the oscillation spectra of some δ Scuti stars, e.g. for CD$-24°7599$, with a regular spacing of around $26\,\mu$Hz (Handler *et al.* 1997), or the well-studied star FG Vir, for which a regular spacing around $46\,\mu$Hz (Breger *et al.* 2009) in its oscillation spectrum composed of 58 frequencies was found. Those works used different techniques, the histogram of frequency differences and the Fourier transform, to detect the regularities.

Later on, similar studies were performed using precise data from space, e.g. the 88 frequencies in the oscillation spectrum of the δ Scuti star HD 209775 (Matthews 2007) observed by the *MOST* satellite, finding a regular spacing of $\sim$50 μHz. More recently, regular patterns were also found in the oscillation spectra of δ Scuti stars observed by *CoRoT* (García Hernández *et al.* 2009, Mantegazza *et al.* 2012, García Hernández *et al.* 2013) and *Kepler* (Hernández *et al.* 2013). The large number of oscillation modes and the wide frequency range cause the regular spacings to arise without making any

Table 1. Ranges of the four parameters used to construct the current model dataset
representative of intermediate-mass stars.

Parameter	Lowest	Highest	Step
$M/M_\odot$	1.25	2.20	0.01
[Fe/H]	−0.52	+0.08	0.20
$\alpha_{\rm ML}$	0.50	1.50	0.50
$d_{\rm ov}$	0.10	0.30	0.10

assumption about the distribution of the modes. Moreover, those regularities were found
to be bounded (within 5 μHz) in the very low radial-order modes, different from the
so-called *asymptotic regime* corresponding to frequencies with radial orders above $n =$
6 approximately (e.g. Antoci *et al.* 2011). The distance between these two regimes is
approximately of the order of the bounded region.

This work intends to understand the physical properties of the periodicities observed in
the low-frequency domain of δ Scuti stars, and also to search for any possible connection
between two predicted periodicity regimes (low- and high-frequency domains).

2. Method

We examine a dense sample of asteroseismic models representative of A-F main-
sequence stars, i.e. covering the corresponding area in the H-R diagram where classical
pulsations for these stars are expected. We constructed a model collection composed of
approximately 5×10^5 models by varying four of the main physical parameters typically
used in the field for the modelling such stars (see Table 1). For the sake of homogeneity
and precision of the asteroseismic mode sample, models were computed following the
prescriptions suggested by the ESTA/*CoRoT*† working group (see Moya *et al.* 2008).

Pre- and post-main sequence evolutionary stages present more complex oscillation
spectra, so for simplicity this work is focused on the main sequence. Although the
work was initially planned to make a multivariable analysis, the first logical step is
to analyse the regularities as large separations (as claimed in previous works, mainly in
García Hernández *et al.* 2009) with respect to the mean density of stars.

3. Rotation and mixed modes

Rotation effects on oscillations are commonly taken into account through the perturbation approximation (Dziembowski & Goode 1992), which is limited to slow-to-moderate
rotation, i.e. small stellar deformations (e.g. Suárez *et al.* 2005, Reese *et al.* 2006). For
moderate-to-rapid rotators, a complete calculation of the oscillation modes on a de-
formed star (Lignières *et al.* 2006) becomes necessary. However, nowadays this calculation is available for polytropic models and for some more realistic 2D stellar models on
the ZAMS based on the self-consistent field (SCF) method (Jackson *et al.* 2005, Mac-
Gregor *et al.* 2007). Furthermore, the latter models together with the calculation of
non-perturbative oscillations require a significant amount of computing resources as well
as time of computation. Therefore, the use of proper modeling for rapidly rotating stars
would be impractical for the present work.

† http://www.astro.up.pt/corot/

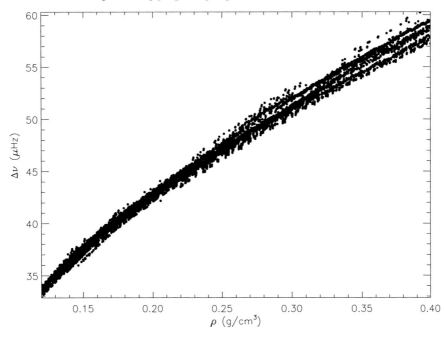

Figure 1. Theoretical large spacing, $\Delta\nu$, as a function of the mean density computed on the model grid. Time evolution is from right to left. The figure was obtained using the graphic tools of a virtual observatory service developed for this work (the tool is presented in Suárez *et al.* 2014).

On the other hand, since periodicities are indeed observed, it might be concluded that rotation effects are not sufficient to break the regularities presumably composed by radial modes and non-radial $m = 0$ modes. Indeed, non-perturbative calculations of the oscillation spectra for rapidly rotating polytropic models indicate that, as rotation increases, the asymptotic structure of the non-rotating frequency spectrum is replaced by a new form of organization (Reese *et al.* 2008, 2009). This new mode-frequency organization also exhibits regular structures, including large separations (Lignières *et al.* 2010) whose variation from the non-rotating case is negligible (Lignières *et al.* 2006). Furthermore, calculations of non-perturbative oscillation frequencies on SCF models show a maximum variation of the large separation of around $2.3\,\mu$Hz for stars rotating up to 40% of the Keplerian velocity (Reese, private comm.), which is small compared with the interval in which the periodicities are observed ($\sim$10 μHz). Considering all the above theoretical arguments, we are allowed to use non-rotating models for the present study.

Another issue that might hamper the detection of periodicities is the presence of mixed modes. This phenomenon is implicitly considered in our work since our models cover the whole main-sequence phase. However, the phenomenon is not fully covered, since both the polytropic and the SCF models in ZAMS are not expected to properly show the avoided crossing phenomenon. More evolved SCF models should be studied in order to better analyse the impact of rotation (and the mixed) modes in the detection of regularities (work in progress).

4. Results and discussion

We found the following theoretical linear relation

$$\Delta\nu/\Delta\nu_\odot = 0.776\,(\rho/\rho_\odot)^{0.46}. \tag{4.1}$$

between the large separation (calculated in the δ Scuti lower frequency range) and the mean density obtained from the grid of models and represented in Fig. 1. This behavior is quite close to $\Delta\nu \propto \rho^{1/2}$ predicted for solar-like stars, opening the way for studying the evolution of regular patterns in pulsating stars, and its relation to stellar evolution.

We present no errors in the coefficients of the fitting because not all the theoretical $(\Delta\nu_i, \bar{\rho}_i)$ are independent from each other, and therefore regression error estimates are meaningless. Nevertheless it is possible to analyse the domain of validity of the method. The minimum error considered comes from the effect of rotation, i.e. $2.3\,\mu$Hz corresponding to the variation of the large separation with the stellar deformation due to rotation. This yields maximum errors predicted for the estimate of the mean density range from 11% to 21% of the total variation of $\bar{\rho}$ in the main sequence (see details in Suárez *et al.* 2014).

The diagnostic applied to real stars (see details in Suárez *et al.* 2014) is consistent with previous results found in the literature, making Eq 4.1 a powerful diagnostic tool for the study of A-F stars, including δ Scuti stars and/or the hybrid phenomenon not yet understood (Uytterhoeven *et al.* 2011). The method also provides an estimate of the frequency of the fundamental radial mode. The strength of this diagnostic tool is that it is almost model independent, since all the models contained in the heterogenous dataset follow the same trend.

In addition, these results extend the characterization (using asteroseismology) of planetary systems to A-F type hosting stars. Most of the planets discovered by direct imaging are found to be orbiting such stars. These are critical to understand the spin-orbital interactions between the planets and the hosting star (e.g. Wright *et al.* 2011).

References

Antoci, V., Handler, G., Campante, T. L., *et al.* 2011, *Nature*, 477, 570

Breger, M., Lenz, P., & Pamyatnykh, A. A. 2009, *MNRAS*, 396, 291

Dziembowski, W. A. & Goode, P. R. 1992, *ApJ*, 394, 670

García Hernández, A., Moya, A., Michel, E., *et al.* 2009, *A&A*, 506, 79

García Hernández, A., Moya, A., Michel, E., *et al.* 2013, *A&A*, 559, A63

Hernández, A. G., Pascual-Granado, J., Grigahcène, A., *et al.* 2013, in: J. C. Suárez, R. Garrido, L. A. Balona, & J. Christensen-Dalsgaard (eds.), *Stellar Pulsations: Impact of New Instrumentation and New Insights*, (Berlin, Heidelberg: Springer), p. 61

Handler, G., Pikall, H., O'Donoghue, D., *et al.* 1997, *MNRAS*, 286, 303

Jackson, S., MacGregor, K. B., & Skumanich, A. 2005, *ApJS*, 156, 245

Lignières, F., Rieutord, M., & Reese, D. 2006, *A&A*, 455, 607

Lignières, F., Georgeot, B., & Ballot, J. 2010, *AN*, 331, 1053

MacGregor, K. B., Jackson, S., Skumanich, A., & Metcalfe, T. S. 2007, *ApJ*, 663, 560

Mantegazza, L., Poretti, E., Michel, E., *et al.* 2012, *A&A*, 542, 24

Matthews, J. M. 2007, *CoAst*, 150, 333

Moya, A., Christensen-Dalsgaard, J., Charpinet, S., *et al.* 2008, *Ap&SS*, 316, 231

Reese, D., Lignières, F., & Rieutord, M. 2006, *A&A*, 455, 621

Reese, D., Lignières, F., & Rieutord, M. 2008, *A&A*, 481, 449

Reese, D. R., Thompson, M. J., MacGregor, K. B., *et al.* 2009, *A&A*, 506, 183

Suárez, J. C., Bruntt, H., & Buzasi, D. 2005, *A&A*, 438, 633

Suárez, J. C., García Hernández, A., Moya, A., *et al.* 2014, *A&A*, in press

Uytterhoeven, K., Moya, A., Grigahcène, A., *et al.* 2011, *A&A*, 534, 125

Wright, D. J., Chené, A. N., De Cat, P., *et al.* 2011, *ApJ*, 728, L20

Precision Asteroseismology
Proceedings IAU Symposium No. 301, 2013
J. A. Guzik, W. J. Chaplin, G. Handler & A. Pigulski, eds.

© International Astronomical Union 2014
doi:10.1017/S1743921313014154

Identification of pulsation modes from photometry

Michel Breger[1,2]

[1]Department of Astronomy, University of Texas, Austin, TX 78712, USA
email: breger@astro.as.utexas.edu

[2]Institut für Astrophysik der Universität Wien, Türkenschanzstr. 17, A–1180, Wien, Austria

Abstract. The identification of the detected pulsation modes in terms of the spherical harmonic quantum numbers is crucial for asteroseismology. Light curves obtained in different passbands have become an important tool for mode identifications, which rely on wavelength-dependent amplitudes and phase shifts. We demonstrate this for different types of pulsators and review recent successes from earth-based measurements, especially in determining the important l values. The extensive amount of accurate data needed to determine small phase shifts and accurate amplitude ratios suggests multicolor measurements using space satellites. This motivated the multicolor *BRITE* satellite project, for which the first two satellites have already been launched successfully. We demonstrate the potential from models computed for the *BRITE* wavelengths. Most of the excellent presently available satellite photometry is not multicolor, although frequencies with amplitudes as small as a few parts-per-million have been detected and confirmed. We briefly discuss mode identifications from frequency patterns, including the use of correlations between phase and amplitude changes.

Keywords. techniques: photometric, space vehicles, stars: oscillations (including pulsations), stars: evolution, stars: early-type, stars: δ Scuti

1. Introduction

The recent advances in high-precision photometry have led to the discovery of several hundreds of simultaneously present pulsation frequencies in a large number of individual stars. For space missions such as *Kepler* and *CoRoT*, the present detection threshold limits are near one part-per-million in amplitude. This represents an improvement of two orders of magnitude over previous earth-based telescope measurements.

To use the frequency data for asteroseismic modeling, the correct astrophysical identification of the detected frequency peaks is essential. This involves the identification of the nature of the frequency peaks (e.g., pulsation and the type of mode, rotation, granulation, nonlinear effects such as combination frequencies, magnetic field effects). In addition, for those peaks identified with pulsation modes, we need to identify the discrete spherical harmonic quantum numbers (l, m, n) for as many of the detected oscillation modes as possible. The latter process is generally referred to as 'mode identification'. However, the recently obtained rich frequency data for many types of pulsators have shown that considerable attention also needs to be paid to the astrophysical identification of the frequency peaks, especially to those with small amplitudes. This is made possible by examining the regularities in the frequency spacings. Especially due to the extremely high frequency resolution of some spacecraft data, these regularities can be determined very precisely and lead to different families of frequencies, each associated with specific astrophysical properties of the star.

2. Mode identification from multicolor photometry

During its pulsation cycle, a star changes in temperature, radius and geometrical cross-section. The relative changes depend on the type of pulsation mode (e.g., gravity and pressure modes) as well as the spherical harmonic quantum numbers. In principle, it is possible to separate the effects by studying the light variability as a function of wavelength (multicolor photometry), radial velocity and changes in the spectroscopic line-profiles. In practice, such information is available for only a few stars. Photometrically, the changes lead to wavelength-dependent amplitudes and times of light maxima. If we observe with two different filters, we will see an amplitude ratio and phase difference between the two passbands.

Already in 1977, in a pioneering paper, Dziembowski (1977) discussed the comparison of light and radial-velocity variations in order to derive the spherical harmonics of nonradial modes. Balona & Stobie (1979) applied the method to the hot β Cep stars to identify radial and quadrupole modes. Watson (1988) applied a 'modern' method of comparing amplitude ratios, A_{B-V}, and phase differences, $\phi_{B-V} - \phi_V$, derived from light curves taken through two passbands to the large-amplitude δ Scuti star AI Vel. He successfully identified the two observed modes to be radial.

During the last ten years, the method has been extended and applied to many types of pulsating stars by a large number of colleagues. Daszyńska-Daszkiewicz et al. (2003, 2005) introduced the nonadiabatic f-parameter, which is the ratio of the amplitude of the bolometric flux perturbations to the radial displacement. While the theoretical values of f are obtained from stellar pulsation calculations, multicolor photometry together with radial-velocity measurements provide the empirical counterparts together with the l value. A number of extensive discussions and summaries of the theoretical problems and results have become available in the last few years and we refer to an excellent review by Daszyńska-Daszkiewicz & Pamyatnykh (2013). The possible sources of error in the theoretical values of amplitude ratios and phase differences of main-sequence B-type stars were discussed by Daszyńska-Daszkiewicz & Szewczuk (2011). Some of the most complex seismic modelling for mode identification of a B-type pulsator (γ Peg) was performed by Walczak et al. (2013).

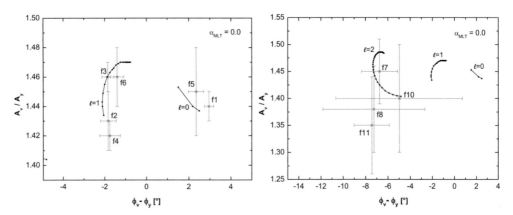

Figure 1. Amplitude ratios and phase differences in the v and y passbands for different pulsation modes of 44 Tau. The observationally well-determined values (for l values of 0 and 1) are shown in the left panel, while the $l = 2$ modes have larger uncertainties (after Lenz et al. 2008).

Let us illustrate the comparison between the amplitude ratio and phase difference observations with models for two extensively studied A/F stars. Figure 1 shows the

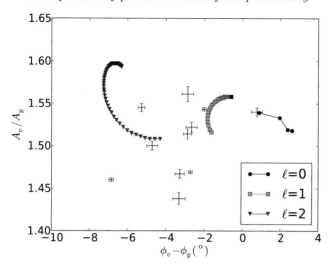

Figure 2. Amplitude ratios and phase differences in the v and y passbands for different pulsation modes of 4 CVn. The error bars are determined by comparing results from 702 nights covering eight different years and the results may be the most precise two-color values obtained so far. The comparison with a preliminary theoretical model (see text) covering the frequency range from 4.5 to 9.5 d^{-1} is also shown.

results for the extremely slow rotator 44 Tau (142 nights, Lenz *et al.* 2008), for which the l values of the dominant pulsation modes could be successfully identified.

The evolved δ Scuti star 4 CVn has been photometrically studied for more than 50 years. From 2005–2012, we obtained more than 700 nights of high-precision photometry of this star with the Vienna Automatic Photoelectric Telescope. The Strömgren v and y passbands were used. Figure 2 presents the amplitude ratios and phase differences for eleven dominant pulsation modes, together with preliminary models computed by P. Lenz and A.A. Pamyatnykh. The models used a mass of 2.1 solar masses, $\alpha = 0.2$ and a metallicity of [M/H] = +0.2 to fit the radial mode. The small error bars of the observational data provide the opportunity to refine the input values used for the theoretical models computed for this star. We note here that the spectroscopic line-profile analyses agree with the photometric mode identifications.

2.1. *Are the formal uncertainties underestimated?*

The extensive multiyear data of 4 CVn also allow us to examine whether the size of the formal errors computed from multiple-least-squares fits to the photometric observations (e.g., with PERIOD04, Lenz & Breger 2005) are possibly underestimated. Such an underestimation could be caused by systematic (as opposed to random) errors in photometry. We have computed amplitude ratios and phase differences with their annual uncertainties for each of the years from 2005–2012. We then averaged the annual values of the amplitude ratios and phase shifts and compared the standard deviation of the averages with those expected from the annual values. The process was applied to the seven modes with the highest amplitudes. For the amplitude ratio we found that the annual results had a standard deviation of 0.82 of the expected value, while for the phase shifts a ratio of 1.12 was obtained.

We conclude that the formal least-squares errors computed for the amplitude ratios and phase shifts are confirmed.

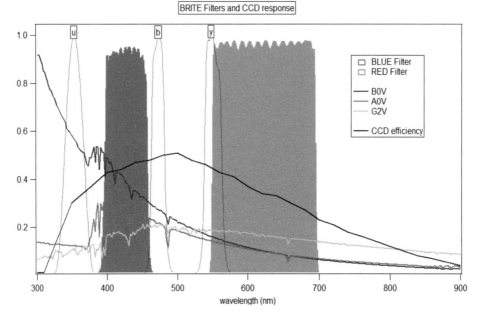

Figure 3. Transmissions of the blue and red interference filters together with the spectral energy distributions of B0 V, A0 V and G2 V stars. The response of the CCD *BRITE* detector is also shown together with the positions of the Strömgren *u*, *b* and *y* passbands. (Diagram courtesy of Werner Weiss.)

3. The multicolor *BRITE* satellites and earth-based radial-velocity measurements

Recent space missions, such as *Kepler*, provide a quantum jump in our observational studies of stellar pulsation. These space observations are single-color, which means that multicolor mode identification is not possible. This provides one of the main motivations for the *BRITE* nanosatellite mission and we refer to an extensive discussion in another paper in this volume. The *BRITE* telescopes have small apertures of 3 cm and are therefore limited to observing bright stars. Based on photon-statistics calculations, the amplitude ratios and phase shifts expected for the *BRITE* nanosatellites should nevertheless be accurate and enable mode identifications.

Figure 3 shows the transmissions of the blue and red interference filters.

The chosen filters provide an excellent separation of the modes with different *l* values in the amplitude ratio, phase difference diagram. This is illustrated in Fig. 4, which shows the results of computations by J. Daszyńska-Daszkiewicz. The SPB model uses $T_{\text{eff}} = 16\,000$ K, $\log L/L_\odot = 3.2$ and a mass of 6 solar masses, while the RR Lyrae star model was computed with a mass of 0.63 solar masses, $T_{\text{eff}} = 6540$ K, and $\log L/L_\odot = 1.68$, and $\alpha_{\text{conv}} = 0.0$. Both models use OP opacities with $X = 0.74$, $Z = 0.0006$.

3.1. *Why earth-based radial-velocity measurements should be simultaneous*

The *BRITE* project uses two (rather than three or more) passbands. Consequently, additional radial velocity measurements from the ground are highly recommended to eliminate many of the remaining uncertainties of the pulsation models used for mode identifications. We would like to stress that such measurements should be nearly simultaneous with the space measurements and not be spaced, say, six months apart due to the limited accessibility of stars observed from the ground. The reason is the fact that

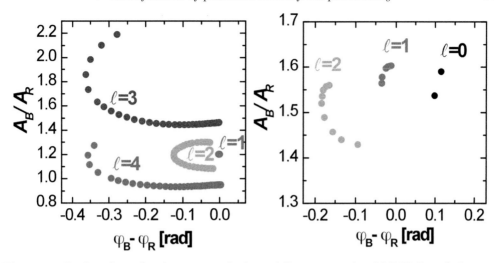

Figure 4. Predicted amplitude ratios and phase differences in the *BRITE B* and *R* pass-bands for different pulsation modes of a SPB (left panel) and a RR Lyrae star model (right panel). Note the clear separations of the different *l* values. (Models and diagrams courtesy of J. Daszyńska-Daszkiewicz.)

Table 1. The case for simultaneous photometric/radial-velocity data: Observed phase shifts in two δ Scuti stars over 6 months.

Star	Mode	Phase shift cycles	Comments
4 CVn	7.38 d^{-1}	0.082 ± 0.006	Steady change 2005–2012, 702 telescope nights
KIC 8054146	25.95 d^{-1}	0.086 ± 0.002	Relatively stable mode, 3 years SC *Kepler* data

nonradial pulsation modes show small, steady amplitude and period changes. These accumulate over several months to such an extent that the correct phase shifts between radial velocities and photometry can no longer be determined.

Let us illustrate this point by using presently unpublished, recent analyses of two well-studied stars: 4 CVn with a total of over 700 nights of ground-based observations from 2005 to 2012, and KIC 8054146, studied continuously by the *Kepler* spacecraft for three years. Both stars are pulsators of spectral types A/F, but the argument should apply to B stars as well. In both stars, all observed modes show slow, systematic period changes. For each star, we pick a typical, stable mode of relatively high amplitude and calculate the phase changes over six months. We note again that these stars were observed for many years and that the results are relatively independent of the choice of six-month time period. The results are shown in Table 1. They indicate that the photometric and spectroscopic measurements should not be spaced months apart.

4. Mode identification from frequency patterns

Recent space measurements of pulsating stars have revealed the existence of hundreds of frequency peaks in individual stars. Since these space missions have been single-color, multicolor mode identification has not been possible. Consequently, the examination of frequency patterns, amplitude variability and lifetime determinations of individual

M. Breger

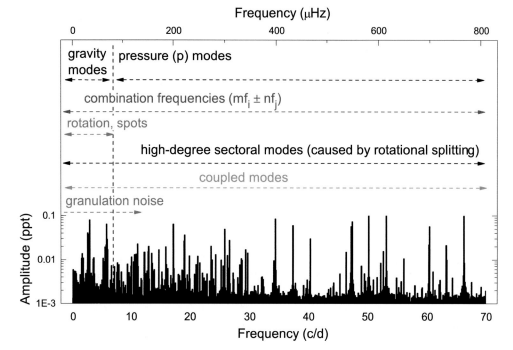

Figure 5. Different astrophysical origins of detected frequencies in main-sequence and slightly evolved A stars. While the frequencies and amplitudes of KIC 8054146 (Breger *et al.* 2012) are given as an example, the interpretations should fit other B, A and F stars as well.

frequencies have become the major tools of analysis. A successful application of this method of mode identification was the discovery of 24 gravity modes of the same l degree through equal period-spacing in CoRoT ID 105733033 by Chapellier *et al.* (2012).

Figure 5 presents a non-exhaustive list of different astrophysical origins in different frequency regions. It demonstrates that the astrophysical origin of each frequency peak needs to be examined carefully.

Combination frequencies and harmonics: Multiple pulsation modes lead to a number of nonlinear interaction terms of the form $(mf_1 + nf_2)$, where f_1 and f_2 are the pulsation frequencies, and m and n are integers. The amplitudes of these combination frequencies are generally much smaller than those of f_1 and f_2. Nevertheless, at the amplitude level of several parts-per-million, as many as hundreds of combination frequencies can be seen. Therefore, it becomes extremely important to separate these from independent pulsation modes by using data with high frequency resolution (i.e., long data sets).

Rotation: *Kepler* observations of A stars, analyzed by Balona (2011), have revealed that the distribution of dominant low frequencies match the expected distribution for rotational frequencies. In fact, for KIC 9700322, the unusually slow rotation suggested by the low-frequency peak at 0.16 d^{-1} was confirmed by an HET spectrum showing $v \sin i = 19 \pm 1\,\mathrm{km\,s^{-1}}$ and the spacing of an $l = 2$ quintuplet (Breger *et al.* 2011). Due to differential rotation, spots can also lead to several low-frequency peaks.

Granulation noise: Kallinger & Matthews (2010) have argued that even in stars of spectral type A, a large number of low-amplitude peaks in the power spectrum are actually granulation noise.

High-degree sectoral gravity modes: Due to rotational splitting, in the observer's frame of reference, high-degree sectoral modes are shifted to high frequencies with equidistant

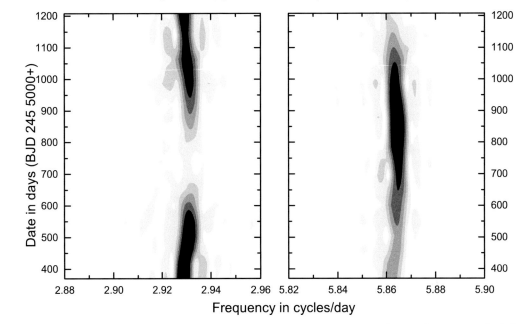

Figure 6. Fourier time plots covering three years for the 2.93 and 5.86 (= 2 × 2.93) d^{-1} frequencies of KIC 8054146. Both frequencies have strongly variable amplitudes of similar sizes. The variations are, however, not connected. This is seen in other main-sequence stars as well and demonstrates that the higher frequency is not a $2f$ harmonic of the lower frequency. The amplitude variations of both frequencies can be correctly predicted by mode coupling of specific high frequencies.

spacing. This was already shown by Kennelly *et al.* (1998) during their spectroscopic MUSICOS multisite project for a number of δ Scuti stars such as τ Peg. Recently, prograde sectoral modes have also been reported for Rasalhague (Monnier *et al.* 2010) and KIC 8054146 (Breger *et al.* 2013).

5. Those pesky f and $2f$ low-frequency peaks

A large number of stars of spectral type B, A and F studied by the *Kepler* spacecraft show low frequencies with an apparent harmonic term, i.e., f and $2f$. Could these peaks be pulsation and what are their mode identifications? A conventional explanation of the $2f$ term would be the Fourier expansion of a nonsinusoidal light curve associated with f. It would then be ignored for mode identification or the determination of the rotational velocity. However, in a number of A/F stars, the $2f$ term has amplitudes as large as those of the frequency f. Furthermore, both the f and $2f$ low frequencies are also preferred spacings of high-frequency p modes. This challenges the explanation in terms of harmonics.

For many of these stars, several years of continuous photometric data are available. This allows us to examine the amplitude and phase variations of the different peaks. For the star KIC 8054146 these variations are considerable. In Fig. 6, we show the Fourier time plots of the 2.93 and 5.86 d^{-1} frequency peaks, which use the same amplitude scale. The amplitude variations are not similar, which does not support the Fourier harmonics theory. In fact, mode-coupling by high-frequency p modes can correctly predict the amplitude variations of these low-frequency peaks to a few parts-per-million. We note here that the star is a rapidly rotating star, for which mode coupling is expected.

6. Conclusion

In this short review, we summarized recent developments of mode identification from photometry. We emphasized that with the advent of extremely high-precision spacecraft data, the first step of mode identification is to separate the independent pulsation modes from the multitude of additional frequency peaks. While the recognition of specific frequency patterns in the amplitude vs. frequency diagram has been successful in a number of cases, we require multicolor photometry in order to determine the discrete spherical harmonic quantum numbers (m, l, n) for as many of the detected oscillation modes as possible.

Such multicolor photometry will be provided by the recently launched *BRITE* satellites. These measurements should be accompanied by near-simultaneous radial-velocity measurements. Since the majority of stars to be studied are fast rotators, the observations will also provide valuable input for the inclusion of fast rotation into the present stellar models.

Acknowledgements

This investigation has been supported by the Austrian Fonds zur Förderung der wissenschaftlichen Forschung through project P21830-N16.

References

Balona, L. A. 2011, *MNRAS*, 415, 1691

Balona, L. A. & Stobie, R. S. 1979, *MNRAS*, 189, 649

Breger, M., Balona, L., Lenz, P., *et al.* 2011, *MNRAS*, 414, 1721

Breger, M., Fossati, L., Balona, L., *et al.* 2012, *ApJ*, 759, 62

Breger, M., Lenz, P., & Pamyatnykh, A. A. 2013, *ApJ*, 773, 56

Chapellier, E., Mathias, P., Weiss, W. W., Le Contel, D., & Debosscher, J. 2012, *A&A*, 540, A117

Daszyńska-Daszkiewicz, J. & Pamyatnykh, A. A. 2013, in: J. C. Suarez, R. Garrido, L. A. Balona & J. Christensen-Dalsgaard (eds.), *Stellar Pulsations*, Astrophysics and Space Science Proceedings, Vol. 31 (Berlin: Springer), p. 179

Daszyńska-Daszkiewicz, J. & Szewczuk, W. 2011, *ApJ*, 728, 2011

Daszyńska-Daszkiewicz, J., Dziembowski, W., & Pamyatnykh, A. A. 2003, *A&A*, 407, 999

Daszyńska-Daszkiewicz, J., Dziembowski, W., & Pamyatnykh, A. A. 2005, *A&A*, 441, 641

Dziembowski, W. 1977, *AcA*, 27, 203

Kallinger, T. & Matthews, J. M. 2010, *ApJ*, 711, L35

Kennelly, E. J., Brown, T. M., Kotak, R., *et al.* 1998, *ApJ*, 495, 440

Lenz, P. & Breger, M. 2005, *CoAst*, 146, 53

Lenz, P., Daszyńska-Daszkiewicz, J., Pamyatnykh, A. A., & Breger, M. 2008, *CoAst*, 153, 40

Monnier, J. D., Townsend, R. H. D., Che, X., *et al.* 2010, *ApJ*, 725, 1192

Walczak, P., Daszyńska-Daszkiewicz, J., Pamyatnykh, A. A., & Zdravkov, T. 2013, *MNRAS*, 432, 822

Watson, R. D. 1988, *Ap&SS*, 140, 255

Precision Asteroseismology
Proceedings IAU Symposium No. 301, 2013   © International Astronomical Union 2014
J. A. Guzik, W. J. Chaplin, G. Handler & A. Pigulski, eds.   doi:10.1017/S1743921313014166

Identification of pulsation modes from spectroscopy

Katrien Uytterhoeven[1,2]

[1]Instituto de Astrofísica de Canarias,
38205 La Laguna, Tenerife, Spain
email: katrien@iac.es

[2]Departamento de Astrofísica, Universidad de La Laguna,
38200 La Laguna, Tenerife, Spain

Abstract. Time-series of high-resolution spectra of massive main-sequence pulsators contain information on the degree l and azimuthal number m of a pulsation mode. I present an overview of existing mode-identification techniques that have been developed to derive l and m from spectroscopic data. I also discuss the data quality needed to perform such a study. Through some examples from the literature I show that the optimal way to identify modes in heat-driven non-radial pulsators is by 1) using multi-site campaign data, 2) combining different spectroscopic mode-identification techniques, and 3) combining results from photometric and spectroscopic mode-identification studies.

Keywords. stars: oscillations, methods: data analysis, techniques: spectroscopic, line: profiles, stars: variables: δ Scuti, stars: variables: other

1. Introduction

Well-defined oscillation frequencies and their associated mode degree l, preferably in combination with constraints on the azimuthal number m, are indispensable for asteroseismic modelling. In case of the Sun, solar-like oscillators, and white dwarfs, the values (l, m) can be derived from equidistant frequency or period spacings. Other methods are needed to determine (l, m) for heat-driven non-radial pulsators such as β Cephei, δ Scuti, γ Doradus, and Slowly Pulsating B (SPB) stars. Generally, their frequency spectra do not show particular frequency patterns, and can be sparse or dense (see, e.g., Uytterhoeven *et al.* 2011; Balona *et al.* 2011). I present here an overview of existing mode-identification (mode-ID) techniques that mainly are applicable to massive ($M > 1\,M_\odot$) main-sequence pulsators. Note that some of the techniques have also been successfully applied to subdwarf B-type (sdB) stars (Telting *et al.* 2008).

Most mode-ID techniques are based on a comparison between observations and oscillation theory. The observables used are generally the frequency of the oscillation, its associated amplitude in a specific passband or in radial velocity, and the phase with respect to a reference epoch. Techniques used for mode identification based on multi-colour photometry are discussed in the overview by Breger (these proceedings). Here I focus on techniques based on high-resolution spectroscopy.

The advantage of spectral time-series is that they contain information on both l and m values through the principle of Doppler Imaging. As there is a one-to-one correlation between the position on the stellar surface and the position in the line profile of a rotating star, stellar pulsation velocity field variations on the stellar surface and photospheric pulsational temperature and gravity variations give rise to so-called line-profile variations (LPVs). Through an analysis of LPVs wave numbers can be identified.

2. Spectroscopic mode-identification techniques

I discuss here the main features, advantages, and disadvantages of the three main mode-ID techniques that have been developed since the 1970s, and refer to other reviews for additional discussion on the different spectroscopic techniques that are in use (e.g. Aerts & Eyer 2000; Mantegazza 2000; Telting 2003).

2.1. *Line Profile Fitting*

The Line Profile Fitting technique was developed following the advent of the first high-resolution spectra in the late 1970s–early 1980s (Smith 1977; Campos & Smith 1980a,b; Baade 1982, 1984; Smith 1983). This technique relies on a comparison between observed LPVs and theoretical profiles, whereby the theoretical line profiles are computed for a large grid of the pulsational and rotational parameters. Disadvantages of the Line Profile Fitting technique include: 1) the calculation of the theoretical profiles is very time-consuming; 2) there are many free parameters involved in the fitting; 3) there is no unique solution as several combinations of (l, m) might give similarly good results; 4) the selection of the *best* set of parameters is subjective, unless the code includes the calculation of a standard deviation per wavelength pixel; 5) the fitting is limited to mono-periodic pulsators. The main advantage of Line Profile Fitting is that it is an excellent tool to fine-tune parameters that are already constrained. The advice is hence to use the Line Profile Fitting technique in combination with other mode-ID techniques.

2.2. *Moment Method*

The Moment Method is based on the analysis of the time variations of the velocity moments of a line profile, which determine the shape of the line profile in a quantitative way. For instance, the first moment $\langle v \rangle$ is a measure for the centroid of the line profile, i.e. the radial velocity, the second moment $\langle v^2 \rangle$ defines the variance of the line profile, i.e. the equivalent width, and the third velocity moment $\langle v^3 \rangle$ measures the skewness of the line profile. This method was first proposed by Balona (1986a,b; 1987), and was further developed by Aerts et al. (1992), Aerts (1996), and Briquet & Aerts (2003). In the latest version by Briquet & Aerts (2003), the observed moments and their amplitudes are compared with theoretically calculated moments using an elaborate weighing function, called discriminant. The lower the value of the discriminant, the better the fit. In addition to an identification of (l, m), the Moment Method provides also constraints on the inclination angle between the stellar rotation axis and the line of sight (i), the projected rotational velocity $(v \sin i)$, and the intrinsic width of the line profile (σ). The expected accuracy on the determined values is $l \pm 1, |m| \pm 1$.

Disadvantages of the Moment Method are: 1) the method is restricted to the identification of low-degree modes $(l \leqslant 4)$; 2) there does not exist a statistical criterion to quantify the significance of the obtained solutions. As a result, several of the highest-ranked solutions need to be considered for further analysis. Further constraints on the best (l, m)-couple can be obtained by confrontation with Line Profile Fitting. Advantages are: 1) the method works well for multi-periodic stars; 2) the method works also for slow rotators.

As a side note I mention a spin-off technique of the Moment Method developed and applied to B-type stars by Cugier et al. (1994), Cugier & Daszyńska (2001), and Daszyńska-Daszkiewicz et al. (2003, 2005). The technique is based on both observed velocity moments and photometric observables in combination with linear non-adiabatic pulsation models, and allows the unambiguous identification of l through the construction of a parameter f that describes the ratio between the relative luminosity variations and the relative radial displacement of the stellar surface. The method has also been successfully applied to a pulsating subdwarf (Baran *et al.* 2008).

2.3. *Doppler Imaging*

The third and most used mode-ID technique is Doppler Imaging, which relies on the Fourier analysis of the observed intensity variations at each position in the line profile. The LPVs in Fourier space can be described by the distributions of the amplitude and phase variations across the line profile for each detected frequency, which are a direct measure for the mode parameters l and m. No modelling is involved in the direct application of this technique. The first concept of the Doppler Imaging technique was outlined by Vogt & Penrod (1983), and since then several versions have been developed (e.g. Gies & Kullavanijaya 1988; Kennelly *et al.* 1992; Telting & Schrijvers 1997; Mantegazza 2000; Zima 2006). I focus here on two of the best known versions of the Doppler Imaging technique, the Intensity Period Search (IPS) method and Fourier Parameter Fit (FPF) method, and compare their performance.

- **Intensity Period Search (IPS) Method** (Telting & Schrijvers 1997): (l, m)-values are determined for each frequency relying on the concept that the phase difference of a frequency ν in the phase diagram is an indicator for the value of the degree l, while the phase difference of its first harmonic (2ν) is a measure for $2|m|$. Also high-degree modes (up to $l \leqslant 20$) can be identified with the IPS method. The expected accuracy on the determined values is $l \pm 1, |m| \pm 2$.

The main disadvantage of the IPS method is that only l and m are constrained, with no information on other parameters such as i, $v \sin i$, or σ. Advantages are: 1) there is no modelling involved; 2) the method works for both mono- and multi-periodic stars; 3) the method can handle fast rotation.

- **Fourier Parameter Fit (FPF) Method** (Zima 2006): (l, m)-values are determined for each frequency by computation of the zero point, the amplitude, and phase for every wavelength bin across the line profile and by comparison with theoretically calculated values using a χ^2-test to assess the significance. For $m \leqslant 2$, the azimuthal number is unambiguously identified. For other values, the expected accuracy is $l \pm 1, |m| \pm 1$. The FPF method works best for slow to moderate rotators ($v \sin i < 100$ km s^{-1}).

The main disadvantage of the FPF method is that it is very time consuming given the computation of the theoretical values. Advantages are: 1) the $|m|$-value is extremely well constrained; 2) the method provides additional constraints on i and $v \sin i$; 3) the method works for both mono- and multi-periodic stars.

2.4. *Comparison between the different methods*

In Table 1 I provide a schematic comparison between the different spectroscopic mode-ID techniques discussed. The main conclusions are that the Moment Method and Doppler Imaging technique are complementary, and that the Line Profile Fitting technique is only efficient when the parameters are already constrained. In that case Line Profile Fitting helps to narrow down the number of possible (l, m) solutions.

Finally, I point out that there exists a freely available software package FAMIAS†, created by Zima (2008), aimed for the frequency analysis and mode identification of main-sequence pulsators, that includes the Moment Method and FPF method.

---

† The software package FAMIAS is freely downloadable from http://www.ster.kuleuven.be/~zima/famias/.

Table 1. Comparison between different spectroscopic mode-ID techniques (+: yes; −: no; .: ?).

	Line Profile Fitting	Moment Method	IPS method	FPF method
can handle fast rotators	+	−	+	−
can handle slow rotators	+	+	−	+
sensitive to sectoral modes ($|m| = l$)	.	−	+	+
sensitive to zonal modes ($m = 0$)	.	+	−	−
modelling involved	+	+	−	+
can handle multiperiodicity	−	+	+	+
constraints on extra parameters	+	+	−	+
detection limit l	.	$\leqslant 4$	$\leqslant 20$	$\leqslant 4$

3. Quality requirements of the spectral time-series and challenges

The mode-ID techniques described above are applied to spectral time-series. Below is a list of the main requirements of the spectra and related challenges.

3.1. *High-resolution spectra*

Spectral mode identification requires time-series of high-resolution spectra. A spectral resolution of at least $R = 40\,000$, preferably $R > 60\,000$, is needed to properly resolve the LPVs caused by the oscillations. Unfortunately, not many specialized spectroscopic instruments are available for long-term monitoring. High-resolution spectrographs are mainly attached to 1 to 2-m class telescopes, with the consequence that currently spectral mode identification is generally limited to bright targets ($V \leqslant 8$ mag for pulsators with pulsation periods of the order of a few hours or shorter).

3.2. *High signal-to-noise spectra*

A second requirement to perform an analysis of LPVs is a signal-to-noise ratio (SNR) per pixel of preferably at least 200. How easy a high-SNR value is obtained in a reasonable exposure time depends on the efficiency of the spectrograph. To have a good phase coverage of the oscillation period, the exposure time should not exceed 10% of the pulsation period. Again, this implies that spectroscopic time-series are only feasable for bright targets with 2-m class telescopes. The use of cross-correlation techniques, such as the Least-Squares Deconvolution (LSD) method (Donati *et al.* 1997), has allowed the analysis of LPVs from low-SNR spectra as the spectral information of all available lines in the spectrum is collapsed into one cross-correlation profile of much higher SNR. The disadvantage of cross-correlation techniques is that information on individual lines formed in different layers of the atmosphere is lost. On the other hand, the main advantage is that mode identification has become possible for slightly fainter stars ($V \leqslant 9.5$ mag for pulsators with pulsation periods of order of a few hours; $V \leqslant 11$ mag for pulsators with pulsation periods of order of a day, when using 2-m-class telescopes). As a side note I want to add that it is extremely difficult to obtain a high-quality spectral time-series for *Kepler* targets as most of them are fainter than $V = 11$ mag.

3.3. *Long time span*

To be able to resolve harmonic, sum and beat frequencies a sufficiently long time span of the spectral time series is needed. This means (continuous) monitoring of the target for several weeks to months. As long-term access to a single telescope is generally not feasible, unless when using a dedicated telescope such as the 1.2-m Mercator telescope with HERMES spectrograph on La Palma, and as single-site observations leave gaps in the time-series, the best way to go are multi-site campaigns. The success of dedicated observing campaigns using spectroscopic instruments at different observing sites for purposes of mode identification is proven by several examples (e.g. ν Eri: De Ridder *et al.* 2004; FG Vir: Zima *et al.* 2006; 12 Lac: Desmet *et al.* 2009; the *CoRoT* ground-based follow-up observations: Uytterhoeven *et al.* 2007). An alternative is the use of a dedicated network of robotic telescopes, such as the upcoming SONG network (Stellar Observations Network Group, Grundahl *et al.*, these proceedings; Uytterhoeven *et al.* 2012).

3.4. *Substantial number of spectra*

To successfully identify some of the observed pulsation modes, the time-series need to consist of a substantial number of spectra. The following empirical rule is valid: 'the more modes you want to identify, the more spectra you need'. There seems to be a correlation between the number of spectra in the time series and the number of modes that can be identified. Be aware that for each mode you want to identify, you need at least 200–300 spectra (accumulative).

4. Successful studies

Next, I present the successful application of spectroscopic mode-ID techniques using examples from the literature. The main aim is to illustrate the importance of, on the one hand, combining different techniques, and, on the other hand, combining photometric and spectroscopic data, to come to a unique mode identification.

4.1. *The power of combining multi-colour photometry and spectroscopy*

The hybrid β Cephei/SPB star 12 Lac was the subject of an extensive and dedicated photometric and spectroscopic multi-site campaign. From the 750 hours of available multi-colour photometric data Handler *et al.* (2006) constrained the degree l of *five* modes. Using the Moment Method and the FPF method on a time-series consisting of 1820 spectra Desmet *et al.* (2009) arrived at an unique identification of (l, m) for *four* of the highest amplitude modes. Subsequently, Daszyńska-Daszkiewicz *et al.* (2013) used amplitudes and phases from the different passbands and the first moment, the so-called non-adiabatic observables, to construct the f parameter (see Sect. 2.2). They arrived at an unique identification of l for *six* modes. This example shows that the maximum amount of information on the mode parameters only can be obtained by combining photometric and spectroscopic methods.

4.2. *The power of combining different spectroscopic mode-identification techniques*

From a photometric and spectroscopic multi-site campaign on the β Cep star HD 180642, resulting in 1234 multi-colour photometric points and 262 high-resolution spectra, Briquet *et al.* (2009) detected a dominant frequency and several low-amplitude frequencies. From photometry the dominant mode was identified as radial, while identification of l was possible for two additional non-radial modes. The spectroscopic mode-ID of the additional modes turned out to be very challenging due to the radial mode with very high amplitude that dominates the LPVs. Customized versions of the Moment Method and the FPF method were needed to arrive at a satisfactory solution. Only thanks to combining the results from the two spectroscopic mode-ID techniques it was possible to unambiguously identify one of the additional modes as a (3, 2)-mode.

4.3. *Recent results*

Finally, I provide a short overview of recent studies on spectroscopic mode identification in main-sequence pulsators.

The *CoRoT* ground-based spectroscopic follow-up observations allowed the identification of pulsation modes in the β Cephei stars HD 180642 (Briquet *et al.* 2009) and HD 43317 (Pápics *et al.* 2012), the Be stars HD 49330 (Floquet *et al.* 2009) and HD 181231 (Neiner *et al.* 2009), the δ Scuti stars HD 50844 (Poretti *et al.* 2009) and HD 50870 (Mantegazza *et al.* 2012), and the hybrid γ Doradus/δ Scuti star HD 49434 (Uytterhoeven *et al.* 2008).

Other spectroscopic multi-site campaigns were carried out on the γ Doradus stars HD 12901 and HD 135825 (see Brunsden *et al.*, this volume; Brunsden *et al.* 2012a,b) and the β Cephei star V2052 Oph (Briquet *et al.* 2012).

5. A view on the future and conclusions

5.1. *Near-infrared spectroscopy*

A new and innovative way of deriving mode parameters from spectra might be the use of near-infrared spectroscopic data. Near-infrared spectra probe different parts of line-forming regions in the atmosphere with respect to visual spectra, and might hence provide extra information for the identification of modes. This technique might be worth exploring for lower mass main-sequence stars, as they are cool enough to produce spectral lines in this wavelength region. A pilot study has already been performed by Amado (2007) for a high-amplitude δ Scuti (HADS) star, and LPVs have been detected in the near-infrared spectral region of CRIRES/VLT spectra. Additional observational and theoretical work is needed to investigate this option further. The future instrument CARMENES (Quirrenbach *et al.* 2012) at the 3.5-m telescope at Calar Alto observatory might be the perfect spectrograph for dedicated observing campaigns in the near-infrared.

5.2. *Network of robotic telescopes*

Upcoming networks of small robotic telescopes equipped with a high-resolution spectrograph such as SONG or LCOGT (Las Cumbres Observatory Global Telescope Network), dedicated to asteroseismic studies in the former case and to all types of variability studies in the latter case, will facilitate the gathering of continuous time-series needed for mode identification of main-sequence pulsators for selected bright targets ($V \leqslant 5$ mag).

5.3. *Spectroscopic mode identification for faint stars?*

Optimally, one needs easy access to 8-m class telescopes equipped with a high-resolution spectrograph for long-term monitoring to broaden the pool of targets that can be studied for mode identification from bright to fainter stars. However, this wish probably will not be granted in the coming few years.

5.4. *To conclude*

To conclude, it has become clear that the best way forward for obtaining a reliable mode identification in heat-driven main-sequence pulsators is the combined use of different techniques, whereby joining photometric and spectroscopic data.

Acknowledgements

KU acknowledges funding by the Spanish National Plan of R&D for 2010, project AYA2010-17803.

References

Aerts, C. 1996, *A&A*, 314, 115

Aerts, C. & Eyer, L. 2000, *ASP-CS*, 210, 113

Aerts, C., de Pauw, M., & Waelkens, C. 1992, *A&A*, 266, 294

Amado, P. J. 2007, *CoAst*, 151, 57

Baade, D. 1982, *A&A*, 105, 65

Baade, D. 1984, *A&A*, 135, 101

Balona, L. A. 1986a, *MNRAS*, 219, 111

Balona, L. A. 1986b, *MNRAS*, 220, 647

Balona, L. A. 1987, *MNRAS*, 224, 41

Balona, L. A., Pigulski, A., De Cat, P., *et al.* 2011, *MNRAS*, 413, 2403

Baran, A., Pigulski, A., & O'Toole, S. J. 2008, *MNRAS*, 385, 255

Briquet, M. & Aerts, C. 2003, *A&A*, 398, 687

Briquet, M., Uytterhoeven, K., Morel T., *et al.* 2009, *A&A*, 506, 269

Briquet, M., Neiner, C., Aerts, C., *et al.* 2012, *MNRAS*, 427, 483

Brunsden, E., Pollard, K. R., Cottrell, P. L., Wright, D. J., & De Cat, P. 2012, *MNRAS*, 427, 2512

Brunsden, E., Pollard, K. R., Cottrell, P. L., Wright, D. J., De Cat, P., & Kilmartin, P. M. 2012, *MNRAS*, 422, 3535

Campos, A. J. & Smith, M. A. 1980a, *ApJ* 238, 250

Campos, A. J. & Smith, M. A. 1980b, *ApJ* 238, 667

Cugier, H. & Daszyńska, J. 2001, *A&A*, 377, 113

Cugier, H., Dziembowski, W. A., & Pamyatnykh, A. A. 1994, *A&A*, 291, 143

Daszyńska-Daszkiewicz, J., Dziembowski, W. A., & Pamyatnykh, A. A. 2003, *A&A*, 407, 999

Daszyńska-Daszkiewicz, J., Dziembowski, W. A., & Pamyatnykh, A. A. 2005, *A&A*, 441, 641

Daszyńska-Daszkiewicz, J., Szewczuk, W., & Walczak, P. 2013, *MNRAS*, 431, 3396

De Ridder, J., Telting, J. H., Balona, L. A., *et al.* 2004, *MNRAS*, 351, 324

Desmet, M., Briquet, M., Thoul, A., *et al.* 2009, *MNRAS*, 396, 1460

Donati, J.-F., Semel, M., Carter, B. D., Rees, D. E., & Collier Cameron, A. 1997, *MNRAS*, 291, 658

Floquet, M., Hubert, A.-M., Huat, A.-L., *et al.* 2009, *A&A*, 506, 103

Gies D. R. & Kullavanijaya A. 1988, *ApJ*, 326, 813

Handler, G., Jerzykiewicz, M., Rodríguez, E., *et al.* 2006, *MNRAS*, 365, 327

Kennelly, E. J., Walker, G. A. H., & Merryfield, W. J. 1992, *ApJ*, 400, L71

Mantegazza, L. 2000, *ASP-CS*, 210, 138

Mantegazza, L., Poretti, E., Michel, E., *et al.*, 2012, *A&A*, 542, A24

Neiner, C., Gutiérrez-Soto, J., Baudin, F., *et al.* 2009, *A&A*, 506, 143

Pápics, P. I., Briquet, M., Baglin, A., *et al.* 2012, *A&A*, 542, 55

Poretti, E., Michel, E., Garrido, R., *et al.* 2009, *A&A*, 506, 85

Quirrenbach, A., Amado, P. J., Seifert, W., *et al.* 2012, in: *Ground-based and Airborne Instrumentation for Astronomy IV*. Proceedings of the SPIE, Vol. 8446, 84460R

Smith, M. A. 1977, *ApJ*, 215, 574

Smith, M. A. 1983, *ApJ*, 265, 338

Telting, J. H. 2003, *Ap&SS*, 284, 85

Telting J. H. & Schrijvers C. 1997, *A&A*, 317, 723

Telting, J. H., Geier, S., Østensen, R. H., *et al.* 2008, *A&A*, 492, 815

Uytterhoeven, K. & Poretti, E., the *CoRoT* SGBOWG 2007, *CoAst*, 150, 371

Uytterhoeven, K., Mathias P., Poretti E., *et al.* 2008, *A&A*, 489, 1213

Uytterhoeven, K., Moya, A., Grigacène, A., *et al.* 2011, *A&A*, 534, A125

Uytterhoeven, K., Pallé, P. L., Grundahl, F., *et al.* 2012, *AN*, 333, 1107

Vogt S. S. & Penrod G. D. 1983, *PASP*, 95, 565

Zima, W. 2006, *A&A*, 455, 227

Zima, W. 2008, *CoAst*, 155, 12

Zima, W., Wright, D., Bentley, J., *et al.* 2006, *A&A*, 455, 235

Precision Asteroseismology
Proceedings IAU Symposium No. 301, 2013
J. A. Guzik, W. J. Chaplin, G. Handler & A. Pigulski, eds.

© International Astronomical Union 2014
doi:10.1017/S1743921313014178

Interpretation of the oscillation spectrum of HD 50230 — a failure of richness

Wojciech Szewczuk[1], Jadwiga Daszyńska-Daszkiewicz[1] and Wojciech Dziembowski[2]

[1] Instytut Astronomiczny, Uniwersytet Wrocławski, Wrocław, Poland
email: szewczuk@astro.uni.wroc.pl, daszynska@astro.uni.wroc.pl

[2] Warsaw University Observatory, Warsaw, Poland
email: wd@astrouw.edu.pl

Abstract. Attempts to interpret the observed oscillation spectrum of the SPB star HD 50230 are reported. We argue that a nearly equidistant period spacing found in the oscillation spectrum of the star is most likely accidental. The observed period distribution requires excitation of modes with the degree $l > 4$. Much more may be learned from the rich oscillation spectrum of the star but most of the work is still ahead of us.

Keywords. stars: oscillations, stars: rotation, stars: individual: HD 50230

1. Introduction

HD 50230 is a star of the B3 V spectral type and a visual brightness of 8.95 mag. The star had been regarded as constant until its first satellite observations were conducted. Degroote *et al.* (2010, 2012) detected more than 500 significant peaks in the *CoRoT* data. With their spectroscopic observations they discovered that HD 50230 is a double-lined spectroscopic binary with projected equatorial velocity of 7 and 117 km s^{-1} for the primary and secondary, respectively. For the primary component, they determined effective temperature of $T_{\rm eff,1} = 18000 \pm 1500$ K and surface gravity of $\log g_1 = 3.8 \pm 0.3$ dex. For the secondary component they found only an upper limit $T_{\rm eff,2} \leqslant 16000$ K and assumed $\log g_2 \approx 4$ dex.

2. Peaks almost uniformly spaced in period

In the rich oscillation spectrum of HD 50230 Degroote *et al.* (2010) extracted eight peaks almost uniformly spaced in period. Invoking the asymptotic theory, they interpreted these peaks as a sequence of modes with the same spherical harmonic degree, l, azimuthal order, m, and consecutive radial orders, n. Assuming $l = 1$ and $m = 0$ they found that main-sequence star models with a mass of $7-8$ $M_\odot$ can reproduce the observed period spacing.

To test Degroote's interpretation we re-analysed the *CoRoT* data. Using the Lomb-Scargle periodogram we found 515 significant frequency peaks comparing to 556 frequencies found by Degroote at al. (2012). Most our frequencies are consistent with the Degroote's determinations. Surprisingly, we found in our set many sequences of peaks nearly uniformly spaced in periods. Unfortunately, these sequences do not yield sufficient clues to mode identification and are likely accidental.

In the left panel of Fig. 1, we present examples of three sequences (filled symbols). For comparison, as open squares, a sequence found by Degroote *et al.* (2012) is also shown. We chose these sequences because if we assume that frequencies with the period spacing

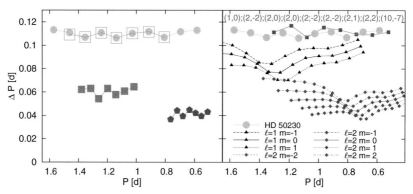

Figure 1. *Left:* Nearly equidistant period spacings found in the *CoRoT* photometric data of the star HD 50230. Filled circles, squares and pentagons denote equidistant series found in the set of periods determined by us whereas the series published by Degroote *et al.* (2012) is marked with open squares. *Right:* Period spacings, in the model with a mass $M = 5.0\,M_\odot$, effective temperature, $\log T_{\mathrm{eff}} = 4.227$, luminosity, $\log L/L_\odot = 2.77$, metallicity, $Z = 0.015$ and rotational velocity, $V_{\mathrm{rot}} = 10$ km s^{-1}. Squares show selected modes with various degrees and azimuthal orders (numbers in brackets) with spacings similar to that found in HD 50230 (dots). For comparison, sequences of consecutive dipolar (triangles) and quadruple (rhombs) modes are shown.

of $\Delta P \approx 0.11$ d are dipole modes then, according to the asymptotic theory, those with $\Delta P \approx 0.06$ and 0.04 d correspond to modes with $l = 2$ and 3, respectively. Unfortunately, they have some common frequencies. Thus, at best, there must be some missing modes.

For further analysis, we selected nine frequencies from the sequence with $\Delta P \approx 0.11$ d. Two frequencies with the shortest periods were omitted because they have too low radial orders and may not follow the asymptotic theory. Next, we tried to reproduce these nine frequencies adopting two approaches. Neglecting the effects of rotation, we reached the best fits for models with the following parameters: $M = 7.7\,M_\odot$, $Z = 0.030$, $\alpha_{\mathrm{ov}} = 0.0$, $\log T_{\mathrm{eff}} = 4.2337$ and $M = 7.1\,M_\odot$, $Z = 0.025$, $\alpha_{\mathrm{ov}} = 0.2$, $\log T_{\mathrm{eff}} = 4.2189$. Here, Z is the metal abundance by mass fraction and α_{ov} is the overshooting parameter from the convective core expressed in the terms of pressure scale height, H_p. However, the corresponding $\chi^2 \approx 2 \times 10^4$ does not allow us to accept these solutions. A better fit was obtained when we took into account rotational splitting. Then, interpreting the sequence as retrograde dipole modes for a model with $M = 5.4\,M_\odot$, $Z = 0.010$, $\alpha_{\mathrm{ov}} = 0.6$, $\log T_{\mathrm{eff}} = 4.2561$, and $V_{\mathrm{rot}} = 25$ km s^{-1} we got $\chi^2 \approx 6 \times 10^3$. These large values of χ^2 result from very small frequency errors.

3. Theoretical modes nearly equally spaced in period

Since oscillation spectra of high- and moderate-order g modes are dense, we may expect that modes with different values of spherical harmonic degree l, and azimuthal order m, can accidentally form period sequences nearly equally spaced in period. To demonstrate such cases, we calculated pulsation modes with $l = 1 - 10$ for models with $M = 5\,M_\odot$ and different rotational velocities. Then, in the same way as we did for the observations, we searched for modes equally spaced in period, considering only unstable modes. We found many sequences with different mean period spacings composed of modes with different pairs (l, m). An example is shown in the right panel of Fig. 1 as squares. The nearly constant period spacing is accidental and in no way is related to the asymptotic property of g modes. The calculated sequences at fixed l and m, depicted in the same figure, show lower mean spacing and much larger deviation from constancy. These results support

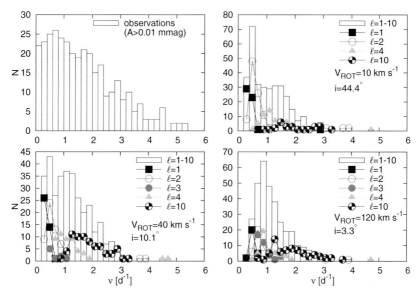

Figure 2. A comparison of histograms for frequencies detected in HD 50230 (upper left) and frequencies of unstable modes in the model with $M = 6.95\,M_\odot$, $\log T_{\rm eff} = 4.255$, $Z = 0.015$ and $\alpha_{\rm ov} = 0$ and three rotational velocities: 10 km s^{-1}, 40 km s^{-1} and 120 km s^{-1}. In the theoretical histograms, N stands for the number of all unstable modes with l from 1 to 10 and $A \geqslant 0.01$ mmag. The values of the inclination angle, i, result from $V_{\rm rot} \sin i = 7$ km s^{-1}.

the conclusion that, in dense oscillation spectra, appearance of equally spaced sequences may be accidental and they should be treated with caution.

4. Comparing histograms of observed and calculated frequencies

When we realized that the observed period spacing cannot be interpreted according to the asymptotic theory we lost the clue to mode identification which is prerequisite for deriving seismic constraints on stellar parameters. Since the prospects for progress looked to us rather grim, we decided to look for a different application of the rich frequency data on HD 50230.

One unsolved problem in the stellar pulsation theory is the amplitude limitation in stars with a large number of unstable modes. The observed distribution of peaks over period ranges may yield an important hint leading us to the solution of this difficult problem. To this aim the observed distribution must be confronted with simulations based on linear nonadiabatic calculations for models constrained by measured values of $\log T_{\rm eff}$, $\log g$ and $V_{\rm rot} \sin i$. The linear calculations do not yield mode amplitudes and this information we want to extract from data. We should search for the best fit to the observed distribution assuming various mode-selection principles. The set of considered modes, which all must be unstable, may be terminated at some sufficiently high l if the observational threshold cannot be reached.

In the histograms shown in Fig. 2, we assume the random distribution of the r.m.s. amplitude of relative variations of the stellar radius, *the intrinsic amplitude of a mode*, and considered modes with the photometric amplitudes $A \geqslant 0.01$ mmag. We included modes with degrees up to $l = 10$, because our simulations showed that for higher l, the amplitudes in the *CoRoT* band do not exceed the value of $A = 0.01$ mmag. We use the same model reproducing the central values of $\log T_{\rm eff}$ and $\log g$ (see Sect. 1) for the primary but with three distinct rotation rates: $V_{\rm rot} = 10$, 40, 120 km s^{-1}. The values of the inclination angle result from $V_{\rm rot} \sin i = 7$ km s^{-1} which was kept constant. These rotation rates are higher than those estimated by Degroote *et al.* (2010, 2012) based on analysis of high-order p modes. However, their estimate is uncertain and refers to the different part of the interior. The effects of rotation were included in the framework of the traditional approximation, e.g., Lee & Saio (1997), Townsend (2003, 2005). For comparison, the observed histogram was shown in the left upper panel of Fig. 2.

As one can see, we were unable to reproduce the observed distribution of frequencies. Independently of rotation, we have a shortage of unstable and "visible" ($A \geqslant 0.01$ mmag) theoretical modes with the shortest as well as with highest frequencies, above about 2 d^{-1}. For higher rotation rate, one could expect that retrograde modes complement the lowest frequency range. But the higher values of $V_{\rm rot}$ imply the lower values of the inclination angle which in turn favor the axisymmetric modes ($m = 0$). We considered also a possibility that some of the low-amplitude peaks may result from combinations and found it unlikely. Our results give certain limits on the intrinsic amplitudes of modes.

5. Conclusions

In dense oscillation spectra such as in the case of HD 50230, equidistant period spacings can be very likely accidental and one should be cautious when interpreting such structures. Although the dream of rich oscillation spectra in the B-type pulsators has come true, we still do not have any clue to identify angular numbers of observed frequencies. Without additional observations which would allow for mode identification, a reliable seismic stellar model of the star cannot be constructed.

We see prospects for gaining new insight into nonlinear mode selection in stars from available data on HD 50230. Our efforts toward explaining the observed distribution of peaks in its oscillation spectrum will continue.

Acknowledgements

WD was supported by Polish NCN grant DEC-2012/05/B/ST9/03932. WS was supported by Polish NCN grant 2012/05/N/ST9/03905. Calculations have been carried out using resources provided by Wrocław Centre for Networking and Supercomputing (http://wcss.pl), grant No. 265.

References

Degroote, P., Aerts, C., Baglin, A., *et al.* 2010, *Nature*, 464, 259
Degroote, P., Aerts, C., Michel, E., *et al.* 2012, *ApJ*, 542, 88
Lee, U. & Saio, H. 1997, *ApJ*, 491, 839
Townsend, R. H. D. 2003, *MNRAS*, 340, 1020
Townsend, R. H. D. 2005, *MNRAS*, 360, 465

Precision Asteroseismology
Proceedings IAU Symposium No. 301, 2013   © International Astronomical Union 2014
J. A. Guzik, W. J. Chaplin, G. Handler & A. Pigulski, eds.   doi:10.1017/S174392131301418X

Spectroscopy of γ Doradus stars

E. Brunsden[1], K. R. Pollard[1], P. L. Cottrell[1], D. J. Wright[2], P. De Cat[3] and P. M. Kilmartin[1]

[1]Department of Physics and Astronomy, University of Canterbury, Private Bag 4800,
Christchurch, New Zealand
email: emily.brunsden@gmail.com

[2]Department of Astrophysics, University of New South Wales, Sydney, Australia
[3]Royal Observatory of Belgium, Ringlaan 3, 1180 Brussel, Belgium

Abstract. The MUSICIAN programme at the University of Canterbury has been successfully identifying pulsation modes in many γ Doradus stars using hundreds of ground-based spectroscopic observations. This paper describes some of the successful mode identifications and emerging patterns of the programme. The hybrid γ Doradus/δ Scuti star HD 49434 remains an enigma, despite the analysis of more than 1700 multi-site high-resolution spectra. A new result for this star is apparently distinct line-profile variations for the γ Doradus and δ Scuti frequencies.

Keywords. line: profiles, techniques: spectroscopic, stars: individual (HD 135825, HD 12901, γ Doradus, HD 49434), stars: variables: other, stars: oscillations (including pulsations)

1. Introduction

Spectroscopic mode identification of individual pulsating stars is a challenging and time-consuming task. The γ Doradus class of pulsators make for particularly awkward targets due to their characteristic $1 - 3$ d^{-1} frequencies. These frequencies have significant aliasing problems due to ground-based observing cycles and long beat periods. To overcome these problems the MUSICIAN programme at the University of Canterbury has been targeting γ Doradus stars in frequent observing runs over several months to years. Typically 100 to 400 spectra are acquired for each star from Mt John University Observatory's 1-m McLellan telescope with the High Efficiency and Resolution Canterbury University Large Échelle Spectrograph (HERCULES). Data are then combined with spectra from other sites where possible for analysis. Successful mode identifications have now been made for a number of stars, four of which are described here.

2. Data reduction and analysis techniques

This section comprises of a brief overview of the currently used methods for data reduction and analysis. More detail can also be found in Brunsden *et al.* (2012b). Spectra acquired are reduced using the standard reduction packages offered for each spectrograph. Cross-correlation of up to 5000 lines per spectrum is performed, avoiding regions of telluric and strong hydrogen lines. The cross-correlation is performed independently for each site, which produces representative line profiles for each observation. An example of the profiles generated for HD 135825 is given in Fig. 1.

Frequency and mode identification using the representative line profiles is performed in FAMIAS (Zima 2008). Frequencies are calculated using the moment method (Balona 1986a,b, Balona 1987, Aerts *et al.* 1992) and using the individual pixels across the line profile (Mantegazza 2000). Fourier spectra of the pixel frequencies detected in HD 135825

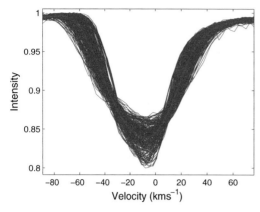

Figure 1. Representative line profile of 295 observations of HD 135825. Pulsations can be clearly seen deforming lines from the mean line profile.

Figure 2. The Fourier spectrum showing frequencies found using the pixel-by-pixel technique and successive prewhitening. The smooth line shows the significance level from FAMIAS.

with each prewhitening stage are shown in Fig. 2. Mode identifications are performed in FAMIAS using the pixel-by-pixel technique (Zima 2009). This method fits the variations of the line profiles to a grid of models to identify the pulsation modes. A χ^2 minimisation routine is used to identify the best modes. As an example, the standard deviation profiles of the four frequencies identified in HD 135825 are shown in Fig. 3 with their mode fits.

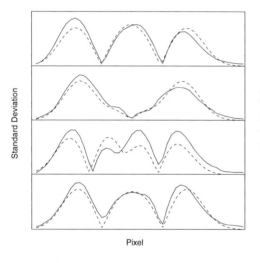

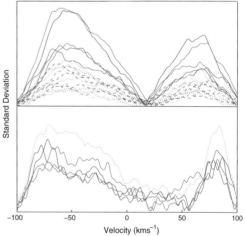

Figure 3. The fit (dashed line) of the mode identification to the standard deviation profile and phase (solid line) of the four identified frequencies in HD 135825.

Figure 4. Standard deviation profiles of the δ Scuti (top) and γ Doradus (bottom) range frequencies detected in HD 49434.

3. Summary of results

The frequencies and mode identifications of several γ Doradus stars are given in Table 1 of Pollard *et al.* (these proceedings), including the three described in this paper.

HD 135825 was found to have a single dominant, high-amplitude frequency and three lower amplitude frequencies. It is confirmed that HD 135825 is a true γ Doradus star.

Considering the prior studies (Aerts 2004, Moya 2005), all of the candidate frequencies and modes found for HD 12901 are well supported. The occurrence of five $(1, 1)$ modes in this star suggests they may be the same low degree (l) for sequential values of n (Tassoul 1980). This led to an investigation into the period spacings of the identified frequencies which showed no consistent spacings.

The frequencies found in the analysis of γ Doradus were almost identical to those found in previous spectroscopic and photometric studies (Balona 1994, Balona *et al.* 1996, Dupret *et al.* 2005a, Tarrant *et al.* 2008). Two frequencies have been shown to be stable over twenty years since their first identification by Cousins (1992). γ Doradus shows an excellent agreement between the frequencies and modes found in photometry and spectroscopy giving independent confirmation of the modes detected in FAMIAS.

A prevalence of $(1, 1)$ modes in γ Doradus stars is beginning to emerge. This could be explained by considering the selection bias of ground-based targets. The stars selected for our ground-based observational work are bright targets ($V < 8$ for spectroscopy) with high-amplitude pulsations. The $l = 1$ modes are expected to have the highest amplitude in these stars due to the large surface area covered in each segment of the mode. In addition, stars with high inclinations (i close to $90°$) will have higher observed amplitudes for sectoral modes and those with low inclinations, higher-amplitude tesseral and zonal modes (Schrijvers *et al.* 1997, Reese *et al.* 2013). For rotating stars, the presence of an equatorial wave-guide (Townsend 2003) would suppress the tesseral modes and thus only the stars with sectoral modes would be classified as γ Doradus stars.

4. The special case of HD 49434

The presence of both γ Doradus and δ Scuti-range frequencies suggests that HD 49434 may be a hybrid star of the two classes. This classification was made by Uytterhoeven *et al.* (2008) based on the frequencies found and the proximity of the star in the intersection of the pulsation groups in the Hertzsprung-Russell diagram. The star was also monitored by the *CoRoT* mission with more than 800 frequencies identified (Chapellier *et al.* 2011). The addition of 1100 spectra to the 700 analysed in Uytterhoeven *et al.* (2008) refined the spectroscopic picture of this star, with 31 frequencies extracted. These frequencies match very poorly those detected by *CoRoT*. Upon inspection of the standard deviation profiles, two distinct shapes were observed (Fig. 4). These two groups corresponded to the frequencies in the δ Scuti range and in the γ Doradus range of expected frequencies.

The classification of the shapes of the line profiles of the frequencies suggests physical differences between the two groups of frequencies and two groups of modes. The shift of the centre of the standard deviation profile with respect to the centre of the line profile is currently being investigated.

Acknowledgements

This work was supported by the Marsden Fund administered by the Royal Society of New Zealand. Mode identification results obtained with the software package FAMIAS developed in the framework of the FP6 European Coordination action HELAS (http://www.helas-eu.org/).

References

Aerts, C., de Pauw, M., & Waelkens, C. 1992 *A&A*, 266, 294

Aerts, C., Cuypers, J., De Cat, P., *et al.* 2004 *A&A*, 415, 1079

Balona, L. A. 1986a, *MNRAS*, 219, 111

Balona, L. A. 1986b, *MNRAS*, 220, 647

Balona, L. A. 1987, *MNRAS*, 224, 41

Balona, L. A., Krisciunas, K., & Cousins, A. W. J. 1994, *MNRAS*, 270, 905

Balona, L. A., Böhm, T., Foing, B. H., *et al.* 1996, *MNRAS*, 281, 1315

Brunsden E., Pollard K. R., Cottrell P. L., Wright D. J. & De Cat P. 2012a, *MNRAS*, 427, 2512

Brunsden E., Pollard K. R., Cottrell P. L., Wright D. J., De Cat P., & Kilmartin P. M. 2012b, *MNRAS*, 422, 3535

Chapellier, E., Rodríguez, E., Auvergne, M., *et al.* 2011, *A&A*, 525, A23

Cousins, A. W. J. 1992, *Observatory*, 112, 53

Dupret, M.-A., Grigahcène, A., Garrido, R., De Ridder, J., Scuflaire, R., & Gabriel, M. 2005a, *MNRAS*, 360, 1143

Dupret, M.-A., Grigahcène, A., Garrido, R., Gabriel, M., & Scuflaire, R. 2005b, *A&A*, 435, 927

Mantegazza, L. 2000, *ASP-CS*, 210, 138

Moya, A., Suárez, J. C., Amado, P. J., Martin-Ruíz, S., & Garrido, R. 2005, *A&A*, 432, 189

Reese, D. R., Prat, V., Barban, C., van't Veer-Menneret, C., & MacGregor, K. B. 2013, *A&A*, 550, A77

Schrijvers, C., Telting, J. H., Aerts, C., Ruymaekers, E., & Henrichs, H. F. 1997, *A&AS*, 121, 343

Tarrant, N. J., Chaplin, W. J., Elsworth, Y. P., Spreckley, S. A., & Stevens, I. R. 2008, *A&A*, 492, 167

Tassoul, M. 1980, *ApJS*, 43, 469

Townsend, R. H. D. 2003, *MNRAS*, 343, 125

Uytterhoeven, K., Mathias, P., Poretti, E., *et al.* 2008, *A&A*, 489, 1213

Zima, W. 2008, *CoAst*, 157, 387

Zima, W. 2009, *A&A*, 497, 827

Precision Asteroseismology
Proceedings IAU Symposium No. 301, 2013
J. A. Guzik, W. J. Chaplin, G. Handler & A. Pigulski, eds.

© International Astronomical Union 2014
doi:10.1017/S1743921313014191

Helioseismology in the 1980s and 1990s

Philip R. Goode

Big Bear Solar Observatory, New Jersey Institute of Technology
40386 N. Shore Lane, Big Bear City, CA, USA
email: pgoode@bbso.njit.edu

Abstract. Over more than twenty years, Wojtek Dziembowski and I collaborated on nearly fifty papers, which were concentrated in helioseismology through the 1980s and 1990s, but extended early into the new century. In this review, I discuss the most significant results of this collaboration and some of the underlying sociology that contributed to the intensity and longevity of our collaboration. Our work began with placing limits on the Sun's buried magnetic field and ended with extracting from the solar-cycle dependent oscillation frequency changes the roles (and net result) of competing dynamical drivers of changes in the solar diameter.

Keywords. Sun: helioseismology, Sun: oscillations, Sun: interior, Sun: photosphere

1. Introduction

In the 1970s, my research was concentrated on the problem of renormalizing the nucleon-nucleon interaction in the presence of a core nucleus. The problem was difficult because the renormalization of the two-nucleon interaction presented a poorly convergent perturbative expansion (see Goode & Koltun 1975). The field was mature and progress was slow, which made the field one of diminished excitement. Then I learned of helioseismology and thought my experience in treating perturbations around complex spherical geometries would be applicable. However, my knowledge of astronomy was limited. Fortunately, Henry Hill introduced me to Wojtek Dziembowski and this began a collaboration that spanned more than twenty years. Wojtek's pioneering paper on nonradial oscillations had already become the handbook for the study of stellar oscillations (Dziembowski 1977) and I have worn out several copies.

Our collaboration began in earnest in 1982 and ended in 2005. It was my most productive, significant, educational and rewarding collaboration. In our collaboration, I typically spent one month each summer in Warsaw where we worked seated at desks facing each other. Then, Wojtek spent one month each winter in New Jersey where we shared an office and meals. I always looked forward to our time together because we would do something worthwhile scientifically, as we had fun and discussed all manner of topics. According to the paper count by A.A. Pamyatnych, our collaboration was Wojtek's most productive. It ended with a sense of fulfillment and sadness as we moved in different directions with his interest returning to the stars, and the wisdom of his choice is reflected in the astounding results that one sees in the proceedings of this conference and the central role of Wojtek in the field of asteroseismology. His work has been pioneering.

2. The 1980s

In 1983, we wrote a little known paper for the proceedings of the European Physical Society Meeting on Solar Oscillations held in Catania, Italy. The paper developed simple formulae describing, asymptotically, the effect of rotation and magnetic fields on solar oscillation frequencies. I still use the formulae in this paper (Dziembowski & Goode 1984) and recommend it to people just learning helioseismology. Figure 1 is a picture of the two of us at that meeting. The picture was taken in the ruins of an ancient Sicilian amphitheatre and shows Wojtek making a scientific argument while I listen, think and learn more astronomy.

This was a truly exciting time in helioseismology. Duvall and Harvey (1984) had measured fine structure in the spectrum of the Sun's five-minute period oscillations from which we (Duvall *et al.* 1984) were able to determine the internal rotation of the Sun in its equatorial plane. We learned that the Sun has surface-like rotation down to its core in the equatorial plane, and so the Sun had "spun-down", which was somewhat of a surprise at the time. Shortly thereafter, Duvall *et al.* (1986) went to the South Pole and made more complete measurements of the fine structure in the oscillation spectrum from which they determined that surface-like differential rotation persisted beneath the surface. In two separate works in 1989, we determined that the surface-like differential rotation persisted to the base of the convection zone where there was a sharp transition to solid-body rotation (Brown *et al.* 1989, Dziembowski *et al.* 1989) at a mean surface rate. These papers used splitting data from Sacramento Peak Observatory (Brown & Morrow 1987) and from Big Bear Solar Observatory (Libbrecht 1989).

3. The 1990s

In the early 1990s, we did our most mathematically pure work on stellar oscillations in which we developed a complete formalism, valid through second order in differential rotation $(\Omega(r,\theta))$, describing the effect of rotation on stellar oscillation frequencies (Dziembowski & Goode 1992). We found that second-order effects on solar oscillation frequencies are dominated by effects of distortion, which must be accounted for in any effort to determine the Sun's internal magnetic field. For solar oscillations, accidental degeneracies can occur, but we found that they cannot lead to large frequency shifts. However for evolved δ Scuti stars, calculated spectra are dense, and due to rotational

Figure 1. Dziembowski and Goode in Catania, 1983.

perturbation members of neighboring multiplets may overlap. Here, we emphasized the seismic potential of modes with mixed p- and g-mode character. The paper was frustrating in that a large effort did not lead to a hoped-for significant second-order effect of rotation on solar oscillations, but was satisfying in its completeness.

The problem of the Sun's neutrino deficit had been the subject of endless controversy beginning with the magnificent work of Ray Davis that ultimately led to his Nobel Prize. After Davis's work was accepted, the question was whether or not there was an astrophysical solution to the Sun's neutrino paradox (aside: with typical incisive humor Wojtek described this as determining who owns the universe – astrophysicists or particle physicists). We inverted p-mode frequencies from Libbrecht's BBSO data, plus BiSON data for the low-l modes, to obtain the run of pressure, density and mass, as well as the fractional surface helium abundance – the *seismic solar model* (Dziembowski *et al.* 1994, 1995). We assumed that gravity and pressure are the only forces acting and that the adiabatic exponent is close to the model value except in the outmost layers. To obtain the run of the temperature, further assumptions would have been required. The results of the seismic model agreed quite well with the standard model. Several variations of the standard model used were considered: a young Sun, p+p enhanced by 1.034 (this reaction occurs at too low an energy to be measured in the laboratory, and so there is some uncertainty in it), the ^{3}He $+^3$He part of the p-p chain enhanced by a factor of nine (under the unlikely assumption of a low-energy resonance in the reaction, which has the effect of circumventing the requisite two-thirds of the p-p chain that produces neutrinos with sufficient energy to be detected in the Davis's deep underground Homestake Mine experiment in which ^{37}Cl is converted to ^{37}Ar in a gigantic tank of cleaning fluid). The deviations from the standard with seismic models determined with the variations led to several conclusions. First, there is no astrophysical solution to the solar neutrino problem (the Sun is prosaic and the particle physicists "own the universe"). Second, the seismic age of the Sun is comparable to that determined from meteorite data.

An aside: During this same time period, we attended a helioseismology meeting at the Tata Institute set to coincide with a total eclipse of the Sun in 1995. During this time, we hired a car and took a two-man (plus driver) tour for several days to many sites of interest. Among other adventures, we found ourselves in a line for a crematorium on the banks of the Ganges in the holy city of Varanasi, and in another city were chased from a hotel room by a "giant" white lizard. The highlight of the trip was the total eclipse where we selected the playing field of a boys school as our place to view it. While I was setting up to watch, I turned and saw Wojtek surrounded by teenage schoolboys listening to him giving an extemporaneous talk about the Sun and solar eclipses. A great, but also typical moment. On that trip, as usual, all topics were open for discussion, and often discussed. We also had a daylong adventure to visit the Taj Mahal (see Fig. 2).

Later in the 1990s, we used MDI/SoHO data on f mode frequencies to determine the seismic radius of the Sun. We found that Kosovichev and Schou had the same idea, so we published the result in a single paper (Schou *et al.* 1998). The principle was simple: we noted that high-l f modes (in this case $l = 88-250$) are surface waves described by a simple formula for the angular frequency in which $\omega \approx \sqrt{GM_\odot/R_\odot{}^3}$, where G is the universal gravitational constant and $M_\odot$ is the mass of the Sun, while $R_\odot$ is photospheric radius of the Sun. With this we determined a value of the photospheric radius of the Sun that was 0.3 Mm smaller than the textbook value (Allen 1973). Christensen-Dalsgaard & Brown (1998) subsequently showed that this difference was due to a 0.3 Mm difference between the true photospheric radius and that measured by transit. Thus, the radius in standard solar models had to be reduced by 0.3 Mm.

Figure 2. 1995 at the Taj Mahal.

4. Ending the collaboration

Early in the new century, after twenty years of close collaboration, global helioseismology had become a mature science and neither of us had an interest in switching to local helioseismology. With *Kepler* and *CoRoT* coming, Wojtek would return to his first love and the wisdom of that choice is written all over this conference. As for me, I wanted to build a solar telescope. So we agreed on one last piece of work to address a number of problems that had vexed us over the years and end the collaboration on a high note.

It had long been known that both f- and p-mode frequencies (after removing mode inertia) closely track the solar cycle and increase with increasing solar activity. However, they do not show the same frequency dependence. In particular, f-mode frequencies abate with increasing mode frequency, while p-mode frequencies increase more strongly with increasing frequency. From this, we (Dziembowski & Goode 2005) were able to demonstrate, in a nearly self-consistent manner, that both do not have the same origin. The f-mode behavior is due to rising field effects primarily about 5 Mm beneath the surface, while the field growth is much smaller near the surface as observed (Lin 1995, Lin & Rimmele 1999), which is the region to which higher frequency f modes are more sensitive. Such field growth near the surface is too weak to directly affect the p-mode frequencies. Rather, the field growth very slightly blocks the turbulent pressure and the heat flow and the resulting cooling shrinks the outer cavity to net the measured p-mode behavior. The final result is a Sun that shrinks ~1 km from activity minimum to activity maximum, far too small to be of climatological significance and further presents no reason to imagine significant changes in the solar diameter over historical timescales.

5. Afterwards

Wojtek has enjoyed a time of great interest in the field of his first love since our collaboration ended. I changed fields to build a new solar telescope (NST).

The NST is the first facility-class telescope built in the US in a generation. Its 1.6 m aperture was chosen because it is the smallest aperture that can resolve features of order

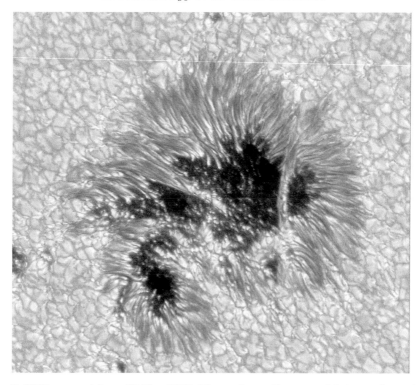

Figure 3. NST sunspot from 22 May 2013. These observations are able to resolve a darkness at the center of umbral dots and fine structure in the light bridge.

100 km (diffraction limit of ∼40 km). This matters because the photon mean free path in the solar photosphere is ∼100 km. After several years of effort, the telescope saw first light in January 2009 and with improving instruments and new generation adaptive optics (AO), the telescope now reaches its diffraction limit in the bluest visible wavelengths. Observations with AO regularly are made with lock of several hours. An image of a sunspot is shown in Fig. 3.

Acknowledgements

P.R.G gratefully acknowledges partial support by NASA (NNX13AG14G), NSF (AGS-1250818 and AFOSR (FA2386-12-1-3018).

References

Allen, C. W. 1973, in *Astrophysical Quantities*, 3rd Ed., p. 169
Brown, T. M., Christensen-Dalsgaard, J., Dziembowski, W. A., Goode, P. R., Gough, D. O., & Morrow, C. A. 1989, *ApJ*, 343, 526
Brown, T. M. & Morrow, C. A. 1987, *ApJ*, 314, L21
Christensen-Dalsgaard, J. & Brown, T. M. 1998, *ApJ*, 500, 195
Duvall, T. L., Dziembowski, W. A., Goode, P. R., Gough, D. O., Harvey, J. W., & Leibacher, J. W. 1984, *Nature*, 310, 22
Duvall, T. L. & Harvey, J. W. 1984, *Nature*, 310, 19
Duvall, T. L., Harvey, J. W., & Pomerantz, M. 1986, *Nature*, 321, 500
Dziembowski, W. A. 1977, *AcA*, 27, 203
Dziembowski, W. A. & Goode, P. R. 1984, *MemSAI*, 55, 185
Dziembowski, W. A. & Goode, P. R. 1992, *ApJ*, 394, 670

Dziembowski, W. A. & Goode, P. R. 2005, *ApJ*, 625, 548

Dziembowski, W. A., Goode, P. R., & Libbrecht, K. G. 1989, *ApJ*, 337, L53

Dziembowski, W. A., Goode, P. R., Pamyatnykh, A. A., & Sienkiewicz, R. 1994, *ApJ*, 432, 417

Dziembowski, W. A., Goode, P. R., Pamyatnykh, A. A., & Sienkiewicz, R. 1995, *ApJ*, 445, 509

Goode, P. R. & Koltun, D. S. 1975, *Nucl. Phys. A*, 243, 44

Libbrecht, K. G. 1989, *ApJ*, 336, 1092

Lin, H. 1995, *ApJ*, 514, 448

Lin, H. & Rimmele, T. R. 1999, *ApJ*, 446, 421

Schou, J., Kosovichev, A. G., Goode, P. R., & Dziembowski, W. A. 1997, *ApJ*, 489, L197

Precision Asteroseismology
Proceedings IAU Symposium No. 301, 2013
J. A. Guzik, W. J. Chaplin, G. Handler & A. Pigulski, eds.

© International Astronomical Union 2014
doi:10.1017/S1743921313014208

Distance determination from the Cepheid and RR Lyrae period-luminosity relations

Chow-Choong Ngeow[1], Wolfgang Gieren[2] and Christopher Klein[3]

[1]Graduate Institute of Astronomy, National Central University, Jhongli 32001, Taiwan
email: cngeow@astro.ncu.edu.tw

[2]Departamento de Astronomia, Universidad de Concepcion, Casilla 160-C, Concepcion, Chile
[3]Astronomy Department, University of California, Berkeley, CA 94720, USA

Abstract. Cepheids and RR Lyrae stars are important pulsating variable stars in distance scale work because they serve as standard candles. Cepheids follow well-defined period-luminosity (PL) relations defined for bands extending from optical to mid-infrared (MIR). On the other hand, RR Lyrae stars also exhibit PL relations in the near-infrared and MIR wavelengths. In this article, we review some of the recent developments and calibrations of PL relations for Cepheids and RR Lyrae stars. For Cepheids, we discuss the calibration of PL relations via the Galactic and the Large Magellanic Cloud routes. For RR Lyrae stars, we summarize some recent work in developing the MIR PL relations.

Keywords. Cepheids, RR Lyrae stars, distance scale

1. Introduction

Classical Cepheids and RR Lyrae stars are pulsating stars that play a vital role in the definition of the distance scale ladder†. This is because they are standard candles in the local Universe that permit the calibration of secondary distance indicators (e.g., the peak brightness of type Ia supernovae). The ultimate goal of the distance scale ladder is to determine the Hubble constant (H_0) with 1% precision and accuracy. The existence of period-luminosity (PL) relations for Cepheids (from optical to infrared wavelengths) makes distance determination using this type of variable star possible. In this article, we review some prospects of the calibration of Cepheid PL relations and their role in the recent distance scale work (Section 2). RR Lyrae stars also obey a PL relation in the infrared, and we review some of the recent developments of such relations in Section 3.

2. The Cepheid period-luminosity relation

The Cepheid PL relation is a 2-D projection of the period-luminosity-color (PLC) relation on the logarithmic period and magnitude plane, where the PLC relation can be derived by combining the Stefan-Boltzmann law, the period-mean density relation for pulsators, and the mass-luminosity relation based on stellar evolution models. Discussion of the physics behind the Cepheid PL relation can be found in Madore & Freedman (1991), and will not be repeated here. The PL relation usually takes the linear form of $M_\lambda = a_\lambda \log P + b_\lambda$, where a_λ and b_λ are the slope and intercept of the relation in bandpass λ, respectively. Once the slopes and intercepts of the multi-band PL relations are determined or calibrated, the distance to a nearby galaxy can be obtained by fitting the calibrated PL relations to the Cepheid data in that galaxy (see Fig. 1).

† For latest version of the distance scale ladder, see http://kiaa.pku.edu.cn/~grijs/distanceladder.pdf

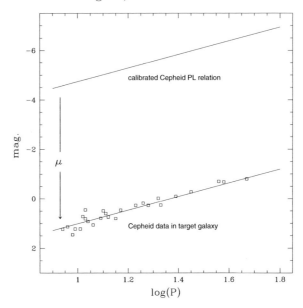

Figure 1. Illustration of using the calibrated Cepheid PL relation to determine the distance modulus to a galaxy. After a calibrated PL relation is adopted, this calibrated PL relation is shifted vertically to fit the observed Cepheids data in a given galaxy, and the vertical offset provides the distance modulus (μ) of the galaxy.

2.1. *Calibration of Cepheid PL relations*

Determining the slope of the PL relation is relatively straightforward. The large number of Cepheids discovered in the Magellanic Clouds permits the determination of the PL slope with $\sim 10^{-2}$ accuracy (Soszyński *et al.* 2008, 2010). The derivation of PL intercepts, on the other hand, is trickier, because distances to a number of Cepheids need to be known or inferred *a priori*. There are two routes to calibrate the Cepheid PL intercepts that are commonly found in literature: the Galactic route and the Large Magellanic Cloud route.

The Galactic route relies on Galactic Cepheids that are located in the solar neighborhood, i.e. those within few kpc. These Cepheids are bright enough that extensive data, both multi-band light curves and radial velocity curves, are available from the literature. However, they suffer from varying extinction and their distances need to be determined independently. A number of Galactic Cepheids is close enough to permit an accurate parallax measurement using *Hipparcos* (van Leeuwen *et al.* 2007) or *Hubble Space Telescope* (*HST*, Benedict *et al.* 2007). In the near future, *Gaia* will provide reliable parallaxes to almost all nearby Galactic Cepheids. Besides parallaxes, distances to Galactic Cepheids can also be determined from the Baade-Wesselink (BW) technique and its variants. The BW technique combines the measurements of radial velocities and angular diameters to derive the distance and mean radius for a given Cepheid. The angular diameter variations can be determined from the infrared surface brightness method (see, for example, Storm *et al.* 2011, and references therein) or the interferometric technique (e.g., as in Gallenne *et al.* 2012). A critical parameter in the BW technique is the projection factor, or p-factor (that converts the observed radial velocity to pulsational velocity), because a 1% error in the p-factor translates to a 1% error in the derived distance. For a Cepheid located in an open cluster, the distance to the Cepheid can be inferred from the distance of its host cluster measured via isochrone fitting (Turner 2010). Finally, the distance to a large number of Cepheids can be obtained from the calibrated Wesenheit function using *HST*

Table 1. Examples of the calibrated multi-band LMC PL relations.

Band	Slope	Fitted Intercept	Calibrated Intercept
V	-2.769 ± 0.023	17.115 ± 0.015	-1.378
I	-2.961 ± 0.015	16.629 ± 0.010	-1.864
J	-3.115 ± 0.014	16.293 ± 0.009	-2.200
H	-3.206 ± 0.013	16.063 ± 0.008	-2.430
K	-3.194 ± 0.015	15.996 ± 0.010	-2.497
3.6 μm	-3.253 ± 0.010	15.967 ± 0.006	-2.526
4.5 μm	-3.214 ± 0.010	15.930 ± 0.006	-2.563
5.8 μm	-3.182 ± 0.020	15.873 ± 0.015	-2.620
8.0 μm	-3.197 ± 0.036	15.879 ± 0.034	-2.614
W	-3.313 ± 0.008	15.892 ± 0.005	-2.601

Note: The PL relations are taken from Ngeow *et al.* (2009), calibrated with $\mu_{\rm LMC} = 18.493$. Extinction corrections have been applied to the data prior to the fitting of PL relations.

parallaxes (Ngeow 2012). Examples of PL relations based on Galactic Cepheids can be found in Tammann *et al.* (2003), Ngeow & Kanbur (2004) and Fouqué *et al.* (2007). It has been argued (see Tammann *et al.* 2003, Kanbur *et al.* 2003 and references therein) that the PL relations calibrated with Galactic Cepheids are preferred in distance scale work, because the spiral galaxies that are used to calibrate the secondary distance indicators have metallicities close to solar value, and hence a metallicity correction to the Cepheid PL relation is not needed to derive distances in this way.

The Large Magellanic Cloud (LMC), located $\sim$50 kpc away, is an irregular galaxy that is far enough to assume that Cepheids in this galaxy lie at the same distance. Yet the LMC is also close enough that stars observed there can be resolved. Therefore, the LMC Cepheids have been commonly used in the previous studies on calibrating the Cepheid PL relations. However, measurements of the LMC distance modulus ($\mu_{\rm LMC}$) show a wide spread, ranging from $\sim$18.0 to $\sim$19.0 mag, with a center around 18.5 ± 0.1 mag (for example, see Freedman *et al.* 2001, Benedict *et al.* 2002, Schaefer 2008)†. This causes the calibration of the PL intercepts to suffer a systematic error of the order of $\sim$5% (Freedman *et al.* 2001). For this reason, some of the PL relations derived from the LMC Cepheids leave the PL intercepts un-calibrated (i.e., the values are taken from fitting only), as shown in Soszyński *et al.* (2008) and Ngeow *et al.* (2009). Nevertheless, this problem is solved with the latest result published by Pietrzyński *et al.* (2013). By using late-type eclipsing binary systems, they determined the distance to the LMC with 2% accuracy, i.e., $\mu_{\rm LMC} = 18.493 \pm 0.048$ (total error). Then, the PL relations for fundamental mode LMC Cepheids given in Soszyński *et al.* (2008) become: $V = -2.762(\pm0.022)\log P - 0.963(\pm0.015)$, $I = -2.959(\pm0.016)\log P - 1.614(\pm0.010)$ (both uncorrected for extinction), and $W = -3.314(\pm0.009)\log P - 2.600(\pm0.006)$. Similarly, the multi-band PL relations from Ngeow *et al.* (2009) can be calibrated, which is summarized in Table 1.

Two additional issues need to be taken into account when calibrating the LMC PL relations: extinction correction and non-linearity of the LMC PL relation. The LMC is known to suffer from differential extinction, hence extinction corrections need to be applied to individual LMC Cepheids by means of extinction maps (e.g., Zaritsky *et al.* 2004, Haschke *et al.* 2011). The LMC PL relation is also known to be non-linear in optical bands: the PL relation can be split into two relations separated at 10 days (for examples, see Sandage *et al.* 2004, Kanbur & Ngeow 2004, Ngeow *et al.* 2005, García-Varela *et al.* 2013). Both these issues, nevertheless, can be remedied by using the Wesenheit function (Madore &

† Also, see the LMC distance moduli compiled in `http://clyde.as.utexas.edu/SpAstNEW/head602.ps`

Freedman 1991, Ngeow & Kanbur 2005, Madore & Freedman 2009, Ngeow *et al.* 2009, Bono *et al.* 2010, Inno *et al.* 2013) or moving to the mid-infrared (MIR, from ~3 μm to ~10 μm, Freedman *et al.* 2008, Ngeow & Kanbur 2008, Madore *et al.* 2009, Ngeow *et al.* 2010, Scowcroft *et al.* 2011) at which extinction is negligible.

2.2. *Examples of distance scale application*

Both the *HST* H_0 Key Project (Freedman *et al.* 2001) and SN Ia *HST* Calibration Program (Sandage *et al.* 2006), two benchmark programs that utilized the Cepheid PL relation in distance scale work, derived a Hubble constant with a 10% uncertainty. Since then, two additional programs, the SH0ES (Supernovae and H_0 for the Equation of State, Riess *et al.* 2011) and the CHP (Carnegie Hubble Program, Freedman *et al.* 2012), aimed to determine the Hubble constant with a 3% uncertainty by reducing or eliminating various systematic errors. Again, the Cepheid PL relation plays an important role in these programs. One of the main differences between the SH0ES program and previous programs is that in the SH0ES program the LMC was replaced with NGC 4258 as an anchoring galaxy in the determination of the distance scale ladder. In NGC 4258, the motions of water masers surrounding its central black hole permit an accurate geometrical distance to be determined (Humphreys *et al.* 2008). To further reduce the systematic errors along the distance scale ladder, the SH0ES program adopted only "ideal" type Ia supernovae in nearby galaxies. They are used to calibrate their peak brightness, using a homogeneous sample of Cepheids, and observed with a single instrument on-board the *HST*. The CHP, on the other hand, recalibrated the *HST* H_0 Key Project distance scale ladder by adopting the MIR PL relation, where the PL slopes are defined by the LMC Cepheids and the PL intercepts are calibrated with Galactic Cepheids that have *HST* parallaxes. Similar to SH0ES, CHP also utilized only a single instrument on-board the *Spitzer Space Telescope* to derive and calibrate the MIR Cepheid PL relations. Both programs derived the Hubble constant with an uncertainty of $\sim3\%$.

3. Period-luminosity relations for RR Lyrae stars

RR Lyrae stars follow PL relations in optical to infrared bands. However, the V-band bolometric correction for RR Lyrae stars is almost independent of temperature, suggesting the slope of their V-band PL relation is zero or very close to it (instead, RR Lyrae stars follow an M_V-[Fe/H] relation in the V-band). In contrast, there is a temperature dependence of the bolometric correction in infrared bands, which translates to an observed K-band PL relation (Bono *et al.* 2001, Bono 2003). The observed K-band PL relation for RR Lyrae stars can be dated back to Longmore *et al.* (1986), who derived the relation based on single-epoch observations of RR Lyrae stars in three globular clusters. Recent calibration of the K-band PL relation, or the PL$_K$-[Fe/H] relation, can be found in, for example, Sollima *et al.* (2006), Borissova *et al.* (2009), Benedict *et al.* (2011) and Dambis *et al.* (2013). When calibrating the K-band PL relation with RR Lyrae stars in globular clusters, one has to be cautious because RR Lyrae stars near the cluster's core may suffer from blending (Majaess *et al.* 2012).

The derivation of the PL relation for RR Lyrae stars can be extended to MIR wavelengths. This is convincingly demonstrated by Klein *et al.* (2011), who derived the MIR PL relations in *Wide-field Infrared Survey Explorer (WISE)* $W1$ (3.4 μm), $W2$ (4.6 μm) and $W3$ (12 μm) bands for 76 field RR Lyrae stars. When deriving these PL relations, Klein *et al.* (2011) employed a Bayesian framework where the posterior distances were based on the data from *Hipparcos*. An updated version of the MIR PL relations with nearly double the sample size is shown in Fig. 2. Independently, Madore *et al.* (2013)

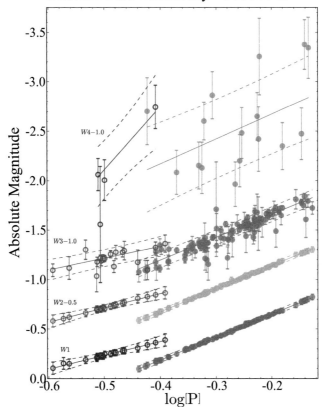

Figure 2. Preliminary RR Lyrae stars PL relations in *WISE*'s bands based on 143 field RR Lyrae stars. Filled and open circles represent the RR Lyrae stars of both Bailey *ab* and *c* type, respectively.

derived similar MIR PL relations based on four Galactic RR Lyrae stars having parallaxes measured by the *HST*.

4. Conclusion

Independent measurements of the Hubble constant via the distance scale ladder are expected to achieve ∼1% uncertainty in the future. This is possible due to a large number of Cepheids and RR Lyrae stars with high-quality data which will become available from various future or on-going projects, such as *Gaia*, the fourth-phase of the Optical Gravitational Lensing Experiment (OGLE-IV), and the VISTA survey of the Magellanic Clouds (VMC). The *James Webb Space Telescope* (*JWST*), which will operate mainly in the MIR, is expected to routinely observe Cepheids beyond 30 Mpc, and it is also expected that data from this satellite will allow to derive a Hubble constant with a 1% uncertainty. Therefore, accurate and independent calibrations of the PL relations for Cepheids and RR Lyrae stars in the MIR are important in the preparation for the *JWST* era.

Acknowledgements

We would like to thank the invitation of the SOC and the LOC for presenting this talk at the conference. CCN acknowledges the support from NSC grant NSC101-2119-M-008-007-MY3.

References

Benedict, G. F., McArthur, B. E., Fredrick, L. W., et al. 2002, AJ, 123, 473

Benedict, G. F., McArthur, B. E., Feast, M. W., et al. 2007, AJ, 133, 1810

Benedict, G. F., McArthur, B. E., Feast, M. W., et al. 2011, AJ, 142, 187

Bono, G. 2003, in: D. Alloin, & W. Gieren (eds.), Stellar Candles for the Extragalactic Distance Scale, Lecture Notes in Physics, 635, 85

Bono, G., Caputo, F., Castellani, V., Marconi, M., & Storm, J. 2001, MNRAS, 326, 1183

Bono, G., Caputo, F., Marconi, M., & Musella, I. 2010, ApJ, 715, 277

Borissova, J., Rejkuba, M., Minniti, D., Catelan, M., & Ivanov, V. D. 2009, A&A, 502, 505

Dambis, A. K., Berdnikov, L. N., Kniazev, A. Y., et al. 2013, MNRAS, 435, 3206

Freedman, W. L., Madore, B. F., Gibson, B. K., et al. 2001, ApJ, 553, 47

Freedman, W. L., Madore, B. F., Rigby, J., Persson, S. E., & Sturch, L. 2008, ApJ, 679, 71

Freedman, W. L., Madore, B. F., Scowcroft, V., et al. 2012, ApJ, 758, 24

Fouqué, P., Arriagada, P., Storm, J., et al. 2007, A&A, 476, 73

Gallenne, A., Kervella, P., Mérand, A., et al. 2012, A&A, 541, A87

García-Varela, A., Sabogal, B. E., & Ramírez-Tannus, M. C. 2013, MNRAS, 431, 2278

Haschke, R., Grebel, E. K., & Duffau, S. 2011, AJ, 141, 158

Humphreys, E. M. L., Reid, M. J., Greenhill, L. J., et al. 2008, ApJ, 672, 800

Inno, L., Matsunaga, N., Bono, G., et al. 2013, ApJ, 764, 84

Kanbur, S. M., Ngeow, C.-C., Nikolaev, S., et al. 2003, A&A, 411, 361

Kanbur, S. M. & Ngeow, C.-C. 2004, MNRAS, 350, 962

Klein, C. R., Richards, J. W., Butler, N. R., & Bloom, J. S. 2011, ApJ, 738, 185

Longmore, A. J., Fernley, J. A., & Jameson, R. F. 1986, MNRAS, 220, 279

Madore, B. F. & Freedman, W. L. 1991, PASP, 103, 933

Madore, B. F., Freedman, W. L., Rigby, J., et al. 2009, ApJ, 695, 988

Madore, B. F. & Freedman, W. L. 2009, ApJ, 696, 1498

Madore, B. F., Hoffman, D., Freedman, W. L., et al. 2013, ApJ, 776, 135

Majaess, D., Turner, D., & Gieren, W. 2012, PASP, 124, 1035

Ngeow, C.-C. 2012, ApJ, 747, 50

Ngeow, C.-C. & Kanbur, S. M. 2004, MNRAS, 349, 1130

Ngeow, C.-C. & Kanbur, S. M. 2005, MNRAS, 360, 1033

Ngeow, C.-C. & Kanbur, S. M. 2008, ApJ, 679, 76

Ngeow, C.-C., Kanbur, S. M., Nikolaev, S., et al. 2005, MNRAS, 363, 831

Ngeow, C.-C., Kanbur, S. M., Neilson, H. R., et al. 2009, ApJ, 693, 691

Ngeow, C.-C., Ita, Y., Kanbur, S. M., et al. 2010, MNRAS, 408, 983

Pietrzyński, G., Graczyk, D., Gieren, W., et al. 2013, Nature, 495, 76

Riess, A. G., Macri, L., Casertano, S., et al. 2011, ApJ, 730, 119

Sandage, A., Tammann, G. A., & Reindl, B. 2004, A&A, 424, 43

Sandage, A., Tammann, G. A., Saha, A., et al. 2006, ApJ, 653, 843

Schaefer, B. E. 2008, AJ, 135, 112

Scowcroft, V., Freedman, W. L., Madore, B. F., et al. 2011, ApJ, 743, 76

Sollima, A., Cacciari, C., & Valenti, E. 2006, MNRAS, 372, 1675

Soszyński, I., Poleski, R., Udalski, A., et al. 2008, AcA, 58, 163

Soszyński, I., Poleski, R., Udalski, A., et al. 2010, AcA, 60, 17

Storm, J., Gieren, W., Fouqué, P., et al. 2011, A&A, 534, A94

Tammann, G. A., Sandage, A., & Reindl, B. 2003, A&A, 404, 423

Turner, D. G. 2010, Ap&SS, 326, 219

van Leeuwen, F., Feast, M. W., Whitelock, P. A., & Laney, C. D. 2007, MNRAS, 379, 723

Zaritsky, D., Harris, J., Thompson, I. B., & Grebel, E. K. 2004, AJ, 128, 1606

Precision Asteroseismology
Proceedings IAU Symposium No. 301, 2013
J. A. Guzik, W. J. Chaplin, G. Handler & A. Pigulski, eds.

© International Astronomical Union 2014
doi:10.1017/S174392131301421X

Pulsating stars as stellar population tracers

Gisella Clementini

INAF - Osservatorio Astronomico,
Via Ranzani n. 1, 40127 Bologna, Italy
email: gisella.clementini@oabo.inaf.it

Abstract. Pulsating stars of different types are in different evolutionary phases, thus allowing one to trace stellar components of different age in the host systems. The light variation caused by the cyclic expansion/contraction of the surface layers makes a pulsating star much easier to identify than constant stars in the same evolutionary phase. Pulsating stars thus offer a powerful tool to disentangle the various stellar generations in systems where stars of different age and metal abundance populate the same regions of the colour-magnitude diagram. An overview is presented of how pulsating stars can be used as tools to study the stellar populations, and the structure and formation process in Local Group galaxies.

Keywords. stars: oscillations, stars: variables: Cepheids, stars: variables: RR Lyrae, stars: variables: other, galaxies: stellar content, galaxies: structure

1. Introduction

A number of virtues make the pulsating stars optimal tools in astrophysics. Their pulsation characteristics (period, amplitudes, pulsation modes) can be determined with great precision, are unaffected by distance and reddening, and are directly linked to the stellar mass, effective temperature, and chemical composition (helium abundance and metallicity), thus allowing one to infer these quantities directly from the pulsation. For instance, the reproduction of the observed light and radial velocity curves of classical Cepheids (CCs; Wood *et al.* 1997, Bono *et al.* 2002), RR Lyrae stars (Di Fabrizio *et al.* 2002, Marconi & Clementini 2005), and δ Scuti variables (McNamara *et al.* 2007) with nonlinear convective pulsation models allows one to predict the variation of the relevant quantities along a pulsation cycle and provides a direct estimate of the star's intrinsic parameters and distance. Recently, Marconi *et al.*'s (2013) application of this technique to the light and radial velocity curves of CEP0227, a CC in a double-lined eclipsing binary in the LMC (Pietrzyński *et al.* 2010), provided intrinsic stellar parameters in excellent agreement with the dynamical estimates. On the other hand, the superb light curves produced by the *CoRoT* and *Kepler* satellites have revealed that an extraordinary large number of frequencies are excited in pulsating stars, thus allowing the determination of the physical parameters through asteroseismology.

The most common types of pulsating variables, the RR Lyrae stars and the Cepheids, are primary stellar distance indicators in the Local Group (LG) and beyond, up to about 20 Mpc. The RR Lyrae stars are standard candles because they obey an absolute magnitude-metallicity relation in the visual band and a period-luminosity-metallicity relation in the K-band; the Cepheids are used as standard candles through the period-luminosity, period-luminosity-colour, and the period-Wesenheit (Madore 1982) relationships. The role of these standard candles in setting the basis of the whole astronomical distance scale will be revolutionized by *Gaia*, the ESA's cornerstone mission scheduled for launch at the end of 2013. The *Gaia* satellite (Lindegren & Perryman 1996, Lindegren 2010, Brown 2013) is expected to discover and measure parallaxes with $\sigma_\pi/\pi <$

10% for about 9 000 CCs in the Milky Way (MW), as well as to observe more than 70 000 RR Lyrae stars in the Galactic halo (Turon *et al.* 2012). By providing a direct measure of distance to local primary distance indicators via parallax, *Gaia* will allow a re-calibration of the secondary indicators and a re-assessment of the whole distance ladder, with significant improvement in our knowledge of the Hubble constant.

Pulsating stars of different types are in different evolutionary phases, thus identifying stellar components of different age within the host system. The spatial distribution of the different types of pulsating stars allows the main structures of the parent galaxy to be revealed, and their pulsation properties can provide hints on the mechanisms through which large galaxies have formed. In the following, examples are presented of the use of pulsating stars as tools to trace the different stellar populations, to study the geometric structure, and to derive information on the formation process of the parent galaxies.

2. Pulsating stars as stellar population tracers

The colour-magnitude diagram (CMD) is the tool commonly used to study the stellar populations and the star formation (SF) history of a galaxy. However, in the case of remote systems for which only the bright portion of the CMD is observable (with present-day facilities, basically systems beyond a few Mpc), it is not easy to prove the existence of a $\geqslant 10$ Gyr-old stellar component. Furthermore, even when the CMD is deep enough to reach the main-sequence turnoff (MSTO) of the oldest population, it may be difficult to disentangle an old component if the system is predominantly composed of young stars and crowding is an issue. Pulsating stars offer a powerful alternative tool to trace stars of different age in a galaxy, as variables of different types arise from parent populations of different age. This property, combined with the advantage of being a pulsating star that is much easier to reveal than a constant star in the same evolutionary phase, makes the pulsating variables invaluable tools specifically in the identification and characterization of the oldest stellar populations in galaxies.

CCs are young ($t \lesssim 100$ Myr), metal-rich, Population I giants ($\log L/L_\odot \sim 3-5$, spectral type F6–K2, $M \sim 3-13\,M_\odot$) that cross the classical instability strip (IS) during the blue-loop phase of the evolutionary tracks. They are preferentially located in gas-rich, dusty, star-forming regions, like galactic bars and spiral arms. Conversely, the RR Lyrae stars are low-mass ($M < 1\,M_\odot$) horizontal-branch (HB) giants ($\log L/L_\odot \sim 1.5-1.8$, A2–F2 spectral type), and along with the Population II Cepheids (P2Cs; low-mass: $M \sim 0.5\,M_\odot$, post-HB giants: $\log L/L_\odot \sim 2$, of F2–G6 spectral type), unambiguously trace the oldest ($t \geqslant 10$ Gyr) stellar generations commonly found in galactic halos. With their mere presence the RR Lyrae stars and the P2Cs provide evidence that a SF episode occurred more than 10 Gyr ago (see, e.g., the case of the NGC 6822, Clementini *et al.* 2003). Furthermore, being roughly 3 magnitudes brighter than the coeval MSTO stars, they are much easier to observe, thus allowing one to trace the SF history back to the first epochs of galaxy formation even for systems where the MSTO is unreachable with present-day facilities. Finally, the ACs ($\log L/L_\odot \sim 1.8-2.3$, spectral type F2–G6, and $M \sim 1.3-2.2\,M_\odot$) are giants evolving form the Zero-Age Horizontal Branch turnover (see Marconi *et al.* 2004, and references therein) for metallicity lower than [Fe/H] $\lesssim -1.7$ dex. They are generally believed to trace an intermediate-age component produced by a relatively recent (a few billion years ago) SF episode, but in principle they could also derive from merging of low-mass, old population stars.

Observations show that RR Lyrae stars are found in all LG galaxies where they are searched for, irrespective of the galaxy morphological type. This provides evidence for a first epoch of SF that is common to all galaxies in the LG, both star-forming and presently quiescent. Observations also show that more massive LG systems (spirals, irregulars)

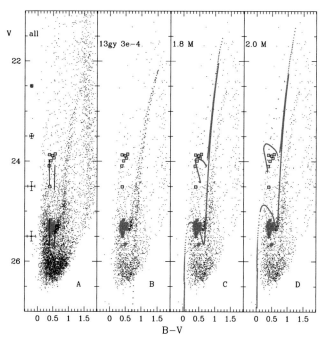

Figure 1. *Panel A:* Colour-magnitude diagram (CMD) of And XIX from Cusano *et al.* (2013). Filled circles and triangles are fundamental-mode (RRab) and first-overtone (RRc) RR Lyrae stars, respectively, filled squares are Anomalous Cepheids (ACs). Black solid lines show the boundaries of the instability strip for RR Lyrae stars and ACs with $Z = 0.0002$, from Marconi *et al.* (2004). *Panel B:* Observed CMD with overlaid stellar isochrones from Padova evolutionary models for $t = 13$ Gyr and $Z = 0.0003$, that well reproduce the galaxy's oldest stellar component traced by the RR Lyrae stars. *Panels C and D:* Stellar evolutionary tracks from the Basti web site for masses of 1.8 $M_\odot$ (*panel C*) and 2.0 $M_\odot$ (*panel D*) that well reproduce the galaxy ACs.

generally contain both CCs (Population I) and RR Lyrae stars (Population II). This suggests that a continuing SF has occurred in these systems. Lower-mass galaxies (mainly dwarf spheroidal galaxies – dSphs) host RR Lyrae stars and ACs (intermediate-age population), thus pointing to mainly early ($\sim$10 Gyr) and intermediate-age (a few Gyrs) SF events (see, e.g., the case of Carina, Dall'Ora *et al.* 2003). Conversely, metal-poor low-mass irregulars host mainly CCs and ACs, hence pointing to prevalently recent SF episodes. These general trends are followed also by the ultra-faint dwarf (UFD) galaxies recently discovered around the MW (see, e.g., Belokurov *et al.* 2006, 2010, and references therein) and by the Andromeda's dSph satellites. Tables 1 and 2 summarize the number and type of pulsating stars identified in the MW UFDs and in the M31 dSphs that have been studied for variability.

Figure 1 shows results from the study of the variable stars in Andromeda XIX (And XIX) by Cusano *et al.* (2013) who have identified 31 RR Lyrae stars (23 RRab and 8 RRc pulsators) and 8 ACs in this new M31 satellite. Panel A shows the galaxy CMD with filled circles and triangles marking RRab and RRc stars, respectively, and filled squares showing the ACs. In Panel B on the CMD are overlaid stellar isochrones from the Padova evolutionary models for $t = 13$ Gyr and $Z = 0.0003$ (from the CMD 2.5 web interface, available at http://stev.oapd.inaf.it/cgi-bin/cmd), that very well reproduce the galaxy's oldest stellar component traced by the RR Lyrae stars. In Panels C and D on the CMD are overlaid stellar evolutionary tracks for masses of 1.8 $M_\odot$ (panel C) and 2.0 $M_\odot$ (panel

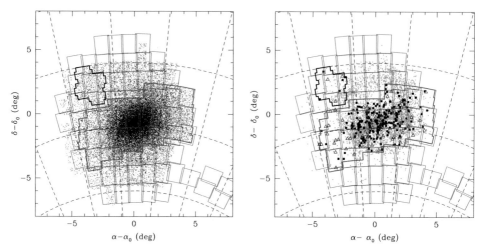

Figure 2. Distribution of RR Lyrae stars (left panel, black points) and Cepheids (classical: grey points, anomalous: open triangles, Population II: filled squares, in the right panel) in the LMC, according to the OGLE III, OGLE IV and EROS-2 surveys. $\alpha_0 = 81.0°$ and $\delta_0 = -69.0°$.

D), from the Basti web site (http://albione.oa-teramo.inaf.it/), that well reproduce the And XIX's ACs.

3. The Large Magellanic Cloud geometry as traced by classical Cepheids and RR Lyrae stars

Different types of pulsating stars are differently located in space within the parent galaxy; thus they probe its three-dimensional structure and also allow the identification of substructures which are the signature of past interactions between the parent galaxy and its satellites. For instance, the local overdensity of RR Lyrae stars detected in the Galactic halo by Vivas *et al.* (2001) turned out to be the northern tidal stream of the Sagittarius dSph dissolving itself into the MW halo. Several substructures in the Galactic halo (e.g., the Pisces Overdensity at about 80 kpc, Watkins *et al.* 2009) have been revealed by overdensities of RR Lyrae stars in the Sloan Digital Sky Survey stripe 82 region (see also Sesar *et al.* 2010). Similarly, evidence for a MW tidal stream reaching beyond 100 kpc and overlapping the Sagittarius stream has been found from the analysis of more than a thousand RR Lyrae stars observed by the Catalina Surveys Mount Lemmon telescope (Drake *et al.* 2013).

An excellent example of how well different types of pulsating stars can trace the geometrical structure of a galaxy is represented by the Large Magellanic Cloud (LMC). The search for microlensing events started at the beginning of the nineties by the MACHO, EROS and OGLE collaborations has led to the discovery of tens of thousands RR Lyrae stars and Cepheids in the LMC (see e.g., Alcock *et al.* 1996, Sasselov *et al.* 1997, Udalski *et al.* 1997). At present, the largest spatial coverage of the LMC is provided by the second generation of the EROS microlensing experiment (EROS-2; Tisserand *et al.* 2007). In Fig. 2 we compare the structure of the LMC as traced by the system RR Lyrae stars (left panel) and Cepheids (right panel), combining results for the LMC variables from the OGLE and the EROS-2 surveys. In each panel of the figure thin (blue in the on-line version) rectangles show the tiles of the VISTA Magellanic Clouds Survey (VMC; Cioni *et al.* 2011), a near-infrared (Y, J, K_s) ESO public survey of the Magellanic System (MS) which is obtaining multi-epoch photometry in the K_s-band to study the MS three-dimensional geometry with the pulsating stars. The thick grey (in red on-line) and black lines delimit instead the regions where identification of the variables comes from the OGLE III (Soszyński *et al.* 2008, 2009) and OGLE IV (Soszyński *et al.* 2012) surveys, respectively. Outside these fields, data for the variables are provided only by the EROS-2 survey. The distributions of RR Lyrae stars (black points in the left panel of Fig. 2) and CCs (grey points in the right panel of Fig. 2) appear to differ significantly. The RR Lyrae stars have a larger density in the central region of the LMC; however, they are present in large numbers also in the peripheral areas covered mainly by EROS-2. Their distribution is smooth and likely traces the LMC halo. Conversely, the CCs are strongly concentrated towards the LMC bar and seem to almost disappear moving outside the region covered by the OGLE III observations (thick grey contours in Fig. 2). However, the larger field covered by EROS-2 reveals the existence of an overdensity of CCs displaced about 2 degrees above the central bar, and running almost parallel to it, to which it connects at its western edge (see, Moretti *et al.* 2013 for details). This overdensity is likely the signature of a northwest spiral arm of the LMC (see also, Schmidt-Kaler 1977, Nikolaev *et al.* 2004). The distribution of CCs in the OGLE IV GSEP field (thick black contours in Fig. 2) also shows an increase in the southeastern part of the field, confirming the overdensity of Cepheids highlighted by the EROS-2 data. This demonstrates how pulsating stars can be used to disentangle the fine structure of the LMC.

4. The Oosterhoff type of the new Milky Way and M31 satellites

Hierarchical merging and accretion of satellites is the mechanism through which large galaxies are predicted to form in the framework of the Λ-cold-dark-matter theory. Leftovers of this cannibalistic process are the streams and disrupted/tidally distorted satellites that we observe today within the halo of the MW and in M31. The pulsation properties of the oldest variables, the RR Lyrae stars, may allow one to identify the "building blocks" that have contributed early on to this merging process. In fact, if the MW and M31 have formed by accretion of systems resembling their present-day dwarf satellites, the RR Lyrae stars in these galaxies should have the same characteristics as the variables observed in the MW and M31 halos. A specific property of the RR Lyrae stars in the Galactic globular clusters (GGCs) is the division into two distinct groups based on the mean period of the RRab pulsators $\langle P_{ab} \rangle$ (Oosterhoff 1939), and the specific number of fundamental to first-overtone pulsators f_c. Oosterhoff I (Oo I) GGCs have $\langle P_{ab} \rangle = 0.55$ d and $f_c \sim 0.17$. Oo II clusters have $\langle P_{ab} \rangle = 0.64$ d and $f_c \sim 0.44$. The GGC distribution is bimodal with probability larger than 99.99% (Catelan 2009, and references therein). An Oosterhoff dichotomy is also observed among the MW field RR Lyrae variables (Kinemuchi *et al.* 2006, Miceli *et al.* 2008, Drake *et al.* 2013), where the RR Lyrae stars belong predominantly to the Oo I group, but with a significant Oo II component. The Oosterhoff types of the bright dSph satellites of the MW are summarized in Table 3 of Clementini (2010), while Table 1 summarizes the Oosterhoff type of the MW UFDs. The more massive dSphs around the MW generally have Oo-Intermediate (Oo-Int) properties, thus they do not resemble the "building blocks" of the MW halo. Conversely, the UFDs generally have Oo II type, hence they may resemble the protogalactic fragments from which the MW formed early on.

We still lack a census of the M31 RR Lyrae stars. Brown *et al.* (2004) identified 55 of these variables in an ACS@*HST* halo field at 11 kpc from the M31 center and found them to have an Oo-Int type. Sarajedini *et al.* (2009) detected about 700 RR Lyrae stars in two ACS fields near M32 about 4-6 kpc from Andromeda's center, and conclude that the M31 spheroid has Oo I properties. Finally, Jeffery *et al.* (2011) discovered 108 RR Lyrae stars in six ACS fields located in the disk, stream and halo of M31 and conclude that the M31 RR Lyrae have Oo I and Oo-Int properties. While none of these studies detected an Oo II

Table 1. Number of pulsating variables and Oosterhoff properties of Milky Way UFDs

Galaxy	$\langle[\text{Fe/H}]\rangle^{(1)}$ dex	N(RR) (ab/c/d)	N(AC)	$f_c^{(2)}$	$\langle P_{ab}\rangle$ day	Oo Type	Reference
Bootes I [3]	−2.55	7/7/1		0.53	0.69	Oo II	Siegel (2006) Dall'Ora et al. (2006)
Canes Venatici I	−1.98	18/5	3	0.22	0.60	Oo Int	Kuehn et al. (2008)
Canes Venatici II	−2.21	1/1			0.74	Oo II	Greco et al. (2008)
Coma [4]	−2.60	1/1			0.67	Oo II	Musella et al. (2009)
Leo IV[5]	−2.54	3			0.66	Oo II	Moretti et al. (2009)
Ursa Major II	−2.47	1			0.66	Oo II	Dall'Ora et al. (2012)
Ursa Major I	−2.18	5/2		0.29	0.63/0.60	Oo Int	Garofalo et al. (2013)
Hercules	−2.41	6/3	1	0.33	0.68	Oo II	Musella et al. (2012)
Leo T	−1.99	1	11		0.60	Oo Int	Clementini et al. (2012)

Notes:
[1] Metallicities are from the compilation of McConnachie (2012).
[2] f_c is the ratio of number of RRc (Nc) to total number of RR Lyrae stars [N(ab+c)]: f_c = Nc/N(ab+c). In the table we list the f_c values only of UFDs containing more than 5 RR Lyrae stars.
[3] Bootes I hosts also a candidate long-period variable.
[4] Coma contains also a short-period variable with $P = 0.125$ d.
[5] Leo IV contains also an SX Phoenicis star.

Table 2. Number of pulsating variables and Oosterhoff properties of M31 dSph satellites

Galaxy	$\langle[\text{Fe/H}]\rangle^{(1)}$ dex	N(RR) (ab/c/d)	N(AC)	f_c	$\langle P_{ab}\rangle$ day	Oo Type	Reference
And I	−1.45	72/26	$1^{(2)}$	0.27	0.575	Oo I/Oo Int	Pritzl et al. (2005)
And II	−1.64	64/8	1	0.11	0.571	Oo I	Pritzl et al. (2004)
And III[3]	−1.78	39/12	4	0.24	0.657	Oo II	Pritzl et al. (2005)
And V	−1.6	7/3		0.30	0.685?	Oo II?	Mancone & Sarajedini (2008) Sarajedini (2010), private com.
And VI	−1.3	91/20	6	0.18	0.588	Oo Int	Pritzl et al. (2002)
And IX	−2.2	> 30					Ripepi et al., in prep.
And X	−1.93	9/6		0.40	0.71	Oo II	Coppola et al., in prep.
And XI	−2.0	10/5		0.33	0.62	Oo Int	Yang & Sarajedini (2011)
And XIII	−1.9	12/5		0.30	0.66	Oo II	Contreras Ramos et al., in prep. Yang & Sarajedini (2011)
And XVI	−2.1	3/6		0.67	0.64	Oo II	Mercurio et al., in prep.
And XIX	−1.9	23/8	8	0.26	0.62	Oo Int	Cusano et al. (2013)
And XXI	−1.8	37/5	8	0.12	0.63	Oo II/Oo Int	Garofalo et al., in prep.

Notes:
[1] Metallicities are from the compilation of McConnachie (2012).
[2] Uncertain classification, this variable could as well be a P2C.
[3] And III contains also a P2C.

component in the M31 halo, first studies of the Oosterhoff dichotomy in the M31 GCs (Clementini *et al.* 2009, Contreras Ramos *et al.* 2013), showed that both Oo-Int and Oo II types seem to occur. Table 2 summarizes the Oosterhoff type of the M31 dSphs for which this information is currently available, and Fig. 3 shows the period-amplitude diagram

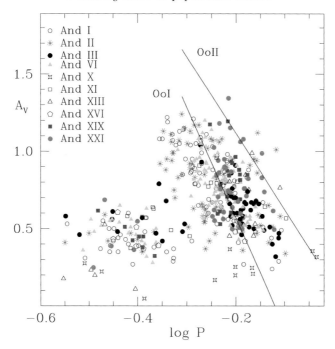

Figure 3. Period-amplitude diagram in the V-band for the RR Lyrae stars in the M31 dwarf satellites And I, And II, And III, And VI, And X, And XI, And XIII, And XVI, And XIX and And XXI. Solid lines show the loci defined by Oo I and Oo II GGCs according to Clement & Rowe (2000).

of the RR Lyrae stars in these systems. According to the $\langle P_{ab} \rangle$ values the M31 dSphs analyzed so far are of Oo II and Oo-Int types; however, in the $\log P - A_V$ diagram (Fig. 3) these systems appear to be preferentially of Oo I type. Clearly, further investigations are needed to shed light on the Oosterhoff properties of the M31 RR Lyrae stars.

Acknowledgements

Results presented in this review are largely based on work by undergraduates, PhDs, post-Docs, and young researchers. For this, special thanks go to (in alphabetic order): M. Cignoni, R. Contreras Ramos, G. Coppola, F. Cusano, A. Garofalo, A. Mercurio, M.I. Moretti, and T. Muraveva. It is a pleasure to thank Marcella Marconi and Vincenzo Ripepi for comments on an earlier version of this review.

References

Alcock, C., Allsman, R. A., Axelrod, T. S., *et al.* 1996, *AJ*, 111, 1146
Belokurov, V., Walker, M. G., Evans, N. W., *et al.* 2010, *ApJ*, 712, L103
Belokurov, V., Zucker, D. B., Evans, N. W., *et al.* 2006, *ApJ*, 647, L111
Bono, G., Castellani, V., & Marconi, M. 2002, *ApJ*, 565, L83
Brown, A. 2013, in: L. Cambresy, F. Martins, E. Nuss, & A. Palacios (eds.), Proc. Soc. Francaise d'Astronomie et d'Astrophysique (SF2 A) 2013, in press, (arXiv: 1310.3485)
Brown, T. M., Ferguson, H. C., Smith, E., *et al.* 2004, *AJ*, 127, 2738
Catelan, M. 2009, *Ap&SS*, 320, 261
Cioni M.-R. L., Clementini G., Girardi L., *et al.* 2011, *A&A*, 527, 116
Clement, C. M. & Rowe, J. 2000, *AJ*, 120, 2579
Clementini, G. 2010, in: C. Sterken, N. Samus, & L. Szabados (eds.), *Variable Stars, the Galactic Halo and Galaxy Formation*, (Moscow: Sternberg Astronomical Institute of Moscow University Publications), p. 107, (arXiv: 1002.1575)

Clementini, G., Held, E. V., Baldacci, L., & Rizzi, L. 2003, *ApJ*, 588, L85

Clementini, G., Contreras, R., Federici, L., *et al.* 2009, *ApJ*, 704, L103

Clementini, G., Cignoni, M., Contreras Ramos, R., *et al.* 2012, *ApJ*, 756, 108

Contreras Ramos, R., Clementini, G., Federici, L., *et al.* 2013, *ApJ*, 765, 71

Cusano, F., Clementini, G., Garofalo, A., *et al.* 2013, *ApJ*, 779, 7

Dall'Ora, M., Ripepi, V., Caputo, F., *et al.* 2003, *ApJ*, 126, 197

Dall'Ora, M., Clementini, G., Kinemuchi, K., *et al.* 2006, *ApJ*, 653, L109

Dall'Ora, M., Kinemuchi, K., Ripepi, V., *et al.* 2012, *ApJ*, 752, 42

Di Fabrizio, L., Clementini, G., Marconi, M., *et al.* 2002, *MNRAS*, 336, 841

Drake, A. J., Catelan, M., Djorgovski, S. G., *et al.* 2013, *ApJ*, 765, 154

Garofalo, A., Cusano, F., Clementini, G., *et al.* 2013, *ApJ*, 767, 62

Greco, C., Dall'Ora, M., Clementini, G., *et al.* 2008, *ApJ*, 675, L73

Jeffery, E. J., Smith, E., Brown, T. M., *et al.* 2011, *AJ*, 141, 171

Kinemuchi, K., Smith, H. A., Woźniak, P. R., & McKay, T. A. 2006, *AJ*, 132, 1202

Kuehn, C., Kinemuchi, K., Ripepi, V., *et al.* 2008, *ApJ*, 674, L81

Lindegren, L. 2010, in: S. A. Klioner, P. K. Seidelmann, & M. H. Soffel (eds.), *Relativity in Fundamental Astronomy: Dynamics, Reference Frames, and Data Analysis*, Proc. IAU Symposium No. 261 (Cambridge: Cambridge University Press), p. 296

Lindegren, L. & Perryman, M. A. C. 1996, *A&AS*, 116, 579

Madore, B. F. 1982, *ApJ*, 253, 575

Mancone, C. & Sarajedini, A. 2008, *AJ*, 136, 1913

Marconi, M. & Clementini, G. 2005, *AJ*, 129, 2257

Marconi, M., Fiorentino, G., & Caputo, F. 2004, *A&A*, 417, 1101

Marconi, M., Molinaro, R., Bono, G., *et al.* 2013, *ApJ*, 768, L6

McConnachie, A. W. 2012, *AJ*, 144, 4

McNamara, D. H., Clementini, G., & Marconi, M. 2007, *AJ*, 133, 2752

Miceli, A., Rest, A., Stubbs, C. W., *et al.* 2008, *ApJ*, 678, 865

Moretti, M. I., Clementini, G., Muraveva, T., *et al.* 2013, *MNRAS*, in press (arXiv:1310.6849)

Moretti, M. I., Dall'Ora, M., Ripepi, V., *et al.* 2009, *ApJ*, 699, L125

Musella, I., Ripepi, V., Clementini, G., *et al.* 2009, *ApJ*, 695, L83

Musella, I., Ripepi, V., Marconi, M., *et al.* 2012, *ApJ*, 765, 121

Nikolaev, S., Drake, A. J., Keller, S. C., *et al.* 2004, *ApJ*, 601, 260

Oosterhoff, P. T. 1939, *Observatory*, 62, 104

Pietrzyński, G., Thompson, I. B., Gieren, W., *et al.* 2010, *Nature*, 468, 542

Pritzl, B. J., Armandroff, T. E., Jacoby, G. H., & Da Costa, G. S. 2002, *AJ*, 124, 1464

Pritzl, B. J., Armandroff, T. E., Jacoby, G. H., & Da Costa, G. S. 2004, *AJ*, 127, 318

Pritzl, B. J., Armandroff, T. E., Jacoby, G. H., & Da Costa, G. S. 2005, *AJ*, 129, 2232

Sarajedini, A., Mancone, C. L., Lauer, T. R., *et al.* 2009, *AJ*, 138, 184

Sasselov, D. D., Beaulieu, J. P., Renault, C., *et al.* 1997, *A&A*, 324, 471

Schmidt-Kaler, T. 1977, *A&A*, 54, 771

Sesar, B., Ivezić, Ž., Grammer, S. H., *et al.* 2010, *ApJ*, 708, 717

Siegel, M. H. 2006, *ApJ*, 649, L83

Soszyński I., Poleski R., Udalski A., *et al.* 2008, *AcA*, 58, 163

Soszyński, I., Udalski, A., Szymanski, M. K., *et al.* 2009, *AcA*, 59, 1

Soszyński, I., Udalski, A., Poleski, R., *et al.* 2012 *AcA*, 62, 219

Tisserand, P., Le Guillou, L., Afonso, C., *et al.* 2007, *A&A*, 469, 387

Turon, C., Luri, X., & Masana, E. 2012, *Ap&SS*, 341, 15

Udalski, A., Kubiak, M., & Szymański, M. 1997, *AcA*, 47, 319

Vivas, A. K., Zinn, R., Andrews, P., *et al.* 2001, *ApJ*, 554, L133

Watkins, L. L., Evans, N. W., Belokurov, V., *et al.* 2009, *MNRAS*, 398, 1757

Wood, P. R., Arnold, A., & Sebo, K. M. 1997, *ApJ*, 485, L25

Yang, S.-C. & Sarajedini, A. 2012, *MNRAS*, 419, 1362

Precision Asteroseismology
Proceedings IAU Symposium No. 301, 2013   © International Astronomical Union 2014
J. A. Guzik, W. J. Chaplin, G. Handler & A. Pigulski, eds.   doi:10.1017/S1743921313014221

Constraints on pre-main-sequence evolution from stellar pulsations

M. P. Casey[1], K. Zwintz[2] and D. B. Guenther[1]

[1] Institute for Computational Astrophysics, Department of Astronomy and Physics,
Saint Mary's University, Halifax, NS B3H 3C3, Canada
email: `mcasey@ap.smu.ca`, `guenther@ap.smu.ca`

[2] Institute of Astronomy, University Leuven, Leuven, Belgium
email: `konstanze@ster.kuleuven.be`

Abstract. Pulsating pre-main-sequence (PMS) stars afford the earliest opportunity in the lifetime of a star to which the concepts of asteroseismology can be applied. PMS stars should be structurally simpler than their evolved counterparts, thus (hopefully!) making any asteroseismic analysis relatively easier. Unfortunately, this isn't necessarily the case. The majority of these stars (around 80) are δ Scuti pulsators, with a couple of γ Doradus, γ Doradus – δ Scuti hybrids, and slowly pulsating B stars thrown into the mix. The majority of these stars have only been discovered within the last ten years, with the community still uncovering the richness of phenomena associated with these stars, many of which defy traditional asteroseismic analysis.

A systematic asteroseismic analysis of all of the δ Scuti PMS stars was performed in order to get a better handle on the properties of these stars as a group. Some strange results have been found, including one star pulsating up to the theoretical acoustic cut-off frequency of the star, and a number of stars in which the most basic asteroseismic analysis suggests problems with the stars' positions in the Hertzsprung-Russell diagram. From this we get an idea of the constraints — or lack thereof — that these results can put on PMS stellar evolution.

Keywords. stars: pre–main-sequence, stars: oscillations, stars: fundamental parameters

---

1. Introduction

Pre-main-sequence (PMS) stars afford the earliest opportunity in the lifetime of a star to apply the techniques of asteroseismology, and therefore the earliest test of stellar-evolution models. PMS stars are stars that have only recently emerged from the cloud from which they were formed, and have yet to start burning hydrogen into helium in their cores. Their primary source of energy is the conversion of gravitational potential energy into light; hence they are slowly contracting to the zero-age main-sequence (ZAMS) of hydrogen burning. Stars with mass between about 1.5 and 5 $M_\odot$ can be shown to spend part of their evolution in the classical instability strip of the Hertzsprung-Russell diagram (HRD), and might be expected to pulsate. However, most of the discoveries have come within the past ten years, and the total number of confirmed or candidate stars is only around 80 objects, with the vast majority being δ Scuti stars, with a couple of γ Doradus stars and δ Scuti-γ Doradus hybrids thrown into the mix. There are also a couple of slowly pulsating B-star candidates. These proceedings will focus on the δ Scuti stars, but please see K. Zwintz's contribution, also in this volume, for the remainder. The study of this class of star is quite a new field!

Like their more-evolved counterparts, PMS δ Scuti stars are generally A stars, pulsating in pressure modes at frequencies ranging from approximately 5 to 80 d^{-1}, or 55 to 925 μHz, corresponding to periods of around 18 minutes to 5 hours. So far, it has been found that all confirmed or candidate members of the class are low-amplitude δ Scuti

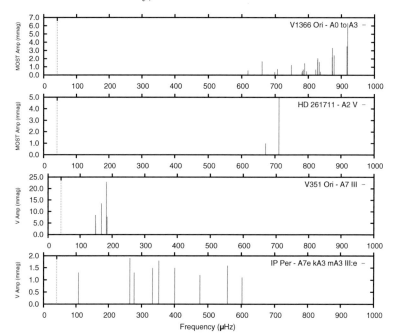

Figure 1. Sample PMS pulsation spectra. Sources of spectra: V1366 Ori, Casey *et al.* (2013); HD 261711, Zwintz *et al.* (2009b); V351 Ori, Ripepi *et al.* (2003); IP Per, Ripepi *et al.* (2006).

stars, with pulsation amplitudes in the millimagnitude range, typically simultaneously pulsating in many independent frequencies at once, both in radial and non-radial modes. Figure 1 shows four sample pulsation spectra.

1.1. *A brief history*

The first discovery of PMS δ-Scuti candidates was of V588 and V589 Mon in the young open cluster NGC 2264 by Breger (1972). In this cluster only the O and B stars are on the main sequence, hence these two A stars (if cluster members) would be PMS stars less than 10 million years of age. Despite this encouraging start, it was 20+ years later before the next member of the class was discovered, V856 Sco (HR 5999) by Kurtz & Marang (1995).

As an Herbig Ae (HAe) star in the Lupus III dark cloud, the discovery of pulsations in V856 Sco partially revealed why discoveries of PMS were so scarce up to this point in time. Herbig Ae stars are thought to be young stars still surrounded by circumstellar material left over from the star-formation process. This circumstellar material carries a number of consequences: 1) It manifests itself in the hydrogen Balmer lines as emission; hence the *Ae* designation for the star. 2) It causes an infrared excess in the spectral energy distribution of the star (e.g. see Figure 8 of van den Ancker *et al.* 1996), which makes the intrinsic properties of the star harder to ascertain. 3) If the circumstellar material is clumpy, aperiodic and unpredictable eclipses of the light from the star of up to 2.5 magnitudes in depth, will result (a UX Ori variable). The depth of each "eclipse" can vary considerably from one extinction event to the next. Fortunately, the time scale for events is on the order of days to weeks, so with the right cadence of observations, UX Ori variability can be readily distinguished from δ Scuti variability. However, it is unlikely that δ Scuti variability will be discovered atop the UX Ori variability unless the δ Scuti variability is specifically sought. 4) The observed colour of a UX Ori variable changes as an eclipse event deepens, becoming redder and redder with increased extinction until the depth reaches 1.5 to 2 magnitudes in V, at which point "blueing" might start to occur — the colour of the star becomes bluer with increased extinction as the observations become contaminated with reflected blue light from the circumstellar material (again, see van den Ancker *et al.* 1996). The above four points make the intrinsic properties of an HAe star particularly hard to determine, and any δ Scuti variability hard to spot compared to (say) a more evolved A star. It is quite easy to misplace an HAe star in the HRD; when this happens even the most basic asteroseismic analysis of the star will fail, as will be demonstrated below.

Theory

The first theoretical asteroseismic analysis of a PMS δ Scuti star was that of V856 Sco by Marconi & Palla (1998), in which an instability strip was calculated for the first three radial orders of pulsation. The calculated strip was roughly coincident with the instability strip of more-evolved counterparts. By the year 2000 a few more stars had been discovered; however it was only with Zwintz (2005) and the discovery of a number of PMS δ Scuti stars in open clusters that this subclass of pulsating star was fully established. Zwintz (2008) then contained a list of all known PMS confirmed or candidate δ Scuti stars to that date, and showed observationally that the PMS δ Scutis occupy the same part of a colour-magnitude diagram as their more-evolved counterparts. Additionally, by comparing observed to modelled frequencies of NGC 6383 170, Zwintz *et al.* (2007) conclusively showed for the first time a PMS δ Scuti star pulsating in non-radial modes. Guenther *et al.* (2007) further successfully fit 5 of 6 observed frequencies to models for NGC 6530 53, 6 of 7 frequencies for star 13, and 9 of 9 frequencies of star 38, suggesting perhaps that the asteroseismic analysis of PMS δ Scuti stars might not be too difficult!

However, up to this point in time, all measurements had been with ground-based instruments, and so the advent of high-quality spaced-based observations from the *Microvariablity and Oscillations of Stars* (*MOST*) satellite mission (Walker *et al.* 2003) proved problematic; suddenly the observations had caught up the theory, and the theory was shown to be wanting. *MOST* observations of V1026 Sco (HD 142666) discovered 12 frequencies, of which only five could be fit to models at any one time (Zwintz *et al.* 2009a). Furthermore, *MOST* observations of the candidate PMS pulsators in NGC 2264 produced results that could sometimes be successfully compared to models, sometimes not (Guenther *et al.* 2009). It was now time for a systematic survey of the collective characteristics of PMS pulsators, e.g. pulsation frequencies (amplitudes, numbers), galactic environment, rotational velocities, etc., and to subject each star to as thorough an asteroseismic analysis as possible. The results of this study can be found in Casey (2011), with the highlights shown below.

2. Results of a systematic study

Collectively, the pulsation spectra of PMS δ Scuti stars reveal many phenomena: some pulsating over a wide range of frequencies that must span many different radial orders of pulsation; some pulsating in a group of low frequencies; some pulsating in a group of high frequencies. The problems associated with an asteroseismic analysis of many of these stars are demonstrated by the case of V1026 Sco (HD 142666), with twelve frequencies detected by *MOST* ranging from 66.8 μHz to 324.4 μHz, including three frequencies, f_5, f_9, and f_{12} that could possibly be a rotationally-split triplet. The best fit to the asteroseismic data (see Figs. 2 and 3), as bad as it is, is nowhere close to the position in the HRD as determined by other means (classification spectra, distance to star, etc.) One or the other or both must be wrong. How do we determine which might be the case?

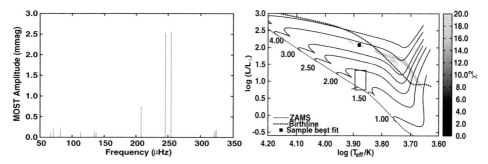

Figure 2. *Left:* V1026 Sco pulsation spectrum. *Right:* HRD position, including χ^2 best-fit values to models. The "birthline" in this (and subsequent) figures refers to the birthline of Palla & Stahler (1990).

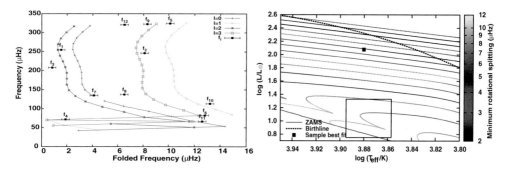

Figure 3. V1026 Sco echelle diagram (left) of best-fit solution (black dot at right), and estimate of expected minimum rotational splitting with $v \sin i = 70$ km s^{-1} (left).

2.1. *Rotational velocity*

A survey of the known $v \sin i$ values for PMS δ Scuti stars reveals that for the most part they rotate at relatively high velocities, generally 50 km s^{-1} or above, with 75 to 150 km s^{-1} being more typical (Casey 2011). As a star rotates, rotational splittings of non-radial l modes appear (called multiplets) compared to a non-rotating model (the rotation breaks the spherical symmetry of the system). At low velocities the splittings are small and symmetric, and relatively easy to spot (if sufficient resolution is obtained in the frequency domain). However, if the velocity is large enough, the splittings become asymmetric, and multiplets of various non-radial l orders can overlap, making it extremely difficult to identify any systematic spacings that may exist without some sort of statistical analysis (e.g. see Deupree & Beslin 2010; Deupree 2011). In the case of V1026 Sco numerous $v \sin i$ values are reported, but the lowest value of 70 ± 2 km s^{-1} is reported by Mora *et al.* (2001). Figure 3 shows an approximate estimate of the *minimum* rotational splitting one might expect for the star given this value for $v \sin i$. For the HRD position given by the box, we would expect splittings of around 10 to 14 μHz, ruling out the high-frequency triplet as members of one multiplet, but more likely to be a chance alignment of overlapping multiplets. However, for the region of the HRD from which the best-fit solution occurs (black square dot), minimum splittings of around 2 μHz are found, indicating the possibility that the triplet (or perhaps maybe two frequencies of the triplet) are members of one rotationally-split multiplet. If the latter is true, then the problem is to figure out why the non-asteroseismic HRD position is so different; perhaps the distance to the star is wrong, or the extinction has been underestimated, either of which might significantly underestimate the luminosity of the star. Regardless, rotation

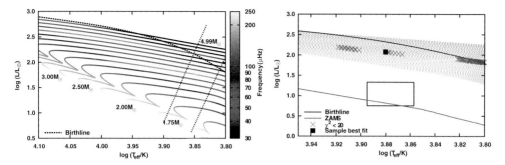

Figure 4. *Left:* fundamental frequencies of PMS models. *Right:* models with the fundamental frequency $< f_{11}$ of V1026 Sco (shaded region), along with location of best-fit models to the observed pulsation spectrum.

seems to be the primary reason for which detailed asteroseismic analysis of PMS stars seems to fail. At least two-dimensional modelling that includes rotation, such as that by Deupree & Beslin (2010), but specifically for PMS stars, is required for progress. However, in the interim there are other attributes of the pulsation spectra that allow us to gain some insight into these stars.

2.2. Lowest pulsation frequency and the fundamental mode of pulsation

If we assume all frequencies in a spectrum are pressure modes (p-modes), then some interesting limits can be put on the luminosities of stars. This is not to say that these frequencies *are* p-modes, but if the assumption is made, what are the consequences of this assumption? In the case of V1026 Sco, the lowest-detected frequency is $f_{11} = 66.8 \ \mu$Hz, or 5.78 d^{-1}. If this is a p-mode, then one would expect this frequency to be *higher* than the frequency of the fundamental mode of star. Therefore, one would eliminate any stellar model from consideration for which the fundamental frequency is *higher* than the star's *lowest* detected frequency. The result is shown in Fig. 4. Given that the fundamental frequency decreases with increasing luminosity, this eliminates all the lower-luminosity models that happen to be in agreement with V1026 Sco's non-asteroseismic position in the HRD, indicating that f_{11} is not a p-mode, or the HRD position is wrong. More work is therefore needed on this star! Through this simple test, lower-luminosity constraints on 32 stars can be placed, with 10 stars found to be "out of position" in the HRD, that is, their asteroseismic and non-asteroseismic positions in the HRD are not in agreement.

2.3. Highest pulsation frequency and the acoustic cut-off frequency

In Casey *et al.* (2013) it was found that the HAe star V1366 Ori (HD 34282), also a δ Scuti star, is most likely pulsating at frequencies just below the theoretical acoustic cut-off frequency (ACF) of the star, a first for a δ Scuti star. The ACF is a theoretical maximum pulsation frequency above which a star should not be expected to pulsate (Aerts *et al.* 2010). If a signal above the ACF *is* detected, then an explanation is needed, such as the modified ACF that seems to exist in some roAp stars (e.g. Saio *et al.* 2012), or the convectively-driven pseudomodes that are observed in the Sun (Jiménez *et al.* 2011). Figure 5 shows that the ACF varies systematically in the HRD, decreasing with increased luminosity for a given effective temperature. Therefore, the *highest* pulsation frequency detected within a pulsation spectrum places *maximum* luminosity constraints on the star if the frequency should be high enough — the detected frequency needs to be below the ACF. In the case of V1026 Sco, with highest-detected frequency of $f_5 = 324 \ \mu$Hz, this constraint eliminated some of the highest-luminosity models from

Table 1. Stars determined to be out of position in the HRD according to asteroseismology. All stars are dimmer than asteroseismology would suggest. For V351 Ori it is the *Hipparcos* distance that causes trouble.

Star	Star formation region or cluster
1) V375 Lac	LkHα 233 DC
2) V856 Sco	Lupus 3 DC
3) V1026 Sco	Sco R1
4) PX Vul	Vul R2
5) IP Per	Per OB2
6) GSC 07380-01173	NGC 6383
7) WW Vul	In Vulpecula
8) VV Ser	Serpens Cloud
9) V351 Ori (Hip)	Orion B
10) CQ Tau	Taurus molecular cloud

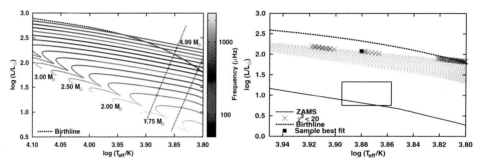

Figure 5. *Left:* ACFs of PMS models (models in the region just above the ZAMS with an ACF greater than 1000 μHz are not shown). *Right:* same as right side of Fig. 4, but also missing models with ACFs > f_5.

consideration, and it then becomes obvious in Fig. 5 why the best-fit models (as horrible as they are) were confined to a narrow horizontal band in the HRD: these are the only models under consideration that are able to even support the observed *range* of the pulsation spectrum. By using this test maximum-luminosity constraints can be placed on 20 of the candidate stars, although none of these constraints result in disagreement with the non-asteroseismic position of a star in the HRD.

2.4. *Out-of-position stars and HRD of PMS δ Scuti stars.*

Table 1 lists 10 candidate stars for which the asteroseismic and non-asteroseismic position in the HRD do not agree. Figure 6 contains the pulsation spectra of four stars and demonstrates how the limits work. A frequency falling within the right-hand-side shaded area puts a lower luminosity limit on the star via the fundamental-mode test — the lower the frequency, the higher the luminosity of the star must be for the frequency to be a *p*-mode. Similarly, a frequency falling within the left-hand-side shaded area puts an upper luminosity limit on the star via the ACF test — the higher the frequency, the lower the luminosity of the star must be for the frequency to be *p*-mode in a regular-atmosphere star (e.g. not a roAp star). For δ Scuti frequency ranges, there is only a very narrow frequency range between about 225 and 250 μHz into which a frequency can fall and *not* put a luminosity constraint on the star.

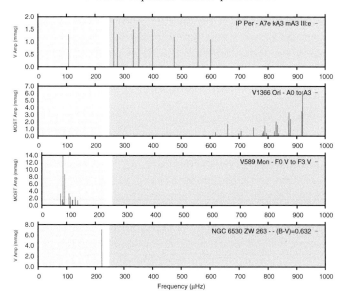

Figure 6. Limits on pulsation spectra: left shaded area, limits imposed by the fundamental frequency; right shaded area, limits imposed by the acoustic cut-off frequency. Sources of spectra: IP Per, Ripepi *et al.* (2006), V1366 Ori, Casey *et al.* (2013); V589 Mon, Guenther *et al.* (2009); NGC 6530 ZW 263, Zwintz & Weiss (2006).

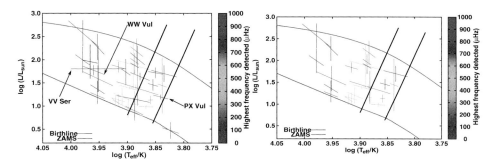

Figure 7. *Left:* HRD of PMS pulsators for which fundamental parameters could be deduced. *Right:* same as left, but with stars "out of position" in the HRD removed.

3. HRD of δ Scuti pulsators

After all this analysis, the left-hand side of Fig. 7 shows an HRD of PMS δ Scuti stars, along with the Marconi and Palla instability strip. Also indicated is the highest frequency detected within the pulsation spectrum of each star. For comparison, the left-hand side of the figure shows the same information except the "out-of-position" have been removed. All of the higher-frequency stars are in the lower left of the distribution, the lower-frequency are stars nearer to the top of the distribution, and the mid-range stars are spread around in the middle to the lower right. The stars to the left of the theoretical instability strip tend to be pulsating in orders higher than the first three radial modes. It will be interesting to see if further detections follow this pattern.

Two of the eliminated candidate stars (for which pulsation detections were too preliminary), PX Vul and WW Vul, have since been determined by *MOST* observations *not* to be pulsating, thus showing the usefulness of the above tests for determining which stars need further observations or theoretical work. Additionally, the lowest frequencies of another out-of-position star, VV Ser, may be *g*-modes (see Ripepi *et al.* 2007; Casey 2011); the fundamental-frequency test tends to flag such candidates fairly readily as being out of position in the HRD, whereas the HRD position may be fine, but it is the *asteroseismic* interpretation of the star that might need further work.

4. Summary

The systematic analysis of PMS δ Scuti pulsators from around 70 members/candidates, with 40 analyzed in greater detail, found examples of stars pulsating at all possible radial *p*-mode orders from the fundamental frequency up to the acoustic cut-off frequency. Rapid rotation was found to be the most probable reason for the failure of the detailed asteroseismic analysis of candidate stars. Using the fundamental mode as a test, minimum luminosity constraints could be placed on 32 stars, whereas the ACF test put maximum luminosity constraints on 20 stars. Ten "problem stars" were found, in which the non-asteroseismic and asteroseismic positions of the stars in the HRD were not in agreement, and for which future work (in particular) should be directed. The results are a new HRD for PMS δ Scuti stars, showing a distinct pattern of pulsation above the first three radial orders of pulsation, as originally calculated by Marconi & Palla (1998). As for specific constraints on PMS evolution from asteroseismology, until the models catch up, it is very difficult to impose anything but the wide constraints demonstrated here. Future work (better HRD positions in particular) are another requirement. See the work of Zwintz *et al.*, also in this volume, for more details.

References

Aerts, C., Christensen-Dalsgaard, J., & Kurtz, D. W. 2010, *Asteroseismology* (Berlin: Springer)
Breger, M. 1972, *ApJ*, 171, 539
Casey, M. P. 2011, Ph.D. thesis, Saint Mary's University, Canada
Casey, M. P., Zwintz, K., Guenther, D. B., *et al.* 2013, *MNRAS*, 428, 2596
Deupree, R. G. 2011, *ApJ*, 742, 9
Deupree, R. G. & Beslin, W. 2010, *ApJ*, 721, 1900
Guenther, D. B., Kallinger, T., Zwintz, K., Weiss, W. W., & Tanner, J. 2007, *ApJ*, 671, 581
Guenther, D. B., Kallinger, T., & Zwintz, K., *et al.* 2009, *ApJ*, 704, 1710
Jiménez, A., García, R. A., & Pallé, P. L. 2011, *ApJ*, 743, 99
Kurtz, D. W. & Marang, F. 1995, *MNRAS*, 276, 191
Marconi, M. & Palla, F. 1998, *ApJ*, 507, L141
Mora, A., Merín, B., & Solano, E., *et al.* 2001, *A&A*, 378, 116
Palla, F. & Stahler, S. W. 1990, *ApJ*, 360, L47
Ripepi, V., Marconi, M., Bernabei, S., *et al.* 2003, *A&A*, 408, 1047
Ripepi, V., Bernabei, S., Marconi, M., *et al.* 2006, *A&A*, 449, 335
Ripepi, V., Bernabei, S., Marconi, M., *et al.* 2007, *A&A*, 462, 1023
Saio, H., Gruberbauer, M., Weiss, W. W., Matthews, J. M., & Ryabchikova, T. 2012, *MNRAS*, 420, 283
van den Ancker, M. E., The, P. S., & de Winter, D. 1996, *A&A*, 309, 809
Walker, G., Matthews, J. M., Kuschnig, R., *et al.*, 2003, *PASP*, 115, 1023
Zwintz, K. 2005, Ph.D. thesis, Universität Wien, Austria
Zwintz, K. 2008, ApJ, 673, 1088
Zwintz, K. & Weiss, W. W. 2006, *A&A*, 457, 237
Zwintz, K. & Guenther, D. B., Weiss W. W. 2007, *ApJ*, 655, 342
Zwintz, K., Kallinger, T., Guenther, D. B., *et al.* 2009a, *A&A*, 494, 1031
Zwintz, K., Hareter, M., Kuschnig, R., *et al.* 2009b, *A&A*, 502, 239

Precision Asteroseismology
Proceedings IAU Symposium No. 301, 2013
J. A. Guzik, W. J. Chaplin, G. Handler & A. Pigulski, eds.

© International Astronomical Union 2014
doi:10.1017/S1743921313014233

The Araucaria Project: the Baade-Wesselink projection factor of pulsating stars

Nicolas Nardetto[1], Jesper Storm[2], Wolfgang Gieren[3], Grzegorz Pietrzyński[4], and Ennio Poretti[5]

[1]Laboratoire Lagrange, UMR7293, Université de Nice Sophia-Antipolis, CNRS, Observatoire
de la Côte d'Azur, Nice, France
email: Nicolas.Nardetto@oca.eu

[2]Leibniz Institute for Astrophysics, An der Sternwarte 16, 14482, Potsdam, Germany
email: jstorm@aip.de

[3]Departamento de Astronomía, Universidad de Concepción, Casilla 160-C, Concepción, Chile
email: wgieren@astro-udec.cl

[4]Warsaw University Observatory, Al. Ujazdowskie 4, 00-478, Warsaw, Poland
email: pietrzyn@astrouw.edu.pl

[5]INAF – Osservatorio Astronomico di Brera, Via E. Bianchi 46, 23807 Merate (LC), Italy
email: ennio.poretti@brera.inaf.it

Abstract. The projection factor used in the Baade-Wesselink method of determining the distance of Cepheids makes the link between stellar physics and the cosmological distance scale. A coherent picture of this physical quantity is now provided based on several approaches. We present the latest news on the expected projection factor for different kinds of pulsating stars in the Hertzsprung-Russell diagram.

Keywords. stars: oscillations (including pulsations), stars: atmospheres

1. Short review on the projection factor of Cepheids

For many decades the Cepheid stars have been used to calibrate the distance scale and the Hubble constant through their well-known period-luminosity (PL) relation (see Riess *et al.* 2011 and Freedman & Madore 2010 for a review). Recently, using the Baade-Wesselink (BW) method to determine distances of Cepheids, Storm *et al.* (2011a) found that the K-band PL relation is nearly universal and can be applied to any host galaxy largely independent of metallicity. The projection factor is a key quantity of the BW methods: it is used to convert the radial velocity variation into the pulsation velocity of the star. There are several ways to study the projection factor. One can use geometrical or static models, hydrodynamical analysis, or even direct observations when the distance of the star is known. In the purely geometric approach, two effects are considered only: the limb-darkening of the star (in the continuum) and the expansion of the atmosphere (at constant velocity). The projection factor is then an integration of the pulsation velocity field (associated with the line-forming region) projected on the line of sight and weighted by the surface brightness of the star, which is defined for instance by $I(\cos\theta) = 1 - u_V + u_V \cos\theta$, where u_V is the limb darkening of the star in the V band and θ is the angle between the normal of the star and the line of sight (Claret *et al.* 2011). In this case, the geometric projection factor can be derived as follows: $p_0 = 3/2 - u_V/6$ (Getting 1934). However, this definition of the projection factor implies a specific method of the radial velocity determination, which is the first moment or centroid method (Burki *et al.* 1982). Depending on the limb-darkening considered for the Cepheid studied, the value of the projection factor is different: $p = 24/17 = 1.415$ ($u_V = 0.60$, Getting 1934), $p = 1.375$

($u_V = 0.75$, Van Hoof & Deurinck 1952) or $p = 1.360$ ($u_V = 0.80$, Burki *et al.* 1982). The latter value has been widely used in spectroscopy. Recently, Neilson *et al.* (2012) derived the geometrical projection factor as a function of the period and for several photometric bands using a radiative transfer in spherical geometry and found a slightly lower value of $p = 1.33$ for δ Cep. Additional studies should be also mentioned like Gray & Stevenson (2007) and Hadrava *et al.* (2009) in which a geometrical model is directly fitted to the observed spectral line profile (the pulsation velocity is then an output). Such approaches are formally consistent with the geometrical method.

The second approach to study the Baade-Wesselink projection factor is to consider a hydrodynamical model, which describes the dynamical structure of the atmosphere of the star (in particular the atmospheric velocity gradient). Using a so-called *piston* model in which the radial velocity curve is used as an input, Sabbey *et al.* (1995) found a mean value of the projection factor of $p = 1.34$. However, this value was derived using the bi-sector method of the radial velocity determination (applied to theoretical line profiles), which unfortunately makes the comparison with other studies quite uncertain. On the other hand, using a *self-consistent* model of the pulsation (requiring few fundamental parameters such as the stellar mass, the luminosity, the effective temperature and the chemical composition), Nardetto *et al.* (2004) found that the atmospheric velocity gradient (and other dynamical effects) reduce the geometric projection factor (found at $p_0 = 1.39$ with the model) by about 9%, leading to a projection factor of $p = 1.27 \pm 0.01$. This value is however consistent with the Gaussian fit method of the radial velocity determination (applied to a spectral line with a typical depth of $D = 0.2$). In this 9% decrease, 5% comes from the dynamical structure of the atmosphere and 4% from using the Gaussian fit method. Indeed later, Nardetto *et al.* (2007) provided a revised value of the projection factor, $p = 1.33 \pm 0.02$, applicable together with the first moment method (and consistent with a plane parallel model atmosphere). It is worth noticing that the projection factor is generally supposed as being constant over the pulsation phase following Nardetto *et al.* (2006b).

If one uses the cross-correlated radial velocity (which includes many lines and also a Gaussian fit of the cross-correlated mean line profile with a typical depth of $D = 0.25$), a lower value of the projection factor is found (by about 11% compared to the initial geometrical projection factor $p_0 = 1.39$), i.e. $p = 1.25 \pm 0.05$ (Nardetto *et al.* 2009). One can say approximatively that in these 11%, 7% comes from the dynamical structure of the atmosphere and 4% from the Gaussian fit.

Mérand *et al.* (2005) applied the *inverse* Baade-Wesselink method using an infrared FLUOR/CHARA interferometric observation of δ Cep. In this approach, the projection factor is fitted, while the distance of δ Cep is known (from the Hubble Space Telescope (*HST*) parallax) at the 4% level (Benedict *et al.* 2007). They found $p = 1.27 \pm 0.05$ (using the cross-correlation method to derive the radial velocity). Then, deriving the infrared surface brightness angular diameters of δ Cep, and applying again the *inverse* BW method, Groenewegen (2007) and Laney & Joner (2009) found similarly a value of the projection factor of $p = 1.27$. Later, Storm *et al.* (2011b) constrained *directly* the period-projection factor (Pp) relation using spectroscopic and photometric observations of Cepheids in the Large Magellanic Cloud (hereafter LMC). In this method, the zero-point of the Pp relation is again based on the *HST* trigonometric parallaxes of Galactic Cepheids, but the slope is derived from the BW distances of LMC Cepheids (all Cepheids in the LMC used by Storm can be assumed to be at the same distance, leading to an extra constraint on the period projection factor relation). The corresponding value for δ Cep itself is $p = 1.41 \pm 0.05$. It has been shown that the metallicity has no impact (at least theoretically) on the projection factor (Nardetto *et al.* 2011). Using a similar

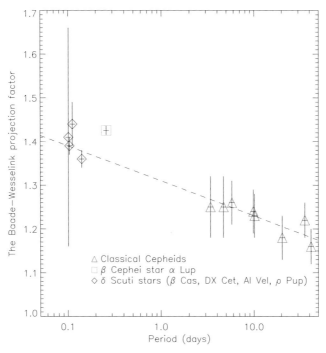

Figure 1. The Baade-Wasselink projection factor as a function of the period for different kinds of pulsating stars. The case of the β Cephei star α Lup is described by Nardetto *et al.* (2013b).

method, Groenewegen (2013) found recently a value of the projection factor which is also quite high ($p = 1.33$). The latest result comes from Pilecki *et al.* (2013), who constrained the projection factor using a short-period Cepheid ($P = 3.8$ days) in an eclipsing binary system. They found $p = 1.21 \pm 0.04$.

This short review shows that a lot of work has been done to constrain the BW projection factor. And even if some discrepancies remain concerning the inverse photometric BW method of determining the projection factor, a consensus is currently emerging. In particular, we emphasise that the fact that the projection factor derived from the surface-brightness techniques is overestimated has no impact on the distances, because at the same time, the amplitude of the photometric angular diameter curve is underestimated. One can say finally that the photometric version of the BW method is *self-consistently* calibrated using the *HST* parallaxes to set the zero point and the distances to LMC Cepheids with a large range of periods to constrain the p-factor relation with pulsation period. However, Ngeow *et al.* (2012) found indeed a significant dispersion in the period-projection factor relation, and this should be also investigated.

2. The projection factor for other types of pulsating stars

One possible way to better understand the dynamical structure of Cepheids, and in particular the k-term† (Nardetto *et al.* 2006a, 2008), the mass loss (Nardetto *et al.* 2008), and the projection factor, is to perform comparison with other kinds of pulsating stars (as soon as they pulsate in a dominant radial model).

---

† The k-term is a residual asymmetry (not related to the pulsation) observed in the spectral line profiles of Cepheids.

In the framework of the Araucaria Project (Gieren *et al.* 2005) of distances determination in the Local Group, we determined the Baade-Wesselink projection factor for four δ Sct stars: ρ Pup ($p = 1.36 \pm 0.02$), DX Cet ($p = 1.39 \pm 0.02$), AI Vel ($p = 1.44 \pm 0.05$), and β Cas ($p = 1.41 \pm 0.25$). (Refer to Nardetto *et al.* (2013a) for ρ Pup and DX Cet and to Guiglion *et al.* (2013) for AI Vel and β Cas.) Figure 1 shows how all these values fit in an excellent way the extension toward short periods of the relation found for Cepheids, i.e., $p = [-0.08 \pm 0.05] \log P + [1.31 \pm 0.06]$ (Nardetto *et al.* 2009). This result seems more robust than the similar one obtained by Laney & Joner (2009) using an indirect method based on the comparison of geometric and pulsation parallaxes. On the other hand, the projection factor of the β Cep star α Lup is 8σ above the relation (Fig. 1). By omitting α Lup we can find a relation common to δ Sct stars and classical Cepheids.

3. Conclusion

The projection factor is a very complex quantity which involves all the physical structure of a Cepheid's atmosphere. Nevertheless, it is now well constrained using geometrical, hydrodynamical modelling and also direct observations (trigonometric parallaxes and interferometry). Thanks to these efforts to better understand the projection factor, the BW technique of distance determination is becoming one of the more robust methods in the path to the Hubble constant.

References

Benedict, G. F., McArthur, B. E., Feast, M. W., *et al.* 2007, *AJ*, 133, 1810
Burki, G., Mayor, M., & Benz, W. 1982, *A&A*, 109, 258
Claret, A. & Bloemen, S. 2011, *A&A*, 529, A75
Freedman, W. L. & Madore, B. F. 2010, *ARAA*, 48, 673
Getting, I. A. 1934, *MNRAS*, 95, 139
Gieren, W., Pietrzyński, G., Bresolin, F., *et al.* 2005, *The Messenger*, 121, 23
Gray, D. F. & Stevenson, K. B. 2007, *PASP*, 119, 398
Groenewegen, M. A. T. 2007, *A&A*, 474, 975
Groenewegen, M. A. T. 2013, *A&A*, 550, A70
Guiglion, G., Nardetto, N., Mathias, P., *et al.* 2013, *A&A*, 550, L10
Hadrava, P., Šlechta, M., & Škoda, P. 2009, *A&A*, 507, 397
Laney, C. D. & Joner, M. D. 2009, *AIP-CP*, 1170, 93
Mérand A., Kervella, P. & Coudé du Foresto, V. 2005, *A&A*, 447, 783
Nardetto, N., Fokin, A., Mourard, D., *et al.* 2004, *A&A*, 428, 131
Nardetto, N., Mourard, D., Kervella, P., *et al.* 2006a, *A&A*, 453, 309
Nardetto, N., Fokin, A., Mourard, D., *et al.* 2006b, *A&A*, 454, 327
Nardetto, N., Mourard, D., Mathias, P. *et al.* 2007, *A&A*, 471, 661
Nardetto, N., Stoekl, A., Bersier, D., *et al.* 2008a, *A&A*, 489, 1255
Nardetto, N., Groh, J. H., Kraus, S., *et al.* 2008b, *A&A*, 489, 1263
Nardetto, N., Gieren, W., Kervella P. *et al.* 2009, *A&A*, 502, 951
Nardetto, N., Fokin, A., Fouqué, P. *et al.* 2011, *A&A*, 534, L16
Nardetto, N., Poretti E., Rainer M., *et al.* 2013a, *A&A*, submitted
Nardetto, N., Mathias, P., Fokin, A., *et al.* 2013b, *A&A*, 553, A112
Neilson, H. R., Nardetto, N., Ngeow, C.-C., *et al.* *A&A*, 541, A134
Ngeow, C.-C., Neilson, H. R., Nardetto, N., *et al.* 2012, *A&A*, 543, A55
Pilecki, D., Graczyk, D., Pietrzyński, G., *et al.* 2013, *MNRAS*, 436, 953
Riess, A. G., Lucas M., Casertano S., *et al.* 2011, *ApJ*, 730, 119
Sabbey, C. N., Sasselov, D. D., Fieldus, M. S., *et al.* 1995, *ApJ*, 446, 250
Storm, J., Gieren, W., & Fouqué, P. 2011a, *A&A*, 534, A94
Storm, J., Gieren, W., & Fouqué, P. 2011b, *A&A*, 534, A95
Van Hoof, A. & Deurinck, R. 1952, *ApJ*, 115, 166

Precision Asteroseismology
Proceedings IAU Symposium No. 301, 2013
J. A. Guzik, W. J. Chaplin, G. Handler & A. Pigulski, eds.
© International Astronomical Union 2014
doi:10.1017/S1743921313014245

The position of confirmed pre-main sequence pulsators in the H-R diagram and an overview of their properties

Konstanze Zwintz[1], Mike Casey[2] and David Guenther[2]

[1] Institute of Astronomy, Katholieke Universitet Leuven,
Celestijnenlaan 200D, 3001 Leuven, Belgium
email: konstanze.zwintz@ster.kuleuven.be

[2] Dept. of Astronomy and Physics, St. Mary's University
Halifax, NS B3H 3C3, Canada

Abstract. Pre-main sequence (PMS) stars can become vibrationally unstable during their evolution to the zero-age main sequence (ZAMS). As they gain their energy from gravitational contraction and have not started nuclear fusion in their cores yet, their inner structures are significantly different to those of (post-) main sequence stars and can be probed by asteroseismology.

Using photometric time series from ground and from space (*MOST, CoRoT & Spitzer*) the number of confirmed pulsating pre-main sequence stars has increased significantly within the last years and allowed to find members of new classes of PMS pulsators. Apart from the well-established group of δ Scuti type PMS stars, members of the groups of PMS γ Doradus, PMS δ Scuti – γ Doradus hybrid and PMS slowly pulsating B (SPB) stars have been discovered. For five PMS δ Scuti candidates, space photometry has revealed that they only show irregular variability, but no pulsations.

The unique high-precision space data were combined with dedicated high-resolution spectra to probe the parameter space in the H-R diagram and study the properties of PMS pulsators in comparison to their evolutionary stage.

Keywords. techniques: spectroscopic, stars: Hertzsprung-Russell diagram, stars: pre-main sequence, stars: variables: δ Scuti, stars: oscillations

1. Introduction

Asteroseismology of pulsating pre-main sequence (PMS) stars has the potential of testing the validity of current models of PMS structure and evolution. Although the first two δ Scuti type pulsating PMS stars were already detected in 1972 in the young cluster NGC 2264 (Breger 1972), it took more than 20 years until the existence of this new group of stars was confirmed by the discovery of pulsation in the Herbig Ae field star HR 5999 (Kurtz & Marang 1995). This first asteroseismic analysis of a PMS δ Scuti star triggered the computation of the theoretical PMS δ Scuti instability strip for the first three radial modes (Marconi & Palla 1998), because at that time it was assumed that PMS stars are purely radial pulsators with only few excited modes.

In the following years numerous studies have been devoted to the search for and investigation of oscillations in PMS stars. Soon it was clear that PMS stars pulsate with radial and non-radial modes (e.g., Zwintz *et al.* 2007) and can show rich pulsation frequency spectra (e.g., Zwintz *et al.* 2009). Hence, the theoretical PMS instability strip for the first three radial modes (Marconi & Palla 1998) has become insufficient for higher radial orders and non-radial pulsators.

149

The comparison of the hot and cool border of the classical instability strip with observations has been an important test for stellar structure and evolution codes. The determination of these borders by dedicated observations of PMS stars is comparably important for the theory. With the 36 known PMS δ Scuti pulsators in 2008, it was possible for the first time to investigate the instability region for PMS stars empirically and compare it with theoretical predictions (Zwintz 2008). The location of all known PMS pulsators in the H-R diagram suggested a PMS δ Scuti instability region that is slightly inclined toward the bluer (i.e. hotter) side relative to the classical instability strip under the plausible assumption that both instability regions coincide on the ZAMS. This analysis (Zwintz 2008) was based on Johnson photometric colors which has been the only common observable for all 36 stars. As young stars are often surrounded by remnants of their birth clouds, the photometric colors were often contaminated by the circumstellar matter and resulted in errors in effective temperature of up to 1000 K. Therefore, it was obvious that spectroscopic observations for all PMS pulsators were required for any further (asteroseismic) study.

2. Observational status of PMS pulsators

53 confirmed PMS δ Scuti stars are known as of September 2013 – their evolutionary stage was assessed either due to the stars' memberships to very young open clusters or because they are Herbig Ae stars; their pulsations have been studied using space or ground based photometric time series. They are radial and non-radial pulsators with periods between about 18 minutes and 6 hours. Additionally several PMS δ Scuti candidates (with ambiguous evolutionary state or low amplitude pulsations) have been found.

One PMS δ Scuti – γ Doradus hybrid (Ripepi et al. 2011) was discovered as well as the first two PMS γ Doradus stars (Zwintz et al. 2013). The presence of PMS SPB stars was suggested with two candidates in NGC 2244 (Gruber et al. 2012).

A large part of these discoveries is based on space photometry obtained by the *MOST* (Walker et al. 2003) and *CoRoT* (Baglin 2006) satellites. Data for a total of 46 PMS pulsators are available from these two space telescopes. Observations were taken also within the CSI2264 project (Cody et al. 2013), which targeted the young cluster NGC 2264 and involved simultaneous observations of the four satellites *MOST, CoRoT, Spitzer* (Werner et al. 2004) and *Chandra* (Weisskopf et al. 2002).

The pulsational characteristics for all these stars have been determined well from the high-precision photometric time series. But PMS pulsators were lacking reliable fundamental parameters that are crucial for an asteroseismic interpretation, for a correct determination of their positions in the H-R diagram and for stellar evolution studies in context with pulsations.

2.1. Non-pulsating PMS stars

From the list of 36 PMS pulsators published in Zwintz (2008), for five stars dedicated *MOST* space photometry revealed that they only show irregular variations but no pulsations. The previously conducted ground based photometric observations on relatively short time bases of a few hours mimicked a periodic signal that in fact originates from large irregular variability caused by circumstellar material. BF Ori, UX Ori and WW Vul have effective temperatures ranging from 8500 K to 9500 K, while HD 35929 and PX Vul are on the cooler side with effective temperatures of 6800 K and 6700 K, respectively. A detailed study will be given in Zwintz et al. (2013, in preparation).

3. Spectroscopic observations

High resolution, high signal-to-noise spectroscopic observations were taken for all PMS pulsators, for several irregular variable PMS stars and some known constant comparison stars using the McDonald 2.7-m telescope and Tull echelle spectrograph, ESO VLT with UVES, ESO 3.6-m telescope with HARPS and CFHT with ESPaDOnS. Literature data were available for additional 15 objects.

The spectroscopic analysis was performed using the LLMODELS model atmosphere code (Shulyak *et al.* 2004), the VALD database (Kupka *et al.* 1999) and SYNTH3 for computation of synthetic spectra (Kochukhov 2007). Use was made of an updated version (N. Piskunov, 2013, priv. comm.) of the SME software package (Valenti & Piskunov 1996) that uses a LLMODELS model atmosphere grid and CM treatment of convection.

4. Results

4.1. *PMS δ Scuti pulsators*

PMS δ Scuti stars basically populate the same instability region as their classical, (post-) main sequence counterparts. The previously suggested slight inclination of the region towards the hotter (i.e., bluer) side might still be seen, while the gap at the lower red corner suggested in 2008 could not be confirmed. Masses of PMS δ Scuti stars range from 1.5 to ∼3.5 $M_\odot$; $v \sin i$ values lie between 10 and 190 km s^{-1}. There is no correlation between the position of the stars in the H-R diagram and their $v \sin i$ values.

4.2. *PMS γ Doradus stars and PMS δ Scuti – γ Doradus hybrid(s)*

CoID 102699796 (Ripepi *et al.* 2011) is the first PMS δ Scuti – γ Doradus hybrid, which is nicely positioned between the δ Scuti and the γ Doradus instability regions. NGC 2264 VAS 20 and NGC 2264 VAS 87 are the first two PMS γ Doradus stars (Zwintz *et al.* 2013). With effective temperatures of 6220 K and 6380 K they are cooler than the predicted cool border of the PMS γ Doradus instability region (Bouabid *et al.* 2011). Several more candidates for both groups discovered in NGC 2264 are currently under investigation.

4.3. *PMS Slowly Pulsating B (SPB) stars*

PMS SPB stars are more massive than the PMS δ Scuti stars, hence have a much shorter PMS evolutionary phase and are located much closer to the ZAMS. The discovery and analysis of PMS SPB type stars allows to investigate the stars' transition phases from the pre- to the main sequence evolutionary stage, i.e. from gravitational contraction to the onset of hydrogen core burning. The first two candidates of SPB type objects among PMS stars, GSC 00154-01871 and GSC 00154-00785, have been found in the young open cluster NGC 2244 (Gruber *et al.* 2012). Additional data for young B type objects are provided by the recent *MOST* observations of the young cluster NGC 2264. At least five new candidate SPB stars close to the ZAMS have been revealed and are currently under investigation.

Acknowledgements

We are grateful to N. Piskunov for providing us with an updated version of the SME software and to L. Fossati, T. Ryabchikova and C. Aerts for fruitful discussions. KZ receives a Pegasus Marie Curie Fellowship of the Research Foundation - Flanders (FWO). This investigation has been supported by the Austrian Fonds zur Förderung der wissenschaftlichen Forschung through project P 21830-N16 (PI: M. Breger). DBG acknowledges the funding support of the Natural Sciences and Engineering Research Council (NSERC) of Canada.

References

Baglin, A. 2006, in: M. Fridlund, A. Baglin, J. Lochard, & L. Conroy (eds.), *The CoRoT mission, pre-launch status, stellar seismology and planet finding*, ESA SP-1306, (Noordwijk: ESA Publication Division), p. 537

Bouabid, M.-P., Montalbán, J., Miglio, A., Dupret, M.-A., Grigahcène, A., & Noels, A. 2011, *A&A*, 531, 145

Breger, M. 1972, *ApJ*, 171, 539

Cody, A. M., Stauffer, J. R., Micela, G., Baglin, A., & CSI 2264 Team 2013, *AN*, 334, 63

Gruber, D., Saio, H. Kuschnig, R., *et al.* 2012, *MNRAS*, 420, 291

Kochukhov, O. 2007, in: I. I. Romanyuk, D. O. Kudryavtsev, O. M. Neizvestnaya, & V. M. Shapoval (eds.), *Physics of Magnetic Stars*, (Special Astrophysical Observatory, RAS), p. 109

Kupka, F., Piskunov, N., Ryabchikova, T. A., Stempels, H. C., & Weiss, W. W. 1999, *A&AS*, 138, 119

Kurtz, D. W. & Marang, F. 1995, *MNRAS*, 276, 191

Marconi, M. & Palla, F. 1998, *ApJ*, 507, 141

Ripepi, V., Cusano, F., di Criscienzo, M., *et al.* 2011, *MNRAS*, 416, 1535

Shulyak, D., Tsymbal, V., Ryabchikova, T., Stütz Ch., & Weiss, W. W. 2004, *A&A*, 428, 993

Valenti, J. A. & Piskunov, N. 1996, *A&AS*, 118, 595

Walker, G., Matthews, J. M., Kuschnig, R., *et al.* 2003, *PASP*, 115, 1023

Weisskopf, M. C., Brinkman, B., Canizares, S., Garmire, G., Murray, S., & Van Speybroeck, L. P. 2002, *PASP*, 114, 1

Werner, M. W., Roellig, T. L., Low, F. J., *et al.* 2004, *ApJS*, 154, 1

Zwintz, K. 2008, *ApJ*, 673, 1088

Zwintz, K., Guenther, D. B., & Weiss, W. W. 2007, *ApJ*, 655, 342

Zwintz, K., Kallinger, T., Guenther, D. B., *et al.* 2009, A&A, 494, 1031

Zwintz, K., Fossati, L., Ryabchikova, T., *et al.* 2013, *A&A*, 550, 121

Precision Asteroseismology
Proceedings IAU Symposium No. 301, 2013
J. A. Guzik, W. J. Chaplin, G. Handler & A. Pigulski, eds.

© International Astronomical Union 2014
doi:10.1017/S1743921313014257

The pulsation-rotation interaction: Greatest hits and the B-side

Rich Townsend

Department of Astronomy, University of Wisconsin-Madison, Madison, WI 53706, USA
email: `townsend@astro.wisc.edu`

Abstract. It has long been known that rotation can have an appreciable impact on stellar pulsation — by modifying the usual p and g modes found in the non-rotating case, and by introducing new classes of modes. However, it's only relatively recently that advances in numerical simulations and complementary theoretical treatments have enabled us to model these phenomena in any great detail. In this talk I'll review highlights in this area (the 'Greatest Hits'), before considering the flip side (or the 'B-side', for those of us old enough to remember vinyl records) of the pulsation-rotation interaction: how pulsation can itself influence internal rotation profiles.

Keywords. stars: oscillations, stars: interiors, stars: rotation, hydrodynamics, waves

1. Introduction

Back at the dawn of civilization, the principal medium for distributing music was the vinyl gramophone record. Those who grew up in the vinyl era will recall that a record had two sides — the 'A' side featuring the hit(s) that usually motivated the initial purchase of the record, and the 'B' side which contained somewhat more esoteric material often destined to languish in obscurity.†

I bring up these facts to draw a strained analogy to the pulsation-rotation interaction in stellar astrophysics. The 'A' side with which we're all familiar comprises the effects of rotation on pulsation; but there's also an accompanying 'B' side which considers the influence that pulsation might have on rotation, and indeed the host star's overall evolution. In this contribution I first review the 'greatest hits' on the 'A' side (Sections 2–6), before highlighting some important developments from the 'B' side (Section 7).

2. Perturbative approaches

Ledoux (1949) and Cowling & Newcomb (1949) first considered the effects of slow rotation on the oscillation frequencies of a star. As seen from an inertial frame, these can be expressed as

$$\omega = \omega_0 + m\Omega(1 - C_{n,l}), \qquad (2.1)$$

where ω_0 is the frequency the mode would have in the absence of rotation, Ω is the rotation angular frequency (for now, assumed uniform), $C_{n,l}$ the Ledoux constant and n, l, m are the usual mode radial order, harmonic degree and azimuthal order, respectively. The Ledoux constant accounts for the effects of the Coriolis force on the mode. It is usually positive, because the Coriolis force tends to counteract the restoring force on displaced fluid elements for prograde modes ($m > 0$), leading to smaller frequencies (vice-versa for

---

† There are, of course, exceptions: *Rock Around the Clock* was first released by Bill Haley & His Comets (a very astronomical band!) on the B-side.

retrograde modes with $m < 0$). (Sometimes, however, phenomena such as mode coupling can produce negative $C_{n,l}$; Takata & Saio, these proceedings, present an example of this). The other term in the parentheses accounts for the Doppler shift in transforming from the co-rotating frame to the inertial frame.

The frequency splitting described by eqn. (2.1) is linear in m. This is entirely analogous to the Zeeman splitting of atomic energy levels in a weak magnetic field, and the same first-order perturbation expansion approach underpins the analysis of both phenomena. Moving to more rapid rotation requires a higher-order perturbation expansion. Simon (1969) and a number of subsequent authors extended the formalism to second order in Ω, and Soufi *et al.* (1998) took it to third order; however, these treatments are significantly more complicated than eqn. (2.1). The value of going to even high orders is moot, because the complexity of the problem becomes unmanageable; equally importantly, there's a point where the effects of rotation can no longer be considered a small perturbation to the non-rotating pulsation equations. Then, so-called 'non-perturbative' approaches are required.

3. Non-perturbative approaches

The pulsation equations in a rotating star comprise a 2-dimensional boundary value problem (BVP), with radius r and co-latitude θ as the independent variables, and the frequency ω serving as an eigenvalue. Non-perturbative approaches to solving these equations fall into four main groups.

3.1. *Direct methods*

Conceptually, the simplest non-perturbative approach is to approximate the pulsation equations using finite differences on a 2-D (r, θ) grid. This leads to a large set of algebraic equations, which can be solved using sparse-matrix algorithms. Clement (1998, plus a number of earlier papers) and Savonije *et al.* (1995) use direct methods, but they have not been more widely adopted for reasons which aren't immediately obvious. (My personal perspective is that the numerical aspects of direct methods are rather daunting).

3.2. *Spectral methods*

Spectral methods expand the angular dependence of solutions as (typically large, but finite) sums of spherical harmonics $Y_l^m(\theta, \phi)$, with the same azimuthal orders but harmonic degrees $l = |m|, |m| + 2, |m| + 4, \ldots$ for even-parity modes, and $l = |m| + 1, |m| + 3, |m| + 5, \ldots$ for odd-parity modes. This reduces the pulsation equations to a 1-D BVP, with the expansion coefficients being the unknowns. This BVP is solved using the same general techniques as in the non-rotating case (see, e.g., Townsend & Teitler 2013, and references therein), although the computational cost is much higher.

Spectral methods are the oldest of the approaches described here, dating back to the pioneering work on the oscillations of rotating polytropes by Durney & Skumanich (1968). They have become increasingly popular in recent years (e.g., Lee & Baraffe 1995; Reese *et al.* 2006; Ouazzani *et al.* 2012), perhaps driven by the advent of inexpensive high-performance computing hardware. One criticism leveled at spectral methods is that the truncation of the spherical-harmonic expansion necessarily makes them approximate. This is technically true, but only inasmuch as *any* numerical solution of a system of differential equations is an approximation. The number of spherical harmonics can always be made sufficiently large to achieve the desired level of accuracy — much as the grid spacing in a finite-difference method can always be made sufficiently small.

3.3. *Ray-tracing methods*

Ray-tracing methods treat the pulsation equations in an asymptotic limit analogous to the geometric limit of optics. The resulting eikonal equation is integrated using the method of characteristics (at a computational cost much smaller than the direct or spectral methods described above), to find the ray paths followed by short-wavelength acoustic waves through a rotating model star. The properties of these rays can be studied using Poincaré surface sections, which mark each passage of a ray through a fixed-radius surface with a point plotted in the θ-k_θ (polar wavenumber) plane.

Lignières & Georgeot (2008) use ray tracing to show that the acoustic properties of rapidly rotating polytropes fall into three main groups: rays bouncing internally between mid-latitude surface regions in the northern and southern hemispheres, rays confined to a surface layer at all latitudes, and rays completely filling the interior. These groups correspond, respectively, to the three classes of global p mode explored by Reese *et al.* (2009) using a spectral method: island modes (small $l - |m|$), whispering gallery modes (large $l - |m|$), and chaotic modes (intermediate $l - |m|$).

3.4. *Traditional approximation*

The traditional approximation neglects the horizontal component of the angular velocity vector when evaluating the Coriolis force. Originating in the geophysical literature (see Eckart 1960), it is a reasonable approximation in radiative regions when both the oscillation frequency and the rotation frequency are much smaller than the local Brunt-Väisälä frequency N — that is, for intermediate- and high-order g-modes.

If the centrifugal force and the gravitational potential perturbations are also neglected, the traditional approximation brings a huge simplification to the pulsation equations: it allows them to be separated in r and θ. Solution then proceeds as in the non-rotating case, with only two substantive changes: spherical harmonics $Y_l^m(\theta, \phi)$ are replaced by so-called Hough functions $\Theta(\theta) \exp(im\phi)$ (Hough 1897; see also Townsend 2003a), which are the eigenfunctions of Laplace's tidal equation, and $l(l+1)$ terms are replaced by the corresponding eigenvalues λ of the tidal equation.

These eigenvalues depend on the 'spin parameter' $\nu = 2\Omega/\omega_{\rm c}$, where

$$\omega_{\rm c} \equiv \omega - m\Omega \qquad (3.1)$$

is the oscillation frequency in the co-rotating frame. In the limit $\nu \to 0$, $\lambda \to l(l+1)$; but for $\nu \gtrsim 1$, λ departs markedly from its non-rotating value. It is this 'inertial regime' which remains inaccessible to perturbative approaches. (As an aside: the significance of ν is that it measures by how much the star turns during one oscillation cycle — and thus to what extent a given mode is 'aware' its frame of reference is rotating).

There's an important caveat to the traditional approximation: it neglects the possibility of resonant coupling between pairs of modes with the same m. Lee & Saio (1989) demonstrated that these resonances are manifested in avoided crossings between the mode frequencies; but in the traditional approximation the avoided crossings are transformed into ordinary crossings, because the resonances are suppressed. This explains why the traditional approximation is unable to reproduce the properties of the rosette modes found in rapidly rotating polytropes by Ballot *et al.* (2012); as Takata & Saio (these proceedings) demonstrate, these modes result from near-degeneracies in the non-rotating frequency spectrum, which are then pushed into resonance by the Coriolis force.

4. The rapidly rotating limit

The complexity of the higher-order perturbative approaches (Section 2) might make us concerned that understanding oscillations in rapidly rotating stars is going to be extremely difficult. Fortunately, however, this often turns out not to be the case; new regularities appear in the frequency spectra, in just the same way that atomic energy levels become regular again in the strong-field limit (the Paschen-Back effect).

4.1. Regularities in p-mode spectra

Based on fitting to spectral-method calculations, but also guided by insights from ray tracing, Reese *et al.* (2009) propose an empirical formula for the regularities seen in the frequency spectra of island acoustic modes:

$$\omega_c \approx \tilde{n}\tilde{\Delta}_{\tilde{n}} + \tilde{l}\tilde{\Delta}_{\tilde{l}} + m^2\tilde{\Delta}_{\tilde{m}} + \tilde{\alpha}. \tag{4.1}$$

Here, the various $\tilde{\Delta}$ terms together with $\tilde{\alpha}$ are obtained from least-squares fitting to the calculated frequency spectra, while $\tilde{n}$ and $\tilde{l}$ are new mode indices which correspond to the number of eigenfunction nodes along and parallel to the ray paths stretching between the two mid-latitude surface endpoints of island modes (see Fig. 3 of Reese *et al.* 2009). These indices can be related to the radial order and harmonic degree of the modes' non-rotating counterparts via

$$\tilde{n} = 2n, \qquad \tilde{l} = \frac{l - |m|}{2}$$

for even-parity modes, and by

$$\tilde{n} = 2n + 1, \qquad \tilde{l} = \frac{l - |m| - 1}{2}$$

for odd-parity modes.

A hand-waving narrative can be used explain the form of eqn. (4.1). The terms proportional to $\tilde{n}$ and $\tilde{l}$ appear by direct analogy to the standard asymptotic expression for p-mode frequencies in a non-rotating star (e.g., Aerts *et al.* 2010, their eqn. 3.216), which contains terms linear in n and l. The term proportional to m^2 accounts for the bulk effects of the centrifugal force, which do not depend on the sign of m. Finally, the $\tilde{\alpha}$ term accounts for the phase of waves at the stellar surface.

4.2. Regularities in g-mode spectra

Ballot *et al.* (2010) explore g modes in rotating stars using a spectral method. Although these authors' focus is primarily on the inadequacies of perturbative approaches in the inertial regime ($\nu > 1$), their Figs. 4 and 5 illustrate quite strikingly that, as with the p modes above, new regularities appear in the g-mode frequency spectrum at rapid rotation rates.

These are a consequence of mode trapping in an equatorial waveguide. When $\nu > 1$ the Coriolis force prevents g modes from propagating outside of the region $|\cos\theta| \leqslant \nu^{-1}$. In the limit $\nu \gg 1$, the trapping can be modeled using an asymptotic treatment of Laplace's tidal equation first developed by Matsuno (1966). The eigenvalue λ (cf. Section 3.4) is found as

$$\lambda \approx \begin{cases} \nu^2(2l_\mu - 1)^2 & l_\mu \geqslant 1, \\ m^2 & l_\mu = 0 \end{cases} \tag{4.2}$$

(e.g., Bildsten *et al.* 1996), where the mode index l_μ counts the number of zonal nodes in the radial displacement eigenfunction. This index is related to the harmonic degree and azimuthal order of the modes' non-rotating counterparts via

$$l_\mu = l - |m|$$

for prograde and axisymmetric ($m = 0$) modes, and

$$l_\mu = l - |m| + 2$$

for retrograde modes.

The co-rotating frequencies of high-order g modes depend on λ via

$$\omega_c \approx \frac{\sqrt{\lambda}}{\pi(n+\alpha)} \int \frac{N}{r} \, \mathrm{d}r.$$

This is just the usual asymptotic expression (e.g., eqn. 3.235 of Aerts *et al.* 2010) with $\sqrt{l(l+1)}$ replaced by $\sqrt{\lambda}$. Combining this with eqn. (4.2) and solving for ω_c yields

$$\omega_c \approx \left[\frac{2\Omega(2l_\mu - 1)}{\pi(n+\alpha)} \int \frac{N}{r} \, \mathrm{d}r \right]^{1/2} \tag{4.3}$$

for modes with $l_\mu \geqslant 1$, and

$$\omega_c \approx \frac{m}{\pi(n+\alpha)} \int \frac{N}{r} \, \mathrm{d}r \tag{4.4}$$

for the $l_\mu = 0$ modes, which are sometimes labeled equatorial Kelvin modes (Townsend 2003a). These Kelvin modes have an azimuthal phase velocity ω_c/m that doesn't depend on m; they may therefore be able to explain the unusual uniformly spaced low-frequency modes seen in some δ Scuti stars (e.g., KIC 8054146; see Breger, these proceedings).

These expressions indicate that a mode multiplet with a given n, l and $-l \leqslant m \leqslant l$ will reorganize itself into $l + 2$ distinct frequencies, corresponding to the permitted indices $0 \leqslant l_\mu \leqslant l+1$. One corresponds to the Kelvin mode $m = l$. Of the remaining $l+1$ distinct frequencies, the lowest corresponds to the $m = l - 1$ mode, the next $l - 1$ are made from pairings between prograde and retrograde modes with the same l_μ, and the final, highest frequency corresponds to the $m = -1$ mode. This is exactly the pattern seen in Figs. 4 and 5 of Ballot *et al.* (2010) in the rapidly rotating limit (note that these authors' sign convention for m is reversed), confirming the analysis here.

4.3. *Inertial modes*

The foregoing discussion focuses on pulsation modes of rapidly rotating stars that have oscillatory counterparts in the non-rotating limit (in other words, ω_c^2 remains greater than zero and deforms continuously as Ω is varied between the two limits). However, the Coriolis force introduces additional classes of 'inertial' mode whose non-rotating counterparts have $\omega^2 \leqslant 0$. Regions with negative N^2 can be stabilized against perturbations with wavenumber $\mathbf{k}$ if

$$N^2 k_\perp^2 > -(2\boldsymbol{\Omega} \cdot \mathbf{k})^2$$

(here, $k_\perp$ is the horizontal component of $\mathbf{k}$). Convective fluid motions are then transformed into oscillatory motions, with the Coriolis force serving as the restoring force.

The Coriolis force likewise transforms the trivial toroidal modes having $\omega^2 = 0$ in the non-rotating limit into r (Rossby) modes with $\omega_c^2 > 0$. The restoring force for r modes does not depend on the stellar structure, but instead comes from conservation of total vorticity (see Saio 1982 for an illuminating discussion). In the slowly rotating limit their frequencies are given by

$$\omega_c = -\frac{2m\Omega}{l(l+1)}.$$

The opposite signs of ω_c and m in this expression tell us that r modes are always retrograde in the co-rotating frame. Toward more rapid rotation the frequencies depart from this formula, acquiring a dependence on the underlying stellar structure. The departures are especially pronounced for the $m = -l$ modes, which behave more like g modes and indeed follow the equatorial waveguide expression (4.3) with $l_\mu = 1$. These 'mixed gravity-Rossby' modes pair up with the $m = l - 1$ modes of frequency multiplets, so the number of distinct frequencies in the multiplets remains unchanged. Due to their g-mode character they can be excited by the κ mechanism, as demonstrated for instance in Townsend (2005).

5. Mode visibilities

Rapid rotation tends to reduce the photometric visibility of oscillations. For p modes with small $l - |m|$ this is a consequence of their transformation into island modes, whose surface amplitudes are appreciable only at mid-latitudes; for g modes, this results from confinement in the equatorial waveguide. As a result, it becomes challenging to detect modes photometrically in rapidly rotating stars — and even if modes *can* be seen, they are subject to strong selection effects (which at low frequencies strongly favor equatorial Kelvin modes; see Townsend 2003b). For an overview of recent developments in this area, see Daszyńska-Daszkiewicz *et al.* (2007) and Reese *et al.* (2013).

6. Differential rotation

So far I've focused on the simple case of uniform rotation. However, there's evidence from a multitude of sources that the internal rotation of stars is differential in r and/or θ (the most well-known example is the Sun; see Thompson *et al.* 2003). Both perturbative and non-perturbative approaches can readily be adapted to handle differential rotation; but here, let's focus on an even simpler analysis.

If the Coriolis and centrifugal forces are neglected, then the only effect of rotation is the Doppler shift in transforming between co-rotating and inertial frames. However, in a differentially rotating star the notion of a global co-rotating frame must be replaced by a continuous sequence of local frames which rotate with angular frequency $\Omega(r, \theta)$. Via equation (3.1), the co-rotating frequency is thus a function of position in the star, and in principle can vanish wherever $m\Omega = \omega$. At these locations, known as 'critical layers' (see Mathis *et al.*, these proceedings), the dispersion relation for low-frequency gravity waves

$$k_r^2 \approx k_\perp^2 \frac{N^2}{\omega_c^2}$$

suggests that the radial component k_r of the wavenumber should diverge. In reality, what will happen is that the wavelength near a critical layer becomes so short that significant radiative dissipation occurs. Thus, critical layers in a differentially rotating star can play a pivotal role in governing mode excitation and damping.

7. Wave transport of angular momentum

Let's now flip the record to the 'B' side, and discuss what impact stellar pulsations might have on their host star. Just as waves transport energy, they also transport momentum. A series of papers by Ando (1981, 1982, 1983) first considered how the angular momentum transported by non-axisymmetric waves alters the internal rotation profile of a star. To model this process we can perform a Reynolds decomposition of the azimuthal momentum equation to find the angle-averaged volumetric torque as

$$\frac{\partial}{\partial t}\left\langle \varpi\overline{\rho v}_\phi \right\rangle = -\frac{1}{4\pi r^2}\frac{\partial}{\partial r}L_J - \frac{\partial}{\partial t}\left\langle \varpi\overline{\rho' v'_\phi} \right\rangle - \left\langle \overline{\rho'\frac{\partial\Phi'}{\partial\phi}} \right\rangle.$$

Here $\varpi \equiv \sin\theta$, while the overline denotes the average over azimuth and the angled brackets the average over co-latitude. The bracketed term on the left-hand side represents the angular momentum in a spherical shell of unit thickness. The first term on the right-hand side is the torque arising from the gradient of the wave angular momentum luminosity L_J; the second term is the rate-of-change of the angular momentum stored in wave motions; and the third term is the gravitational torque.

Focusing on the first term, the angular momentum luminosity (that is, the net amount of angular momentum flowing through a spherical surface in unit time) is given by

$$L_J = 4\pi r^2 \left\langle \varpi\left(\overline{\rho v'_r v'_\phi} + \overline{v}_\phi\overline{\rho' v'_r}\right) \right\rangle, \tag{7.1}$$

to second order in the pulsation amplitude (e.g., Lee & Saio 1993). The first term in the parentheses is the Reynolds stress generated by the radial and azimuthal fluid motions. It vanishes in the case of pure standing waves, because v'_r and v'_ϕ are exactly 90 degrees out of phase. However, departures from this strict phase relation arise when waves acquire a propagative component — either due to non-adiabatic effects, or from wave leakage at the outer boundary. In the non-adiabatic case, the Reynolds stress term for prograde modes transports angular momentum from excitation regions to dissipation regions (vice-versa for retrograde modes; see Ando 1986).

The second term in the parentheses of eqn. (7.1) is the eddy mass flux. Again, this term vanishes for pure standing waves, but becomes non-zero when waves acquire a propagative component. Shibahashi (these proceedings) proposes the intriguing hypothesis that the eddy mass flux of g modes, leaking through the surface layers of Be stars, can transport the angular momentum necessary to build a quasi-Keplerian disk. The transport is particularly effective in the outer layers because the Eulerian pressure perturbation ρ' is large due to the steep density gradient there.

Stellar evolution calculations which include wave transport of angular momentum have so far focused primarily on stochastically excited modes (see, e.g., Talon 2008 and references therein). However, simple estimates of transport by overstable global modes suggest that they can have a significant impact on rotation profiles over timescales which are evolutionarily short (Townsend 2009). Interest in this topic is steadily growing; just this year a number of new papers have appeared exploring topics such as wave transport in massive stars (Rogers *et al.* 2013) and pre-main sequence stars (Charbonnel *et al.* 2013), and the interaction between wave transport and critical layers (Alvan *et al.* 2013).

8. Summary

To summarize this review, I'd like to highlight an encouraging trend. Much of the recent progress in understanding the pulsation-rotation interaction has been driven by numerical simulations. However there have been multiple parallel efforts to develop complementary theoretical narratives for the interaction. These have allowed us to retain a firm grasp on what's really going on in the simulations, and also reassured us that the rapidly rotating limit might not be as difficult to understand as we once thought. Let's ensure this trend does not disappear in the future, by always remembering the wonderful adage by Hamming (1987): *'the purpose of computing is insight, not numbers'*.

Acknowledgements

I acknowledge support from NSF awards AST-0908688 and AST-0904607 and NASA award NNX12AC72G. I'd also like to thank the IAU, and the SOC/LOC, for helping support my travel to this meeting.

References

Aerts, C., Christensen-Dalsgaard, J., & Kurtz, D. W. 2010, *Asteroseismology* (Dordrecht: Springer)

Alvan, L., Mathis, S., & Decressin, T. 2013, *A&A*, 553, A86

Ando, H. 1981, *MNRAS*, 197, 1139

Ando, H. 1982, *A&A*, 108, 7

Ando, H. 1983, *PASJ*, 35, 343

Ando, H. 1986, *A&A*, 163, 97

Ballot, J., Lignières, F., Reese, D. R., & Rieutord, M. 2010, *A&A*, 518, A30

Ballot, J., Lignières, F., Prat, V., Reese, D. R., & Rieutord, M. 2012, *ASP-CS*, 462, 389

Bildsten, L., Ushomirsky, G., & Cutler, C. 1996, *ApJ*, 460, 827

Charbonnel, C., Decressin, T., Amard, L., Palacios, A., & Talon, S. 2013, *A&A*, 554, A450

Cowling, T. G. & Newing, R. A. 1949, *ApJ*, 109, 149

Daszyńska-Daszkiewicz, J., Dziembowski, W. A., & Pamyatnykh, A. A. 2007, *AcA*, 57, 11

Durney, B. & Skumanich, A. 2008, *ApJ*, 152, 255

Eckart, C. 1960, *Hydrodynamics of Oceans and Atmospheres* (Oxford: Pergamon Press)

Hamming, R. W. 1987, *Numerical Methods for Scientists and Engineers* (Dover, New York)

Hough, S. S. 1898, *Phil. Trans. Roy. Soc. Lon.*, 191, 139

Ledoux, P. 1949, *Mem. Soc. R. Sci. Liège*, 4(9), 3

Lee, U. & Baraffe, I. 1995, *A&A*, 301, 419

Lee, U. & Saio, H. 1993, *MNRAS*, 261, 415

Lignières, F. & Georgeot, B. 2008, *Phys. Rev. E*, 78, 016215

Matsuno, T. 1966, *J. Meteorol. Soc. Japan*, 44, 25

Ouazzani, R.-M., Dupret, M.-A., & Reese, D. R. 2012, *A&A*, 547, A75

Reese, D. R., Lignières, F., & Rieutord, M. 2006, *A&A*, 455, 621

Reese, D. R., MacGregor, K. B., Jackson, S., Skumanich, A., & Metcalfe, T. S. 2009, *A&A*, 506, 189

Reese, D. R., Prat, V., Barban, C., van 't Veer-Menneret, C., & MacGregor, K. B. 2013, *A&A*, 550, A77

Rogers, T. M., Lin, D. N. C., McElwaine, J. N., & Lau, H. H. B. 2013, *ApJ*, 772, 21

Saio H. 1982, *ApJ*, 256, 717

Savonije, G. J., Papaloizou, J. C. B., & Alberts, F. 1995, *MNRAS*, 277, 471

Simon, R. 1969, *A&A*, 2, 390

Soufi, F., Goupil, M.-J., & Dziembowski, W. A. 1998, *A&A*, 334, 911

Talon, S. 2008, in: C. Charbonnel & J.-P. Zahn (eds.), *Stellar Nucleosynthesis: 50 years after B2FH*, EAS Publ. Ser., 32, 81

Thompson, M. J., Christensen-Dalsgaard, J., Miesch, M. S., & Toomre, J. 2003, *ARAA*, 41, 599

Townsend, R. H. D. 2003a, *MNRAS*, 340, 1020

Townsend, R. H. D. 2003b, *MNRAS*, 344, 125

Townsend, R. H. D. 2005, *MNRAS*, 364, 573

Townsend, R. H. D. 2009, *AIP-CS*, 1170, 355

Townsend, R. H. D. & Teitler, S. A. 2013, *MNRAS*, 435, 3406

Precision Asteroseismology
Proceedings IAU Symposium No. 301, 2013
J. A. Guzik, W. J. Chaplin, G. Handler & A. Pigulski, eds.

© International Astronomical Union 2014
doi:10.1017/S1743921313014269

Transport of angular momentum in solar-like oscillating stars

Mariejo Goupil[1], Sébastien Deheuvels[2], Joao Marques[3], Yveline Lebreton[4,5], Benoit Mosser[1], Rafa García[6], Kevin Belkacem[1], and Stéphane Mathis[6]

[1] Observatoire de Paris, LESIA, CNRS UMR 8109, F-92190 Meudon, France
email: mariejo.goupil@obspm.fr

[2] Université de Toulouse, UPS-OMP IRAP, CNRS, F-31400 Toulouse, France;
[3] Institut d'Astrophysique Spatiale, UMR 8617, CNRS, Université Paris XI,
91405 Orsay Cedex, France;
[4] Observatoire de Paris, GEPI, CNRS UMR 8109, F-92190 Meudon, France;
[5] Institut de Physique de Rennes, Université de Rennes 1, CNRS UMR 6251,
F-35042 Rennes, France;
[6] Laboratoire AIM Paris-Saclay, CEA/DSM-CNRS-Université Paris Diderot, IRFU/SAp
Centre de Saclay, 91191 Gif-sur-Yvette, France

Abstract. Our current understanding and modeling of angular momentum transport in low-mass stars are briefly reviewed. Emphasis is set on single stars slightly younger that the Sun and on subgiants and red giants observed by the space missions *CoRoT* and *Kepler*.

Keywords. stars: solar-type, stars: oscillations, stars: rotation, stars: evolution, stars: interiors

---

1. Introduction and background

All stars rotate to some extent from their birth throughout their lifetimes. The evolution of angular momentum (AM) therefore is an essential feature of stellar evolution. For a star rotating as a solid body, the total specific AM is $j = k^2 R^2 \Omega$, where Ω is the rotation rate, R is the stellar radius and k is a constant of order unity. When no external torque is applied, the AM of the system is conserved and one must expect an increase of Ω when the star contracts – during the pre-main sequence (PMS) for instance – and a slowing down of expanding stellar regions – such as during the subgiant and red giant phase. Observations however indicate that low-mass stars slow down during most of their lifetime, and develop differential rotation in radius. Contracting regions that are expected to rotate quite fast do not, while expanding regions which ought to rotate slowly rotate faster than expected. This means that AM transport occurs within stars. The issue then is to identify and model the braking processes that cause the internal transport and evolution of stellar AM of low -mass stars. This review is limited to 'isolated' stars (no binaries or effects from planets), and focuses on low-mass stars of $\sim 0.9 - 1.5\, M_\odot$.

The convective regions of one-dimensional stellar models are assumed instantaneously chemically homogenized and in solid-body rotation (but see Potter 2012 and Brun & Palacios 2009). In radiative regions, one needs to solve a time-dependent equation for the local specific AM, $j(r) = r^2 \Omega(r)$ of the form:

$$\frac{\partial j}{\partial t} + \dot{r}\, \frac{\partial j}{\partial r} = -\frac{1}{\rho r^2} \frac{\partial (r^2 \mathcal{F})}{\partial r} + \left(\frac{dj}{dt}\right)_{\text{ext}}, \qquad (1.1)$$

where $r(m)$ is the radius enclosing the mass m, $\dot{r}$ the time derivative of the radius and $\mathcal{F}(r)$ is the angular momentum flux. In order to solve Eq. (1.1), one must specify $\mathcal{F}(r)$, the surface AM losses $(dj/dt)_{ext}$ and the initial condition $j_0(r) = j(r, t = 0)$. The AM flux is the result of several AM transport processes (see Zahn 2007, Talon 2008a, Maeder 2009, and Mathis 2013 for reviews). It is given by $\mathcal{F}(r) = \mathcal{F}_{MC}(r) + \mathcal{F}_{turb}(r) + \mathcal{F}_{IGW}(r) + \mathcal{F}_{B}(r)$ where the currently identified transport processes are:

• Turbulent transport: Turbulence in radiative regions is generated by instabilities of various types (Endal & Sofia 1978, Talon 2008a). Their combined effect is modeled as a diffusive process and contributes to a total turbulent viscosity ν_v. The AM flux then is $\mathcal{F}_{turb} = -\rho r^2 \nu_v (\partial \Omega / \partial r)$. This process transports AM from inner fast rotating layers to outer slower regions and can be dominant in regions with sharp rotation gradients. Several evolutionary codes include the AM transport only as a turbulent diffusion description (see for instance Denissenkov *et al.* 2010, Paxton 2013). The diffusion coefficients for the AM and chemical elements are then calibrated by fitting appropriate observations such as those for the Sun or open clusters. The validity of these calibrated values cannot be general. Besides, turbulent transport as a diffusion process cannot account for all possible types of AM transport.

• Transport by meridional circulation: Large-scale motions are driven by internal stresses, surface AM losses and structural changes. Zahn (1992) proposed to describe the combined effect of advective transport by meridional circulation and the diffusive (differential rotation) shear-induced turbulence as a 1D process (here CMST approach). The basic assumption is that a shear instability, acting on dynamical timescales leads to a large horizontal turbulent viscosity. This causes rapid homogenization on horizontal surfaces, and the rotation rate is a function of radius only (shellular rotation). The meridional circulation AM flux then is $\mathcal{F}_{MC} = -\rho r^2 \Omega U_r / 5$, where U_r is the vertical meridional circulation velocity. Zahn (1992), Maeder & Zahn (1998), and Mathis & Zahn (2004) derived prescriptions for the transport coefficients for the coupled evolutions of AM and chemical elements (see Mathis 2013 for a recent review). Several codes have implemented this approach (Palacios 2013). It does however suffer from several uncertainties in the prescriptions of the transport coefficients (Meynet 2013, Maeder *et al.* 2013).

• Transport by internal gravity waves (IGW) in stellar radiative regions was first discussed by Press (1981) and Schatzman (1993). IGW transport has been proven to be efficient to transport AM and influence the chemical mixing. It is able to make the solar rotation rigid and to reproduce the cool side of the Li dip. The AM flux in that case results in successive AM-extraction propagating fronts from the inner to the outer layers. See reviews such as Talon (2008a) and Palacios (2013). Several open issues on the wave generation and propagation still cast some uncertainties on the quantitative efficiency of this transport for determining the rotation profile in radiative regions (Lecoanet & Quataert 2013, Alvan *et al.* 2013; see also Mathis 2013 and references therein).

• Fossil magnetic fields and hydromagnetic instabilities. A fossil magnetic field is able to make the rotation nearly uniform (Mestel 1953). Its dynamics and its interactions with meridional currents, differential rotation and turbulence are complex (Mathis 2011 and references therein). The result as to whether such a process is able to enforce a rigid rotation in the solar radiative region, for instance, is still debated (Zahn 2009, Strugarek *et al.* 2011, Garaud *et al.* 2013).

• A magnetized wind has long been identified as an important process responsible for AM surface losses (Schatzman 1962, Skumanich 1972). Its description led to a prescription for $(dj/dt)_{ext}$ to be used in stellar models (Kawaler 1988) which involves a proportionality constant, K. The K value is usually set by imposing the rotation period of the present Sun at the age of the Sun. However a realistic description of magnetized wind-driven AM losses remains quite complex and theoretical work is still ongoing to improve the prescription for $(dj/dt)_{ext}$ (Reiners & Mohanty 2012, Matt *et al.* 2012).

• The AM evolution must be coupled to that of the chemical elements. The evolution of the chemical elements is described as a diffusive process with a transport coefficient including impact of the meridional circulation, turbulence, etc. (Zahn 1992, Maeder & Zahn 1998, Pinsonneault 1997, Talon 2008a, Palacios 2013).

Studies of the interactions of these different processes in terms of rotation profiles and mixing processes benefit nowadays from the results of several 3D numerical simulations (for instance Brun & Rempel 2009, Rogers *et al.* 2006).

2. Evolution of stellar AM

The evolution of AM in stellar interiors of low-mass stars from the PMS up to advanced stages has been the subject of many papers, reviews and lectures, for instance Talon (2008a), Maeder (2009), Pinsonneault (2010), Bouvier (2013), Palacios (2013), Mathis (2013), and Montalbán & Noels (2013).

Early stages to TAMS: For a typical dense molecular cloud, the specific AM is $j_{cloud} \sim 10^{17} - 10^{18}$ m^2 s^{-1}. With a radius $R \sim 2 - 3\,R_\odot$ and a rotation period in the range 1 day to 20 days, a typical T Tauri star (age $\sim$10 Myr) has a specific angular momentum $\sim 10^{12} - 10^{13}$ m^2 s^{-1} (assuming rigid rotation), a decrease of 5 orders of magnitude. The AM loss occurs through complex hydromagnetic star-disk interactions (Ferreira 2013). Early on the PMS (at a few Myr), the temperature increase due to contraction in the center of the star causes the appearance of a radiative core, some core-envelope decoupling, and the development of differential rotation with radius.

Gallet & Bouvier (2013) showed the evolution of the surface rotation of stars in several open clusters with ages running from early PMS to mid-MS. The velocity dispersion observed for low-mass stars on the PMS is maintained on the ZAMS ($\sim$10 km s^{-1} up to $150 - 200$ km s^{-1}). This can be explained by the coupling between the magnetized surface of the star and its surrounding disk in the early phase of PMS. This interaction acts as a braking torque on the surface of the star which is forced to rotate at the constant disk rotation rate (disk-locking) as long as the disk does not dissipate, and the disk rotation is taken as the initial condition for the AM evolution of stars when including the PMS evolution. When the disk disappears (at age 5 to 10 Myr), the gravitational contraction of the star leads to an increase of the uniform rotation rate. Once on the ZAMS (roughly $25 - 100$ Myr), the star evolution slows down and angular momentum can be carried away efficiently by magnetized winds. The spin-down timescale is of order of a few tenths of a Gyr, shorter than the nuclear evolutionary timescale. Between 100 Myr and $1 - 2$ Gyr, the rotation depends on the parameters of the disk-locking description and the description of the magnetized wind AM loss. Later, the rotation follows the Skumanich (1972) spin-down law when surface magnetic braking is operating.

At arrival on the ZAMS, the star does not rotate as a solid body, although this is often taken as an initial condition. Using a simplified two-zone model, Gallet & Bouvier (2013) suggest that the slow rotators develop a high degree of differential rotation between the radiative core and the convective envelope whereas the faster ones are nearly in solid-body rotation. This is also found when assuming AM transport by MCST (see Marques *et al.* 2013, Palacios 2013 and references therein). MCST and surface magnetic braking are not able to enforce rigid rotation on the MS (Pinsonneault *et al.* 1989, Matias & Zahn 1998) and the core still rotates faster than the surface. On the other hand, AM transport by IGW has been shown to be very efficient in flattening the rotation profile in radiative zones of solar -type stars at the age of the Sun (Charbonnel & Talon 2005).

Direct seismic probes of the internal rotation profile of low-mass main sequence stars are not currently possible. Apart from rotation information from open clusters, what we have at our disposal are surface rotation periods for thousands of stars from high quality photometric light curves provided by *Kepler*'s and *CoRoT*'s observations (for instance Affer *et al.* 2012, McQuillan *et al.* 2013).

The variations of surface Li abundance and rotation with age can serve to constrain the AM internal transport (Pinsonneault 1989, Talon 2008b). The surface lithium abundance is too depleted for stars on the MS when using standard models with no transport processes other than atomic diffusion and convection. On the other hand, rotationally-induced mixing accounts for the observed destruction of Li on the blue edge of the Li gap. However, the core-envelope coupling of the MCST approach generates a too-large Li depletion for stars on the cool side of the Li dip ($T_{\rm eff} < 6700$ K). The AM transport by IGW, by reducing the meridional circulation, limits the mixing, and hence the Li destruction. IGW transport is efficient for stars with a deep enough outer convective region. It can account very well for the cool side of the Li dip.

It would be desirable to have a few stars that, like the Sun, can serve as calibrators for the various prescriptions used to describe the AM surface loss and transport. *CoRoT* (Michel *et al.* 2008) and *Kepler* (Gilliland 2011) provided high-quality photometric data which resulted in very precise high radial order p-mode oscillation frequencies for a large number of low-mass main sequence stars. From the measurement of individual frequencies and their spacings, and careful stellar modelling, the mass and age of a star can be precisely derived. Together with a precise determination of the surface rotation period, such a star can then be used to calibrate the rotation period–age empirical relation, as has been done for the Sun, and to constrain magnetic braking processes for early MS stars. An example is the star HD 52265 which has been observed with *CoRoT*. The rotation period was also derived to be $P_{\rm rot} = 12.3\pm0.15$ days from the *CoRoT* light-curve variation (Ballot *et al.* 2011). Using seismic constraints (Roxburgh 2005 and references therein), Lebreton & Goupil (in preparation) found a precise age of 2.35 ± 0.25 Gyr for a mass in the range $1.22-1.27\ M_{\odot}$.

Figure 1 displays the surface rotation rate, Ω, as a function of age for HD 52265 and for three types of evolutionary models which correspond to different assumptions about the magnetic braking and AM internal transport prescriptions. The models are computed with the CESTAM code (Marques *et al.* 2013). The solar value of the magnetic braking constant is obtained for a calibrated solar model with a surface rotation period of 27 days at the solar age (here $K_{\odot} = 6.5 \times 10^{47}$ in cgs units). The solar K value is too large for the models to agree with HD 52265 observed values. The seismic constraints impose a very narrow range of possible values for $K \sim (4.06\text{--}4.85) \times 10^{47}$ (cgs) as shown in the inset of Fig. 1. Other parameters such as disk lifetime and period have also some influence (not shown here). The result depends on the physical description of the stellar models, in particular the description of AM transport. For instance, the solar value for K becomes acceptable if one assumes an additional viscosity, $\nu_{\rm add}$, to the canonical prescription for ν_v of the MCST approach (see Sect. 2). The additional viscosity can be seen as mimicking the diffusive transport due to a fossil magnetic field for instance. The adopted value here $\nu_{\rm add} = 3 \times 10^4$ cm^2 s^{-1} is similar to that found by Decressin *et al.* (2009) for MS stars and Eggenberger *et al.* (2012) for a red giant star.

Seismic Constraints Beyond the MS: When the star leaves the main sequence, structural changes occur with the contraction of the inner regions and a huge extension of the envelope. If one assumes local conservation of AM, one must expect a rotation profile increasing toward the interior in the radiative region. When rotationally induced mixing in the radiative region is described by the MCST, it is found that the structural

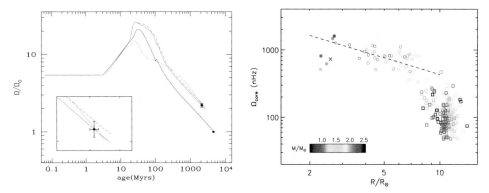

Figure 1. *Left*: Evolution of the surface rotation normalized to the solar surface rotation versus age for $1.265\,M_\odot$ models including MCST and for different values of the magnetic braking constant K (solid magenta highest peak, $K = 4.54 \times 10^{47}$ (in cgs units), dashed magenta, $K = 4.06 \times 10^{47}$). The blue curve (with second-highest peak) represents the evolution for a $1\,M_\odot$ model with MCST and $K = 6.5 \times 10^{47}$ (cgs); the dot indicates the location of the Sun. The cross indicates the position of HD 52265. The orange curve (with the lowest peak) corresponds to $1.265\,M_\odot$ models with $K_\odot$ and additional viscosity (see text). A zoom-in on the location of HD 52265 (black crosses) is shown in the inset. *Right*: Core rotation rate as a function of the seismic stellar radius for subgiants and red giants. The open symbols correspond to the stars that were studied by Mosser *et al.* (2012a) (circles: RGB stars, squares: clump stars). The filled symbols correspond to late subgiants and early red giants (Deheuvels *et al.* 2013) and the cross to a young giant (Deheuvels *et al.* 2012). (Credit: Deheuvels *et al.* 2013.)

changes remain dominant over the MCST due to the short evolution timescales (Palacios *et al.* 2013). Marques *et al.* (2013) show the evolution of the central and surface rotations for a $1.3\,M_\odot$ model computed according to MCST from the PMS to the RGB: beyond the TAMS, the surface rotation decreases while the central rotation sharply increases. A steep rotation gradient at the edge of the H-burning shell develops due to rapid core contraction at the TAMS (see also Palacios *et al.* 2006, Eggenberger *et al.* 2012). While the surface rotation decreases to $\sim 0.1 - 1\,\mu$Hz, the central rotation reaches $100\,\mu$Hz for the subgiant phase and $200\,\mu$Hz for the RG phase (Marques *et al.* 2013).

Before 2006, observational constraints on the internal AM transport of low-mass RG stars were coming from the observed surface abundance anomalies. The red giants have long been known to be pulsating variables. However, it is only in the early 2000s that lower mass RG were suspected to oscillate with solar-like oscillations because of their outer convective region. Nonradial solar-like oscillations for red giants were confirmed by the *CoRoT* (De Ridder *et al.* 2009) and later *Kepler* (Bedding *et al.* 2010).

Subgiants and red giants oscillate with mixed modes (Bedding *et al.* 2010, Mosser *et al.* 2011, 2012a). Mixed modes are trapped in two resonant cavities: they behave as gravity modes in the deep interior where both the Brünt-Väisälä and Lamb frequencies are high. In the outer part the modes are trapped as p modes. In between the modes are evanescent (Unno *et al.* 1989, Dziembowski *et al.* 2001). In stars where they can be excited, mixed modes can probe the inner structure (for instance the convective cores of δ Scuti stars (Dziembowski & Pamyatnykh 1991) or of RG stars (Montalbán *et al.* 2013). Mixed modes can also serve to probe the inner rotation profile (Goupil *et al.* 1996).

When the star evolves off the MS, its change of structure modifies the p and g cavities and therefore modifies the respective number of p- and g-dominated modes in the observed frequency spectrum (Montalbán *et al.* 2010). Observed mixed modes of subgiants are not fully dominated by their g nature and therefore remain sensitive to the surface properties. On the other hand, the red giants are oscillating mostly with g-dominated mixed modes which probe the core. Those properties are responsible for specific frequency patterns in subgiant and RG power spectra.

Seismic information provides mass and radius as for MS stars. In addition, due to the presence of g-dominated modes, one is also able to discriminate between clump stars and giant stars (Bedding *et al.* 2011, Mosser *et al.* 2011). This information is extremely valuable to investigate the evolution of the rotation profile and to provide constraints on the dominant AM transport during evolution.

For slowly rotating stars, the information on the rotation profile comes from the rotational splittings, $\delta = \int_0^R K(r)\Omega(r)dr$, where the rotational kernel $K(r)$ involves the associated eigenfunction for the fluid displacement, and $\Omega(r)$ is the rotation profile at radius r in the star. The (p- or g-dominated) nature of the modes strongly influences the behavior of the rotational splittings as a function of frequency (Goupil *et al.* 2013). The observed rotational splittings of hundreds of giant stars revealed that the rotation of their cores is $5-10$ times larger than the surface rotation (Beck *et al.* 2012, Mosser *et al.* 2012b, Deheuvels *et al.* 2012). For the subgiants we have access to both $\Omega_{\rm core}$ and $\Omega_{\rm env}$, the surface rotation averaged over the acoustic cavity which corresponds mostly to the convective envelope (Deheuvels *et al.* 2013). Figure 1b (from Deheuvels *et al.* 2013) compiles the seismically measured averaged core rotation of subgiants and red giants as a function of the seismically determined stellar radius of the star. The core of the subgiants appears to spin up with evolution, which would mean that the core contraction prevails over AM losses to the envelope. In contrast, at the bottom of the RGB and up the ascent of the RGB, the core rotation of the red giants clearly decreases with evolution. Some efficient AM transport mechanism must operate during these phases (Mosser *et al.* 2012b). The averaged core rotation of the clump stars has decreased further by a factor of six compared to that of RG stars. However some care must be taken when interpreting the rotational splittings in terms of spin up or down. Indeed the rotational splittings provide an average rotation on central layers which might rotate differentially in radius.

The rotational splittings indicate a mean core rotation of the order of one μHz to be compared to the $\sim 100-200\,\mu$Hz typical for subgiant and red giant stellar models at the bottom of the ascending RGB, when assuming AM transport according to MCST. The theoretical rotational splittings are then larger than the observed ones by two orders of magnitude (Eggenberger *et al.* 2012, Marques *et al.* 2013, Ceillier *et al.* 2013).

What could be the origin of the discrepancy? Varying several parameters entering the physical description of stellar models or the chemical composition can decrease the core rotation rate by at most one order of magnitude (Ceillier *et al.* 2013), the most efficient one being an increase of the horizontal turbulent diffusion coefficient by two orders of magnitude (Marques *et al.* 2013). In order to mimic an additional AM transport in the stellar models, Eggenberger *et al.* (2012) add an ad hoc viscosity (representing a yet unknown braking mechanism) in the diffusion coefficient for the AM transport. They found that in order to reproduce the observations for one red giant studied by Beck *et al.* (2012), the value of the additional viscosity is similar to the value required to account for the observed spin-down of MS slowly rotating solar like stars (Denissenkov *et al.* 2010). In an attempt to identify some properties of the missing AM transport processes, Ceillier *et al.* (2013) showed that imposing solid-body rotation on the main sequence only and MCST on the subgiant phase is able to decrease the core rotation by one order of magnitude; again this is not enough to reproduce the observed rotation profiles. The process must also be efficient during the short evolutionary subgiant phase. Two possible missing AM transport processes have so far been identified: a fossil magnetic field associated with hydromagnetic instabilities and internal gravity waves. Both have been shown to be efficient in reducing the sharp rotation gradient in stellar models.

3. Conclusions

We have briefly reviewed our current understanding of internal AM transport in low-mass stars based on recently collected high-quality data and concomitant theoretical developments. The main result is that the cores of red giants rotate much more slowly than expected. This is in agreement with the seismic results for hot white dwarfs which are found to rotate slowly and must then have lost their AM before the white-dwarf phase (Charpinet *et al.* 2009). This nevertheless highlights the need for including additional AM transport processes in current stellar models. Many open questions remain. On the theoretical side, more work is needed to model better the impact of the interaction of the star with its circumstellar environment, the impact of instabilities on AM transport, the prescriptions for dynamical effects, the quantitative importance of magnetic fields and IGW. On the observational side, the goals should be the detection of solar-like oscillations for PMS stars, the detection of seismic individual splittings for MS stars and early post-MS subgiants. One must then go beyond *CoRoT* and *Kepler* results and support the *PLATO* project (Rauer *et al.*, submitted).

References

Affer, L., Micela, G., Favata, F., & Flaccomio, E. 2012, *MNRAS*, 424, 11
Alvan, L., Mathis, S., & Decressin, T. 2013, *A&A*, 553, A86
Beck, P. G., Montalbán, J., Kallinger, T., *et al.* 2012, *Nature*, 481, 55
Ballot, J., Gizon, L., Samadi, R., *et al.* 2011, *A&A*, 530, 97
Bedding, T. R., Huber, D., Stello, D., *et al.* 2010, *ApJ*, 713, 176
Bedding, T. R., Mosser, B., Huber, D., *et al.*, 2011, *Nature*, 471, 608
Bouvier, J. 2013, *EAS Publ. Series*, 62, 143
Brun, A. S. & Palacios, A. 2009, *ApJ*, 702, 1078
Brun, A. S. & Rempel, M. 2009, *Space Sci. Revs*, 144, 151
Ceillier, T., Eggenberger, P., García, R. A., & Mathis, S. 2013, *A&A*, 555, A54
Charbonnel, C. & Talon, S. 2005, *Science*, 309, 2189
Charpinet, S., Fontaine, G., & Brassard, P. 2009, *Nature*, 461, 501
Decressin, T., Mathis, S., Palacios, A., Siess, L., Talon, S., Charbonnel, C., & Zahn, J.-P. 2009, *A&A*, 495, 271
Denissenkov, P. A., Pinsonneault, M., Terndrup, D. M., & Newsham, G. 2010, *ApJ*, 716, 1269
De Ridder, J., Barban, C., Baudin, F., *et al.* 2009, *Nature*, 459, 398
Deheuvels, S., García, R. A., Chaplin, W. J., *et al.* 2012, *ApJ*, 756, 19
Deheuvels, S., *et al.* 2013, *ApJ*, submitted
Dziembowski, W. A. & Pamyatnykh, A. A. 1991, *A&A*, 248, L11
Dziembowski, W. A., Gough, D. O., Houdek, G., & Sienkiewicz, R. 2001, *MNRAS*, 328, 601
Eggenberger, P., Montalbán, J., & Miglio, A. 2012, *A&A*, 544, L4
Endal, A. S. & Sofia, S. 1978, *ApJ*, 220, 279
Ferreira, J. 2013, *EAS Publ. Series*, 62, 169
Garaud, P., Meru, F., Galvagni, M., & Olczak, C. 2013, *ApJ*, 764, 146
Gallet, F. & Bouvier, J. 2013, *A&A*, 556, A36
Gilliland, R. L. 2011, *ASP-CS*, 448, 167
Goupil, M.-J., Dziembowski, W. A., Goode, P. R., & Michel, E. 1996, *A&A*, 305, 487
Goupil, M.-J., Mosser, B., Marques, J. P., *et al.* 2013, *A&A*, 549, A75
Kawaler, S. D. 1988, *ApJ*, 333, 236
Lecoanet, D. & Quataert, E. 2013, *MNRAS*, 430, 2363

Maeder, A. 2009, *Physics, Formation and Evolution of Rotating Stars*, Astronomy and Astrophysics Library (Berlin, Heidelberg: Springer)

Maeder, A. & Zahn, J.-P. 1998, *A&A*, 334, 1000

Maeder, A., Meynet, G., Lagarde, N., & Charbonnel, C. 2013, *A&A*, 553, A1

Màrques, J. P. & Goupil, M. J. 2013, in: M. J. Goupil, K. Belkacem, C. Neiner, F. Lignières, & J. J. Green (eds.), *Studying Stellar Rotation and Convection*, Lecture Notes in Physics, 865, 75

Marques, J. P., Goupil, M. J., Lebreton, Y., *et al.* 2013, *A&A*, 549, A74

Mathis, S. 2011, in: J.-P. Rozelot, & C. Neiner (eds.), *The Pulsations of the Sun and the Stars*, Lecture Notes in Physics, 832, 275

Mathis, S. 2013, in: M. J. Goupil, K. Belkacem, C. Neiner, F. Lignières, & J. J. Green (eds.), *Studying Stellar Rotation and Convection*, Lecture Notes in Physics, 865, 23

Mathis, S. & Zahn, J.-P. 2004, *A&A*, 425, 229

Matias, J. & Zahn, J.-P. 1998, in: J. Provost, & F.-X. Schmieder (eds.), *Sounding solar and stellar interiors*, Proc. IAU Symposium No. 181 (Kluwer Academic), poster volume

Matt, S. P., MacGregor, K. B., Pinsonneault, M. H., & Greene, T. P. 2012, *ApJ*, 754, L26

McQuillan, A., Aigrain, S., & Mazeh, T. 2013, *MNRAS*, 432, 1203

Meynet, G., Ekström, S., Maeder, A., *et al.* 2013, in: M. J. Goupil, K. Belkacem, C. Neiner, F. Lignières, & J. J. Green (eds.), *Studying Stellar Rotation and Convection*, Lecture Notes in Physics, 865, 3

Mestel, L. 1953, *MNRAS*, 113, 716

Michel, E., Baglin, A., Auvergne, M., *et al.* 2008, *Science*, 322, 558

Montalbán, J. & Noels, A. 2013, in: J. Montalbán, A. Noels, & V. Van Grootel (eds.), *Ageing Low Mass Stars: From Red Giants to White Dwarfs*, European Physical Journal Web of Conferences, 43, id. 03002

Montalbán, J., Miglio, A., Noels, A., Scuflaire, R., & Ventura, P. 2010, *ApJ*, 721, L182

Montalbán, J., Miglio, A., Noels, A., Dupret, M.-A., Scuflaire, R., & Ventura, P. 2013, *ApJ*, 766, 118

Mosser, B., Barban, C., Montalbán, J., *et al.* 2011, *A&A*, 532, A86

Mosser, B., Goupil, M. J., Belkacem, K., *et al.* 2012a, *A&A*, 540, A143

Mosser, B., Goupil, M. J., Belkacem, K., *et al.* 2012b, *A&A*, 548, A10

Palacios, A. 2013, *EAS Publ. Series*, 62, 227

Palacios, A., Charbonnel, C., Talon, S., & Siess, L. 2006, *A&A*, 453, 261

Paxton, B., Cantiello, M., Arras, P., *et al.* 2013, *ApJS*, 208, 4

Pinsonneault, M. 1997, *ARAA*, 35, 557

Pinsonneault, M. H. 2010, in: C. Charbonnel, M. Tosi, & F. Primas (eds.), *Light Elements in the Universe*, Proc. IAU Symposium No. 268 (Cambridge, UK: Cambridge University Press), p. 375

Pinsonneault, M. H., Kawaler, S. D., Sofia, S., & Demarque, P. 1989, *ApJ*, 338, 424

Potter, A. T. 2012, Ph.D. Thesis, University of Cambridge

Press, W. H. 1981, *ApJ*, 245, 286

Reiners, A. & Mohanty, S. 2012, *ApJ*, 746, 43

Rogers, T. M., Glatzmaier, G. A., & Jones, C. A. 2006, *ApJ*, 653, 765

Roxburgh, I. W. 2005, *A&A*, 434, 665

Schatzman, E. 1993, *A&A*, 279, 431

Schatzman, E. 1962, *Annales d'Astrophysique*, 25, 18

Skumanich, A. 1972, *ApJ*, 171, 565

Strugarek, A., Brun, A. S., & Zahn, J.-P. 2011, *A&A*, 532, A34

Talon, S. 2008a, *EAS Publ. Series* 32, 8

Talon, S. 2008b, *MemSAIt*, 79, 569

Unno, W., Osaki, Y., Ando, H., Saio, H., & Shibahashi, H. 1989, *Nonradial oscillations of stars* (Tokyo: University of Tokyo Press)

Zahn, J.-P. 1992, *A&A*, 265, 115

Zahn, J.-P. 2007, *EAS Publ. Series*, 26, 147

Zahn, J.-P. 2009, *CoAst*, 158, 27

Precision Asteroseismology
Proceedings IAU Symposium No. 301, 2013
J. A. Guzik, W. J. Chaplin, G. Handler & A. Pigulski, eds.

© International Astronomical Union 2014
doi:10.1017/S1743921313014270

Pulsations of rapidly rotating stars with compositional discontinuities

Daniel R. Reese[1], Francisco Espinosa Lara[2,3], Michel Rieutord[2,3]

[1]Institut d'Astrophysique et Géophysique de l'Université de Liège,
Allée du 6 Août 17, 4000 Liège, Belgium
email: daniel.reese@ulg.ac.be

[2]Université de Toulouse, UPS-OMP, IRAP, Toulouse, France

[3]CNRS, IRAP, 14 avenue Edouard Belin, 31400 Toulouse, France

Abstract. Recent observations of rapidly rotating stars have revealed the presence of regular patterns in their pulsation spectra. This has raised the question as to their physical origin, and, in particular, whether they can be explained by an asymptotic frequency formula for low-degree acoustic modes, as recently discovered through numerical calculations and theoretical considerations. In this context, a key question is whether compositional/density gradients can adversely affect such patterns to the point of hindering their identification. To answer this question, we calculate frequency spectra using two-dimensional ESTER stellar models. These models use a multi-domain spectral approach, allowing us to easily insert a compositional discontinuity while retaining a high numerical accuracy. We analyse the effects of such discontinuities on both the frequencies and eigenfunctions of pulsation modes in the asymptotic regime. We find that although there is more scatter around the asymptotic frequency formula, the semi-large frequency separation can still be clearly identified in a spectrum of low-degree acoustic modes.

Keywords. stars: oscillations, stars: rotation, stars: interiors

1. Introduction

Recent observations of pulsation spectra in rapidly rotating stars have revealed the presence of frequency patterns. For instance, García Hernández *et al.* (2009, 2013) found recurrent frequency spacings in two δ Scuti stars observed by *CoRoT*, thereby allowing the construction of an *echelle* diagram in the latter case. Similarly, Breger *et al.* (2012, 2013) found multiple sequences of very uniformly spaced frequencies in a δ Scuti star observed by *Kepler*. These observations show that although the pulsation spectra of δ Scuti stars lack the *simple* frequency patterns present in solar-type pulsators, regular patterns do exist in such stars and need to be explained.

Among the various possible explanations, one particularly interesting option is the asymptotic frequency pattern for low-degree acoustic modes (i.e. island modes) in rapidly rotating stars, recently discovered through numerical (Lignières *et al.* 2006; Reese *et al.* 2008, 2009) and theoretical considerations (Lignières & Georgeot 2008, 2009; Pasek *et al.* 2011, 2012). Identifying such a pattern in rapidly rotating stars could yield useful information such as the mean density (Reese *et al.* 2008; García Hernández *et al.* 2013). However, an open question is to what extent the pattern is affected by strong gradients or glitches (such as μ gradients, ionisation zones, or boundaries of convective regions), and whether this can hinder its identification. In order to answer this question, we investigate the pulsation spectra of rapidly rotating models with sharp discontinuities.

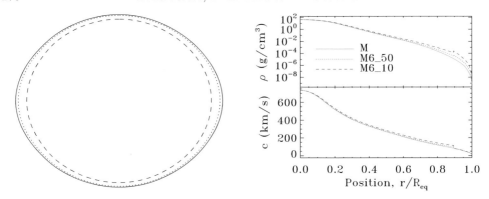

Figure 1. *Left:* Meridional cross-sections of discontinuities in models M6_xx (dashed line) and M7_xx (dotted line), and stellar surface (solid line). *Right:* Density and sound velocity profiles for three of the models.

2. Numerical calculations

We worked with various 3 $M_\odot$ models, where the surface is rotating at 70 % of the Keplerian break-up rotation rate ($v_{eq} = 340 - 350$ km s^{-1}). These models were produced by the 2D multi-domain spectral code ESTER, which self-consistently calculates the rotation profile, Ω (Rieutord & Espinosa Lara 2009, 2013; Espinosa Lara & Rieutord 2013). Its multi-domain approach is well-suited to introducing discontinuities without sacrificing numerical accuracy, since these can be made to coincide with domain boundaries. In what follows, we worked with five different models: M which is smooth, M6_50, M6_10, M7_50, and M7_10. Models M6_xx have a discontinuity deeper within the star (see Fig. 1, left panel). In all cases, the discontinuities follow isobars. The surface hydrogen content is decreased by 50% and 90% in models Md_50 and Md_10, and corresponds to a 17% and 39% jump in the speed of sound, respectively (see Fig. 1, right panel).

Adiabatic calculations of acoustic pulsation modes were carried out thanks to the TOP code which fully takes into account the effects of rotation (Reese *et al.* 2009). Regularity conditions were applied in the centre, the simple mechanical condition $\delta p = 0$ was enforced at the stellar surface, and the perturbation to the gravity potential was made to vanish at infinity. Various matching conditions were needed to ensure that the perturbation of the pressure, the gravity potential, and its gradient, remain continuous across the *perturbed* discontinuity. Furthermore, the fluid domain had to be kept continuous by making sure that the deformation caused by the fluid displacement is the same below and above the discontinuity. Similar calculations had previously been carried out in Reese *et al.* (2011). However, these calculations did not take into account the fact that the matching conditions apply across the *perturbed* discontinuity, and the results were less conclusive because the discontinuity was located deeper within the star, where acoustic island modes are less sensitive.

3. Results

We first turn our attention to the effects of discontinuities on the eigenfunctions. Figure 2 shows the meridional cross-section of an island mode as well as the sound velocity and mode profile along a heuristically determined path. As can be seen in the right panel, the discontinuity modifies the wavelength as well as the amplitude of the oscillations. Further tests confirm that the wavelength scales with the sound velocity.

Another effect which has already been pointed out by Reese *et al.* (2011) is a slight deviation of the mode at the discontinuity.

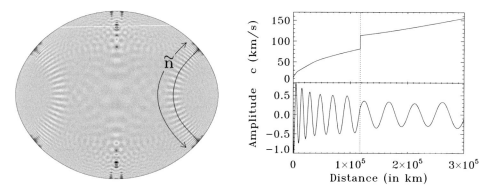

Figure 2. *Left:* Meridional cross-section of an island mode. *Right:* Sound velocity and mode profile along the path shown in the left panel (only part of the profile is shown for legibility).

At low rotation rates, discontinuities affect the frequencies by superimposing an oscillatory pattern over the usual frequency spectrum (e.g. Monteiro *et al.* 1994). A similar effect takes place here, as illustrated by the semi-large frequency separations shown in Fig. 3, although the oscillatory pattern is less regular. One can also calculate the scatter between the numerical frequencies and a simplified version of the asymptotic formula (see Eq. (27) of Reese *et al.* 2009). The scatter, $\langle(\nu_{\mathrm{asymp.}} - \nu)^2\rangle^{1/2}/\Delta_{\tilde{n}}$, ranges from 0.0143 for model M to 0.0436 for model M7_10. Even in the best case, the scatter is more than an order of magnitude larger than the scatter obtained around the main sequence of equidistant frequencies found in Breger *et al.* (2012), thereby supporting the conclusion that this sequence is not caused by an asymptotic behaviour.

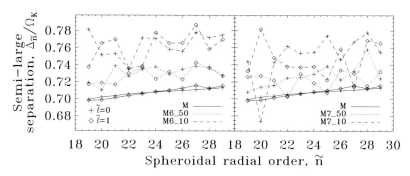

Figure 3. Semi-large frequency separation, $\Delta_{\tilde{n}} = \nu_{\tilde{n}+1} - \nu_{\tilde{n}}$, for axisymmetric ($m = 0$) modes, as a function of $\tilde{n}$, the spheroidal radial order (see Fig. 2, left panel).

One can then investigate whether it is possible to recover the semi-large frequency separation, $\Delta_{\tilde{n}}$. Figure 4 shows histograms of frequencies differences for three models. The lightly shaded areas show all frequency differences, whereas the dark areas show the frequency differences from modes with adjacent $\tilde{n}$ values and the same $(\tilde{l}, m)$ values. The upper row is based on the original numerical frequencies. In all cases, the semi-large frequency separation $\Delta_{\tilde{n}}$ shows up clearly. However, it turns out that the rotation rate is close to $\Delta_{\tilde{n}}$ thereby amplifying the signal, due to island mode multiplets (Pasek *et al.* 2012). This can be seen by comparing the light and dark regions in the histograms. In

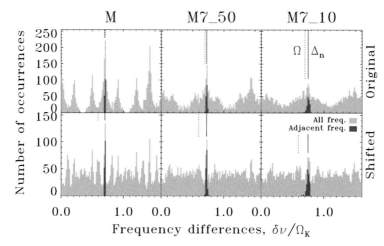

Figure 4. Histograms of frequency differences for three models (see text for details).

the lower row, the frequencies were shifted by $0.1m$, thereby mimicking a lower rotation rate. Even in this situation, a peak remains at $\Delta_{\tilde{n}}$ for all three models.

In conclusion, although discontinuities lead to more scatter around the asymptotic behaviour of island modes and may complicate mode identification, they are unable to mask features such as the semi-large frequency separation. Hence, the asymptotic formula remains a viable explanation for stars such as those observed by García Hernández et al. (2009, 2013).

Acknowledgements

DRR is financially supported through a postdoctoral fellowship from the "Subside fédéral pour la recherche 2012", University of Liège. FEL and MR acknowledge the support of the French Agence Nationale de la Recherche (ANR), under grant ESTER (ANR-09-BLAN-0140).

References

Breger, M., Fossati, L., Balona, L. A., et al. 2012, ApJ, 759, 62
Breger, M., Lenz, P., & Pamyatnykh, A. A. 2013, ApJ, 773, 56
Espinosa Lara, F. & Rieutord, M. 2013, A&A, 552, A35
García Hernández, A., Moya, A., Michel, E., et al. 2009, A&A, 506, 79
García Hernández, A., Moya, A., Michel, E., et al. 2013, A&A, 559, A63
Lignières, F. & Georgeot, B. 2008, Phys. Rev. E, 78, 016215
Lignières, F. & Georgeot, B. 2009, A&A, 500, 1173
Lignières, F., Rieutord, M., & Reese, D. 2006, A&A, 455, 607
Monteiro, M. J. P. F. G., Christensen-Dalsgaard, J., & Thompson, M. J. 1994, A&A, 283, 247
Pasek, M., Georgeot, B., Lignières, F., & Reese, D. R. 2011, Phys. Rev. Letters, 107, 121101
Pasek, M., Lignières, F., Georgeot, B., & Reese, D. R. 2012, A&A, 546, A11
Reese, D. R., Lignières, F., & Rieutord, M. 2008, A&A, 481, 449
Reese, D. R., MacGregor, K. B., Jackson, S., Skumanich, A., & Metcalfe, T. S. 2009, A&A, 506, 189
Reese, D. R., Espinosa Lara, F., & Rieutord, M. 2011, in: C. Neiner, G. Wade, G. Meynet, & G. Peters (eds.), Active OB stars: structure, evolution, mass loss, and critical limits, Proc. IAU Symposium No. 272 (Cambridge: Cambridge University Press), p. 535
Rieutord, M. & Espinosa Lara, F. 2009, CoAst, 158, 99
Rieutord, M. & Espinosa Lara, F. 2013, Lecture Notes in Physics, 865, 49

Precision Asteroseismology
Proceedings IAU Symposium No. 301, 2013
J. A. Guzik, W. J. Chaplin, G. Handler & A. Pigulski, eds.

© International Astronomical Union 2014
doi:10.1017/S1743921313014282

A working hypothesis about the cause of Be stars: Episodic outward leakage of low-frequency waves excited by the iron-peak κ-mechanism

Hiromoto Shibahashi

Department of Astronomy, University of Tokyo, Tokyo 113-0033, Japan
email: `shibahashi@astron.s.u-tokyo.ac.jp`

Abstract. Observations indicate that a circumstellar disk is formed around a Be star while the stellar rotation is below the break-up velocity. I propose a working hypothesis to explain this mystery by taking account of the effect of leaky waves upon angular momentum transfer.

In B-type stars near the main sequence, low-frequency nonradial oscillations are excited by the κ-mechanism in the iron bump. They transport angular momentum from the driving zone to the surface. As a consequence, the angular momentum is gradually deposited near the stellar surface. This results in a gradual increase in the "critical frequency for g-modes", and g-modes eventually start to leak outward, long before the surface rotation reaches the break-up velocity. This leads to a substantial amount of angular momentum loss from the star, and a circumstellar disk is formed. The oscillations themselves will be soon damped owing to kinetic energy loss. Then the envelope of the star spins down and angular momentum loss stops soon. The star returns to being quiet and remains calm until nonradial oscillations are newly built up by the κ-mechanism to sufficient amplitude and a new episode begins.

According to this view, the interval of episodic Be-star activity corresponds to the growth time of the oscillation, and it seems in good agreement with observations.

Keywords. stars: emission-line, Be, stars: oscillations, stars: rotation

---

1. Motivation

Some B-type stars near the main sequence show Balmer line emissions, and they are classified as "Be stars" (Collins 1987). Sharp absorption lines often appear in what is known as a "shell" spectrum, indicating the presence of a cool thin disk surrounding the star. The line shapes and intensities, or even the presence of emission lines, vary with a timescale of the order of decades. It is widely accepted that this kind of activity is preceded by episodic mass loss, occurring quasi-periodically at intervals ranging from several years to some decades, from the equatorial region of the star, and forming a cool gas disk around the star (see, e.g., the review by Porter & Rivinius (2003) and references therein). In the active mass-eruption phase, Be stars show Balmer line emissions. In the quiescent phase, they show normal B-type spectra.

The mechanism of the episodic mass loss in Be stars is, however, as yet unknown. Previously, Be stars were presumed to rotate at their critical break-up velocity, over which the centrifugal force exceeds the gravity at the equator, and to eject their mass from the equatorial zone (Struve 1931). Indeed, the spectroscopically observed rotation velocities of Be stars are higher than those of normal B-type stars. However, careful analyses indicate that their velocities are not high enough to reach the break-up limit (Porter 1996; Frémat *et al.* 2005), though this conclusion is still controversial (e.g., Townsend *et al.* 2004). Some additional mechanisms for the angular momentum transport have been examined to bring the already rapid stellar rotation to its critical value at the surface, and allows the star to eject material. Such attempts are nevertheless incompatible with the observational fact that the rotational Doppler velocities of Be stars are lower than the break-up limit.

The view that rapid rotation of Be stars is likely to be responsible for episodic mass loss presumes that the rapid rotation is an intrinsic characteristic of these stars. It does not attempt to explain the essential question of why Be stars are more rapidly rotating than normal B-type stars. It sidesteps the issue and is then far from a logical explanation. We should go behind the outward form to grasp the inner meaning of the phenomenon. It is uncertain yet even whether rapid rotation of Be stars is the cause of the Be phenomenon or rather the consequence of them.

2. Mystery of angular momentum transfer

Be stars eject their mass, while their surface rotation is not high enough to reach the break-up limit. The essential problem in Be stars is how to transfer angular momentum from the inside of the star to the outer atmosphere by keeping the stellar rotation below the critical break-up velocity. At first glance, this appears to be a quandary. The problem is then how to overcome this dilemma. Nonradial g-modes have been regarded as a promising mechanism to redistribute the angular momentum in stars in general (Unno *et al.* 1989, see also recent more comprehensive reviews, e.g., Lee 2008, Mathis & Alvan 2013). In fact, non-axisymmetric nonradial oscillation modes work to transfer angular momentum in a star from the region where the oscillations are excited to other regions where dissipation works to damp the oscillations. As for Be stars, some attempts were made to see whether the nonradial g-modes efficiently work to accelerate the stellar surface rotation up to the break-up velocity (Ando 1983, 1986; Osaki 1986, 1999; Lee & Saio 1993; Lee 2008, 2013; Cranmer 2009; Neiner 2013). However, as far as we accept the aforementioned observational fact, we think that it is beside the point to consider the case that the stars rotate at the break-up limit.

3. Angular momentum transfer by leaky gravity waves

3.1. *Outwardly leaky gravity waves*

In this paper, we propose a mechanism of angular momentum transfer to the circumstellar environment without much deposition of angular momentum into the layers below the photosphere. Such a situation becomes possible when the nonradial gravity waves become leaky outward from the photosphere. However, this possibility has not been taken into account, while the presence of g-modes with leakage in B-type stars was discussed in detail by Townsend (2000a,b). Instead, in previous studies concerning the angular momentum transfer, the waves were thought to be reflected near the stellar surface to make the oscillation modes a standing wave. Such a treatment of the reflective surface boundary is justified if the oscillation frequency is higher than the critical frequency for g-modes at the surface. However, the critical frequency for g-modes is dependent on the stellar rotation speed, and the reflective boundary condition is not necessarily applicable.

3.2. *Pulsations and Be stars*

First of all, we suppose that non-axisymmetric nonradial g-modes are excited in B type stars in a certain effective temperature range due to the iron-peak κ-mechanism.

Pulsations in Be stars were often thought to be a key factor to understand the physical cause of Be stars. Many Be stars show photometric variability which is interpreted as low-degree g-modes, while some others show line-profile variation which is interpreted to be caused by intermediate-degree g-modes. Those g-modes of either low or intermediate degrees are most probably excited by the iron-peak κ-mechanism. It was often claimed, however, that pulsations were not seen in all the Be stars and then pulsations were unlikely to be responsible for the Be phenomena (e.g., Balona 2003).

In our view, any non-axisymmetric, nonradial, low-frequency modes, whether they are detectable or not (in other words, whether or not they are low- or intermediate-degree modes), can transport the angular momentum into the circumstellar environment. The leaky wave conditions are more easily realized in the case of high-degree modes, which are hard to detect. Hence, lack of detected pulsations is off the point.

3.3. *Gravity modes in a rotating star*

These modes transfer the angular momentum to the stellar surface, and the rotation at the surface is gradually accelerated, though still below the break-up speed. In a rotating star, the Coriolis force modifies significantly the oscillation characteristics of the star, particularly those of low-frequency modes. It is known that, with the increase of the rotation frequency, the eigenfunctions are concentrated in low latitudinal zones. This means that the horizontal wavelength becomes gradually shorter with the increase of rotation frequency, that is, the horizontal wavenumber becomes higher near the stellar surface.

Propagation features of nonradial oscillations can be understood with the local dispersion relation, in which the Lamb frequency and the Brunt-Väisälä frequency appear, to a good approximation, as the critical frequencies. A gravity wave does not propagate in radial directions and is evanescent if the frequency is between these two critical frequencies. If waves become evanescent at the stellar surface, they are reflected inward and can become standing waves. The Brunt-Väisälä frequency is high at the stellar surface except within convective zones, while the Lamb frequency being proportional to the sound speed is quite low due to the low temperature near the surface, and then the gravity waves are usually standing waves.

However, this situation is different when the star begins to rotate substantially due to acceleration caused by angular momentum transfer of g-modes. The horizontal wavelength becomes short, and, as a consequence, the Lamb frequency, being proportional to the horizontal wavenumber, becomes high. Then high-order g-modes begin to propagate and leak outward (Shibahashi & Ishimatsu 2012). For more detailed analyses, see Ishimatsu & Shibahashi (2013).

3.4. *On-off mechanism of the valve for the leaky waves*

With the increase of stellar rotation frequency, the critical frequency for g-modes becomes higher, and then high-order g-modes eventually become leaky outward, though the stellar rotation is still below the break-up speed. Once g-modes become leaky, the angular momentum is transferred into the circumstellar environment to form a disk around the star. Since angular momentum is not deposited at the photosphere by leaky waves but transferred through it, the aforementioned apparent dilemma is solved. On the other hand, the wave energy is suddenly lost and then the oscillations are damped soon. The stellar surface rotation slows down, and this shuts the valve for the leaky wave and the reflective boundary is recovered. The star remains quiet until new nonradial oscillations are built up by the κ-mechanism to sufficient amplitude. The circumstellar disk, once formed, gradually loses its angular momentum and the matter will eventually fall back to the star. All the above processes recur, and the timescale of this cycle is governed by the growth rates of g-modes and it is of the order of a decade. In this view, the fairly rapid rotation of Be stars is not a direct cause for the Be phenomena, but an inevitably induced consequence of presence of non-axisymmetric nonradial g-modes excited by the iron-peak κ-mechanism in the process of Be phenomena.

3.5. *Is rapid rotation the main cause of Be phenomena?*

It is certain that Be stars rotate more rapidly than normal B-type stars. It was suspected that mass loss would be a result of the star rotating at critical rotational velocity since Struve (1931). In the view described in this paper, the rapid rotation of Be stars is neither an intrinsic feature nor the cause of Be phenomena, but an inevitable consequence of angular momentum transfer by g-modes excited by the iron bump κ-mechanism.

The amount of angular momentum transferred into the upper atmosphere is highly dependent on the pulsation amplitude. Even if any single mode would not be enough to carry the angular momentum above the photosphere, the total sum of the contributions of many self-excited modes would be substantial. This implies that, during the process of accelerating the stellar rotation by low-frequency g-modes, the angular momentum would be transferred into the upper atmosphere above the photosphere to form a circumstellar disk long before the surface rotation reaches the break-up limit.

References

Ando, H. 1983, *PASJ*, 35, 343

Ando, H. 1986, *A&A*, 163, 97

Balona, L. A. 2003, *ASP-CS*, 305, 263

Collins, G. W., II 1987, in: A. Slettebak & T. P. Snow (eds.), *Physics of Be stars*, Proc. IAU Colloqium No. 92 (Cambridge & New York: Cambridge University Press), p. 3

Cranmer, S. R. 2009, *ApJ*, 701, 396

Frémat, Y., Zorec, J. Hubert, A.-M., & Floquet, M. 2005, *A&A*, 440, 305

Ishimatsu, H. & Shibahashi, H. 2013, *ASP-CS*, 479, 325

Lee, U. 2008, *CoAst*, 157, 203

Lee, U. 2013, *ASP-CS*, 479, 311

Lee, U. & Saio, H. 1993, *MNRAS*, 261, 415

Mathis, S. & Alvan, L. 2013, *ASP-CS*, 479, 295

Neiner, C., Mathis, S., Saio, H., & Lee, U. 2013, *ASP-CS*, 479, 319

Osaki, Y. 1986, *PASP*, 98, 30

Osaki, Y. 1999, in: B. Wolf, O. Stahl, & A. W. Fullerton (eds.), *Variable and Non-spherical Stellar Winds in Luminous Hot Stars*, Proc. IAU Colloqium No. 169, Lecture Notes in Physics, 523, 329

Porter, J. M. 1996, *MNRAS*, 280, L31

Porter, J. M. & Rivinius, T. 2003, *PASP*, 115, 1153

Shibahashi, H. & Ishimatsu, H. 2013, *Astrophys. Space Sci. Proc.*, 31, 49

Struve, O. 1931, *ApJ*, 73, 94

Townsend, R. H. D. 2000a, *MNRAS*, 318, 1

Townsend, R. H. D. 2000b, *MNRAS*, 319, 289

Townsend, R. H. D., Owocki, S. P., & Howarth, I. D. 2004, *MNRAS*, 350, 189

Unno, W., Osaki, Y., Ando, H., Saio, H., & Shibahashi, H. 1989, *Nonradial Oscillations of Stars* (2nd Edition) (Tokyo: University of Tokyo Press)

Precision Asteroseismology
Proceedings IAU Symposium No. 301, 2013
J. A. Guzik, W. J. Chaplin, G. Handler & A. Pigulski, eds.

© International Astronomical Union 2014
doi:10.1017/S1743921313014294

Pulsation – convection interaction

F. Kupka, E. Mundprecht, H. J. Muthsam

Faculty of Mathematics, University of Vienna, Oskar-Morgenstern-Platz 1,
A-1090, Vienna, Austria

Abstract. A lot of effort has been devoted to the hydrodynamical modelling of Cepheids in one dimension. While the recovery of the most basic properties such as the pulsational instability itself has been achieved already a long time ago, properties such as the observed double-mode pulsation of some objects and the red-edge of the classical instability strip and their dependence on metallicity have remained a delicate issue. The uncertainty introduced by adjustable parameters and further physical approximations introduced in one-dimensional model equations motivate an investigation based on numerical simulations which use the full hydrodynamical equations. In this talk, results from such two-dimensional numerical simulations of a short period Cepheid are presented. The importance of a carefully designed numerical setup, in particular of sufficient resolution and domain extent, is discussed. The problematic issue of how to reliably choose fixed parameters for the one-dimensional model is illustrated. Results from an analysis of the interaction of pulsation with convection are shown concerning the large-scale structure of the He II ionization zone. We also address the influence of convection on the atmospheric structure. Considering the potential of hydrodynamical simulations and the wealth of ever improving observational data an outlook on possible future work in this field of research is given.

Keywords. convection, stars: oscillations, stars: variables: Cepheids, stars: variables: RR Lyrae

1. Motivation

Astrophysics has enormously benefitted from the advent of large scale surveys and from ever improving high precision measurements. The data on Cepheids observed in the OGLE-III field of the LMC (Soszyński *et al.* 2008), for instance, comprises a unified sample of the observed distribution of various types of Cepheid pulsators. This set of data is capable of challenging theoretical models well beyond what had been possible when discrepancies between first generation models and observational data on the famous pulsational mass problem had become evident (Stobie 1969). Likewise, the identification of non-radial modes in the classical pulsator V445 Lyr, an RR Lyr type star (Guggenberger *et al.* 2012), widens future possibilities to constrain the models of such types of stars.

Although these developments are intriguing from the viewpoint of hydrodynamics and stellar physics, the non-specialist might be tempted to ask: how is it possible that classical pulsators such as Cepheids, after all those years of research, are still of interest to astrophysics in general? The most straightforward answer is that Cepheids are one of the most crucial parts of the cosmic distance ladder (cf. de Grijs 2011). Their precise quantitative description holds the key to them becoming an instrument of simultaneous precision measurement of distance and chemical composition in galaxies beyond the Local Group.

Well-known problems of research on Cepheids that keep reappearing include the already introduced discrepancies found when comparing masses from stellar evolution with those resulting from stellar pulsation modelling (Keller 2008, Pietrzyński *et al.* 2010, Cassisi & Salaris 2011) and explanations to the location of double-mode (and even triple-mode) pulsators (cf. the data discussed in Soszyński *et al.* 2008). 1D models that have previously been claimed to have resolved this question (Kolláth *et al.* (1998) and Feuchtinger (1998) for RR Lyr stars; cf. the critical discussion in Buchler (2009)) eventually turned out only to have done so by chance, lacking a solid physical justification for their success (see the general discussion in Smolec & Moskalik (2008a) and the work on double-mode Cepheids in Smolec & Moskalik (2008b) as well as the discussion of beat Cepheids in Smolec & Moskalik (2010)). This is unfortunate also since the period ratios of such objects are sensitive to metallicity Z (cf. Buchler & Szabó 2007) which would allow their use for simultaneous metallicity and distance determinations in more distant galaxies. This is even more so since similar investigations exist now, e.g., for short-period classical Cepheids (Klagyvik *et al.* 2013). It thus appears natural to question whether we can trust one dimensional (1D, averaged over spherical shells) models to interpret the relations of period P, luminosity L, colour, and metallicity Z when convection is important.

The main challenges to previous modelling of 1D hydrodynamical pulsation and stellar evolution originate from their dependence on variants of non-local mixing-length theory (MLT) which entails up to 8 free parameters (Buchler & Kolláth 2000, Smolec & Moskalik 2008b) that are either calibrated or guessed while their values determine the model predictions themselves such as the location of red edge of the instability strip of Cepheids. The latter, given here as an example, depends on turbulent pressure and thus its modelling which requires the specification of closure parameters. Likewise, the parameter determination with spectroscopy relies mostly on the assumption of static, homogeneous atmospheres (e.g. Luck *et al.* 2013, Takeda *et al.* 2013). One might wonder whether that at least approximately holds and whether any systematic differences might be introduced this way.

All these problems motivate the attempt to perform numerical hydrodynamical simulations of classical pulsators (see also the work of Gastine & Dintrans 2011a and Gastine & Dintrans 2011b on such simulations for idealized microphysics and the work of Geroux & Deupree 2011 and Geroux & Deupree 2013 on simulations of RR Lyrae stars in this context). Initially, such work may be restricted to two spatial dimensions (2D), while in the long-term three dimensional (3D) simulations probably will prevail.

2. The ANTARES code

A recently developed tool to perform such simulations is ANTARES (A Numerical Tool for Astrophysical RESearch), a general purpose Fortran95 code for hydrodynamical simulations (for details see Muthsam *et al.* 2010). Its development has been initiated by H.J. Muthsam and has mostly been done at the Faculty of Mathematics at Univ. of Vienna, Austria and during earlier stages also at the Max-Planck-Institute for Astrophysics in Garching, Germany. More recently, development work has also occurred at the BTU Cottbus, Germany, and at the Lab. d'Astrophys. Toulouse, France. Presently, the code is available to developers and direct collaborators and consists of about 150,000 lines of code in active modules. It is a modular, fully MPI parallelized program (with optional OpenMP parallelization for load balance) and has been demonstrated to scale up to more than 1 000 CPU cores (see Happenhofer *et al.* 2013). Post-processing of simulation results is mostly done with the Paraview system, available at http://www.paraview.org/, but also with separate statistics programs operating on the output data.

3. Numerical challenges

The numerical challenges ensuing from any attempt to pursue a 2D or 3D hydrodynamical simulation of a classical pulsator are best illustrated by an example. To this end we model a Cepheid with the following basic stellar (and simulation) parameters. We assume a $T_{\text{eff}} = 5125$ K, a mass of $M = 5\,M_\odot$ as well as a radius of $R \sim 38.5\,R_\odot$. This results in a luminosity of $L \sim 913\,L_\odot$ and a surface gravity of $\log(g) \sim 1.97$. For the chemical composition we assume $X = 0.7$, $Y = 0.29$, $Z = 0.01$ and the GN93 mixture (Grevesse & Noels 1993). In the simulations the LLNL equation of state (Rogers *et al.* 1996, Rogers & Nayfonov 2002) is used in combination with OPAL opacities (Iglesias & Rogers 1996) to compute a model with ANTARES for realistic microphysical conditions. Since the surface layers, for which a radiative transfer solver is used, extend to a temperature range below the limits of the OPAL tables, the latter are supplemented by opacity data from Ferguson *et al.* (2005).

For a Cepheid with these parameters pulsations in the fundamental mode with a period of $P = 3.85$ d are eventually found in the model simulations discussed in the following. In these simulations only the outer 42% of the radius is modelled with a typical vertical grid spacing of 0.47 Mm near the surface and 124 Mm in the interior (modelling of only the outer 42% implies that P is somewhat too short in comparison with 1D models that extend into the stellar core region). For these conditions, a 1D stellar structure model has kindly been provided to us by G. Houdek which can be used to start the simulations.

The computational concept underlying the following work is based on the idea to simulate the flow in a wedge with a fixed opening angle. Azimuthally this wedge is open for in- and outflow and periodic boundary conditions are assumed in that direction. An integer multiple of such wedges constitutes a closed ring located at the stellar equator (thus, there is no flow in polar direction, but the geometry assumed for the radiative transfer is that of a fully three dimensional configuration (cf. Mundprecht 2011, Mundprecht *et al.* 2013 for further details). Vertically (i.e. along the radial direction) the boundaries are considered to be closed (a very recent development is an open boundary at the top though this is not used for the simulations discussed in the following). Due to self-gravity of the star there is an extreme stratification with respect to density (and pressure) along the radial direction. This is particularly pronounced in the stellar atmosphere layers. To handle this problem it has hence been necessary to introduce a radially stretched grid which is co-moving with the mean radial, pulsational velocity (Mundprecht 2011, Mundprecht *et al.* 2013).

Several grid and geometry configurations have been considered more closely: a model with an opening angle of 1° and 800×300 points (radially and azimuthally, respectively, here and in the following). This model is suitable to resolve the zone of hydrogen ionization and thus also the photospheric layers. Secondly, a model with an opening angle of 10° and 510×800 points has been constructed to study the dynamical behaviour of the He II ionization zone. For a satisfactory simulation of both regions simultaneously a model with 3° opening angle and grid refinement has been constructed as well (further details on these models are given in Mundprecht *et al.* 2013).

We summarise first some results of this work on the numerical requirements. One important realization has been that the number of grid points cannot be scaled from solar simulations. For the similar case of A-type stars this has already been discussed in Kupka *et al.* (2009). Instead, a much higher resolution is required to resolve the divergence of the radiative flux, Q_{rad}, and thus the radiative cooling in the H I ionization zone. This was concluded from 1D models of various resolutions in which case an 8 to 16 times finer grid spacing is needed. Even if the smoothing effect of a convective rather than radiative structure is considered, the resolution has to be at the very least four times higher which consequently places more restrictive conditions on the permitted time step. An increased number of grid points is inevitably required in both the radial and the azimuthal direction to avoid artificial numerical viscosity along the azimuthal direction and to resolve layers temporarily inclined with respect to the radial direction which altogether increases the computational costs. To illustrate the practical implications of a poorly resolved calculation Mundprecht *et al.* (2013) show how light curves computed from an insufficient default resolution and from a four times higher resolution differ in the case of a 1D radiative Cepheid model (no non-local MLT used to model the convective flux and turbulent pressure). Indeed, artifacts (a high frequency noise and a shift of the intensity level) are introduced in the lightcurve computation from this simulation. They are caused by the insufficient resolution and cannot be removed by simply averaging the predicted lightcurve in time.

As can be illustrated with a model with a 3×4 increased resolution obtained through a grid refinement zone (triple resolution radially and quadruple resolution azimuthally), which ranges from the top region of the model to optical depths slightly above 100, the photosphere becomes much more dynamical. The maximum convective flux in the H I ionization zone increases by factors of five to eight. The price to pay in this case is a much smaller radiative time scale where the most stringent restrictions are due to layers just underneath the H I ionization zone. The grid refinement zone thus has to be placed such that it ends closely to the bottom of this zone to reduce the computational costs.

4. Pulsation–convection interaction

The models with different opening angles and thus different resolution can be used for a variety of investigations. High resolution models such as the 1° model by Mundprecht *et al.* (2013) are suited for studying the temporal development of the H I ionization zone and the dynamics of the photosphere. Models with large opening angle such as the 10° model by the same authors can be used to investigate the development of convection in the He II ionization zone, in particular with respect to phase shifts and the interaction between convection and pulsation. Moreover, they can be used to probe models for the computation of F_{conv} and determine, for instance, one of the model parameters of the Kuhfuß and Stellingwerf models. The simulations are also useful to directly compute work integrals and study their behaviour as a function of phase and location within and outside the convection zone.

In the following subsection we will discuss global properties of the He II convection zone by considering a whole annulus instead of a sector of some relatively small opening angle. Subsequently, we will provide some results regarding the structure of the atmosphere of our model Cepheid.

4.1. *The ring model*

One might wonder whether a full ring instead of a wedge with finite opening angle is an affordable simulation domain for models of Cepheids. Indeed, this turns out to be feasible. Reducing the radial solution somewhat, a grid of 277×13000 points (rather than just 510×800) can be set up such that the domain azimuthally covers a full 360° angle. With physical parameters otherwise identical to the 10° case and with $500,000$ CPU-core hours available, it has been possible to conduct such a simulation for no less than 14 pulsation cycles.

The aim in setting up this model has been to perform a study of the natural azimuthal width of the flow structures. Figure 1 provides an insight to some of the properties. At some instance of time, the convective vortices are visualized by displaying the convective flux F_{conv}. It turns out that during the evolution of the model the vortices had some tendency to group together so that in effect the distribution of vortices is quite irregular.

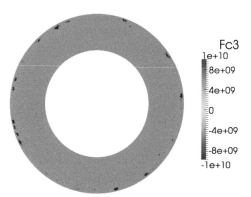

Figure 1. Convective flux in the ring model (in erg s^{-1} cm^{-2}). This model serves to investigate the He II ionization zone. In order to better represent its flux carrying vortices, the colour scale for the has been limited to approximately the central ten percent of the total range of the convective flux. For discussion see text.

This is in contrast to what does, or even can, happen when working with a sector just 10° wide.

Naturally, such findings provoke thoughts whether or not such mechanisms may ultimately lead to the excitation of nonradial pulsations. While the present 360° simulation is not suitable yet to study the excitation of nonradial modes, this is certainly an exciting perspective of future simulations of this kind.

4.2. *The atmosphere*

As already mentioned, interpretation of spectroscopic observations overwhelmingly relies on the use of one-dimensional, static atmospheres. As a matter of fact, however, one has to deal with pulsation and convection and to consider, in addition, the interaction of these two phenomena which posess characteristic times which are, moreover, not too dissimilar. There are observations which cast doubt on the simple static picture usually assumed in analysis. Even setting aside standard spectroscopic investigations showing, for example, a varying degree of microturbulence as a function of phase, other hints to complications have popped up. Analyzing an eclipsing binary, one component of which is actually a Cepheid, it has been noticed in Pilecki *et al.* (2013) that the limb darkening is in strong disagreement with theoretical predictions from static model atmospheres. Obviously, this indicates a wrong temperature structure of the static atmospheric models.

Let us provide a short discussion of atmospheric properties based on our 1° model. Figure 2 shows the temperature structure at an instance of time. The steep rise of temperature in the hydrogen ionization front is easily visible. (The top of the star is to the left side of the figure.) As is clearly seen in the temperature, an outgoing ray hits several shock fronts in the atmosphere. Details can be discerned more closely when moving to Fig. 3. Here, the course of several quantities extracted along the line visible in Fig. 2 is displayed, namely, the density of gas, the temperature (labelled by t) and the vertical velocity vx. All those quantities are normalized so as to range from 0 to 1 along the probing line. The shock visible in the middle has a density contrast of about 3.9 and a temperature jump of about 270 K, the temperature there being about 5800 K. The difference in Mach numbers across the shock is 1.6. In contrast to the case of solar granulation, where shocks appear only rather high in the atmosphere, our model also exhibits shocks throughout all atmospheric depth ranges. Often, they dissolve from the hydrogen ionization zone and move upwards. We have observed shocks of up to Mach 4 strength.

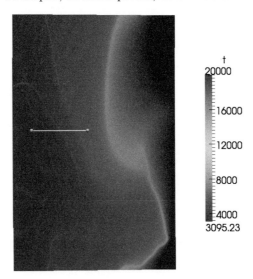

Figure 2. Temperature (labelled t and in units of K) in regions near and above the hydrogen ionization zone. The top of the atmosphere is to the left in the figure. Data are probed along the line inserted and displayed in the next figure.

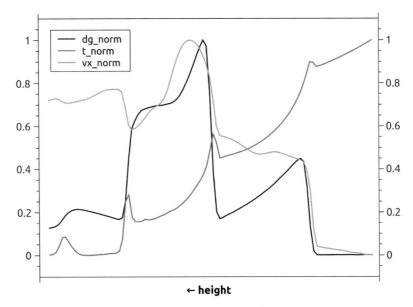

Figure 3. Various physical quantities (all normalized to be located in the range from 0 to 1) extracted along the probing line visible in the preceding figure). dg: density of gas; t: temperature; vx: vertical velocity.

In addition, when looking into the properties of the atmospheric convection at various phases a rather strong phase dependency is obvious.

5. Recent work and outlook

Since simulations of Cepheids and related stars are so demanding numerically, we have ongoing work regarding numerical methodology.

Significant progress has recently also been made on the subject of upper open boundary conditions, which are required for stable long-term runs of high resolution models. Such conditions, similar to what has recently been discussed by Grimm-Strele *et al.* (2013), have been ported to the scenario of a polar, co-moving grid and are currently being tested.

Implicit time integration to further ease the restrictions imposed by radiative diffusion are not only necessary for a more efficient model computation, but are even a prerequisite for the simulation of long time series. A new, strong stability-preserving, implicit-explicit Runge-Kutta (SSP IMEX RK) method has been developed for this purpose (Higueras *et al.* 2012). The method is already operational and working for 1D simulations with ANTARES. It has now been implemented and is currently being tested for the 2D case. A parallelized, non-linear multigrid solver has been developed to perform such simulations on suitable hardware (see Happenhofer 2013 for these developments). The simulation of wedges at moderate opening angle but very high resolution and the computation of full 360° models will become much more accessible by these improvements.

Of course, the long-term goal remains to be 3D simulations for the same scenarios and with similar resolution and model accuracy. However, in the near future, the astrophysical interpretation of existing results and their detailed comparisons with observational data will have highest priority.

Acknowledgements

This research has been supported by the Austrian Science Fund, FWF grant P18224. FK acknowledges support through FWF grants P21742 and P25229. We are thankful to G. Houdek for supplying us with one-dimensional starting models for the simulations performed with the ANTARES code. The simulations have been run on the VSC-1 and VSC-2 clusters of the Vienna Universities.

References

Buchler, J. R. 2009, *AIP-CP*, 1170, 61
Buchler, J. R. & Kolláth, Z. 2000, *Annals New York Academy of Sciences*, 898, 39
Buchler, J. R. & Szabó, R. 2007, *ApJ*, 660, 723
Cassisi, S. & Salaris, M. 2011, *ApJ*, 728, L43
de Grijs, R. 2011, *An Introduction to Distance Measurement in Astronomy* (New York: Wiley)
Ferguson, J. W., Alexander, D. R. Allard, F., *et al.* 2005, *ApJ*, 623, 585
Feuchtinger, M. U. 1998, *A&A*, 337, L29
Gastine, T. & Dintrans, B. 2011a, *A&A*, 528, A6
Gastine, T. & Dintrans, B. 2011b, *A&A*, 530, L7
Geroux, C. M. & Deupree, R. G. 2011, *ApJ*, 731, 18
Geroux, C. M. & Deupree, R. G. 2013, *ApJ*, 771, 113
Grimm-Strele, H., Kupka, F., Löw-Baselli, B., Mundprecht, E., Zaussinger, F., & Schiansky, P. 2013, arXiv: 1305.0743
Grevesse, N. & Noels, A. 1993, in: N. Prantzos, E. Vangioni-Flam & M. Casse (eds.), *Origin and Evolution of the Elements*, p. 15
Guggenberger, E., Kolenberg, K., Nemec, J. M., *et al.* 2012, *MNRAS*, 424, 649
Happenhofer, N., Grimm-Strele, H., Kupka, F., Löw-Baselli, B., & Muthsam, H. 2013, *J. Comput. Phys.*, 236, 96
Happenhofer, N. 2013, Ph.D. thesis, University of Vienna, to be submitted
Higueras, I., Happenhofer, N., Koch, O., & Kupka, F. 2012, *ASC Report 14/2012* (Vienna University of Technology, Vienna) (http://www.asc.tuwien.ac.at/preprint/2012/asc 14x2012.pdf)
Iglesias, C. A. & Rogers, F. J. 1996, *ApJ*, 464, 943

Keller, S. C. 2008, *ApJ*, 677, 683

Klagyivik, P., Szabados, L., Szing, A., Leccia, S., & Mowlavi, N. 2013, *MNRAS*, 434, 2418

Kolláth, Z., Beaulieu, J. P., Buchler, J. R., & Yecko, P. 1998, *ApJ*, 502, L55

Kupka, F., Ballot, J., & Muthsam, H. J. 2009, *CoAst*, 160, 30

Luck, R. E., Andrievsky, S. M., Korotin, S. N., & Kovtyukh, V. V. 2013, *AJ*, 146, 18

Mundprecht, E. 2011, Ph.D. Thesis, University of Vienna

Mundprecht, E., Muthsam, H. J., & Kupka, F. 2013, *MNRAS*, 435, 3191

Muthsam, H. J., Kupka, F., Löw-Baselli, B., Obertscheider, C., Langer, M., & Lenz, P. 2010, *New Astron.*, 15, 460

Pietrzyński, G., Thompson, I. B., Gieren, W., *et al.* 2010, *Nature*, 486, 542

Pilecki B., Graczyk, D. Pietrzyński, G., *et al.* 2013, *MNRAS*, 436, 953

Rogers, F. J., Swenson, F. J., & Iglesias, C. A. 1996, *ApJ*, 456, 902

Rogers, F. J. & Nayfonov, A. 2002, *ApJ*, 576, 1064

Smolec, R. & Moskalik, P. 2008a, *AcA*, 58, 193

Smolec, R. & Moskalik, P. 2008b, *AcA*, 58, 233

Smolec, R. & Moskalik, P. 2010, *A&A*, 524, A40

Soszyński, I., Poleski, R., Udalski, A., *et al.* 2008, *AcA*, 58, 163

Stobie, R. S. 1969, *MNRAS*, 144, 485

Takeda, Y., Kang, D.-I., Han, I., Lee, B.-C., & Kim, K.-M. 2013, *MNRAS*, 432, 769

Precision Asteroseismology
Proceedings IAU Symposium No. 301, 2013
J. A. Guzik, W. J. Chaplin, G. Handler & A. Pigulski, eds.

© International Astronomical Union 2014
doi:10.1017/S1743921313014300

Atomic diffusion and element mixing in pulsating stars

Georges Alecian

LUTH (Observatoire de Paris - CNRS), Observatoire de Meudon, F-92190 Meudon, France,
email: georges.alecian@obspm.fr

Abstract. Stellar plasmas are multicomponent anisotropic gases. Each component (chemical element) of these gases experiences specific forces related to its properties, which leads each element to diffuse with respect to the others. There is no reason why a stellar plasma should remain homogeneous except if mixing motions enforce homogeneity. Because atomic diffusion is a very slow process, the element separation only occurs in places where mixing motions are weak enough not to erase the effect of the ineluctable tendency of chemical elements to migrate. In this talk, I will present how atomic diffusion and mixing processes compete in stars (interiors as well as atmospheres), and I will show various cases where atomic diffusion is believed to have noticeable effects. This concerns several types of stars throughout the H-R diagram, including pulsating ones.

Keywords. atomic processes, diffusion, turbulence, stars: mass loss, stars: abundances

1. Introduction

In multicomponent gases, diffusion is usually considered as a physical process which tends to homogenize mixtures by smoothing concentration gradients. Therefore, it is familiar to consider the second spatial derivative of the particle number density n_k in the continuity equation for the species k (in the plane-parallel case):

$$\partial_t n_k + \ldots + D_k \cdot \partial_z^2 n_k = 0, \qquad (1.1)$$

where D_k is the diffusion coefficient for k. No advective term appears in this expression. However, in modeling atomic diffusion in stars, one often speaks about diffusion velocity of elements, which may sound odd to some physicists. Actually, following Chapman & Cowling (1970) one can define in the framework of the kinetic theory of gases the relative velocities V_{kt} of the species k with respect to each of the other components t of the mixture (see Alecian & Michaud 2005) by:

$$\sum_t \frac{p_k p_t}{p D_{kt}} V_{kt} = A_k, \qquad (1.2)$$

where p_k, p_t and p are the partial and total pressures respectively, D_{kt} is the diffusion coefficient of k with respect to t (related to the collision rate between these two species), and A_k a term gathering all the forces acting on particle k (gravity, radiation force, etc...). The relative velocities V_{kt} have a statistical meaning in the sense that they are obtained from the average deviation of each species from their thermal velocity. In a star, if one considers that metals are trace elements in a gas dominated by protons and electrons, one can solve for the diffusion velocities V_{D_k} of each metal with respect to the stellar plasma. This velocity is then an average velocity of particles belonging to the population of type k, and it may be introduced in the form of an additional advective term due to atomic diffusion in Eq. 1.1. However, no fluid motion can be related to this velocity; it is the velocity of the *displacement* of the population of k inside the medium, which could be at rest or moving. Speaking about diffusion velocities of metals in the context of stellar modeling, one refers generally to this picture. Notice that, in advanced stellar modeling, metals (including He) are generally not considered as trace elements, which complicates significantly the equations describing the process.

In Sect. 2, I will discuss in more detail the diffusion velocity, and I will show how one can formulate the competition of atomic diffusion and mixing processes. In Sect. 3, I will present the case of AmFm stars where atomic diffusion is known to act in internal radiative zones. Sect. 4 will be devoted to hotter main-sequence stars.

2. Mixing and mass loss vs. atomic diffusion

We consider here the simple case of the diffusion of a trace element in ionization state i (charge Z_i), with respect to protons in a ternary mixture (ions-protons-electrons). In a formulation inspired from Chapman & Cowling (1970), the diffusion velocity may be written as:

$$V_{ip} \approx \underbrace{-\left(D_{ip}+D_{\mathrm{turb}}\right)\partial_r \ln \frac{n_i}{n_p}}_{(1)} + D_{ip}\left[\underbrace{A_i \frac{m_p}{kT}\left(g_i^{\mathrm{rad}}-g\right)}_{(2)} - \underbrace{\left(\frac{Z_i+1}{2}\right)\frac{m_p}{kT}g}_{(3)} + ...\right], \qquad (2.1)$$

where m_p is the proton mass, A_i the atomic mass, k the Boltzmann constant, and T the temperature.

One can highlight three main terms as indicated in the expression above (the term due to thermal diffusion is not shown here). The first term (1) may be called the *pure diffusion* term. This term corresponds to the usual smoothing process mentioned in commenting Eq. 1.1. In this first term, a turbulent diffusion coefficient (D_{turb}) has been added to the microscopic diffusion coefficient by Schatzman (1969) to account for turbulent mixing. The second term (2) corresponds to the competition of gravity against the radiative force, as introduced by Michaud (1970). The third term (3) was introduced by Aller & Chapman (1960) and corresponds to the effect of the electric field due to diffusion of electrons. The diffusion velocity of an element k may be estimated by computing a weighted average of the ion velocities.

The continuity equation for the k particle number density may be written as:

$$\partial_t n_k + \nabla \left[n_k \cdot \left(\mathbf{V}_{D_k}+\mathbf{V}_M\right)\right] = 0, \qquad (2.2)$$

where $\mathbf{V}_M$ is a macroscopic velocity of the plasma (e.g., stellar wind, large scale circulation). So, both processes (atomic diffusion and mixing) are gathered in the expression of the diffusion velocity. Outside radiative zones, mixing processes are so strong and D_{turb} so large in term (1) of Eq. 2.1, that all the other terms are negligible. In that case, one recovers the continuity equation in the form of Eq. 1.1 with just the smoothing term (with $D_k = D_{\mathrm{turb}}$). On the contrary, in radiative zones, $D_{\mathrm{turb}} = 0$, and the smoothing term (pure diffusion) can have some efficiency only after strong abundance gradients have been built up by atomic diffusion dominated by the competition between gravity and radiative acceleration in term (2).

The wind velocity $\mathbf{V}_M$ may be modeled as a radial flow of matter. It may be simply obtained assuming the conservation of a constant mass flux in spherical geometry:

$$V_M(r) = \dot{M}_{\mathrm{star}}\left(4\pi \rho r^2\right)^{-1}, \qquad (2.3)$$

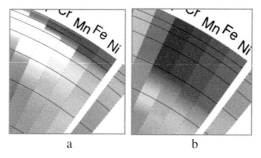

a   b

Figure 1. This figure is a magnified part of Fig. 5 in Michaud *et al.* (2011) (the reader is encouraged to see the original legend in their paper for color code, scales, and model details). It shows the concentration of some iron-peak elements in the upper layers of a main-sequence A star with the same fundamental parameters as Sirius A (see Landstreet 2011). In the left panel (the case assuming mass loss), metals are concentrated in high layers compared to the right panel (without mass loss) where overabundances extend deeper.

where $\dot{M}_{\text{star}}$ is the mass-loss rate and ρ the mass density at radius r. The stellar mass loss has a special status because it is not a mixing process, but it contributes to homogenize the medium. Indeed, if $\mathbf{V}_{\text{M}} \gg \mathbf{V}_{\text{D}_k}$, the local mixture is replaced by fresh homogeneous mixture (brought by the mass-loss flux from deep layers) faster than atomic diffusion can build up abundance stratifications.

3. The case of AmFm stars

The case of AmFm stars may illustrate both the role of turbulent mixing and mass loss in modeling the stellar interiors. AmFm stars constitute the cooler group of chemically peculiar stars of the main sequence (in the T_{eff} range 7000 – 10 000 K). To explain AmFm stars, one usually considers a first phase when, because of the low rotation of these stars, atomic diffusion of helium (gravitational settling) leads to a decrease of the depth of the outer convection zone. A second phase, when heavier elements diffuse in a radiative zone, follows. After the first phase, this radiative zone has a higher upper boundary where atomic diffusion is more efficient (see Charbonneau & Michaud 1988). Alecian (1996) has proposed that, to explain the systematic underabundances of Ca and Sc in AmFm stars, it is necessary to assume a small mass loss in addition to atomic diffusion. Some years later, Richer *et al.* (2000) who developed the Montreal evolution code (with no mass loss), had shown that during the second phase, the iron accumulation may trigger a new convection zone that makes deeper the upper boundary of the radiative zone. Such a new mixing zone in the higher part of the radiative zone appears to be still compatible with most (but not all, see below the case of Sc) of the abundance anomalies observed in AmFm stars.

A new version of the Montreal evolution code (Vick *et al.* 2010) has been applied by Michaud *et al.* (2011) to the case of Sirius A, which is a hot Am star (see Landstreet 2011). Figure 1 shows two computations realized by Michaud *et al.* (2011). The left panel shows the result assuming mass loss, while the right panel corresponds to the case with no mass loss. Both models give surface abundance pattern close to the observed one, but quite different abundance stratifications in the interior. To know which one of these two models is the closest to real stars, Michaud *et al.* (2011) suggest that asteroseismology could help to distinguish between mass loss and turbulence as dominant process for abundance stratifications in AmFm stars. On the other hand, Alecian *et al.* (2013) discussed the case of scandium in AmFm stars and have instead confirmed the result of Alecian (1996) and Leblanc & Alecian (2008), who find that underabundances of Sc in AmFm stars requires that diffusion of Sc occur just below the superficial convection of H (without a convection zone due to iron), with a small mass loss. So, the presence of mass loss in AmFm stars (and for the hotter HgMn stars) remains a subject of debate which could benefit from asteroseismology.

Another interesting aspect concerning AmFm stars is the fact that they share the same region of the H-R diagram as δ Scuti stars. The classical model has considered that there was a dichotomy of the two groups: slowly rotating stars become AmFm because of helium settling, and fast rotators become δ Scuti stars because helium remains in the outer layers and can trigger the κ mechanism. One knows that the situation is not so simple. Apart from the coexistence with, for instance, λ Boo stars, and δ Del stars in the same region of the H-R diagram, some Am stars pulsating like δ Scuti stars have been observed at the red edge of the instability strip. However, one can still consider that the main features of the classical model remain valid, and that evolution models need to be refined to understand intermediate cases, or stars transiting from one subgroup to the other.

4. Hotter stars

4.1. *The case of chemically peculiar stars*

Considering hotter main-sequence stars leads us first to look at the ApBp chemically peculiar stars (we will not consider here He-weak or He-rich type stars). This group is composed by three subgroups (roAp, magnetic ApBp, HgMn), which share a main characteristic: atomic diffusion occurs in their atmospheres. Because atomic diffusion is more efficient when particle density is weak, abundance anomalies are much stronger in these stars than in AmFm stars.

As discussed by Michaud (1970), several properties of these stars allow one to assume that either there are no mixing processes in their atmospheres, or that these processes are not strong enough to compete with atomic diffusion (slow rotation, not enough He and too hot atmospheres to have convection, very strong magnetic fields, etc.). During several decades, modeling of atomic diffusion in ApBp stars relied on this assumption of stability. Even if this assumption was justified by several serious arguments, there were no observational measurements to confirm it. This is no longer the case, since many recent spectroscopic observations show evidence of abundance stratifications in the atmospheres of ApBp stars (Ryabchikova *et al.* 2002; Ryabchikova 2005; Shulyak *et al.* 2009; Thiam *et al.* 2010). In addition to these *classical* approaches to detect abundance stratifications, one has some observations (spectroscopic variabilities) of propagating waves in roAp stars, which unveil high altitude clouds of heavy elements (Mkrtichian *et al.* 2008). Clouds are also observed through rotational modulation in some spectral lines (see, for instance, Adelman *et al.* 2002; Ryabchikova *et al.* 2002; Freyhammer *et al.* 2009; Briquet *et al.* 2010). Because such abundance stratifications or clouds cannot exist in presence of mixing, one can say now that stable atmospheres exist.

Among ApBp stars, only roAp stars are well known to pulsate. The situation is less clear for HgMn stars (in the T_{eff} range $10\,000 - 16\,000$ K), which may be considered as a continuation of the AmFm family in the upper main sequence. According to the instability strip for SPB type pulsators calculated by Miglio *et al.* (2007), HgMn stars may be inside it. However, their calculations are based on homogeneous distribution of metal abundances, which is not the case for HgMn stars. Models calculated by Turcotte & Richard (2003), with partially stratified abundances, give the same result: HgMn

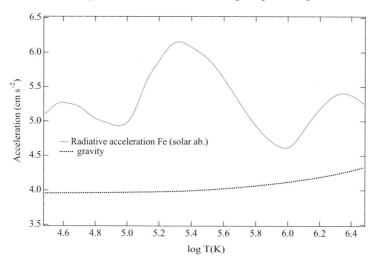

Figure 2. Radiative acceleration on Fe in a 10 $M_\odot$ star. The solid (red) line is the logarithm of the radiative acceleration (in cm s^{-2}) plotted vs. the logarithm of the temperature (in K). The dotted line is the local gravity.

stars could be SPB-type pulsators, but these evolution models need to be improved. Observations with *CoRoT* of some HgMn stars show that at least some of them present mono-periodic photometric variations (Alecian *et al.* 2009; Morel *et al.* 2013); however, it is not yet established whether these variations are pulsations or rotational modulation.

4.2. *About β Cep stars*

There are no chemically peculiar stars hotter than $T_{\text{eff}} \approx 18\,000$ K (except stars with helium abundance anomalies), because mass-loss rates become too large. However, even if one cannot identify superficial strong abnormal abundances, one cannot exclude that atomic diffusion may affect abundances inside radiative zones of hot stars. To illustrate this possibility, we have computed the radiative acceleration of Fe in a 10 $M_\odot$ star (Fig. 2) and the corresponding diffusion velocity (Fig. 3). These computations have been carried out using the same code as the one used by Alecian *et al.* (2013) for Sc, and a model (for an age of about 10 million years, with solar abundances) obtained with the stellar evolution code CLES (Scuflaire 2005). Figure 2 shows that the radiative acceleration is 100 times larger than gravity for layers with $\log T \approx 5.3$ (position of the iron bump). This means that Fe is strongly supported by the radiation field, and so the diffusion velocity is relatively large in these layers (we have neglected here the effect of the thermal diffusion, which will moderate the efficiency of the radiative acceleration). The radiative acceleration on iron at solar abundance (as shown in Fig. 2) is strong enough to support large overabundances of Fe (about 10^4 times the solar value, if one does a simple extrapolation!). Of course, such huge overabundance will never be reached, first because the diffusion time is too long compared with the evolution time of the star, and second because atomic diffusion competes with other processes. In Fig. 3 the diffusion velocity is compared to a mass-loss velocity computed for this model using Eq. 2.3, for a mass-loss rate of 10^{-12} $M_\odot \text{yr}^{-1}$. Indeed, the diffusion velocity is large enough to change significantly the Fe abundance in less than 10^5 years†. According to these calculations, we see that iron can accumulate inside the radiative zone of a 10 $M_\odot$ star, provided that the

---

† The diffusion timescale is defined as the time needed for atoms of the species under consideration to diffuse through a distance of one pressure scale height.

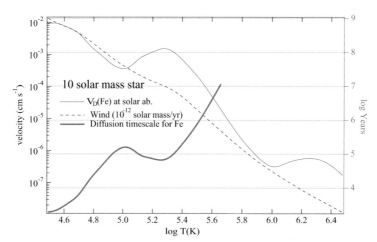

Figure 3. Diffusion velocity, diffusion timescale of Fe, and mass-loss velocity in a 10 $M_\odot$ star. The diffusion velocity (red solid line) is obtained using the radiative acceleration shown in Fig. 2, the mass-loss velocity (dashed line) corresponds to a mass-loss rate of 10^{-12} $M_\odot \mathrm{yr}^{-1}$. Left axis corresponds to the velocity (in $\mathrm{cm\,s}^{-1}$); right axis is the diffusion timescale (blue heavy solid line). The timescale curve ends at about 10^7 years, which is the age of the star for this model.

mass-loss rate is not much larger than 10^{-12} $M_\odot \mathrm{yr}^{-1}$ (10^{-11} $M_\odot \mathrm{yr}^{-1}$ seems still acceptable), and that the velocity due to large scale circulations (due to rotation for instance) is not significantly larger than 10^{-3} $\mathrm{cm\,s}^{-1}$ in layers with $\log T \approx 5.3$. A detailed estimate of iron accumulation would require detailed calculation with an evolution code.

Such results could apply to the case of some β Cep stars, for which models need larger opacities in the iron bump layers to account for the observed pulsation pattern (Bourge *et al.* 2006). This is especially interesting for β Cep stars with low metallicity because the final local overabundances built up by atomic diffusion in iron-bump layers do not depend on the initial metallicity (except through the internal structure and details of the evolution history of a low-metallicity star). In stars with low metallicity, the diffusion velocities could even be larger, and diffusion timescale shorter than in stars with solar metallicity.

5. Conclusion

Because atomic diffusion is a very slow process, and because the abundance inhomogeneities are fragile when faced with the mixing processes usually considered in stellar modeling, the efficiency of atomic diffusion has been often considered with caution, and too often neglected.

Nowadays, it is possible to say that we have a coherent picture of chemically peculiar stars in the framework of atomic diffusion, even if the modeling of individual stars remains a challenge. The important requirement to make this process efficient (the absence of mixing processes) is supported by several observations.

There is no reason to consider that atomic diffusion has visible effects only in chemically peculiar stars. Chemically peculiar stars should be considered as extreme cases, and mild abundance stratifications should exist in all outer radiative zones at least for main-sequence stars heavier than about 1.5 $M_\odot$ (and at least for layers with $\log T \lesssim 6$. where partial ionization is large enough, and diffusion timescales short enough). By mild abundance stratifications, we mean variations of local abundances by a factor around 2 or 3 for instance. Such a departure from homogeneous abundances, for a given metallicity, may have significant effects on models when accurate opacities are required. There are large efforts to make available accurate opacity tables, and a 10% change in the opacity of iron is considered significant. So, what if the uncertainties in the local abundances are around a factor of two at the iron-bump position? Asteroseismology is certainly a fantastic tool to probe such local properties of stellar interiors, and the case of β Cep stars with low metallicity is a perfect illustration of what could be addressed by refining the physics implemented in models.

References

Adelman, S. J., Gulliver, A. F., Kochukhov, O. P., & Ryabchikova, T. A. 2002, *ApJ*, 575, 449
Alecian, G. 1996, *A&A*, 310, 872
Alecian, G. & Michaud, G. 2005, *A&A*, 431, 1
Alecian, G., Gebran, M., Auvergne, M., *et al.* 2009, *A&A*, 506, 69
Alecian, G., LeBlanc, F., & Massacrier, G. 2013, *A&A*, 554, A89
Aller, L. H. & Chapman, S. 1960, *ApJ*, 132, 461
Bourge, P., Alecian, G., Thoul, A., Scuflaire, R., & Théado, S. 2006, *CoAst*, 147, 105
Briquet, M., Korhonen, H., González, J. F., Hubrig, S., & Hackman, T. 2010, *A&A*, 511, A71
Chapman, S. & Cowling, T. G. 1970, *The Mathematical Theory of non-uniform Gases* (3rd ed.; Cambridge: Cambridge University Press)
Charbonneau, P. & Michaud, G. 1988, *ApJ*, 327, 809
Freyhammer, L. M., Kurtz, D. W., Elkin, V. G., *et al.* 2009, *MNRAS*, 396, 325
Landstreet, J. D. 2011, *A&A*, 528, A132
Leblanc, F. & Alecian, G. 2008, *A&A*, 477, 243
Michaud, G. 1970, *ApJ*, 160, 641
Michaud, G., Richer, J., & Vick, M. 2011, *A&A*, 534, A18
Miglio, A., Montalbán, J., & Dupret, M.-A. 2007, *MNRAS*, 375, L21
Mkrtichian, D. E., Hatzes, A. P., Saio, H., & Shobbrook, R. R. 2008, *A&A*, 490, 1109
Morel, T., Briquet, M., Auvergne, M., *et al.* 2013, *A&A*, in press
Richer, J., Michaud, G., & Turcotte, S. 2000, *ApJ*, 529, 338
Ryabchikova, T. 2005, in: G. Alecian, O. Richard, & S. Vauclair (eds.), *EAS Publications Series*, Vol. 17, p. 253
Ryabchikova, T., Piskunov, N., Kochukhov, O., *et al.* 2002, *A&A*, 384, 545
Schatzman, E. 1969, *A&A*, 3, 331
Scuflaire, R. 2005, *Institut d'Astrophysique et de Géophysique*, Université de Liège, Belgium, CLES, Tech. rep.
Shulyak, D., Ryabchikova, T., Mashonkina, L., & Kochukhov, O. 2009, *A&A*, 499, 879
Thiam, M., Leblanc, F., Khalack, V., & Wade, G. A. 2010, *MNRAS*, 405, 1384
Turcotte, S. & Richard, O. 2003, *Ap&SS*, 284, 225
Vick, M., Michaud, G., Richer, J., & Richard, O. 2010, *A&A*, 521, A62

Precision Asteroseismology
Proceedings IAU Symposium No. 301, 2013
J. A. Guzik, W. J. Chaplin, G. Handler & A. Pigulski, eds.

© International Astronomical Union 2014
doi:10.1017/S1743921313014312

Radiative hydrodynamic simulations of turbulent convection and pulsations of *Kepler* target stars

Irina N. Kitiashvili[1,2,3]

[1]Hansen Experimental Physics Laboratory, Stanford University, Stanford, CA 94305, USA
email: irinasun@stanford.edu

[2]Center for Turbulence Research, Stanford University, Stanford, CA 94305, USA

[3]Kazan Federal University, Kazan, 420008, Russia

Abstract. The problem of interaction of stellar pulsations with turbulence and radiation in stellar convective envelopes is central to our understanding of excitation mechanisms, oscillation amplitudes and frequency shifts. Realistic ("ab initio") numerical simulations provide unique insights into the complex physics of pulsation-turbulence-radiation interactions, as well as into the energy transport and dynamics of convection zones, beyond the standard evolutionary theory. 3D radiative hydrodynamics simulations have been performed for several *Kepler* target stars, from M- to A-class along the main sequence, using a new 'StellarBox' code, which takes into account all essential physics and includes subgrid scale turbulence modeling. The results reveal dramatic changes in the convection and pulsation properties among stars of different mass. For relatively massive stars with thin convective envelopes, the simulations allow us to investigate the dynamics the whole envelope convection zone including the overshoot region, and also look at the excitation of internal gravity waves. Physical properties of the turbulent convection and pulsations, and the oscillation spectrum for two of these targets are presented and discussed in this paper. In one of these stars, with mass 1.47 $M_\odot$, we simulate the whole convective zone and investigate the overshoot region at the boundary with the radiative zone.

Keywords. stars: Hertzsprung-Russell diagram, oscillations, convection, plasmas, turbulence, methods: numerical

1. Introduction

It is known that oscillations of a star contain important information about the interior structure and physical and dynamical properties of turbulent convection. Recent progress in observational capabilities and theory opens new challenges for understanding the dynamical evolution of stars. Identification of high angular degree (high-l) modes is very tricky because of complicated interactions of turbulent convection, radiation, effects of differential rotation and magnetic field (e.g. Benomar *et al.* 2012a). The unprecedented quality of observations made by *Kepler* (Chaplin *et al.* 2011; Uytterhoeven *et al.* 2011) allows us to estimate physical and dynamical properties of stars more accurately (e.g. Verner *et al.* 2011; Benomar *et al.* 2012b), thus providing additional constraints to theoretical models (e.g. Kjeldsen *et al.* 2008; Bonaca *et al.* 2012).

Numerical modeling of the convective interior dynamics can provide important clues on how to connect theoretical models and observations. Recent advances in computational capabilities have allowed modelers to develop 3D models, which revealed significant deviations from previous 2D models (Guzik 2011). In this paper I present initial results of new realistic modeling of stellar convection.

2. Realistic modeling of stellar convection

It is known that helioseismology measurements and surface observations of the solar radius and abundances have discrepancies, which can be related to highly turbulent properties of the near-surface layers (Lefebvre *et al.* 2006; Guzik & Mussack 2010). Therefore accurate modeling of subsurface turbulent layers is critically important for our understanding of sources of the observed oscillations, and of properties of stellar convection. The simulations of the stellar convection were performed using a 3D radiative MHD code 'StellarBox'. The code was carefully tested and compared with other similar codes (e.g. Jacoutot *et al.* 2008b). It has been used for the investigation of oscillatory and dynamical properties of solar magnetoconvection (e.g. Jacoutot *et al.* 2008a; Kitiashvili *et al.* 2011a,b), and compared with high-resolution observations (e.g. Kitiashvili *et al.* 2010, 2013).

We perform realistic hydrodynamic simulations of stellar convection, where initial interior structure models are chosen to be consistent with observed surface properties of stars. The initial background models of the stellar interior structure were obtained using the standard 1D stellar evolution code 'CESAM' (Morel 1997; Morel & Lebreton 2008). These models of the interior structure were used as initial conditions in our numerical model for selected *Kepler* target stars with masses ranging from 1.01 to 1.52 $M_\odot$ (Fig. 1). The stellar masses were approximately estimated from a grid of the calculated stellar models to match observed spectroscopic parameters. The simulations were performed for a small volume of the upper convective zone and the lower atmosphere, confined in a 'box domain', with laterally periodic boundary conditions, and top and bottom boundary conditions open for the radiation energy flux.

Our numerical simulations showed that the dynamics of convective flows in these stars strongly deviate from the solar convection dynamics. Figure 2 shows a horizontal cut of the temperature distribution in the photosphere (panel a) and a vertical cut of the

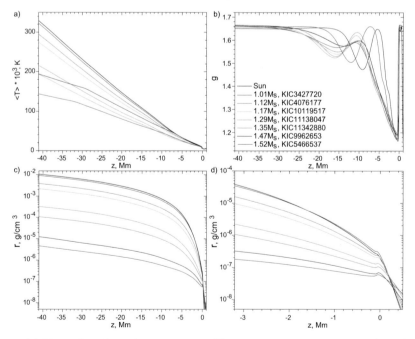

Figure 1. (a) Vertical profiles of temperature, (b) adiabatic exponent, γ, and ($c - d$) density for *Kepler* target stars with masses from 1.01 to 1.52 $M_\odot$.

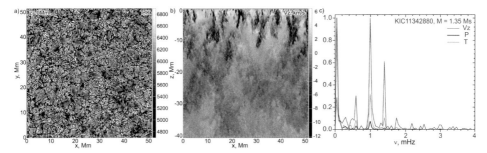

Figure 2. (*a*) Snapshot of the temperature distribution for the *Kepler* target star KIC 11342880 ($1.35 M_\odot$, $\log T_{\rm eff} = 3.822$, $\log L/L_\odot = 0.566$) at the photosphere; (*b*) vertical cut of the vertical velocity distribution; and (*c*) the power spectrum of low-degree stellar oscillations for the mean vertical velocity V_z, pressure P, and temperature near the stellar surface.

vertical velocity (panel *b*). In the case of KIC 11342880 ($1.35 M_\odot$) two basic scales of stellar granulation can be identified: solar-like granulation with a characteristic size of $1-2$ Mm, and larger-scale granules, ~ 3 Mm and larger in size (Fig. 2*a*). Such double-scale structuring disappears below the photosphere, and in the deeper layers the scale of convective patterns increases. Convective downflows in the intergranular lanes are substantially stronger than the downflows in the solar convection; they can reach up to 16 km s^{-1}, and extend up to 40 Mm below the photosphere (Fig. 2*b*). The realistic numerical simulations make it possible to obtain synthetic oscillation spectra and investigate contributions of various oscillation sources to the observed spectra. This investigation is currently underway. Figure 2*c* shows an example of the power spectrum for a model of the star KIC 11342880 obtained from the mean vertical velocity, gas pressure and temperature time-series (corresponding to low-*l* oscillation modes).

Because the stellar convective zone becomes thinner for more massive stars we were able to model the whole convective zone for our second *Kepler* target KIC 9962653 ($1.47 M_\odot$) and investigate the dynamics of the interface between the radiative and convective zones. Figure 3 shows a snapshot of the simulated stellar surface intensity (panel *a*), density fluctuations (panel *b*) and the vertical velocity (panel *c*). In this case, the scale separation between the two populations of small and large granules becomes even stronger, the downdrafts penetrate through the whole convective zone, and overshoot into the radiative zone, reaching speeds up to 30 km s^{-1}. Figure 3*b* shows that in these stars strong downflows overshoot and strongly perturb the upper radiative zone, thus causing excitation of *g* modes.

3. Conclusion

Modern observations of stellar oscillations make it possible to significantly advance understanding of the stellar structure, dynamics and evolution. However, due to the complexity of multi-scale interactions in the highly turbulent stellar convection zones an important step is in building realistic numerical models, where the effects of radiative transfer, ionization, turbulence, internal structure and composition are taken into account from first principles.

We have performed 3D radiative hydrodynamic simulations for a number of selected *Kepler* target stars (Fig. 1), with masses in the range from 1.01 to $1.52 M_\odot$. According to our results, with the increase of stellar mass the convective flows become more and more vigorous, and are characterized by high-speed downflows reaching supersonic velocities (e.g. up to 30 km s^{-1} for $1.47 M_\odot$) and penetrating into deep convective layers (Fig. 2*b*). The vigorous convection at the photosphere tends to organize granulation in

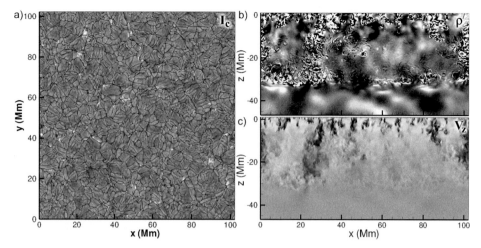

Figure 3. Stellar convection for the *Kepler* target star KIC 9962653 (M = 1.47 $M_\odot$, $\log T_{\rm eff}$ = 3.849, $\log L/L_\odot$ = 0.675): (*a*) intensity, (*b*) density fluctuations (vary in range $\pm 2 \cdot 10^{-7}$ g cm^{-3}), and (*c*) vertical velocity (range: -24 to $+10$ km s^{-1}).

two main populations (Figs. 2*a*, 3*a*). In more massive F- and A–type stars (examples are KIC 9962653 and KIC 5466537 discussed here) the convective zones become shallow. For these stars we are able to model the whole convective zones including the interface with the radiative zone (Fig. 3), which allows us to investigate the mechanism of *g*-mode excitation and characterize the overshooting region.

These preliminary simulations are a first attempt to apply the realistic 'ab-initio' approach, previously used for modeling solar magnetoconvection. This approach allows us to gain insight into the complicated turbulent dynamics of stellar envelopes and use these simulation results for verification and validation of asteroseismology models used for interpretation and analysis of stellar oscillations.

References

Benomar, O., Baudin, F., Chaplin, W. J., *et al.* 2012a, *MNRAS*, 420, 2178
Benomar, O., Bedding, T. R., Stello, D., *et al.* 2012b, *ApJ*, 745, L33
Bonaca, A., Tanner, J. D., Basu, S., *et al.* 2012, *ApJ*, 755, L12
Chaplin, W. J., Kjeldsen, H., Bedding, T. R., *et al.* 2011, *ApJ*, 732, 54
Guzik, J. A. 2011, *Ap&SS*, 336, 95
Guzik, J. A. & Mussack, K. 2010, *ApJ*, 713, 1108
Jacoutot, L., Kosovichev, A. G., Wray, A. A., & Mansour, N. N. 2008a, *ApJ*, 684, L51
Jacoutot, L., Kosovichev, A. G., Wray, A. A., & Mansour, N. N. 2008b, *ApJ*, 682, 1386
Kitiashvili, I. N., Bellot Rubio, L. R., Kosovichev, A. G., *et al.* 2010, *ApJ*, 716, L181
Kitiashvili, I. N., Kosovichev, A. G., Mansour, N. N., & Wray, A. A. 2011a, *ApJ*, 727, L50
Kitiashvili, I. N., Kosovichev, A. G., Mansour, N. N., & Wray, A. A. 2011b, *Solar Phys.*, 268, 283
Kitiashvili, I. N., Abramenko, V. I., Goode, P. R., *et al.* 2013, *Physica Scripta*, 155, 014025
Kjeldsen, H., Bedding, T. R., & Christensen-Dalsgaard, J. 2008, *ApJ*, 683, L175
Lefebvre, S., Kosovichev, A. G., & Rozelot, J. P. 2006, in: D. Barret, F. Casoli, G. Lagache, A. Lecavelier, & L. Pagani (eds.), *SF2A-2006: Proceedings of the Annual meeting of the French Society of Astronomy and Astrophysics*, p. 551
Morel, P. 1997, *A&AS*, 124, 597
Morel, P. & Lebreton, Y. 2008, *Ap&SS*, 316, 61
Uytterhoeven, K., Moya, A., Grigahcène, A., *et al.* 2011, *A&A*, 534, A125
Verner, G. A., Chaplin, W. J., Basu, S., *et al.* 2011, *ApJ*, 738, L28

Precision Asteroseismology
Proceedings IAU Symposium No. 301, 2013
J. A. Guzik, W. J. Chaplin, G. Handler & A. Pigulski, eds.

© International Astronomical Union 2014
doi:10.1017/S1743921313014324

Pulsation of magnetic stars

Hideyuki Saio

Astronomical Institute, Graduate School of Science, Tohoku University,
Aoba-ku, Sendai, Japan
email: `saio@astr.tohoku.ac.jp`

Abstract. Some Ap stars with strong magnetic fields pulsate in high-order p modes; they are called roAp (rapidly oscillating Ap) stars. The p-mode frequencies are modified by the magnetic fields. Although the large frequency separation is hardly affected, small separations are modified considerably. The magnetic field also affects the latitudinal amplitude distribution on the surface. We discuss the properties of axisymmetric p-mode oscillations in roAp stars.

Keywords. stars: magnetic fields, stars: oscillations

1. Observational properties of roAp stars

The group of rapidly oscillating Ap (roAp) stars was discovered by Kurtz (1982) more than thirty years ago (it consisted of five members then). Since then, the number of known roAp stars has increased to around $40-50$ (still increasing); a recent list is given by Kurtz *et al.* (2006). They are chemically-peculiar main-sequence stars having masses ranging from ~ 1.5 to $\sim 2\,M_\odot$, and effective temperatures from ~ 6400 to $\sim 8400\,\mathrm{K}$. On the H-R diagram (Fig 1, left panel), they lie in and around the δ Sct instability strip. However, the oscillation frequencies are much higher than those of δ Sct variables as seen in the right panel of Fig. 1, which plots dominant frequencies of roAp and δ Sct stars with respect to effective temperature. The roAp stars pulsate in high-order p modes whose frequencies range from ~ 1 to $\sim 3\,\mathrm{mHz}$ (periods; $\sim 6-21\,\mathrm{min}$), while δ Sct stars pulsate in low-order p modes. The pulsations are generally multi-periodic, and some stars show nearly equally spaced frequencies; e.g., HR 1217 (Kurtz *et al.* 2005).

Another important property of roAp stars is the presence of strong global (mainly dipole) magnetic fields. Fig. 2 shows the surface magnetic field moduli of some roAp stars as a function of the dominant pulsation frequency. The field strength ranges from ~ 1 to $24\,\mathrm{kG}$ (mostly between 2 and $6\,\mathrm{kG}$), which shows no appreciable correlation with the pulsation frequency. The lack of a correlation might be indicating that the magnetic field itself does not play a direct role in exciting high-order p modes in roAp stars.

2. Excitation of high-order p modes in roAp stars

The κ mechanism in the He II ionization zone, which excites low-order p modes in δ Sct stars, is ineffective in exciting high-order p modes in roAp stars, because the thermal time there is much longer than the periods. Rather, the H ionization zone is more important for roAp stars, although the convective flux weakens the effect of the κ mechanism. Balmforth *et al.* (2001) found that if a strong magnetic field suppresses the convection in the H ionization zone in the polar regions, high-order p modes are excited. Based on this assumption, Cunha (2002) obtained an instability region for high-order p modes, which is largely consistent with the positions of roAp stars in the H-R diagram.

The comparison on the H-R diagram, however, does not tell whether the mechanism works for the frequencies of actual roAp stars (Fig. 1). A better comparison might be

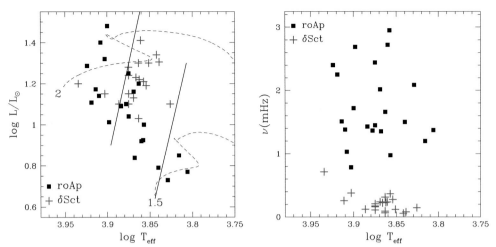

Figure 1. *Left:* The positions of roAp stars (filled squares) in the H-R diagram compared with the positions δ Sct stars (pluses). Two solid lines indicate the stability boundaries of radial fundamental modes obtained by Dupret *et al.* (2005) including time-dependent convection. *Right:* Main pulsation frequencies of the stars in the left panel are plotted with respect to effective temperature.

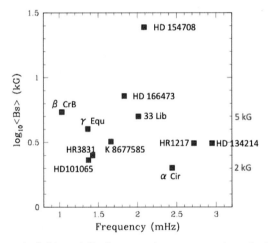

Figure 2. Surface magnetic field moduli of some roAp stars are plotted with respect to the dominant oscillation frequencies. The magnetic field strengths were taken from Shulyak *et al.* (2013), Ryabchikova *et al.* (2008), Mathys *et al.* (1997), Hubrig *et al.* (2005), Kochukhov *et al.* (2004), Balona *et al.* (2013).

possible on the $\log T_{\rm eff} - \nu L/M$ plane (ν = oscillation frequency). In this plane, the minimum and maximum frequencies of excited high-order p modes at various main-sequence evolution stages of various masses are located, respectively, on two lines, forming an instability range. Fig. 3 shows the instability region and the positions of some roAp stars whose parameters are relatively well determined. Obviously, some of the well studied roAp stars are far from the instability region, and their oscillation frequencies seem to be above the acoustic cutoff frequency. In fact, pulsation phase variations in the atmosphere of some of these stars indicate the presence of running waves (e.g., HR 1217, Saio *et al.* 2010; HD 134214, Saio *et al.* 2012). We still do not fully understand the excitation of high-order p modes in roAp stars.

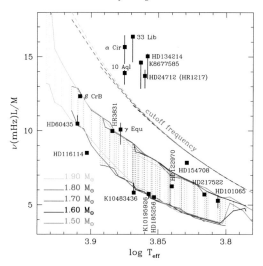

Figure 3. The instability region of high-order p modes (shaded area) in the $\log T_{\rm eff} - \nu L/M$ plane, where ν is oscillation frequency, and luminosity L and mass M are in solar units. Also plotted are some of the roAp stars whose parameters are relatively well known. Dashed lines are acoustic cutoff frequencies along evolutionary tracks of models with various masses.

3. Effect of magnetic fields on high-order p-mode pulsations

3.1. *Linearized basic equations*

In the presence of a magnetic field, the motion of the ionized gas is affected by the Lorentz force $(\nabla \times \boldsymbol{B}) \times \boldsymbol{B}/(4\pi\rho)$ (in the MHD approximation with cgs units) with ρ being the gas density. The ratio of the Lorentz force to the pressure-gradient force, $|(\nabla p)/\rho|$, is estimated roughly as $(C_{\rm A}/C_{\rm S})^2$, where the Alfvén speed $C_{\rm A}$ and sound speed $C_{\rm S}$ are defined as

$$C_{\rm A} = \frac{B}{\sqrt{4\pi\rho}} \quad \text{and} \quad C_{\rm S} = \sqrt{\Gamma_1 \frac{p}{\rho}} \tag{3.1}$$

where Γ_1 is the first adiabatic index. The runs of the ratio $C_{\rm A}/C_{\rm S}$ for $B = 2$ and $25\,{\rm kG}$ shown in Fig. 4 indicate that the Lorentz force is important in the layers as deep as the He II ionization zone in the roAp stars. Geometrically, however, the layers are thinner than $\sim$5% of the stellar radius.

Including the Lorentz force, a linearized momentum equation for nonradial pulsations may be written as

$$\frac{d\boldsymbol{v}}{dt} = \frac{\rho'}{\rho^2}\frac{dp}{dr}\boldsymbol{e}_r - \frac{1}{\rho}\nabla p' + \frac{1}{4\pi\rho}(\nabla \times \boldsymbol{B}') \times \boldsymbol{B}_0, \tag{3.2}$$

where $\boldsymbol{v}$ is pulsation velocity, and $\boldsymbol{e}_r$ is the unit vector in the radial direction. The prime ($'$) indicates the Eulerian perturbation of the quantity, and the subscript 0 on $\boldsymbol{B}$ means its equilibrium value. The equilibrium magnetic field is assumed to be force free ($\nabla \times \boldsymbol{B}_0 = 0$; a dipole field is assumed). We have used the Cowling approximation, in which the Eulerian perturbation of gravitational potential is neglected.

Assuming the ideal MHD condition (neglecting magnetic diffusivity), the linearized magnetic induction equation is given as

$$\frac{\partial \boldsymbol{B}'}{\partial t} = \nabla \times (\boldsymbol{v} \times \boldsymbol{B}_0). \tag{3.3}$$

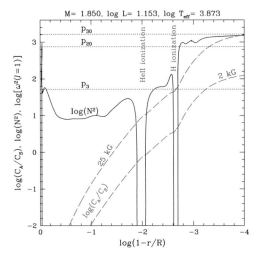

Figure 4. Runs of Brunt-Väisälä frequency, N, and the ratio of the Alfvén speed to the sound speed, C_A/C_S, for magnetic field strengths of 2 and 25 kG. Horizontal dotted lines indicate the square of pulsation frequencies of some dipole p modes. The squares of frequencies, ω^2 and N^2 are normalized by GM/R^3, with G being the gravitational constant.

In addition to these equations, the linearized mass-conservation equation

$$\frac{\partial \rho'}{\partial t} + \nabla \cdot (\rho \boldsymbol{v}) = 0, \tag{3.4}$$

and the adiabatic relation $\delta p/p = \Gamma_1 \delta\rho/\rho$ form a closed set of equations for adiabatic oscillations, where δ means the Lagrangian perturbation.

3.2. Local analysis – Magneto-acoustic waves

To understand the basic properties of oscillations, we use here a local analysis, where perturbations are assumed to be proportional to $\exp(i\sigma t - ik_z z - ik_x x)$ with the plane-parallel approximation; z and x are distances in radial and latitudinal directions. This form corresponds to axisymmetric modes, in which the Alfvén wave (torsional) is excluded. Substituting this form of perturbation into eqs. (3.2) – (3.4) and using the adiabatic relation we have a dispersion relation,

$$\sigma^4 - \sigma^2[N^2 + k^2(C_S^2 + C_A^2)] + C_S^2 C_A^2 k^2 k_{\shortparallel}^2 + N^2\left(C_S^2 k_x^2 + C_A^2 k^2 \frac{B_z^2}{B^2}\right) = 0 \tag{3.5}$$

(Appendix A in Saio & Gautschy 2004a), where $k^2 = k_z^2 + k_x^2$, $k_{\shortparallel}$ is the wavenumber parallel to the magnetic field (i.e., $k_{\shortparallel} = \boldsymbol{k} \cdot \boldsymbol{B}/|B|$), and N is the Brunt-Väisälä frequency.

Since, in most of the cavity of high-order p modes (Fig. 4), N^2 is not important, we neglect N^2 in eq. (3.5). Solving the equation for σ^2 we obtain

$$\sigma^2 = \frac{1}{2}k^2(C_S^2 + C_A^2)\left(1 \pm \sqrt{1 - \frac{4V_C^2}{C_S^2 + C_A^2}\frac{k_{\shortparallel}^2}{k^2}}\right), \tag{3.6}$$

where V_C (cusp velocity) is defined as

$$V_C^2 = \frac{C_S^2 C_A^2}{C_S^2 + C_A^2}. \tag{3.7}$$

If either $C_S \gg C_A$ or $C_S \ll C_A$, $4V_C^2/(C_S^2 + C_A^2) \ll 1$. Then, from eq. (3.6) we obtain two types of waves

$$\sigma^2 \approx k^2(C_S^2 + C_A^2) \quad \text{(fast wave)}, \qquad \sigma^2 \approx k_\parallel^2 V_C^2 \quad \text{(slow wave)}. \qquad (3.8)$$

In these conditions, in which $(C_S^2 + C_A^2) \gg V_C^2$, fast waves decouple from slow waves, because the wave number of fast waves is much smaller than that of slow waves for a given frequency. In the condition of $C_S \sim C_A$, however, the two waves have similar wave numbers and a coupling occurs.

In the outermost layers where $C_A \gg C_S \sim V_C$ (Fig. 4, eq. 3.7), slow waves correspond to p-mode acoustic oscillations, while in the deep interior where $C_S \gg C_A \sim V_C$, fast waves correspond to p-mode pulsations. In these layers p-mode pulsations decouple from magnetic waves. In the intermediate layers, in which $C_S \sim C_A \sim V_C$, the coupling occurs; i.e., p-mode pulsation generates a slow wave which propagates downwards. Roberts & Soward (1983) argued that the slow wave will be dissipated before it reaches the stellar center because the wavelength of the slow wave decreases rapidly as the ratio C_A/C_S decreases. This means that the slow wave carries away a fraction of the pulsation energy, so that the pulsation damps even in the adiabatic analysis. This damping is incorporated in later works solving pulsation equations for magnetized stars.

3.3. *Comparison of frequency shifts calculated by different methods*

Three different methods have been developed to calculate oscillation frequencies taking into account the generation of the slow wave and its damping effect; the first one is by Dziembowski & Goode (1996), which is extended by Bigot *et al.* (2000); the next one is based on a variational principle developed by Cunha & Gough (2000); and the third one (Saio & Gautschy 2004a) is somewhat similar to the first one, expanding the eigenfunction by a sum of terms associated with spherical harmonics. These three methods are summarized by Saio (2008).

Figure 5 compares results obtained by the three different methods. Plotted are real and imaginary parts of $\nu - \nu_0$, where ν is the frequency of a p mode at $B_p = 1\,\mathrm{kG}$ and ν_0 is the frequency of the mode without a magnetic field. The left panel compares the results of Cunha & Gough (2000) for $l = 1$ and 3 axisymmetric modes of the polytropic model having $(M, R) = (2\,M_\odot, 2\,R_\odot)$ with the results of Bigot *et al.* (2000) for a 1.8 $M_\odot$ ZAMS model. The results of Cunha & Gough (2000) show that the frequency shift by the magnetic field increases with frequency, but at a certain frequency it jumps down and then starts increasing again. This property, which is governed by a parameter, $\nu B_p^{0.25}$ (for a polytrope of index 3), is confirmed by Saio & Gautschy (2004a). Although the calculations by Bigot *et al.* (2000) do not extend beyond the jump, the results agree with the results of Cunha & Gough (2000) up to the jump.

The right panel of Fig. 5 compares the results of Bigot *et al.* (2000) for $l = 0, 2$ with those obtained by the method of Saio & Gautschy (2004a) for a 1.8 $M_\odot$ ZAMS model, which is similar to the model of Bigot *et al.* (2000). Obviously, both results agree well with each other. Those comparisons show that the results from the three different methods agree well up to the jump, although some discrepancy in the amount of the jump is indicated by Saio (2008).

4. Comparison with KIC 8677585

Among the three roAp stars found by the *Kepler* satellite, KIC 8677585 has the largest number of frequencies (Balona *et al.* 2013), which are more or less regularly spaced, but some frequencies form tight groups as seen in the left panel of Fig. 6. Such a group of

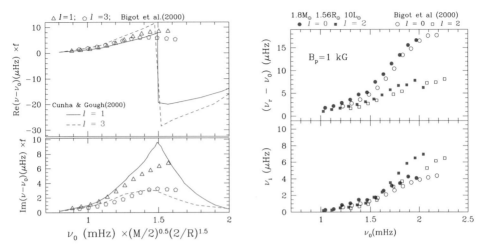

Figure 5. *Left:* Magnetic frequency shifts of high-order p modes at $B_{\rm p} = 1\,$kG obtained by Cunha & Gough (2000) (solid lines for $l = 1$ and dashed lines for $l = 3$) for a $2\,M_\odot$ (and $R = 2\,R_\odot$) polytropic (index 3) model are compared with the results for a $1.8\,M_\odot$ ZAMS model obtained by Bigot *et al.* (2000). To normalize the difference in mass and radius, a factor $f = (M/2)^{0.5}(2/R)^{1.5}$ is multiplied to the frequencies. *Right:* Comparison between the results obtained by Bigot *et al.* (2000) and by the method of Saio & Gautschy (2004a) for even modes ($l = 0, 2$).

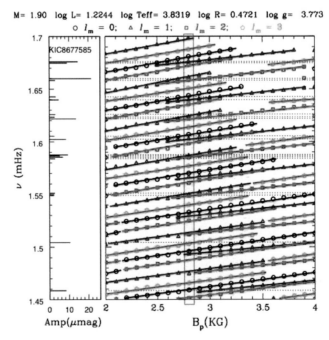

Figure 6. *Left:* An amplitude spectrum of KIC 8677585 drawn using the data from Balona *et al.* (2013). *Right:* Frequencies of various modes calculated by the method of Saio (2005) (a nonadiabatic extension of Saio & Gautschy 2004) are plotted as a function of $B_{\rm p}$ (magnetic strength at a pole). The model with the parameters shown in the top line gives a best fit with the frequencies of KIC 8677585 at $B_{\rm p} = 2.8\,$kG (the right panel of Fig. 7). Frequencies belonging to the same mode are connected by solid lines. Generally the frequency of a mode increases with $B_{\rm p}$, but at some $B_{\rm p}$ the effective latitudinal degree l_m changes.

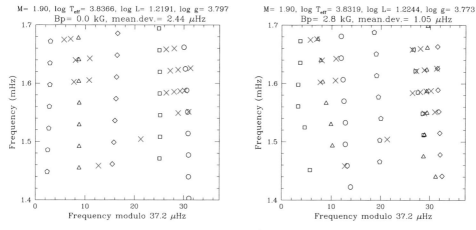

Figure 7. Oscillation frequencies of KIC 8677585 (crosses; Balona *et al.* 2013) from the *Kepler* satellite are compared with a model that does not include magnetic field (left panel), and with a model that does include the effect of a dipole magnetic field of 2.8 kG.

high-order p modes is one of the properties which can be explained by the effect of a magnetic field.

As seen in the right panel of Fig. 6, the frequency of a mode generally increases with B_p. But the rate of the increase (i.e., $d\nu/dB_p$) depends slightly on the (effective) latitudinal degree (l_m). Therefore, at some value of B_p, frequencies of two (sometimes three) modes get very close. Such a character can produce tight groupings seen in KIC 8677585. An echelle diagram of the model produced at $B_p = 2.8$ kG is compared with the observed frequencies in the right panel of Fig. 7. Compared with the left panel for a model with $B_p = 0$, it is obvious that including the magnetic effect improves the fitting with the observed frequencies, indicating the theoretical prediction for the magnetic effect on the p-mode frequencies to be in the right direction. However, we note that KIC 8677585 shows, in addition to the high-frequency oscillations, some low-frequency oscillations (34 – 74 μHz; Balona *et al.* 2013). We do not understand how they are excited.

5. Magnetic effects on the amplitude distribution at the surface

Because of the Lorentz force, the latitudinal dependence of a pulsation mode cannot be expressed by a single spherical harmonic. It changes gradually with the strength of the magnetic field. Generally speaking, the amplitude on the surface is confined more strongly to the polar regions, which corresponds to the increasing contribution from higher l components (Saio & Gautschy 2004a). This phenomenon was confirmed observationally by Kochukhov (2004), who obtained the velocity amplitude distribution of the roAp star HR 3831 and found that the velocity amplitude of the "dipole" mode is more concentrated toward the magnetic axis than the dependence of the Legendre function $P_1(\cos\theta)$; i.e., there is a considerable contribution from $P_3(\cos\theta)$.

The modification of the amplitude distribution on the surface also affects the line-profile variations caused by the oscillations. Contribution from higher l components makes the "wavelength" of the profile variation shorter (Saio & Gautschy 2004b), in agreement with the observed ones (e.g., HR 3831, Kochukhov 2006).

6. Effect of rotation – Oblique pulsator model

We discuss here the effects of rotation, which we have disregarded so far. Most of the roAp stars rotate slowly with an axis generally inclined to the magnetic axis (oblique rotator). To explain the observational property that the pulsation amplitude and phase modulate with the rotation phase, Kurtz (1982) invented the *oblique pulsator model*, in which the pulsation axis is aligned with the magnetic axis which is inclined to the rotation axis. In this model, the alignment between the pulsation and the magnetic axes is assumed, but it is not strictly possible because the rotation effect, whose symmetry axis is inclined to the magnetic axis, modifies the pulsation. Bigot & Dziembowski (2002) investigated this difficult problem and found that the pulsation axis of a nearly axisymmetric mode does not stay on one direction but draws an ellipse during a pulsation cycle. Later, Bigot & Kurtz (2011) applied this modified oblique pulsator model to various roAp stars to fit observed amplitude and phase modulations better than the fits by the original model.

From a different point of view, Gough (2012) argued that even if the pulsation axis is once aligned to the magnetic axis, the alignment would be lost because it should precess due to the Coriolis force. To avoid this difficulty, he proposed that pulsation should be excited only when the axis is closely aligned to the magnetic axis.

References

Balmforth, N. J., Cunha, M. S., Dolez, N., Gough, D. O., & Vauclair, S. 2001, *MNRAS*, 323, 362
Balona, L. A., Catanzaro, G., Crause, L., *et al.* 2013, *MNRAS*, 432, 2808
Bigot, L. & Dziembowski, W. A. 2002, *A&A*, 391, 235
Bigot, L. & Kurtz, D. W. 2011, *A&A*, 536, A73
Bigot, L., Provost, J., Berthomieu, W., Dziembowski, W. A., & Goode, P. R. 2000, *A&A*, 356, 218
Cunha, M. S. 2002, *MNRAS*, 333, 47
Cunha, M. S. & Gough, D. 2000, *MNRAS*, 319, 1020
Dupret, M.-A., Grigahcène, A., Garrido, R., Gabriel, M., & Scuflaire, R. 2005, *A&A*, 435, 927
Dziembowski, W. A. & Goode, P. R. 1996, *ApJ*, 458, 338
Gough, D. 2012, *Geo. Ast. Fluid Dyn.*, 106, 429
Hubrig, S., Nesvacil, N., Schöller, M., *et al.* 2005, *A&A*, 440, L37
Kochukhov, O. 2004, *ApJ*, 615, L149
Kochukhov, O. 2006, *A&A*, 446, 1051
Kochukhov, O., Drake, N. A., Piskunov, N., & de la Reza, R. 2004, *A&A*, 424, 935
Kurtz, D. W. 1982, *MNRAS*, 200, 807
Kurtz, D. W., Cameron, C., Cunha, M. S., *et al.* 2005, *MNRAS*, 358, 651
Kurtz, D. W., Elkin, V. G., Cunha, M. S., *et al.* 2006, *MNRAS*, 372, 286
Mathys, G., Hubrig, S., Landstreet, J. D., Lanz, T., & Manfroid, J. 1997, *A&AS*, 123, 353
Roberts, P. H. & Soward, A. M. 1983, *MNRAS*, 205, 1171
Ryabchikova, T., Kochukhov, O., & Bagnulo, S. 2008, *A&A*, 480, 811
Saio, H. 2005, *MNRAS*, 360, 1022
Saio, H. 2008, *JPhCS*, 118, 12018
Saio, H. & Gautschy, A. 2004a, *MNRAS*, 350, 485
Saio, H. & Gautschy, A. 2004b, *ASP-CS*, 310, 478
Saio, H., Ryabchikova, T., & Sachkov, M. 2010, *MNRAS*, 403, 1729
Saio, H., Gruberbauer, M., Weiss, W. W., Matthews, J. M., & Ryabchikova, T. 2012, *MNRAS*, 420, 283
Shulyak, D., Ryabchikova, T., & Kochukhov, O. 2013, *A&A*, 551, A14

Precision Asteroseismology
Proceedings IAU Symposium No. 301, 2013   © International Astronomical Union 2014
J. A. Guzik, W. J. Chaplin, G. Handler & A. Pigulski, eds.   doi:10.1017/S1743921313014336

Pulsation and mass loss across the H-R diagram: From OB stars to Cepheids to red supergiants

Hilding R. Neilson

Department of Physics & Astronomy, East Tennessee State University,
PO Box 70300, Johnson City, TN 37614, USA
email: `neilsonh@etsu.edu`

Abstract. Both pulsation and mass loss are commonly observed in stars and are important ingredients for understanding stellar evolution and structure, especially for massive stars. There is a growing body of evidence that pulsation can also drive and enhance mass loss in massive stars and that pulsation-driven mass loss is important for stellar evolution. In this review, I will discuss recent advances in understanding pulsation-driven mass loss in massive main-sequence stars, classical Cepheids and red supergiants and present some challenges remaining.

Keywords. circumstellar matter, stars: evolution, stars: mass loss, Cepheids

1. Introduction

Stellar winds are ubiquitous in massive stars, whether they be hot O-type stars or cool red supergiants and are an crucial ingredient for stellar and galactic evolution. However, the underlying physics of stellar winds and mass loss is still not well-understood across the Hertzsprung-Russell diagram (HRD).

Mass loss in massive stars plays an important role in determining their evolution and how they end as supernovae. The most obvious examples of this evolution are Wolf-Rayet stars that eject their envelopes, exposing their helium cores. Because of that mass loss, these stars appear to evolve at hot effective temperatures, $\geqslant 30\,000$ K, and explode as Type Ib/c supernovae. Georgy (2012) showed that changing the mass-loss rates of red supergiant stars forces them to evolve blueward and potentially explode during a yellow-supergiant phase of evolution, consistent with observations (Maund *et al.* 2011). Similarly, understanding the wind-driving mechanism and mass-loss rates of red supergiant stars would provide insights into questions about the progenitor masses of Type II-P and IIn supernovae (Smartt 2009).

Similarly, understanding stellar mass loss provides insight into the circumstellar medium of massive stars and feedback into the interstellar medium. Evidence for this is found across the HRD, such as bow shocks observed about the O-type star ζ Oph (Gvaramadze *et al.* 2012), the Galactic center star IRS 8 (Rauch *et al.* 2013), the prototype Cepheid, δ Cep (Marengo *et al.* 2010) and the red supergiant Betelgeuse (Ueta *et al.* 2008; Cox *et al.* 2012). Further, changes in the wind due to stellar evolution could lead to the formation of multiple bow shocks (Mohamed *et al.* 2012; Mackey *et al.* 2012; Decin *et al.* 2012). Observations of stellar wind bow shocks can be used to infer mass-loss rates and wind velocities.

The challenge for understanding stellar winds is that there are many driving mechanisms possible. In Fig. 1, I show a HRD outlining regions for different mass-loss mechanisms and regions where mass-loss rates are significant but not understood, such as

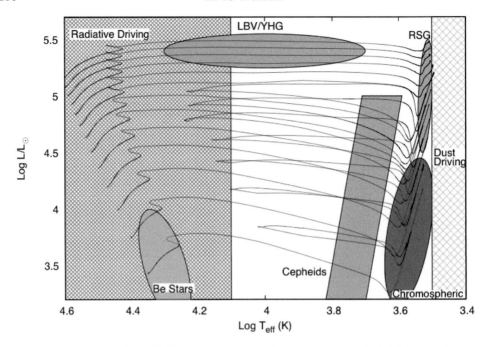

Figure 1. Hertzsprung-Russell diagram showing regions of different wind-driving mechanisms. For regions denoting Cepheids, luminous blue variables and red supergiants the wind-driving mechanism is poorly understood.

in Cepheids, luminous blue variable stars and red supergiant stars. In hot stars, with $T_{\rm eff} > 20\,000$ K, radiative line-driving is the dominant mechanism for accelerating a stellar wind (Lamers & Cassinelli 1999; Vink 2011). In these stars, radiation accelerates Fe III and IV ions and these ions interact with other atoms in the photosphere, driving the wind. In the coolest stars, the mechanism is similar, but instead of accelerating ions, radiation accelerates dust that forms in the stellar photosphere (e.g., Mattsson *et al.* 2010; Mattsson & Höfner 2011). For lower mass stars, like the Sun and red giant stars, winds are driven by Alfenic waves and/or turbulence generating a chromosphere and corona that also accelerates a wind (e.g., Cranmer & Saar 2011). Rotation is yet another mechanism that can drive mass loss, especially in Be stars (Porter & Rivinius 2003).

Even with a plethora of diverse wind-driving mechanisms, there still exist stars for which stellar mass loss is not understood. However, a potential mechanism for driving mass loss is stellar pulsation, especially in stars such as luminous blue variable stars, Cepheids and red supergiant stars. In this review, I will discuss progress in understanding how pulsation can drive and influence stellar winds. However, research on the topic of pulsation-driven mass loss has focused on specific stellar types and not the physics of the pulsation-driven mass loss in general. As such, I will present recent results as a tour of stars on the HRD where pulsation-driven mass loss is important.

2. Massive OB stars

The first step in the tour of the HRD is massive OB stars, in particular OB supergiants. Radiative line-driving is important, but there is evidence that radiative line-driven theory is insufficient to explain observed mass-loss rates in many of these stars. Aerts *et al.* (2010) presented photometric and spectroscopic observations of the hot supergiant HD 50064 and found variation of P Cygni line profiles consistent with periodic mass loss. The period of the mass-loss variation is the same as the period of brightness variations measured from photometry. The authors suggested that the mass loss must be driven by or is modulated by pulsation, in this case strange-mode oscillations (Glatzel & Kiriakidis 1993). Similarly, Kraus *et al.* (these proceedings) show similar line-profile variations in OB supergiants consistent with pulsation-driven mass loss. This adds further evidence for the connection between pulsation and mass loss in massive stars; however, these works do not discuss how pulsation can drive a wind.

Sanyal & Langer (2013) presented preliminary stellar evolution models of a massive star that develops an inflated envelope with a density inversion near the surface. The model also is unstable to pulsation, in which the radius varies by about 10%, but the luminosity amplitude is only about ten milli-magnitudes. This variation is consistent with strange-mode oscillations. The authors also find that the pulsation instability is connected to the mass loss, but more research is required.

Massive blue supergiants may also pulsate due to the ϵ mechanism. Moravveji *et al.* (2012) presented evidence for this by modeling the blue supergiant Rigel and finding that the ϵ mechanism drives gravity-mode oscillations. It is possible that these gravity modes might contribute to the stellar wind and enhance the mass-loss rate.

3. Luminous Blue Variable stars

Luminous blue variables (LBV) stars may be the prototype for pulsation-driven and eruptive mass loss and can be subdivided into two groups: the η Carinae-like LBVs and the S Doradus stars. The former stars are primarily eruptive variables, named for η Car which is famous for observed eruptions in the 19th and early 20th-century (Humphreys & Koppelman 2005; Davidson & Humphreys 2012). The latter stars are defined by more regular pulsation and microvariations (e.g., Saio *et al.* 2013).

The cause of the observed eruptions in η Car are still a mystery. Smith (2013) modeled the great 19th century eruption as a strong wind with mass-loss rates, $\dot{M} = 0.33\ M_{\odot}\ \mathrm{yr}^{-1}$ lasting 30 years followed by an explosion, ejecting about 10 $M_{\odot}$ of material. The model is similar to that expected for a Type IIn supernova. However, the source of the explosion was not described. Guzik *et al.* (2005) computed hydrodynamic models of massive stars that undergo pulsation and found that pulsation could cause the radiative luminosity in the star to increase with a delay before the convective luminosity compensates. The radiative luminosity surpasses the Eddington luminosity, hence driving an eruption. In this scenario, pulsation in stars near the Eddington limit is a potential mass-loss mechanism.

The S Doradus stars appear to pulsate more regularly than the η Carinae stars where pulsation is driven by either the iron-bump opacity or the strange-mode instability. The stars also straddle the effective temperature $T_{\mathrm{eff}} \approx 22\,000$ K, which defines the boundary of the bi-stability mechanism for radiative-driven winds (e.g., Vink *et al.* 2013). For $T_{\mathrm{eff}} > 22\,000$ K, the dominant opacity for driving the wind is the Fe IV ions, while for cooler stars the dominant opacity is Fe III. The change in opacity causes significant differences in mass-loss rates, with the cooler stars having greater mass-loss rates than those on the hotter side of the bi-stability region. Pulsation could drive a star to oscillate between the hot and cool side of the S Doradus instability strip and change the mass-loss rate (Smith *et al.* 2004). Pulsation can also directly drive mass loss in these stars, possibly by strange-mode oscillations (Grott *et al.* 2005). A combination of the two mechanisms might generate a pseudo-photosphere, making an S Dor star appear as a yellow hypergiant, as suggested by Vink (2012). As before, stellar pulsation can enhance the wind, but the underlying physics is uncertain.

4. β Cephei stars

β Cephei stars are massive stars where pulsation is driven by the iron opacity bump that oscillate both radially and non-radially with pulsation periods of hours. However, these stars appear to have the opposite problem as LBV stars; the observed mass-loss rates are smaller than expected from radiative line-driven theory. Oskinova *et al.* (2011) presented observations of seven β Cephei stars plus a number of non-pulsating, magnetic B-type stars, all of which have smaller mass-loss rates than expected, i.e., the weak-wind problem (e.g. Crowther *et al.* 2006).

One phenomenon common to all these stars is X-ray emission, which affects the ionization structure in the photosphere, hence changing the observed mass-loss rate. Classical Cepheids also display X-ray emission that is correlated with pulsation (Engle & Guinan 2012); this hints at the possibility that X-ray emission in β Cep stars is also pulsation related. There is no evidence for phase-dependent X-ray emission (Raassen *et al.* 2005) but the pulsation period is short relative to the integration time required to detect X-ray photons. If pulsation is generating X-ray emission, then pulsation can be acting to stall the stellar wind.

5. Be stars

The Be stars are an interesting class of stars displaying emission lines consistent with the presence of a circumstellar mass loss disk generated by rapid rotation (Struve 1931). However, it is not obvious how a disk is generated if Be stars rotate at less than their critical velocity (e.g., Marsh Boyer *et al.* 2012), although this is still an issue of contention (Rivinius 2013). One potential resolution is stellar pulsation and pulsation-driven mass loss (Baade 1983; Ando 1986).

Be stars are well-known non-radial pulsating stars, where gravity modes have been observed (Diago *et al.* 2009; Neiner *et al.* 2009; Semaan *et al.* 2013). Pulsation provides a number of potential avenues to help generate mass loss, either by pulsation-rotation interactions (e.g., Townsend, these proceedings; Goupil, these proceedings), or directly as suggested by Cranmer (2009) and Shibahashi (these proceedings).

For the latter case, Cranmer (2009) presented an analytic prescription for driving a disk outflow in Be stars with rotation velocities as slow as 60% of the critical velocity, where non-radial pulsations transfer angular momentum and accelerate that material to a Keplerian velocity. This model is somewhat ad hoc as there is no obvious mechanism for transferring the angular momentum. However, Shibahashi does build upon models of wave leakage in the upper photosphere (e.g., Townsend 2000a,b), providing a plausible connection. The Cranmer (2009) mass-loss mechanism depends on non-radial pulsations instead of radial pulsation, but still suffers the same issue of not being able to detail the connection between pulsation and mass loss beyond some prescription.

6. Classical Cepheids

From the hot star side of the HRD, I consider cooler effective temperatures with a discussion of Classical Cepheids. These stars are arguably the archetype for stellar pulsation; they are bright radially pulsating stars that are ideal standard candles for cosmology (Freedman *et al.* 2001; Ngeow *et al.* 2009; Freedman *et al.* 2012). However, recent infrared and radio observations show that the prototypical Cepheid, δ Cephei, is undergoing mass loss (Marengo *et al.* 2010; Matthews *et al.* 2012). This result was surprising and raises important questions about the role of pulsation and mass loss in these stars.

There is evidence for mass loss in other Cepheids based on infrared observations. Barmby *et al.* (2011) presented *Spitzer* infrared observations of Galactic Cepheids and found, at best, tentative evidence for infrared excess in these stars. On the other hand, Neilson *et al.* (2009, 2010) analyzed optical and infrared observations of Large Magellanic Cloud Cepheids and found evidence that infrared excess is common, suggesting mass-loss rates of the order $\dot{M} = 10^{-8} - 10^{-7} \ M_\odot \ \mathrm{yr}^{-1}$.

However, infrared excess is not the only indicator for Cepheid mass loss. Neilson *et al.* (2012a) showed that the observed rate of period change and other fundamental parameters of the Cepheid Polaris cannot be fit by stellar evolution models unless Polaris is losing mass at a rate of the order $10^{-7} - 10^{-6} \ M_\odot \ \mathrm{yr}^{-1}$. This result was extended to a population of almost 200 Galactic Cepheids (Turner *et al.* 2006). For that case, Neilson *et al.* (2012b) compared the observed fraction of Cepheids with positive and negative period change with predictions from population synthesis models. A negative rate of period change is consistent with a Cepheid evolving to hotter effective temperatures and a positive rate is consistent with redward evolution. Mass loss acts to increase the rate of period change in a positive direction, but it also decreases the timescale of blueward evolution. The observed fraction of Cepheids with positive period change is about 67%, and Neilson *et al.* (2012b) predicted the fraction is more than 80% if mass loss is not included. When stellar evolution models are computed with Cepheid mass-loss rates, $\dot{M} = 10^{-7} \ M_\odot \ \mathrm{yr}^{-1}$, then the predicted fraction is reduced to about 70%, consistent with observations. This is the first evidence that all Cepheids undergo significant mass loss.

While mass loss is important in Cepheids, there is no obvious mass-loss theory. Neilson & Lester (2008, 2009) developed a prescription for pulsation-driven mass loss in Cepheids, in which a wind is driven by pulsation-generated shocks. Previous calculations suggest there are multiple shocks propagating in a Cepheid photosphere at various phases (Fokin *et al.* 1996) and it was hypothesized that these shocks add momentum to an outflow, similar to the suggestion of Bowen (1988). Mass-loss rates for a sample of Galactic Cepheids were predicted to be $\dot{M} = 10^{-10} - 10^{-7} \ M_\odot \ \mathrm{yr}^{-1}$, smaller than those suggested by infrared observations and rates of period change. However, when the theory was incorporated with stellar-evolution models, the mass-loss rates were sufficient to resolve the Cepheid mass discrepancy (Keller 2008; Neilson *et al.* 2011). The pulsation-driven mass-loss prescription is interesting, but insufficient to describe observations.

7. Red supergiant stars

A Cepheid evolves to cooler effective temperatures and becomes a red supergiant (RSG) star. Mass loss in RSGs is typically understood in terms of the Reimer's relation or a similar type of mass-loss prescription (e.g., Reimers *et al.* 1975; Schröder & Cuntz 2005). However, the Reimer's relation is a measure of average mass-loss rates, and RSG winds can vary significantly (Willson 2000). While mass-loss rates have been measured, the driving mechanism is not understood even though mass loss is a crucial ingredient for stellar evolution and supernova progenitors (Langer 2012).

Radial pulsation was suggested as one mechanism for driving superwinds in RSG stars. Yoon & Cantiello (2010) found that RSG stellar-evolution models are pulsationally unstable near the end of their lives. Pulsation amplitudes were found to increase, and the authors assumed that this pulsation drives a superwind. Mass-loss rates were assumed to are function of the amplitude growth rate. The enhanced mass-loss rates are significant enough for RSG stars to lose three or more solar masses; hence Yoon & Cantiello (2010) hypothesized that pulsation-driven mass loss could explain the observed dearth of Type IIP supernovae progenitors with mass $M \geqslant 16.5 \ M_\odot$, as noted by Smartt (2009).

Georgy (2012) also computed stellar-evolution models with enhanced mass loss. He found that by increasing mass-loss rates by factors of three to five would cause RSG stars to evolve blueward to hotter effective temperatures and end up as yellow supergiants when they explode as supernovae. This model also assumes that RSG stars undergo pulsation-driven mass loss, but there is no defined theory and almost no observational evidence for pulsation-driven mass loss.

However, the observed bow shock structure of Betelgeuse provides a tantalizing hint for pulsation-driven mass loss. Decin *et al.* (2012) found multiple arc structures about Betelgeuse's bow shock. These multiple arcs are too close to be a result of stellar evolution, but may be caused by pulsation changing the structure of the stellar wind. This hypothesis is speculative and needs to be tested by hydrodynamic models. Further work is required to begin to understand the role of pulsation-driven mass loss in RSG stars.

8. AGB stars

Asymptotic giant branch stars are perhaps the best understood examples of stars undergoing pulsation-driven mass loss. These stars are particularly cool, allowing dust to form in the photosphere. Radiation then accelerates the dust in a wind, analogous to radiative-driven winds in hot stars. However, this dust-driven wind is efficient only in the coolest and/or carbon-rich stars (e.g., Wachter *et al.* 2008). In other AGB stars, such as the M-type AGB stars, dust does not form easily, hence another wind-driving mechanism is necessary (Woitke 2006).

Stellar pulsation is the most likely candidate for driving mass loss in these stars. Höfner & Andersen (2007) suggested that pulsation in M- and S-type AGB stars generate shocks that levitate material, extending the photosphere. The extended, cooler photosphere allows for silicate dust to form that would not form otherwise. Radiation can then take over and accelerate a wind. Freytag & Höfner (2008) computed three-dimensional simulations of an AGB atmosphere that include convection and pulsation and verified the hypothesis that pulsation could levitate material in the photosphere. Furthermore, Bladh & Höfner (2011) and Bladh *et al.* (2013) presented radiation-hydrodynamic atmosphere models, again, showing that the combination of the dust formation in the photosphere and pulsation drives an outflow. Currently, understanding AGB mass loss is limited more by knowledge of which dust species form in the photosphere than the role of pulsation.

9. Outlook

In this review, I described how stellar pulsation can influence mass loss in a number of different stellar types ranging from massive OB stars to coolest AGB stars. There has been significant progress in understanding pulsation-driven mass loss in the past decade, including some of the first observational hints.

While observations are beginning to hint at pulsation-driven mass loss, there is not yet a proverbial smoking gun. The most compelling observations include spectral line variations in massive OB stars and the large mass-loss rates measured for Cepheids. For the most part, evidence for pulsation-driven mass loss is the inability for other driving mechanisms to explain observed mass-loss rates, even in LBV and AGB stars. An ideal observation connecting pulsation and mass loss would be variations of P Cygni profiles in Cepheids and AGB stars. In hot stars, other observations are necessary to confirm pulsation-driven mass loss since P Cygni profiles measure mass-loss rates from radiative line-driving. Variations of those profiles measure changes of mass-loss rates and/or changes due to pulsation velocities.

Observational evidence is still circumstantial partly because there is also no general theory for pulsation-driven mass loss. Current theories are targeted toward specific stellar types and specific challenges, such as explosive events in η Carinae LBV stars and dust formation in AGB stars. For instance, the theory of radiative line-driven winds is general (Castor *et al.* 1975) and can be applied to all stars; that theory is only limited to hot stars because of how it depends on the ionization structure of the photosphere. The one distinct challenge for developing a general theory for pulsation-driven mass loss is, in most cases, that pulsation is not the sole driving mechanism. In hot stars line driving is the dominant mechanism while in cool stars dust driving is the dominant mechanism. Pulsation enhances mass loss in these stars; thus any pulsation-driven mass-loss theory must couple with other wind-driving mechanisms.

Developing a theory of pulsation-driven mass loss is a difficult challenge, but the future is bright. Hydrodynamic modeling, such as that by Guzik & Lovekin (2012) and Bladh *et al.* (2013) is progressing and new insights are being discovered. New models will shed further light on mass loss in LBV and AGB stars as well as explore the physics for pulsation and mass loss in β Cephei, RSG, massive blue supergiant, and Be stars as well as Classical Cepheids. Greater insight in pulsation-driven mass loss will also provide deeper understanding of the detailed stellar evolution and precision stellar astrophysics.

References

Aerts, C., Lefever, K., Baglin, A., *et al.* 2010, *A&A*, 513, L11

Ando, H. 1986, *A&A*, 163, 97

Baade, D. 1983, *A&A*, 124, 283

Barmby, P., Marengo, M., Evans, N. R., *et al.* 2011, *AJ*, 141, 42

Bladh, S. & Höfner, S. 2012, *A&A*, 546, A76

Bladh, S., Höfner, S., Nowotny, W., Aringer, B., & Eriksson, K. 2013, *A&A*, 553, A20

Bowen, G. H. 1988, *ApJ*, 329, 299

Castor, J. I., Abbott, D. C., & Klein, R. I. 1975, *ApJ*, 195, 157

Cox, N. L. J., Kerschbaum, F., van Marle, A.-J., *et al.* 2012, *A&A*, 537, A35

Cranmer, S. R. 2009, *ApJ*, 701, 396

Cranmer, S. R. & Saar, S. H. 2011, *ApJ*, 741, 54

Crowther, P. A., Lennon, D. J., & Walborn, N. R. 2006, *A&A*, 446, 279

Davidson, K. & Humphreys, R. M. 2012, *Nature*, 486, E1

Decin, L., Cox, N. L. J., Royer, P., *et al.* 2012, *A&A*, 548, A113

Diago, P. D., Gutiérrez-Soto, J., Auvergne, M., *et al.* 2009, *A&A*, 506, 125

Engle, S. G. & Guinan, E. F. 2012, *JASS*, 29, 181

Fokin, A. B., Gillet, D., & Breitfellner, M. G. 1996, *A&A*, 307, 503

Freedman, W. L., Madore, B. F., Gibson, B. K., *et al.* 2001, *ApJ*, 553, 47

Freedman, W. L., Madore, B. F., Scowcroft, V., *et al.* 2012, *ApJ*, 758, 24

Freytag, B. & Höfner, S. 2008, *A&A*, 483, 571

Georgy, C. 2012, *A&A*, 538, L8

Glatzel, W. & Kiriakidis, M. 1993, *MNRAS*, 263, 375

Grott, M., Chernigovski, S., & Glatzel, W. 2005, *MNRAS*, 360, 1532

Guzik, J. A., Cox, A. N., & Despain, K. M. 2005, *ASP-CS*, 332, 263

Guzik, J. A. & Lovekin, C. C. 2012, *The Astronomical Review*, 7, 030000

Gvaramadze, V. V., Langer, N., & Mackey, J. 2012, *MNRAS*, 427, L50

Höfner, S. & Andersen, A. C. 2007, *A&A*, 465, L39

Humphreys, R. M. & Koppelman, M. 2005, *ASP-CS*, 332, 159

Keller, S. C. 2008, *ApJ*, 677, 483

Lamers, H. J. G. L. M. & Cassinelli, J. P. 1999, *Introduction to Stellar Winds* (Cambridge, UK: Cambridge University Press)

Langer, N. 2012, *ARAA*, 50, 107

Mackey, J., Mohamed, S., Neilson, H. R., Langer, N., & Meyer, D. M.-A. 2012, *ApJ*, 751, L10

Marengo, M., Evans, N. R., Barmby, P., *et al.* 2010, *ApJ*, 725, 2392

Marsh Boyer, A. N., McSwain, M. V., Aragona, C., & Ou-Yang, B. 2012, *AJ*, 144, 158

Matthews, L. D., Marengo, M., Evans, N. R., & Bono, G. 2012, *ApJ*, 744, 53

Mattsson, L. & Höfner, S. 2011, *A&A*, 533, A42

Mattsson, L., Wahlin, R., & Höfner, S. 2010, *A&A*, 509, A14

Maund, J. R., Fraser, M., Ergon, M., *et al.* 2011, *ApJ*, 739, L37

Mohamed, S., Mackey, J., & Langer, N. 2012, *A&A*, 541, A1

Moravveji, E., Moya, A., & Guinan, E. F. 2012, *ApJ*, 749, 74

Neilson, H. R. & Lester, J. B. 2008, *ApJ*, 684, 569

Neilson, H. R. & Lester, J. B. 2009, *ApJ*, 690, 1829

Neilson, H. R., Cantiello, M., & Langer, N. 2011, *A&A*, 529, L9

Neilson, H. R., Engle, S. G., Guinan, E., *et al.* 2012a, *ApJ*, 745, L32

Neilson, H. R., Langer, N., Engle, S. G., Guinan, E., & Izzard, R. 2012b, *ApJ*, 760, L18

Neilson, H. R., Ngeow, C.-C., Kanbur, S. M., & Lester, J. B. 2009, *ApJ*, 692, 81

Neilson, H. R., Ngeow, C.-C., Kanbur, S. M., & Lester, J. B. 2010, *ApJ*, 716, 1136

Neiner, C., Gutiérrez-Soto, J., Floquet, M., *et al.* 2009, *CoAst*, 158, 319

Ngeow, C.-C., Kanbur, S. M., Neilson, H. R., Nanthakumar, A., & Buonaccorsi, J. 2009, *ApJ*, 693, 691

Oskinova, L. M., Todt, H., Ignace, R., *et al.* 2011, *MNRAS*, 416, 1456

Porter, J. M. & Rivinius, T. 2003, *PASP*, 115, 1153

Raassen, A. J. J., Cassinelli, J. P., Miller, N. A., Mewe, R., & Tepedelenlioğlu, E. 2005, *A&A*, 437, 599

Rauch, C., Mužić, K., Eckart, A., *et al.* 2013, *A&A*, 551, A35

Reimers, D. 1975, *Mémoires of the Société Royale des Sciences de Liège*, 8, 369

Rivinius, T. 2013, in: J. C. Suárez, R. Garrido, L. A. Balona, & J. Christensen-Dalsgaard (eds.), *Stellar Pulsations: Impact of New Instrumentation and New Insights*, ASSP, Vol. 31, 253

Saio, H., Georgy, C., & Meynet, G. 2013, *MNRAS*, 433, 1246

Sanyal, D. & Langer, N. 2013, in: *Massive Stars: From Alpha to Omega*, http://www.a2omega-conference.net/Posters/SessionII_47_Sanyal.pdf

Schröder, K.-P. & Cuntz, M. 2005, *ApJ*, 630, L73

Semaan, T., Gutiérrez-Soto, J., Frémat, Y., *et al.* 2013, in: J. C. Suárez, R. Garrido, L. A. Balona, & J. Christensen-Dalsgaard (eds.), *Stellar Pulsations: Impact of New Instrumentation and New Insights*, ASSP, Vol. 31, 261

Smartt, S. J. 2009, *ARAA*, 47, 63

Smith, N. 2013, *MNRAS*, 429, 2366

Smith, N., Vink, J. S., & de Koter, A. 2004, *ApJ*, 615, 475

Struve, O. 1931, *ApJ*, 73, 94

Townsend, R. H. D. 2000a, *MNRAS*, 318, 1

Townsend, R. H. D. 2000b, *MNRAS*, 319, 289

Turner, D. G., Abdel-Sabour Abdel-Latif, M., & Berdnikov, L. N. 2006, *PASP*, 118, 410

Ueta, T., Izumiura, H., Yamamura, I., *et al.* 2008, *PASJ*, 60, 407

Vink, J. S. 2011, *Ap&SS*, 336, 163

Vink, J. S. 2012, in: K. Davidson & R. M. Humphreys (eds.), *η Carinae and the Supernova Impostors*, Astrophysics and Space Science Library, Vol. 384, 221

Vink, J. S. 2013, *ASP-CS*, 470, 121

Wachter, A., Winters, J. M., Schröder, K.-P., & Sedlmayr, E. 2008, *A&A*, 486, 497

Willson, L. A. 2000, *ARAA*, 38, 573

Woitke, P. 2006, *A&A*, 460, L9

Yoon, S.-C. & Cantiello, M. 2010, *ApJ*, 717, L62

Precision Asteroseismology
Proceedings IAU Symposium No. 301, 2013
J. A. Guzik, W. J. Chaplin, G. Handler & A. Pigulski, eds.

© International Astronomical Union 2014
doi:10.1017/S1743921313014348

Asteroseismic signatures of magnetic activity variations in solar-type stars

Travis S. Metcalfe[1,2]

[1] Space Science Institute, Boulder CO 80301 USA;
email: travis@spsci.org

[2] Stellar Astrophysics Centre, Aarhus University, DK-8000 Aarhus C, Denmark

Abstract. Observations of magnetic activity cycles in other stars provide a broader context for our understanding of the 11-year sunspot cycle. The discovery of short activity cycles in a few stars, and the recognition of analogous variability in the Sun, suggest that there may be two distinct dynamos operating in different regions of the interior. Consequently, there is a natural link between studies of magnetic activity and asteroseismology, which can characterize some of the internal properties that are relevant to dynamos. I provide a brief historical overview of the connection between these two fields (including prescient work by Wojtek Dziembowski in 2007), and I highlight some exciting results that are beginning to emerge from the *Kepler* mission.

Keywords. stars: activity, stars: interiors, stars: magnetic fields, stars: oscillations

1. Background

In early 2005, I attended a seminar given by David Salabert in which he described his work documenting subtle shifts in the solar oscillation frequencies throughout the 11-year sunspot cycle (Salabert *et al.* 2004). The measurements relied on data from the IRIS network, and they clearly showed that even the low-degree ($l \leqslant 3$) solar oscillation frequencies were shifted by a few tenths of a μHz between magnetic minimum and maximum. If such frequency shifts were detectable in the Sun observed as a star, I realized that it might be possible to see similar behavior in other stars.

The high-degree oscillation frequency shifts in the Sun through the solar cycle were first characterized by Libbrecht & Woodard (1990). Using the thousands of oscillation modes then available from helioseismology they showed that the magnitude of the shift depended on both the geometry (spherical degree, l) and the frequency of the oscillation, with the largest shifts observed for higher degrees and at higher frequencies. The initial interpretation of these observations was given by Goldreich *et al.* (1991), who matched the frequency dependence of the shifts by considering a direct magnetic perturbation to the near-surface propagation speed of the acoustic waves.

Dziembowski & Goode (2005) developed a similar formalism to explain modern space-based observations of the solar acoustic oscillations (p modes) as well as the surface gravity waves (f modes). They identified some secondary effects that were needed to explain the shifts observed in both sets of modes: a decrease in the radial component of the turbulent velocity, and the associated changes in temperature. Shortly after this work was published, I contacted Wojtek Dziembowski to ask whether I could use his code to calculate the expected shifts in low-degree p modes for other solar-type stars.

At the time there were very few stars with detections of solar-like oscillations, but Fletcher *et al.* (2006) would soon publish evidence of a marginally significant shift in the p-mode frequencies of α Cen A by comparing ground-based observations with earlier data from the *WIRE* satellite, and Bedding *et al.* (2007) would see a similar (but statistically insignificant) shift when comparing two ground-based asteroseismology campaigns on β Hyi. Wojtek happily sent me a copy of his code, and with his help I spent more than a year trying to figure out how to adapt it for other stars before he generously invited me to come to Warsaw for a week and work on it together. During that week we kick-started a project on β Hyi, and we submitted the paper six months later (Metcalfe *et al.* 2007).

2. Predictions

As the activity level of the Sun rises from minimum to maximum, the p-mode oscillations are gradually shifted to higher frequencies. The magnitude of the shift is proportional to the change in activity level, so the simplest prediction for other stars is to assume that the mean shift in the p-mode frequencies scales with activity level. In other words, the largest shifts would be expected in the most active stars. This was the approach taken by Chaplin *et al.* (2007), who were working contemporaneously.

Wojtek took a slightly different approach to predicting the frequency shifts in other stars. He parametrized the shifts as $\Delta\nu \propto A_0 \, [R/M] \, Q_j(D_c)$, where A_0 scales with the activity level as in Chaplin *et al.* (2007), R and M are the radius and mass, and Q_j is a function of D_c which is the depth of the source of the perturbation below the photosphere. He demonstrated that this parametrization could remove all of the dependence on spherical degree and most of the frequency dependence in MDI data of the Sun with D_c fixed at 0.3 Mm. To extend the relation to other stars, he assumed that D_c would scale with the pressure scale height H_p in the outer layers: $D_c \propto H_p \propto L^{1/4} R^{3/2}/M$, which can be expressed in terms of the luminosity L, radius and mass.

At the time we were doing this work, β Hyi was the only star with a known activity cycle and multiple asteroseismic observing campaigns. The activity cycle had been observed in the Mg II h and k lines by the *IUE* satellite, and several years of additional observations were available in the archive after the initial characterization by Dravins *et al.* (1993). Phil Judge did a complete reanalysis of the *IUE* data, and determined a cycle period of 12 years with a maximum at 1986.9. Marty Snow produced a comparable record of solar Mg II h and k flux so we could scale the observed change in β Hyi to predict a mean shift in the oscillation frequencies. Just by luck, β Hyi was near magnetic maximum during the first detection of solar-like oscillations by Bedding *et al.* (2001), and close to magnetic minimum during the subsequent asteroseismic campaign by Bedding *et al.* (2007). The ground-based data were insufficient for a quantitative test, but Wojtek's relation qualitatively reproduced the observed shifts (Metcalfe *et al.* 2007).

Two years later, Christoffer Karoff led a project to define an optimal sample of asteroseismic targets in the *Kepler* field that would also be monitored for stellar activity variations from ground-based Ca II H and K measurements obtained throughout the mission (Karoff *et al.* 2009, 2013). When trying to determine which stars would show the largest frequency shifts, he was confronted with two conflicting predictions (see Fig. 1). The relation proposed by Chaplin *et al.* (2007) suggested that the largest shifts would be expected in the young active K stars. Wojtek's relation in Metcalfe *et al.* (2007) predicted that the hotter F stars would exhibit the largest shifts, which would grow even larger as the stars evolved. To be safe, the sample covered the full range of temperatures.

3. Confirmation

At the Beijing SONG workshop in March 2010, I gave a contributed talk on monitoring stellar magnetic activity cycles with SONG. During the coffee break Rafa García told me about some asteroseismic observations of the F star HD 49933, and he invited me to participate. The activity level of the star was not known, so I added it to a time-domain

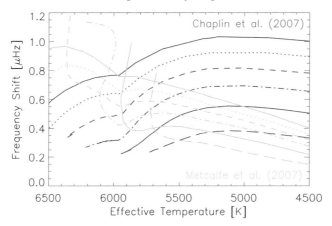

Figure 1. Predictions of the cycle-induced frequency shift as a function of effective temperature and age, with the scaling proposed by Chaplin *et al.* (2007) shown in blue and that proposed by Metcalfe *et al.* (2007) shown in green. Solid lines are Padova isochrones at an age of 1 Gyr, and the sets of dashed lines show progressively older isochrones at 1.6, 2.5, 4, 6.3, and 10 Gyr (adapted from Karoff *et al.* 2009).

survey of Ca II H and K emission that I was conducting for a sample of bright stars in the southern hemisphere (Metcalfe *et al.* 2009).

García *et al.* (2010) discovered anti-correlated changes in the frequencies and amplitudes of the oscillations in HD 49933 during 150 days of continuous monitoring by *CoRoT*. The frequency shifts were positive, and passed through a minimum while the amplitudes increased and passed through a maximum—the same pattern of changes that occurs in the Sun as it passes through a magnetic minimum. Convection stochastically excites and intrinsically damps solar-like oscillations. Magnetic fields inhibit convection, suppressing the oscillation amplitudes while simultaneously shifting the frequencies.

Salabert *et al.* (2011) pushed the analysis further, and examined the frequency dependence of the shifts. Just as in the Sun, the shifts grew steadily larger toward higher frequencies. It seemed clear that *CoRoT* had made the first asteroseismic detection of a stellar magnetic cycle, but there was one striking difference between HD 49933 and the Sun. The frequency shifts observed in this ~2 Gyr-old F star at 6600 K were 4–5 times larger than solar, providing the first confirmation of Wojtek's relation.

4. Future prospects

The archive of *Kepler* data represents an unprecedented opportunity to study the short-period magnetic cycles that have been observed in some rapidly rotating F stars. The high precision time-series photometry collected every 30 minutes over the past four years can be used to measure rotation periods from spot modulation and to monitor the longer-term brightness changes associated with the stellar cycle. Furthermore, for targets that have been observed in short cadence (1-minute sampling), asteroseismology allows a characterization of the star including key dynamo ingredients such as the depth of the surface convection zone (Mazumdar *et al.* 2012) and radial differential rotation (Deheuvels *et al.* 2012). The asteroseismic data can also be used to monitor the solar-like oscillations over time, allowing a search for the same pattern of changes that have been seen for the Sun and HD 49933 in response to their magnetic cycles. Rapidly rotating F stars are the ideal targets because they show the shortest cycle periods (Metcalfe *et al.* 2010), and the frequency shifts are significantly larger than in the Sun (see Fig. 1). Young, rapidly rotating K stars can also show relatively short cycles (Metcalfe *et al.* 2013), but the asteroseismic signatures are expected to be smaller.

Mathur *et al.* (2013) recently examined the archive of *Kepler* observations for a sample of 22 rapidly rotating F stars. Wavelet analysis of the long light curves revealed clear signatures of latitudinal differential rotation and evidence for short magnetic cycles in a few stars. The best target in the sample has three years of continuous asteroseismic data spanning what appears to be a complete magnetic cycle, so additional tests of Wojtek's relation should soon be possible.

Acknowledgements

This long story has gradually unfolded with the collaboration and support of Tim Bedding, Phil Judge, Christoffer Karoff, Bill Chaplin, Rafa García, Savita Mathur, and David Salabert. I would particularly like to thank Wojtek Dziembowski for graciously inviting me to work with him at the Copernicus Astronomical Center for a week in September 2006. This work was partially supported by NASA grant NNX13AC44G.

References

Bedding, T. R., Butler, R. P., Kjeldsen, H., *et al.* 2001, *ApJ*, 549, L105
Bedding, T. R., Kjeldsen, H., Arentoft, T., *et al.* 2007, *ApJ*, 663, 1315
Chaplin, W. J., Elsworth, Y., Houdek, G., & New, R. 2007, *MNRAS*, 377, 17
Deheuvels, S., García, R. A., Chaplin, W. J., *et al.* 2012, *ApJ*, 756, 19
Dravins, D., Linde, P., Fredga, K., & Gahm, G. F. 1993, *ApJ*, 403, 396
Dziembowski, W. A. & Goode, P. R. 2005, *ApJ*, 625, 548
Fletcher, S. T. Chaplin, W. J., Elsworth Y., *et al.* 2006, *MNRAS*, 371, 935
García, R. A., Mathur, S., Salabert, D., *et al.* 2010, *Science*, 329, 1032
Goldreich, P., Murray, N., Willette, G., & Kumar, P. 1991, *ApJ*, 370, 752
Karoff, C., Metcalfe, T. S., Chaplin, W. J., *et al.* 2009, *MNRAS*, 399, 914
Karoff, C., Metcalfe, T. S., Chaplin, W. J., *et al.* 2013, *MNRAS*, 433, 3227
Libbrecht, K. G. & Woodard, M. F. 1990, *Nature*, 345, 779
Mathur, S., García, R. A., Ballot, J., *et al.* 2013, *A&A*, submitted
Mazumdar, A., Monteiro, M. J. P. F. G., Ballot, J., *et al.* 2012, *AN*, 333, 1040
Metcalfe, T. S., Dziembowski, W. A., Judge, P. G., & Snow, M. 2007, *MNRAS*, 379, L16
Metcalfe, T. S., Judge, P. G., Basu, S., *et al.* 2009, *Solar Analogs II workshop* (arXiv: 0909.5464)
Metcalfe, T. S., Basu, S., Henry, T. J., *et al.* 2010, *ApJ*, 723, L213
Metcalfe, T. S., Buccino, A. P., Brown, B. P., *et al.* 2013, *ApJ*, 763, L26
Salabert, D., Fossat, E., Gelly, B., *et al.* 2004, *A&A*, 413, 1135
Salabert, D., Régulo, C., Ballot, J., *et al.* 2011, *A&A*, 530, A127

Precision Asteroseismology
Proceedings IAU Symposium No. 301, 2013
J. A. Guzik, W. J. Chaplin, G. Handler & A. Pigulski, eds.

© International Astronomical Union 2014
doi:10.1017/S174392131301435X

Pulsations as a mass-loss trigger in evolved hot stars†

Michaela Kraus[1], Dieter H. Nickeler[1], Maximiliano Haucke[2], Lydia Cidale[2,3], Roberto Venero[2,3], Marcelo Borges Fernandes[4], Sanja Tomić[5] and Michel Curé[6]

[1] Astronomický ústav, Akademie věd České Republiky,
Fričova 298, 251 65 Ondřejov, Czech Republic
email: kraus@sunstel.asu.cas.cz

[2] Departamento de Espectroscopía Estelar, Facultad de Ciencias Astronómicas y Geofísicas,
Universidad Nacional de La Plata, Paseo del Bosque s/n, B1900FWA, La Plata, Argentina

[3] Instituto de Astrofísica de La Plata, CCT La Plata, CONICET-UNLP,
Paseo del Bosque s/n, B1900FWA, La Plata, Argentina

[4] Observatório Nacional,
Rua General José Cristino 77, 20921-400 São Cristovão, Rio de Janeiro, Brazil

[5] Department of Astronomy, Faculty of Mathematics, University of Belgrade,
Studentski trg 16, 11000 Belgrade, Serbia

[6] Departmento de Física y Astronomía, Facultad de Ciencias, Universidad de Valparaíso,
Av. Gran Bretaña 1111, Casilla 5030, Valparaíso, Chile

Abstract. During their post-main sequence evolution, massive stars pass through several short-lived phases, in which they experience enhanced mass loss in the form of clumped winds and mass ejection events of unclear origin. The discovery that stars populating the blue luminous part of the Hertzsprung-Russell diagram can pulsate suggests that stellar pulsations might influence or trigger enhanced mass loss and eruptions. We present recent results for two objects in different phases: a B[e] star at the end of the main sequence and a B-type supergiant.

Keywords. stars: early-type, stars: emission-line, stars: Be, stars: mass loss, stars: oscillations

1. Introduction

The post-main sequence evolution of massive stars is one of the major unsolved problems in massive star research. Massive stars can pass through several short-lived phases, in which they lose tremendous amounts of mass via enhanced mass loss and eruptive mass ejection events of yet unknown origin. During the classical Blue Supergiant (BSG) stage, mass loss occurs via line-driven winds. The mass-loss rates involved are still uncertain and strongly depend on whether the winds emerge smoothly from the stellar surface or whether instabilities occur at the base of the wind, which result in the formation of clumpy structures. The cause of such instabilities is still unclear, but stellar pulsations, which were recently found in a few BSGs (e.g., Saio *et al.* 2006, Lefever *et al.* 2007, Kraus *et al.* 2012), might influence, and maybe even trigger, enhanced mass loss from evolved hot stars. Here we present recent results for a B[e] star and a BSG.

† Based on observations acquired at the Ondřejov Observatory, Czech Republic, the Dominion Astrophysical Observatory, Herzberg Institute of Astrophysics, National Research Council of Canada, and with the HERMES spectrograph, which is supported by the Fund for Scientific Research of Flanders (FWO), Belgium, the Research Council of K.U.Leuven, Belgium, the Fonds National Recherches Scientific (FNRS), Belgium, the Royal Observatory of Belgium, the Observatoire de Genève, Switzerland and the Thüringer Landessternwarte Tautenburg, Germany.

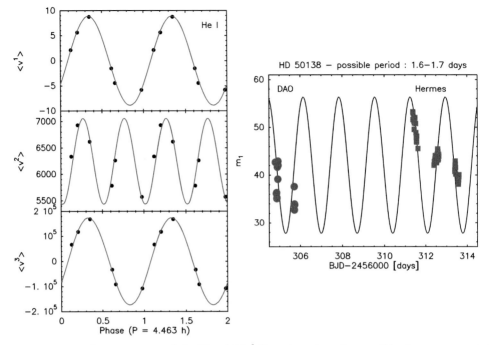

Figure 1. First three moments of the He I 4026 Å line phased to the possible short-term period of 4.463 h (left). Radial velocity (first moment) variations in the combined data sets from DAO and HERMES imply the existence of an additional longer period with larger amplitude (right).

2. The Galactic B[e] star HD 50138

HD 50138, as a member of the B[e] stars, is surrounded by high-density material giving rise to strong Balmer and forbidden emission lines, and a circumstellar dusty disk ($i = 56 \pm 4°$) resolved by interferometry (Borges Fernandes *et al.* 2011). It experienced outbursts and shell-ejection phases in the past, and its location at the end of (or slightly beyond) the main sequence (Borges Fernandes *et al.* 2009), close to confirmed pulsating Be stars, suggests pulsations as possible trigger for the outbursts.

First indications for pulsational activity in the atmosphere of HD 50138 were found from a sample of high-resolution spectra obtained during different observing runs at the 1.2-m Mercator (HERMES) and DAO telescopes. The data showed strong night-to-night variability in all photospheric and wind lines (Borges Fernandes *et al.* 2012). In addition, the photospheric lines displayed a large broadening component of $30 - 40 \, \mathrm{km \, s^{-1}}$ in excess of the stellar rotational broadening ($v \sin i = 74.7 \pm 0.8 \, \mathrm{km \, s^{-1}}$). Such high values of excess broadening (referred to as 'macroturbulence') are well known from BSGs and are attributed to pulsational activity (Aerts *et al.* 2009). Application of the moment method (Aerts *et al.* 1992, North & Paltani 1994) to the photospheric He I and Si II lines suggests the presence of two possible periods (4.463 h and $1.6-1.7$ d). The results for the He I 4026 Å line are depicted in Fig. 1. If these periods are confirmed, HD 50138 will be the first pulsating B[e] star, providing a very important milestone for our understanding of the triggering mechanism leading to mass ejection events in B[e] stars.

3. The blue supergiant star 55 Cyg = HD 198478

A second group of evolved hot and massive stars discussed during this meeting (see Godart *et al.*, this volume) are BSGs. Members of this class were long known to display

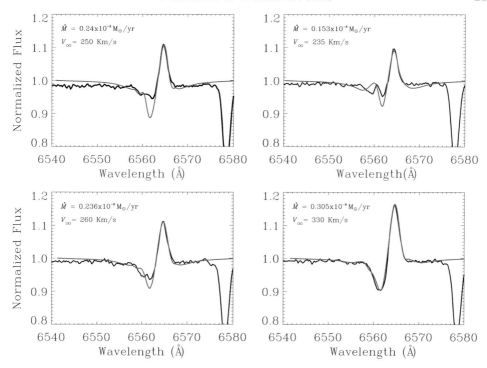

Figure 2. Fits to the Hα profile observed within four consecutive nights (Sep 21 – 24, 2010). Note the strong increase in $\dot{M}$ and v_∞ from the 2nd to the last night.

strong photometric and spectroscopic variability, and the profiles of their photospheric lines contain large contributions from macroturbulent broadening in excess of rotational broadening (e.g., Markova & Puls 2008), indicating stellar pulsational activity. In fact, recent theoretical investigations by Saio *et al.* (2006) revealed the presence of a new instability domain in the HRD covering the location of the BSGs.

We study a sample of bright northern BSGs located within this instability domain using the Perek 2-m telescope at Ondřejov Observatory. One of the objects we are surveying is the early B-type supergiant 55 Cyg (HD 198478). While its stellar parameters have been determined accurately from optical spectroscopy ($T_{\mathrm{eff}} = 17\,500 \pm 500\,\mathrm{K}$; $\log L/L_\odot = 5.1 \pm 0.2$; $v\sin i = 37 \pm 2\,\mathrm{km\,s^{-1}}$; $v_{\mathrm{macro}} = 53\,\mathrm{km\,s^{-1}}$; Markova & Puls 2008), the situation is less clear regarding the wind parameters. Mass-loss rates and terminal wind velocities are typically obtained from the emission component of the Hα line. However, the Hα line displays strong night-to-night variability. From our long-term observations, we found a zoo of profile shapes ranging from P Cygni, to pure single emission, to almost complete disappearance, to double- or multiple-peaked, and no cyclic variation was found over a total of 25 consecutive observing nights. Consequently, modeling the emission component of observations taken in different nights delivered different sets of wind parameters.

So far, we have collected a total of 339 spectra in the Hα region distributed over 59 nights between August 2009 and August 2013. The spectral coverage is 6270 – 6730 Å with a resolution of $R \simeq 13\,000$. Of these, we modeled the Hα profile from 32 different nights using the NLTE code FASTWIND (Puls *et al.* 2005) to obtain the wind parameters. We found that the mass-loss rate, $\dot{M}$, and terminal wind velocity, v_∞, change simultaneously with large night-to-night variability (Fig. 2). The value in both parameters spreads over more than a factor of three: $\dot{M} = (1.4 - 4.3) \cdot 10^{-7}\,\mathrm{M_\odot/yr}$ and $v_\infty = 180 - 700\,\mathrm{km\,s^{-1}}$.

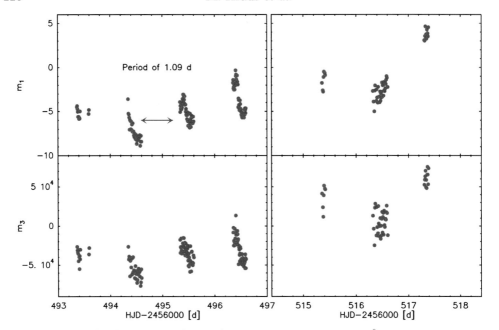

Figure 3. First (top) and third (bottom) moments of the He I 6678 Å line showing identical variations typical for pulsations. The moments were computed from time series within four (left) and three (right) consecutive nights. Both observing epochs suggest a possible period of 1.09 d. The shift in radial velocity between the two epochs indicates additional superimposed period(s).

From the moment analysis of time series within four and three consecutive nights of the He I 6678 Å line, we obtained a possible pulsation period of 1.09 d (Fig. 3). The shift in radial velocity between the first and second set of time series suggests that the 1.09 d period is superimposed on a second (and probably more) period(s). A proper period and mode analysis (work in progress) is necessary to confirm the identifications.

Acknowledgements

M.K. and D.H.N. acknowledge financial support from GAČR under grant number P209/11/1198. Financial support for International Cooperation of the Czech Republic (MŠMT, 7AMB12AR021) and Argentina (Mincyt-Meys, ARC/11/10) is acknowledged. The Astronomical Institute Ondřejov is supported by the project RVO:67985815. L.C. acknowledges financial support from CONICET (PIP 0300) and the Agencia (préstamo BID, PICT 2011/0885).

References

Aerts, C., De Pauw, M., & Waelkens, C. 1992, *A&A*, 266, 294
Aerts, C., Puls, J., Godart, M., & Dupret, M.-A. 2009, *A&A*, 508, 409
Borges Fernandes, M., Kraus, M., Chesneau, O., *et al.* 2009, *A&A*, 508, 309
Borges Fernandes, M., Meilland, A., Bendjoya, P., *et al.* 2011, *A&A*, 528, A20
Borges Fernandes, M., Kraus, M., Nickeler, D. H., *et al.* 2012, *A&A*, 548, A13
Kraus, M., Tomić, S., Oksala, M. E., & Smole, M. 2012, *A&A*, 542, L32
Lefever, K., Puls, J., & Aerts, C. 2007, *A&A*, 463, 1093
Markova, N. & Puls, J. 2008, *A&A*, 478, 823
North, P. & Paltani, S. 1994, *A&A*, 288, 155
Puls, J., Urbaneja, M. A., Venero, R., *et al.* 2005, *A&A*, 435, 669
Saio, H., Kuschnig, R., Gautschy, A., *et al.* 2006, *ApJ*, 650, 1111

Precision Asteroseismology
Proceedings IAU Symposium No. 301, 2013
J. A. Guzik, W. J. Chaplin, G. Handler & A. Pigulski, eds.

© International Astronomical Union 2014
doi:10.1017/S1743921313014361

Testing microphysics data

Przemysław Walczak and Jadwiga Daszyńska-Daszkiewicz

Instytut Astronomiczny, Uniwersytet Wrocławski, ul. Kopernika 11, 51-622 Wrocław, Poland
email: walczak@astro.uni.wroc.pl, daszynska@astro.uni.wroc.pl

Abstract. High precision asteroseismic data provide a unique opportunity to test input microphysics such as stellar opacities, chemical composition or equation of state. These tests are possible because pulsational frequencies as well as amplitudes and phases of the light variations are very sensitive to the internal structure of a star. We can therefore compute pulsation models and compare them with observations. The agreement or differences should tell us whether some models are adequate or not, and which input data need to be improved.

Keywords. stars: variables: β Cep, stars: individual: θ Oph, γ Peg, 12 Lac, ν Eri

1. Introduction

One of the most important ingredients of the stellar input physics are opacities. The value of the opacity coefficient affects the pulsational properties of models. There are two basic asteroseismic tools: frequencies and the corresponding values of the complex non-adiabatic f-parameter. The f-parameter is defined as the ratio of the radiative flux change to the radial displacement at the photospheric level (Daszyńska-Daszkiewicz *et al.* 2003, 2005). Its value determines amplitudes and phases of light variations. In the case of B-type stars, the empirical value of the f-parameter can be determined only for modes that are visible both in multicolor photometry and spectroscopy. It is important to add that the empirical f-parameter depends slightly on the input from model atmospheres. A discussion of these effects can be found in Daszyńska-Daszkiewicz & Szewczuk (2012) and Daszyńska-Daszkiewicz *et al.* (2013). Here, in all computations we adopt the LTE models of stellar atmospheres by Kurucz (2004) with microturbulence velocity of $\xi_t = 2\,\mathrm{km\,s^{-1}}$.

We tested three available opacity tables: OP (Seaton 2005), OPAL (Iglesias & Rogers 1996) and new data from Los Alamos (LA) (Magee *et al.* 1995). In Fig. 1, we show a comparison of the Rosseland mean opacity, κ, plotted as a function of the temperature within stellar models. The metallicity parameter was set to $Z = 0.02$. In the left and right panels we considered stellar models of $5\,M_\odot$, $\log T_{\mathrm{eff}} \sim 4.196$ and $10\,M_\odot$, $\log T_{\mathrm{eff}} \sim 4.373$, respectively.

We can easily notice two high opacity bumps. The first one, occurring at the lower temperature, is connected with ionization of He II. The second one, the so-called Z bump, is caused by a large number of transition lines of iron-group elements. Although there are some small differences near the Z bump, the OP and OPAL data are quite similar. On the other hand, the LA opacity coefficient is in general smaller than the OP and OPAL ones, especially in the region of the Z bump. As one can expect, this fact has huge consequences on pulsational instability in models of B-type stars.

2. Inferring constraints on opacities

We have chosen four β Cephei-type stars for our tests: θ Ophiuchi (HD 157056), γ Pegasi (HD 886), 12 Lacertae (HD 214993) and ν Eridani (HD 29248). In Fig. 2, we plot their

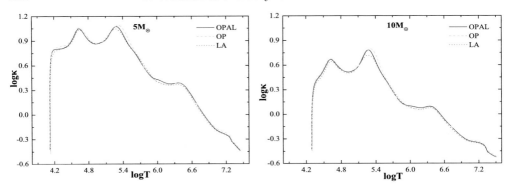

Figure 1. The Rosseland mean opacity, κ, as a function of the temperature, T, inside of the 5 $M_\odot$ (left panel) and 10 $M_\odot$ (right panel) stellar models with effective temperatures $\log T_{\rm eff} \sim 4.196$ and $\log T_{\rm eff} \sim 4.373$, respectively. Three sources of opacity data were considered: the OP, OPAL and LA tables.

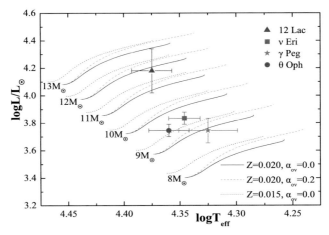

Figure 2. The H-R diagram with positions of four β Cephei stars: θ Oph, γ Peg, 12 Lac and ν Eri. The theoretical evolutionary tracks for masses from 8 to 13 $M_\odot$ were calculated for different values of metallicity, Z, and overshooting parameter, $\alpha_{\rm ov}$. Only the main-sequence part of evolution is shown.

positions in the Hertzsprung-Russell diagram. We added also evolutionary tracks from ZAMS to TAMS for masses from 8 up to 13 $M_\odot$ in steps of 1 $M_\odot$. The tracks were calculated with the OP opacity tables, two values of metallicity ($Z = 0.015$ and 0.020) and two values of the overshooting parameter ($\alpha_{\rm ov} = 0.0$ and 0.2). Unless otherwise noted, in all computations we assumed the chemical composition by Asplund *et al.* (2009) and the initial hydrogen abundance $X = 0.7$.

In Fig. 2 we can easily see that the masses of the stars are from about 9 $M_\odot$ to about 12 $M_\odot$. All stars are most likely in the core hydrogen-burning evolution phase. We can also notice the huge impact of metallicity and the overshooting parameter on the theoretical tracks. The lower the metallicity, the higher the mass that can be derived from the H-R diagram. A high value of the overshooting parameter prolongs the duration of the main-sequence phase.

2.1. θ Ophiuchi

θ Oph is a β Cephei-type star that pulsates in at least seven frequencies (Handler *et al.* 2005). Three of them were also found in radial velocity measurements (Briquet

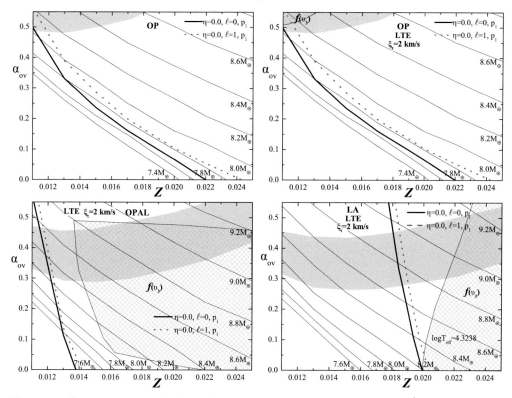

Figure 3. Seismic models of θ Oph fitting two frequencies ($\nu_3 = 7.4677$ d^{-1} as an $l = 0$, p$_1$ mode and $\nu_6 = 7.8742$ d^{-1} as $l = 1$, p$_1$ mode) on the overshooting ($\alpha_{\rm ov}$) *vs.* metallicity (Z) planes. Grey areas indicate models lying inside of the observational error box. We show also lines of constant mass (thin solid lines) and instability borders for the radial mode (thick solid line) and dipole mode (thick dashed line). In the upper panels we used the OP opacity tables. In the upper right panel we marked models fitting additionally the empirical value of the f-parameter for ν_3 (hatched area). The bottom panels are the same as the upper right panel, except that we used the OPAL (left panel) and LA (right panel) opacities.

et al. 2005). The effective temperature of θ Oph, $\log T_{\rm eff} = 4.360 \pm 0.018$, was determined by Handler *et al.* (2005). The luminosity, $\log L/L_\odot = 3.746 \pm 0.045$, was calculated taking into account the *Hipparcos* parallax $\pi = 7.48 \pm 0.17$ mas and the bolometric correction from the calibration by Flower (1996).

In our modelling we used only axisymmetric modes (with azimuthal number $m = 0$). In the case of θ Oph, mode identification indicates two centroid modes: $\nu_3 = 7.4677$ d^{-1} (radial p$_1$) and $\nu_6 = 7.8742$ d^{-1} (dipole p$_1$) (Daszyńska-Daszkiewicz & Walczak 2009). We have found models fitting these two frequencies for different values of metallicity, Z, and the overshooting parameter, $\alpha_{\rm ov}$. The results for the OP opacity tables are shown in the upper left panel of Fig. 3 on the overshooting parameter ($\alpha_{\rm ov}$) *vs.* metallicity (Z) plane. We marked lines of constant masses (thin lines) and instability borders for the radial (thick solid line) and dipole (thick dashed line) modes. Models that lie above these instability borders are excited. Instability borders were defined as the zero value of the instability parameter, $\eta = W / \int_0^R \left| \frac{dW}{dr} \right| dr$, where W is the work integral and R is the stellar radius. It can be seen that there exist a lot of models fitting the ν_3 and ν_6 frequencies. Only for the low values of metallicity and the overshooting parameter were we unable to find seismic models (bottom left corner of the panel). The grey area indicates models lying inside of the observational errors of the effective temperature and luminosity of θ Oph. Models below the grey area are cooler and less bright than the error box.

The radial mode, ν_3, was detected both in photometry and spectroscopy. Therefore we were able to derive the empirical value of the non-adiabatic f-parameter for this mode. We compared this value with its theoretical equivalent and found models fitting it (within the errors). In the upper right panel of Fig. 3 we showed the same figure as in the upper left panel, but in addition we marked models which also fit the empirical value of the f-parameter for the ν_3 mode (hatched area labeled with $f(\nu_3)$).

As we can see, models fitting two frequencies (ν_3 and ν_6) and the f-parameter of the radial mode are located inside the observational error box. On the other hand, these models have very efficient overshooting from the convective core, with $\alpha_{ov} \sim 0.5$, which is not expected in a rather slowly rotating star like θ Oph ($V_{rot} \approx 30$ km s^{-1}).

A different situation arises in the case of the OPAL and LA opacities (bottom left and right panel of Fig. 3, respectively). For a given value of Z and α_{ov}, OPAL and LA models have much higher masses than OP models. Also, the effective temperature and luminosity are larger, and models that are inside of the error box appear for less effective core overshoot ($\alpha_{ov} \sim 0.3 - 0.5$).

With the OPAL opacities we were able to find quite a large number of models fitting the f-parameter of the radial mode ν_3. Some of these models lie inside of the observational error box. For the case of the LA opacities, we managed to fit the f-parameter only for models with metallicities larger that about 0.02. Moreover, for the LA opacity models with Z lower than 0.02, the modes considered are stable.

In the case of θ Oph, the models turned out to be very sensitive to the differences between opacities. As we could see, the value of κ has also a very large impact on the f-parameter.

2.2. γ Pegasi

γ Peg is a B2 spectral-type star that pulsates in at least 14 modes (Handler *et al.* 2009). Six of the modes have very low frequencies (< 0.9 d^{-1}), typical for the Slowly Pulsating B-type stars. The remaining 8 modes are of the β Cep-type. Because of this, the star is a hybrid pulsator of the β Cep/SPB type.

The effective temperature ($\log T_{eff} = 4.325 \pm 0.026$) as well as the luminosity ($\log L/L_\odot = 3.744 \pm 0.09$) of γ Peg, shown in Fig. 2 were adopted from Walczak *et al.* (2013). We chose two well-identified β Cep modes: the radial p$_1$ ($\nu_1 = 6.58974$ d^{-1}) and dipole g$_1$ ($\nu_5 = 6.01616$ d^{-1}), and constructed models fitting them.

The models are shown in Fig. 4 on the α_{ov} vs. Z plane in three panels corresponding to computations with the OP, OPAL and LA opacities. Here, the unstable modes are below the drawn instability borders. For γ Peg we were also able to determine the empirical values of the f-parameter for the ν_1 and ν_5 modes. They are marked in Fig. 4 as hatched areas labeled as $f(\nu_1)$ and $f(\nu_5)$. Unfortunately, we could not find a single model that would fit the empirical values of the f-parameter for these two modes simultaneously. The OP and OPAL models are quite similar. There is only a difference in the position of models fitting the f-parameter. In the case of the OPAL data the hatched areas are shifted to a higher value of metallicity. With the LA data, however, we did not find models fitting the empirical value of the f-parameter for the ν_5 mode (in the metallicity range considered, $Z \in 0.007 - 0.025$). The LA models fitting the f-parameter for ν_1 mode have high metallicities, $Z \approx 0.023$, and are outside of the observational error box.

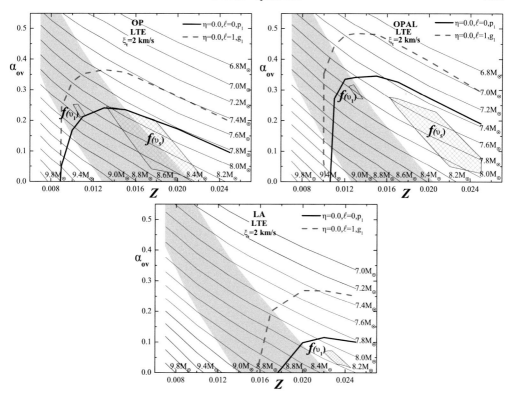

Figure 4. Seismic models of γ Peg fitting two frequencies ($\nu_1 = 6.58974$ d^{-1} as $l = 0$, p$_1$ mode and $\nu_5 = 6.01616$ d^{-1} as $l = 1$, g$_1$ mode) on the $\alpha_{\rm ov}$ *vs.* Z plane. In the upper left panel we used the OP opacities, in the upper right – OPAL and in the bottom – LA.

2.3. *12 Lacertae*

12 Lac is a well known pulsating β Cep/SPB type star. It pulsates in 11 modes (Handler *et al.* 2006). One mode is SPB-type. In Fig. 2, we showed the error box of 12 Lac. The values of the effective temperature ($\log T_{\rm eff} = 4.375 \pm 0.018$) and luminosity ($\log L/L_\odot = 4.18 \pm 0.16$) were taken from Handler *et al.* (2006).

Based on the mode identification (Daszyńska-Daszkiewicz *et al.* 2013), we know that at least two modes are axisymmetric: $\nu_2 = 5.066346$ d^{-1} ($l = 1$, g$_1$) and $\nu_4 = 5.334357$ d^{-1} ($l = 0$, p$_1$). Models fitting these two frequencies are plotted in Fig. 5 on the $\alpha_{\rm ov}$ vs. Z plane. We managed to derive the value of the empirical f-parameter for these two modes. We were also able to find models which fit the f-parameter for these two modes simultaneously (hatched regions in Fig. 5).

As we can see, in the case of the OP data (upper left panel of Fig. 5), there are plenty of models fitting two frequencies (ν_2 and ν_4) and their f-parameters that are inside of the error box (grey area). The OPAL models fitting additionally the f-parameter for ν_2 and ν_4 are outside of the error box (upper right panel of Fig. 5). A similar situation occurred in the case of the LA opacities. Models fitting the f-parameters are outside of the error box.

2.4. *ν Eridani*

ν Eri is one of the most extensively studied β Cep/SPB pulsators. As is the case γ Peg, this star pulsates in at least 14 modes (Handler *et al.* 2004; Jerzykiewicz *et al.* 2005). Two of them are SPB-type. This star pulsates in three well-identified centroid modes (e.g.,

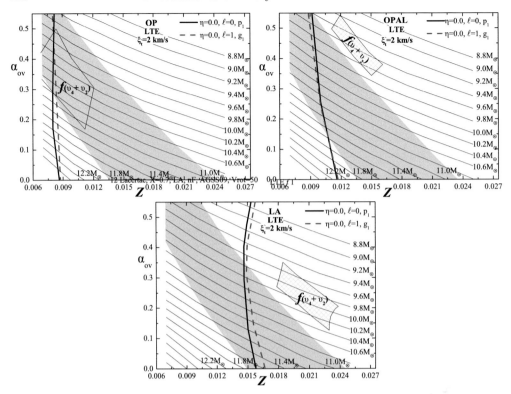

Figure 5. Seismic models of 12 Lac fitting two frequencies ($\nu_2 = 5.066346$ d^{-1} as $l = 1$, g$_1$ mode and $\nu_4 = 5.334357$ d^{-1} as $l = 0$, p$_1$ mode) on the $\alpha_{\rm ov}$ *vs.* Z plane. In the upper left panel we used the OP opacities, in the upper right – OPAL and in the bottom – LA.

Daszyńska-Daszkiewicz & Walczak 2010): one radial p$_1$ mode ($\nu_1 = 5.7632828$ d^{-1}), and two dipoles: g$_1$ ($\nu_4 = 5.6372470$ d^{-1}) and p$_1$ ($\nu_6 = 6.243847$ d^{-1}). There is also a $\nu_9 = 7.91383$ d^{-1} mode, which could be the centroid of the dipole p$_2$ mode.

The effective temperature of ν Eri, $\log T_{\rm eff} = 4.346 \pm 0.014$, was adopted from Daszyń-ska-Daszkiewicz *et al.* (2005). The luminosity, $\log L/L_\odot = 3.835 \pm 0.045$, was calculated with the *Hipparcos* parallax $\pi = 483 \pm 19$ mas (van Leeuwen 2007). We used also the Flower (1996) bolometric correction corresponding to the effective temperature of ν Eri.

In the left panel of Fig. 6, we plotted models fitting three frequencies of ν Eri (ν_1, ν_4 and ν_6). The results are presented on the Z *vs.* $\alpha_{\rm ov}$ plane. The large dots correspond to models fitting additionally also the ν_9 frequency. Because we used three frequencies, we have lines of models instead of a plane, like in the case of θ Oph, γ Peg or 12 Lac. Four frequencies reduce the number of models to one point in this kind of figure. We marked also the direction of increasing mass and effective temperature.

We can easily notice that seismic models calculated with different opacity tables are well separated in metallicity. The highest values of Z occur for the OP data ($Z = 0.016 - 0.018$). With OPAL opacities the metallicity is in the range from 0.0135 to 0.0145. The lowest metallicity was found with the LA tables: $Z = 0.013 - 0.0145$. In the right panel of Fig. 6, we marked seismic models of ν Eri on the H-R diagram. We showed also the ν Eri error box. Unfortunately, seismic models are only partially inside the error box. It is interesting that the LA models have somewhat higher values of the effective temperature and fit the observational parameters slightly better.

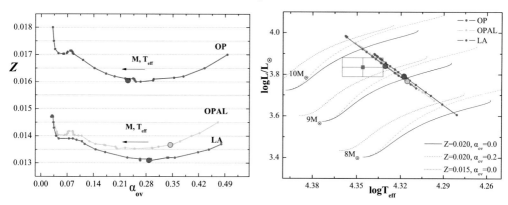

Figure 6. *Left panel*: seismic models of ν Eri fitting three frequencies ($\nu_1 = 5.7632828$ d^{-1} as $l = 0$, p$_1$ mode, $\nu_4 = 5.6372470$ d^{-1} as $l = 1$, g$_1$ mode and $\nu_6 = 6.243847$ d^{-1} as $l = 1$, p$_1$ mode) on the Z *vs.* $\alpha_{\rm ov}$ plane. *Right panel*: H-R diagram with position of seismic models of ν Eri.

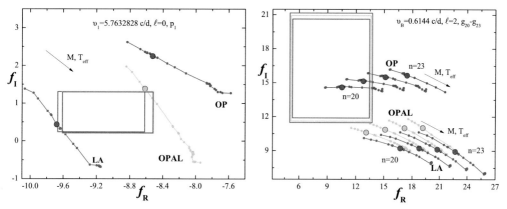

Figure 7. Comparison of the empirical (box) and theoretical (line) values of the f-parameter for the ν_1 mode (left panel) and $\nu_{\rm B}$ mode (right panel).

In Fig. 7, we show a comparison of the empirical and theoretical values of the non-adiabatic f-parameter for two modes: radial ν_1 (left panel) and quadrupole $\nu_{\rm B} = 0.6144$ d^{-1} (right panel) which is of SPB type. We plotted the imaginary part of the f-parameter as a function of its real part. The boxes represent empirical values and the lines - theoretical. The large dots mark models fitting also the ν_9 frequency.

We see that the agreement is rather poor. In the case of the radial mode, we have some marginal agreement for the LA and OPAL models. For the $\nu_{\rm B}$ frequency we plotted a few modes with different radial orders (from $n = 20$ to $n = 23$). These models do not fit exactly the $\nu_{\rm B}$ frequency, but are very close to it. In this case, the OP models fit the empirical value of the f-parameter much better than the LA or OPAL models.

The low metallicity of the LA models as well as the low value of the opacity coefficient itself cause large problems with excitation of modes. In Fig. 8 we plotted the instability parameter, η, as a function of frequency for three seismic models of ν Eri computed with the OP and OPAL opacities (left panel) and LA tables (right panel). The short vertical lines correspond to frequency spectrum of ν Eri. We can see that the LA model is almost entirely stable. The OP and OPAL models cannot excite the high frequency modes. Also the very low frequencies are stable, especially for OPAL opacities.

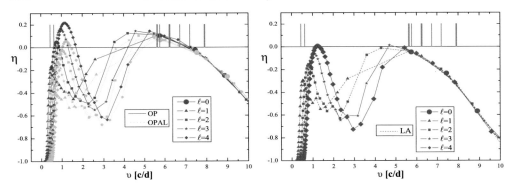

Figure 8. Instability parameter, η, as a function of the frequency for three seismic models of ν Eri calculated with the OP and OPAL data (left panel) and LA opacities (right panel).

3. Summary

The B-type pulsators are very suitable for testing the opacity tables because a small difference in κ results in quite large differences in seismic models. We could see that both frequencies and the f-parameter are sensitive to the opacities.

We found that, in case of θ Oph, the OPAL tables are the best. 12 Lac prefers OP data, while models of γ Peg are rather similar with the OP and OPAL opacities. The LA data seem not to be good for γ Peg. For ν Eri, the f-parameter of the radial mode prefers the LA or OPAL opacities, but the SPB-type mode favors instead the OP tables. It seems that, in some parameter space, the OP opacities are better, in others – the OPAL data. The LA opacity table values are definitely too low; they are much smaller than OP or OPAL, especially in the region of the Z bump, where the differences reach 9 to 10%.

Although the presented results are not unambiguous, they show that further improvements and corrections in the opacity computations are needed.

Acknowledgment. Calculations have been carried out using resources provided by the Wroclaw Centre for Networking and Supercomputing (http://wcss.pl), grant No. 265.

References

Asplund, M., Grevesse, N., Sauval, A. J., & Scott, P. 2009, *ARAA*, 47, 481
Briquet M., Lefever K., Uytterhoeven K., & Aerts C. 2005, *MNRAS*, 362, 619
Daszyńska-Daszkiewicz, J. & Szewczuk, W. 2011, *ApJ*, 728, 17
Daszyńska-Daszkiewicz, J. & Walczak, P. 2009, *MNRAS*, 398, 1961
Daszyńska-Daszkiewicz, J., Dziembowski, W. A., & Pamyatnykh, A. A. 2003, *A&A*, 407, 999
Daszyńska-Daszkiewicz, J., Dziembowski, W. A., & Pamyatnykh, A. A. 2005, *A&A*, 441, 641
Daszyńska-Daszkiewicz, J., Szewczuk, W., & Walczak, P. 2013, *MNRAS*, 431, 3396
Flower P. J. 1996, *ApJ*, 469, 355
Handler, G., Shobbrook, R. R., Jerzykiewicz, M., *et al.* 2004, *MNRAS*, 347, 454
Handler, G., Shobbrook, R. R., & Mokgwetsi T. 2005, *MNRAS*, 362, 612
Handler, G., Jerzykiewicz, M., Rodríguez, E., *et al.* 2006, *MNRAS*, 365, 327
Handler, G., Matthews, J. M., Eaton, J. A., *et al.* 2009, *ApJ*, 698, 56
Iglesias, C. A. & Rogers, F. J. 1996, *ApJ*, 464, 943
Jerzykiewicz, M., Handler, G., Shobbrook, R. R., *et al.* 2005, *MNRAS*, 360, 619
Kurucz, R. L. 2004, http://kurucz.harvard.edu
Magee, N. H., Abdallah, J., Jr., Clark, R. E. H., *et al.* 1995, *ASP-CS*, 78, 51
Seaton, M. J. 2005, *MNRAS*, 362, 1
van Leeuwen, F. 2007, *A&A*, 474, 653
Walczak, P., Daszyńska-Daszkiewicz, J., Pamyatnykh, A. A., & Zdravkov, T. 2013, *MNRAS*, 432, 822

Precision Asteroseismology
Proceedings IAU Symposium No. 301, 2013
J. A. Guzik, W. J. Chaplin, G. Handler & A. Pigulski, eds.

© International Astronomical Union 2014
doi:10.1017/S1743921313014373

Iron-group opacities in the envelopes of massive stars

Maëlle Le Pennec and Sylvaine Turck-Chièze

CEA/DSM/IRFU/SAp, CE Saclay, 91191 Gif-sur-Yvette, France
email: maelle.le-pennec@cea.fr; Sylvaine.Turck-Chieze@cea.fr

Abstract. β Cephei and SPB stars are pulsating stars for which the excitation of modes by the κ mechanism, due to the iron-group opacity peak, seems puzzling. We have first investigated the origins of the differences noticed between OP and OPAL iron and nickel opacity calculations (up to a factor 2), a fact which complicates the interpretation. To accomplish this task, new well-qualified calculations (SCO-RCG, HULLAC and ATOMIC) have been performed and compared to values of these tables, and most of the differences are now well understood. Next, we have exploited a dedicated experiment on chromium, iron and nickel, conducted at the LULI 2000 facilities. We found that, in the case of iron, detailed calculations (OP, ATOMIC and HULLAC) show good agreement, contrary to all of the non-detailed calculations. However, in the case of nickel, OP calculations show large discrepancies with the experiments but also with other codes. Thus, the opacity tables need to be revised in the thermodynamical conditions corresponding to the peak of the iron group. Consequently we study the evolution of this iron peak with changes in stellar mass, age, and metallicity to determine the relevant region where these tables should be revised.

Keywords. opacity, atomic data, atomic processes, stars: oscillations

1. Introduction

The κ mechanism is responsible for the pulsation of stars between 1.6 to 20 $M_\odot$. For massive stars (2.5 to 20 $M_\odot$), this mechanism is due to M-shell transitions for the elements of the iron group (chromium, iron, nickel and copper) which induce an opacity bump. SPB and β Cephei stars are examples of such pulsating stars. β Cephei (M > 8 $M_\odot$) are particularly interesting because they will evolve into supernovae and thus are linked to our understanding of the interstellar medium enrichment. However, they are particularly poorly understood. Indeed, there are, for instance, some difficulties interpreting the pulsations of these stars, as one observes modes which were calculated to be stable in theoretical predictions using OP or OPAL opacity tables (Pamyatnykh 1999, Zdravkov & Pamyatnykh 2009). Furthermore, depending on the mass of the star, some of the modes seem better predicted using OP (Seaton & Badnell 2004) or OPAL (Rogers & Iglesias 1992) tables. This fact suggests that some of these opacities could be inaccurately determined for both tables (Daszyńska-Daszkiewicz & Walczak 2010) or that some hydrodynamic process plays an important role not yet understood. We are studying in this paper the first possibility: an inaccurate determination of the iron opacity bump. To deal with this problem, several activities have been developed at CEA in France to improve the present situation: new calculations have been developed and compared to understand the differences, and two XUV campaigns of experiments have been conducted at the LULI 2000 facility for the different elements at temperature around 25 eV and density of about 2 mg cm^{-3}.

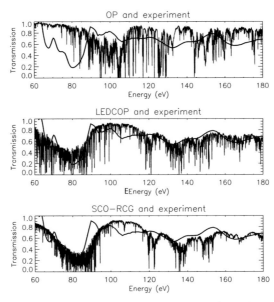

Figure 1. Comparison between nickel transmission spectrum (linked to the opacity through an exponential) taken at LULI 2000 (thick continuous line) and respectively OP, LEDCOP and SCO-RCG. The OP and SCO-RCG calculations are given at 27 eV, 3.4 mg cm^{-3} and LEDCOP at 26 eV, 2 mg cm^{-3}. The domain 60 – 180 eV (700 000 K – 2×10^6 K) is the domain where the Rosseland mean is the most important. From Turck-Chièze *et al.* (2013).

2. Calculations and experiments

Opacity codes are based on different approaches (Turck-Chièze *et al.* 2011): statistical (STA, SCO), detailed (OP, HULLAC, Bar-Shalom *et al.* 2001; ATOMIC, Magee *et al.* 2004; LEDCOP, OPAS) or mixed (SCO-RCG). The two major contributors to the iron bump are iron and nickel, so the calculations have been performed for these two elements as highest priority. Comparisons have been made for tabulated temperature and density values near the experimental ones. These comparisons show that detailed calculations tend to agree, except for the OP results (Gilles *et al.* 2011). The interaction of configuration plays an important role for iron in this domain of temperature and density, and largely explains the difference from the statistical calculations (Gilles *et al.* 2012). For OP, in the case of iron, the Rosseland mean values show differences of around 6 – 7% with ATOMIC and HULLAC, but up to 40% with statistical calculations. In the case of nickel, OP differs clearly from the other codes, showing discrepancies of at least 50% (Turck-Chièze *et al.* 2013).

Figure 1 presents the first analysis of the experiment on nickel (Turck-Chièze *et al.* 2013) compared to different code results (OP, LEDCOP and SCO-RCG). The OP calculations, in fact extrapolated from iron, disagree strongly with the experiments and other calculations. This result confirms some conclusions of Salmon *et al.* (2012). New OP, HULLAC and ATOMIC calculations are in progress, but take a very long time to perform. One concludes that the origin of the differences between OP and OPAL is of different nature for iron and nickel.

3. Study of the iron bump

We have calculated numerous stellar models to determine the domain where new calculations need to be performed. For this task, we have explored how the iron peak varies

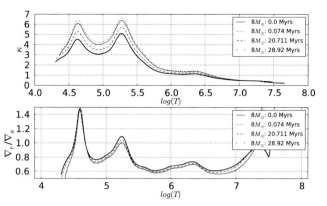

Figure 2. Age variation of the amplitude of the iron bump. Top panel: Opacity in $cm^2 g^{-1}$ versus $\log T$ in the case of a $8\,M_\odot$ star. The first bump at $\log T = 4.6$ is linked to partially ionized helium, the second at 5.25 is the iron bump and the third at 6.3 is the deep iron bump, linked to L-shell bound-free transitions of iron. Bottom panel: ratio ∇_r/∇_a versus $\log T$.

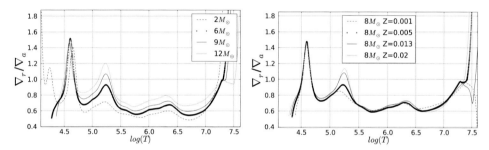

Figure 3. *Left*: Influence of the mass on the amplitude of the iron bump. *Right*: Influence of the metallicity on the iron bump for an $8\,M_\odot$ star. The ratio of the two gradients is always around one so, depending on the properties of stars, convective instability can appear.

with mass, age and metallicity during the main sequence of SPB and β Cep stars. All models were calculated using the stellar evolution code MESA (Paxton *et al.* 2011) and the OPAL opacity tables with the AGSS09 (Asplund *et al.* 2009) abundance mixture. Figure 2 shows that opacity uncertainties during the stellar lifetime have large consequences on the stability of the acoustic modes. Adopting the Schwarzschild criterion to see the onset of the convective instability (ratio of the radiative gradient to the adiabatic gradient greater than 1), one observes that this ratio is very near to 1 at the iron-bump region from one source of opacity calculation to another, and so the resulting structure of the star can be different. In less than 10 Myr, a small convective zone appears in the iron bump region of an $8\,M_\odot$. Precise knowledge of the age of the star is needed to correctly predict the observed modes.

Figure 3 shows the influence of mass and metallicity on the iron bump. The gradient ratio varies rapidly with mass and metallicity, and is around 1 for these types of stars, so a precise knowledge of the opacity is required to properly determine the theoretical frequencies.

This study allows us to determine the thermodynamical conditions in which the iron-peak opacity must be carefully estimated for stars between 2 to $20\,M_\odot$. It is known that opacities vary rapidly with temperature, so one can notice that this peak appears always at the same position independently of the chosen conditions between 100 000 K to 320 000 K. However, the free electron density, N_e, principally due to the totally ionized

Table 1. Domain of investigation of the iron bump for stars from 2.5 to 20 $M_\odot$ and $Z = 0.02$. The free electron density decreases with stellar mass. The present domain will be reduced for specific analyses, and the study extended to the Magellanic Clouds.

Mass	$N_{\rm e,min}$ (cm^{-3})	$N_{\rm e,max}$ (cm^{-3})
2.5 $M_\odot$	$7.87 \cdot 10^{16}$	$1.57 \cdot 10^{19}$
6 $M_\odot$	$1.93 \cdot 10^{16}$	$3.94 \cdot 10^{18}$
10 $M_\odot$	$8.63 \cdot 10^{15}$	$1.98 \cdot 10^{18}$
14 $M_\odot$	$4.63 \cdot 10^{15}$	$1.22 \cdot 10^{18}$
20 $M_\odot$	$2.43 \cdot 10^{15}$	$7.69 \cdot 10^{17}$

helium and hydrogen at these temperatures, varies with mass:

$$N_{\rm e} = \rho \sum_i \frac{Q_i \chi_i}{A_i}.$$

Q_i is the ionization charge, χ_i the relative contribution in mass, A_i the atomic weight of species i and ρ is the density.

4. Conclusion

The OPAC consortium studies (new calculations and new experiments) help to understand the discrepancies between OP and OPAL in the iron bump which excites the modes of β Cephei and SPB stars. Iron opacities are better estimated by OP calculations at relatively high temperature but OP nickel opacities are not correct for all the considered cases. As the iron group peak varies strongly with age, mass and composition of the star, the properties of the stars need to be correctly known and the opacities of the different elements of the iron group must be precisely calculated for a range of T and $N_{\rm e}$ that we are determining. New tables are in construction thanks to the HULLAC, ATOMIC and SCO-RCG codes, and we hope for results in 2014. At that time, radiative acceleration and non LTE-conditions will also be investigated.

References

Asplund, M., Grevesse, N., Sauval, A. J., & Scott, P. 2009, *ARAA*, 47, 481
Bar-Shalom, A., Klapisch, M., & Oreg, J. 2001, *J. Quant. Spectrosc. Radiat. Transf.*, 71, 169
Daszyńska-Daszkiewicz, J. & Walczak, P. 2010, *MNRAS*, 403, 496
Gilles, D., Turck-Chièze, S., Loisel, G., *et al.* 2011, *High Energy Density Physics*, 7, 312
Gilles, D., Turck-Chièze, S., Busquet, M., *et al.* 2012, in *ECLA 2011*, vol. 58, EAS Pub. Ser., p. 51
Magee, N. H., Abdallah, J., Colgan, J., *et al.* 2004, *AIP-CS*, 730, 168
Paxton, B., Bildsten, L., Dotter, A., *et al.* 2011, *ApJS*, 192, 3
Pamyatnykh, A. A. 1999, *AcA*, 49, 119
Rogers, F. J. & Iglesias, C. A. 1992, *ApJS*, 79, 507
Salmon, S., Montalbán, J., Morel, T., *et al.* 2012, *MNRAS*, 422, 3460
Seaton, M. J. & Badnell, N. R. 2004, *MNRAS*, 354, 457
Turck-Chièze, S., The OPAC Consortium 2011a, *J. Phys. Conf. Ser.*, 271, 012035
Turck-Chièze, S., The OPAC Consortium 2011b, *Ap&SS* 336, 103
Turck-Chièze, S., Gilles, D., Le Pennec, M., *et al.* 2013, *High Energy Density Phys.*, 9, 473
Zdravkov, T. & Pamyatnykh, A. A. 2009, *AIP-CS*, 1170, 338

Precision Asteroseismology
Proceedings IAU Symposium No. 301, 2013
J. A. Guzik, W. J. Chaplin, G. Handler & A. Pigulski, eds.

© International Astronomical Union 2014
doi:10.1017/S1743921313014385

Asteroseismology, standard candles and the Hubble Constant: what is the role of asteroseismology in the era of precision cosmology?

Hilding R. Neilson[1], Marek Biesiada[2], Nancy Remage Evans[3], Marcella Marconi[4], Chow-Choong Ngeow[5], and Daniel R. Reese[6]

[1] Dept. of Physics & Astronomy, East Tennessee State University, , PO Box 70300, Johnson City, TN 37614, USA
email: `neilsonh@etsu.edu`
[2] Department of Astrophysics and Cosmology, Institute of Physics, University of Silesia, Uniwersytecka 4, 40-007 Katowice, Poland
[3] Smithsonian Astrophysical Observatory, MS 4, 60 Garden St., Cambridge, MA 02138, USA
[4] INAF-Osservatorio astronomico di Capodimonte, Via Moiariello 16, I-80131 Napoli, Italy
[5] Graduate Institute of Astronomy, National Central University, Jhong-Li 32001, Taiwan
[6] Institut d'Astrophysique et Géophysique de l'Université de Liège, Allée du 6 Août 17, 4000 Liège, Belgium

Abstract. Classical Cepheids form one of the foundations of modern cosmology and the extra-galactic distance scale; however, cosmic microwave background observations measure cosmological parameters and indirectly the Hubble Constant, H_0, to unparalleled precision. The coming decade will provide opportunities to measure H_0 to 2% uncertainty thanks to the *Gaia* satellite, *JWST*, ELTs and other telescopes using Cepheids and other standard candles. In this work, we discuss the upcoming role for variable stars and asteroseismology in calibrating the distance scale and measuring H_0 and what problems exist in understanding these stars that will feed back on these measurements.

Keywords. Cepheids, cosmological parameters, distance scale, stars: horizontal-branch, stars: oscillations

1. Introduction

The *Planck* satellite, launched May 14th 2009, is another milestone in the cosmic microwave background radiation experiments. It carries an array of 74 detectors aboard, covering the frequency range from 25 to 1000 GHz, and with an angular resolution ranging from $30'$ in low frequency bands to $5'$ in high frequencies. This resolution is a considerable improvement over the previous *WMAP* experiment. Recently, the *Planck* team released 29 papers summarizing the first year of data analysis. With *Planck* (and with some of its forerunners) we have begun the era of CMB experiments which are more limited by systematics and foregrounds than by the noise. Removing the foregrounds like: thermal and anomalous dust emission, free-free emission, synchrotron emission, cosmic infrared background, CO rotational emission or secondary Sunyaev-Zel'dovich (SZ) anisotropies resulted in producing unique maps (for unresolved components) and catalogues (for point sources) of these foreground components. The *Planck* data are in agreement with a six-parameter $(\Omega_c h^2, \Omega_b h^2, \tau, \theta_A, A_s, n_s)$ ΛCDM model.

However, there are tensions. First of all, the inferred Hubble constant $H_0 = 67 \pm 1.2\,\mathrm{km\,s^{-1}Mpc^{-1}}$ (Planck Collaboration *et al.* 2013a) is much lower than recently measured on local calibrators: $H_0 = 73.8 \pm 2.4\,\mathrm{km\,s^{-1}\,Mpc^{-1}}$ in the SH0ES survey (Riess *et al.* 2011) or $H_0 = 74.3 \pm 1.5 \pm 2.1\,\mathrm{km\,s^{-1}\,Mpc^{-1}}$ in the Carnegie Hubble Program (Freedman *et al.* 2012) based on calibrations of the infrared Leavitt Law (e.g., Ngeow *et al.* 2009, Freedman *et al.* 2011). A new measurement, independent of cosmic ladder calibration (Suyu *et al.* 2013) obtained from time delays of two strong gravitational lensing systems RXJ 1131−1231 and B1608 + 656 gave an even higher result: $H_0 = 75.2 \pm 4.2\,\mathrm{km\,s^{-1}\,Mpc^{-1}}$. Therefore it is really a sort of tension between *Planck* and other evidence concerning the Hubble constant. There is also another tension in the value of the matter density parameter, Ω_m, which is higher than other independent measurements, and even worse it is internally incompatible with *Planck* Sunyaev-Zeldovich data analysis (Planck Collaboration *et al.* 2013b).

Is this a reason for worry? Every inconsistency is. The strength of modern cosmology lies not in single precise experiments but in consistency across independent and unrelated pieces of evidence. On the other hand it is only the first year of *Planck* mission data which was analyzed – the rest of the mission awaits the data/results release. At last if we are sure the systematics are all accounted for and tension remains, there could be a signal to go beyond the simplest ΛCDM model.

Independent local, precise measurements of H_0 are needed. From a cosmological perspective, in the context of the Dark Energy problem, where the most important goal is to establish details of the expansion and large-scale-structure formation history, reliably determining H_0 in other dedicated experiments would provide invaluable prior information. In particular, knowing H_0 with 1% accuracy would improve the figure of merit in future Dark Energy experiments by 40% (Weinberg *et al.* 2012). Freedman & Madore (2010) also noted that at the precision from *WMAP* observations, one requires better than 2% precision measurements of the Hubble Constant to constrain cosmological parameters.

The cause of the difference between Cepheid and CMB measurements is unclear but in the coming era of *Gaia* and the James Webb Space Telescope (*JWST*), it is possible that Cepheids can be employed to measure H_0 to better than 2% precision. When combined with CMB observations, we can expect to also constrain cosmological parameters (e.g., Planck Collaboration *et al.* 2013a). While *Gaia* and *JWST* will allow for unprecedented precision measurements of the Hubble Constant using Cepheids and other standard candles, there are a number of potential challenges as well as opportunities to be considered using precision asteroseismology.

2. The role of Cepheids

2.1. *The mid-infrared Cepheid Leavitt Law*

The Cepheid Leavitt Law or PL relation plays an essential role in the extra-galactic distance scale ladder (for reviews of the Cepheid PL relation and its role in distance scale work, see Madore & Freedman 1991, Ngeow 2012, Ngeow *et al.*, these proceedings). The Cepheid PL relation has been well developed in the optical bands, as well as in the near infrared JHK bands. Recently, the calibration of the Cepheid PL relation has moved to the mid infrared (MIR, mainly for the 3.6 μm- and 4.5 μm-band) for the LMC Cepheids (Freedman *et al.* 2008, Ngeow & Kanbur 2008, Scowcroft *et al.* 2011), the SMC Cepheids (Ngeow & Kanbur 2010), and the Galactic Cepheids (Marengo *et al.* 2010a, Monson *et al.* 2012). Advantages of using MIR PL relations in distance scale work include: (a) extinction corrections in the MIR are much smaller than in optical bands, and can be safely ignored; (b) the metallicity effect is expected to be minimal in the MIR; (c) the intrinsic dispersion of the MIR PL relation is 2-3 times smaller than for optical counterparts; and most importantly (d) *JWST* operated in MIR, is ideal to measure a Hubble constant with $\sim 1\%$ error from Cepheid observations out to 100 Mpc.

The metallicity dependency of the Cepheid MIR PL relation merits further discussion. The slopes of the 3.6 μm PL relation based on $\sim 10^3$ LMC and SMC Cepheids are almost identical: -3.25 ± 0.01 (Ngeow *et al.* 2009) vs. -3.23 ± 0.02 (Ngeow & Kanbur 2010), indicating metallicity does not affect the slope in this wavelength. Synthetic 3.6 μm PL relations based on pulsation models, however, suggest the opposite (Ngeow *et al.* 2012). More observations of Cepheids in nearby galaxies that span a wide range in metallicity are needed to resolve this issue, similar to analyses done for optical wavelengths (Bono *et al.* 2010). The 4.5 μm PL relation, on the other hand, is affected by CO absorption (Marengo *et al.* 2010a) and could be potentially used as a metallicity indicator (Scowcroft *et al.* 2011).

2.2. *Classical Cepheids and metallicity dependence*

In order to predict and interpret the observed pulsation properties of Classical Cepheids and their role as primary distance indicators, nonlinear convective hydrodynamical models are required (e.g.,Wood *et al.* 1997, Bono *et al.* 1999). In particular, extensive sets of nonlinear convective pulsation models at various chemical compositions have been computed (e.g., Bono *et al.* 1999, Fiorentino *et al.* 2002, Marconi *et al.* 2005, 2010) to provide a theoretical calibration of the extragalactic distance scale and an accurate investigation of metallicity and helium content effects (Caputo *et al.* 2000, Marconi *et al.* 2005, 2010, Fiorentino *et al.* 2007, Bono *et al.* 2008, 2010). The main result of these investigations is that optical PL relations have significant spread (the intrinsic dispersion is about 0.2-0.3 mag) and depends on the chemical composition of the host galaxy, reflecting the topology and finite width of the instability strip. Moreover, they are affected by nonlinearity at the longest periods or a break around 10 days. All these systematic effects are reduced when moving toward the near infrared bands or when introducing the color term in a Period-Luminosity-Colour or Wesenheit relation (e.g., Caputo *et al.* 2000, Marconi 2009, Bono *et al.* 2010 and references therein). In particular both observational and theoretical investigations find that the $V, (V - I)$-Wesenheit relation has a negligible dependence on chemical composition (Bono *et al.* 2010, Marconi *et al.* 2010).

By transforming the theoretical scenario into the *HST* filters, Fiorentino *et al.* (2013) compared their model predictions with the observed properties of Cepheids in the *HST* sample galaxies adopted by Riess *et al.* (2011), who provided an estimate of the Hubble constant at the 3% level of uncertainty. Fiorentino *et al.* (2013) concluded that when adopting the predicted PL relation in the *HST* infrared (160W) filter the inferred extragalactic distance scale, and corresponding Hubble constant, is in agreement with that obtained by Riess *et al.* (2011).

2.3. *The Ultra Long Period Cepheids*

The Ultra Long Period Cepheids (ULPs) are fundamental-mode Cepheid-like pulsators with $P \geqslant 80$ days (see Bird *et al.* 2009 for an extensive discussion), and were first identified in the Magellanic Clouds (Freedman *et al.* 1985). They are much brighter (M_I from -7 to -9 mag) than short-period Cepheids, hence *HST* is able to observe them up to distances of 100 Mpc. These pulsators have been identified both in metal rich (Riess *et al.* 2009, 2011, Gerke *et al.* 2011) and metal poor (Pietrzyński *et al.* 2004, 2006, Gieren *et al.* 2004, Fiorentino *et al.* 2010) galaxies. The two longest-period ULPs have been discovered in IZW18 (Fiorentino *et al.* 2010), the most metal poor galaxy containing Cepheids, and their behaviour seems to be consistent with the extrapolation of the Cepheid Wesenheit PL Law to higher masses and luminosities (Marconi *et al.* 2010). Computation of hydrodynamical models of ULP Cepheids is in progress, but new extensive observations allowing the sampling of more than one pulsation cycle are needed in order to confirm their nature. Understanding these pulsators is crucial because, thanks to their brightness, these pulsators are in principle able to reach cosmologically interesting distances in one step (Fiorentino *et al.* 2012). Finally we note that the *Gaia* satellite will provide a direct calibration of LMC ULPs with parallaxes at μarcsec accuracy.

2.4. *Convective core overshoot & mass loss*

One of the key issues in understanding classical Cepheids from the perspective of stellar evolution is the Cepheid mass discrepancy (Keller 2008), for which there are two likely solutions. The first is convective core overshoot during main sequence evolution (Huang & Wiegert 1983) and the second is Cepheid mass loss (Neilson & Lester 2008, Marengo *et al.* 2010b). Neilson *et al.* (2011) suggested that the mass discrepancy could be solved by a combination of the two possibilities and Neilson *et al.* (2012a,b) showed that observed rates of period change provide an observational avenue to distinguish the two phenomena and that mass loss is a ubiquitous property of Galactic Cepheids. By understanding the role of mass loss in Cepheids, we can better constrain Cepheid properties, i.e., mass-luminosity relation, and how infrared excess from a stellar wind might contaminate the Cepheid Leavitt Law (Neilson *et al.* 2010).

2.5. *Cepheid masses & binarity*

Cepheids are of considerable value to the extragalactic distance scale. Their value depends on how well we understand them, and that understanding is linked to how well we know their masses. Measured masses, of course, come from binary systems. The foundation is the ground-based orbits of Cepheids in spectroscopic binaries. These lack the inclination, however there are four ways in which Cepheid masses are determined in the Milky Way, in addition to the exciting results for the double-lined eclipsing binaries in the LMC.

(*a*) High resolution ultraviolet spectra from satellites (*HST* and *IUE*) have revealed several double-lined spectroscopic binaries. Measuring the orbital velocity amplitude of the hot companion, and combining it with the orbital velocity amplitude of the Cepheid from the ground-based orbit and the mass of the main sequence companion from the spectral energy distribution provides the Cepheid mass (S Mus and V350 Sgr with *HST*; SU Cyg with *IUE*; e.g., Evans *et al.* 2011 and references therein).

(*b*) Astrometric motion has been detected in *HST* FGS observations of Cepheids (Benedict *et al.* 2007). Combining that with the spectroscopic orbit of the Cepheid and the mass of the secondary (Evans *et al.* 2009) provides the mass of the Cepheid (W Sgr and FF Aql).

(*c*) Astrometric motion of both stars in the Polaris system combined with the spectroscopic orbit provides a fully dynamical mass (Evans *et al.* 2008).

(*d*) Interferometry of the V1334 Cyg system (Gallenne *et al.* 2013) has detected the companion and demonstrated orbital motion. This approach, particularly when applied to a number of Cepheids promises to increase the number of measured masses in the Milky Way significantly.

Continuing work is in progress on improving the accuracy of the masses and extending the list using interferometric measurements. Comparison of masses with evolutionary tracks provided by G. Bono favors moderate main sequence core convective overshoot.

2.6. *Beyond the LMC*

Another avenue to more precisely calibrate the Cepheid Leavitt Law is to use M31 Cepheids as an anchor instead of LMC Cepheids. The distance modulus of the LMC is about 18.5 ± 0.1 (e.g., Wagner-Kaiser & Sarajedini 2013, Dambis *et al.* 2013, Marconi *et al.* 2013). The uncertainty in the LMC distance is one of the largest sources of uncertainty for calibrating the Leavitt Law. An alternative calibration source is M31 Cepheids. The distance to M31 has been measured to 3% precision (Riess *et al.* 2012, Valls-Gabaud *et al.* 2013). This increased precision allows for an independent calibration of the Leavitt Law and in the era of *JWST*, M31 Cepheids will help measure H_0 to better precision.

3. Beyond classical Cepheids

RR Lyrae variable stars are arguably another of the most powerful standard candles (Caputo 2012, Marconi 2012, Cacciari 2013). Current calibrations of the RR Lyrae PL relation have allowed for precision measurements of the distance and structure of the LMC (e.g., Haschke *et al.* 2012), the structure of the Galactic halo (e.g., Sesar *et al.* 2013) and the properties of globular clusters as far as M31 (e.g., Contreras Ramos *et al.* 2013). Further, the RR Lyrae PL relation has been calibrated to similar precision as the Cepheid Leavitt Law, especially at infrared wavelengths (e.g., Cáceres & Catelan 2008, Klein *et al.* 2011, Madore *et al.* 2013).

However, RR Lyrae stars also suffer from similar uncertainties as Cepheids as standard candles. One such challenge is the lack of measured distances to field RR Lyrae stars; *HST* astrometric parallaxes have been measured for only five RR Lyrae stars (Benedict *et al.* 2011), meaning most PL relations must be calibrated using distances measured by alternative methods. There is also an observed metallicity dependence (e.g., Dambis 2013). Furthermore, RR Lyraes also undergo the Blazhko effect (Buchler & Kolláth 2011) and strong photospheric shocks have been observed (Chadid & Preston 2013).

If one considers stars with multiple oscillation frequencies, then it is possible to apply more sophisticated tools from asteroseismology in order to estimate various stellar parameters, which in turn may lead to more precise estimates of their luminosities, hence distances. The frequency spectra of solar-type oscillators, including both main sequence stars such as the Sun and red giants, follow well-defined patterns. From these spectra it is possible to extract $\Delta\nu$, the large frequency separation (i.e. the separation between modes of consecutive radial order), and ν_{max}, the frequency at maximum amplitude. These quantities then intervene in scaling relations (Kjeldsen *et al.* 1995, Belkacem *et al.* 2011) that have been used to obtain masses and radii of hundreds of red giants with quoted error bars of 7% and 3%, respectively (e.g. Kallinger *et al.* 2010, Mosser *et al.* 2010). Such relations are subject to various uncertainties. For instance, the outer portions of solar-type stars are convective and give rise to surface effects which are poorly modeled. Kjeldsen *et al.* (2008) recently proposed a recipe for correcting/removing such effects from the frequencies. Another source of uncertainty is the structural differences between the reference star/model (typically the Sun) used in a scaling relation and the stars to which the relation is applied. In order to minimize such effects, one can choose reference models which are closer to the observed stars, and/or apply seismic inversions. Recently, Reese *et al.* (2012) showed how to directly invert for the mean density of a star. Such an approach seeks the frequency combination which is least sensitive to structural differences between the star and the reference model, yet minimizes the effects of observational errors. Applying such an approach may decrease the error by 0.5 to 1% compared to the more traditional scaling relation between $\Delta\nu$ and the mean density.

Pulsating variable stars and Cepheids will anchor the cosmic distance scale, but there are a number of other standard candles that can be employed to measure H_0, including the tip of the red giant branch (Lee *et al.* 1993, the Tully-Fisher relation (Tully & Fisher 1977) and Type Ia supernovae (Phillips 1993).

The tip of the RGB is a powerful standard candle for calibrating primary distance indicators such as Cepheids. This method is based on the observation that the the tip of the RGB occurs at a constant luminosity and has been calibrated for the LMC (Salaris & Girardi 2005) and globular clusters (Bellazzini *et al.* 2001). The tip of the RGB, however, depends on the age and metallicity of clusters as well as calibrations from either stellar evolution calculations (Valle *et al.* 2013) or empirical measurements (Bellazzini *et al.* 2001). These uncertainties need to be understood in greater detail for the tip of the RGB to be used as a distance indicator to measure H_0 to unprecedented precision.

The Tully-Fisher relation is another important standard candle, where the luminosity of a spiral galaxy is related to the rotational velocity of a galaxy. Because the Tully-Fisher relation can be applied at much greater distances than Cepheids, it is an important complement for measuring H_0. Recently, Sorce *et al.* (2013) calibrated the relation at 3.6 μm using the same thirteen galaxy clusters as described by Tully & Courtois (2012), who calibrated the relation in the I-band. In both works, the authors measured H_0 to be approximately $75 \, \text{km} \, \text{s}^{-1} \text{Mpc}^{-1}$. However, the Tully-Fisher relation has a number of uncertainties and corrections, such as a color correction (Sorce *et al.* 2013) and depends on the calibration of the Cepheid Leavitt Law.

Type Ia supernovae were instrumental for the discovery of dark energy (Riess *et al.* 1998, Perlmutter *et al.* 1999) and have been exploited for measuring H_0 to unprecedented precision (Riess *et al.* 2011, Barone-Nugent *et al.* 2012). The challenge for employing Type Ia SNe is observing them in nearby galaxies with independently-determined distances. At large redshift, these standard candles can be calibrated using photometric and spectroscopic redshifts, but in the local group, the calibration requires using Cepheids or other standard candles (Phillips *et al.* 2006, Riess *et al.* 2011). While Type Ia SNe are one of best standard candles, they are still not ideal. For instance, the luminosities of Type Ia SNe differ as a function of host galaxy properties (Sullivan *et al.* 2011).

4. Summary

The next decade is bright for measuring H_0 to less than 2% precision and maybe even 1% thanks to new observatories and satellites such as *Gaia*, *JWST*, LSST and the upcoming extremely large telescopes. The Cepheid PL relation will be calibrated by measuring distances to thousands of Cepheids by *Gaia* (Dennefeld 2011), while *JWST* and LSST will observe Cepheids as far as 100 Mpc (Freedman & Madore 2012).

While the future is bright, there are many current opportunities as well. Surveys such as VVV, VMC and OGLE are constraining Cepheids and variable star properties and physics. For instance, the discovery of Cepheids in eclipsing binary systems are constraining the Cepheid mass discrepancy (Cassisi & Salaris 2011, Neilson & Langer 2012, Prada Moroni *et al.* 2012, Marconi *et al.* 2013) as well as the structure of the Leavitt Law (e.g., Ngeow *et al.* 2009). These observations also allow for constraining the metallicity and the helium abundances of Cepheids, hence constraining the effects of variations in the chemical composition on the Leavitt Law (Marconi *et al.* 2010). We are further able to employ long-term observations of rates of period change to explore Cepheid mass loss and circumstellar media (Neilson *et al.* 2012b). As such, some key ingredients for better understanding the Cepheid Leavitt Law include the metallicity dependence, convective core overshoot, mass loss and stellar masses.

There are a number of complementary standard candles that can be calibrated by asteroseismic and radial pulsation measurements. RR Lyrae variable stars form one such distance indicator for measuring distances to galaxies in the Local group such as the LMC and M31, hence calibrating the Cepheid Leavitt law. The precision of RR Lyrae standard candles will be improved by *Gaia* parallax measurements as well as better understanding of metallicity dependencies and their evolution from the red giant branch, i.e., the color dispersion. Asteroseismic observations of red giant stars provide other potential standard candles, in particular seismic inversions measure stellar mean densities, suggesting the potential for calibrating red giants as standard candles.

Because of the significant improvements in the past decade for calibrating standard candles, it is possible to measure H_0 to better than 3% precision and it is likely that the measured uncertainty will be reduced to less than 2% in the coming decade. At that precision, measurements of H_0 using standard candles will complement cosmic microwave background determination of cosmological parameters. Furthermore, measuring more precise values of H_0 will also require precision understanding of stellar astrophysics.

References

Barone-Nugent, R. L., Lidman, C., Wyithe, J. S. B., *et al.* 2012, *MNRAS*, 425, 1007
Belkacem, K., Goupil, M. J., Dupret, M. A., Samadi, R. *et al.* 2011, *A&A* 530, A142
Bellazzini, M., Ferraro, F. R., & Pancino, E. 2001, *ApJ*, 556, 635
Benedict, G. F., McArthur, B. E., Feast, M. W., *et al.* 2011, *AJ*, 142, 187
Bird, J. C., Stanek, K. Z., & Prieto, J. L. 2009, *ApJ*, 695, 874
Bono, G., Marconi, M., & Stellingwerf, R. F. 1999, *ApJS*, 122, 167
Bono, G., Caputo, F., Fiorentino, G., Marconi, M., & Musella, I. 2008, *ApJ*, 684, 102
Bono, G., Caputo, F., Marconi, M., & Musella, I. 2010, *ApJ*, 715, 277
Buchler, J. R. & Kolláth, Z. 2011, *ApJ*, 731, 24
Cacciari, C. 2013, in: R. de Grijs (ed.), *Advancing the Physics of Cosmic Distances*, Proc. IAU Symposium No. 289 (Cambridge: Cambridge University Press), p. 101
Cáceres, C. & Catelan, M. 2008, *ApJS*, 179, 242
Caputo, F., Marconi, M., & Musella, I. 2000, *A&A*, 354, 610
Caputo, F. 2012, *Ap&SS*, 341, 77
Cassisi, S. & Salaris, M. 2011, *ApJ*, 728, L43
Chadid, M. & Preston, G. W. 2013, *MNRAS*, 434, 552
Contreras Ramos, R., Clementini, G., Federici, L., *et al.* 2013, *ApJ*, 765, 71
Dambis, A. K., Berdnikov, L. N., Kniazev, A. Y., *et al.* 2013, *MNRAS*, 435, 3206
Dennefeld, M. 2011, *ASP-CS*, 451, 317
Evans, N. R., Schaefer, G. H., Bond, H. E., *et al.* 2008, *AJ*, 136, 1137
Evans, N. R., Massa, D., & Proffitt, C. 2009, *AJ*, 137, 3700
Evans, N. R., Berdnikov, L., Gorynya, N., Rastorguev, A., & Eaton, J. 2011, *AJ*, 142, 87
Fiorentino, G., Caputo, F., Marconi, M., & Musella, I. 2002, *ApJ*, 576, 402
Fiorentino, G., Marconi, M., Musella, I., & Caputo, F. 2007, *A&A*, 476, 863
Fiorentino, G., Contreras Ramos, R., Clementini, G., *et al.* 2010, *ApJ*, 711, 808
Fiorentino, G., Clementini, G., Marconi, M., *et al.* 2012, *Ap&SS*, 341, 143
Fiorentino, G., Musella, I., & Marconi, M. 2013, *MNRAS*, 434, 2866
Freedman, W. L. & Madore, B. F. 2010, *ARAA*, 48, 673
Freedman, W. L., Grieve, G. R., & Madore, B. F. 1985, *ApJS*, 59, 311
Freedman, W. L., Madore, B. F., Rigby, J., Persson, S. E., & Sturch, L. 2008, *ApJ*, 679, 71
Freedman, W. L., Madore, B. F., Scowcroft, V., *et al.* 2011, *AJ*, 142, 192
Freedman, W. L., Madore, B. F., Scowcroft, V., *et al.* 2012, *ApJ*, 758, 24
Gallenne, A., Monnier, J. D., Mérand, A., *et al.* 2013, *A&A*, 552, A21
Gerke, J. R., Kochanek, C. S., Prieto, J. L., Stanek, K. Z., & Macri, L. M. 2011, *ApJ*, 743, 176
Gieren, W., Pietrzyński, G., Walker, A., *et al.* 2004, *AJ*, 128, 1167
Haschke, R., Grebel, E. K., & Duffau, S. 2012, *AJ*, 144, 106
Huang, R. Q. & Weigert, A. 1983, *A&A*, 127, 309
Kallinger, T., Mosser, B., Hekker, S., *et al.* 2010, *A&A*, 522, A1
Kjeldsen, H. & Bedding, T. R. 1995, *A&A*, 293, 87

Kjeldsen, H., Bedding, T. R., & Christensen-Dalsgaard, J. 2008, *ApJ*, 683, L175
Keller, S. C. 2008, *ApJ*, 677, 483
Klein, C. R., Richards, J. W., Butler, N. R., & Bloom, J. S. 2011, *ApJ*, 738, 185
Lee, M. G., Freedman, W. L., & Madore, B. F. 1993, *ApJ*, 417, 553
Madore, B. F. & Freedman, W. L. 1991, *PASP*, 103, 933
Madore, B. F., Hoffman, D., Freedman, W. L., *et al.* 2013, *ApJ*, 776, 135
Marconi, M. 2009, *MemSAIt Suppl.*, 80, 141
Marconi, M. 2012, *MemSAIt Suppl.*, 19, 138
Marconi, M., Musella, I., & Fiorentino, G. 2005, *ApJ*, 632, 590
Marconi, M., Musella, I., Fiorentino, G., *et al.* 2010, *ApJ*, 713, 615
Marconi, M., Molinaro, R., Bono, G., *et al.* 2013, *ApJ*, 768, L6
Marengo, M., Evans, N. R., Barmby, P., *et al.* 2010a, *ApJ*, 709, 120
Marengo, M., Evans, N. R., Barmby, P., *et al.* 2010b, *ApJ*, 725, 2392
Monson, A. J., Freedman, W. L., Madore, B. F., *et al.* 2012, *ApJ*, 759, 146
Mosser, B., Belkacem, K., Goupil, M., *et al.* 2010, *A&A* 517, A22
Neilson, H. R. & Lester, J. B. 2008, *ApJ*, 684, 569
Neilson, H. R., Ngeow, C.-C., Kanbur, S. M., & Lester, J. B. 2010, *ApJ*, 716, 1136
Neilson, H. R., Cantiello, M., & Langer, N. 2011, *A&A*, 529, L9
Neilson, H. R., Engle, S. G., Guinan, E., *et al.* 2012a, *ApJ*, 745, L32
Neilson, H. R., Langer, N., Engle, S. G., Guinan, E., & Izzard, R. 2012b, *ApJ*, 760, L18
Ngeow, C.-C. 2012, *Journal of Tapei Astronomical Museum*, 10, 1
Ngeow, C.-C. & Kanbur, S. M. 2008, *ApJ*, 679, 76
Ngeow, C.-C. & Kanbur, S. M. 2010, *ApJ*, 720, 626
Ngeow, C.-C., Marconi, M., Musella, I., Cignoni, M., & Kanbur, S. M. 2012, *ApJ*, 745, 104
Ngeow, C.-C., Kanbur, S. M., Neilson, H. R., Nanthakumar, A., & Buonaccorsi, J. 2009, *ApJ*,
    693, 691
Perlmutter, S., Aldering, G., Goldhaber, G., *et al.* 1999, *ApJ*, 517, 565
Phillips, M. M. 1993, *ApJ*, 413, L105
Phillips, M. M., Feldmeier, J. J., & Jacoby, G. H. 2006, *Rev. Mexicana AyA*, 27, 196
Pietrzyński, G., Gieren, W., Udalski, A., *et al.* 2004, *AJ*, 128, 2815
Pietrzyński, G., Gieren, W., Soszyński, I., *et al.* 2006, *AJ*, 132, 2556
Planck Collaboration, Ade, P. A. R., Aghanim, N., *et al.* 2013a, arXiv: 1303.5076
Planck Collaboration, Ade, P. A. R., Aghanim, N., *et al.* 2013b, arXiv: 1305.5080
Prada Moroni, P. G., Gennaro, M., Bono, G., *et al.* 2012, *ApJ*, 749, 108
Reese, D. R., Marques, J. P., Goupil, M. J., Thompson, M. J., & Deheuvels, S. 2012, *A&A* 539,
    A63
Riess, A. G., Filippenko, A. V., Challis, P., *et al.* 1998, *AJ*, 116, 1009
Riess, A. G., Macri, L., Casertano, S., *et al.* 2009, *ApJ*, 699, 539
Riess, A. G., Macri, L., Casertano, S., *et al.* 2011, *ApJ*, 730, 119
Riess, A. G., Fliri, J., & Valls-Gabaud, D. 2012, *ApJ*, 745, 156
Salaris, M. & Girardi, L. 2005, *MNRAS*, 357, 669
Scowcroft, V., Freedman, W. L., Madore, B. F., *et al.* 2011, *ApJ*, 743, 76
Sesar, B., Ivezić, Ž., Stuart, J. S., *et al.* 2013, *AJ*, 146, 21
Sorce, J. G., Courtois, H. M., Tully, R. B., *et al.* 2013, *ApJ*, 765, 94
Sullivan, M., Conley, A., Howell, D. A., *et al.* 2010, *MNRAS*, 406, 782
Suyu, S. H., Auger, M. W., Hilbert, S., *et al.* 2013, *ApJ*, 766, 70
Tully, R. B. & Courtois, H. M. 2012, *ApJ*, 749, 78
Tully, R. B. & Fisher, J. R. 1977, *A&A*, 54, 661
Valle, G., Dell'Omodarme, M., Prada Moroni, P. G., & Degl'Innocenti, S. 2013, *A&A*, 554, A68
Valls-Gabaud, D. 2013, in: R. de Grijs (ed.), *Advancing the Physics of Cosmic Distances*, Proc.
    IAU Symposium No. 289 (Cambridge: Cambridge University Press), p. 235
Wagner-Kaiser, R. & Sarajedini, A. 2013, *MNRAS*, 431, 1565
Weinberg, D. H., Mortonson, M. J., Eisenstein, D. J., *et al.* 2012, *Phys. Rep.*, 530, 87
Wood, P. R., Arnold, A., & Sebo, K. M. 1997, *ApJ*, 485, L25

Precision Asteroseismology
Proceedings IAU Symposium No. 301, 2013
J. A. Guzik, W. J. Chaplin, G. Handler & A. Pigulski, eds.

© International Astronomical Union 2014
doi:10.1017/S1743921313014397

Blazhko effect in Cepheids and RR Lyrae stars

Róbert Szabó

Konkoly Observatory, Research Center for Astronomy and Earth Sciences
of the Hungarian Academy of Sciences
Konkoly Thege Miklós út 15-17.
H-1121, Budapest, Hungary
email: szabo.robert@csfk.mta.hu

Abstract. The Blazhko effect is the conspicuous amplitude and phase modulation of the pulsation of RR Lyrae stars that was discovered in the early 20th century. The field of study of this mysterious modulation has recently been invigorated thanks to the space photometric missions providing long, uninterrupted, ultra-precise time-series data. In this paper I give a brief overview of the new observational findings related to the Blazhko effect, such as extreme modulations, irregular modulation cycles and additional periodicities. I argue that these findings together with dedicated ground-based efforts provide us now with a fairly complete picture and a good starting point to theoretical investigations. Indeed, new, unpredicted dynamical phenomena have been discovered in Blazhko RR Lyrae stars, such as period doubling, high-order resonances, three-mode pulsation and low-dimensional chaos. These led to the proposal of a new explanation to this century-old enigma, namely a high-order resonance between radial modes. Along these lines I present the latest efforts and advances from the theoretical point of view. Lastly, amplitude variations in Cepheids are discussed.

Keywords. stars: variables: RR Lyrae, stars: variables: Cepheids, Blazhko effect, stars: individual: RR Lyr, V445 Lyr, V473 Lyr, α UMi, pulsation, hydrodynamics, *Kepler*

1. Introduction

Although representatives of several pulsating variable types show amplitude modulation, the Blazhko effect of RR Lyrae stars is unique in many respects. First, the percentage of modulated fundamental mode RR Lyrae stars can be as high as 50% (Konkoly Blazhko Survey, Jurcsik *et al.* 2009). The modulation time scale ranges from a few days to several years. Second, besides amplitude modulation, simultaneous phase modulation (or, equivalently period variation) is always present in modulated RRab stars (Benkő *et al.* 2010). The third point is a freshly distilled lesson from a series of recent discoveries: the modulation of the high-amplitude nonlinear pulsational modes are accompanied by several dynamical phenomena (period doubling, resonances, chaos) and this feature makes the Blazhko effect in RR Lyrae stars a unique feature among the classes of pulsating variables. We note in passing that these dynamical effects have a special role in RR Lyrae stars, even the Blazhko effect itself might have a dynamical origin (i.e. resonance between radial modes, see later).

The Blazhko effect was discovered by S. Blažko more than a century ago (Blažko 1907) when he could not fit the period of RW Dra with a constant value (phase modulation). A few years later H. Shapley (Shapley 1916) noticed the different heights of observed maxima of RR Lyrae, the prototype, discovering the amplitude modulation.

Only a few years have passed since the latest major review on the topic (Kovács 2009). By coincidence, the launch of *Kepler* (Borucki *et al.* 2010) preceded this excellent review only by a few months. This space telescope has opened a completely new window to the intricate behavior of the Blazhko-modulated RR Lyrae stars. Therefore I focus on the enormous progress that has been made since Kovács' (2009) review.

A very important aspect is that without assuming any special physical mechanism, the mathematical description of the modulation of a carrier wave (the pulsation in our case) can be elegantly derived from simple considerations (Benkő et al. 2011, Szeidl & Jurcsik 2009). This kind of description admittedly does not convey any knowledge about the underlying physical mechanism, but helps to discriminate between physical interpretations. In particular, instead of detecting only one side frequency (doublet) or symmetric side-peaks (triplet) (Alcock et al. 2003) or sometimes quintuplets (Hurta et al. 2008) around the dominant frequency and its harmonics from ground-based observations, with space data we can find a large (in theory infinite) number of side peaks in Blazhko RR Lyrae stars. For example, in the Fourier-spectrum of V1127 Aql up to 6-8 side peaks on both sides could be detected around the dominant frequency and its harmonics with *CoRoT* (Chadid et al. 2010). The side peak structure is only one aspect of the modulation. Many more unexpected results came from space photometric missions, and I continue with highlighting some of the spectacular *Kepler* observational results.

2. The *Kepler* revolution – observational results

The homogeneous photometric data for more than 200 000 stars delivered by *Kepler* is unprecedented and proved to be a treasure trove both for transiting exoplanet finding and stellar astrophysics. The ultra-high precision (e.g. 8 ppm per point precision for the bright, heavily saturated RR Lyr, the prototype, Kolenberg et al. 2011) exceeds by orders of magnitude what has been possible from the ground. In addition, the 4-year-long, quasi-uninterrupted monitoring allowed the exploration of a completely new region of parameter space. For many discoveries not the precision, but the continuity was the most important factor. The continuous observations were interrupted only for a few hours per month for data downloading or a few days in case of shorter and up to two weeks for longer technical problems. Some stars that fell on the dead Module 3 show regular, 3-month-long gaps annually.

With this extraordinary instrumental setup apart from a few RRc stars (Moskalik et al., in preparation) forty-four RRab stars were observed, 17 of them being Blazhko-modulated, which is close to 40%. A few more RR Lyrae stars have been discovered either serendipitously (background contamination) or by meticulous investigations by dedicated citizen scientists (PlanetHunters) and will be added to the list soon. Unfortunately, after four years of operations *Kepler* has lost its second reaction wheel, which degrades the pointing stability, hence photometric precision, as well.

2.1. *Extreme modulations – modulation extremes*

One of the most intriguing features that *Kepler* was able to show us is the huge variety of modulation shapes, and in some cases its cycle-to-cycle variations. There are extreme cases with large variability in the modulation cycles, while in other stars the modulation seems to be more or less regular. The poster child of the rapid variations in modulation characteristics is V445 Lyr (Guggenberger et al. 2012), where significant variations are observed from cycle to cycle. In addition, drastic light curve shape variation is evident: in Blazhko-maxima the light curve is that of a normal RRab star, but in Blazhko-minima the amplitude is an order of magnitude less, and it transforms into a double-maxima, irregular light curve, that does not resemble any RRab light variations. A sibling of this object is *CoRoT* 105288363 (Guggenberger et al. 2011).

Another important revelation was the multiperiodic nature of the modulations. One of the best-observed examples found by ground-based observations is CZ Lac (Sódor *et al.* 2011), which was followed by many more examples from *Kepler*. In fact, based on the four-year long *Kepler* data, it seems that most of the modulated RRab stars show multiple periodicities in their modulations, except the longest-period ones, where more data would be necessary to establish any multi-periodicity (Benkő *et al.*, in preparation). In CZ Lac two simultaneous modulation periods were detected and both changed its period from one observing season to another. This fact, along with the multiperiodic/irregular behavior of the modulation, immediately raises the question whether the underlying dynamics is chaotic, and the changing, seemingly multiperiodic modulations are simply the manifestations of a chaotic behavior. There are indications that this scenario might indeed be plausible (see below).

2.2. *The unexpected beauty: period doubling*

One of the most remarkable discoveries on the Blazhko-front has been the detection of a well-known dynamical phenomenon, the period doubling (hereafter PD), which has never been observed in RR Lyrae stars before *Kepler*. PD was easily noticed in the first long-cadence (29.5-min) *Kepler* light curves (Kolenberg *et al.* 2010, Szabó *et al.* 2010) as alternating maxima in some of the Blazhko RR Lyrae stars. PD (or period-2 state) manifests itself in the frequency domain as the presence of half-integer frequencies between the dominant pulsational frequency and its harmonics. After the first detection, it turned out that the majority of Blazhko-modulated stars shows PD, at least temporarily. In each case, the strength of PD varies with time, and it can vanish for long time intervals, which partly explains why it could remain unnoticed before the space photometry era. Anyhow, the fact that PD is present in most of the Blazhko RRab stars, and has never been seen in non-modulated RR Lyrae stars, even with the precision allowed by *Kepler* (Szabó *et al.* 2010, Nemec *et al.* 2011), demonstrates boldly that there should be a strong connection between PD and the Blazhko effect itself. In Fig. 1 we see a characteristic and very strong PD phase of RR Lyrae, the eponym of its class.

2.3. *Additional periodicities in modulated RR Lyrae stars*

Interestingly, in addition to the traditional frequency solution of the Blazhko RR Lyrae stars (dominant mode, its harmonics and modulation side peaks), extra frequencies can be frequently found in Blazhko RR Lyrae stars if space photometric data of sufficiently high precision are available. These frequencies are present in non-Blazhko stars as well (Moskalik 2013 and Moskalik, these proceedings) and their origin (if pulsation) can be either radial (Benkő *et al.*, these proceedings) or non-radial modes. However, in Blazhko RRab stars most of these frequencies tend to show up around the radial low-order overtones, namely around the predicted frequencies of the first (O1) and second (O2) overtones (see e.g. Benkő *et al.* 2010). As has been shown by Wojtek Dziembowski, nonradial modes can be unstable in RR Lyrae star models (Dziembowski 1977). Such modes are most easily excited in the vicinity of radial overtones (Van Hoolst *et al.* 1998), so the radial/nonradial dilemma cannot be readily solved. It is worth noting in this context that in at least one case (RR Lyrae itself) we were able to prove that the radial O1 mode is excited (see Sect. 3.2).

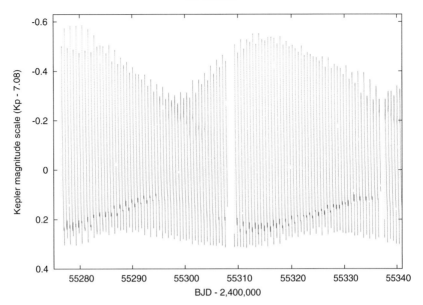

Figure 1. Q5 short-cadence (1-min sampling) *Kepler* flux data of RR Lyrae, the prototype converted to the *Kepler* magnitude scale and shifted by an arbitrary number. Period doubling is persistent throughout this 2-month section. Note the large difference between the heights of consecutive maxima at the beginning. The plot contains more than 91 000 individual data points.

3. The *Kepler* revolution – theoretical aspects

How can these recent findings fit into a coherent theoretical framework? Will eventually all the pieces of the Blazhko jigsaw puzzle fall into place? In what follows I am presenting the latest theoretical advancements inspired by the discoveries discussed above.

3.1. *Resonances, resonances, resonances*

Resonances play a crucial role in Cepheids, but because of the narrower region in the parameter space occupied by RR Lyrae stars, resonances were thought to play a negligible role in these high-amplitude horizontal branch pulsators. Surprisingly, Szabó *et al.* (2010) and Kolláth *et al.* (2011) were able to prove by computing hydrodynamic models that the origin of the PD is undoubtedly a high-order (9:2) resonance between the fundamental mode and a high-order *radial* (strange) overtone. This resonance is at work for a relatively large portion of the mass - luminosity - effective temperature - metallicity parameter space. Although the 8th and the 10th radial overtones can also be trapped in the outer regions, hence not so heavily damped or even excited, a large survey of models demonstrated unambiguously that the ninth overtone is coupled very strongly to the fundamental mode and causes the period doubling. A series of period-doubling bifurcations can lead to chaos. We were able to find such bifurcation cascades in RR Lyrae models. Further full hydro computations showed that our models often approach other high-order resonant states, like 14:19, 20:27, etc., between the fundamental and the first overtone modes, creating a huge variety of complex dynamical behavior.

3.2. *Triple-mode states*

One such dynamical state is a three-mode condition. Molnár *et al.* (2012) investigated the hydro models with PD that showed period-6 behavior as well (where the height of maxima repeats after 6 pulsation periods), indicating that the model can temporarily be close to the 3:4 resonance between the fundamental and first overtone. This period-6 characteristic was found in the *Kepler* observations of RR Lyr, the prototype. Based on the models, the detection of the first overtone was predicted with low amplitude, and indeed Q5 – Q6 *Kepler* data of RR Lyrae showed the O1 frequency with high significance (Molnár *et al.* 2012). Based on this result, the presence of the first radial overtone itself is well established, but additional, non-radial modes cannot be completely ruled out either. This three-mode state (fundamental, O1 and O9) on the one hand represents a new pulsational behavior dissimilar to the well-known double-mode (RRd) pulsators, and on the other hand enables an even more diverse bonanza of complex dynamical states, including chaos (Plachy *et al.* 2013).

3.3. *Radial resonance as an explanation of the Blazhko effect*

Buchler & Kolláth (2011) moved forward and showed using the amplitude equation formalism that if the 9:2 resonance between the fundamental mode and the 9th overtone is present, then in a large part of parameter space regular, irregular, and even chaotic modulations occur naturally. This result is a very important step towards the understanding of the Blazhko effect. The final step of proving the concept with full hydrodynamics models still remains to be done, but it is worth mentioning here that Smolec & Moskalik (2012) were able to produce modulated hydrodynamic models of BL Her stars in the presence of period doubling. The radial resonance paradigm is currently the most tenable, and the only one that is backed up by full hydrodynamic models. Additional resonances and the presence of nonradial modes are also highly probable, and finding them should be of prime priority. Much work is required to fully validate the resonance model, e.g. to reconcile the models with occurrence rates and observed quantities, and to work out a similar mechanism for Blazhko RRc stars.

3.4. *Low-dimensional chaos in the Blazhko modulation*

Based on recent work of Plachy *et al.* (in preparation), there are indications that the modulation of some of the *Kepler* RRab stars with shorter modulation periods may be the result of low-dimensional chaos. The work is based on advanced mathematical methods, like global flow reconstruction and compares various return maps of the observed data to synthetic, chaos-generated data sets. The key factor for the successful application of the method is the observed number of modulation cycles. RR Lyrae stars having the shortest period (20–30 days) modulations accumulated enough cycles during *Kepler*'s 4-year operation to be suitable for such analyses. For the long-period Blazhko stars more observational data would be essential. If *Kepler*'s operation is extended and it continues to observe the same field in two-wheel mode (e.g. to detect transit-timing variation of the already discovered planetary systems), then there is a good chance that a few *Kepler* Blazhko stars will be monitored as well (Molnár *et al.* 2013a). We emphasize here again that the radial resonance model (Buchler & Kolláth 2011) is able to predict chaotic modulation cycles, hence observations seem to confirm the theoretical predictions.

4. Amplitude variations in Cepheids

Although several flavors of amplitude variation have been detected in Cepheids, these do not always resemble the Blazhko effect seen in RR Lyrae stars. One such example is the large number (19%) of first overtone - second overtone (O1/O2) double-mode Cepheids in the Large Magellanic Cloud that show long-period ($P_{\mathrm{mod}} > 700$ d), anti-correlated amplitude modulation discovered by Moskalik & Kołaczkowski (2009). The true occurrence rate can be higher, considering the finite length of the OGLE-II observations and

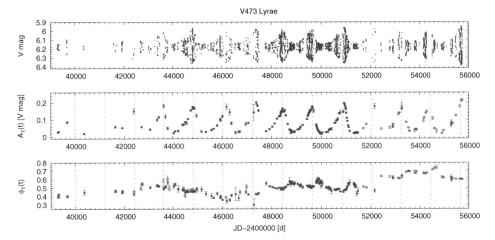

Figure 2. Blazhko-like variations of V473 Lyr. Upper panel: light curve, middle panel: amplitude variation (A_1 term of the Fourier-decomposition), bottom panel: phase variation. The vertical dashed lines denote the 1205-day modulation cycles. The full circles are V data, open symbols are scaled Strömgren v observations to match the amplitudes in V. In the bottom plot the triangles are O-C data taken from the literature (Molnár *et al.* 2013b and references therein.)

a significant number of O1/O2 double-mode Cepheids showing amplitude variations on even longer time scales. The origin of this phenomenon is currently unknown.

Another prominent example is the amplitude change seen in the famous Cepheid α UMi or Polaris (Turner *et al.* 2005). The light as well as the radial velocity amplitude of this overtone Cepheid had been decreasing until the mid-nineties, then it rebounded and it has been increasing since then (Bruntt *et al.* 2008, Spreckley & Stevens 2008). It is not clear whether the variation is secular or cyclic. Obviously, more observations are required to answer this question. The cause of the variation is also a matter of debate, the suggested mechanisms range from variations in the stellar structure due to evolution through the presence of an additional (presumably) nonradial mode. Unless it is an extremely long-period modulation, there is no indication that the amplitude variation of Polaris would be related to the Blazhko effect in RR Lyrae stars.

Contrary to Polaris, V473 Lyrae (Burki *et al.* 1986) is a possible example for a Blazhko Cepheid. The characteristics of the light variation of this presumably second overtone Cepheid are almost identical to the Blazhko-modulated RRab stars (Molnár *et al.*, in preparation). The argument is based on a recent analysis of new photometric data (Molnár *et al.* 2013b) showing Blazhko-like amplitude modulation and simultaneous period variation, a distinctive characteristic of Blazhko RR Lyrae stars (see Fig. 2).

As a side note we add that the variations we discussed in this section fit the trend that was found on shorter time scales (Evans *et al.*, these proceedings and references therein), namely a more pronounced tendency for instability in the oscillations of overtone pulsators compared to fundamental-mode Cepheids.

5. The future of the Blazhko research field

From the observational side the success story of space photometry is expected to be continued. NASA's *TESS* (Transiting Exoplanet Survey Satellite, Ricker *et al.* 2010) mission with a launch scheduled for 2017 will monitor the whole sky (it will observe a given field for 27 days) searching for short-period planets around bright stars. The design of four large field-of-view cameras will allow good photometric precision. A small fraction of the sky around the ecliptic poles will be observed continuously, which will be an excellent hunting ground for longer period Blazhko stars. The European counterpart of *TESS* is the *PLATO* (PLAnetary Transits and Oscillations of Stars) mission (Rauer *et al.* 2013). If selected, the arrangement of 32 normal and 2 fast telescopes on a common platform and the judiciously designed overlap in the monitored fields between the telescopes will provide large field-of-views and high precision. During the nominal 5-yr mission half of the sky could be covered with long (from months to up to a year) staring phases peppered with shorter step-and-stare intervals. The earliest launch date for *PLATO* is 2024. In both cases the missions will most probably avoid the overcrowded low galactic latitude regions. Needless to say, both space photometric projects can continue the pioneering work of *CoRoT* and *Kepler* and contribute to the understanding of the long-standing Blazhko-problem.

There is great potential in spectroscopic observations as well. Systematic, dedicated studies covering several Blazhko cycles are rare. Promising results have already been published recently, e.g. the detection of He I and He II lines in Blazhko stars (Preston 2009, 2011, Gillet *et al.* 2013, see also the contributions of Guggenberger and Kolenberg, these proceedings). Understanding the complicated dynamics of the pulsating atmosphere and the study of any non-spherical asymmetries (presence of nonradial modes for instance) would greatly benefit from such investigations.

On the theoretical front, one-dimensional hydrocodes, e.g., the Budapest-Florida code (Kolláth *et al.* 2002), or the Warsaw code (Smolec & Moskalik 2008) still have a role in understanding physical concepts and mechanisms, especially in the light of the radial resonance paradigm. It has become increasingly clear that multi-dimensional codes (Kupka, these proceedings, Geroux & Deupree 2011, 2013, and these proceedings) will greatly advance our understanding of stellar pulsation in general and of the Blazhko effect in particular in the near future. When long simulations with many different initial conditions become feasible in two or three-dimensions, the self-consistent treatment of convection, the ability of modeling nonradial motions (or even modes) will make these hydrocodes excellent tools in helping to convey a better picture of the mode selection mechanism, the interaction between pulsation and convection, and the modulation mechanism itself. Clearly, the modeling of the these inherently three-dimensional objects and dynamical processes in multi-dimensions requires an enormous amount of work, but I strongly believe that a better understanding of RR Lyrae stars and the Blazhko effect is well worth the effort.

Acknowledgements

The author gratefully acknowledges the Lendület-2009 Young Researchers' Program and the János Bolyai Research Scholarship of the Hungarian Academy of Sciences, the HUMAN MB08C 81013 grant of the MAG Zrt., the Hungarian OTKA grant K83790, the KTIA URKUT_10-1-2011-0019 grant, the European Community's FP7/2007-2013 programme under grant agreement no. 269194 (IRSES/ASK), and the IAU for the travel grant. The inspiring discussions and influential work of Wojtek Dziembowski in the field of stellar pulsation are also thankfully acknowledged. Fruitful discussions with Z. Kolláth, L. Molnár, E. Plachy, J. M. Benkő, K. Kolenberg, P. Moskalik, and R. Smolec are appreciated. The author wishes to dedicate this review to the memory of J. Robert Buchler.

References

Alcock, C., Alves, D. R., Becker, A., *et al.* 2003, *ApJ*, 598, 597
Benkő, J. M., Kolenberg, K., Szabó, R., *et al.* 2010, *MNRAS*, 409, 1585

Benkő, J. M., Szabó R. & Paparó, M. 2011, *MNRAS*, 417, 974

Borucki, W. J. Koch, D., Basri, G., *et al.* 2010, *Science*, 327, 977

Bruntt, H., Evans, N. R., Stello, D., *et al.* 2008, *ApJ*, 683, 433

Blažko, S. 1907, *AN*, 175, 327

Buchler, J. R. & Kolláth, Z. 2011, *ApJ*, 731, 24

Burki, G., Schmidt, E. G., Arellano Ferro, A., *et al.* 1986, *A&A*, 168, 139

Chadid, M., Benkő, J. M., Szabó, R., *et al.* 2010, *A&A*, 510, A39

Dziembowski, W. 1977, *AcA* 27, 95

Geroux, C. M. & Deupree, R. G. 2011, *ApJ*, 731, 18

Geroux, C. M. & Deupree, R. G. 2013, *ApJ*, 771, 113

Gillet, D., Fabas, N., & Lèbre, A. 2013, *A&A* 553, A59

Guggenberger, E., Kolenberg, K., Poretti, E., *et al.* 2011, *MNRAS*, 415, 1577

Guggenberger, E., Kolenberg, K., Nemec, J. M., *et al.* 2012, *MNRAS*, 424, 649

Hurta, Zs., Jurcsik, J., Szeidl, B., & Sódor, Á. 2008, *AJ*, 135, 957

Jurcsik, J., Sódor, Á., Szeidl, B., *et al.* 2009, *MNRAS*, 400, 1006

Kolenberg, K., Szabó, R., Kurtz, D. W., *et al.* 2010, *ApJ*, 713, L198

Kolenberg, K., Bryson, S., Szabó, R., *et al.* 2011, *MNRAS*, 411, 878

Kolláth, Z., Buchler, J. R., Szabó, R., & Csubry, Z. 2002, *A&A*, 385, 932

Kolláth, Z., Molnár, L., & Szabó, R. 2011, *MNRAS*, 414, 1111

Kovács, G. 2009, *AIP-CP*, 1170, 261

Molnár, L., Kolláth, Z., Szabó, R., *et al.* 2012, *ApJ*, 757, L13

Molnár, L., Szabó, R., Kolenberg, K., *et al.* 2013a, arXiv: 1309.0740

Molnár, L., Szabados, L., Dukes, R. J. Jr, Győrffy, Á., & Szabó, R. 2013b, *AN*, in press (arXiv: 1309.2108)

Moskalik, P. 2013, in: J. C. Suárez, R. Garrido, L. A. Balona, & J. Christensen-Dalsgaard (eds.), *Stellar Pulsations, Astrophysics and Space Science Proceedings*, Vol. 31 (Berlin, Heidelberg: Springer-Verlag), p. 103

Moskalik, P. & Kołaczkowski, Z. 2009, *MNRAS*, 394, 1649

Nemec, J. M., Smolec, R., Benkő, J. M., *et al.* 2011, *MNRAS*, 417, 1022

Plachy, E., Kolláth, Z., & Molnár, L. 2013, *MNRAS*, 433, 3590

Preston, G. W. 2009, *A&A*, 507, 1621

Preston, G. W. 2011, *AJ*, 141, 6

Rauer, H., Catala, C., Aerts, C., *et al.* 2013, *Experimental Astronomy*, submitted (arXiv: 1310.0696)

Ricker, G. R., Latham, D. W., Vanderspek, R. K., *et al.* 2010, *BAAS*, 42, 459

Shapley, H. 1916, *ApJ*, 43, 217

Smolec, R. & Moskalik, P. 2008, *AcA*, 58, 193

Smolec, R. & Moskalik, P. 2012, *MNRAS*, 426, 108

Spreckley, S. A. & Stevens, I. R. 2008, *MNRAS*, 388, 1239

Sódor, Á., Jurcsik, J., Szeidl, B., *et al.* 2011, *MNRAS*, 411, 1585

Szabó, R., Kolláth, Z., Molnár, L., *et al.* 2010, *MNRAS*, 409, 1244

Szeidl, B. & Jurcsik, J. 2009, *CoAst*, 160, 17

Turner, D. G., Savoy, J., Derrah, J., Abdel-Sabour, A.-L. M., & Berdnikov, L. N. 2005, *PASP*, 117, 207

Van Hoolst, T., Dziembowski, W. A., & Kawaler, S. D. 1998, *MNRAS*, 297, 536

Precision Asteroseismology
Proceedings IAU Symposium No. 301, 2013
J. A. Guzik, W. J. Chaplin, G. Handler & A. Pigulski, eds.

© International Astronomical Union 2014
doi:10.1017/S1743921313014403

Multi-mode oscillations in classical Cepheids and RR Lyrae-type stars

Paweł Moskalik

Copernicus Astronomical Centre, ul. Bartycka 18, 00-716 Warsaw, Poland
email: pam@camk.edu.pl

Abstract. I review different types of multi-mode pulsations observed in classical Cepheids and in RR Lyrae-type stars. The presentation concentrates on the newest results, with special emphasis on recently detected nonradial oscillations.

Keywords. stars: oscillations, stars: variables: Cepheids, stars: variables: RR Lyr

1. Introduction

Most of classical (Pop. I) Cepheids and RR Lyrae-type stars are periodic, single-mode radial pulsators. The long-term lightcurve modulation, observed in some RR Lyrae-type variables, was for many years the only known exception from this simple picture. This phenomenon, known as the Blazhko effect, was discovered a century ago (Blazhko 1907, Shapley 1916) and still lacks a satisfactory explanation. Current understanding of the Blazhko modulation has been reviewed by Kovács (2009) and by Szabó (these proceedings) and will not be discussed here. Instead, I will focus on various forms of multi-mode pulsations, i.e. pulsations with several different oscillation modes simultaneously excited.

2. Classical double-mode radial pulsators

F+1O double-mode Cepheids. The first double-mode pulsators, U TrA and TU Cas, were identified more than fifty years ago (Oosterhoff 1957a,b). These two Cepheids pulsate simultaneously in the two lowest radial modes – the fundamental mode (F) and the first overtone (1O). Double-mode Cepheids of this class have period ratios of $P_1/P_0 = 0.694 - 0.746$. Almost 200 such stars are currently known in the Galaxy and the Magellanic Clouds (Soszyński *et al.* 2008b, 2010a, 2011b, 2012, Marquette *et al.* 2009, Smolec & Moskalik 2010 and references therein). A few have also been discovered in M33 (Beaulieu *et al.* 2006) and most recently in M31 (Poleski 2013a).

F+1O double-mode RR Lyrae-type stars (RRd stars). These variables are RR Lyrae-type analogs of the F+1O double-mode Cepheids. AQ Leo, the prototype of the class, was discovered by Jerzykiewicz & Wenzel (1977). Since then, nearly 2000 RR Lyrae stars of this type have been identified. They are observed in the Galactic field (Wils 2010, Poleski 2013b), in many Galactic globular clusters (e.g. Walker 1994, Walker & Nemec 1996, Corwin *et al.* 2008), in the Magellanic Clouds (Soszyński *et al.* 2009, 2010b, 2012) and in nearby dwarf spheroidal galaxies (e.g. Cseresnjes 2001, Clementini *et al.* 2006, Bernard *et al.* 2009). In most stellar populations, their period ratios are in a very narrow range of $P_1/P_0 = 0.742 - 0.748$. The only exception is the Galactic Bulge, where variables with P_1/P_0 as low as 0.726 are found (Soszyński *et al.* 2011a). These low period ratios indicate higher metallicities, up to $Z \sim 0.008$.

1O+2O double-mode Cepheids. Another type of double-mode pulsations was detected a few years later in the short period Cepheid CO Aur (Mantegazza 1983). In this star the period ratio is very different, $P_2/P_1 = 0.801$, implying pulsations in the first (1O) and the second (2O) radial overtones. Currently, about 500 double-mode Cepheids of this type are known, most in the Magellanic Clouds (Soszyński *et al.* 2008b, 2010a, 2012, Marquette *et al.* 2009). Only 19 are identified in the Galaxy (Smolec & Moskalik 2010 and references therein; Soszyński *et al.* 2011b). One such variable has also been found in the spiral galaxy M31 (Poleski 2013a).

3. New types of multi-mode radial pulsators

In the next 20 years following the discovery of CO Aur, no new types of multi-mode radial pulsators were identified. This started to change only with the advent of microlensing surveys, which collected photometric data for thousands of Cepheids and RR Lyrae-type stars. Such data allowed finding rare forms of oscillations. In just a few years, the inventory of multi-mode radial pulsators was enlarged by several new classes. Further discoveries came from analysis of ultra-precise photometry obtained with *Kepler* and CoRoT planet-hunting space telescopes.

1O+3O double-mode Cepheids. Cepheids pulsating in the first (1O) and the third (3O) radial overtones are extremely rare. Only two examples are known, both in the Large Magellanic Cloud (hereafter LMC, Soszyński *et al.* 2008a). Both variables have period ratio of $P_3/P_1 = 0.677$. The amplitude of the third overtone is very low, in both cases below 0.03 mag in the *I*-band.

F+2O double-mode RR Lyrae-type stars. This type of pulsation is also very rare and so far has been detected only in 9 stars (see Moskalik 2013 and references therein). All but one have been discovered with photometry obtained from space. The ratio of the two pulsation periods falls in a very narrow range of $P_2/P_0 = 0.582 - 0.593$, implying pulsations in the radial fundamental mode (F) and the second radial overtone (2O). In most stars of this class the fundamental mode is modulated, i.e. displays a Blazhko effect. However, this is not a rule. In two variables with the longest pulsation periods (V350 Lyr and KIC 7021124) the amplitudes are constant.

Triple-mode Cepheids. The first two variables of this type were identified by Moskalik *et al.* (2004). Currently, only 10 triple-mode Cepheids are known in both Magellanic Clouds and in the Galaxy (Soszyński *et al.* 2008a, 2010a, 2011b). They can be divided into two classes: those which pulsate in the fundamental mode and the first two radial overtones (F+1O+2O type; 4 objects) and those which pulsate in the first three radial overtones (1O+2O+3O type; 6 objects). Properties of all known triple-mode Cepheids are summarized in Table 1. These rare objects are very important and interesting targets for asteroseismology. Modeling of their three pulsation periods strongly constrains parameters of these stars and, consequently, imposes tight constraints on Cepheid evolutionary tracks (Moskalik & Dziembowski 2005).

Blazhko effect in 1O+2O double-mode Cepheids. Long-term periodic amplitude modulation was detected in many 1O+2O double-mode Cepheids of the LMC (Moskalik *et al.* 2004, 2006, Moskalik & Kołaczkowski 2009). This behaviour is quite common and occurs in at least 20% (but possibly in as many as 35%) of these stars. Both overtones vary with the common period ($P_B > 700$ d), but variations are anticorrelated – the maximum amplitude of one mode coincides with minimum amplitude of the other. The phases (or equivalently periods) of both modes are modulated, too. The phenomenon closely resembles the Blazhko effect in the RR Lyrae-type stars. The only difference is that not one, but two radial modes are being modulated. Within the accuracy of the data, this modulation is periodic, and the modulation period is the same for both overtones.

Table 1. Triple-mode Cepheids.

star	P_0 [day]	P_1 [day]	P_2 [day]	P_3 [day]	P_1/P_0	P_2/P_1	P_3/P_2	ref.
1O+2O+3O Cepheids								
OGLE-LMC-CEP-1847		0.5795	0.4666	0.3921		0.8052	0.8404	1,2
OGLE-LMC-CEP-2147		0.5413	0.4360	0.3663		0.8056	0.8401	1,2
OGLE-LMC-CEP-3025		0.5687	0.4582	0.3850		0.8057	0.8402	2
OGLE-SMC-CEP-3867		0.2688	0.2174	0.1824		0.8086	0.8392	3
OGLE-BLG-CEP-16		0.2955	0.2340	0.1951		0.7919	0.8337	4
OGLE-BLG-CEP-30		0.2304	0.1830	0.1522		0.7943	0.8316	4
F+1O+2O Cepheids								
OGLE-LMC-CEP-0857	1.4631	1.0591	0.8552		0.7239	0.8075		2
OGLE-LMC-CEP-1378	0.5150	0.3849	0.3094		0.7474	0.8037		2
OGLE-SMC-CEP-1077	0.8973	0.6565	0.5289		0.7316	0.8056		3
OGLE-SMC-CEP-1350	0.9049	0.6660	0.5379		0.7360	0.8076		3

References: 1 – Moskalik *et al.* (2004), 2 – Soszyński *et al.* (2008a), 3 – Soszyński *et al.* (2010a), 4 – Soszyński *et al.* (2011b).

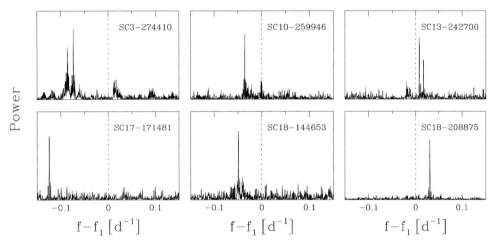

Figure 1. Nonradial modes in LMC first-overtone Cepheids. Prewhitened frequency spectra after subtracting the dominant radial mode (indicated with a dashed line) are displayed.

4. Nonradial modes in Cepheids and RR Lyrae-type stars

Excitation of nonradial modes is very common in pulsating stars. Modes of this type are observed in all classes of main-sequence pulsators (Gautschy & Saio 1996) as well as in oscillating white dwarfs (Winget & Kepler 2008, Fontaine, these proceedings), subdwarfs (Randall, these proceedings) and in stochastically driven subgiants and red giants (Hekker, these proceedings). But classical pulsating variables – Cepheids and RR Lyrae-type stars – seemed to be different. They seemed to avoid excitation of any nonradial oscillations at all. This picture has changed only in the last decade. New observations prove that nonradial modes are present also in the classical pulsators.

4.1. *Nonradial modes in LMC Cepheids*

The first convincing detection of nonradial modes in classical Cepheids came with the analysis of LMC photometry of the OGLE-II survey (Moskalik *et al.* 2004, Moskalik & Kołaczkowski 2008, 2009). In 37 first overtone Cepheids, low amplitude secondary frequencies have been found in close proximity to the dominant radial mode ($\Delta f < 0.125\,\mathrm{d^{-1}}$). Six of these variables are displayed in Fig. 1. Similar secondary frequencies

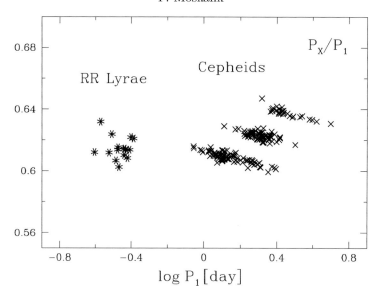

Figure 2. Petersen diagram for Cepheids and RR Lyrae-type stars with period ratio of
$P_X/P_1 = 0.60 - 0.64$.

have also been detected in two F+1O double-mode Cepheids, where they are found close to the first radial overtone. In most cases only one secondary peak is present, but sometimes two peaks are seen. In the latter case, both peaks always appear on the same side of the radial mode. The observed frequency pattern cannot be explained by any form of modulation of the radial mode. Modulation should always produce an equally spaced frequency multiplet centered on the primary peak, and such a structure is not found in any of these stars. Because secondary frequencies are too close to the frequency of the radial first overtone, they cannot be explained by any radial mode, either. Consequently, they must correspond to nonradial modes of oscillations.

4.2. Double-mode pulsators with period ratios of 0.60 – 0.64

Cepheids. In some Magellanic Cloud Cepheids a different type of multiperiodicity is found – a single secondary mode is detected with a frequency much above that of the radial mode. The ratio of the secondary and the primary period falls in a narrow range of $P_X/P_1 = 0.60 - 0.64$. The first eight Cepheids of this class have been discovered in the LMC by Moskalik & Kołaczkowski (2008). Currently, more than 170 such pulsators are known in both Magellanic Clouds (Soszyński *et al.* 2008b, 2010a, Moskalik & Kołaczkowski 2009). The phenomenon is restricted to the first overtone Cepheids and F+1O double-mode Cepheids (only one star). When plotted on the period ratio vs. period diagram (so called Petersen diagram), these variables form three very tight parallel sequences (see Fig. 2). They are as tight as in case of the radial double-mode pulsators. However, none of the sequences fits theoretical prediction for the two radial modes (Dziembowski & Smolec 2009, Dziembowski 2012). This implies that the secondary mode is nonradial. Considering that this mode has to be unstable, the theoretical analysis suggests that it must be an f-mode of high angular degree of $l = 42 - 50$ (Dziembowski 2012).

RR Lyrae-type stars. Secondary modes with the period ratio of $0.60 - 0.64$ are found in RR Lyrae-type stars as well. 16 such variables have been identified so far, including four stars observed by the *Kepler* space telescope (Moskalik *et al.* 2013, 2014 and references

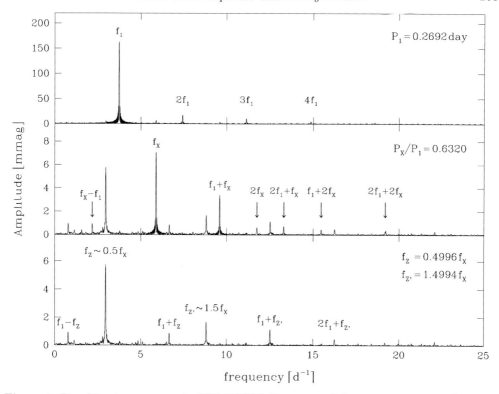

Figure 3. Prewhitening sequence for KIC 5520878. Upper panel: frequency spectrum of original data. Middle panel: frequency spectrum after subtracting f_1 and its harmonics. Bottom panel: frequency spectrum after subtracting f_1, f_X and their harmonics and linear combinations.

therein). Most of these variables are dominated by the first radial overtone (RRc stars), but two are double-mode pulsators (RRd stars). The amplitude of the secondary mode is always extremely low, in the mmag range. Interestingly, this mode is found in all four RRc stars observed by *Kepler*. This suggests that such double-mode pulsations are rather common and should be detected in many more RRc variables, provided high-quality photometry is available. Just like in the case of Cepheids, comparison of observed period ratios with theoretical models implies that the secondary frequency must correspond to a nonradial mode (Moskalik *et al.* 2014).

The RR Lyrae-type stars displaying the puzzling period ratio of $0.60 - 0.64$ are very similar to their Cepheid siblings. In both cases this phenomenon occurs only in the first-overtone pulsators and in the F+1O double-mode pulsators. The period ratios fall in the same narrow range. This is shown in Fig. 2, where we plot both groups of variables on a common Petersen diagram. The only difference between the two groups is that classical Cepheids split into three separate sequences, whereas the RR Lyrae-type stars do not. Finally, in both types of stars the secondary mode is nonradial. These similarities lead to the conclusion that Cepheids and RR Lyrae-type stars with the period ratio of $P_X/P_1 = 0.60 - 0.64$ form a very homogenous group, constituting a new, well defined class of multimode pulsators.

Period doubling of the secondary mode in RR Lyrae-type stars. In Fig. 3 we display a prewhitening sequence for KIC 5520878, one of the RRc stars in the *Kepler* field (Moskalik *et al.* 2014). After subtracting the dominant radial pulsation (f_1) and its harmonics, the

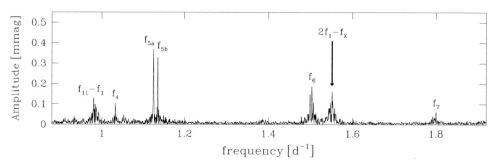

Figure 4. Low frequency low amplitude modes in KIC 5520878. Fourier transform of Q1 – Q10 lightcurve is computed after prewhitening the data of the dominant frequency f_1, f_X, their harmonics, subharmonics and 5 highest combination peaks. Nonradial modes appear at f_4, f_5 (doublet), f_6 (triplet) and f_7. $f_{11} - f_1$ is a combination peak associated with another low amplitude mode. The frequency of the radial fundamental mode is estimated to be $f_0 \sim 2.75\,\mathrm{d}^{-1}$.

frequency spectrum of the residuals (middle panel) reveals a secondary frequency at $f_X = 5.879\,\mathrm{d}^{-1}$, which yields a period ratio of $P_X/P_1 = 0.632$. A harmonic of f_X and several linear combinations with f_1 are also present. But ultra-precise *Kepler* photometry reveals even more. After prewhitening the data with f_1, f_X and their harmonic and combination frequencies (bottom panel), the highest residual peak appears at $f_Z = 2.937\,\mathrm{d}^{-1}$, that is at $\sim f_X/2$. In other words, f_Z is not an independent frequency, but a subharmonic of f_X. Another subharmonic is found at $f_{Z'} \sim (3/2)f_X$. The detection of subharmonic frequencies is highly significant. Their presence is a characteristic signature of a period doubling of the secondary mode, f_X. The same signature of the period doubling behaviour is seen in all four RRc variables in the *Kepler* field (Moskalik *et al.* 2013, 2014). It is also detected in the RRd variable AQ Leo (Gruberbauer *et al.* 2007). Thus, RR Lyrae-type stars with the period ratio of 0.60 – 0.64 are yet another class of pulsators in which period doubling can occur. We recall, that this phenomenon has recently been discovered in two other classes of pulsating variables – in Blazhko RRab stars (Szabó *et al.* 2010) and in the BL Herculis-type stars (Smolec *et al.* 2012). Its origin can be traced to a half-integer resonance between the pulsation modes (Moskalik & Buchler 1990).

4.3. *Other nonradial modes in RRc stars*

With the benefit of high-quality *Kepler* data it is possible to search for modes which are even weaker than f_X and its subharmonic peaks. Such modes are discovered in three *Kepler* RRc variables (Moskalik *et al.* 2014). All but one of these oscillations have amplitudes below 0.4 mmag. The majority of them generate combination peaks with the dominant radial pulsation (f_1), which proves that they indeed originate in the RRc star and do not come from blending with another variable.

The richest harvest of low amplitude modes is found in KIC 5520878, where 15 such oscillations are detected. Based on the period ratios, one of these modes can be identified with the second radial overtone, but all others must be nonradial. Interestingly, several modes have frequencies significantly lower than the radial fundamental mode (see Fig. 4). This implies that these nonradial oscillations are not *p*-modes, but must be of either gravity or mixed mode character. Low-frequency modes of this type are discovered also in the other two RRc stars.

4.4. *Nonradial modes in RRab stars*

Low-amplitude modes other than the radial second overtone are detected in six RRab variables. They are listed in Table 2. For each secondary mode, instead of its period, we

Table 2. Secondary periodicities in RRab stars

star	P_0 [day]	P/P_0	P/P_0	ref.
V 1127 Aql	0.3560	0.7271	0.6966	1,2
RR Lyr	0.5669	0.7557		5
V 354 Lyr	0.5617	0.7295	0.8555	3
V 360 Lyr	0.5576	0.7210		3
V 445 Lyr	0.5131	0.7306	0.7031	4
CoRoT 105288363	0.5674	0.7216	0.7764	4
		0.7407		

References: 1 – Chadid *et al.* (2010), 2 – Poretti *et al.* (2010), 3 – Benkő *et al.* (2010), 4 – Guggenberger *et al.* (2012), 5 – Molnár *et al.* (2012).

provide the period ratio of this mode and the dominant fundamental radial pulsation, P/P_0. The period ratios given in the column 4 of the Table are clearly incompatible with those of the radial modes. Therefore, these secondary oscillations must be nonradial. The situation is somewhat more complicated for modes listed in column 3. Their period ratios are more or less similar to that observed in the RRd stars. On this ground, the secondary modes of column 3 are usually identified with the radial first overtone. However, the double-mode RR Lyrae-type stars do not populate the P/P_0 vs. P_0 diagram in a random way. Instead, they form a rather narrow, well defined sequence on this diagram (Soszyński *et al.* 2009, 2011a). Putting this differently, there is an empirical relation between the period ratio and the period. It is easy to check that V1127 Aql is the only variable of Table 2 which follows this relation. For all the remaining variables, the secondary modes listed in column 3 deviate considerably from the empirical trend. Therefore, identification of these modes with the radial first overtone should be treated with great caution. These periodicities most likely correspond to nonradial modes, the frequencies of which are just close to that of the radial overtone. We note, that a small number of similar outliers from the empirical P/P_0 vs. P_0 relation have also been identified in the LMC and in the Galactic bulge (Soszyński *et al.* 2009, 2011a). The secondary periods in those objects might correspond to nonradial modes, too.

The tendency of nonradial oscillations to appear preferentially around the frequency of the radial overtone is not a surprise. In fact, it is consistent with the theory. Linear nonadiabatic calculations show that nonradial modes can be excited in the RR Lyrae-type stars, and their growth rates are highest in the vicinity of the radial modes (Van Hoolst *et al.* 1998, Dziembowski & Cassisi 1999). The growth rates reach maximum values at frequencies slightly above that of the radial mode. This is true both for modes of $l = 1, 2$ and for the strongly trapped modes (STU modes) of higher spherical degrees. Thus, nonradial oscillations which are most likely to be excited should yield period ratios somewhat lower than the radial first overtone. This prediction is in good agreement with observational picture summarized in Table 2.

Acknowledgements

This research is supported by the Polish National Science Centre through grant DEC-2012/05/B/ST9/03932. Generous IAU support in the form of a travel grant is also acknowledged.

References

Beaulieu, J.-P., Buchler, J. R., Marquette, J.-B., *et al.* 2006, *ApJ*, 653, L101
Benkő, J. M., Kolenberg, K., Szabó, R., *et al.* 2010, *MNRAS*, 409, 1585
Bernard, E. J., Monelli, M., Gallart, C., *et al.* 2009, *ApJ*, 699, 1742

Blazhko, S. 1907, *AN*, 175, 325

Chadid, M., Benkő, J. M., Szabó, R., *et al.* 2010, *A&A*, 510, A39

Clementini, G., Greco, C., Held, E. V., *et al.* 2006, *MemSAIt*, 77, 249

Corwin, T. M., Borissova, J., Stetson, P. B., *et al.* 2008, *AJ*, 135, 1459

Cseresnjes, P. 2001, *A&A*, 375, 909

Dziembowski, W. A. 2012, *AcA*, 62, 323

Dziembowski, W. A. & Cassisi, S. 1999, *AcA*, 49, 371.

Dziembowski, W. A. & Smolec, R. 2009, *AIP-CP*, 1170, 83

Gautschy, A. & Saio, H. 1996, *ARAA*, 34, 551

Gruberbauer, M., Kolenberg, K., Rowe, J. F., *et al.* 2007, *MNRAS*, 379, 1498

Guggenberger, E., Kolenberg, K., Nemec, J. M., *et al.* 2012, *MNRAS*, 424, 649

Jerzykiewicz, M. & Wenzel, W. 1977, *AcA*, 27, 35

Kovács, G. 2009, *AIP-CP*, 1170, 261

Mantegazza, L. 1983, *A&A*, 118, 321

Marquette, J.-B., Beaulieu, J.-P., Buchler, J. R., *et al.* 2009, *A&A*, 495, 249

Molnár, L., Kolláth, Z., Szabó, R. *et al.* 2012, *ApJ*, 757, L13

Moskalik, P. 2013, in: J. C. Suárez, R. Garrido, L. A. Balona, & J. Christensen-Dalsgaard (eds.), *Stellar Pulsations: Impact of New Instrumentation and New Insights*, Astrophysics and Space Sci. Proc., Vol. 31 (Berlin, Heidelberg: Springer-Verlag), p. 103

Moskalik, P. & Buchler, J. R. 1990, *ApJ*, 355, 590

Moskalik, P. & Dziembowski, W. A. 2005, *A&A*, 434, 1077

Moskalik, P. & Kołaczkowski, Z. 2008, *CoAst*, 157, 343

Moskalik, P. & Kołaczkowski, Z. 2009, *MNRAS*, 394, 1649

Moskalik, P., Kolenberg, K., Smolec, R., *et al.* 2014, *MNRAS*, to be submitted

Moskalik, P., Kołaczkowski, Z., & Mizerski, T. 2004, *ASP-CS*, 310, 498

Moskalik, P., Kołaczkowski, Z., & Mizerski, T. 2006, *MemSAIt*, 77, 563

Moskalik, P., Smolec, R., Kolenberg, K., *et al.* 2013, in: J. C. Suárez, R. Garrido, L. A. Balona, & J. Christensen-Dalsgaard (eds.), *Stellar Pulsations: Impact of New Instrumentation and New Insights*, Astrophysics and Space Sci. Proc., Vol. 31 (Berlin, Heidelberg: Springer-Verlag), Poster No. 34 (arXiv: 1208.4251)

Oosterhoff, P. T. h. 1957a, *BAN*, 13, 317

Oosterhoff, P. T. h. 1957b, *BAN*, 13, 320

Poleski, R. 2013a, *ApJ*, 778, 147

Poleski, R. 2013b, A&A, submitted (arXiv: 1309.1168)

Poretti, E., Paparó, M., Deleuil, M., *et al.* 2010, *A&A*, 520, A108

Shapley, H. 1916, *ApJ*, 43, 217

Smolec, R. & Moskalik, P. 2010, *A&A*, 524, A40

Smolec, R., Soszyński, I., Moskalik, P., *et al.* 2012, *MNRAS*, 419, 2407

Soszyński, I., Dziembowski, W. A., Udalski, A., *et al.* 2011a, *AcA*, 61, 1

Soszyński, I., Poleski, R., Udalski, A., *et al.* 2008a, *AcA*, 58, 153

Soszyński, I., Poleski, R., Udalski, A., *et al.* 2008b, *AcA*, 58, 163

Soszyński, I., Poleski, R., Udalski, A., *et al.* 2010a, *AcA*, 60, 17

Soszyński, I., Udalski, A., Pietrukowicz, P., *et al.* 2011b, *AcA*, 61, 285

Soszyński, I., Udalski, A., Poleski, R., *et al.* 2012, *AcA*, 62, 219

Soszyński, I., Udalski, A., Szymański, M. K., *et al.* 2009, *AcA*, 59, 1

Soszyński, I., Udalski, A., Szymański, M. K., *et al.* 2010b, *AcA*, 60, 165

Szabó, R., Kolláth, Z., Molnár, L., *et al.* 2010, *MNRAS*, 409, 1244

Van Hoolst, T., Dziembowski, W. A., & Kawaler, S. D. 1998, *MNRAS*, 297, 536

Walker, A. R. 1994, *AJ*, 108, 555

Walker, A. R. & Nemec, J. M. 1996, *AJ*, 112, 2026

Wils, P. 2010, *IBVS*, 5955

Winget, D. E. & Kepler, S. O. 2008, *ARAA*, 46, 157

Precision Asteroseismology
Proceedings IAU Symposium No. 301, 2013
J. A. Guzik, W. J. Chaplin, G. Handler & A. Pigulski, eds.
© International Astronomical Union 2014
doi:10.1017/S1743921313014415

RR Lyrae studies with *Kepler*: showcasing RR Lyr

Katrien Kolenberg[1,2] Robert L. Kurucz[1], Robert Stellingwerf[3], James M. Nemec[4], Paweł Moskalik[5], Luca Fossati[6], and Thomas G. Barnes[7]

[1] Harvard-Smithsonian Center for Astrophysics, 60 Garden Street, Cambridge MA 02138, USA
email: kkolenberg@cfa.harvard.edu

[2] Institute of Astronomy, Leuven University, Celestijnenlaan 200d, 3001 Heverlee, Belgium

[3] Stellingwerf Consulting, 11033 Mathis Mountain Rd SE Huntsville AL 35803-2813, USA

[4] Dept. of Physics & Astronomy, Camosun College, Victoria, Br. Columbia, V8P 5J2, Canada

[5] Copernicus Astronomical Centre, ul.Bartycka 18, 00-716, Warsaw, Poland

[6] Argelander-Inst. für Astronomie der Univ. Bonn, Auf dem Hügel 71, D-53121, Bonn, Germany

[7] The University of Texas at Austin, McDonald Observatory, 82 Mt. Locke Rd.
McDonald Observatory, Texas, 79734, USA

Abstract. Four years into the *Kepler* mission, an updated review on the results for RR Lyrae stars is in order. More than 50 RR Lyrae stars in the *Kepler* field are observed with *Kepler* and each one of them can provide us with new insight into this class of pulsating stars. Ground-based spectroscopy of the *Kepler* targets allows us to narrow down their physical parameters. Previously, we already reported a 50% occurrence rate of modulation in the RRab stars, a large variety of modulation behavior, period doubling in several Blazhko stars, the detection of higher-overtone radial modes, probable non-radial modes and new types of multiple-mode RR Lyrae pulsators, among both the RRab and the RRc stars. In addition, the quasi-continuous photometry obtained over several years with *Kepler* allows one to observe changes in Blazhko behavior and additional longer cycles. These observations have sparked new theoretical modelling efforts. In this short paper we showcase RR Lyr itself. The star has been observed with *Kepler* in short cadence, and some remarkable features of its pulsation behavior are unveiled in this long-studied prototype, through the *Kepler* photometry and additional spectroscopic data.

Keywords. stars: variables: RR Lyrae, stars: individual: RR Lyr, techniques: photometric, techniques: spectroscopic

1. Introduction

Kepler had been operational at full capacity for four years until the second of its four reaction wheels failed in May 2013, thereby compromising its capacity to point accurately. Towards the end of *Kepler*'s nominal mission time, over 40 RR Lyrae stars were known in the observed 115-square-degree field and all of them were included in the target list for the extended mission, which started in November 2012. More recently, several RR Lyrae stars have been found serendipitously, e.g., through the citizen science program "PlanetHunters" (www.planethunters.org), bringing the total of known *Kepler* RR Lyrae stars to about 50.

To complement the wide-band filter observations of the *Kepler* targets, Jeon *et al.* (these proceedings) have obtained extensive multicolor photometry for all the known RR Lyrae stars in the *Kepler* field. In addition, high-resolution spectroscopy has been obtained for 41 RR Lyrae stars in the *Kepler* field, leading to an accurate [Fe/H] determination for each one (Nemec *et al.* 2013). An overview of published studies (reviews excluded) on the

Table 1. RR Lyrae studies with *Kepler*, and spin-off studies/results published thus far.

Topic	Authors	N_{stars}	Remark[a]
First results (period doubling)	Kolenberg *et al.* (2010a)	2	A
Period doubling in *Kepler* data	Szabó *et al.* (2010)	3+4	A, M
Flavors of variability	Benkő *et al.* (2010)	29	A
Mathematical description of modulation	Benkő *et al.* (2011)		M, S
RR Lyr long-cadence data	Kolenberg *et al.* (2011)	1	A
Testing Stothers model on RR Lyr	Smolec *et al.* (2011)	1	A, M
Non-modulated RR Lyrae stars	Nemec *et al.* (2011)	19	A
Modeling of period doubling	Kolláth *et al.* (2011)		M
Radial resonance Blazhko model	Buchler & Kolláth (2011)		M, S
RRc multiperiodicity	Moskalik *et al.* (2012)	4	A
Nonlinear asteroseismology for RR Lyr	Molnár *et al.* (2012)	1	A, M
Multiperiodicity in V445 Lyr	Guggenberger *et al.* (2012)	1	A
(Quasi-)periodic modulation in BL Her stars	Smolec & Moskalik (2012)		M, S
Stitching together *Kepler* data of a Blazhko Star	Çelik *et al.* (2012)	1 (example)	A
Comparison of non-repetitive Blazhko cycles	Guggenberger (2012)	1	A
Metallicity [Fe/H], v_{rad}, etc. from spectra	Nemec *et al.* (2013)	41	A, G
RR Lyr in short cadence	Stellingwerf *et al.* (2013)	1	A
Additional RR Lyrae stars?	Kinemuchi (2013)	TBD	A, G
Chaotic pulsation models	Plachy *et al.* (2013)		M, S
Shock Blazhko model	Gillet (2013)	1 (example)	M
Ground-based multicolor data	Jeon *et al.* (these proc.)	41	G
Nonradial modes or not?	Benkő (these proc.)	41	A
Interchange of alternating amplitudes	Plachy *et al.* (these proc.)	3	A
Multi-mode resonances in RR Lyrae stars	Molnár *et al.* (these proc.)	4	A, M

[a] A: Analysis, M: Modelling, G: Ground-based data, S: Spin-off.

Kepler data and their spin-offs is given in Table 1. In this paper we focus on the brightest RR Lyrae star in the *Kepler* field, RR Lyr.

2. Showcasing RR Lyr

The analysis of the long-cadence *Kepler* data of RR Lyr (quarters Q1–Q2, May – September 2009) was published by Kolenberg *et al.* (2011). As of Q5 (March 2010) the star was observed in short cadence. Molnár *et al.* (2012) found evidence for the first radial overtone in the Q5 and Q6 short-cadence data. A frequency analysis performed on the Q5–Q15 short-cadence data (March 2010–January 2013) leads to the following conclusions (see also Stellingwerf *et al.* 2013):

• There is clear evidence for a longer cycle in RR Lyr. A cycle of duration of about 4 years was reported by Detre & Szeidl (1973). The Blazhko amplitude variation of about 40% at the beginning of the *Kepler* measurements decreases significantly (to as low as 0-10% in light amplitude) and then increases again (see Fig. 1).

• There is a variation of the 0.5667-day pulsation period (from 0.5655 to 0.5685 day) within each Blazhko cycle, with the shorter period corresponding to the larger amplitude phases in the Blazhko cycle. This variation of half a percent is huge considering that the mean period has barely changed since the observations by Shapley (1916).

• There is a variation of the Blazhko period itself (from 39.2 days at maximum Blazhko amplitude to 38.4 days at the amplitude minimum) over the 4-year modulation cycle, i.e., about 2%.

• There is a small but detectable variation (a few mmag) of the mean brightness over the Blazhko cycle.

Spectroscopic data allow an in-depth analysis of the pulsations of RR Lyrae stars (see also Guggenberger, these proceedings). A spectroscopic data set of RR Lyr, obtained with the high-resolution ($R = 60\,000$) spectrograph attached to the 2.7-m telescope at McDonald Observatory was described by Kolenberg *et al.* (2010b). Integration times were

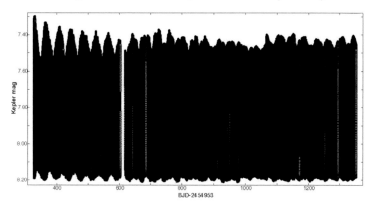

Figure 1. *Kepler* short-cadence data of RR Lyr (Q5–Q15). Individual cycles are not visible, but the envelope of the Blazhko modulation is. There is clear evidence for a variation in the strength of the Blazhko effect.

typically 20 minutes. A detailed abundance analysis was performed on the spectrum in the most quiescent phase in RR Lyr's pulsation cycle. This is the phase when the star reaches its maximum radius, around pulsation phase ∼0.3 for RR Lyr.

We fitted a static Kurucz atmosphere model to the spectrum in the most quiescent phase. Using an instrumental profile of 5 km s^{-1} (corresponding to the instrumental resolution), the best fit is obtained for a macroturbulent velocity $v_{\mathrm{mac}} = 5 \pm 2$ km s^{-1} and a projected rotational velocity $v \sin i = 5 \pm 2$ km s^{-1}.

Short-cadence spectroscopy was obtained simultaneously with the short cadence *Kepler* data over three nights in 2010 (Kolenberg *et al.*, in preparation). With a 1-minute spectrum taken every few minutes the photospheric layers can be followed in detail over time. In Fig. 2 we show the radial-velocity variations for two spectral lines, the Hγ line at 4340.462 Å and a Cr line at 4351.811 Å, over a few hours of spectra coinciding with the descending branch (declining light) of the star (pulsation phase [0.0 – 0.35]). We can detect shifts of 1 km s^{-1} for spectra taken only four minutes apart, even in this "quieter" phase interval of the pulsation cycle. As a consequence, spectra taken over 20 minutes, an integration time necessary to obtain a sufficient signal-to-noise ratio for a detailed line profile analysis on a 3-m class telescope, show smearing of the order of 5 km s^{-1}. That

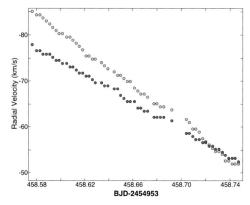

Figure 2. Radial velocities derived from two lines for the short-cadence spectra. The steeper radial-velocity curve corresponds to the Hγ line, the other one to the Cr line.

implies that the macroturbulent velocity parameter $v_{mac} = 5$ km s^{-1} adopted in our fit to the observed spectrum is probably a representative lower limit.

Assuming $v \sin i = 5$ km s^{-1} implies, for a star with a radius of about $5 R_\odot$, that $P_{rot} \sim 50$ d $\times \sin i$. As a consequence, rotation period and Blazhko period could correspond, which is an essential component in some models for the Blazhko effect (Shibahashi & Takata 1995; Dziembowski & Mizerski 2004), though currently not the most favored explanation. The observation of changing or multiple modulation periods does pose a problem for the models linking the modulation directly to the rotation, but one could think of a scenario in which features at different latitudes on the star rotate at different speeds, translated into the light variation. However, often the observed modulation periods for a given star are sometimes too far apart (e.g., 40 days and 120 days) to make this a plausible scenario.

Acknowledgements

We are particularly grateful to Wojtek Dziembowski for his important contributions to the longstanding puzzle of the Blazhko effect in RR Lyrae stars. KK also is very indebted to him for his encouragement and enthusiasm. KK acknowledges support from Marie Curie Fellowship 255267 SAS-RRL within the 7th European Community Framework Program. Some of her work on the *Kepler* data was funded through research from the Austrian Research Foundation (FWF projects P19962 and T359).

References

Benkő, J., Kolenberg, K., Szabó, R., *et al.* 2010, *MNRAS*, 409, 1585
Benkő, J, Szabó, R. & Paparó, M. 2011, *MNRAS*, 417, 974
Buchler, J. R. & Kolláth, Z. 2011, *ApJ*, 731, 24
Çelik, L., Ekmekçi, F., Nemec, J., *et al.* 2012, arXiv: 1202.3607
Detre, L. & Szeidl, B. 1973, *IBVS*, 764
Dziembowski, W. A. & Mizerski, T. 2004, *AcA*, 54, 363
Gillet, D. 2013, *A&A*, 554, A46
Guggenberger, E. 2012, *AN*, 333, 1044
Guggenberger, E., Kolenberg, K., Nemec, J. M., *et al.* 2012, *MNRAS* 424, 649
Kinemuchi, K., 2013, arXiv: 1310.0544
Kolenberg, K., Szabó, R., Kurtz, D. W. *et al.*, 2010a, *ApJ*, 713, L198
Kolenberg, K., Fossati, L., Shulyak, D., *et al.*, 2010b, *A&A* 519, 64
Kolenberg, K., Bryson, S., Szabó, R., *et al.*, 2011, *MNRAS* 411, 878
Kolláth, Z., Molnár, L., & Szabó, R., 2011, *MNRAS* 414, 1111
Molnár, L., Kolláth, Z., Szabó, R., *et al.*, 2012, *ApJ*, 757, L13
Moskalik, P., Smolec, R., Kolenberg, K., *et al.*, 2012, arXiv: 1208.4251
Nemec, J. M., Smolec, R., Benkő, J. M., *et al.*, 2011, *MNRAS*, 417, 1022
Nemec, J. M., Cohen, J. G., Ripepi, V., *et al.*, 2013, *ApJ*, 773, 181
Plachy, E., Kolláth, Z., & Molnár, L., 2013, *MNRAS* 433, 3590
Shibahashi, H. & Takata, M., 1995, *ASP-CS*, 83, 42
Smolec, R. & Moskalik, P., 2012, *MNRAS* 426, 108
Smolec, R., Moskalik, P., Kolenberg, K., *et al.*, 2011, *MNRAS*, 414, 2950
Stellingwerf, R. F., Nemec, J. M., & Moskalik, P., 2013, arXiv: 1310.0543
Szabó, R., Kolláth, Z., Molnár, L., *et al.*, 2010, *MNRAS*, 409, 1244
Szeidl, B., 1988, in: G. Kovács, L. Szabados, & B. Szeidl (eds.), *Multimode Stellar Pulsations* (Budapest: Konkoly Observatory, Kultura), p. 45

Precision Asteroseismology
Proceedings IAU Symposium No. 301, 2013
J. A. Guzik, W. J. Chaplin, G. Handler & A. Pigulski, eds.

© International Astronomical Union 2014
doi:10.1017/S1743921313014427

Bisector analysis of RR Lyrae: atmosphere dynamics at different phases

Elisabeth Guggenberger[1], Denis Shulyak[2], Vadim Tsymbal[3] and Katrien Kolenberg[4,5]

[1]Institut für Astrophysik, Universität Wien, Türkenschanzstrasse 17, 1180 Vienna, Austria
email: elisabeth.guggenberger@univie.ac.at

[2]Institute of Astrophysics, Georg-August University, Friedrich-Hund-Platz 1, 37077, Göttingen, Germany

[3]Tavrian National University, Vernadskiys Avenue 4, Simferopol, Crimea, 95007, Ukraine

[4]Harvard-Smithsonian Center for Astrophysics, 60 Garden Street, Cambridge, MA 02138, USA

[5]Instituut voor Sterrenkunde, University of Leuven, Celestijnenlaan 200D, B 3001 Heverlee, Belgium

Abstract. This article reports some preliminary results on an analysis of line bisectors of metal absorption lines of RR Lyrae, the prototype of its class of pulsators. The extensive data set used for this study consists of a time series of spectra obtained at various pulsation phases as well as different Blazhko phases. This setup should allow a comparison of the atmospheric behaviour, especially of the function of radial velocity versus depth at differing Blazhko phases, but (almost) identical pulsation phase, making it possible to investigate whether the modulation causes a change in the atmospheric motion of RR Lyrae. While the nature of the Blazhko modulation has often been investigated photometrically and described as a change in the light curve, studies on time series of high resolution spectra are rare. We present for the first time work on line bisectors at different phases of RR Lyr.

Keywords. line: profiles, techniques: spectroscopic, stars: atmospheres, stars: variables: RR Lyrae

1. Introduction

Dedicated photometric surveys such as for example the Konkoly survey (Jurcsik *et al.* 2009) as well as high-precision satellite data for example from the *Kepler* mission (Benkő *et al.* 2010) have recently been very successful at revealing many detailed characteristics of RR Lyrae stars and their major unsolved problem: the Blazhko effect. Both from modeling and from spectroscopic studies it is known, however, that violent dynamic phenomena happen during the pulsation in the atmosphere of RR Lyrae stars. An important question is how the motion of the atmosphere changes when the Blazhko phase changes. As the pulsational motion of the atmosphere can best be studied spectroscopically, especially with the help of line asymmetries, we have chosen a method for our study that has never before been applied to RR Lyrae stars: a bisector analysis.

Spectral line bisectors (which are defined as the midpoint of a horizontal line connecting the two sides of the line profile) are a powerful tool to quantitatively study line asymmetries, allowing to extract information about velocity fields in the stellar atmosphere. The most well-known application of line bisectors is probably the study of solar and stellar granulation and convection (see for example Gray 2005). A very different application of the analysis of bisectors is the detection of false positives in the hunt for extrasolar planets, where a change of the bisector during the observed radial velocity variation indicates that the signature in the radial velocity most likely does not originate from a planet but from stellar activity.

Another application of spectral line bisectors – the one used in this study – is the derivation of pulsational velocity fields, which relies on the fact that different parts of the line profile are formed at different depths of the stellar atmosphere. Using model atmospheres and synthetic spectra, it is possible to associate an observed part of the line profile with a specific layer in the star, allowing to derive atmospheric dynamics from the velocity information contained in the bisectors.

2. The spectra, the models and the analysis

The data set used for this analysis consists of 55 spectra with a resolution of $R = 60\,000$ taken with the Robert G. Tull Coudé Spectrograph at the 2.7-m telescope of McDonald Observatory. 45 spectra were obtained near a Blazhko phase of 0.3 covering a full pulsation cycle, and 10 spectra were taken around a Blazhko phase of 0.8. A first analysis of these spectra as well as the technical details were presented by Kolenberg *et al.* (2010) who selected the spectrum obtained at the most quiescent phase (near maximum radius) to perform an abundance analysis. During their analysis they also found that the microturbulence velocity (v_{mic}) is depth-dependent. The fundamental parameters, and specifically the element abundances, obtained by them for the "quiet" phase serve as the basis for any future analysis of the remaining spectra.

In a second step, all the other spectra were subjected to a self-consistent detailed analysis. Fundamental parameters were determined for each spectrum, and the function of the depth-dependent v_{mic} was found for each phase in an iterative process. The results of this extensive analysis will soon be published (Fossati *et al.*, in preparation). The LLMODELS code (Shulyak *et al.* 2004) was used to calculate static model atmospheres.

For the bisector analysis, we selected all Fe I, Fe II, Ti II, Cr II, Ca II and Mg I lines that were unblended and therefore suitable for a reliable analysis, resulting in a sample of about 100 lines in total. Synthetic spectra were then calculated for all phases and for the complete wavelength range based on the final models obtained by Fossati *et al.* (in preparation), using a new version of the SYNTHV code by Vadim Tsymbal (Tsymbal 1996). The bisector analysis code then extracts the "classical" bisector (i.e., intensity versus velocity) as well as the bisector with formation depth (i.e. $\log(\tau)$ versus velocity) for the selected lines and the selected spectra. The resulting motion of a single line bisector during the pulsation can be seen in Fig. 1 where an Fe II line is shown as an example. It is obvious that strong tilts and distortions of the bisector take place at some phases, and the wavelength shifts due to the strong pulsational motions can be seen. Figure 2 gives a zoom-in revealing the curvature of a bisector of a Ti II line. In the subsequent steps, a mean bisector will be calculated for each spectrum including all suitable lines to improve the accuracy, and a comparison to the predictions of pulsation models will be made.

3. Velocity curves and the Van Hoof effect

As a side result of our analysis we also obtain accurate velocity curves for every line that has been selected for the bisector analysis. These allow to study the Van Hoof effect, a phase lag between different layers in a pulsating star which has originally been discovered in β Cephei stars by Van Hoof & Struve (1953) and which has been shown to also exist in RR Lyrae stars by Mathias *et al.* (1995). Following the method proposed by Mathias & Gillet (1993) who used the Van Hoof effect as a tool to study wave propagation in the atmosphere, we directly plot velocities of different lines against each other. As an

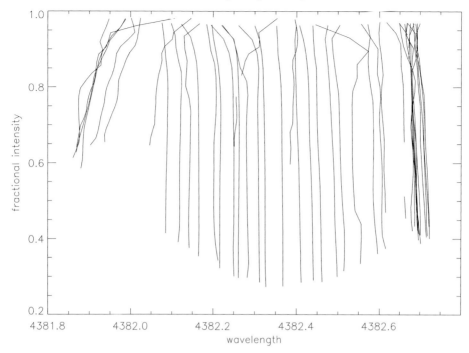

Figure 1. The motion of a line bisector (the Fe II line at 4383.545 Å is shown as an example) during the pulsation. Motion is counterclockwise. The phase of maximum radius is the more or less straight bisector line in the middle of the plot when the line is very deep. Starting from there, the star contracts, resulting in the redshifted lines in the right half of the figure. Some tilted bisectors occur during the contraction. When the minimum radius is reached, lines become shallow, and distorted and tilted bisectors occur during expansion.

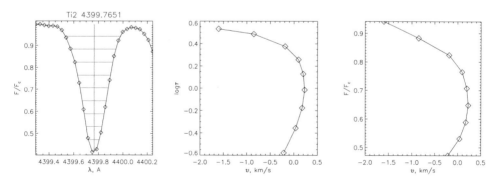

Figure 2. Example of the bisectors derived for a Ti III line. The left panel shows the observed spectral line (points) as well as the interpolation that was done to obtain the bisector points (line). Horizontal lines indicate the intensity values where bisector points were calculated. The middle panel shows the bisector with the optical depth on the y-axis, and the right panels shows the classical bisector as fractional intensity versus velocity.

example, the velocity of the Hγ line versus the velocity of a metal line is shown in Fig. 3. It can be seen that our results so far agree very well with the findings of Mathias & Gillet (1993). As we have a huge number of unblended lines available and will automatically produce velocity curves for each of them, we expect that we can study this effect in some more detail. So far we have used the center of gravity to find the velocity of the line, but other methods will also be tested later in this project.

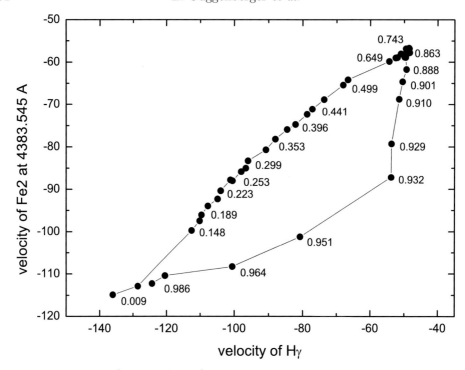

Figure 3. A phase lag (Van Hoof effect) between the Hγ line and an Fe II line. The big area enclosed by the curve indicates a strong phase delay between the hydrogen line and the metal line, consistent with a running wave crossing the different line forming regions in the stellar atmosphere. Numbers in the plot indicate the pulsation phase.

Acknowledgements

The first author acknowledges support from the Austrian Science Fund (FWF), project number P19962-N16. KK acknowledges support from Marie Curie Fellowship 255267 SAS-RRL within the 7th European Community Framework Program.

References

Benkő, J. M., Kolenberg, K., Szabó, R., *et al.* 2010, *MNRAS* 409, 1585
Gray, D. F. 2005, *PASP*, 117, 711
Jurcsik J., Sódor Á., Szeidl B., *et al.* 2009, *MNRAS*, 400, 1006
Kolenberg, K. Fossati, L., Shulyak, D., *et al.* 2010, *A&A*, 519, 64
Mathias, P. & Gillet, D. 1993, *A&A*, 278, 511
Mathias, P., Gillet D., Fokin A. B. & Chadid, M. 1995, *A&A*, 298, 843
Shulyak, D., Tsymbal, V., Ryabchikova, T., Stütz, Ch., & Weiss, W. W. 2004, *A&A*, 428, 993
Tsymbal, V. 1996, *ASP-CS*, 108, 198
Van Hoof, A. & Struve, O. 1953, *PASP*, 65, 158

Precision Asteroseismology
Proceedings IAU Symposium No. 301, 2013
J. A. Guzik, W. J. Chaplin, G. Handler & A. Pigulski, eds.

© International Astronomical Union 2014
doi:10.1017/S1743921313014439

Mode selection in pulsating stars

Radosław Smolec

Nicolaus Copernicus Astronomical Centre,
ul. Bartycka 18, 00-716 Warszawa, Poland
email: smolec@camk.edu.pl

Abstract. In this review we focus on non-linear phenomena in pulsating stars: mode selection and amplitude limitation. Of many linearly excited modes, only a fraction is detected in pulsating stars. Which of them are excited, and why (the problem of mode selection), and to what amplitude (the problem of amplitude limitation) are intrinsically non-linear and still unsolved problems. Tools for studying these problems are briefly discussed and our understanding of mode selection and amplitude limitation in selected groups of self-excited pulsators is presented. We focus on classical pulsators (Cepheids and RR Lyrae stars) and main-sequence variables (δ Scuti and β Cephei stars). Directions of future studies are briefly discussed.

Keywords. stars: oscillations, Cepheids, δ Scuti stars, white dwarfs

1. Introduction

Models of pulsating stars typically predict more unstable modes than are observed. Which of the linearly unstable modes are excited and why – the problem of mode selection – is a difficult non-linear problem, still lacking a satisfactory solution. Closely related is the problem of amplitude limitation, which is non-linear as well. Intrinsic non-linearity of these two problems is a major challenge. Our tools to analyse non-linear pulsation are either restricted to large amplitude radial pulsation, e.g. in Cepheids and RR Lyrae stars (hydrodynamical modelling) or are based on simplified assumptions and depend on unknown parameters (amplitude equation formalism). Therefore these problems received only scant theoretical attention in the past, and the dated but excellent review of Dziembowski (1993) is still mostly up-to-date.

For ground-based observations, the basic mode selection mechanism is of an observational nature. Because of geometric cancellation, modes of degree $l > 2$ are hard to detect from the ground (Dziembowski 1977). For space-based photometry, mode degrees above 10 are reported (e.g. Poretti *et al.* 2009) and since geometrical cancellation is very similar for large l it is hard to point to any obvious limit for maximum l, except that even-l modes are less affected by cancellation and thus more likely to be detected. In this review we focus on intrinsic non-linear mechanisms acting in pulsating stars.

In the next section the tools for studying the non-linear phenomena are briefly discussed. Then we discuss the mode-selection mechanisms: mode trapping (Section 3), non-resonant and resonant mode interaction (Section 4). We next turn to the discussion of amplitude-limiting effects: collective saturation of the driving mechanism (Section 5) and resonant mode coupling (Section 6). A discussion and outlook for the future studies end this review.

2. Tools for mode selection analysis

Linear stability analysis. A linear stability analysis tells us nothing about the mode selection or amplitudes of the excited modes – these are non-linear problems. It is, however, a necessary starting point, as it provides the information on mode eigenfunctions,

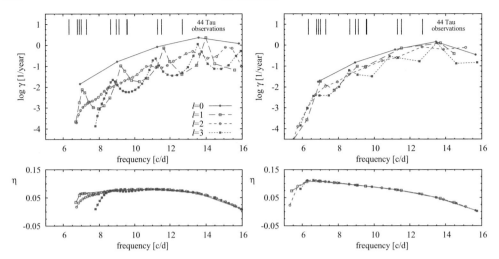

Figure 1. Linear growth rates, γ (top) and Stellingwerf's growth rates, η (bottom) for two models of δ Sct stars at different evolutionary stages: post-main sequence (MS) expansion (left) and post-MS contraction phase (right). In the top panels, frequencies detected in 44 Tau are marked. Models from Lenz *et al.* (2008, 2010).

mode frequencies, σ, and on mode stability through the linear growth rate, γ:

$$\gamma = \frac{\int dW}{2\sigma I} , \qquad (2.1)$$

where

$$dW = \Im\left[\delta P\left(\frac{\delta\rho}{\rho}\right)^{*}\right], \quad I = \int_{M} |\boldsymbol{\xi}|^{2} dm, \qquad (2.2)$$

are local contribution to the work integral (dW, with pressure, δP, and density, $\delta\rho$, perturbations) and mode inertia (I, with radial displacement, $\boldsymbol{\xi}$). Plots of growth rates for low degree modes in δ Sct-type models are shown in the top panels of Fig. 1. The growth rates of non-radial modes are not smooth, but exhibit maxima, particularly pronounced for the evolved model (left panel) and modes of $l=1$. This peculiar frequency dependence of the growth rates reflects the behaviour of mode inertia. Modes trapped in the external layers of the model with small amplitudes in the interior have the smallest inertia and the largest growth rates. The inference that the most unstable, trapped modes will be most easily driven to high amplitude is precarious, however (see Section 3). The maxima of the growth rates do not reflect the properties of the driving region, which is clear if Stellingwerf's (1978) growth rates are considered instead:

$$\eta = \frac{\int dW}{\int |dW|} , \qquad (2.3)$$

with $\eta \in (-1, 1)$. These growth rates are plotted in the bottom panels of Fig. 1 and they smoothly vary with the mode frequency.

Hydrodynamic models. Realistic non-linear hydrodynamic models of radially pulsating stars have been computed for nearly 50 years now (e.g. Christy 1966). Computations are done with direct time-integration, one-dimensional codes. The initial static structure is perturbed with the scaled velocity eigenfunction (initial *kick*) and the time evolution of the model is followed until finite-amplitude steady pulsations are reached (limit cycle). As different initializations may lead to different limit cycles for the same static model (hysteresis), mode selection analysis is time-consuming and requires computation of tens of models. The convergence may be sped-up with the use of a relaxation technique (Stellingwerf 1974), which in addition provides the information about stability of the limit cycles (even unstable ones) through the Floquet exponents.

The early codes were purely radiative. Currently several codes that include convective energy transfer are in use. Two prescriptions for turbulent convection are commonly adopted, either by Stellingwerf (1982) (e.g. in the Italian code, Bono & Stellingwerf 1992) or by Kuhfuß (1986) (e.g. in the Warsaw codes of Smolec & Moskalik (2008a) or in the Florida-Budapest code, e.g. Kolláth *et al.* (2002), with the modified Kuhfuß model). Both models include several free parameters, values of which must be adjusted to match the observational constraints.

Non-linear pulsation codes were successfully used to model the light and radial velocity curves in single-periodic classical pulsators. Understanding of the dynamical phenomena shaping these curves would not be possible however without the insight provided by the analysis of amplitude equations.

Amplitude equations (AEs). If the growth rates of the dominant modes are small compared to their frequencies (weak non-adiabaticity), and assuming weak non-linearity, the hydrodynamic equations governing the stellar pulsation may be reduced to ordinary differential equations for the amplitudes of the excited modes, A_i (e.g. Dziembowski 1982, Buchler & Goupil 1984). In the case when no resonances are present among pulsation modes, the form of the AEs (usually truncated at the cubic terms) is the following:

$$\frac{dA_i}{dt} = \gamma_i \left(1 + \sum_j \alpha_{ij} A_j^2 \right) A_i, \tag{2.4}$$

where α_{ii} and α_{ij} are negative self- and cross-saturation coefficients, respectively.

In the case of resonant mode coupling, the exact form of the amplitude equations depends on the resonance considered. Here we present the complex equations for the parametric resonance $\sigma_a = \sigma_b + \sigma_c + \Delta\sigma$:

$$\frac{dA_a}{dt} = \gamma_a A_a - i\frac{C}{2} A_b A_c e^{-i\Delta\sigma t},$$
$$\frac{dA_{b,c}}{dt} = \gamma_{b,c} A_{b,c} - i\frac{C}{2} A_a A_{c,b}^* e^{i\Delta\sigma t}. \tag{2.5}$$

C is a resonant coupling coefficient.

With reasonable approximations the amplitude equations may be solved analytically. Of particular interest are time-independent solutions (fixed points) which correspond to limit cycles in hydrodynamic computations. Analysis of a fixed point's stability provides direct insight into mode selection. For non-resonant AEs the single-mode fixed points are given by $A_i = 1/\sqrt{-\alpha_{ii}}$ and are stable if $\alpha_{ji}/\alpha_{ii} > 1$ for each j. For the interesting case of non-resonant two-mode interaction, the double-mode solution, with finite amplitude of the two excited modes, is possible once $\alpha_{00}\alpha_{11} - \alpha_{01}\alpha_{10} > 0$ (analysis of cubic AEs), i.e. when the self-saturation exceeds the cross-saturation. A detailed discussion of mode selection scenarios for both non-resonant and resonant mode interaction may be found e.g. in Dziembowski & Kovács (1984) or Buchler & Kovács (1986).

The described mode selection analysis is possible only when the values of the saturation/coupling coefficients are known. These, however, are very difficult to compute. Only with simplistic approximations some analytical estimates are possible. Therefore, most of the work on AEs has been parametric studies. This problem may be overcome for large-amplitude radial pulsators for which hydrodynamic computations may be coupled with the analysis of AEs. For time integration of the same model, but initialized with different initial conditions, the evolution of mode amplitudes may be followed with the help of the analytical signal method (e.g. Kolláth *et al.* 2002). The resulting trajectories are then fitted with the appropriate AEs and the resulting saturation/coupling coefficients may be used to compute all the fixed points and their stability, i.e. to analyse the mode selection. Repeating the procedure for a discrete set of models located at different parts of the HR diagram, and interpolating in-between, yields a consistent picture of mode selection in the full instability strip (Szabó *et al.* 2004). Results of such analysis for Cepheids are reported in Section 4.

3. Mode trapping

Mode trapping as a mode selection mechanism was first proposed by Winget *et al.* (1981) in the context of white-dwarf (ZZ Ceti) pulsations. Trapping in the strongly stratified models of white dwarfs is caused by the resonance between the wavelength of the g-mode and the thickness of one of the compositional layers. The trapped modes have low amplitude in the core, with most of the mode energy confined in the outer regions, where pulsation driving takes place. The mode inertia is low and the growth rate is high (Eq. 2.1). Winget *et al.* (1981) concluded that the trapped modes are much more likely to be excited than adjacent, non-trapped modes. The mode trapping is also present in the models of evolved δ Sct stars, as is clearly visible in Fig. 1 (top, left). In this case, the frequency separation between the trapped modes corresponds to the separation between the consecutive radial overtones. Dziembowski & Królikowska (1990) proposed that mode trapping might be a mode selection mechanism in evolved δ Sct stars. They commented however, that this selection rule relies only on linear non-adiabatic theory, and since the high mode growth rates are not indicators of large amplitude, justification must come from non-linear theory.

The observations themselves invalidate mode trapping as a mode selection rule. In case of white dwarfs mode trapping is clearly detected through the characteristic wave shape of the period spacing vs. period diagram, which allows the identification of the trapped modes. It turns out that the trapped modes are not the ones that have the highest amplitudes. As an example we use the observations of PG 1159-035 – a hot pulsator, but evolved and stratified enough to show the effects of mode trapping (Costa *et al.* 2008). In Fig. 2 we show the frequency spectrum of the star from the 1983 season. The dashed lines mark the location of the trapped modes derived from the period spacing diagrams constructed using six seasons of observations. Although the two highest-frequency trapped modes have high amplitudes, the neighbouring non-trapped mode has the highest amplitude. For the three low-frequency trapped modes, no signal was detected in 1983 season. For two of these modes, a significant detection was made only during one out of the six observing seasons.

In the case of δ Sct stars, regularities observed in the frequency spectra are also interpreted as a result of mode trapping. Breger *et al.* (2009) show that, in the cases of FG Vir, BL Cam, and 44 Tau, there is a preferred frequency spacing between the excited modes, which corresponds to the spacing between radial modes. The asteroseismic model of Lenz *et al.* (2008) showed that mode trapping may be indeed operational in 44 Tau. The growth rates for their best asteroseismic model, located in the post-MS expansion phase, are reproduced in Fig. 1 (top left). Observed modes seem to cluster around the growth rate maxima. However, this model fails to reproduce all of the observable parameters satisfactorily. In their later analysis, Lenz *et al.* (2010) constructed asteroseismic models at an earlier evolutionary phase, the post-MS contraction, and obtained a better model with an excellent fit to all the observed modes (Fig. 1, top right). The mode

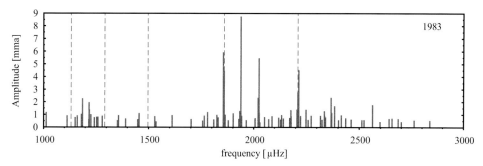

Figure 2. Frequency spectrum of PG 1159-035 (Costa *et al.* 2008) from the 1983 season. Frequencies of trapped modes are marked with dashed lines.

trapping is only barely noticeable for this model and cannot represent a valid mode selection mechanism.

4. Resonant and non-resonant mode coupling

The analysis of mode selection and amplitude limitation is most feasible for large-amplitude radial pulsators, Cepheids and RR Lyrae stars, as direct hydrodynamic models may be computed and complementary analysis of amplitude equations is feasible. Most of these stars are singly-periodic, pulsating either in the fundamental (F) mode or in the first overtone (1O). In many stars double-mode pulsation either of the F+1O or 1O+2O type is detected (see Moskalik 2013 and these proceedings for a review). Already, with the first purely radiative hydrocode, Christy (1966) showed that in the single-periodic models amplitude growth is limited by saturation of the driving mechanism. The selection between fundamental and first overtone pulsation is however still an unsolved problem. The analysis of radiative models yielded the following picture (e.g. Stellingwerf 1975): If only one mode is linearly unstable, then this mode reaches a finite amplitude: first overtone at the blue side of the instability strip and fundamental mode at the red side. If two modes are simultaneously unstable, then either only one limit cycle is stable (F-only or 1O-only domains) or two limit cycles are simultaneously stable, and which is selected depends on the initial conditions (E/O, either-or domain). A star entering the E/O domain from the blue side will continue to pulsate in the 1O mode, while a star entering from the red side will continue to pulsate in the F mode. No double-mode pulsation was found in realistic radiative models of Cepheids and RR Lyrae stars.

Inclusion of turbulent convection in the models seemed to solve the problem. Feuchtinger (1998) reported one double-mode RR Lyrae model and the Florida-Budapest group found double-mode Cepheids and RR Lyrae models in their surveys (Kolláth *et al.* 1998, 2002, Szabó *et al.* 2004, Buchler 2009). Regrettably, how the inclusion of turbulent convection caused the stable double-mode pulsation in these models was not analysed. Smolec & Moskalik (2008b) were able to show that the double-mode pulsation was caused by unphysical neglect of buoyant forces in convectively stable regions of the model. In the absence of a restoring force, the turbulence is not damped effectively below the envelope convective zone and resulting strong eddy-viscous dissipation reduces the amplitudes of fundamental and first overtone modes differentially, favouring the occurrence of double-mode pulsation. With the correct treatment of the buoyant forces, Smolec & Moskalik (2008b) were not able to find satisfactory double-mode Cepheid models. Also computations with the Italian code, adopting Stellingwerf's model of convection, yielded a null result (see Smolec & Moskalik 2010).

Although resonant mode interaction cannot explain the double-mode pulsation for most of the observed variables, it may be operational in some limited parameter ranges. In particular the 2:1 resonance between the fundamental mode and the linearly damped second overtone may decrease the amplitude of the former mode, allowing the growth of the first overtone, as pointed out by Dziembowski & Kovács (1984). Some hydrodynamic resonant double-mode models were in fact found (Smolec 2009, Buchler 2009) and the two long period double-periodic Cepheids discovered recently in M31 by Poleski (2013) are the first good candidates for resonant double-periodic pulsation. For the majority of double-periodic pulsators, an explanation is still missing.

5. Collective saturation of the driving mechanism

Among δ Sct stars and β Cep stars there are variables with one or two dominant radial modes, with amplitudes of the order of 0.1 mag. The attempts to model these stars with hydrodynamic codes failed however. Stellingwerf (1980) computed δ Sct models and got pulsation amplitudes exceeding 1 mag (which he called the *main-sequence catastrophe*). Clearly, the instability cannot be saturated with a single pulsation mode in these stars, as is the case for Cepheids or RR Lyr stars, occupying the high luminosity part of the same instability strip. Similar results were obtained for models of singly-periodic β Cep pulsators computed by Smolec & Moskalik (2007). Linear stability computations predict that many non-radial acoustic modes are unstable in these stars. Smolec & Moskalik (2007) assumed that the instability is collectively saturated not by a single mode, but by tens (n) of acoustic modes simultaneously (and that because of the assumed large l these modes are not detected). Using amplitude equations and assuming that the properties of acoustic modes (saturation coefficients) are the same, one may show that in this case the amplitude drops by a factor $\sqrt{n}$ as compared to the single-mode saturation amplitude predicted by a hydrodynamic model. The agreement with the pulsation amplitudes of multi-periodic β Cep stars is obtained with only a fraction of the available (linearly unstable) modes. In principle, the collective saturation of the instability mechanism explains the amplitudes of β Cep stars (and of δ Sct stars as well); however, there is a serious difficulty with such an explanation. The resulting macroturbulence velocities (and hence the line widths) are too high compared to observations, indicating that other amplitude-limiting mechanisms must be operational.

6. Amplitude limitation in δ Scuti stars

An alternative scenario to collective saturation of the driving mechanism is resonant mode coupling investigated by Dziembowski (1982), who considered the coupling of an unstable acoustic mode to a pair of stable g modes. The AEs appropriate for this case were given in Section 2. The parametric excitation of the g modes starts once the amplitude of the acoustic mode exceeds a critical value, $A_a > A_{\rm crit}$. A steady-state solution is then possible in a limited range of mismatch parameter, and provided that damping of the gravity-mode pair exceeds the acoustic-mode driving. The amplitude of the acoustic mode is then close to the critical amplitude, while amplitudes of the g-modes are much lower, making their detection from the ground impossible. The exact formulae for the stability condition and critical/equilibrium amplitudes may be found in Dziembowski (1982). Dziembowski & Królikowska (1985) applied the formalism for realistic δ Sct models. Strong coupling arises only if the radial orders of the gravity modes are similar, which implies $\sigma_b \approx \sigma_c \approx \sigma_a/2$ and for close and large l values. Because of a large number of potential resonant pairs, Dziembowski & Królikowska (1985) computed the probability distribution for the critical amplitude. Their results showed that, indeed, resonant mode coupling may be a promising amplitude-limitation mechanism. The typical critical amplitude they found is of order of 0.01 mag. Moreover, if rotation is taken into account the critical amplitude drops even further, as the denser g-mode spectrum allows for fine tuning of the resonance condition (Dziembowski *et al.* 1988). Thus, resonant mode coupling nicely explains the observational fact that high amplitude δ Sct stars are slow rotators.

The resonant mode coupling scenario has serious shortcomings, however. Only for low order acoustic modes, which couple to strongly damped global g-modes, there is a large probability that the equilibrium is stable. Higher order p-modes couple preferentially to weakly damped inner g-modes which are not able to halt the amplitude growth. The excitation of many g-mode pairs is then expected, and has to be analysed numerically, which was done by Nowakowski (2005). His results are disappointing however. A static multi-mode solution is not possible then, and strong amplitude variability on a γ_a^{-1} timescale is expected. Moreover, computations for realistic model of the δ Sct star XX Pyx show that resonant mode coupling cannot be a dominant amplitude-limiting effect, as the predicted amplitudes are higher than observed. Whether saturation of the driving mechanism plays a role in δ Sct models has not been investigated in detail yet.

7. Discussion and conclusions

The problems of mode selection and amplitude limitation are some of the most stubborn, still unsolved problems of stellar pulsation theory. They are important for all groups of pulsating stars, and for no group do we have a satisfactory solution. Even for large-amplitude radial pulsators, we do not understand the mechanisms behind the simplest form of multi-mode pulsation, i.e. double-mode pulsation. Triple-mode pulsation and excitation of non-radial modes are even more challenging problems. It seems that their solution must await the development of full 3D hydrodynamic models. Fortunately, such codes are now being developed (Geroux & Deupree 2013, Mundprecht *et al.* 2013), but have not been applied for modelling double-periodic pulsations yet.

For low-amplitude non-radial pulsators our understanding is even poorer, as the use of the amplitude equation formalism, the only available tool to study non-linear and non-radial pulsation, is strongly limited by its complexity and unknown saturation/coupling coefficients.

The most interesting quantities that observations can provide are intrinsic amplitudes of pulsation modes. For their determination, the mode identification and inclination angle are needed. The robust determination of these quantities is however difficult and requires combination of multi-band photometric and spectroscopic observations (Uytterhoeven, these proceedings). Only for a limited number of main-sequence pulsators and usually only for a few detected modes robust mode identifications are available. In the case of stars studied with space telescopes, with hundreds of detected modes, the task seems even more challenging, if possible at all. The space observations, in particular their statistical analyses, are however of great importance for our understanding of mode selection (Balona & Dziembowski 1999). The frequency distribution of amplitudes of the excited modes may be used to infer the probability distribution of intrinsic amplitudes, assuming some knowledge of the l values, and then compared with model computations. *Kepler* observations of thousands of δ Sct stars (Balona & Dziembowski 2011) make such an analysis feasible.

We stress the need for systematic spectroscopic observations of targets of current and future space missions. The aim is not only the mode identification, but also precise determination of basic stellar parameters, $\log g$ and $\log T_{\text{eff}}$, which are additional important constraints for seismic models and are necessary to study the intriguing problem of the significant contamination of the δ Sct instability strip with apparently non-pulsating stars (Balona & Dziembowski 2011).

Acknowledgements

I am grateful to Wojtek Dziembowski, Paweł Moskalik and Alosha Pamyatnykh for many fruitful discussions and to Patrick Lenz for providing data for Fig. 1. I acknowledge the IAU grant for the conference. This research is supported by the Polish National Science Centre through grant DEC-2012/05/B/ST9/03932.

References

Balona, L. A. & Dziembowski, W. A. 1999, *MNRAS*, 309, 221
Balona, L. A. & Dziembowski, W. A. 2011, *MNRAS*, 417, 591
Bono, G. & Stellingwerf, R. F. 1992, *MemSAIt.*, 63, 357
Breger, M., Lenz, P., & Pamyatnykh, A. A. 2009, *MNRAS*, 396, 291
Buchler, J. R. 2009, *AIP-CP*, 1170, 51
Buchler, J. R. & Goupil M.-J. 1984, *ApJ*, 279, 394
Buchler, J. R. & Kovács, G. 1986, *ApJ*, 308, 661
Christy, R. F. 1966, *ApJ*, 144, 108
Costa, J. E. S., Kepler, S. O., Winget, D. E., *et al.* 2008, *A&A*, 477, 627
Dziembowski, W. 1977, *AcA*, 27, 203
Dziembowski, W. 1982, *AcA*, 32, 147
Dziembowski, W. 1993, *ASP-CS*, 40, 521
Dziembowski, W. & Kovács, G. 1984, *MNRAS*, 206, 497
Dziembowski, W. & Królikowska, M. 1985, *AcA*, 35, 5
Dziembowski, W. & Królikowska, M. 1990, *AcA*, 40, 19
Dziembowski, W., Królikowska, M., & Kosovichev, A. 1988, *AcA*, 38, 61
Feuchtinger, M. U. 1998, *A&A*, 337, 29
Geroux, C. M. & Deupree, R. G. 2013, *ApJ*, 771, 113
Kolláth, Z., Beaulieu, J. P., Buchler, J. R., & Yecko, P. 1998, *ApJ*, 502, L55
Kolláth, Z., Buchler, J. R., Szabó R., & Csubry, Z. 2002, *A&A*, 385, 932
Kuhfuß, R. 1986, *A&A*, 160, 116
Lenz, P., Pamyatnykh, A. A., Breger, M., & Antoci, V. 2008, *A&A*, 478, 855
Lenz, P., Pamyatnykh, A. A., Zdravkov, T., & Breger, M. 2010, *A&A*, 509, A90
Moskalik, P. 2013, *Ap&SS*, 31, 103
Mundprecht, E., Muthsam, H. J., & Kupka, F. 2013, *MNRAS*, 435, 3191
Nowakowski, R. 2005, *AcA*, 55, 1
Poleski, R. 2013, *ApJ*, 778, 147
Poretti, E., Michel, E., Garrido, R., *et al.* 2009, *A&A*, 506, 85
Smolec, R. 2009, *EAS Publ. Ser*, 38, 175
Smolec, R. & Moskalik, P. 2007, *MNRAS*, 377, 645
Smolec, R. & Moskalik, P. 2008a, *AcA*, 58, 193
Smolec, R. & Moskalik, P. 2008b, *AcA*, 58, 233
Smolec, R. & Moskalik, P. 2010, *A&A*, 524, A40
Stellingwerf, R. F. 1974, *ApJ*, 192, 139
Stellingwerf, R. F. 1975, *ApJ*, 195, 441
Stellingwerf, R. F. 1978, *AJ*, 83, 1184
Stellingwerf, R. F. 1980, *Lecture Notes in Physics*, 125, 50
Stellingwerf, R. F. 1982, *ApJ*, 262, 330
Szabó, R., Kolláth, Z., & Buchler, J. R. 2004, *A&A*, 425, 627
Winget, D. E., Van Horn, H. M., & Hansen, C. J. 1981, *ApJ*, 245, L33

Precision Asteroseismology
Proceedings IAU Symposium No. 301, 2013   © International Astronomical Union 2014
J. A. Guzik, W. J. Chaplin, G. Handler & A. Pigulski, eds.   doi:10.1017/S1743921313014440

Pulsations in white dwarf stars

G. Fontaine[1], P. Bergeron[1], P. Brassard[1], S. Charpinet[2], P. Dufour[1], N. Giammichele[1], S. K. Randall[3], and V. Van Grootel[4]

[1] Département de Physique, Université de Montréal,
Montréal, Québec, Canada H3C 3J7
email: fontaine@astro.umontreal.ca

[2] CNRS, Université de Toulouse, UPS-OMP, IRAP,
14 av. E. Belin, 31400 Toulouse, France

[3] European Southern Observatory, Karl-Schwarzschild-Str. 2
85748 Garching bei München, Germany

[4] Institut d'Astrophysique et de Géophysique de l'Université de Liège
Allée du 6 Août 17, B-4000 Liège, Belgique

Abstract. We first present a brief description of the six distinct families of pulsating white dwarfs that are now known. These are all opacity-driven pulsators showing low- to mid-order, low-degree gravity modes. We then discuss some recent highlights that have come up in the field of white dwarf asteroseismology.

Keywords. stars: oscillations, white dwarfs

---

1. The families of pulsating white dwarfs

A general view of the distribution of known pulsating stars lying at the bottom of the spectroscopic H-R diagram is provided by Fig. 1. Note that only those pulsators with reliable estimates of their atmospheric parameters have been plotted. There exist many more pulsating stars of the kinds, particularly of the ZZ Cet and V777 Her type, but those have yet to be well characterized spectroscopically speaking. One can identify at least six different categories of pulsating white dwarfs, and three different types of pulsating hot subdwarfs. The properties of the latter are reviewed by S. Randall *et al.* in these proceedings, and will no longer be referred to in this article.

The ZZ Cet pulsators are H-atmosphere (or DA) white dwarfs found in a very narrow instability strip in the lower right corner of the H-R diagram as indicated in Fig. 1. The DA white dwarfs themselves constitute some 80% of post-AGB stars and, hence, form the majority of white dwarfs. A representative evolutionary track for a 0.593 $M_\odot$ DA star connects the post-AGB domain with the ZZ Cet instability strip in the figure. The pulsations detected in ZZ Cet stars correspond to low- to mid-order, low-degree g modes. The ZZ Cet stars are opacity-driven pulsators. The exact excitation mechanism is sometimes referred to as convective driving, but the ultimate culprit is the recombination of H in the envelope as a cooling DA white dwarf enters and then transits across a narrow strip.

A closely related family is that of the V777 Her stars which involves, this time, He-atmosphere (or DB) white dwarfs. The latter are the cool descendants of the other 20% of post-AGB objects. That smaller fraction corresponds to stars that experience a very late He flash in their evolution, which causes a violent mixing episode that destroys the remaining H in these objects. Thus, about 20% of all white dwarfs are born devoid of hydrogen. The DB stars descend from those H-deficient objects (a representative case is illustrated by the dashed evolutionary track) and constitute a spectral type based

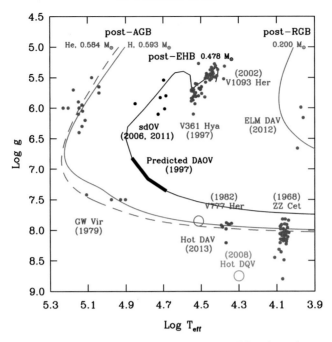

Figure 1. Regions of the log g-log T_{eff} plane where the several families of compact pulsators are found. The year of the discovery of the prototype of each class is indicated. Five of the distinct classes are identified by their official IAU names, while the others, more recently discovered, have yet to be given IAU variable star names. To guide the eye, four representative evolutionary tracks have been plotted. Further details are provided in the text.

on He I lines visibility, concentrated between about 30 000 K and 12 000 K in effective temperature. The pulsations detected in V777 Her DB white dwarfs are of the same nature as those found in ZZ Cet stars. Convective driving is again at work, this time associated with the recombination of He in the envelope. The V777 Her instability strip is found at higher effective temperatures than the ZZ Cet strip because the ionization potentials of He I and He II are larger than that of H I.

The hottest pulsating stars known are the GW Vir white dwarfs found on the left side of Fig. 1. They span a huge domain in effective temperature and surface gravity. These pulsators are white dwarfs that belong to the PG1159 spectral type, characterized by mixed He-C-O atmospheres. The PG1159 white dwarfs are the immediate, very hot descendants of that fraction of 20% of post-AGB stars that underwent a late He flash. Their atmospheres/envelopes still reflect a composition (a mixture of He, C, and O in roughly comparable proportions, but which varies from star to star) that resulted from that mixing event. Ultimately, gravitational settling of C and O produces almost pure He atmospheres, and PG1159 stars evolve into cooler white dwarfs of the DO and DB spectral types. The pulsations observed in GW Vir stars are again low- to mid-order, low-degree g modes. The excitation mechanism is now a classical κ-mechanism (convection has little impact in these very hot stars) associated with the presence of opaque high ions of C and O in the envelope.

An extremely rare type of white dwarfs, the so-called Hot DQ stars, has been discovered relatively recently by Dufour *et al.* (2007). These are C-atmosphere white dwarfs that bunch around 20 000 K in effective temperature and have relatively high surface gravities by white dwarf standards (see Fig. 1). Their exact origin is uncertain and they are still being characterized. They appear to be highly magnetic (with field strengths exceeding

1 MGauss) and half of them have been found to pulsate. The pulsations observed would correspond to mid- to low-order, low-degree g modes. Convective driving associated with the recombination of C in the envelope is the mechanism responsible for the excitation of pulsation modes in these Hot DQV variables.

Gravity-mode pulsations have been detected recently in three extremely low-mass (ELM) DA white dwarfs (Hermes *et al.* 2012, 2013a). The locations of these variable stars are indicated in Fig. 1, and a representative 0.2 $M_\odot$ evolutionary track is also plotted. The ELM white dwarfs are relatively rare objects produced by binary (common envelope) evolution. They are post-RGB remnants and are constituted of a He core surrounded by a relatively thick H envelope. The pulsations detected correspond to low- to mid-order, low-degree g modes. Given their much lower surface gravities (and considerably reduced average densities), the periods involved are much longer (thousands of seconds versus hundreds of seconds) compared to other pulsating white dwarfs.

An even more recent development has been the report of Kurtz *et al.* (2013) concerning the probable discovery of yet another class of pulsating white dwarfs. The newly-uncovered variable stars are DA white dwarfs that cluster around 30 000 K in effective temperature (see Fig. 1), and should be referred to as Hot DAV white dwarfs. The tentative explanation is that these stars have very thin H outer layers sitting on top of a He mantle in which the driving/damping region resides. In effect, these objects are DBs "disguised" as DAs. Low-order, low-degree g-mode pulsations would be driven by the same opacity mechanism that operates in V777 Her stars.

Finally, we recall a prediction made by Charpinet *et al.* (1997) to the effect that low-mass, post-EHB DAO white dwarfs should show very low-order, low-degree g-mode pulsations driven by the ϵ-mechanism associated with H-shell burning at the base of the H envelope in these stars. The heavy curve segment along the illustrated post-EHB track in Fig. 1 identifies the domain where ϵ-driving should be in action. It should be pointed out here that post-EHB evolution contributes at most 2% of the total white dwarf population. In addition, the evolutionary phase indicated by the heavy curve segment is relatively rapid, meaning that these hot, low-mass DA white dwarfs are quite rare. And indeed, a real pulsator of the kind has yet to be found, but the prediction has been made.

2. Recent highlights in white dwarf seismology

In the rest of this contribution, we focus on some recent highlights of interest in the field of white dwarf seismology. The reader will find an exhaustive review of the progress made in this field up to 2008 in Fontaine & Brassard (2008).

2.1. *An enlightening discussion*

Saio (2013) has recently presented a brief but, in our view, particularly interesting discussion on the driving mechanisms in pulsating white dwarfs. For those interested in a deeper physical understanding of these mechanisms as well as a first acquaintance with the effects of large magnetic fields in pulsating white dwarfs, this is highly recommended reading.

2.2. *The total angular momentum of white dwarfs*

Seismology has been successfully used by Charpinet *et al.* (2009) to map the internal rotation profile of PG 1159−035, the prototype of the PG1159 spectral type for white dwarfs, and also the prototype of the GW Vir class of pulsating white dwarfs. This is depicted in Fig. 2, which shows that some 99% of the mass of the star has been covered, thus allowing one to compute essentially the total angular momentum and total

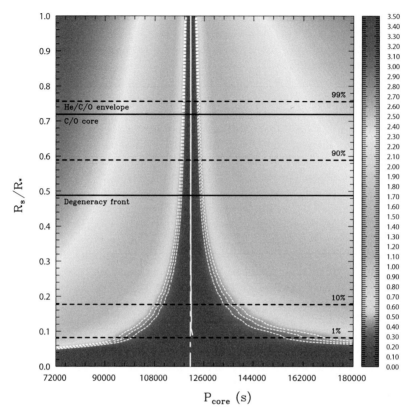

Figure 2. Internal rotation profile of the GW Vir star PG 1159−035 illustrated as depth (on a radius scale) versus the local rotation period. The dashed horizontal lines labelled 99%, 90%, 10%, and 1% correspond to the levels in the seismic model below which 99%, 90%, 10%, and 1% of the total mass is concentrated. The sounding method loses its sensitivity over the inner ∼10% of the radius (as indicated by the diverging confidence curves depicting the 1, 2, and 3 σ levels), but this inner region contains only about 1% of the total mass. This implies that the rotation profile is secure over some 99% of the mass of the model.

rotation energy of that object. The rotation profile in PG 1159−035 corresponds to solid body rotation. Much more importantly, Fig. 2 reveals that PG 1159−035 rotates in bulk extremely slowly by white dwarf standards, with $P_{rot} = 33.67 \pm 0.24$ h. While it has been known for several decades that isolated white dwarfs are slow rotators at their surfaces (through spectroscopic or polarization measurements), one could never exclude the possibility that their cores could spin quite rapidly, thus "hiding" most of their angular momentum.

We have recently carried out similar asteroseismic analyses for the three additional GW Vir stars with rotation data available in the literature. The results are qualitatively the same as that obtained for PG 1159−035, i.e., these objects all rotate solidly and extremely slowly. In fact the ratio of the total rotation energy to the thermal energy in these objects varies from 10^{-6} in our fastest rotator to 10^{-8} in our slowest one, meaning that, for all practical purposes, these stars have lost all of their angular momentum. Given that these four objects are representative of single star evolution, this finding has very important implications for theories of angular momentum transfer between the radiative core and the convective envelope in red giant phases leading to the final white dwarf phase in stellar evolution.

2.3. *An updated view of the ZZ Cet instability strip*

Considerable efforts have been invested at Université de Montréal over the last two decades to provide a reliable and homogeneous description of the empirical ZZ Cet instability strip, starting with the pioneering efforts of Bergeron *et al.* (1995). Emphasis has been put on gathering high S/N spectroscopic data using the same experimental setup, making sure that the exposure times exceed the dominant periods (in order to obtain meaningful time-averaged spectra), and deriving atmospheric parameters in a strict homogeneous way with the help of state-of-the-art model atmospheres. Gianninas *et al.* (2005) demonstrated that without using these stringent criteria, the view of the empirical ZZ Cet strip remains blurred, in particular when using the low S/N spectra coming out of the SDSS survey. In contrast, Fig. 3 (built by adhering stricly to the above approach) provides the best available view of the empirical ZZ Cet instability domain in the log g-$T_{\rm eff}$ diagram.

Considerable progress has also been made in recent years on the theory front, thanks mainly to the introduction of time-dependent convection (TDC) into nonadiabatic calculations as applied to ZZ Cet pulsators at Université de Liège (Van Grootel *et al.* 2012). Assuming that convection can be described by the so-called ML2/$\alpha = 1.0$ version of the mixing-length theory, convective driving described with TDC leads to a theoretical blue edge given by the solid curve on the left in Fig. 3. However, TDC still fails at explaining the empirical red edge, so Van Grootel *et al.* (2013) have proposed a semi-analytic criterion based on the comparison between the thermal timescale at the base of the convection zone (where the driving region is located in ZZ Cet models) and the cutoff g-mode period (due to energy leakage through the atmosphere). This criterion leads to a theoretical red edge (for dipole modes) given by the solid curve on the right in Fig. 3. Given that the red edge location is sensitive to the assumed convective efficiency used in the equilibrium model, and given that it is obtained independently of the TDC nonadiabatic calculations carried out for obtaining the blue edge, Fig. 3 provides a most satisfying view of the ZZ Cet instability strip. This is the first time that such a good agreement has been obtained between theory and observations. Furthermore, the fact that the three recently-discovered pulsating ELM DA white dwarfs fall into the extension of the theoretical strip into the low-gravity regime is a convincing proof that these are genuine ZZ Cet stars, driven by the same mechanism, but with very low masses. Further details are provided in Van Grootel *et al.* (2013).

2.4. *Discovery of an ultramassive ZZ Cet pulsator*

Hermes *et al.* (2013b) have reported the discovery of *g*-mode pulsations in GD 518, a DA white dwarf (a ZZ Cet star) massive enough to be partly solidified in its core. This is the second pulsating star of the kind, the first one being BPM 37093 discussed, for instance, by Metcalfe *et al.* (2004) and Brassard & Fontaine (2005). GD 518 is the most massive pulsating white dwarf ever found and it is likely to have a core made of neon-oxygen instead of the more standard carbon-oxygen composition. Asteroseismology may yet reveal its true nature. In Fig. 3, GD 518 appears as the lowest (highest gravity) data point, sitting at the red edge of the instability strip, a position consistent with the observed periods since there exist period-effective temperature relationships sensitive on the mass for ZZ Cet stars (see, e.g., Fig. 24 of Brassard & Fontaine 2008). In comparison, BPM 37093 sits just above and slightly to the right of GD 518 in Fig. 3.

2.5. *Finally making sense of the GW Vir instability domain*

The puzzle of the GW Vir instability strip has finally been resolved by Quirion, Fontaine, & Brassard (2012) who presented a first coherent picture of the phenomenon by invoking

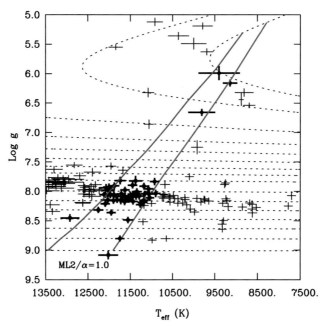

Figure 3. Comparison of the empirical with the theoretical instability domain in the log g-T_{eff} diagram for pulsating DA white dwarfs. Only the stars with reliable and homogeneous determinations of their spectroscopic parameters are plotted here. The heavy symbols indicate the pulsators, while the light crosses correspond to stars that have been monitored for variability but found not to vary. The theoretical boundaries of the strip have been obtained from models computed with the so-called ML2/α = 1.0 version of the mixing-length theory. With this convective efficiency, the predicted $l = 1$ blue edge is given by the solid curve on the left as obtained from detailed nonadiabatic calculations using TDC. For its part, the predicted $l = 1$ red edge is given by the solid curve on the right and was obtained from an independent semi-analytic argument involving the local thermal timescale in the driving region and the cutoff period for dipole g modes.

the competing actions of a stellar wind and of gravitational settling. As indicated above, PG1159 stars are the immediate descendants of "born-again" post-AGB stars that have undergone a late He flash. Their atmosphere/envelope compositions are mixtures of He, C, and O in very roughly comparable proportions, but varying from one object to another. Those PG1159 stars with compositions dominated by C and O over He (about half of them) pulsate as GW Vir variables as they contain enough opaque ions of C and O in the driving region (a pure κ-mechanism is at work here). The problem has been how to explain and maintain such mixed compositions in high-gravity objects, and how to account for the empirical blue and red edges of the instability domain.

Figure 4 illustrates what happens if one artificially maintains a representative GW Vir atmosphere/envelope composition in an evolving model. As indicated by the filled circles, the instability domain then extends all the way down to about ∼30 000 K in effective temperature. However, mixed-atmosphere white dwarfs of the sort do not exist below the GW Vir empirical red edge which is located between ∼70 000 K and ∼80 000 K, depending on the surface gravity (mass). Hence, the initial composition cannot be maintained over long evolutionary timescales and the most obvious cause for this state of affair is the action of diffusion, which cannot be turned off. And the process is particularly efficient in very high-gravity stars such as white dwarfs.

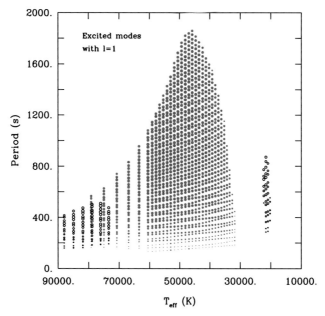

Figure 4. Period spectra of excited dipole modes in two post-PG1159 evolutionary models. The filled circles refer to a model with a fixed envelope chemical composition specified by $X(\mathrm{He}) = 0.38$, $X(\mathrm{C}) = 0.40$, $X(\mathrm{O}) = 0.20$, and $Z = 0.02$. The open circles refer to a similar $0.6\,M_\odot$ model defined by the same initial conditions but for which diffusion and mass loss have been turned on in the calculations. In the latter sequence, a GW Vir red edge is naturally found around $T_{\mathrm{eff}} = 75\,000$ K, and a V777 Her instability phase is also recovered near the end of the evolution. Each circle gives the period of a mode, and its size represents a logarithmic measure of the modulus of the imaginary part σ_I of the complex eigenfrequency. The bigger the circle, the more unstable the mode.

On the other hand, if left unimpeded, diffusion will transform the atmosphere of a PG1159 star into a pure He atmosphere almost instantly, and will lead to the settling of the opaque C and O ions out of the driving region in less than a few hundred years, thus stopping pulsational driving. In other words, white dwarfs of spectral type PG1159 (and their associated GW Vir pulsators) would not exist in that scenario. Thus, clearly, some mechanism must slow down the effects of gravitational settling over the timespan of the PG1159 evolutionary phase, which lasts about 2×10^6 yr, typically. Note that radiative levitation can only maintain small amounts of C and O in the atmosphere/envelope of a PG1159 star, certainly not enough to feed a κ mechanism. Instead, Quirion *et al.* (2012) have proposed the competing effect of a stellar wind, which appears a natural candidate in such very hot stars (and indeed, direct evidence for the presence of a wind exists for some of the most luminous PG1159 stars). Details are provided in the paper.

An animation summarizing these considerations has been presented at the conference. The interested reader may look up that animation in the .ppt presentation of G.F. available on the official web site of the meeting: http://iaus301.astro.uni.wroc.pl.

References

Bergeron, P., Wesemael, F., Lamontagne, R., Fontaine, G., Saffer, R. A., & Allard, N. F. 1995, *ApJ*, 449, 258
Brassard, P. & Fontaine, G. 2005, *ApJ*, 622, 572
Charpinet, S., Fontaine, G., & Brassard, P. 1997, *ApJ*, 489, L149

Charpinet, S., Fontaine, G., & Brassard, P. 2009, *Nature*, 561, 501

Dufour, P., Liebert, J., Fontaine, G., & Behara, N. 2007, *Nature*, 450, 522

Fontaine, G. & Brassard, P. 2008, *PASP*, 120, 1043

Gianninas, A., Bergeron, P., & Fontaine, G. 2005, *ApJ*, 631, 1100

Hermes, J. J., Montgomery, M. H., Winget, D. E., Brown, W. R., Kilic, M., & Kenyon, S. C. 2012, *ApJ*, 750, L28

Hermes, J. J., Montgomery, M. H., Winget, D. E., *et al.* 2013a, *ApJ*, 765, 102

Hermes, J. J., Kepler, S. O., Castanheira, B., *et al.* 2013b, *ApJ*, 771, L2

Kurtz, D. W., Shibahashi, H., Dhillon, V. S., *et al.* 2013, *MNRAS*, 432, 1632

Metcalfe, T. S., Montgomery, M. H., & Kanaan, A. 2004, *ApJ*, 605, L133

Quirion, P.-O., Fontaine, G., & Brassard, P. 2012, *ApJ*, 755, 128

Saio, H. 2013, in: J. Montalbán, A. Noels, & V. Van Grootel (eds.), *Ageing Low Mass Stars: From Red Giants to White Dwarfs*, EPJ Web of Conferences, 43, 05005

Van Grootel, V., Dupret, M.-A., Fontaine, G., Brassard, P., Grigahcène, A., & Quirion, P.-O. 2012, *A&A*, 539, A87

Van Grootel, V., Fontaine, G., Brassard, P., & Dupret, M.-A. 2013, *ApJ*, 762, 57

Precision Asteroseismology
Proceedings IAU Symposium No. 301, 2013
J. A. Guzik, W. J. Chaplin, G. Handler & A. Pigulski, eds.

© International Astronomical Union 2014
doi:10.1017/S1743921313014452

The most massive pulsating white dwarf stars

Barbara G. Castanheira[1] and S. O. Kepler[2]

[1] Department of Astronomy and McDonald Observatory, University of Texas
Austin, TX 78712, USA
email: barbara@astro.as.utexas.edu

[2] Instituto de Física, Universidade Federal do Rio Grande do Sul
91501-970 Porto Alegre, RS, Brazil
email: kepler@if.ufrgs.br

Abstract. Massive pulsating white dwarf stars are extremely rare, because of their small size and because they are the final product of high-mass stars, which are less common. Because of their intrinsic smaller size, they are fainter than the normal size white dwarf stars. The motivation to look for this type of stars is to be able to study in detail their internal structure and also derive generic properties for the sub-class of variables, the massive ZZ Ceti stars. Our goal is to investigate whether the internal structures of these stars differ from the lower-mass ones, which in turn could have been resultant from the previous evolutionary stages.

In this paper, we present the ensemble seismological analysis of the known massive pulsating white dwarf stars. Some of these pulsating stars might have substantial crystallized cores, which would allow us to probe solid physics in extreme conditions.

Keywords. stars: oscillations (including pulsations), stars: white dwarfs

1. Introduction

According to the best current evolutionary models, all single stars with masses below $9-10\,M_\odot$ will end up their lives as white dwarf stars. The final product contains important information about their previous phases, which might be difficult to model due to some more complicated physical processes, such as nuclear reactions and mass loss. As white dwarf stars are basically just cooling, their structure is the simplest in comparison to all previous phases.

The white dwarf interior is a Coulomb plasma of ions and degenerate electron gas, which at lower temperatures will crystallize. The crystallization process emits latent heat, delaying cooling (Winget at al. 2009). For massive white dwarf stars, because of their smaller radii, crystallization is expected to start at higher temperature.

As white dwarf stars cool down, they cross four distinct instability strips (Castanheira *et al.* 2010, Nitta *et al.* 2009, Montgomery *et al.* 2008, Werner & Herwig 2006), depending on their temperature and atmospheric composition. The ZZ Ceti stars (or DAVs) are stars with atmospheres dominated by hydrogen. They are observed to pulsate with a few independent modes in a narrow temperature range of about 2000 K (e.g. Bergeron *et al.* 2004, Mukadam *et al.* 2004). Within the instability strip, as the ZZ Ceti stars cool, the pulsation modes change both in amplitude (energy) and period (thermal timescale). Close to the blue edge of the strip, when the stars start to pulsate, we observe low-amplitude (few mma) short-period ($\sim$200 s) modes, while close to the red edge, the mode periods are much longer (up to $\sim$1500 s) and amplitudes are higher (several mma). The position of the instability strip also depends on mass (Giovannini *et al.* 1998, Hermes *et al.* 2012): the higher the mass, the hotter the stars that will start to pulsate.

All these effects have been well studied for the stars with masses close to the main value of the mass distribution, around 0.6 $M_\odot$. With the recent discoveries of the extreme low-mass white dwarf stars (Hermes *et al.* 2012), these dependencies are starting to be investigated for the lower-mass end. Our motivation is to continue the work started by Castanheira *et al.* (2013) and study the high-mass pulsating white dwarf stars. Our ultimate goal is to be able to constrain their internal structure and determine the differences and similarities between high- and normal-mass white dwarf stars.

One of the most exciting reasons to study massive white dwarf stars is that they are potential precursors of SN Ia. These SN explosions occur when a nearby companion fills its Roche lobe and transfers mass to the white dwarf, which then exceeds the Chandrasekhar mass limit, or by the merging of two white dwarf stars. The luminosity of SN Ia, when used as standard candles, led to the important important discovery of the accelerated expansion of the Universe (Riess *et al.* 1998, Perlmutter *et al.* 1999).

2. Properties of the massive pulsating white dwarf stars

Because of the intrinsically small number of pulsation modes, we have performed a seismological study of massive white dwarf stars using their ensemble properties. We made the assumption that, if similar sets of modes are observable in several stars, their internal structure should be similar. The larger number of massive pulsators allows us to probe the ensemble internal structure of the high-mass end of the ZZ Ceti instability strip. The results from our study can even be extended to the non-variable hydrogen-dominated atmosphere white dwarf stars.

To our knowledge, this is the first attempt to study the ensemble properties of massive pulsating white dwarf stars. With almost 30 stars more massive than 0.8 $M_\odot$ now known, within the known ZZ Cetis, we have a sample large enough to explore some characteristics. We chose to study only the stars with SDSS spectra (see Table 1) in order to have a homogeneous sample in terms of atmospheric determinations, i.e., T_{eff} and $\log g$ (Kleinman *et al.* 2004, 2013).

The current best models predict that pulsations start when the partial ionization zone of hydrogen (H) deepens into the envelope. The base of the partial ionization zone masks the bottom of the convection zone. As the stars cool down, the depth of the convective zone increases, as well as its size. When the thermal timescale at the bottom of the convection zone reaches the timescale of the of g modes, pulsations are detected. Around 11 500 K, there is a sudden deepening of the convection zone in the models; the observed pulsations change in character, with more modes with long periods and high amplitudes excited. For normal-mass white dwarf stars at temperatures slightly lower than 11 000 K, the stars stop pulsating, defining the red edge of the ZZ Ceti instability strip. The observed amplitudes decrease towards the cool part of the strip, consistent with an increase of the depth of the convective zone (Brickhill 1991, Mukadam *et al.* 2006). These were the aspects investigated for the massive pulsating white dwarf stars.

In Fig. 1, we plot T_{eff} as a function of the main observed period (largest amplitude). We can clearly see that there are two families of solutions. The best fits for these two families are, for periods smaller than 500 s:

$$P = 895.53\,[\text{s}] - 0.06\,[\text{s/K}]\,T_{eff} \tag{2.1}$$

and for periods larger than 500 s:

$$P = 3186.54\,[\text{s}] - 0.21\,[\text{s/K}]\,T_{eff} \tag{2.2}$$

Table 1. List of pulsating white dwarf stars with masses above $0.8\,M_\odot$. Temperature and mass were determined from the SDSS spectra (Kleinman *et al.* 2004, 2013). In the last column, we give the reference regarding variability.

Star	T_{eff} (K)	Mass ($M_\odot$)	Main period (s)	Amplitude (mma)	Reference
WD J0000−0046	10831 ± 224	0.83 ± 0.11	611.42	23.0	Castanheira *et al.* (2006)
WD J0048+1521	11260 ± 139	0.82 ± 0.04	615.3	24.8	Mulally *et al.* (2005)
WD J0111+0018	11765 ± 92	0.81 ± 0.03	292.97	22.13	Mukadam *et al.* (2004)
WD J0303−0808	11408 ± 280	0.94 ± 0.08	707	4.1	Castanheira *et al.* (2006)
WD J0349+1036	11715 ± 41	0.86 ± 0.01	184.5	3.76	Castanheira *et al.* (2013)
WD J0825+4119	11830 ± 324	0.92 ± 0.08	653.4	17.1	Mukadam *et al.* (2004)
WD J0855+0653	11075 ± 108	0.88 ± 0.04	849.88	39.09	Castanheira *et al.* (2006)
WD J0925+0509	10922 ± 51	0.86 ± 0.02	1378.93	5.25	Castanheira & Kepler (2009)
WD J0940+0050	10731 ± 125	0.92 ± 0.06	254.67	17.70	Castanheira *et al.* (2013)
WD J1200−0251	11986 ± 143	0.82 ± 0.04	304.78	23.69	Castanheira *et al.* (2013)
WD J1216+0922	11346 ± 268	0.81 ± 0.08	823	45.2	Kepler *et al.* (2005)
WD J1222−0243	11451 ± 105	0.80 ± 0.03	396	22.0	Kepler *et al.* (2005)
WD J1257+0124	11546 ± 335	0.85 ± 0.10	905.8	46.7	Castanheira *et al.* (2006)
WD J1323+0103	11821 ± 277	0.94 ± 0.06	612.23	11.7	Kepler *et al.* (2012)
WD J1337+0104	11511 ± 407	0.97 ± 0.08	715	10.0	Kepler *et al.* (2005)
WD J1612+0830	12026 ± 126	0.90 ± 0.03	114.97	5.17	Castanheira *et al.* (2013)
WD J1641+3521	11277 ± 203	0.84 ± 0.09	809.3	27.3	Castanheira *et al.* (2006)
WD J1650+3010	11057 ± 163	1.02 ± 0.06	339.06	14.71	Castanheira *et al.* (2006)
WD J1711+6541	11311 ± 94	1.02 ± 0.08	606.3	5.2	Mukadam *et al.* (2004)
WD J2128−0007	11460 ± 239	0.81 ± 0.08	302.2	17.1	Castanheira *et al.* (2006)
WD J2159+1322	11676 ± 308	1.05 ± 0.07	801	15.1	Mulally *et al.* (2005)
WD J2208+0654	11104 ± 29	0.92 ± 0.02	757.23	4.46	Castanheira *et al.* (2013)
WD J2209−0919	11546 ± 184	0.82 ± 0.06	894.71	43.94	Castanheira & Kepler (2009)
WD J2214−0025	11560 ± 195	0.83 ± 0.05	255.2	13.1	Mulally *et al.* (2005)
WD J2350−0054	10394 ± 140	0.83 ± 0.08	304.3	17.0	Mukadam *et al.* (2004)

One very interesting feature of our study is that the massive ZZ Ceti stars seem to avoid pulsating with the largest amplitude with periods around 500 s, indicating a mode-selection mechanism or mode trapping. Trapped modes in a compositionally stratified white dwarf model were discovered by Winget *et al.* (1981) for stellar masses around $0.6\,M_\odot$.

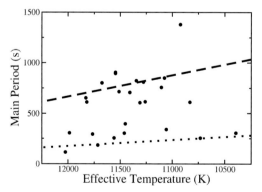

Figure 1. Main observed period as a function of effective temperature. We can identify two separate families of pulsating massive white dwarf stars.

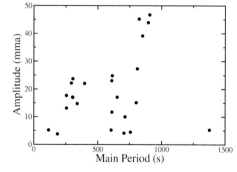

Figure 2. Observed main period versus its respective amplitude. The larger scatter for longer periods indicate that the intrinsic amplitudes are larger than for shorter pulsation periods.

In another analysis, we looked for relations between main period and observed amplitude, as plotted in Fig. 2. There is a scatter in the observed periods versus amplitude. However, we note that the scatter increases for longer periods. Because we do not know the inclination angle of the pulsation axis a priori, the observations of the changes in the scatter indicate that the amplitudes are intrinsically smaller for shorter periods and higher for longer periods. The fact that the longest observed period has a very small amplitude is consistent with theory: just before ceasing to pulsate, these stars should pulsate with small amplitudes and long periods.

3. Conclusions and final remarks

We have performed the first ensemble study of massive pulsating white dwarf stars, finding that these stars seem to avoid pulsating with a dominant mode around 500 s. We have active observing programs at McDonald Observatory and the SOAR Telescope to look for variability in white dwarf stars. In the past years, we have focused our searches on stars that have spectroscopic masses above $0.8 \, M_\odot$. Our goal is to increase the sample of these rare pulsators to further study their ensemble and individual properties.

References

Bergeron, P., Fontaine, G., Billères, M., Boudreault, S., & Green, E. M. 2004, *ApJ*, 600, 404

Brickhill, A. J. 1991, *MNRAS*, 252, 334

Castanheira, B. G. & Kepler, S. O. 2009, *MNRAS*, 396, 1709

Castanheira, B. G., Kepler, S. O., Mullally, F., *et al.* 2006, *A&A*, 450, 227

Castanheira, B. G., Kepler, S. O., Kleinman, S. J., Nitta, A., & Fraga, L. 2010, *MNRAS*, 405, 2561

Castanheira, B. G., Kepler, S. O., Kleinman, S. J., Nitta, A., & Fraga, L. 2013, *MNRAS*, 430, 50

Giovannini, O., Kepler, S. O., Kanaan, A., Wood, A., Claver, C. F., & Koester, D. 1998, *Baltic Astronomy*, 7, 131

Hermes, J. J., Montgomery, M. H., Winget, D. E., Brown, W. R., Kilic, M., & Kenyon, S. J. 2012, *ApJ*, 750, L28

Kepler, S. O., Castanheira, B. G., Saraiva, M. F. O., *et al.* 2005, *A&A*, 442, 629

Kepler, S. O., Pelisoli, I., Peçanha, V., *et al.* 2012, *ApJ*, 757, 177

Kleinman, S. J., Harris, H. C., Eisenstein, D. J., *et al.* 2004, *ApJ*, 607, 426

Kleinman, S. J., Kepler, S. O., Koester, D., *et al.* 2013, *ApJS*, 204, 5

Montgomery, M. H., Williams, K. A., Winget, D. E., Dufour, P., De Gennaro, S., & Liebert, J. 2008, *ApJ*, 678, L51

Mukadam, A. S., Mullally, F., Nather, R. E., *et al.* 2004, *ApJ*, 607, 982

Mukadam, A. S., Montgomery, M. H., Winget, D. E., Kepler, S. O., & Clemens, J. C. 2006, *ApJ*, 640, 956

Mullally, F., Thompson, S. E., Castanheira, B. G., *et al.* 2005, *ApJ*, 625, 966

Nitta, A., Kleinman, S. J., Krzesinski, J., *et al.* 2009, *ApJ*, 690, 560

Perlmutter, S., Aldering, G., Goldhaber, G., *et al.* 1999, *ApJ*, 517, 565

Riess, A. G., Filippenko, A. V., Challis, P., *et al.* 1998, *AJ*, 116, 1009

Werner, K. & Herwig, F. 2006, *PASP*, 118, 183

Winget, D. E., van Horn, H. M., & Hansen, C. J. 1981, *ApJ*, 245, L33

Winget, D. E., Kepler, S. O., Campos, F., *et al.* 2009, *ApJ*, 693, L6

Precision Asteroseismology
Proceedings IAU Symposium No. 301, 2013
J. A. Guzik, W. J. Chaplin, G. Handler & A. Pigulski, eds.

© International Astronomical Union 2014
doi:10.1017/S1743921313014464

The internal structure of ZZ Cet stars using quantitative asteroseismology: The case of R548

N. Giammichele[1], G. Fontaine[1], P. Brassard[1], and S. Charpinet[2]

[1] Département de physique, Université de Montréal,
C.P. 6128, Succursale Centre-Ville, Montréal, QC H3C3J7, Canada
email: noemi@astro.umontreal.ca

[2] Université de Toulouse, UPS-OMP, IRAP, Toulouse, France;
CNRS, IRAP, 14 avenue Edouard Belin, 31400, Toulouse, France

Abstract. We explore quantitatively the low but sufficient sensitivity of oscillation modes to probe both the core composition and the details of the chemical stratification of pulsating white dwarfs. Until recently, applications of asteroseismic methods to pulsating white dwarfs have been far and few, and have generally suffered from an insufficient exploration of parameter space. To remedy this situation, we apply to white dwarfs the same double-optimization technique that has been used quite successfully in the context of pulsating hot B subdwarfs. Based on the frequency spectrum of the pulsating white dwarf R548, we are able to unravel in a robust way the unique onion-like stratification and the chemical composition of the star. Independent confirmations from both spectroscopic analyses and detailed evolutionary calculations including diffusion provide crucial consistency checks and add to the credibility of the inferred seismic model. More importantly, these results boost our confidence in the reliability of the forward method for sounding white dwarf internal structure with asteroseismology.

Keywords. stars: oscillations, white dwarfs

---

1. Introduction

Asteroseismology is a unique tool that allows us to probe the internal structure of a pulsating star. Until recently however, applications of the method to pulsating white dwarfs were few, and greatly suffered from an insufficient exploration of parameter space. The use of fully evolutionary models tend to propagate errors, and by the end of the calculation, uncertainties are greatly boosted. Moreover, such calculations are lenghty and it is impossible to thoroughly explore parameter space. To remedy this situation, we have begun to apply to white dwarfs the same double-optimization technique based on the forward approach that has been successfully used in the context of pulsating hot B subdwarfs (see, e.g., the review of Charpinet *et al.* 2009a). Targeted white dwarfs were chosen lying close to the blue edge, where the simplest pulsators are found. Even if only a few modes are excited in such stars, they still contain individually quite a bit of complementary information since they are low-order modes, far away from the asymptotic regime. We took advantage of a previously unexploited data set of very high S/N, obtained with the CFHT/LAPOUNE combination to extract pulsation periods in the first part of this venture (Giammichele *et al.* 2013a). We carried out our seismic analysis on the basis of five very certain periods observed in R548 (an improved extraction exercise was performed since Giammichele *et al.* 2013a and slightly different pulsation periods were obtained and used hereinafter). Results are presented in Giammichele *et al.* (2013b). This work contains the final validations of this asteroseismic exercise.

285

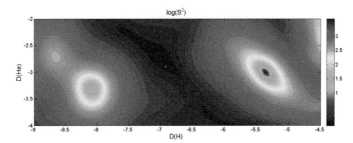

Figure 1. Contour map of the merit function as a function of the top of the transition zones $D(\mathrm{H})$ and $D(\mathrm{He})$, while the remaining parameters are fixed at their optimized values. $D(\mathrm{H})$ and $D(\mathrm{He})$, natural parameters of the optimization exercise, are directly related to the more familiar parameters $\log(M_\mathrm{H}/M_*)$ and $\log(M_\mathrm{He}/M_*)$ through: $\log(M_\mathrm{H}/M_*) = D(\mathrm{H}) + C$ (similar for the helium layer) where C is a small positive term, slightly dependent on the model parameters, determined by the mass of hydrogen or helium present inside the transition zones.

2. Internal structure

2.1. *Optimization procedure*

Given the high quality of our data, our goal is now to apply the same forward approach that has been successfully performed on hot pulsating subdwarfs. The technique relies on a first optimization leading to the best possible fit between the five observed periods (in the present case) with five periods out of the theoretical spectrum of a given stellar model. The second optimization step next leads to the best model in parameter space. To fully define a static white dwarf model, we have to specify the surface gravity (alternatively, the mass), the effective temperature, the chemical stratification in the envelope, and the core composition. Since low radial-order modes are quite insensitive to the prescribed convective efficiency, we fixed the latter to the so-called ML2/$\alpha = 1.0$ prescription and did not vary it in parameter space. The quality of the match is then quantified through a second optimization by a merit function S^2 (an unweighted χ^2), which is minimized as a function of T_eff, $\log g$, $\log(M_\mathrm{H}/M_*)$, and $\log(M_\mathrm{He}/M_*)$, given a core composition. Thus, for every set of T_eff, $\log g$, $\log(M_\mathrm{H}/M_*)$, and $\log(M_\mathrm{He}/M_*)$, we get a unique period spectrum to compare with the observed frequencies.

The forward procedure as implemented in Montréal (a full grid approach) necessitates the computation of many millions of models and their respective period spectra in a reasonable amount of time, and the use of our dedicated cluster of 320 nodes is a necessity. The same exercise with fully evolutionary sequences, including diffusion would be impracticable. It is worth repeating that there is no a priori guarantee to find an optimal model and, moreover, to find an optimal model that would be consistent with independent spectroscopic determinations of the atmospheric parameters.

The resulting optimum asteroseismic model presents a very low $S^2 = 0.83$, with $T_\mathrm{eff} = 12\,095$ K, $\log g = 8.070$, $\log(M_\mathrm{H}/M_*) = -4.54$, and $\log(M_\mathrm{He}/M_*) = -2.42$. Details on the mode identification, given automatically by the solution, are presented in Table 1. Consecutive values of the radial order k, are perfectly consistent with the prediction from nonadiabatic theory that pulsation modes are excited in consecutive modes. Moreover, the nice agreement between the newest spectroscopic determination of $T_\mathrm{eff} = 11\,980$ K, $\log g = 8.06$ (kindly provided to us by P. Bergeron) and our seismic determination of atmospheric parameters boosts our confidence in the validity of the solution found.

2.2. *Chemical stratification*

Figure 1 unambiguously suggests that R548 is composed of a thick hydrogen and thick helium layer as indicated by the deeper minimum. A much less convincing "thin" envelope

Table 1. Mode identification for the best-fitting model with a mixed C-O core composition (mass fraction: 60-40).

Observed periods	Fitted periods	l	k
212.96	212.98	1	2
274.53	274.60	1	3
318.43	318.18	1	4
334.12	334.96	1	5
186.89	187.14	2	4

possibility also appears in the figure, but can readily be ruled out on the grounds of its significantly degraded merit function. There is no degeneracy in the optimized grid of seismic solutions. The chemical stratification with thick hydrogen and helium layers found for R548 ought to be reliable.

A first validation of the use of the forward method in this context is illustrated in Figure 2. We explore the importance of the imposed steepness of the chemical profile at the interface between the H/He and He/C-O layers in our static models. While the transition between the H/He layers leaves a stronger signature on the period spectrum than the deeper He/C-O interface, we still find a very narrow and well-defined minimum for a(H/He) = 1.3 and a(He/C-O) = 3.3, the parameters defining steepness. Here again, these two parameters do not lead to any degeneracy in the asteroseismic solution. More importantly, evolutionary calculations, including diffusion, confirm transition steepness values of a(H/He) = 1.31 and a(He/C-O) = 3.20, in almost perfect agreement with our asteroseismic values.

2.3. Core composition

The distribution of the Brunt-Väisälä frequency in the deepest layers is slightly but sufficiently perturbed by the different possible core compositions to allow us to probe the bulk composition. The best quality fit in terms of minimal $S^2 = 0.83$ corresponds to a chemical composition of the core made of a mixture of C-O of mass fraction 60/40 as can be seen in Fig. 1 of Giammichele *et al.* (2014, in press). Cores composed of pure carbon and oxygen can be unambiguously ruled out from the poor matches of the observed frequency spectrum. If we consider the newest spectroscopic determination of T_{eff} and $\log g$ for R548, the mixed C-O core is the solution that agrees best with spectroscopic

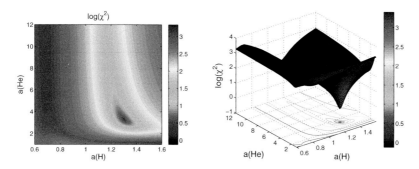

Figure 2. Contour and 3-D map of the merit function S^2 as a function of the values of the chemical profile steepness at the interface between the H/He and He/C-O layers, for the optimal effective temperature and surface gravity.

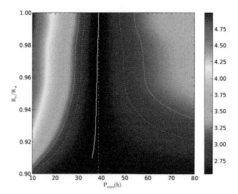

Figure 3. Contour map showing the behaviour of the normalized merit function S^2 in terms of the radius and in terms of the rotation period of the inner region for solid body rotation in a two-zone approach defined by Charpinet *et al.* (2009b). The solution is illustrated by the nearly vertical white curve and the dotted white curves depict the 1, 2 and 3σ contours. In comparison, the vertical dot-dashed white curve gives the exact solution for solid-body rotation.

parameters while displaying the best (lowest) value of the merit function S^2 of the seismic solution. This contributes in giving credibility to the proposed method.

It must be pointed out that it is the imposed smoothness of the core profile (assumed homogeneously mixed) that has allowed us to carry out successfully our asteroseismic exercise. Indeed, through numerous tests, we found that detailed post-AGB models contain too much noise in the core for pursuing quantitative seismology. For example, we implemented in some of our test models the detailed core chemical profiles computed by Salaris *et al.* (2010). Even if the grid of optimized parameters is showing the same global geometry, the surface turns out to be too noisy to identify a global minimum. We believe that current post-AGB models leave core structures that are most probably not physical, rendering the seismic exercise difficult if not impossible.

2.4. Rotation profile structure

Based on the rotational splitting (Giammichele *et al.* 2013a) shown in four out of the five modes listed in Table 1 and on the optimal model of R548, we determined whether or not its internal rotation profile corresponds to solid-body rotation, as seen in Figure 3. Unfortunately, evolution has pushed the *g*-modes to the outer part of the star because of the increase of degeneracy along the cooling track. Unlike the case of GW Vir stars (see Charpinet *et al.* 2009b), rotation kernels have large amplitudes in the outermost layers only. Therefore, we can only probe about 10% of the radius, and less than 1% of the total mass. But within the outer 10% radius, we can assert that R548 is rotating as a solid body. The relatively low period of 39 h is consistent with the suggestion of Charpinet *et al.* (2009b) to the effect that isolated white dwarfs must have lost essentially all of their angular momentum during evolutionary phases prior to the white dwarf stage.

References

Charpinet, S., Fontaine, G., & Brassard, P. 2009a, *A&A*, 493, 595
Charpinet, S., Fontaine, G., & Brassard, P. 2009b, *Nature*, 461, 501
Giammichele, N., Fontaine, G., & Brassard, P. 2013a, *European Physical Journal Web of Conferences*, 43, 5007
Giammichele, N., Fontaine, G., & Brassard, P. 2013b, *ASP-CS*, 469, 49
Salaris, M., Cassisi, S., Pietrinferni, A., Kowalski, P. M., & Isern, J. 2010, *ApJ*, 716, 1241

Precision Asteroseismology
Proceedings IAU Symposium No. 301, 2013
J. A. Guzik, W. J. Chaplin, G. Handler & A. Pigulski, eds.
© International Astronomical Union 2014
doi:10.1017/S1743921313014476

Origin and pulsation of hot subdwarfs

S. K. Randall[1], G. Fontaine[2], S. Charpinet[3], V. Van Grootel[4], and P. Brassard[2]

[1]ESO, Karl-Schwarzschild-Str. 2, 85748 Garching bei München, Germany
email: srandall@eso.org

[2]Département de Physique, Université de Montréal, C.P. 6128, Succ. Centre-Ville, Montréal, QC H3C 3J7, Canada

[3]CNRS, Université de Toulouse, UPS-OMP, IRAP, 14 av. E. Belin, 31400, Toulouse, France

[4]Institut d'Astrophysique et de Géophysique de l'Université de Liège, Allée du 6 Août 17, B-4000 Liège, Belgium

Abstract. We briefly introduce hot subdwarfs and their evolutionary status before discussing the different types of known pulsators in more detail. Currently, at least six apparently distinct types of variable are known among hot subdwarfs, encompassing p- as well as g-mode pulsators and objects in the Galactic field as well as in globular clusters. Most of the oscillations detected can be explained in terms of an iron opacity mechanism, and quantitative asteroseismology has been very successful for some of the pulsators. In addition to helping constrain possible evolutionary scenarios, studies focussing on stellar pulsations have also been used to infer planets and characterize the rotation of the host star.

Keywords. stars: oscillations, subdwarfs

1. Origin of hot subdwarfs

Hot subdwarfs are compact, evolved stars located on the Extreme Horizontal Branch (EHB) of the Hertzsprung-Russell diagram. They are low-mass stars of about $0.5\,M_\odot$ and comprise the spectral types sdB ($T_{\rm eff} \sim 20\,000 - 36\,000$ K) and sdO ($T_{\rm eff} \sim 36\,000 - 80\,000$ K). While most sdB stars are He-poor, the majority of sdO stars are He-rich. The chemical peculiarity observed in hot subdwarfs is caused by diffusion, the competitive action of radiative levitation and gravitational settling.

It is commonly accepted that hot subdwarfs are post red-giant stars that were stripped of too much of their H-envelope before or at the He-flash to sustain H-shell burning. In this canonical scenario, the sdB stars are in the core He-burning phase, while the H-rich sdOs are their short-lived He-shell burning progeny already evolving away from the EHB (Dorman *et al.* 1993). He-rich subdwarfs on the other hand have been suggested to result from the merger of two He-rich white dwarfs (Saio & Jeffery 2000) or a late He-core flash (Brown *et al.* 2001). One of the main problems in our understanding of the formation of hot subdwarfs is the question of what caused the necessary mass loss. Given that more than half of sdB stars in the field appear to reside in binary systems (e.g. Maxted *et al.* 2001), it has been proposed that the mass loss could have occurred as a consequence of binary formation channels involving stable Roche lobe overflow, a common envelope phase, or a combination of the two (Han *et al.* 2002, 2003). In this context, single sdB stars were suggested to have formed via the merger of two He white dwarfs, but this scenario was more recently superseded by the idea of a He white dwarf and main sequence star merger (Clausen & Wade 2011). In globular clusters the EHB may have formed from a He-enriched sub-population (D'Antona *et al.* 2005). Another theory is that the mass loss was enhanced by the presence of planets (Soker 1998).

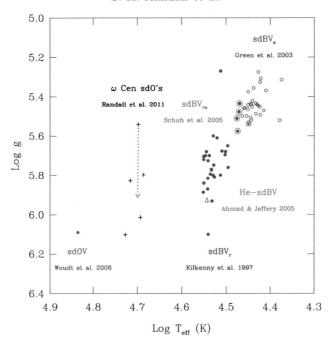

Figure 1. Currently known types of hot subdwarf pulsators. The references refer to the discovery paper of the prototype. The arrow pointing downwards for one of the ω Cen pulsators indicates that the $\log g$ value derived for this object is likely underestimated due to a significant contamination of its spectra by nearby stars.

The validity and relative importance of the proposed evolutionary scenarios as formation channels for the different populations of hot subdwarfs is still unclear. However, first trends are beginning to emerge based on observations of the binary properties of these stars, atmospheric abundance analyses and, more recently, the exploitation of pulsating hot subdwarfs through asteroseismology. The latter technique has the great advantage that not only the mass, but also the internal structure of the star can be determined to high accuracy. Moreover, pulsations can be used to infer substellar companions and planets, to test spin-orbit synchronization in binary systems and to characterise the internal rotation profile of the host star. In the following sections we describe the different types of known hot subdwarf pulsators and give a brief overview of our current understanding of these objects.

2. Hot subdwarf pulsators

The location of the different classes of subdwarf pulsators in the $\log g - T_{\rm eff}$ diagram is displayed in Fig. 1. We now describe each class in more detail, working chronologically based on the time of their discovery.

2.1. *sdBV$_r$ stars (also known as EC 14026 or V361 Hya stars)*

This first class of hot subdwarf pulsators was discovered (Kilkenny *et al.* 1997) independently, but almost at the same time as the theoretical prediction of such objects (Charpinet *et al.* 1996). Clustered in an instability strip between $\sim$29 000 and 36 000 K, sdBV$_r$ stars show short periods on a 100–300 s timescale that are interpreted as low-degree, low-order p modes. The pulsations are driven by a κ-mechanism associated with a local overabundance of iron-peak elements in the driving region, which in turn is achieved

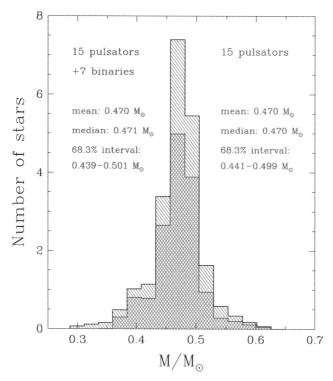

Figure 2. Empirical mass distribution based on the asteroseismological values for 15 sdBV stars as well as on estimates derived from the observed binary properties of 7 eclipsing binaries. The two distributions are indistinguishable from a statistical point of view.

by radiative levitation. Static envelope models incorporating the diffusion of traces of iron in a pure H-envelope (the so-called Montréal second-generation models, e.g. Charpinet *et al.* 1997) can very accurately reproduce the observed pulsation properties of these stars. Not only is the instability strip recovered almost perfectly (see Fig. 4), but quantitative asteroseismology via the forward method (e.g. Brassard *et al.* 2001) has also been very successful. For a discussion of the accuracy and precision achieved from the asteroseismology of sdBV$_r$ stars see Van Grootel *et al.*, these proceedings.

To date, asteroseismic analyses have been carried out for 15 sdBV stars, allowing a first characterization of the mass distribution (Fig. 2; for more details see Fontaine *et al.* 2012). While we are still limited by small number statistics, there is tentative evidence that the high-mass ($\sim 0.5 - 0.7\ M_\odot$) tail arising primarily from the white dwarf merger scenario (Han *et al.* 2003) is much less pronounced than predicted. The empirical mass distribution peaks very close to the canonically expected value of $0.47\ M_\odot$, and shows a similar behaviour to that predicted by the binary formation channels.

2.2. *sdBV$_s$ stars (also known as V1093 Her or Betsy stars)*

The second type of pulsating hot subdwarf lies at the cool end of the sdB star regime ($T_{\rm eff} \sim 22\,000 - 29\,000$ K) and shows high radial-order g-mode oscillations on a timescale of $2000 - 8000$ s (Green *et al.* 2003). While the driving mechanism was quickly identified to be the same iron opacity mechanism as for the sdBV$_r$ stars (Fontaine *et al.* 2003), standard models incorporating a non-uniform abundance profile only of iron are not able to reproduce the blue edge of the observed instability strip. A good agreement between the predicted and observed sdBV$_s$ instability regions was reached only recently using models including nickel as well as iron in the diffusion calculations (Hu *et al.* 2011).

Compared to the sdBV$_r$ stars, the slow pulsators are much more challenging to study, both from an observational and a modeling point of view. Ground-based observations are severely hampered by aliasing, atmospheric contamination and the length of the dataset needed, while the fact that g modes penetrate much deeper into the stellar interior compared to the shallower p modes means that the second-generation envelope models are no longer sufficient for quantitative modeling. Real progress on the asteroseismic exploitation of these objects was made only with the advent of space-bourne missions such as *CoRoT* and in particular *Kepler*, which delivered months of uninterrupted time-series photometry revealing typically hundreds of pulsation frequencies down to amplitudes of a few ppm. In parallel, progress was made on the computational front with the development of the Montréal third-generation models, which are complete static stellar models in thermal equilibrium that accurately describe both the stellar core and the envelope (for a detailed description see Van Grootel *et al.* 2013a). This has allowed quantitative asteroseismology to be carried out for a small number of sdBV$_s$ stars, and the mass and composition of the core to be constrained for the first time (Van Grootel *et al.* 2010a,b, Charpinet *et al.* 2011a).

Another interesting finding to come out of the *Kepler* data is that many of the g-mode pulsators also show a small number of very low-amplitude periodicities in the p-mode regime. Moreover, it appears that the sdBV$_s$ instability strip is not pure as was previously expected, but also contains a fraction ($\sim$25%, based on small-number statistics) of non-variable stars. It is not yet clear why this is, or why the pulsator fraction is so much higher than for the sdBV$_r$ stars (where only $\sim$10% of stars in the instability strip pulsate). One explanation is that the apparently non-pulsating stars have not yet been able to build up the necessary reservoir of heavy elements in the driving region (Fontaine *et al.* 2006), but this remains to be investigated in more detail.

2.3. *sdBV*$_{rs}$ *stars (also known as hybrid pulsators)*

These stars lie at the intersection of the sdBV$_s$ and sdBV$_r$ instability strips and exhibit both rapid p-mode and slow g-mode oscillations at appreciable amplitudes. First discovered by Schuh *et al.* (2005), a handful of such objects are now known. Hybrid pulsators are of great interest to asteroseismology, since both the inner and outer regions of the star can in principle be probed. A potential Rosetta stone, the hybrid pulsator and eclipsing binary 2M 1938 + 4603 was recently observed with *Kepler* (Østensen *et al.* 2010). Asteroseismology however has so far been foiled, both by the very fast rotation of this object and the likely post-EHB status of this star.

2.4. *A He-sdBV: LS IV* $-14°116$

The unusual He-sdBV LS IV $-14°116$ (Ahmad & Jeffery 2005) is located well within the sdBV$_r$ instability strip at $T_{\rm eff} \sim 35\,000$ K and $\log g \sim 5.9$, but quite puzzlingly exhibits g-mode pulsations in the same (2000–5000 s) period range as the sdBV$_s$ stars (Green *et al.* 2011). It is only moderately helium enhanced with $\log q(\mathrm{He/H}) \sim -0.6$, but shows a strange abundance pattern with Ge, Sr, Y and Zr enhanced by factors of up to 10 000 compared to solar values (Naslim *et al.* 2011). It has been proposed that these strange atmospheric abundances are indicative of a strong magnetic field, which may influence the driving of pulsation modes via the κ-mechanism and potentially shift the g-mode instability strip to higher temperatures. Alternatively, non-adiabatic pulsation calculations involving hot-flasher and post-merger evolutionary models suggest that periodicities on a similar time scale as those observed in LS IV $-14°116$ may be driven by the ϵ-mechanism

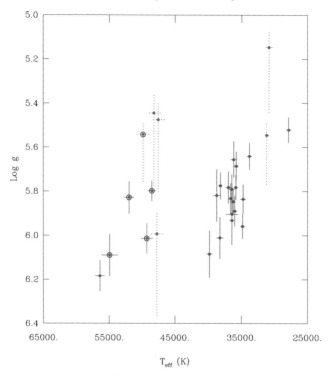

Figure 3. The instability strip in ω Cen. The five known pulsators are marked by large open circles, while the targets that were found to be constant down to a detection threshold of $< 1\%$ are denoted by small points. The continuous error bars refer to targets whose spectra were uncontaminated by nearby stars and for which reliable atmospheric parameters could be determined. Dotted error bars indicate stars with contaminated spectra, where in particular the derived $\log g$ value is likely to have been underestimated. The He-rich stars (blue in the online version) are clustered around $35\,000 - 40\,000$ K while the remaining stars are H-rich.

acting on the unstable He-burning shells present just before the star settles on the EHB (Miller Bertolami *et al.* 2011, 2013).

2.5. *An sdOV star among the field population: SDSS J160043.6 + 074802.9*

To date the only confirmed sdO pulsator in the Galactic field is the very hot ($\sim$68 500 K), moderately helium-enhanced ($\log q(\mathrm{He/H}) \sim -0.64$, Latour *et al.* 2011) star SDSS J160043.6 + 074802.9, which exhibits very rapid p-mode pulsations with periods in the $60 - 120$ s range (Woudt *et al.* 2006). These pulsations can be driven by the same iron opacity mechanism as the pulsations in the sdBV$_\mathrm{r}$ and sdBV$_\mathrm{s}$ stars (Fontaine *et al.* 2008, see also Fig. 4).

2.6. *The ω Cen sdOV stars: the first EHB pulsators detected in a globular cluster*

Members of this most recently discovered class of EHB pulsator (Randall *et al.* 2011) have so far only been found in the unusual globular cluster ω Cen. The five currently known variables (Randall *et al.* 2013) appear to form a homogeneous class in terms of pulsation properties and atmospheric parameters. They are all H-rich sdO stars, all exhibit pulsation periods in the $80 - 125$ s range, and all fall within a nicely defined $T_{\mathrm{eff}} \sim 48\,000 - 54\,000$ K empirical instability strip (see Fig. 3). It is not yet clear whether or not all ω Cen sdOs within this strip are pulsating. However, it is becoming increasingly likely that these stars do not exist among the field sdO population in a similar relative

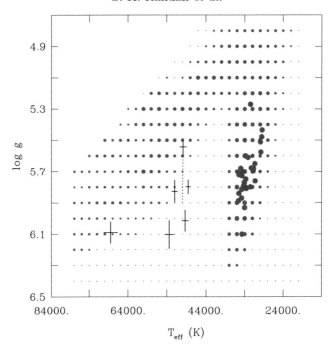

Figure 4. p-mode instability strip predicted by the Montréal second generation models. Each large grid point refers to a model showing unstable radial modes in the p-mode regime, while models indicated by small grid points do not exhibit instabilities. The observed sdBV$_r$ stars are denoted by big (blue in the online version) points superposed on the grid, and the sdO variables are shown by crosses. The five crosses at ~50 000 K refer to the ω Cen pulsators (the dotted error bar extension to lower values of log g for one of the stars implies that the latter is likely underestimated due to the significant contamination of the extracted spectrum), whereas the isolated cross at higher temperatures indicates the field sdOV SDSS J160043.6 + 074802.9.

proportion to the entire EHB star population. In fact, a dedicated study monitoring 36 sdO stars (Johnson *et al.* 2013) failed to uncover a single pulsator. However, there is one object, PB 8783, traditionally accepted as an sdBV$_r$ star by the community, that may in fact turn out to be a field counterpart to the ω Cen pulsators (Østensen 2012). Due to the strong contamination of its optical spectrum by an F-type main sequence companion there is a degeneracy when iteratively fitting the hot and the cool components during the atmospheric analysis, and both solutions giving a temperature of ~36 000 K (sdB) and ~50 000 K (sdO) are possible. An asteroseismic exploration of parameter space in the vicinity of the two spectroscopic solutions (Van Grootel *et al.* 2013b) indicated a preference for the hotter of the two. However, UV spectra will be needed in order to confirm the true nature of this rapid variable.

With the aim of understanding the non-adiabatic properties of the ω Cen variables, we extended the grid of Montréal second-generation models in order to cover the required region of $T_{\mathrm{eff}} - \log g$ parameter space. The results of our instability calculations for low-degree p-mode pulsations for hot subdwarfs between 20 000 and 78 000 K can be seen in Fig. 4. It becomes apparent that the sdOV and sdBV$_r$ star instability strips are in fact connected at low surface gravities. However, while the observed sdBV$_r$ instability strip is perfectly reproduced by the models, the ω Cen pulsators lie for the most part beyond the red edge of the theoretical sdO star instability region. Moreover, the periods excited by the models are too short compared to those observed. This could be due to shortcomings in the models (e.g. the fact that elements such as Ni are not incorporated in the opacity bump computations) or to problems with the accurate spectroscopic determination of the very high temperatures of these objects from optical spectra alone. Again, we need UV spectra to derive reliable temperatures for these objects.

2.7. *EHB pulsators in NGC 2808*

Recently, HST time-series photometry revealed 6 EHB pulsators in the central regions of the globular cluster NGC 2808 (Brown *et al.* 2013). While the intent of these observations had been to uncover counterparts to the ω Cen sdOVs in another globular cluster, the variables that were finally uncovered appear to be a rather inhomogeneous bunch, and none match the ω Cen oscillators in terms of atmospheric parameters and pulsation properties. The one NGC 2808 pulsator that has $T_{\mathrm{eff}} \sim 50\,000$ K is extremely He-rich and has a periodicity of 147 s, longer than any of the ω Cen pulsators. The other two variables for which UV spectroscopy is available lie in the sdBV$_\mathrm{r}$ and sdBV$_\mathrm{s}$ regimes respectively, but both show very short periodicities around 110 s. These findings are extremely puzzling, and present a significant challenge to our understanding of hot subdwarfs.

3. Nice "side-effects" of pulsation studies

The pulsations in hot subdwarfs have been used not only to determine the fundamental parameters of the stars themselves, but also to detect planets orbiting them. For several sdBV$_\mathrm{r}$ stars that have been monitored extensively over several years, small sinusoidal frequency variations of the dominant pulsation modes have been measured and attributed to an orbital wobble caused by sub-stellar companions, either brown dwarfs or giant planets (Silvotti *et al.* 2007, Lutz *et al.* 2012). Two smaller, likely Earth-sized planets were inferred from extremely low-amplitude, low-frequency luminosity modulations detected by *Kepler* for the sdBV$_\mathrm{s}$ star KPD 1943 + 4058 (Charpinet *et al.* 2011b). These planets (or at least their cores) must have somehow survived the red giant stage, and perhaps they played a role in the formation of their host EHB star by enhancing the mass loss.

Another benefit of pulsations is that the rotation rate of the host star can be constrained by the measurement (or absence of) rotational splitting. While most single sdB stars are known to be slow rotators, the situation is different for those stars residing in close binaries. Depending on the proximity and nature of the companion, the rotation rate of the individual binary components will become synchronized with the orbital period. For example, in the 0.57-d sdBV$_\mathrm{s}$+white dwarf binary PG 0101 + 039, a tiny ellipsoidal variation detected at exactly half the orbital period was used to confirm spin-orbit synchronization (Geier *et al.* 2008). On the other hand, the rotational splitting of pulsation frequencies measured by *Kepler* in two sdBV$_\mathrm{s}$ + dM binaries with binary periods $\sim$0.4 d indicates that the rotation period is much longer ($\sim$7 – 10 days) than the binary period (Pablo *et al.* 2012). With asteroseismology it is possible to go even further and measure not just the rotation at the stellar surface, but also in the interior. To date, such measurements are consistent with solid-body rotation (e.g. Charpinet *et al.* 2008).

References

Ahmad, A. & Jeffery, C. S. 2005, *A&A*, 437, L51

Brassard, P., Fontaine, G., Billères, M., *et al.* 2001, *ApJ*, 563, 1013

Brown, T. M., Sweigart, A. V., Lanz, T., Landsman, W. B., & Hubeny, I. 2001, *ApJ*, 562, 368

Brown, T. M., Landsman, W. B., Randall, S. K., Sveigart, A. V., & Lanz, T. 2013, *ApJ*, 777, L22

Charpinet, S., Fontaine, G., Brassard, P., & Dorman, B. 1996, *ApJ*, 471, L103

Charpinet, S., Fontaine, G., Brassard, P., *et al.* 1997, *ApJ*, 483, L123
Charpinet, S., Van Grootel, V., Reese, D., *et al.* 2008, *A&A*, 489, 377
Charpinet, S., Van Grootel, V., Fontaine, G., *et al.* 2011a, *A&A*, 530, A3
Charpinet, S., Fontaine, G., Brassard, P., *et al.* 2011b, *Nature*, 480, 496
Clausen, D. & Wade, R. A. 2011, *ApJ*, 733, L42
D'Antona, F., Bellazzini, M., Caloi, V., *et al.* 2005, *ApJ*, 631, 868
Dorman, B., Rood, R. T., & O'Connell, R. W. 1993, *ApJ*, 419, 596
Fontaine, G., Brassard, P., Charpinet, S., *et al.* 2003, *ApJ*, 597, 518
Fontaine, G., Brassard, P., Charpinet, S., & Chayer, P. 2006, *MemSAIt*, 77, 49
Fontaine, G., Brassard, P., Green, E. M., *et al.* 2008, *A&A*, 486, L39
Fontaine, G., Brassard, P., Charpinet, S., *et al.* 2012, *A&A*, 539, A12
Geier, S., Nesslinger, S., Heber, U., *et al.* 2008, *A&A*, 477, L13
Green, E. M., Fontaine, G., Reed, M. D., *et al.* 2003, *ApJ*, 583, L31
Green, E. M., Guvenen, B., O'Malley, C. J., *et al.* 2011, *ApJ*, 734, 59
Han, Z., Podsiadlowski, P., Maxted, P. F. L., Marsh, T. R., & Ivanova, N. 2002, *MNRAS*, 336, 449
Han, Z., Podsiadlowski, P., Maxted, P. F. L., & Marsh, T. R. 2003, *MNRAS*, 341, 669
Hu, H., Tout, C. A., Glebbeek, E., & Dupret, M.-A. 2011, *MNRAS*, 418, 195
Johnson, C. B., Green, E. M., Wallace, S., *et al.* 2013, arXiv: 1308.1373
Kilkenny, D., Koen, C., O'Donoghue, D., & Stobie, R. S. 1997, *MNRAS*, 285, 640
Latour, M., Fontaine, G., Brassard, P., *et al.* 2011, *ApJ*, 733, 100
Lutz, R., Schuh, S., & Silvotti, R. 2012, *AN*, 333, 1099
Maxted, P. F. L., Heber, U., Marsh, T. R., & North, R. C. 2001, *MNRAS*, 326, 1391
Miller Bertolami, M. M., Córsico, A. H., & Althaus, L. G. 2011, *ApJ*, 741, L3
Miller Bertolami, M. M., Córsico, A. H., Zhang, X., Althaus, L. G., & Jeffery, C. S. 2013, in: J. Montalbán, A. Noels, & V. Van Grootel (eds.), *Ageing Low Mass Stars: From Red Giants to White Dwarfs*, European Physical Journal Web of Conferences, 43, 4004
Naslim, N., Jeffery, C. S., Behara, N. T., & Hibbert, A. 2011, *MNRAS*, 412, 363
Østensen, R. H. 2012, *ASP-CS*, 452, 233
Østensen, R. H., Silvotti, R., Charpinet, S., *et al.* 2010, *MNRAS*, 409, 1470
Pablo, H., Kawaler, S. D., Reed, M. D., *et al.* 2012, *MNRAS*, 422, 1343
Randall, S. K., Calamida, A., Fontaine, G., Bono, G., & Brassard, P. 2011, *ApJ*, 737, L27
Randall, S. K., Calamida, A., Fontaine, G., *et al.* 2013, in: J. Montalbán, A. Noels, & V. Van Grootel (eds.), *Ageing Low Mass Stars: From Red Giants to White Dwarfs*, European Physical Journal Web of Conferences, 43, 4006
Saio, H. & Jeffery, C. S. 2000, *MNRAS*, 313, 671
Schuh, S., Huber, J., Green, E. M., *et al.* 2005, *ASP-CS*, 334, 530
Silvotti, R., Schuh, S., Janulis, R., *et al.* 2007, *Nature*, 449, 189
Soker, N. 1998, *AJ*, 116, 1308
Van Grootel, V., Charpinet, S., Fontaine, G., *et al.* 2010a, *ApJ*, 718, L97
Van Grootel, V., Charpinet, S., Fontaine, G., Green, E. M., & Brassard, P. 2010b, *A&A*, 524, A63
Van Grootel, V., Charpinet, S., Brassard, P., Fontaine, G., & Green, E. M. 2013a, *A&A*, 553, A97
Van Grootel, V., Charpinet, S., Fontaine, G., Brassard, P., & Green, E. M. 2013b, *ASP-CS*, in press
Woudt, P. A., Kilkenny, D., Zietsman, E., *et al.* 2006, *MNRAS*, 371, 1497

Precision Asteroseismology
Proceedings IAU Symposium No. 301, 2013
J. A. Guzik, W. J. Chaplin, G. Handler & A. Pigulski, eds.

© International Astronomical Union 2014
doi:10.1017/S1743921313014488

The origin and pulsations of extreme helium stars†

C. Simon Jeffery

Armagh Observatory, College Hill, Armagh BT61 9DG, Northern Ireland, UK

Abstract. Stars consume hydrogen in their interiors but, generally speaking, their surfaces continue to contain some 70% hydrogen (by mass) throughout their lives. Nevertheless, many types of star can be found with hydrogen-deficient surfaces, in some cases with as little as one hydrogen atom in 10 000. Amongst these, the luminous B- and A-type extreme helium stars are genuinely rare; only ∼ 15 are known within a very substantial volume of the Galaxy.

Evidence from surface composition suggests a connection to the cooler R CrB variables and some of the hotter helium-rich subdwarf O stars. Arguments currently favour an origin in the merger of two white dwarfs; thus there are also connections with AM CVn variables and Type Ia supernovae. Pulsations in many extreme helium stars provide an opportune window into their interiors. These pulsations have unusual properties, some being "strange" modes, and others being driven by Z-bump opacities. They have the potential to deliver distance-independent masses and to provide a unique view of pulsation physics.

We review the evolutionary origin and pulsations of these stars, and introduce recent progress and continuing challenges.

Keywords. stars: early-type, stars: chemically-peculiar, stars: supergiants, stars: white dwarfs, stars: evolution, stars: oscillations, stars: variable: other, stars: individual: V652 Her, FQ Aqr

---

1. Extreme helium stars

Observing from McDonald Observatory at a maximum altitude of 14°, Popper (1942) reported the B2 star HD 124448 to "show no hydrogen lines, either in absorption or in emission, although the helium lines are sharp and strong ... The abundance of hydrogen appears to be very low in the atmosphere of this star". Indeed, HD 124448 turned out to be the first of around 20 B- and early A- supergiants, with apparent magnitude $9.3 < V < 12.6$, in which hydrogen comprises less than 1 part per thousand of the atmosphere. In terms of kinematics, metallicity and galactic distribution, they have the properties of the Galactic bulge (Jeffery *et al.* 1987). With luminosities approximately ten thousand times solar, the number count is complete for the observable parts of the Galaxy (fainter stars would lie beyond). In addition to their low hydrogen abundances, the true "extreme helium stars" (EHes) show atmospheres which are enriched in nitrogen by a factor ten, carbon by between one and three parts per hundred, and sometimes oxygen by a similar amount. The combination of extreme surface composition and extreme rarity makes these stars interesting, and presents a challenge for the theory of stellar evolution.

The general properties of EHes have prompted suggestions of a connection to the cooler R Coronae Borealis stars (Schönberner 1975), which are better known for their spectacular and unpredictable light variability (Pigott & Englefield 1797). Figure 1 shows the distribution by surface gravity and effective temperature of several classes of hydrogen-

---

† The full version of this paper, including additional sections on the asteroarchæology and origin of EHes, is available online at *ArXiv: 1311.1635*.

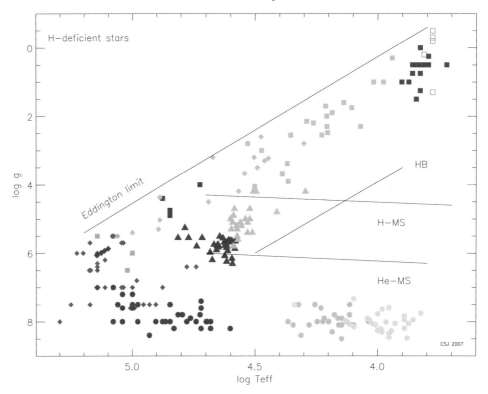

Figure 1. The $\log g - \log T_{\mathrm{eff}}$ diagram for several classes of low-mass hydrogen-deficient stars described in detail by Jeffery (2008a). EHe stars are denoted by squares (light grey, green online), as are RCB (dark grey, red online), HdC (open squares), HesdO^{+} (back, blue online) and O(He) (mid-grey, violet online) stars.

deficient star; stars falling on an imaginary line parallel to the Eddington limit would have the same luminosity-to-mass ratio, corresponding to the evolutionary path of a giant contracting to become a white dwarf. Links between non-variable hydrogen-deficient carbon (HdC), R Coronae Borealis (RCB), extreme helium (EHe), luminous helium-rich subdwarf O (HesdO^{+}) and O(He) stars have all been suggested at one time or another.

In order to establish the origin of EHes and related objects, temperatures and gravities are required. Additional observables are provided by the surface composition and by the fact that many EHes are pulsating. Pulsations provide an opportunity to make direct measurements of mass and radius. The talk of which this report is a summary reviewed the spectroscopic data on surface composition, summarized the principal theories for the evolutionary origin, and discussed the major pulsation properties of EHes. There is only space within these proceedings to present the latter, together with a synopsis of the extraordinary pulsating EHe V652 Herculis. The complete version of this paper is published online (Jeffery 2013). It includes a summary of the spectroscopic data, and a more detailed review of arguments which currently favour the origin of EHes as the product of a double white dwarf merger.

2. Pulsation

2.1. *Discovery and classification*

Landolt (1975) made the first detection of variability in an EHe, the hot star HD 160641 = V2076 Oph which showed a brightening by 0.1 mag over seven hours. The discovery of a

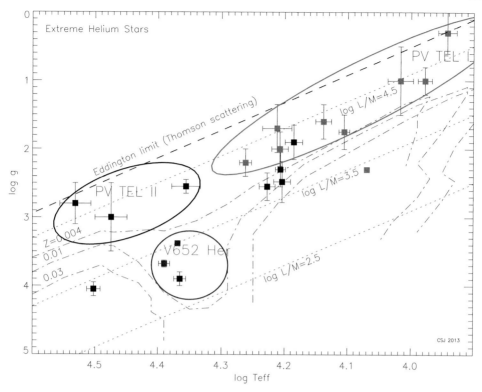

Figure 2. The $\log g - \log T_{\mathrm{eff}}$ diagram for EHe variables, adapted from Jeffery (2008c), including the position of the Eddington limit (assuming Thomson scattering: dashed) and the loci of stars with given luminosity-to-mass ratios (solar units: dotted). Stars above the boundaries shown for metallicities $Z = 0.004, 0.01, 0.03$ (dot-dash) are predicted to be unstable to pulsations (Jeffery & Saio 1999). Ellipses (coloured in electronic version) identify three groups of pulsating helium stars. In the electronic version, PV TEL I variables are shown in purple, PV TEL II variables in blue, and V652 Her variables in green. Non-variables are black.

0.108 d pulsation period in V652 Her (Landolt 1975) prompted searches for short-period variability in other EHes. Walker & Hill (1985) reported irregular small-amplitude variations of weeks to months in HD 168476 = PV Tel, thus codifying the PV Tel variables as a class. Bartolini *et al.* (1982) reported a short-period variation in BD + 10°2179 = DN Leo which could not be verified (Hill *et al.* 1984; Grauer *et al.* 1984). From 1983 – 1988, a St Andrews/SAAO campaign made discoveries of variability in six EHe stars, namely FQ Aqr, NO Ser, V2205 Oph, V2244 Oph and V1920 Cyg (Jeffery & Malaney 1985; Jeffery *et al.* 1986, 1985; Morrison 1987; Morrison & Willingale 1987). As with DN Leo, variability was not confirmed in Popper's star HD 124448 = V821 Cen (Jeffery & Lynas-Gray 1990). In the following decade, variability was discovered in V4732 Sgr, V5541 Sgr and V354 Nor (Lawson *et al.* 1993; Lawson & Kilkenny 1998), and most recently in the enigmatic MV Sgr (Percy & Fu 2012). Following a prediction by Saio (1995), Kilkenny & Koen (1995) discovered 0.1 d pulsations in BX Cir. The distribution of variable and non-variable EHes is shown in Fig. 2. Jeffery (2008c) gives a complete list of properties including original catalogue numbers, GCVS variable star designations, approximate periods, temperatures, and gravities.

The first discovery of the St Andrews/SAAO campaign concerned FQ Aqr, which showed a 21.2 d sinusoidal oscillation with an amplitude of 0.4 mag in V and 0.05 mag in $b - y$. Jeffery & Malaney (1985) associated this with pulsation. Together with periods for other EHes, a pulsation-period effective-temperature relation of the form $\Pi \propto T_{\text{eff}}^3$ was apparent. For stars of the same luminosity, this corresponds to the period mean-density relation for classical radial pulsators. On the basis of this inference, Saio & Jeffery (1988) calculated a series of linear non-adiabatic pulsation models for low-mass high-luminosity stars in the temperature range 7000 to 30000 K. They showed that, for sufficiently high values of L/M, opacity-driven strange-mode radial pulsations would be excited, consistent with the periods observed in the cooler PV Tel variables (Fig. 2). Jeffery (2008c) calls these "Type I" PV Tel variables.

At $T_{\text{eff}} > 20\,000$ K, both V2205 Oph and V2076 Oph were apparently multiperiodic on timescales longer than that consistent with a fundamental radial pulsation, and had been inferred to be g-mode non-radial pulsators (Jeffery et al. 1985; Lynas-Gray et al. 1987). Together with V5541 Sgr, Jeffery (2008c) calls these "Type II" PV Tel variables.

The short-period variability in the lower luminosity star V652 Her was not explained until the introduction of OPAL opacities (Rogers & Iglesias 1992), when Saio (1993) showed that, in the absence of hydrogen, The Z-bump opacity mechanism could easily drive radial pulsations with normal (solar-like) metallicities. Jeffery (2008c) puts BX Cir and V652 Her in the same class, but they would be better labelled V652 Her variables (Fig. 2).

2.2. *Pulsation properties*

What promised to be a class of simply periodic variables with a strict period-temperature relation and the possibility of measuring direct radii using Baade's method turned out to be a chimæra. A second season of observations of FQ Aqr did not reveal a unique period, with 21.5 d and 23.0 d being possible (Jeffery et al. 1986). Kilkenny et al. (1999) monitored both FQ Aqr and NO Ser for five consecutive seasons and, from the Fourier power spectrum, reported the "apparent presence of several periods but, if real, none seems to persist for more than one season." Data from SuperWASP (unpublished), and a wavelet analysis of the Kilkenny et al. data are similarly ambiguous. The only recurring signal occurs at a period of around 20 d, but is not coherent over a long period of time.

The absence of a regular period is not fatal for extracting stellar radii, so long as the oscillation is radial and angular radii and radial velocities can be measured simultaneously. Radial-velocity amplitudes of a few km s^{-1} were measured in PV Tel, FQ Aqr and V2244 Oph at the same time as angular radius measurements were obtained with *IUE* (Jeffery et al. 2001a). Relative phases were consistent with a radial pulsation. In two cases, the resulting radius and mass measurements were consistent with theoretical expectation, but the relative mass errors did not provide a strong constraint for establishing an evolutionary origin.

The aperiodicity of FQ Aqr is reflected in observations of other EHe stars (cf. Walker & Hill 1985). Wright et al. (2006) considerably extended the photometry and spectroscopy of V2076 Oph. Instead of recovering the periods of 0.7 and 1.1 d reported by Lynas-Gray et al. (1987), they reported that "conventional Fourier analysis . . . fails to reveal coherent frequencies" and suggest that the light curve "could be a result of random variations". On the other hand, the high-resolution spectrum of V2076 Oph shows prominent line-profile variability (LPV) on a timescale of a few hours which is symptomatic of a non-radial pulsation. Of note is that, although LPVs are visible in all He I lines, they are only seen in the He II 4686 Å line, and not in other He II lines (Fig. 2: Jeffery 2008b). Such intriguing results are almost impossible to interpret in isolation from photometric monitoring and over such a short time interval.

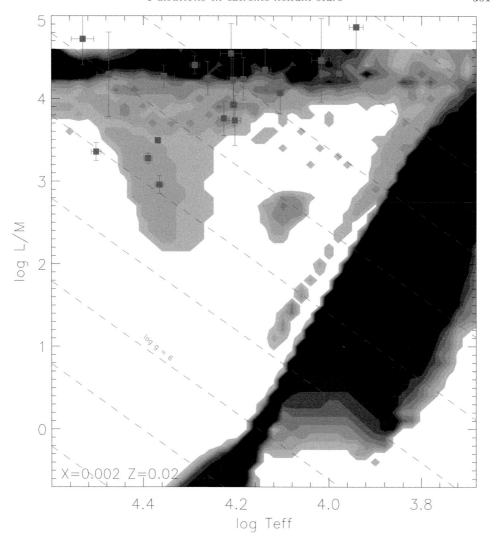

Figure 3. Contour plot showing the number of unstable radial modes with $n < 15$ in linear non-adiabatic pulsation analyses of hydrogen-deficient stellar envelopes. The ordinate is $\log L/M$ in solar units, and the abscissa is $\log T_{\mathrm{eff}}$. The plot is virtually invariant to the mass, at least in the range $0.2 - 1.0\ M_\odot$. The solid squares correspond to EHe stars. White means no unstable modes except for $\log L/M > 4.6$ where model envelopes are difficult to compute. Surface gravity contours for $\log g = 8, 7, 6, \ldots$ are represented as broken lines.

2.3. *Theoretical considerations*

Jeffery & Saio (2013) extended their investigations of pulsation stability as a function of hydrogen abundance (mass fraction: X) to a larger range of effective temperature (T_{eff}), luminosity-to-mass ratio (L/M) and X than before (Jeffery & Saio 1999). Models for $0.9 > X > 0.002$ were computed. Figure 2 shows the instability domain for $X = 0.002$ and solar metallicity. At large X, the classical Cepheid instability strip is clearly seen. At high L/M strange modes are excited. As X is reduced, the Z-bump instability finger starts to develop, and a strip of intermediate-order radial modes develops blueward of the classical instability strip. The strange modes are less sensitive to X, and at low X correspond to the observed locations of the variable EHe stars (Fig. 3). At low metallicity, the Z-bump finger is much diminished, and the lower edge of the strange-mode domain moves to higher L/M ratio (Fig. 2), so the absence of variability in metal-poor EHes is anticipated (Saio & Jeffery 1988; Saio 1993). With no hydrogen, the morphology of the classical strip again changes and other pockets of instability appear, in particular one at $T_{\rm eff} \approx 13\,000\,{\rm K}$ and $L/M \approx 500\,L_\odot/M_\odot$. It remains to be seen whether any real hydrogen-deficient stars can be identified with such models.

Strange-mode oscillations in stars with high L/M ratios have been identified for some time (Wood 1976). They appear theoretically in non-adiabatic pulsation analyses and have no corresponding modes in the adiabatic approximation. In particular, they appear to be associated with stellar envelopes where high opacities in the ionisation zones also lead to a density inversion – effectively creating a radiation-pressure dominated cavity in the stellar interior. They are discussed at greater length by, *inter alia*, Gautschy & Glatzel (1990) and Saio *et al.* (1998).

Evidence that non-radial oscillations may be present in the hottest EHes led Guzik *et al.* (2006) to find a large number of unstable opacity-driven g-modes in models for V2076 Oph. It will be interesting to learn whether extreme non-adiabacity and strange-mode characteristics or frequent switching between closely-spaced modes is responsible for the absence of long-lived coherent periods in these stars.

3. The shocking case of V652 Herculis

BD+13°3224 = V652 Her is an extreme helium star and a B-type giant discovered by Berger & Greenstein (1963). With a nitrogen-rich, carbon-poor surface, it pulsates with a period of 0.108 d which is decreasing at a substantial rate (cf. Kilkenny & Lynas-Gray 1982; Kilkenny *et al.* 2005). This period decrease implies a radius contraction that, together with a radius $2.31 \pm 0.02\,R_\odot$ and mass $0.59 \pm 0.18\,M_\odot$ (Jeffery *et al.* 2001b), in turn implies that V652 Her is evolving to become a hot subdwarf within $\leqslant 10^5$ yr (Jeffery 1984). Single-star models cannot explain the origin of V652 Her, which has an almost completely CNO-processed atmosphere (Jeffery *et al.* 1999). On the other hand, models for the post-merger evolution of two helium white dwarfs very successfully match nearly all of the observational properties (Saio & Jeffery 2000), as well as demonstrating that such mergers will become helium-rich hot subdwarfs (Zhang & Jeffery 2012b).

Saio (1993) demonstrated that Z-bump opacity instability drives fundamental-mode radial pulsations in V652 Her. The radial-velocity curve shows that, for nine tenths of the cycle, the surface layers are nearly ballistic (Hill et al. 1981; Lynas-Gray *et al.* 1984; Jeffery & Hill 1986). Close to minimum radius, the surface acceleration is so large that the atmosphere may be shocked (Jeffery *et al.* 2001b). Since hydrogen-deficient atmospheres are in general more transparent than hydrogen-rich ones, V652 Her provides a unique opportunity to study the dynamical behaviour of a pulsating atmosphere at greater depths (or densities) than is usually the case. Suitable observations could test non-linear hydrodynamic models for pulsation (Fadeyev & Lynas-Gray 1996; Montañés Rodríguez & Jeffery 2002). Jeffery *et al.* (2013a) present a more extended account.

High-speed spectroscopy of V652 Her was obtained with the Subaru High Dispersion Spectrograph. 2 nights of observations cover over six pulsation cycles with a temporal resolution of 174 seconds. These observations aim to identify structure in the line cores around minimum radius, to resolve the passage of the wave through the photosphere using ions with different ionisation potentials, to determine whether this generates a shock front, or whether the photosphere adjusts subsonically, and to establish how close the "free-fall" phase is to a true ballistic trajectory. The interpretation of the observations requires coupling a hydrodynamic model of the pulsation to a radiative transfer code, so that the emergent spectrum can be computed realistically. A summary of progress is given by Jeffery *et al.* (2013a,b).

4. Conclusion

Extreme helium stars form a group of some 15 low-mass supergiants of spectral types A and B. Their helium-dominated atmospheres are extremely hydrogen poor ($< 0.1\%$) and carbon-rich ($1 - 3\%$). Evidence from kinematics, surface composition and distribution in effective temperature and surface gravity points to a strong link with the cooler R CrB stars, and to their origin in the merger of a helium white dwarf with a carbon-oxygen white dwarf.

Most EHe stars are photometric variables with amplitudes around one tenth of a magnitude. Most are also radial-velocity variables with amplitudes of a few $\mathrm{km\,s^{-1}}$. The timescales of these variations as a function of effective temperature are consistent with the stellar dynamical timescales and hence with being due to pulsations. Theoretical models show most EHes to be unstable to opacity-driven radial pulsations. However, in all but two cases (V652 Her and BX Cir), the observed variations are *not strictly* periodic. Additional work is necessary to better characterise the periods and amplitudes and to obtain radii using the Baade-Wesselink method.

V652 Her and BX Cir are less luminous than the majority of EHes. V652 Her can be explained by the merger of two helium white dwarfs evolving to become a helium-rich subdwarf. The carbon-rich BX Cir is harder to explain, but could be similar. Both stars pulsate radially with periods of 0.1 d. The radial-velocity curve of V652 Her is nearly ballistic, with a very steep acceleration phase. New observations and models provide a unique dataset and toolkit for exploring the physics of these pulsations.

Bibliography

Bartolini, C., Bonifazi, A., Fusi Pecci, F., *et al.* 1982, *Ap&SS*, 83, 287
Berger, J. & Greenstein, J. L. 1963, *PASP*, 75, 336
Fadeyev, Y. A. & Lynas-Gray, A. E. 1996, *MNRAS*, 280, 427
Gautschy, A. & Glatzel, W. 1990, *MNRAS*, 245, 597
Grauer, A. D., Drilling, J. S., & Schönberner, D. 1984, *A&A*, 133, 285
Guzik, J. A., Peterson, B. R., Cox, A. N., & Bradley, P. A. 2006, *MemSAIt*, 77, 131
Hill, P. W., Kilkenny, D., Schönberner, D., & Walker, H. J. 1981, *MNRAS*, 197, 81
Hill, P. W., Lynas-Gray, A. E., & Kilkenny, D. 1984, *MNRAS*, 207, 823
Jeffery, C. S. 1984, *MNRAS*, 210, 731
Jeffery, C. S. 2008a, *ASP-CS*, 391, 3
Jeffery, C. S. 2008b, *ASP-CS*, 391, 53
Jeffery, C. S. 2008c, *IBVS*, 5817
Jeffery, C. S. 2013, arXiv: 1311.1635
Jeffery, C. S. & Hill, P. W. 1986, *MNRAS*, 221, 975
Jeffery, C. S. & Lynas-Gray, A. E. 1990, *MNRAS*, 242, 6
Jeffery, C. S., Malaney, R. A. 1985, *MNRAS*, 213, 61P
Jeffery, C. S. & Saio, H. 1999, *MNRAS*, 308, 221
Jeffery, C. S. & Saio, H. 2013, *MNRAS*, 435, 885
Jeffery, C. S., Skillen, I., Hill, P. W., Kilkenny, D., Malaney, R. A., & Morrison, K. 1985, *MNRAS*, 217, 701
Jeffery, C. S., Drilling, J. S., & Heber, U. 1987, *MNRAS*, 226, 317
Jeffery, C. S., Hill, P. W., & Morrison, K. 1986, in: K. Hunger, D. Schönberner, N. Kameswara Rao (eds.), *Hydrogen-Deficient Stars and Related Objects*, Proc. IAU Colloqium No. 87, Astrophysics and Space Sci. Library, Vol. 128 (Dordrecht: D. Reidel Publishing Co.) p. 95

Jeffery, C. S., Hill, P. W., & Heber, U. 1999, *A&A*, 346, 491

Jeffery, C. S., Starling, R. L. C., Hill, P. W., & Pollacco, D. 2001a, *MNRAS*, 321, 111

Jeffery, C. S., Woolf, V. M., & Pollacco, D. L. 2001b, *A&A*, 376, 497

Jeffery, C. S., Shibahashi, H., Kurtz, D., Elkin, V., Montañés-Rodríguez, P., & Saio, H. 2013a, in: H. Shibahashi, A. E. Lynas-Gray (eds.), *Progress in Physics of the Sun and Stars: A New Era in Helio- and Asteroseismology, ASP-CS*, in press

Jeffery, C. S., Shibahashi, H., Kurtz, D., Elkin, V., Montañés- Rodríguez, P., & Saio, H. 2013b, in: V. Van Grootel, E. Green, G. Fontaine, S. Charpinet (eds.), *Hot Subdwarf Stars and Related Objects, ASP-CS*, in press

Kilkenny, D. & Koen, C. 1995, *MNRAS*, 275, 327

Kilkenny, D. & Lynas-Gray, A. E. 1982, *MNRAS*, 198, 873

Kilkenny, D., Lawson, W. A., Marang, F., Roberts, G., & van Wyk, F., 1999, *MNRAS*, 305, 103

Kilkenny, D., Crause, L. A., & van Wyk, F. 2005, *MNRAS*, 361, 559

Landolt, A. U. 1975, *ApJ*, 196, 789

Lawson, W. A. & Kilkenny, D. 1998, *Observatory*, 118, 1

Lawson, W. A., Kilkenny, D., van Wyk, F., Marang, F., Pollard, K., & Ryder, S. D. 1993, *MNRAS*, 265, 351

Lynas-Gray, A. E., Schönberner, D., Hill, P. W., & Heber, U. 1984, *MNRAS*, 209, 387

Lynas-Gray, A. E., Kilkenny, D., Skillen, I., & Jeffery, C. S. 1987, *MNRAS*, 227, 1073

Montañés Rodríguez, P. & Jeffery, C. S. 2002, *MNRAS*, 384, 433

Morrison, K. 1987, *MNRAS*, 224, 1083

Morrison, K. & Willingale, G. P. H. 1987, *MNRAS*, 228, 819

Percy, J. R. & Fu, R. 2012, *Journal AAVSO*, 40, 900

Pigott, E. & Englefield, H. C. 1797, *Phil. Trans. Ser. I*, 87, 133

Popper, D. M. 1942, *PASP*, 54, 160

Rogers, F. J. & Iglesias, C. A. 1992, *ApJS*, 79, 507

Saio, H. 1993, *MNRAS*, 260, 465

Saio, H. 1995, *MNRAS*, 277, 1393

Saio, H. & Jeffery, C. S. 1988, *ApJ*, 328, 714

Saio, H. & Jeffery, C. S. 2000, *MNRAS*, 313, 671

Saio, H., Baker, N. H., & Gautschy, A. 1998, *MNRAS*, 294, 622

Schönberner, D. 1975, *A&A*, 44, 383

Walker, H. J. & Hill, P. W. 1985, *A&AS*, 61, 303

Wood, P. R. 1976, *MNRAS*, 174, 531

Wright, D. J., Lynas-Gray, A. E., Kilkenny, D., *et al.* 2006, *MNRAS*, 369, 2049

Zhang, X. & Jeffery, C. S. 2012, *MNRAS*, 419, 452

Precision Asteroseismology
Proceedings IAU Symposium No. 301, 2013
J. A. Guzik, W. J. Chaplin, G. Handler & A. Pigulski, eds.

© International Astronomical Union 2014
doi:10.1017/S174392131301449X

Reaching the 1% accuracy level on stellar mass and radius determinations from asteroseismology

V. Van Grootel[1]†, S. Charpinet[2,3], G. Fontaine[4], P. Brassard[4], and E. M. Green[5]

[1]Institut d'Astrophysique et de Géophysique de l'Université de Liège, Allée du 6 Août 17, B-4000 Liège, Belgium
email: valerie.vangrootel@ulg.ac.be

[2]Université de Toulouse, UPS-OMP, IRAP, Toulouse, France

[3]CNRS, IRAP, 14 avenue Edouard Belin, 31400 Toulouse, France

[4]Université de Montréal, Pavillon Roger-Gaudry, Département de Physique, CP 6128, Succ. Centre-Ville, Montréal QC, H3C 3J7, Canada

[5]Steward Observatory, University of Arizona, 933 North Cherry Avenue, Tucson, AZ, 85721, USA

Abstract. Asteroseismic modeling of subdwarf B (sdB) stars provides measurements of their fundamental parameters with a very good precision; in particular, the masses and radii determined from asteroseismology are found to typically reach a precision of 1% containing various uncertainties associated with their inner structure and the underlying microphysics (composition and transition zones profiles, nuclear reaction rates, etc.). Therefore, the question of the accuracy of the stellar parameters derived by asteroseismology is legitimate. We present here the seismic modeling of the pulsating sdB star in the eclipsing binary PG 1336−018, for which the mass and the radius are independently and precisely known from the modeling of the reflection/irradiation effect and the eclipses observed in the light curve. This allows us to quantitatively evaluate the reliability of the seismic method and test the impact of uncertainties in our stellar models on the derived parameters. We conclude that the sdB star parameters inferred from asteroseismology are precise, accurate, and robust against model uncertainties.

Keywords. stars: binaries: eclipsing, stars: subdwarfs, stars: oscillations (including pulsations), stars: individual: PG 1336−018 (NY Virginis).

1. Introduction to subdwarf B stars

Subdwarf B (sdB) stars are hot and compact objects with effective temperatures $T_{\rm eff}$ between 20 000 – 40 000 K and surface gravities $\log g$ in the range 5.0 to 6.2. They occupy the so-called extreme horizontal branch (EHB), burning helium in the core and having a very thin residual H-rich envelope. This extreme thinness of the hydrogen envelope is the original feature of sdB stars, and is the main difficulty when explaining the formation of such objects. They are thought to be post-RGB (Red Giant Branch) stars that went though the He-flash and that have lost most of their envelope through binary interaction. While about half of sdB stars reside in binaries with a stellar companion, the recent discoveries of planets around single sdB stars (e.g. Charpinet *et al.* 2011) also support the idea that planets could influence the evolution of their host star, by triggering the mass loss necessary for the formation of an sdB star.

† Chargé de recherches, Fonds de la Recherche Scientifique, FNRS, rue d'Egmont 5, B-1000 Bruxelles, Belgium.

The sdB stars exhibit pulsation instabilities driving both acoustic modes of a few minutes and gravity modes with 1–4 h periods. To date, 15 sdB stars have been modeled by asteroseismology (Fontaine *et al.* 2012), allowing for seismic determinations of their global and structural parameters (total mass M_*, surface gravity log g, radius R_*, thickness of the H-rich envelope, mass and composition of the He-burning core, etc.). The mass distribution of sdB stars, although still based on small number statistics, is consistent with the idea that sdB stars are post-RGB stars (Van Grootel *et al.* 2013a). The masses, surface gravities and radii of sdB stars are typically determined from asteroseismology with a very good precision of ∼1%, 0.1%, and 0.6%, respectively. We demonstrate here that these numbers are reliable, and the sdB parameters inferred from asteroseismology are precise, accurate, and robust against model uncertainties. We used for this purpose PG 1336−018, one of the only two known eclipsing binaries where the sdB component is a pulsating sdB star. This allows us to compare results obtained from the two independent techniques of asteroseismology and orbital light curve and eclipse modeling.

2. PG 1336−018, the Rosetta stone of sdB asteroseismology

PG 1336−018 (NY Virginis) is an eclipsing binary made of an sdB pulsator and an M dwarf. Orbital solutions, including the mass and radius of the sdB component, are provided by Vučković *et al.* (2007) from eclipses and light curve modeling. From the pulsational point of view, the sdB component of PG 1336−018 exhibits 28 p-mode periodicities in the 96 to 205 s range (Kilkenny *et al.* 2003). The atmospheric parameters of the sdB component of PG 1336−018 are (Van Grootel *et al.* 2013b): $T_{\rm eff} = 32807 \pm 82$ K, log $g = 5.771 \pm 0.015$, and log N(He)/N(H) = −2.918 ± 0.089.

3. Method for asteroseismology of hot subdwarfs

The forward modeling approach developed to perform objective asteroseismic modeling of subdwarf pulsators has been described in detail by Charpinet *et al.* (2008). We fit directly and simultaneously all observed independent pulsation periods with theoretical ones calculated from sdB models, in order to minimize a merit function defined by

$$S^2 = \sum_{i=1}^{N_{\rm obs}} \left(\frac{P_{\rm obs}^i - P_{\rm th}^i}{\sigma_i} \right)^2, \tag{3.1}$$

where $N_{\rm obs}$ is the number of observed independent periodicities and σ_i a weight, here equal to σ_d, the density of theoretical modes in the considered range. The method performs a double-optimization procedure in order to find the minima of the merit function, under the external constraints from spectroscopy and from mode identification (if available). The optimal model(s) define the asteroseismic solution(s). Error estimates are provided using probability distributions defined for each input parameter of a stellar model, from a likelihood function defined by

$$\mathcal{L}(a_1, a_2, a_3, a_4) \propto e^{-\frac{1}{2}S^2}. \tag{3.2}$$

From there, the probability density function for parameter a_1 (for example the mass) is

$$\mathcal{P}(a_1)\mathrm{d}a_1 \propto \mathrm{d}a_1 \iiint \mathcal{L}(a_1, a_2, a_3, a_4)\mathrm{d}a_2\mathrm{d}a_3\mathrm{d}a_4. \tag{3.3}$$

This density of probability function is normalized assuming that the probability of the value of a_1 to be in the specified range is equal to 1. Finally, the errors associated to each parameter are defined by the 1σ range of the corresponding probability distribution.

4. Seismic modeling of PG 1336–018

The optimization procedure is launched in a vast parameter space where sdB stars are found (see Van Grootel *et al.* 2013b for details). The parameters of the optimal model are presented in Table 1. The quoted errors are the 1σ range of the corresponding probability density function, shown on Fig. 1 for the case of the stellar mass. Similar probability functions are obtained for the other model parameters. The stellar parameters inferred from asteroseismology are found to be remarkably consistent, within the 1σ errors of each method, with both the preferred orbital solution obtained from the binary light curve modeling (model II, see Vučković *et al.* 2007) and the spectroscopic estimate for the surface gravity of the star. We can therefore affirm that the stellar parameters determined from asteroseismology are accurate. We can also note that the optimal model is not an outlier of the probability distribution, but is well within the 1σ range of each distribution (see Fig. 1 for the mass). In summary, stellar models for asteroseismology of sdB stars allow for both precise and accurate determinations of the stellar parameters, in particular mass and radius. But we can wonder how the model uncertainties impact on this result. There are indeed three main sources of uncertainties in sdB models:

• The nonuniform envelope iron profile. In our sdB models, the equilibrium is assumed between radiative levitation of iron (in the H-rich envelope) and gravitational settling, ignoring competing processes such as stellar winds, thermohaline convection, etc.;

• The core/envelope transition profile, not smoothed by diffusion in our models;

• The He-burning nuclear reaction rates, from uncertainties in nuclear physics.

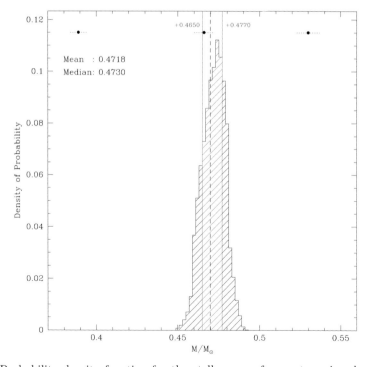

Figure 1. Probability density function for the stellar mass from asteroseismology. The filled circles with the dotted lines are the three orbital solutions for the mass of the sdB component proposed by Vučković *et al.* (2007) with their 1σ uncertainties. The red hatched part between the two vertical solid red lines defines the 1σ range, containing 68.3% of the mass distribution. The blue vertical dashed line indicates the mass of the optimal model solution of lowest S^2-value.

Table 1. Structural parameters of PG 1336−018 derived from asteroseismology, spectroscopy, and orbital light curve analysis.

Quantity	Asteroseismology	Spectroscopy	Orbital light curve modeling		
	This study		Vučković *et al.* (2007)		
			Model I	Model II	Model III
$M_*/M_\odot$	0.471 ± 0.006 (1.3%)	...	0.389 ± 0.005	0.466 ± 0.006	0.530 ± 0.007
$R_*/R_\odot$	0.1474 ± 0.0009 (0.6%)	...	0.14 ± 0.01	0.15 ± 0.01	0.15 ± 0.01
$\log g$	5.775 ± 0.007 (0.1%)	5.771 ± 0.015	5.74 ± 0.05	5.77 ± 0.06	5.79 ± 0.07
$T_{\rm eff}$ (K)	32850 ± 175 (0.5%)	32 807 ± 82	...	...	...
$\log(M_{\rm env}/M_*)$	−3.83 ± 0.06 (1.6%)	...	...	...	...
$\log(1 - M_{\rm core}/M_*)$	unconstrained	...	...	...	...
$X_{\rm core}$(C+O)	0.58 ± 0.06 (10%)	...	...	...	...
$\log(L/L_\odot)$	22.9 ± 0.6 (2.6%)	23.3 ± 1.5	...	...	...

In order to quantitatively appreciate the impact of these uncertainties on the stellar parameters, we built new stellar models with modified physics and carried out again seismic analyses of PG 1336−018 following the same procedure as before. First, we built stellar models with our standard equilibrium envelope iron profile divided by two and by four, and we also built models with a uniform solar abundance. We therefore re-did three seismic analyses of PG 1336−018 with the modified models. As a result, despite significant changes in the iron abundance profiles, the derived parameters are mostly unaffected (e.g. the mass) or only subject to very small systematic drifts compared to the standard seismic solution. Secondly, for the core/envelope transition profile, we built modified models with smoothed transition profiles, and carried out again a seismic analysis. No significant drift on stellar mass, radius and $\log g$ was observed. Finally, we multiplied by two the $^{12}C(\alpha,\gamma)^{16}O$ rate and increased by 10% the triple-α rate. The seismic analysis with these modified models led to almost unchanged stellar parameters. Details on these experiments and on sdB models can be found in Van Grootel *et al.* (2013b).

5. Conclusion

We presented here the seismic modeling of the p-mode sdB pulsator in the eclipsing system PG 1336−018. This very rare configuration allowed us to test some of the stellar parameters inferred from asteroseismology with the values obtained independently from the orbital light curve analysis of Vučković *et al.* (2007). We also tested the impact of the uncertainties of the input physics in the stellar models. We conclude that seismic parameters determined from asteroseismology for sdB stars are both precise, accurate, and robust against model uncertainties. We can indeed achieve ∼1% accuracy for mass and radius determinations from asteroseismology. We also demonstrated that the best-fit (optimal) model is not an outlier of the statistical distribution of potential solutions, and can therefore safely be considered as the most representative model of the star.

References

Charpinet, S., Van Grootel, V., Reese, D., *et al.* 2008, *A&A*, 489, 377
Charpinet, S., Fontaine, G., Brassard, P. *et al.* 2011, *Nature*, 480, 496
Fontaine, G., Brassard, P., Charpinet, S., *et al.* 2012, *A&A*, 539, 12
Kilkenny, D., Reed, M. D., O'Donoghue, D., *et al.* 2003, *MNRAS*, 345, 834
Van Grootel, V., Fontaine, G., Charpinet, S., *et al.* 2013a, *EPJ Web of Conferences*, 43, 04007
Van Grootel, V., Charpinet, S., Fontaine, G., *et al.* 2013b, *A&A*, 553, 97
Vučković, M., Aerts, C., Østensen, R., *et al.* 2007, *A&A*, 471, 605

Precision Asteroseismology
Proceedings IAU Symposium No. 301, 2013
J. A. Guzik, W. J. Chaplin, G. Handler & A. Pigulski, eds.

© International Astronomical Union 2014
doi:10.1017/S1743921313014506

Being rich helps –
the case of the sdBV KIC 10670103

Jurek Krzesinski and Szymon Bachulski

Pedagogical University of Cracow,
ul. Podchorazych 2, 30-084 Cracow, Poland
email: jk@astro.as.up.krakow.pl

Abstract. We present a study of KIC 10670103, a pulsating hot subdwarf in the *Kepler* field. By means of Fourier analysis, we investigate periodic signals associated with pulsations. Using asymptotic relationships and rotational multiplets we identify degrees of modes. The Fourier spectrum appears to be rich in $l = 1$ and $l = 2$ multiplets allowing derivation of a ∼90-day rotation period of the star from rotational splittings. Comparing the identified gravity-mode period spacing pattern with theoretical models we show that KIC 10670103 has to be a thick-envelope sdBV.

Keywords. stars: subdwarfs, oscillations, rotation

1. Introduction

Pulsating stars are a rich source of knowledge about stellar interiors and evolution. Thanks to asteroseismology, we have the tools to study their internal compositions, envelope masses, rotation and other parameters by interpreting their frequency spectra (FS). In this work, we present the properties of a pulsating hot subdwarf (sdBV) star, KIC 10670103, in the *Kepler* field (Borucki *et al.* 2010, Koch *et al.* 2010, Jenkins *et al.* 2010).

The sdBVs are extended horizontal branch stars with masses of about $0.5\,M_\odot$ and effective temperatures between 20 000 and 40 000 K (Saffer *et al.* 1994). Having lost too much of their hydrogen envelope mass prior to the helium flash, they cannot sustain H-shell burning. Therefore, they will not become asymptotic giant branch stars; instead, they will simply cool down as white dwarfs.

The sdBVs come in two brands: short-period (1.5 – 5 minutes) p-mode, V361 Hya-type pulsators (Kilkenny *et al.* 1997) and long-period (45 – 120 minutes) g-mode, V1093 Her-type pulsators (Green *et al.* 2003). There are also a few hybrid sdBV stars known; they show both types of modes simultaneously.

KIC 10670103 is a g-mode pulsator, but has some extreme properties which make it a unique star for seismic analysis. It has the longest observed pulsation periods (up to 4.5 hours) and the lowest effective temperature, $T_{\rm eff} \approx 20\,900$ K (Reed *et al.* 2010, 2011) of all sdBV stars. Using one month of data from the *Kepler* exploratory phase, Reed *et al.* (2010) identified 28 modes, which made KIC 10670103 one of the richest pulsators among the sdBV stars. Since then, *Kepler* has observed KIC 10670103 for over 2.5 years and it is time to see what can be done with 30 months (Q5 – Q14) of new data. Figure 1 presents the pulsation pattern of the star, i.e. the FS of 928 days (∼30 months) of KIC 10670103 time-series photometry. The pulsation pattern is dominated by two large-amplitude modes with frequencies between 130 and 150 μHz. These are responsible for the ∼2.5-day beat present in the light curve of the star.

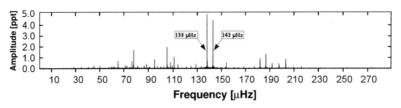

Figure 1. The FS of the 2.5-year *Kepler* data of KIC 10670103. The pulsation pattern of the star is dominated by two large-amplitude modes (marked with arrows).

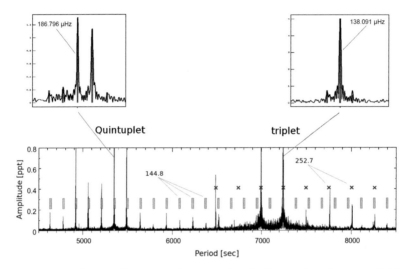

Figure 2. Mode identification from period spacing (at the bottom) and from multiplets (upper part of the figure). Vertical rectangles mark positions of $l = 2$ modes while crosses denote $l = 1$ modes. The numbers 144.8 and 252.7 denote the period spacings (in s) for these modes.

2. Mode identification

Having found the frequencies of the modes, one can perform mode identification in either the period or frequency domains. For g modes, the former allows identification of the degree l using period spacings in the asymptotic limit (Unno *et al.* 1979, Smeyers & Tassoul 1987). A neat and brief description of the procedure can be found in Reed *et al.* (2011).

As one can see in Fig. 2 (bottom panel), there are two distinct spacings in the FS. The shorter, 144.8 s (marked with a set of vertical rectangles) corresponds to the $l = 2$ mode period spacing. The longer one, 252.7 s (marked by a set of crosses), corresponds to $l = 1$ modes. In Fig. 2, some crosses appear at the same positions as rectangles and one may not be able to distinguish between $l = 1$ and $l = 2$ modes, unless the identification can be done via multiplets.

An example of such a mode can be found near the period of 7250 s. A closer look at its structure (upper right panel of Fig. 2) allows identification of the mode as $l = 1$ since a triplet is clearly visible. The decision whether it is indeed a triplet or in fact a partial quintuplet can be made on the basis of the spacings of the components of the multiplet. Multiplets are created when the rotation of a star lifts the azimuthal node degeneracy, and the spacing between multiplet components depends on the angular velocity of the star and on l. Therefore, the spacing can be used not only to identify the modes/multiplets, but also to determine the rotation period of the star. In the case of KIC 10670103, the spacing for quintuplets ($l = 2$ modes) is 0.107 μHz, while it is 0.069 μHz for triplets

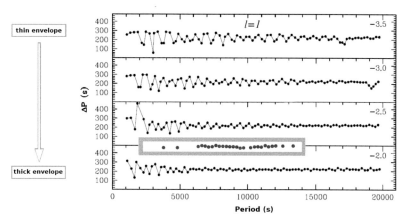

Figure 3. A part of Fig. 4 by Charpinet *et al.* (2013) presenting the deviations of trapped gravity $l = 1$ mode periods from equal period spacing (ΔP) for different thicknesses of the hydrogen envelope. The envelope mass ($\log q(\mathrm{H}) = \log M_{\mathrm{env}}/M_*$) is shown in the right upper corner of each panel. The rectangle inserted between two bottom panels, with a sequence of solid circles, represents $l = 1$ modes identified in this work. The sdBV model with the thickest envelope fits the trapping pattern of KIC 10670103 best.

($l = 1$ modes). The rotation periods derived from the triplet and quintuplet spacings are equal to $\sim$84 and $\sim$90 days, respectively.

Using both the period spacing and multiplets, one can double check mode identification whenever possible and prepare a diagram of period spacings vs. period, to look for a gravity mode trapping pattern. It is crucial to identify long continuous chains of modes, since some conclusions about the stellar structure can be made right from the scatter of period spacing diagrams, such as those presented by Charpinet *et al.* (2013) (their Fig. 4). It appears that the chemical transition between the envelope and mantle of an sdBV star results in a distinct pattern of trapped modes in the period spacing diagrams. As a result, the mean period spacing between consecutive modes of the same degree l is cyclically perturbed. For thin envelopes, the deviation from equal period spacing is severe for all periods. For thick envelopes, mode trapping causes large scatter of period spacings for periods shorter than 5000 s and becomes small for longer periods, where the period spacing is virtually constant. In Fig. 3 we present a partial set of models from Charpinet *et al.* (2013) with an inset of the observed g-mode trapping pattern for KIC 10670103. The observed pattern is best fit by the sdBV model with the thickest envelope.

3. Summary

Thanks to the long time-series photometry from *Kepler*, we were able to extract over 300 frequencies from the light curve of KIC 10670103. Using period spacings and multiplet patterns we identified $l = 1$ and $l = 2$ modes of KIC 10670103 and we used them to determine the stellar rotation period and the mass of its hydrogen envelope. It appears that the star is a very slow rotator with a rotation period of about 90 days. Since some white dwarfs are also slow rotators (Fontaine, these proceedings), slowly rotating sdBs may be an evolutionary link between red giants and slowly rotating white dwarfs.

Comparing our period spacing pattern of $l = 1$ (and $l = 2$ modes, not shown here) with Charpinet *et al.* (2013) (their Fig. 3), we conclude that the weak gravity mode trapping observed in KIC 10670103 is due to the thick envelope of the star. The mass of the hydrogen envelope defined as $\log q(\mathrm{H}) = \log(M_{\mathrm{env}}/M_*)$ was estimated to be between -2.5

J. Krzesinski & S. Bachulski

and −2. This probably makes KIC 10670103 the star with the most massive hydrogen envelope among all known sdBVs. This conclusion requires verification against a model computed specifically for KIC 10670103 but the grid of models presented by Charpinet *et al.* (2013) allows for such a preliminary conclusion.

Acknowledgement

This project was supported by Polish National Science Center grant 2011/03/D/ST9/01914.

References

Borucki, W. J., Koch, D., Basri, G., *et al.* 2010, *Science*, 327, 977

Charpinet, S., Van Grootel, V., Brassard, P., Fontaine, G., Green, E. M., & Randall, S. K. 2013, in: J. Montalbán, A. Noels, & V. Van Grootel (eds.), *Ageing Low Mass Stars: From Red Giants to White Dwarfs*, European Physical Journal Web of Conferences, 43, 4005

Green, E. M., Fontaine, G., Reed, M. D., *et al.* 2003, *ApJ*, 583, L31

Jenkins, J., Caldwell, D. A., Chandrasekaran, H., *et al.* 2010, *ApJ*, 713, L87

Kilkenny, D., Koen, C., O'Donoghue, D., & Stobie, R. S. 1997, *MNRAS*, 285, 640

Koch D. G., Borucki, W. J. Basri, G., *et al.* 2010, *ApJ*, 713, L79

Reed, M. D., Kawaler, S. D., Østensen, R. H., *et al.* 2010, *MNRAS*, 409, 1496

Reed, M. D., Baran, A., Quint, A. C., *et al.* 2011, *MNRAS*, 414, 2885

Safer, R. A., Bergeron, P., Koester, D., & Liebert, J. 1994, *ApJ*, 432, 351

Smeyers, P. & Tassoul, M. 1987, *ApJS*, 65, 429

Unno, W., Osaki, Y., Ando, H., & Shibahashi, H. 1979, *Nonradial Oscillations of Stars* (Tokyo: Univ. of Tokyo Press)

Precision Asteroseismology
Proceedings IAU Symposium No. 301, 2013
J. A. Guzik, W. J. Chaplin, G. Handler & A. Pigulski, eds.

© International Astronomical Union 2014
doi:10.1017/S1743921313014518

Pulsations in hot supergiants

Melanie Godart[1], Arlette Grotsch-Noels[2] and Marc-Antoine Dupret[2]

[1] Dept. of Astronomy, University of Tokyo, Japan
email: melanie.godart@gmail.com

[2] Dept. of Astrophysics, Geophysics and Oceanography, University of Liège, Belgium

Abstract. Massive stars are the cosmic engines that shape and drive our Universe. Many issues such as their formation, their stability and the mass loss effects, are far from being completely understood. Recent ground-based and space observations have shown pulsations in massive MS and post-MS stars, such as acoustic and gravity modes excited by the κ-mechanism and even solar-like oscillations. Theoretical studies emphasized the presence of strange modes in massive models, and recent theoretical analyses have shown that hot supergiants can pulsate in oscillatory convective modes. We review the instability domains of massive stars as well as their excitation mechanisms and present the latest results.

Keywords. stars: early-type, stars: oscillations, supergiants, Wolf-Rayet

1. Introduction

Hot supergiants comprise massive stars in different evolutionary states, including O and B supergiants, Luminous Blue Variables (LBV) and Wolf-Rayet (WR) stars. Periodic variability has been detected in such stars with a possible link to pulsations. This review summarizes the theoretical efforts in the field of massive star pulsations. Massive stars pulsate in *regular* and *strange* modes. Regular modes are excited with periods around hours for the β Cephei type modes (low-order p and g modes) and with longer periods ($\sim$ days) for Slowly Pulsating B (SPB) type modes (high-order g modes). Strange modes are excited in a larger range of periods, of the order of hours or days. Several driving mechanisms can be responsible for the pulsations: the ϵ-mechanism, the κ-mechanism and strange-mode instabilities. Moreover, Belkacem *et al.* (2009) reported the presence of solar-like oscillations in β Cephei stars. Both the convective region associated with the metal (Z) opacity bump and the convective core are found to efficiently drive acoustic modes in $10\,M_\odot$ models (Belkacem *et al.* 2010). Their results have been reinforced by observations since Degroote *et al.* (2010) have detected solar-like oscillations in a massive main-sequence (MS) star observed by *CoRoT*.

2. Regular modes in massive stars

Though the instabilities due to the ϵ-mechanism have been extensively studied in the framework of massive stars (Simon & Stothers 1969, Ziebarth 1970), this mechanism was later neglected in favor of stronger instabilities produced by the κ-mechanism and strange-mode instabilities. However, recently 19 significant frequencies have been observed in the B supergiant Rigel (Moravveji *et al.* 2012a), with periods ranging from 1 to 75 days. Moravveji *et al.* (2012b) suggest that these modes are excited by the ϵ-mechanism occurring in the hydrogen-burning shell.

Less massive stars such as SPB and β Cephei stars are pulsating due to the κ-mechanism activated by the Z opacity bump at $\log T \sim 5.2$. Their instability domains are well known to this date thanks to the works of Pamyatnykh (1999) and later with the

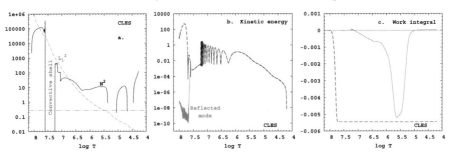

Figure 1. *a*) Dimensionless Brunt-Väisälä (solid line) and Lamb frequencies (dashed line) as a function of $\log T$ in an $18\,M_\odot$ post-MS model. *b*) Kinetic energy of 2 modes of $\omega = 0.52$. *c*) Work integral for the same modes. The reflected mode is excited (solid line) while the mode entering the radiative core is damped (dashed line) (Godart 2011).

new OP opacities (Seaton *et al.* 1994) and the new abundances (Asplund *et al.* 2005), thanks to Miglio (2007), Miglio *et al.* (2007), and Pamyatnykh (2007). The computations were limited to the MS and/or to the lower masses (up to $20\,M_\odot$). However, the instability domain of these stars could be extended to larger masses (e.g. Kiriakidis *et al.* 1992, Moskalik & Dziembowski 1992). One reason for the limitation of the computation is perhaps the following: Post-MS massive stars consist of a dense pure helium core surrounded by a large hydrogen envelope. The high density contrast between both regions causes a large Brunt-Väisälä frequency, producing (1) a large number of oscillations around the large Brunt-Väisälä frequency, generating possible numerical issues; and (2) a strong radiative damping in the very dense core, which should cancel the occurrence of g modes (see e.g. Eq. 14 of Godart *et al.* 2009). However, in the mid-2000s, new observations contradicted these assumptions. Indeed, periodic variabilities were detected in B supergiant stars (Saio *et al.* 2006, Lefever *et al.* 2007) attributed to non-radial g mode pulsations. In addition, Saio *et al.* (2006) provided an explanation for the presence of g modes in such post-MS massive stars: an intermediate convective zone (ICZ) surrounds the pure helium core and prevents the g modes from entering the core. The authors showed that in that case, the κ-mechanism activated by the Z opacity bump is sufficient to excite the g modes. This effect was later investigated by Gautschy *et al.* (2009) and Godart *et al.* (2009). These authors showed that the presence of an ICZ depends on the past history of the star, since it is produced thanks to the occurrence of a region of neutrality of the temperature gradients (semi-convection) during the MS. Moreover, some processes, such as mass loss and overshooting or the Ledoux criterion for convection, are able to prevent the formation of the semi-convective region on the MS and, therefore, to suppress the ICZ and the occurrence of g modes on the post-MS (Godart *et al.* 2009, Lebreton *et al.* 2009). We show in Figs. 1a-c the study (propagation diagram, kinetic energy and work integral) of two modes of close frequencies for an $18\,M_\odot$ post-MS model. The stable mode crosses the convective barrier and presents a high amplitude in the stellar core, illustrated by a large kinetic energy (Fig. 1b, black dashed line). Hence, this mode suffers a strong damping in the core ($\log T \sim 7.8$), much more efficient than the κ-mechanism in the superficial layers (see Fig. 1c, representing the work integral of the mode). On the other hand, the unstable mode (grey solid line) is reflected on the convective shield: the amplitude is very small in the stellar core. This mode is excited by the κ-mechanism at $\log T \sim 5.3$ (see the work integral).

Figure 2 shows the instability domains for massive stars of Saio (2011) (left panel) and Godart *et al.* (2011) (right panel). Low-order p and strange modes (black dashed lines)

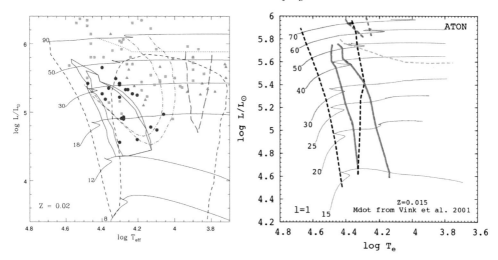

Figure 2. *Left:* Instability domains for various types of modes (Saio 2011, fig. 1, by permission of Oxford University Press on behalf of The Royal Astronomical Society). The low-order p and strange modes are shown in black dashed lines, the boundaries becoming horizontal due to strange-mode instability. Solid red and blue lines stand for the g modes ($l = 1$ and $l = 2$, respectively). *Right:* κ-mechanism instability domains in the HR diagram (Godart 2011). The $l = 1$ instability domains for low-order p and g modes (15 to 30 $M_\odot$) and for low-order p and g modes and adiabatic strange modes (40 to 70 $M_\odot$) are shown in the black dashed lines. Solid red lines stand for the high-order g modes.

are found to be excited on the MS and the post-MS (the strange modes are excited in the higher luminosity models only), while high-order g modes are excited in the post-MS phase. Recently, Ostrowski *et al.* (2012) and Daszyńska-Daszkiewicz *et al.* (2013) reported that the ICZ is not a necessary condition for the g mode detection. They point out that a minimum in the Brunt-Väisälä frequency is sufficient for forcing the trapping of the g mode outside the radiative damping core. The computations were performed with the Ledoux criterion for convection and the mean molecular weight gradient prevents therefore the ICZ to form in the 16 $M_\odot$ model. The minimum in the Brunt-Väisälä frequency is associated with the change of the temperature gradient from the adiabatic to the radiative one above the semi-convective region. Note that for larger masses, the ICZ should appear, though it will stay smaller than when taking the Schwarzschild's convection treatment.

3. Strange modes

In addition to the usual spectrum of regular modes, other instabilities are also responsible for the pulsations in massive stars. The term *strange modes* was put forward by Cox *et al.* (1980) because of the *strange* behaviour of these modes in a modal diagram: it consists in the plot of the mode frequency (normalized by the dynamical time $\sqrt{R^3/GM}$) as a function of a stellar parameter, e.g. $T_{\rm eff}$ or M, for a homogeneous set of models, e.g. ZAMS models or pure He stars. The variation of the regular p-mode frequency scales as $\sqrt{GM/R^3}$. Nonetheless, due to their confinement in a small cavity, strange modes present a very different behaviour. Indeed, the dimensionless frequency, ω, depends on the size of the mode propagation cavity: $\omega \sim n\pi \sqrt{(R^3/GM)}(\int_{\rm cavity} c_s^{-1} dr)$, which means that the dimensionless frequency of a strange mode decreases compared to that of a regular mode if the size of the cavity increases more rapidly than the radius of

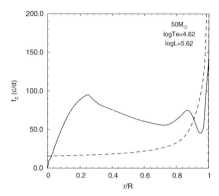

Figure 3. Critical frequency of radial modes (solid line) in $50\,M_\odot$ MS model. The dashed line stands for f_t, the inverse of the time needed for an acoustic wave to travel from r to the surface and back (Godart 2011).

the star. This is the case on the MS, i.e. with decreasing $T_{\rm eff}$ or increasing age, or for a homogeneous set of models of different masses. Figure 2 of Glatzel & Kiriakidis (1993a) shows a modal diagram in which adiabatic and non-adiabatic strange modes (see below for the definitions) are visible, with decreasing frequencies. Reviews of strange modes may be found in, e.g., Saio *et al.* (1998), Noels *et al.* (2008), Glatzel (2009), Saio (2009) and Godart (2011). Strange modes are found to be present in various types of stars and mass ranges, such as the pure He stars, low mass supergiants, massive main-sequence stars, evolved massive stars, central stars of planetary nebulae, classical Cepheids, and Wolf-Rayet stars. The phenomenon of strange modes is not even restricted to stellar objects: accretion disks can also be unstable due to strange modes (Glatzel & Mehren 1996).

Adiabatic and nonadiabatic strange modes: By limiting our discussion to the radial adiabatic strange modes (see below), we can easily understand their origin. Figure 3 shows the angular critical frequency for radial modes, roughly given by $c_s/2H_p$, as a function of the stellar radius. In low- and intermediate-mass stars, the cavity created by this critical frequency extends over the whole star. However, with increasing mass, a local minimum appears in the superficial layers, and therefore a new propagating cavity. This cavity (which appears for masses above $\sim 40\,M_\odot$) comprises a new set of modes: the strange modes. A large contribution of radiation pressure to the total pressure in addition to the proximity of opacity bumps are necessary to produce the cavity. Indeed, these conditions create a density and hence a sound-speed inversion. The information which could be obtained from strange modes is thus related to that superficial cavity, in particular to the sound speed and thus the temperature and the density profiles. For more details and for the determination of this critical frequency see, e.g., Saio *et al.* (1998) and Godart (2011). Though the critical frequency for non-adiabatic strange modes is different, these modes also propagate in a narrow superficial cavity. The confinement of the modes leads to two major consequences: (1) the dimensionless angular frequencies decrease with increasing mass of the star, explaining the peculiar behaviour of strange modes in a modal diagram, and (2) the modes have a small kinetic energy and therefore a large imaginary part of the eigenfrequency. The imaginary part of an adiabatic strange mode (labelled S1u) is visible in fig. 2 of Glatzel & Kiriakidis (1993a): it is a few orders of magnitude larger than the imaginary part of regular modes.

Saio *et al.* (1998) show in their fig. 2 (left panel) two modal diagrams for a set of ZAMS homogeneous models: the adiabatic dimensionless frequency distribution is given

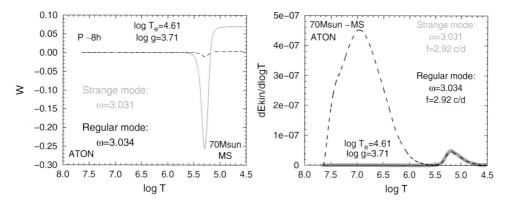

Figure 4. *Left:* Work integral for two unstable modes: regular mode (dashed line) and strange mode (solid line). The imaginary part of the strange mode is much larger because of the mode trapping. *Right:* Kinetic energy (surface under the curve) of the same modes. The amplitude confinement of the strange mode is clearly visible (thick grey line) (from Godart 2011).

as a function of the mass in the top panel while the real part of the non-adiabatic frequency is given in the middle panel. A quick look at the top panel allows one to notice the presence of a strange mode (decreasing frequency). Since this panel displays the adiabatic frequencies, *the strange modes have an adiabatic counterpart*. Regular modes and adiabatic strange modes interact with an avoided crossing.

On the other hand, the middle panel of the same figure presents one or several additional spectra superimposed on the normal mode spectrum: the *non-adiabatic strange modes*. These modes do not just cross the regular ones but an interaction may occur with an accompanying *unfolding* of the crossings. The modes coalesce, forcing one of the participating modes to become overstable and the other one to become stable. The real parts of the frequencies of the two interacting modes are essentially identical while their imaginary parts are approximately of the same magnitude but with opposite signs.

Excitation mechanism: Two mechanisms are responsible for the strange mode pulsations: the enhanced κ-mechanism and the strange-mode instabilities (SMI). The κ-mechanism (activated by the Z or the He opacity bump) is enhanced by the confinement of the mode in a small region. This driving mechanism accounts for the strange modes with an adiabatic counterpart. The work integral (positive at the surface for unstable modes) and the kinetic energy of a regular and a strange mode are shown in Fig. 4: both modes are excited by the κ-mechanism due to the Z opacity bump. Figure 4 (right panel) emphasizes the confinement of the strange mode in a superficial region which induces a large inertia and hence a larger growth rate for the strange mode ($\eta = W\sigma/[4\pi E_k]$).

The SMI derives directly from the definition of the strange mode without adiabatic counterpart: one component of the pair is always unstable. Hence, if the conditions for the existence of strange modes are fulfilled (high L/M, high radiation pressure), they can always be excited. Glatzel *et al.* (1993) represent in their fig. 2 the real (top panel) and imaginary (bottom panel) part of the frequencies for a set of pure helium stars. It is evident that the coalescence of two modes coincides with the appearance of two nearly complex conjugate eigenfrequencies.

Prospects: With the addition of a model atmosphere, for example the FASTWIND model atmosphere (Santolaya-Rey *et al.* 1997, Puls *et al.* 2005), a temperature inversion may appear in some models generating an atmospheric cavity in which *atmospheric strange modes* are found excited with periods of the order of days (Godart *et al.* 2010, 2011). In addition, *oscillatory convection modes* (Saio 2011) in the g mode period range have been found to be present in massive star models with very large growth rates. Saio (2011) computed instability domains for these modes (Fig. 2, left panel, dot-dashed lines) by assuming that they could be observable if the ratio of the photospheric amplitude to the maximum amplitude in the interior is larger than 0.2. Moreover, our preliminary results for the effect of convection on the driving show that the use of time-dependent convection rather than frozen convection still allows these modes to be excited, though the instability is weaker. Sonoi & Shibahashi (these proceedings) studied the instability of oscillatory convective modes in $60\,M_\odot$ models, and suggest a link between these instabilities and the lack of stars above the Humphreys-Davidson (HD) limit (Humphreys & Davidson 1979).

4. Hot pulsators

O and B supergiants: Models of O and B supergiants show pulsation of low-order p and g modes (β Cephei type modes) on the MS, high-order g modes on the post-MS (SPB type modes) and strange modes on the MS and the post-MS. Variations have been detected in OB and A supergiant stars (Burki *et al.* 1978, Lovy *et al.* 1984, Kaufer *et al.* 1997, Waelkens *et al.* 1998), and have been suggested to be related to pulsations (Saio *et al.* 2006, Lefever *et al.* 2007). Saio *et al.* (2006) proposed to name this new type of pulsating stars the SPB supergiant stars (SPBsg). Instabilities due to the ϵ-mechanism have been detected (Moravveji *et al.* 2012a, 2012b) and stochastic oscillations have been observed also (Degroote *et al.* 2010). Periodic variabilities have been detected in O-type MS stars (e.g. Blomme *et al.* 2011, Mahy *et al.* 2011). Furthermore, Aerts *et al.* (2010) have detected a strange-mode candidate in the B supergiant star HD 50064 from *CoRoT* photometry. The candidate mode has a period of 37 d, with a sudden amplitude increase of a factor 1.6 occurring once on a timescale of 137 d. Models of this kind of star show strange-mode pulsations with periods of the order of days, though such periods could result from g mode pulsations (Godart 2011).

Luminous Blue Variables (LBV): The limits of SMIs were investigated in post-MS models (Glatzel & Kiriakidis 1993b, Kiriakidis *et al.* 1993). The instability limits of modes associated with He ionization appear to be similar to the HD limit, while the modes associated with the Z bump give strongly metallicity-dependent instability domains (Kiriakidis *et al.* 1993). Hence strange modes could be responsible for the violent mass-loss outbursts in highly luminous stars (LBV phenomenon), which was previously explained by a dynamical instability (Stothers & Chin 1993). However, this dynamical instability could not reproduce the HD limit: it appears at much lower effective temperature and for high metallicity. Glatzel & Kiriakidis (1993b) then suggested that strange modes could trigger the violent mass-loss event. Their results show good agreement with the HD limit, though this limit is not proven to be related to strange modes yet.

Wolf-Rayet Stars (WR): After an LBV event, massive stars become WR (e.g. Chiosi & Maeder 1986), also subject to pulsations. Short periods (of order or less than an hour) were attributed to the radial fundamental mode driven by the ϵ-mechanism (Noels & Gabriel 1981). Since strange modes were theoretically found in pure He stars, these were suggested to be present in WR stars (Glatzel *et al.* 1993). Longer periods (order of hours) have later been suggested (e.g. Rauw *et al.* 1996 and references therein), with a possible link to non-radial pulsations (Scuflaire & Noels 1986, Noels & Scuflaire 1986). Recently, Lefèvre *et al.* (2005) observed a longer pulsation period of about 9.8 hours in WR 123 with *MOST*. Based on the results of Glatzel *et al.* (1993) who found strange modes with periods shorter than 30 minutes in He-star models for WR, Lefèvre *et al.* (2005) concluded that the period they discovered was too long to be excited by

SMI. However, Dorfi *et al.* (2006) showed that this instability is in fact consistent with a strange-mode pulsation by adding a H-rich envelope to the model in order to increase the radius and therefore the pulsation period. This is however in contradiction with the results of Crowther *et al.* (1995), who found a very small H abundance in the envelope of WR 123. Townsend & MacDonald (2006) suggested that the 9.8 hour period may be attributed to g mode instability occurring in the *deep opacity bump*, at $\log T \sim 6.3$. In that case, the excited periods range from 10 to 21 hours for late WN (WNL) models and from 3 to 13 hours for early WN (WNE) models, in agreement with Lefèvre *et al.*'s observed period.

5. Conclusions

Massive stars are nowadays surrounded by many mysteries whose implications extend far beyond MS and post-MS phases. With asteroseismology, we have a tool to attempt to unravel these mysteries and improve our understanding of stellar evolution. Observational proof of pulsations in the most massive stars is increasing, and new strategies are emerging, for instance, the use of the *macroturbulent* broadening affecting the line profiles of massive stars. This extra broadening has been suggested to be produced by pulsation modes (e.g. Lucy 1976, Aerts *et al.* 2009), and recent observational work seems to support this hypothesis (Simón-Díaz *et al.* 2010, Simón-Díaz 2011). But the wealth and diversity of new observations require a solid theoretical framework to understand and interpret the physical processes at play. High-order g modes probe the deepest layers and bring information about the mixed regions along with the processes affecting the mixing such as mass loss and overshooting, or even the convective criterion. Strange modes probe the superficial layers and could bring information on the density inversion region and the opacity profile while oscillatory convection modes are related to the convective region properties and the convection treatment to adopt.

Acknowledgements

MG thanks the science organizing committee of the conference for the invitation. This research has been funded by the Japanese Society for Promotion of Science (JSPS).

References

Aerts, C., Puls, J., Godart, M., & Dupret, M.-A. 2009, *A&A*, 508, 409
Aerts, C., Lefever, K., Baglin, A., *et al.* 2010, *A&A*, 513, L11
Asplund, M., Grevesse, N., & Sauval, A. J. 2005, *ASP-CS*, 336, 25
Belkacem, K., Samadi, R., Goupil, M., *et al.* 2009, *Science*, 324, 1540
Belkacem, K., Dupret, M. A., & Noels, A. 2010, *A&A*, 510, A6
Blomme, R., Mahy, L., Catala, C., *et al.* 2011, *A&A*, 533, A4
Burki, G., Maeder, A., & Rufener, F. 1978, *A&A*, 65, 363
Chiosi, C. & Maeder, A. 1986, *ARA&A*, 24, 329
Crowther, P. A., Smith, L. J., & Hillier, D. J. 1995, *A&A*, 302, 457
Daszyńska-Daszkiewicz, J., Ostrowski, J., & Pamyatnykh, A. A. 2013, *MNRAS*, 432, 3153
Degroote, P., Briquet, M., Auvergne, M., *et al.*, 2010, *A&A*, 519, A38
Dorfi, E. A., Gautschy, A., & Saio, H. 2006, *A&A*, 453, L35
Gautschy, A. 1992, *MNRAS*, 259, 82
Gautschy, A. 2009, *A&A*, 498, 273
Glatzel, W. 2009, *CoAst*, 158, 252
Glatzel, W. & Kiriakidis, M. 1993a, *MNRAS*, 262, 85
Glatzel, W. & Kiriakidis, M. 1993b, *MNRAS*, 263, 375

Glatzel, W. & Mehren, S. 1996, *MNRAS*, 282, 1470

Glatzel, W., Kiriakidis, M., & Fricke, K. J. 1993, *MNRAS*, 262, L7

Godart, M. 2011, Ph.D. thesis, University of Liège

Godart, M., Noels, A., Dupret, M., & Lebreton, Y. 2009, *MNRAS*, 396, 1833

Godart, M., Dupret, M.-A., Noels, A., Aerts, C., & Simón-Díaz, S. 2010, *AN* 331, P52

Godart, M., Dupret, M.-A., Noels, A., *et al.* 2011, in: C. Neiner, G. Wade, G. Meynet & G. Peters (eds.), *Active OB stars: Structure, Evolution, Mass-Loss, and Critical Limits*, IAU Symposium No. 272, (Cambridge: Cambridge University Press), p. 503

Grevesse, N. & Noels, A. 1993, in: B. Hauck, S. Paltani & D. Raboud, (eds.), *Perf. de l'Assoc. Vaud. des Cherch. en Phys.*, p. 205

Humphreys, R. M. & Davidson, K. 1979, *ApJ*, 187, 871

Kaufer, A., Stahl, O., Wolf, B., *et al.* 1997, *A&A*, 320, 273

Kiriakidis, M., Fricke, K. J., & Glatzel, W. 1993, *MNRAS*, 264, 50

Kiriakidis, M., Glatzel, W., & Fricke, K. J. 1996, *MNRAS*, 281, 406

Lebreton, Y., Montalbán, J., Godart, M., Morel, P., Noels, A., & Dupret, M. 2009, *CoAst*, 158, 277

Lefever, K., Puls, J., & Aerts, C. 2007, *A&A*, 463, 1093

Lefèvre, L., Marchenko, S. V., Moffat, A. F. J., *et al.* 2005, *ApJ*, 634, L109

Lovy, D., Maeder, A., Noels, A., & Gabriel, M. 1984, *A&A*, 133, 307

Lucy, L. B. 1976, *ApJ*, 206, 499

Mahy, L., Gosset, E., Baudin, F., *et al.* 2011, *A&A*, 525, A101

Miglio, A. 2007, Ph.D. thesis, Université de Liège

Miglio, A., Montalbán, J., & Dupret, M.-A. 2007, *CoAst*, 151, 48

Moravveji, E., Moya, A., & Guinan, E. F. 2012b, *ApJ*, 749, 74

Moskalik, P. & Dziembowski, W. A. 1992, *A&A*, 256, L5

Noels, A. & Gabriel, M. 1981, *A&A*, 101, 215

Noels, A. & Scuflaire, R. 1986, *A&A*, 161, 125

Noels, A., Dupret, M.-A., & Godart, M. 2008, *Journal of Physics, Conf. Ser.*, 118, 012019

Ostrowski, J., Daszyńska-Daszkiewicz, J., & Pamyatnykh, A. A. 2012, *AN*, 333, 946

Pamyatnykh, A. A. 1999. *AcA*, 49, 119

Pamyatnykh, A. A. 2007, *CoAst*, 150, 207

Puls, J., Urbaneja, M. A., Venero, R., Repolust, T., Springmann, U., Jokuthy, A., & Mokiem, M. R. 2005, *A&A*, 435, 669

Rauw, G., Gosset, E., Manfroid, J., Vreux, J.-M., & Claeskens, J.-F. 1996, *A&A*, 306, 783

Rogers, F. J. & Iglesias, C. A. 1992, *ApJS*, 79, 507

Saio, H. 2009, *CoAst*, 158, 245

Saio, H. 2011, *MNRAS*, 412, 1814

Saio, H., Baker, N. H., & Gautschy, A. 1998, *MNRAS*, 294, 622

Saio, H., Kuschnig, R., Gautschy, A., *et al.* 2006, *ApJ*, 650, 1111

Santolaya-Rey, A. E., Puls, J., & Herrero, A. 1997, *A&A*, 323, 488

Scuflaire, R. & Noels, A. 1986, *A&A*, 169, 185

Seaton, M. J., Yan, Y., Mihalas, D., & Pradhan, A. K. 1994, *MNRAS*, 266, 805

Shibahashi, H. 1979, *PASJ*, 31, 87

Simon, N. R. & Stothers, R. 1969, *ApJ*, 156, 377

Simón-Díaz, S. 2011, *Bulletin de la Société Royale des Sciences de Liège*, 80, 86

Simón-Díaz, S., Herrero, A., Uytterhoeven, K., Castro, N., Aerts, C., & Puls, J. 2010, *ApJ*, 720, L174

Stothers, R. B. & Chin, C.-W. 1993, *ApJ*, 408, L85

Townsend, R. H. D. & MacDonald, J. 2006, *MNRAS*, 368, L57

Waelkens, C., Aerts, C., Kestens, E., Grenon, M., & Eyer, L. 1998, *A&A*, 330, 215

Ziebarth, K. 1970, *ApJ*, 162, 947

Precision Asteroseismology
Proceedings IAU Symposium No. 301, 2013
J. A. Guzik, W. J. Chaplin, G. Handler & A. Pigulski, eds.

© International Astronomical Union 2014
doi:10.1017/S174392131301452X

Pulsations of blue supergiants before and after helium core ignition

Jakub Ostrowski and Jadwiga Daszyńska-Daszkiewicz

Instytut Astronomiczny, Uniwersytet Wrocławski, ul. Kopernika 11, 51-622 Wrocław, Poland
email: ostrowski@astro.uni.wroc.pl

Abstract. We present results of pulsation analyses of B-type supergiant models with masses of $14-18\,M_\odot$, considering evolutionary stages before and after helium core ignition. Using a non-adiabatic pulsation code, we compute instability domains for low-degree modes. For selected models in these two evolutionary phases, we compare properties of pulsation modes. Significant differences are found in oscillation spectra and the kinetic energy density of pulsation modes.

Keywords. stars: early-type, stars: supergiants, stars: oscillations, stars: evolution

1. Introduction

Slowly Pulsating B-type supergiants (SPBsg) are a new class of pulsating variable stars. They have been discovered by Saio *et al.* (2006) who found 48 frequencies in the light variations of the blue supergiant HD 163899 (B2 Ib/II, Klare & Neckel 1977, Schmidt & Carruthers 1996) and attributed them to g- and p-mode pulsations. This was unexpected because it was believed that g modes cannot propagate in stars beyond the main sequence due to very strong radiative damping in the helium core.

The discovery has prompted a few groups (Godart *et al.* 2009, Daszyńska-Daszkiewicz *et al.* 2013) to reanalyse pulsation stability in models of B-type stars after the Terminal Age Main Sequence (TAMS) and to further studies of SPBsg variables. The presence of g-mode pulsations in B-type post-main-sequence stars has been explained by a partial reflection of some modes at an intermediate convective zone (ICZ) related to the hydrogen-burning shell or at a chemical gradient zone surrounding the radiative core. However, all studies of these objects published so far are based on the assumption that HD 163899 has not reached the phase of helium core ignition, i.e., it is in the phase of hydrogen shell burning. This assumption does not have to be made, because the blue loop can reach temperatures of early B spectral types. In this paper we investigate this possibility and compare two SPBsg models: before and after He core ignition.

2. Instability domains

Our evolutionary models were calculated with the MESA evolution code (Modules for Experiments in Stellar Astrophysics, Paxton *et al.* 2011, Paxton *et al.* 2013). We adopted a hydrogen abundance at ZAMS of $X = 0.7$, metal abundance of $Z = 0.015$ and OPAL opacity tables (Iglesias & Rogers 1996) with the AGSS09 metal mixture (Asplund *et al.* 2009). We took into account convective overshooting from the hydrogen and helium core and inward overshooting from non-burning convective zones, using the exponential formula (Herwig 2000):

$$D_{\mathrm{ov}} = D_{\mathrm{conv}} \exp\left(-\frac{2z}{f\lambda_P}\right), \qquad (2.1)$$

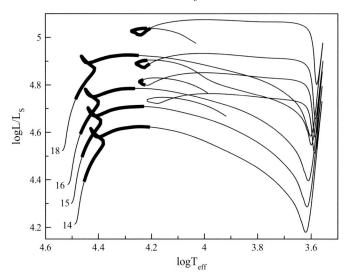

Figure 1. The H-R diagram with instability domains, marked as thick lines, for modes of degree $l = 0, 1, 2$ excited in models with masses of $14 - 18\,M_\odot$.

where D_{conv} is the mixing-length-theory-derived diffusion coefficient at a user-defined location near the core boundary, λ_P is the pressure scale height at that location, z is the distance in the radiative layer away from that location, and f is an adjustable parameter, which we set to 0.01. All effects of rotation and mass loss were neglected. We performed non-adiabatic pulsation analyses using the code of Dziembowski (1977).

In Fig. 1, we present instability domains for the modes of the degree $l = 0, 1, 2$ excited in models with masses of $14 - 18\,M_\odot$. There is an instability strip beyond the TAMS which is very similar to the one from previous calculations (e.g. Daszyńska-Daszkiewicz *et al.* 2013). The main difference is the presence of unstable non-radial modes on the blue loops for more massive models ($M \gtrsim 14\,M_\odot$ and $\log T_{\mathrm{eff}} \gtrsim 4.2$) whereas radial modes are stable in this evolutionary stage. The existence of pulsation instability on the blue loop, as well as the blue loops themselves, depend critically on the metallicity, Z. For example, for $Z = 0.02$ there are no unstable modes on the blue loops in the considered range of masses. More details will be given by Ostrowski & Daszyńska-Daszkiewicz (in preparation).

3. Pulsation modes before and after He core ignition

We compare pulsation properties of two models with similar positions in the H-R diagram: one during the hydrogen shell-burning phase ($16\,M_\odot$, $\log T_{\mathrm{eff}} = 4.343$, $\log L/L_\odot = 4.705$, Model 1) and the second during core helium burning ($15\,M_\odot$, $\log T_{\mathrm{eff}} = 4.244$, $\log L/L_\odot = 4.815$, Model 2). The main difference between these models is the presence of a convective core in the model on the blue loop. Beyond the core, the propagation diagrams of these two models are qualitatively very similar; in particular, both have a fully developed intermediate convective zone.

In Fig. 2 we depict the instability parameter, η, as a function of frequency for Model 1 (top panel) and Model 2 (bottom panel). This parameter tells us whether the pulsation mode is unstable ($\eta > 0$) or not ($\eta \leqslant 0$). In the case of Model 1, we can see a regular structure controlled by mode trapping, and two instability regions can be identified

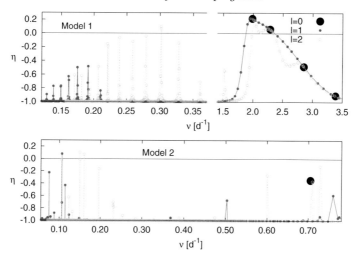

Figure 2. Instability parameter, η, for Model 1 (top panel) and Model 2 (bottom panel).

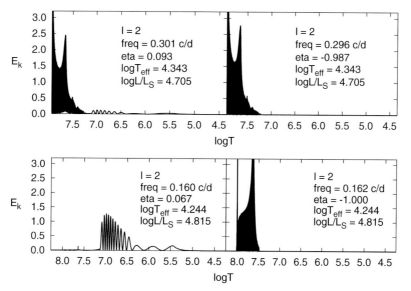

Figure 3. *Top panel*: the kinetic energy density of unstable (left panel) and stable (right panel) $l = 2$ modes with close frequencies from Model 1. The values of ν and η are listed in the panels. *Bottom panel*: the same as in the top panel, except for Model 2.

(Daszyńska-Daszkiewicz *et al.* 2013) whereas in Model 2 only a single very low frequency mode is excited.

In the top panel of Fig. 3, we present the kinetic energy density for two very close frequency quadrupole high-order g modes for the hydrogen shell-burning model. One of them (left panel) is unstable and the other stable (right panel). Their properties are described in the panels. For both modes, most of their energy is confined to the radiative helium core, where very strong damping occurs. The main difference between them is that due to a partial reflection at the ICZ, the unstable mode has some small amount of energy confined in the outer envelope. This energy is crucial for instability of pulsation modes in SBPsg stars and it is sufficient for the κ mechanism operating in the Z-bump region to efficiently drive the mode.

A similar plot for the blue-loop model is shown in the bottom panel of Fig. 3. There are also two very close frequency quadrupole high-order g modes, one unstable (left panel) and one stable (right panel). The behavior of E_k is different for these modes than for the modes from the model before core helium burning. The entire energy of the stable mode is confined to the radiative zone in-between the convective core and the ICZ (there is strong damping in this area), whereas the unstable mode has almost all of its energy trapped in the outer radiative zone. That means that unstable modes on the blue loops have to be almost entirely reflected at the ICZ. This could explain why we observe in our models that undergo core helium burning many fewer unstable modes than in models before helium core ignition.

4. Conclusions

Our work has shown that blue loops can reach temperatures of B spectral types and there are unstable modes during this phase of evolution. It means that SPBsg stars might undergo core helium burning but whether they actually do or not is still an open question. We found that there is a huge difference in the behavior of the kinetic energy density of pulsation modes between models before and after helium core ignition. On the blue loop, the pulsation modes have to be almost entirely reflected at the ICZ in order to be unstable whereas for the models that undergo hydrogen shell burning beyond the TAMS, a partial reflection at the ICZ is sufficient.

References

Asplund, M., Grevesse, N., Sauval, A. J., & Scott, P. 2009, *ARAA*, 47, 481
Daszyńska-Daszkiewicz, J. & Ostrowski, J., Pamyatnykh A. A. 2013, *MNRAS*, 432, 3153
Dziembowski, W. 1977, *AcA*, 27, 95
Godart, M., Noels, A., Dupret, M.-A., & Lebreton, Y. 2009, *MNRAS*, 396, 1833
Herwig, F. 2000, *A&A*, 360, 952
Iglesias, C. A. & Rogers, F. J. 1996, *ApJ*, 464, 943
Klare, G. & Neckel, T. 1977, *A&AS*, 27, 215
Paxton, B., Bildsten, L., Dotter, A., *et al.* 2011, *ApJS*, 192, 3
Paxton, B., Cantiello, M., Arras, P., *et al.* 2013, *ApJS*, 208, 4
Saio, H., Kuschnig, R., Gautschy, A., *et al.* 2006, *ApJ*, 650, 1111
Schmidt, E. G. & Carruthers, G. R. 1996, *ApJS*, 104, 101

Precision Asteroseismology
Proceedings IAU Symposium No. 301, 2013
J. A. Guzik, W. J. Chaplin, G. Handler & A. Pigulski, eds.

© International Astronomical Union 2014
doi:10.1017/S1743921313014531

Solar-like oscillations in subgiant and red-giant stars: mixed modes

S. Hekker[1,2] and A. Mazumdar[3]

[1] Max Planck Institute for Solar System Research, Katlenburg-Lindau, Germany
email: Hekker@mps.mpg.de

[2] Astronomical Institute "Anton Pannekoek", University of Amsterdam, Amsterdam, the Netherlands

[3] Homi Bhabha Centre for Science Education, TIFR, Mumbai, India

Abstract. Thanks to significant improvements in high-resolution spectrographs and the launch of dedicated space missions *MOST*, *CoRoT* and *Kepler*, the number of subgiants and red-giant stars with detected oscillations has increased significantly over the last decade. The amount of detail that can now be resolved in the oscillation patterns does allow for in-depth investigations of the internal structures of these stars. One phenomenon that plays an important role in such studies are mixed modes. These are modes that carry information of the inner radiative region as well as from the convective outer part of the star allowing to probe different depths of the stars.

Here, we describe mixed modes and highlight some recent results obtained using mixed modes observed in subgiants and red-giant stars.

Keywords. stars: oscillations (including pulsations), stars: evolution

---

1. Introduction

Solar-like oscillations are oscillations stochastically excited in the outer convective layers of low-mass stars on the main sequence (such as the Sun), subgiants and red-giant stars (e.g. Goldreich & Keeley 1977, Goldreich & Kumar 1988). Effectively, some of the convective energy is converted into energy of global oscillations. The main characteristics of these oscillations are that a) essentially all modes are excited albeit with different amplitudes; b) the stochastic driving and damping results in a finite mode lifetime.

In Fourier space the solar-like oscillation characteristics yield a regular pattern of oscillation frequencies with a roughly Gaussian envelope, where each frequency peak has a width inversely proportional to the mode lifetime. Hence, resolved individual frequency peaks can be fitted using a Lorentzian profile to determine height, width and frequency of the oscillation mode (see inset in Fig. 1). The regular pattern of the frequencies in the Fourier spectrum can be described asymptotically (Tassoul 1980, Gough 1986):

$$\nu_{n,l} \simeq \Delta\nu \left(n + \frac{l}{2} + \epsilon \right) - \delta\nu_{n,l}, \tag{1.1}$$

where ν is frequency and ϵ is a phase term. $\Delta\nu$ is the regular spacing in frequency of oscillations with the same angular degree l and consecutive orders n (large frequency separation, horizontal dashed-dotted line in Fig. 1). $\delta\nu_{n,l}$ is the so-called small frequency separation between pairs of odd or even modes. The large frequency separation, $\Delta\nu$, is proportional to the square root of the mean density of the star. Another characteristic of the Fourier spectrum is the frequency of maximum oscillation power, $\nu_{\max}$, which depends on the surface gravity and effective temperature of the star.

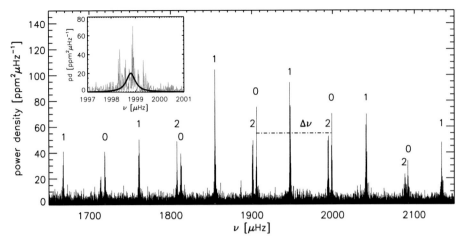

Figure 1. Fourier power density spectrum of a main-sequence star (KIC 3656479, Hekker 2013). The angular degree (l) of the modes are indicated. The horizontal dashed-dotted line indicates $\Delta\nu$ (see Eq. 1.1). The inset shows a zoom of the radial mode at $\sim$1999 μHz with a Lorentzian profile overplotted.

For a full overview of solar-like oscillations we refer to Aerts *et al.* (2010), Chaplin & Miglio (2013) and Hekker (2013).

2. Mixed modes

2.1. *Cavities*

The oscillations described above are pure acoustic pressure (p) modes. Non-radial p modes reside in a cavity in the outer parts of the star bound at the bottom by the Lamb frequency (S_l), where the horizontal phase speed of the wave equals the local sound speed. At the top the cavity is limited by the cut-off frequency $\nu_{\mathrm{ac}} \propto g/\sqrt{T_{\mathrm{eff}}}$ (Brown *et al.* 1991) above which the atmosphere is not able to trap the modes and the oscillations become traveling waves, so called high-frequency or pseudo-modes (e.g., Karoff 2007). When stars evolve the p-mode frequencies decrease, mostly due to the decrease in surface gravity (increase in radius) and hence decrease in cut-off frequency. At the same time oscillations that reside in the inner radiative region of the star have increasing frequencies with evolution. These so-called gravity (g) modes have buoyancy as their restoring force. These modes reside in a cavity defined by finite values of the Brunt-Väisälä or buoyancy frequency N. This is the frequency at which a vertically displaced parcel will oscillate within a statically stable environment. The peak of the Brunt-Väisälä frequency increases with evolution due to the increase in the core gravity. Oscillations can only be sustained when $\nu < S_l, N$ (g mode) or $\nu > S_l, N$ (p mode). A region where either of these conditions is not satisfied is an evanescent zone for the respective mode at a particular frequency. See Fig. 2 for the dipole Lamb frequency and Brunt-Väisälä frequency as a function of stellar radius defining the respective p- and g-mode cavities of a main-sequence star (top panel), subgiant (middle panel) and a red giant (bottom panel).

2.2. *Avoided crossings*

In subgiants and red giants the frequencies in both p- and g-mode cavities have similar values and a coupling between these frequencies can persist if the evanescent zone is narrow, i.e., the oscillation is not damped out. In that case a p and g mode with

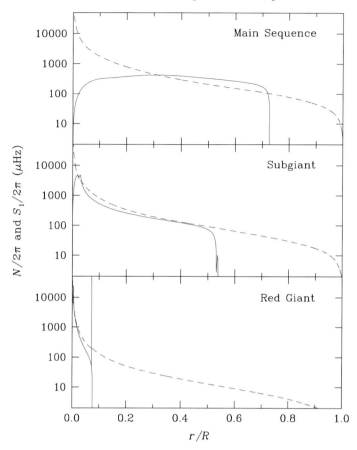

Figure 2. The Brunt-Väisälä (N, red solid line) and the Lamb frequency for $l = 1$ modes (S_1, blue dashed line) are shown as functions of fractional radius (r/R) for models of a $1\,M_\odot$ star in the main sequence (top panel), subgiant (middle panel) and red-giant (bottom panel) phases.

similar frequencies and same spherical degree undergo a so-called avoided crossing. The interactions between the modes will affect (or bump) the frequencies. This bumping can be described as a resonance interaction of two oscillators (e.g. Aizenman *et al.* 1977, Deheuvels & Michel 2010, Benomar *et al.* 2013).

In short the avoided crossings can be viewed using a system of two coupled oscillators $y_1(t)$ and $y_2(t)$ with a time dependence:

$$\frac{d^2 y_1(t)}{dt^2} = -\omega_1^2 y_1 + \alpha_{1,2} y_2 \tag{2.1}$$

$$\frac{d^2 y_2(t)}{dt^2} = -\omega_2^2 y_1 + \alpha_{1,2} y_1 \tag{2.2}$$

where $\alpha_{1,2}$ is the coupling term between the two oscillators, and ω_1, ω_2 are the eigenfrequencies of the uncoupled oscillators ($\alpha_{1,2} = 0$). In the case of uncoupled oscillators the eigenfrequencies can cross at ω_0 where $\omega_1 = \omega_2$.

If the coupling term $\alpha_{1,2}$ is very small compared to the difference between the eigenfrequencies ($\alpha_{1,2} \ll |\omega_1^2 - \omega_2^2|$), then the eigenfrequencies of the system are hardly perturbed and have values close to ω_1 and ω_2. However, if the difference between the eigenfunctions is small compared to the coupling term ($|\omega_1^2 - \omega_2^2| \ll \alpha_{1,2}$), then the eigenfrequencies can

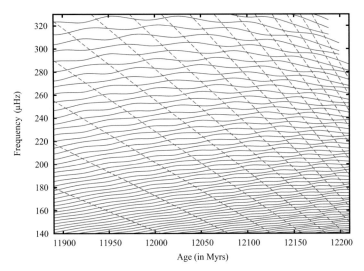

Figure 3. The evolution of frequencies of a $1 M_\odot$ star with age in the red-giant phase. The blue continuous lines depict the $l = 1$ modes while the red dashed lines represent the $l = 0$ modes of different radial orders.

be approximated by $\omega^2 = \omega_0^2 \pm \alpha_{1,2}$. These two eigenmodes behave as mixed modes, one with dominant features from ω_1 and the other one with dominant features from ω_2.

For a $1 M_\odot$ star, Fig. 3 shows the variation of frequencies with evolution. The general trend of the p-mode frequencies is to decrease with age, while g-mode frequencies increase with age (see Section 2.1). Therefore at a particular age, a g mode and a p mode of the same angular degree and similar frequencies can exist, which would — instead of crossing each other to continue these opposite trends — interact to produce a pair of mixed modes with close frequencies, as explained above. This is visible in Fig. 3 as a series of bumps in the $l = 1$ modes, where each bump is located at an avoided crossing. One of these mixed modes will have dominant features from the underlying p mode, while the other mixed mode will have dominant features from the underlying g mode. Figure 4 shows a Fourier power density spectrum of a subgiant with a mixed-mode pair.

As evident from Fig. 3, an avoided crossing can occur at a specific frequency only at a specific age. Thus observations of mixed modes allows for a precise estimate of the age. However, the age determination is model dependent and the actual value of the stellar age might differ as a function of physical processes included in the model.

In more evolved stars the density of g modes around a given frequency can be high and multiple g modes can interact with different coupling terms with a single p mode to produce multiple mixed modes. Figure 5 shows a Fourier power density spectrum of a red-giant star with multiple mixed dipole modes.

2.3. Period spacing

The high-order g modes in the inner radiative region have (in an asymptotic approximation) a typical spacing in period ($\Delta\Pi$, e.g. Tassoul 1980). Since the regular pattern of the frequencies is broken by the avoided crossings, the period spacing will also be affected. For red giants with multiple mixed modes per p-mode order, this results in smaller observed $\Delta\Pi$ for pressure-dominated mixed modes (lying close to the underlying p mode) and increasing values of $\Delta\Pi$ for mixed modes with more g-dominated character, i.e., with frequencies further away from the underlying p mode. The least coupled, i.e. most g-dominated, modes have a $\Delta\Pi$ close to the asymptotic value. However these modes are

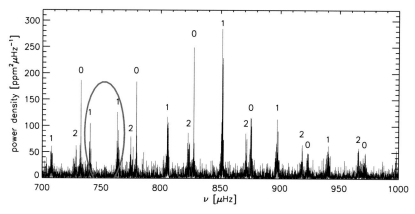

Figure 4. Fourier power density spectrum of a subgiant (KIC 11395018, Hekker 2013). The degree (*l*) of the modes is indicated. A mixed mode pair is present at 740.3 and 764.0 μHz highlighted by the (red) oval.

the hardest to detect due to their low heights in the Fourier power spectrum (Dupret *et al.* 2009).

3. Recent results

3.1. *Subgiants*

For subgiants a recent highlight is the detection of radial differential rotation (Deheuvels *et al.* 2012). Due to the different sensitivities of mixed modes to different parts of the star (either the internal or the outer region of the star depending on their predominant g- or p-mode nature) it is possible to study stellar properties, such as rotation, as a function of radius. Deheuvels *et al.* (2012) were able to detect rotational splittings in 17 $l = 1$ mixed modes in a subgiant, which have different pressure-gravity mode sensitivity. They concluded that the core rotates approximately five times faster than the surface.

3.2. *Red giants*

Gravity-dominated mixed modes in red-giant stars have recently been discovered observationally for $l = 1$ modes by Beck *et al.* (2011). From these mixed modes the period spacings can be derived. Bedding *et al.* (2011) and Mosser *et al.* (2011) showed that the observed period spacings of red giants in the hydrogen-shell-burning phase ascending the red-giant branch are significantly different from the period spacing for red giants also burning helium in the core. This provides a clear separation into these two groups of stars that are superficially very similar. This effect was explained by Christensen-Dalsgaard (2011) as being due to convection in the central regions of the core in the helium-burning stars. The buoyancy frequency is nearly zero in the convective region. As the period spacing is inversely proportional to the integral over N, the period spacing of stars with convection in the core is higher compared to the period spacing of stars without a convective region in the core.

Similar to the subgiant mentioned in the previous subsection, the rotational splittings of dipole mixed modes with different p/g nature have been measured for a red-giant star (Beck *et al.* 2012), which revealed that the core rotates approximately ten times faster than the surface. Subsequently, Mosser *et al.* (2012) showed that stars ascending the red-giant branch experience a small increase of the core rotation followed by a significant slow-down in the later stages of the red-giant branch resulting in slower rotating cores in

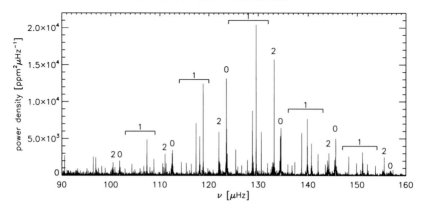

Figure 5. Fourier power density spectrum of a red-giant star (KIC 9145955, Hekker 2013). The degree (l) of the modes is indicated. For the dipole modes the approximate range of the observed mixed modes is indicated.

red-clump (or horizontal-branch) stars compared to faster rotating cores in stars on the red-giant branch.

4. Future

With the long-term datasets currently available from *Kepler*, mixed modes can be detected in many stars. This allows for further studies of the radial differential rotation (see Section 3 and Di Mauro *et al.*, these proceedings). Furthermore, mixed modes will also allow for further studies of the internal structure of subgiants and red-giant stars, possibly including core overshoot, and the presence of secondary helium flashes. The latter are present in stellar evolution models of low-mass stars which ignite helium in a degenerate core, but it is still unclear whether this is a realistic representation. Therefore, it is evident that mixed modes have great diagnostic potential especially for subgiants and red-giant stars.

Acknowledgements

SH acknowledges financial support from the Netherlands Organisation for Scientific Research (NWO). AM acknowledges support from the NIUS programme of HBCSE.

References

Aerts, C., Christensen-Dalsgaard, J., & Kurtz, D. W., 2010, *Asteroseismology*, Astronomy and Astrophysics Library (Springer Science+Business Media B.V.)
Aizenman, M., Smeyers, P., & Weigert, A. 1977, *A&A*, 58, 41
Beck, P. G., Bedding, T. R., Mosser, B., *et al.* 2011, *Science*, 332, 205
Beck, P. G., Montalbán, J., Kallinger, T., *et al.* 2012, *Nature*, 481, 55
Bedding, T. R., Mosser, B., Huber, D., *et al.* 2011, *Nature*, 471, 608
Benomar, O., Bedding, T. R., Mosser, B., *et al.* 2013, *ApJ*, 767, 158
Chaplin, W. J. & Miglio, A. 2013, *ARAA*, 51, 353
Christensen-Dalsgaard, J. 2011, arXiv: 1106.5946
Deheuvels, S., García, R. A., Chaplin, W. J., *et al.* 2012, *ApJ*, 756, 19
Deheuvels, S. & Michel, E. 2010, *Ap&SS*, 328, 259
Dupret, M.-A., Belkacem, K., Samadi, R., *et al.* 2009, *A&A*, 506, 57
Goldreich, P. & Keeley, D. A. 1977, *ApJ*, 212, 243
Goldreich, P. & Kumar, P. 1988, *ApJ*, 326, 462

Gough, D. O. 1986, *Highlights of Astronomy*, 7, 283

Hekker, S. 2013, *Adv. Sp. Res.*, 52, 1581

Karoff, C. 2007, *MNRAS*, 381, 1001

Mosser, B., Barban, C., Montalbán, J., *et al.* 2011, *A&A*, 532, A86

Mosser, B., Goupil, M. J., Belkacem, K., *et al.* 2012, *A&A*, 548, A10

Tassoul, M. 1980, *ApJS*, 43, 469

Precision Asteroseismology
Proceedings IAU Symposium No. 301, 2013
J. A. Guzik, W. J. Chaplin, G. Handler & A. Pigulski, eds.

© International Astronomical Union 2014
doi:10.1017/S1743921313014543

Stochastically excited oscillations in the upper main sequence

Victoria Antoci

Stellar Astrophysics Centre, Department of Physics and Astronomy, Aarhus University
Ny Munkegade 120, DK-8000 Aarhus C, Denmark
email: antoci@phys.au.dk

Abstract. Convective envelopes in stars on the main sequence are usually connected only with stars of spectral types F5 or later. However, observations as well as theory indicate that the convective outer layers in hotter stars, despite being shallow, are still effective and turbulent enough to stochastically excite oscillations. Because of the low amplitudes, exploring stochastically excited pulsations became possible only with space missions such as *Kepler* and *CoRoT*. Here I review the recent results and discuss among others, pulsators such as δ Scuti, γ Doradus, roAp, β Cephei, Slowly Pulsating B and Be stars, all in the context of solar-like oscillations.

Keywords. stars: oscillations, stars: variables: δ Scuti, stars: variables: γ Doradus, stars: variables: Sun-like stars, stars: variables: β Cephei, stars: variables: SPB stars, stars: variables: Be stars

1. Introduction

There are several mechanisms driving oscillations in pulsating variables. The oldest known is the opacity (κ) mechanism acting like a heat engine, converting thermal energy into mechanical energy (e.g. Eddington 1919, Cox 1963). The layers where the thermal energy is stored are the zones connected to (partial) ionisation of abundant elements, which can take place only at specific temperatures. The zone where neutral hydrogen (H) and helium (He) are ionised is at about $14\,000$ K and close to the surface, whereas the second ionisation zone of He II is at $\sim 50\,000$ K. The driving in the He II ionisation zone is the main source of excitation in stars placed in the classical instability strip such as the δ Scuti (δ Sct) stars. The pulsations of the more massive β Cephei (β Cep) and Slowly Pulsating B stars (SPB) are triggered by the κ mechanism operating on the iron-group elements (located at $\sim 200\,000$ K).

The oscillations in γ Doradus stars (γ Dor) are driven by a similar mechanism, in this case, however, it is the bottom of the outer convection zone blocking the flux from the interior (Guzik *et al.* 2000) and is therefore called *convective blocking* (see Fig. 1 for exact location in the Hertzsprung-Russell diagram).

Another important mechanism, especially for the present review, is stochastic driving. In stars like the Sun the modes are intrinsically stable, damped by the turbulent convection. Nevertheless the acoustic energy of the convective motion, which is comparable to the sound speed, is sufficient to cause resonance at a star's natural frequencies where a part of the energy is transferred into global oscillation modes. Because of the large number of convective cells the excitation is random, hence stochastic. This is very different from the κ mechanism, which excites pulsation coherently. This contrast is a very important way of distinguishing between these two types of driving mechanisms in the signal processing. Another important attribute of stochastic driving is that all the modes in a certain frequency range are excited to observable amplitudes, allowing mode

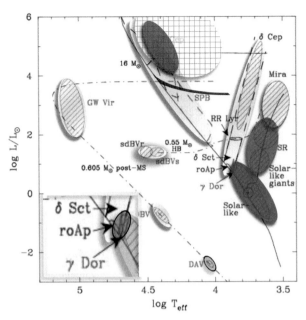

Figure 1. The asteroseismic H-R diagram displays the major types of known pulsating stars. This modified version shows their excitation mechanisms labeled as follows: the κ mechanism is depicted in light grey, the stochastic mechanism in dark grey (red in the electronic form) and the grey area surrounded by a black ellipse (blue area in the electronic form) indicates the convective blocking mechanism. The inset is a close-up of the lower part of the classical instability strip. Note the overlap of the different instability strips. The upper hashed region depicts the region where massive stars have a sub-surface convection zone with convective velocities $\geqslant$ 2.5 km s^{-1}. The black line, just above the SPB stars shows the division line for surface convection in the Fe ionisation zone. Both regions indicating convection in massive stars are from Cantiello et al. (2009). Adapted from J. Christensen-Dalsgaard and G. Handler.

identification from pattern recognition. This is again in absolute contrast to the heat engine where the mechanism selecting which modes are excited to observable amplitudes is not understood (for a review see Smolec, these proceedings).

However, a star will only pulsate if the conditions inside are just right for a mechanism to work. For the κ mechanism, a mode will only be excited if its period of oscillation corresponds to the thermal time scale of the driving zone (e.g. Pamyatnykh 2000), which means that the opacity bump needs to be at a favourable depth in the envelope. Similarly for the convective blocking mechanism, the depth of the convective envelope should be between 3 and 9% (Guzik et al. 2000). As for the stochastic excitation, what matters is not the depth of the convective layer, but how vigorous the convective motions are.

Stochastically excited pressure modes as observed in the Sun are nearly equally spaced in frequency and give rise to a clear comb-like structure. The almost equidistant spacing in frequency of consecutive modes of the same degree l is the so called *large frequency separation* $\Delta\nu$ and measures the mean density of a star. The oscillation spectrum in Sun-like stars as well as in red giants is described by an envelope with a frequency of maximum power, for the Sun $\nu_{\max}$ is at $3090 \ \mu\text{Hz}$ (Huber et al. 2011). The $\nu_{\max}$ and the shape of the oscillation envelope, determined by damping and driving, are related to the acoustic cutoff frequency, $\nu_{\text{ac}} \propto gT_{\text{eff}}^{-1/2}$ (e.g. Brown et al. 1991). From that, Kjeldsen & Bedding (1995) derived the following scaling relation for $\nu_{\max}$ for any other star: $\nu_{\max} = \nu_{\max_\odot}(M/M_\odot)(R/R_\odot)^{-2}(T_{\text{eff}}/T_{\text{eff}_\odot})^{-1/2}$. Kjeldsen & Bedding (1995) also show that $\Delta\nu$, when scaled to the Sun, can also be written as $\Delta\nu = (M/M_\odot)^{1/2}(R/R_\odot)^{-3/2}\Delta\nu_\odot$. From observations with the *CoRoT* and *Kepler* satellites it was demonstrated that there is a clear empirical relation between $\nu_{\max}$ and $\Delta\nu$ of the form: $\Delta\nu = \alpha(\nu_{\max}/\mu\text{Hz})^\beta$, where α and β can have slightly different values, summarised by e.g. Huber *et al.* (2011). All these relations are powerful tools to study solar-like oscillators.

The timescale on which a solar-like mode is excited, the mode lifetime, depends on the damping rate. This can be determined by fitting a Lorentzian profile to an observed mode and measuring its line width, where the inverse of the FWHM times π gives the mode lifetime. Because the power of the mode is spread over a certain frequency range, the amplitudes and frequencies are also derived by fitting Lorentzian profiles.

Detecting solar-like oscillations in Sun-like stars is easy, however identifying stochastically excited modes in other types of pulsators is complicated and requires a lot of data. In particular discriminating between different excitation mechanisms is sometimes impossible. Temporal variability does not necessarily mean stochastic excitation; it can simply be that there are many closely spaced, unresolved frequencies. The good news is that even in the case where coherent and non-coherent signals are present, one can subtract the coherent peaks without destroying the stochastic signal (Antoci *et al.* 2013).

Studying the excitation mechanisms of stars with different temperatures, masses and evolutionary stages allows us to determine the structure of stars as well as understand several physical processes. The existence of convective layers plays a major role in the generation of magnetic fields and stellar activity, in the transport of angular momentum, in mixing processes and diffusion, and even in the alignment of planetary systems.

2. A- and early F-type stars

The A- and early F-type stars are very complex objects, because the transition from deep and effective to shallow convective envelopes takes place within this spectral region. Figure 2 shows how the outer layers change as a function of temperature for stars located on the zero-age main sequence, from theory (Christensen-Dalsgaard 2000). While the outer 30% are fully convective for the Sun, it changes dramatically for slightly higher temperatures and separates into the known ionisation zones (H, He I and He II). For a late A type star, for example, only the outer 2% of the radius is still convective (e.g. Kallinger & Matthews 2009). The right panel of Fig. 2 shows how much of the energy is transported by convection, demonstrating again the drastic change in the convective properties. The grey (electronic version blue) region in the same figure indicates the location where the κ mechanism operates in the He II ionisation zone, as assumed to be the case in the δ Sct stars. Interestingly, Théado *et al.* (2012) find that if diffusion is implemented in the models, heavy elements will gravitationally settle. Due to the strong accumulation of Fe in the Z bump, this region eventually becomes convective, mixing again the material into the higher layers connected by overshooting, hence destroying the steep gradient. Once this layer is homogeneous enough the convection will cease and the cycle repeats. The authors find several convective episodes during the main sequence lifetime of a 2 $M_\odot$ star.

From an observational point of view, Gray & Nagel (1989) found from bisector measurements that the granulation boundary crosses the classical instability strip. Also related to the afore-mentioned transition is the so-called rotation boundary. Stars later than F5 have slow rotation rates, on average $v\sin i < 10$ km s^{-1}, with a sharp increase between F0 and F5 (Royer 2009). The measured rotational velocities for the cool stars inside the instability strip are typically higher than 100 km s^{-1}. Royer (2009) also reports observational evidence for differential rotation in A type stars, which is related to magnetic

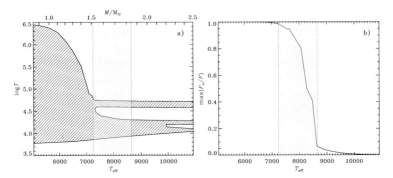

Figure 2. Envelope structure for zero-age main sequence stars as a function of effective temperature (lower abscissa) and mass (upper abscissa). *Left:* the dashed area indicates the convective regions in the outer layers. It can be seen that the depth of the convective envelope changes dramatically within the classical instability strip (grey area, blue in electronic form). While for stars like the Sun, all three convection zones are combined, for hotter more massive stars the H, He I and He II zones start to separate. *Right:* F_c/F indicates how much of the energy is transported by convection. Again a clear transition is present within the instability strip. Courtesy of J. Christensen-Dalsgaard.

dynamos like in the Sun. Chromospheric activity disappears only for stars hotter than 8300 K (Simon *et al.* 2002), demonstrating that the δ Sct and γ Dor pulsators are still affected by the presence of convective envelopes. Landstreet *et al.* (2009) and Kallinger & Matthews (2009) find strong evidence for the signatures of convective motions.

Houdek *et al.* (1999) and Samadi *et al.* (2002) predicted that surface convection in the envelopes of δ Sct stars is still vigorous enough to excite solar-like oscillations. There is no reason why the same prediction should not be generalised for all the stars occupying the same location in the H-R diagram. Solar-like oscillations, if detected, would allow one to perform mode identification from pattern recognition and therefore to also identify the modes of pulsation excited by the κ mechanism. As a result we can then reproduce the deep interior of the star by using the δ Sct oscillations and constrain the overshooting parameter $\alpha_{\rm ov}$, which is the primary factor to influence the lifetime on the main sequence. The solar-like oscillations on the other hand are most sensitive to outer (convective) envelope, which so far is poorly understood in stars with temperatures higher than 6500 K.

2.1. *(Rapidly oscillating) Ap stars*

Rapidly oscillating (ro)Ap stars are stars with magnetic fields of the order of kG, with periods of oscillations similar to the solar ones (4 to 20 min). The regular patterns in the oscillation spectrum of roAp stars are best explained by the oblique pulsator model, where the pulsation axis is inclined with respect to the magnetic and rotation axes (Kurtz 1982, Bigot & Dziembowski 2002). The excitation mechanism is not entirely understood but is interpreted as the opacity mechanism acting in the H ionisation zone. In Ap stars the magnetic field suppresses convection at the magnetic poles reducing the damping (Balmforth *et al.* 2001), triggering the observed p modes. So far there are no solar-like oscillations observed in roAp stars, but also none expected.

2.2. γ *Dor stars*

The generally accepted mechanism triggering the high radial order g mode pulsations in γ Dor stars is the convective blocking mechanism. Interestingly, the instability domain of the γ Dor stars overlaps the one of the δ Sct stars, suggesting that both types of pulsation should be excited simultaneously in stars located there (Fig. 1). Missions like the NASA *Kepler* satellite established that the δ Sct and γ Dor hybrids are a common phenomenon, rather than an exception (Grigahcène *et al.* 2010). Intriguingly, from *Kepler* observations, Uytterhoeven *et al.* (2011) and Tkachenko *et al.* (2012) find hybrid pulsators not only in the region where the two instability domains overlap but distributed all over the δ Sct strip. The convective blocking mechanism, as it is currently understood, cannot operate unless the convective envelope has a certain depth between approximately 3 and 9% (Guzik *et al.* 2000); this is not expected to be the case at the observed effective temperatures. This implies that either the hybrids observed at such high $T_{\rm eff}$ values have still a non-negligible subsurface convection zone or we are missing basic physics. As far as stochastic excitation is concerned, these stars are the perfect targets to search for solar-like oscillations. Nevertheless, up to now, no γ Dor/solar-like hybrids were detected, which is very puzzling. The non-detection of stochastic oscillations in these stars becomes even more mysterious when considering the significant number of solar-like oscillators identified within the γ Dor instability strip. In other words, there are several F type stars with high temperatures, higher than some of the γ Dor pulsators, showing solar-like oscillations but no g modes. Note that here I consider only targets with reliable temperature determinations (W.J. Chaplin, priv. comm.). To summarise, there are no solar-like oscillations detected in γ Dor stars, but from the theoretical point of view we do not understand why. One has to keep in mind, however, that γ Dor are usually observed with a sampling cadence not rapid enough to detect high radial-order p modes.

2.3. δ Sct stars

δ Sct stars are one of the first known groups of pulsators, yet also one of the least understood. It is generally accepted that the complex pulsational behaviour is the result of the κ mechanism acting in the He II ionisation zone. Previously, models predicted more modes to be unstable than observations showed; however, with *CoRoT* and *Kepler*, the opposite seems to be the case (e.g. Poretti *et al.* 2009, García-Hernández *et al.* 2010, Balona & Dziembowski 2011, Uytterhoeven *et al.* 2011). The NASA *Kepler* spacecraft observed hundreds of δ Sct stars at a precision where solar-like oscillations, if present, should be detected. One of these stars was HD 187547 (Antoci *et al.* 2011), at first glance a typical δ Sct star. However, at high frequencies, the authors detected modes approximately equidistantly spaced, as expected for high radial-order p modes, which are not combination frequencies. The κ mechanism, as it is known, cannot excite a continuous frequency region as observed in HD 187547 (Pamyatnykh 2000). Spectroscopic observations exclude that the peaks at high frequencies originate from a possible companion; such a star would be an A- or F-type star and would be visible in the spectrum, which is not. The measured large separation $\Delta\nu$ predicts $\nu_{\rm max}$ to be exactly where the mode with the highest amplitude in the supposed stochastic region is detected. New data, however showed that the mode lifetimes are longer than the observing run, suggesting coherence over more than 450 days (from scaling relations the mode lifetimes are suggested to decrease with increasing $T_{\rm eff}$). This is not compatible with the interpretation of 'pure' solar-like oscillations. The word 'pure' in this context should emphasise that there are no studies on how the κ mechanism interacts with the stochastic excitation, especially in the case where the periods of pulsations have similar time scales.

Very recent results, in collaboration with Margarida Cunha and Günter Houdek, suggest the presence of a new excitation mechanism for at least the $T_{\rm eff}$ domain of HD 187547, i.e. 7500 K. The models, using a non-local and time-dependent treatment of convection (Houdek *et al.* 1999 and references therein), reproduce the major part of the observed modes in HD 187547, as being excited by the turbulent pressure and not the κ mechanism.

The turbulent pressure can be understood as the dynamical component of convection, associated with the transport of momentum (details will be published in a dedicated article). This proves that convection in the envelope of these stars still plays an important role and can excite pulsations even if not stochastically.

3. O- and B-type stars

When it comes to stellar structure, O- and B-type stars are usually described as having a convective core and a purely radiative envelope. Cantiello *et al.* (2009), however, show that this is not necessarily true (Fig. 1). Even though it might comprise a negligible part of the stellar mass, subsurface convection occurring predominantly in the Z bump can not only lead to variability in the observed microturbulent velocity but also influence the stellar evolution of massive stars significantly. This is because the photospheric motion caused by convection can strongly influence mass loss by affecting the stellar winds known to exist in these stars (see Cantiello *et al.* 2009). Furthermore, a convection zone close to the surface can also induce magnetic fields, which may favour mass loss and more importantly, loss of angular momentum.

Samadi *et al.* (2010) investigated from a theoretical point of view whether the turbulent convection can stochastically excite g modes in massive stars. They find that, while low radial-order g modes might be excited by the convection in the core, the convective envelope is primarily exciting high radial-order g modes. However, the same authors predict the amplitudes to be too low to be observed even with *CoRoT* or *Kepler*. Belkacem *et al.* (2010), on the other hand, find in their exploratory models of a 10 $M_\odot$ that both the core and the convective outer layers associated with the Fe bump can efficiently drive stochastic oscillations with detectable amplitudes. However, the authors also comment that the convective properties in these temperature domains are poorly understood. Shiode *et al.* (2013) arrive at a similar conclusion, namely that the convection in the core can stochastically excite g modes to observable amplitudes. They expect the amplitudes to be highest for stars with masses $\geqslant 5$ $M_\odot$. Also from 3D MHD and numerical simulations Browning *et al.* (2004) and Rogers *et al.* (2012) find that convection in the core excites internal gravity modes in stars more massive than the Sun. From the observational point of view there is strong evidence for the presence of stochastically excited oscillations in massive O-type stars observed with *CoRoT*. HD 46149 (O8.5 V) appears to oscillate in stochastic modes fulfilling the scaling laws as expected for solar-type stars and also showing ridges for modes of equal degree but consecutive radial orders as those observed for Sun-like stars (Degroote *et al.* 2010). The κ mechanism is not expected to drive the observed oscillations. Besides that, the temporal variability argues in favour of their stochastic nature. Blomme *et al.* (2011) find 300 significant frequencies in HD 46966 (O8 V), all displaying temporal variability which might be due to granulation noise. The same authors find similar behaviour in other two O type stars.

3.1. *SPB stars*

SPB stars pulsate in high radial-order g modes excited by the κ mechanism associated with the opacity in the Z bump (Fig. 1). Their instability region overlaps that of the β Cep stars, suggesting the presence of both type of pulsations. As in the case of γ Dor and δ Sct stars, observations are consistent with theory (e.g. Handler *et al.* 2009). According to Cantiello *et al.* (2009), the only subsurface convective layer present in these stars is in the He II ionisation zone, but it is shallow and inefficient. No solar-like oscillations driven by the convection in the envelope are expected in SPB stars. If there would be any, these would be excited by the convection in the core but none were observed so far.

3.2. *β Cep stars*

The β Cep stars pulsate in p and g modes excited by the κ mechanism in the Z bump. From Cantiello *et al.* (2009) and the theoretical work summarised above, these stars are expected to have a substantial convection layer in the Z bump. Belkacem *et al.* (2009) reported the detection of solar-like oscillations and opacity driven modes in the β Cep star V1449 Aql. The authors argue that the peaks observed at high frequencies are consistent with stochastic oscillations, because of the temporal variability and the detected spacing interpreted as the large frequency separation. Aerts *et al.* (2011), on the other hand, can seismically reproduce the pulsational behaviour of V1449 Aql without invoking the presence of solar-like oscillations but only κ mechanism excitation. They ascribe the power at higher frequencies to non-linear resonant mode coupling between the dominant radial fundamental mode and many other low-order p modes. Degroote (2013) delivers a third possible explanation for the observed frequency spectrum of V1449 Aql, arguing that chaotic behaviour of the very dominant mode can qualitatively reproduce the observed frequency spectrum. As no other β Cep pulsators have been found to show solar-like oscillations, the case of V1449 Aql is still a matter of debate.

3.3. *Be stars*

Be stars are stars showing emission in their spectra originating from a circumstellar disk. These can be found spreading over several spectral classes (O, B and early A). So far the Be phenomenon is still not fully understood but can be related to rapid rotation, and meanwhile to pulsations (of SPB and β Cep type); the latter were found to be present in all Be stars examined so far (Neiner *et al.* 2013). Other types of variability attributed to rotation, magnetic fields, stellar winds and outbursts were also observed in these stars (for a complete review on Be stars see Porter & Rivinius 2003). A very fortunate case is the Be star HD 49330 (B0 IVe) which was observed during an outburst with *CoRoT* as well as from ground (Huat *et al.* 2009). In this star the authors find a clear correlation between the outburst and the presence of g modes. While during the quiescent phase the pulsations are predominantly in the p-mode regime, during the outburst the g modes are enhanced to higher amplitudes and are also very unstable indicating short mode lifetimes. The modelling of this star, even considering rapid rotation, shows that the κ mechanism cannot excite the observed g modes. A very recent study of another Be star (HD 51452, Neiner *et al.* 2012) suggests that the g modes observed in the two afore-mentioned Be stars are in fact stochastically driven gravito-inertial modes. These oscillations are excited by the convection in the core and enhanced to observable amplitudes by the rapid rotation (see Neiner *et al.* 2012 and Mathis *et al.* 2013 for details).

4. Conclusions

The potential to detect stochastically excited oscillations in stars significantly more massive than the Sun became possible only with the space missions *CoRoT* and *Kepler*. It is not only the quality of the data which is outstanding, but also the uninterrupted and long observing seasons which demonstrated that we need to revisit our knowledge on pulsation mechanisms (Fig. 1). Convection is one of the most complex processes in astrophysics and therefore often neglected with the argument that convection zones in stars hotter than early F are negligible. However, the latest results reviewed in this article demonstrate that the impact of convection needs to be taken into account.

Acknowledgements

Funding for the Stellar Astrophysics Centre (SAC) is provided by The Danish National Research Foundation. The research is supported by the ASTERISK project

(ASTERoseismic Investigations with SONG and *Kepler*) funded by the European Research Council (Grant agreement no.: 267864).

References

Aerts, C., Briquet, M., Degroote, P., Thoul, A., & van Hoolst, T. 2011, *A&A*, 534, A98

Antoci, V., Handler, G., Campante, T. L., *et al.* 2011, *Nature*, 477, 570

Antoci, V., Handler, G., Grundahl, F., *et al.* 2013, *MNRAS*, 435, 1563

Balmforth N. J., Cunha, M. S., Dolez, N., Gough, D. O., & Vauclair, S. 2001, *MNRAS*, 323, 362

Balona, L. A. & Dziembowski, W. A. 2011, *MNRAS*, 417, 591

Belkacem, K., Samadi, R., Goupil, M.-J., *et al.* 2009, *Science*, 324, 1540

Belkacem, K., Dupret, M.-A., & Noels, A. 2010, *A&A*, 510, A6

Bigot, L. & Dziembowski, W. A. 2002, *A&A*, 391, 235

Blomme, R., Mahy, L., Catala, C., *et al.* 2011, *A&A*, 533, A4

Brown, T. M., Gilliland, R. L., Noyes, R. W., & Ramsey, L. W. 1991, *ApJ*, 368, 599

Browning, M. K., Brun, A. S., & Toomre, J. 2004, in: J. Zverko, J. Žižňovský, S. J. Adelman, & W. W. Weiss (eds.), *The A-Star Puzzle*, Proc. IAU Symposium No. 224 (Cambridge: Cambridge University Press), p. 149

Cantiello, M., Langer, N., Brott, I., *et al.* 2009, *A&A*, 499, 279

Cox, J. P. 1963, *ApJ*, 138, 487

Christensen-Dalsgaard, J. 2000, *ASP-CS*, 210, 454

Degroote, P. 2013, *MNRAS*, 431, 255

Degroote, P., Briquet, M., Auvergne, M., *et al.* 2010, *A&A*, 519, A38

Eddington, A. S. 1919, *MNRAS*, 79, 177

García Hernández, A., Moya, A., Michel, E., *et al.* 2009, *A&A*, 506, 79

Grigahcène, A., Antoci, V., Balona, L. A., *et al.* 2010, *ApJ*, 713, L192

Guzik, J. A., Kaye, A. B., Bradley, P. A., Cox, A. N., & Neuforge, C. 2000, *ApJ*, 542, L57

Gray, D. F. & Nagel, T. 1989, *ApJ*, 341, 421

Handler, G., Matthews, J. M., Eaton, J. A., *et al.* 2009, *ApJ*, 698, 56

Houdek, G., Balmforth, N. J., Christensen-Dalsgaard, J., & Gough, D. O. 1999, *A&A*, 351, 582

Huat, A.-L., Hubert, A.-M., Baudin, F., *et al.* 2009, *A&A*, 506, 95

Huber, D., Bedding, T. R., Stello, D., *et al.* 2011, *ApJ*, 743, 143

Kallinger, T. & Matthews, J. M. 2010, *ApJ*, 711, L35

Kjeldsen, H. & Bedding, T. R. 1995, *A&A*, 293, 87

Kurtz, D. W. 1982, *MNRAS*, 200, 807

Landstreet, J. D., Kupka, F., Ford, H. A., *et al.* 2009, *A&A*, 503, 973

Mathis, S., *et al.*, 2013, *A&A*, in press

Neiner, C., Floquet, M., Samadi, R., *et al.* 2012, *A&A*, 546, A47

Neiner, C., *et al.* 2013, H. Shibahashi and A. E. Lynas-Gray (eds.), *Progress in Physics of the Sun and Stars: A New Era in Helio- and Asteroseismology*, *ASP-CS*, in press

Pamyatnykh, A. A. 2000, *ASP-CS*, 210, 215

Poretti, E., Michel, E., Garrido, R., *et al.* 2009, *A&A*, 506, 85

Porter, J. M. & Rivinius, T. 2003, *PASP*, 115, 1153

Rogers, T. M., Lin, D. N. C., & Lau, H. H. B. 2012, *ApJ*, 758, 6

Royer, F. 2009, *Lecture Notes in Physics*, 765, 207

Samadi, R., Goupil, M.-J., & Houdek, G. 2002, *A&A*, 395, 563

Samadi, R., Belkacem, K., Goupil, M.-J., Dupret, M.-A., Brun, A. S., & Noels, A. 2010, *Ap&SS*, 328, 253

Shiode, J. H., Quataert, E., Cantiello, M., & Bildsten, L. 2013, *MNRAS*, 430, 1736

Simon, T., Ayres, T. R., Redfield, S., & Linsky, J. L. 2002, *ApJ*, 579, 800

Théado, S., Alecian, G., LeBlanc, F., & Vauclair, S. 2012, *A&A*, 546, A100

Tkachenko, A., Lehmann, H., Smalley, B., Debosscher, J., & Aerts, C. 2012, *MNRAS*, 422, 2960

Uytterhoeven, K., Moya, A., Grigahcène, A., *et al.* 2011, *A&A*, 534, A125

Precision Asteroseismology
Proceedings IAU Symposium No. 301, 2013
J. A. Guzik, W. J. Chaplin, G. Handler & A. Pigulski, eds.

© International Astronomical Union 2014
doi:10.1017/S1743921313014555

Energy of solar-like oscillations in red giants

Mathieu Grosjean[1]†**, Marc-Antoine Dupret**[1]**, Kevin Belkacem**[2]**,
Josefina Montalbán**[1]**, and Reza Samadi**[2]

[1]Institut d'astrophysique et de géophysique, Université de Liège,
Allée du 6 Août 17, B-4000 Liège, Belgium
email: grosjean@astro.ulg.ac.be

[2]LESIA Observatoire de Paris-Meudon,
5 place Jules Janssen, F-92195 Meudon, France

Abstract. *CoRoT* and *Kepler* observations of red giants reveal a large variety of spectra of nonradial solar-like oscillations. So far, we understood pretty well the link between the global properties of the star (radius, mass, evolutionary state) and the properties of the oscillation spectrum ($\Delta\nu$, $\nu_{\max}$, period spacing). We are interested here in the theoretical predictions of two other components of a power spectrum, the mode linewidths and heights. The study of the energy of the oscillations is of great importance to predict the peak parameters in the power spectrum. We will discuss circumstances under which mixed modes are detectable for red-giant stellar models from 1 to 2 $M_\odot$, with emphasis on the effect of the evolutionary status of the star along the red-giant branch on theoretical power spectra.

Keywords. red giants, solar-like oscillations, mixed-mode lifetimes

1. Introduction

We consider three stellar models of 1.5 $M_\odot$ on the red-giant branch (Fig. 1). These models have been investigated in the adiabatic case by Montalbán & Noels (2013). We have also selected models of 1.0, 1.7 and 2.1 $M_\odot$ which have the same number of mixed modes in a large separation as in model B of 1.5 $M_\odot$ (Fig. 1). All of these models (see details in Table 1) were computed with the code ATON (Ventura *et al.* 2008) using the mixing-length theory (MLT) for the treatment of the convection with $\alpha_{\mathrm{MLT}} = 1.9$. The initial chemical composition is $X = 0.7$ and $Z = 0.02$.

Table 1. Characteristics of the models studied

Model	Mass [$M_\odot$]	Radius [$R_\odot$]	T_{eff} [K]	$\log g$
A	1.5	5.17	4809	3.19
B	1.5	7.31	4668	2.88
C	1.5	11.9	4455	2.46
E	1.0	6.29	4549	2.84
F	1.7	8.06	4682	2.85
G	2.1	10.6	4665	2.72

To compute the mode lifetimes, we used the nonadiabatic pulsation code MAD (Dupret *et al.* 2002) with a nonlocal time-dependent treatment of convection (TDC, Grigahcène *et al.* 2005 (G05), Dupret *et al.* 2006 (D06)). The amplitudes were computed using a stochastic excitation model (Samadi & Goupil 2001 (SG01)) with solar parameters for the description of the turbulence in the upper part of the convective envelope.

† This work is supported through a PhD grant from the F.R.I.A.

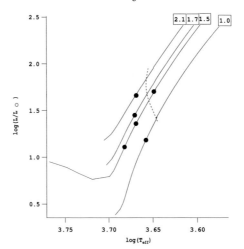

Figure 1. Evolutionary tracks for stars of 1.0, 1.5, 1.7, and 2.1 $M_\odot$ (from right to left). The models studied here are indicated by black dots. The dashed line represents the detectability limit we have found.

2. Energetic aspects

The damping rate η of a mode is given by the expression:

$$\eta = -\frac{\int_V dW}{2\sigma I |\xi_r(R)|^2 M}, \tag{2.1}$$

where σ is the angular frequency of the mode, I the dimensionless mode inertia, ξ_r the radial displacement and R and M the total radius and mass of the star. In deep radiative zones we can write the following asymptotic expression for the work integral (Dziembowski 1977, Van Hoolst *et al.* 1998, Godart *et al.* 2009):

$$-\int_{r_0}^{r_c} \frac{dW}{dr} dr \simeq \frac{K[l(l+1)]^{3/2}}{2\sigma^3} \int_{r_0}^{r_c} \frac{\nabla_{\rm ad} - \nabla}{\nabla} \frac{\nabla_{\rm ad} N g L}{p r^5} dr. \tag{2.2}$$

The factor Ng/r^5 in this expression indicates that the radiative damping increases with the density contrast.

For solar-like oscillations in red giants, in the upper part of the convective envelope, the thermal time scale, the oscillation period and the timescale of most energetic turbulent eddies are of the same order. Hence it is important for the estimation of the damping to take into account the interaction between convection and oscillations. This is made using a nonlocal, time-dependent treatment of the convection which takes into account the variations of the convective flux and of the turbulent pressure due to the oscillations (see G05, D06). The TDC treatment involves a complex parameter β in the closure term of the perturbed energy equation. It is adjusted so that the depression of the damping rates occurs at the frequency of maximum oscillation power, $\nu_{\rm max}$, as suggested by Belkacem *et al.* (2012).

The estimation of the power injected into the oscillations by the turbulent Reynolds stresses is made through a stochastic excitation model (SG01). To compute the height (H) of a mode in the power spectrum we have to distinguish between:
— resolved modes, i.e. those with $\tau < T_{\rm obs}/2$: $H = V(R)^2 \times \tau$,
— unresolved modes, i.e. those with $\tau \geqslant T_{\rm obs}/2$: $H = V(R)^2 \times T_{\rm obs}/2$,
where $V(R)$ is the amplitude of the oscillation at the surface, τ the lifetime of the mode and $T_{\rm obs}$ the duration of observations. In this study, we used $T_{\rm obs} = 1$ year.

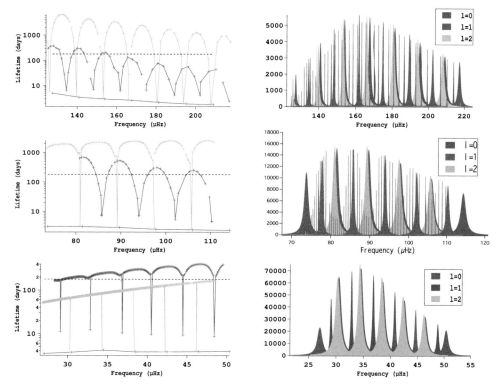

Figure 2. Lifetimes (left) and theoretical power spectra (right) for the model A (top), B (middle) and C (bottom).

3. Results

From models A to C (Fig. 2), we see a progressive attenuation of the modulation of the lifetimes due to an increase of the radiative damping. Because of the high radiative damping, mixed modes are not detectable in model C. These results indicate a theoretical detectability limit for mixed modes on the red-giant branch. For a $1.5\,M_\odot$ star, with one year of observations, this limit occurs around $\nu_{\max} \simeq 50\ \mu\text{Hz}$ and $\Delta\nu \simeq 4.9\ \mu\text{Hz}$.

The models with the same number of mixed modes in one large separation present similar lifetime patterns and similar power spectra (Fig. 3). Since the number of mixed modes in a large separation is approximately given by

$$\frac{n_g}{n_p} \simeq \frac{\Delta\nu}{\Delta P \nu_{\max}^2} \quad \propto \quad \left[\int \frac{N}{r}dr\right] M^{3/2} R^{5/2} T_{\text{eff}} \tag{3.1}$$

we find that this expression provides a good theoretical proxy for the detectability of mixed modes.

4. Conclusions

These results extend to lower masses those found by Dupret *et al.* (2009). On the red-giant branch, mixed modes are detectable until the radiative damping becomes too important. For a $1.5\,M_\odot$ star with 1 year of observations, mixed modes are detectable for stars with $\nu_{\max} \gtrsim 50\ \mu\text{Hz}$ and $\Delta\nu \gtrsim 4.9\ \mu\text{Hz}$. Whatever the mass, models with the same number of mixed modes in a large separation will present similar power spectra

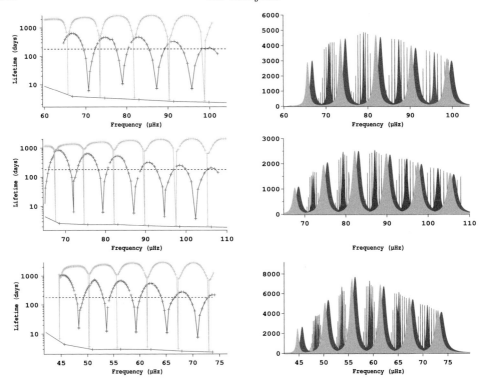

Figure 3. Lifetimes (left) and theoretical power spectra (right) for the model E (top), F (middle) and G (bottom).

and the same detectability of mixed modes. This criterium can thus be used to predict the detectability limit of mixed modes in red giants of different masses.

References

Belkacem, K., Dupret, M.-A., Baudin, F., Appourchaux, T., Marques, J. P., & Samadi, R. 2012, *A&A*, 540, L7

Dupret, M.-A., De Ridder, J., Neuforge, C., Aerts, C., & Scuflaire, R. 2002, *A&A*, 385, 563

Dupret, M.-A., Goupil, M.-J., Samadi R., & Grigahcène, A., Gabriel M. 2006, in: K. Fletcher (ed.), Proceedings of SOHO 18/GONG 2006/HELAS I, *Beyond the Spherical Sun*, ESA SP-624, p. 78.1

Dupret, M.-A., Belkacem, K., Samadi, R., *et al.* 2009, *A&A*, 506, 57

Dziembowski, W. A. 1977, *AcA*, 27, 95

Godart, M., Noels, A., Dupret, M.-A., & Lebreton, Y. 2009, *MNRAS*, 396, 1833

Grigahcène, A., Dupret, M.-A., Gabriel, M., Garrido, R., & Scuflaire, R. 2005, *A&A*, 434, 1055

Montalbán, J. & Noels, A. 2013, in: J. Montalbán, A. Noels, & V. Van Grootel (eds.), 40th Liège International Astrophysical Colloquium. *Ageing Low Mass Stars: From Red Giants to White Dwarfs*, EPJ Web of Conferences, 43, 03002

Samadi, R. & Goupil, M.-J. 2001, *A&A*, 370, 136

Van Hoolst, T., Dziembowski, W. A., & Kawaler, S. D. 1998, *MNRAS*, 297, 536

Ventura, P., D'Antona F. & Mazzitelli, I. 2008, *Ap&SS*, 316, 93

Precision Asteroseismology
Proceedings IAU Symposium No. 301, 2013
J. A. Guzik, W. J. Chaplin, G. Handler & A. Pigulski, eds.

© International Astronomical Union 2014
doi:10.1017/S1743921313014567

The evolution of the internal rotation of solar-type stars

Maria Pia Di Mauro[1], Rita Ventura[2], Daniela Cardini[1], Jørgen Christensen-Dalsgaard[3], Wojciech A. Dziembowski[4] and Lucio Paternò[2]

[1]INAF-IAPS Istituto di Astrofisica e Planetologia Spaziali, Via del Fosso del Cavaliere 100, 00133 Roma, Italy email: maria.dimauro@inaf.it

[2]INAF-Astrophysical Observatory of Catania, Via S. Sofia 78, 95020, Catania, Italy

[3]Stellar Astrophysics Centre, Department of Physics and Astronomy, Aarhus University, Ny Munkegade 120, DK-8000 Aarhus C, Denmark

[4]Warsaw University Observatory, Al. Ujazdowskie 4, 00-478 Warszawa, Poland

Abstract. We discuss the potential of asteroseismic inversion to study the internal dynamics of solar-type stars and to reconstruct the evolution of the internal rotation from the main sequence to the red-giant phase. In particular, we consider the use of gravity and mixed modes and the application of different inversion methods.

Keywords. stars: oscillations, stars: rotation, stars: evolution, Sun: helioseismology, Sun: evolution, Sun: rotation

1. Introduction

The knowledge of the variation of the stellar rotation with radius and time is essential for a complete understanding of the evolution of the internal properties of the Sun and other solar-type stars and for explaining the loss of angular momentum occurring from the main sequence to late evolution stages.

In addition, the determination of the internal angular velocity would help to constrain dynamo theories and the connection between rotation and convection and to establish the role played by the mixing in the stellar core.

Until recently, these and similar issues could not be addressed, since the internal rotation was inaccessible to observation. In the 90's, the development of helioseismology and the high quality of oscillation data provided by the satellite *SOHO* (Scherrer *et al.* 1995) and ground-based networks demonstrated that the internal rotation of the Sun could be investigated through the measurement of the splittings of the oscillation frequencies. In fact, the rotation breaks the spherical symmetry of the solar structure and splits the frequency of each oscillation mode by an amount whose magnitude is related to the angular velocity (Cowling & Newing 1949).

More recently, the unprecedented frequency resolution supplied by the space missions *Kepler* (Borucki *et al.* 2010) and *CoRoT* (Baglin *et al.* 2006) has allowed accurate measurements of rotational splittings in stars in different evolutionary stages, and in particular in red-giant phase (Beck *et al.* 2012, Mosser *et al.* 2012, Deheuvels *et al.* 2012). The results indicate a fast core rotation in stars ascending the red-giant branch and a spin-down of the core in the red-clump stars. This implies that a significant angular momentum transport takes place during the last stages of red-giant branch evolution.

In this article we consider the use of inversion techniques, applied with success to the case of the Sun, in order to reconstruct rotation history in solar-type stars from the main sequence to the red-giant phase of evolution.

2. Asteroseismic inversion

Asteroseismic inversion is a powerful tool that allows one to estimate the physical properties of the stars, by solving integral equations expressed in terms of the experimental data.

Inversion techniques are well known and applied with success to several branches of physics. Applications to helioseismic data have been studied extensively and inversion methods and techniques have been reviewed and compared by several authors, leading to extraordinary results about the internal rotation of the Sun (e.g., Di Mauro 2003).

In order to quantify the internal rotation it is possible to invert the following equation, relating the set of the observed splittings $\delta\nu_{n,l}$ to the internal rotation $\Omega(r)$:

$$\delta\nu_{n,l} = \int_0^R \mathcal{K}_{n,l}(r)\frac{\Omega(r)}{2\pi}\,dr + \sigma_{n,l} \tag{2.1}$$

where $\sigma_{n,l}$ are the errors in the measured splittings, and $\mathcal{K}_{n,l}(r)$ are the kernel functions, calculated on the stellar model for each mode $i \equiv (n,l)$. The inverse problem consists of finding the function $\Omega(r)$ by using a finite set of observed splittings $i = 1,\dots,M$ and their associated errors σ_i. This is an 'ill-posed' problem which can be solved by employing an appropriate technique.

The present article deals with the application of the Optimally Localized Averaging (OLA) method based on the original idea of Backus & Gilbert (1970), which allows one to estimate a localized weighted average of the angular velocity $\bar{\Omega}(r_0)$ at selected target radii '$\{r_0\}$' by means of a linear combination of all the data:

$$\bar{\Omega}(r_0) = \sum_i^M c_i(r_0)2\pi\delta\nu_i = \sum_{i=1}^M c_i(r_0)\int_0^R \mathcal{K}_i(r)\Omega(r)dr \tag{2.2}$$

where $c_i(r_0)$ are the inversion coefficients and $K(r_0,r) = \sum_{i=1}^M c_i(r_0)\mathcal{K}_i(r)$ are the so-called averaging kernels. We look for the coefficients $c_i(r_0)$ which minimize the propagation of the errors and the spread of the averaging kernels.

The errors of the solutions are given by $\delta\bar{\Omega}(r_0) = \left[\sum_{i=1}^M c_i^2(r_0)\sigma_i^2\right]^{1/2}$, while the radial spatial resolution is assumed to be the half-width at half-maximum of the averaging kernels.

The code, developed for the solar rotation (Paternò *et al.* 1996), has been adapted here to be applied to any evolution phase. We also considered the method in the variant form proposed by Pijpers & Thompson (1992) and known as SOLA (Subtractive Optimally Localized Averaging), making attempts to fit the averaging kernel to a Gaussian function of appropriate width, centred at the target radius (Di Mauro & Dziembowski 1998).

3. Inversion results

The properties of the inversion depend both on the mode selection and on the observational errors, which characterize the mode set to be inverted. The large set of data available for the Sun, including modes with harmonic degree $l = 1 - 100$, allows to sound the rotational profile from the core to the upper layers. However, inferences of the interior

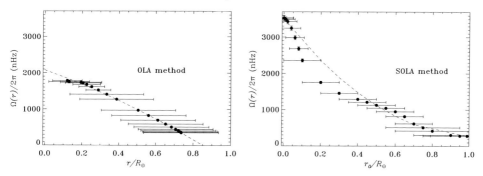

Figure 1. Internal rotation of the Sun as deduced by inversion of $l = 1$ rotational splittings observed with GOLF/*SOHO* with the OLA (left panel) and the SOLA (right panel) methods. The radial resolution is equal to the width of the averaging kernels, while the error in the solution is the standard deviation calculated on the observational uncertainties of the data. Dashed lines are polynomial fitting functions. Data have been kindly provided by the GOLF/*SOHO* team.

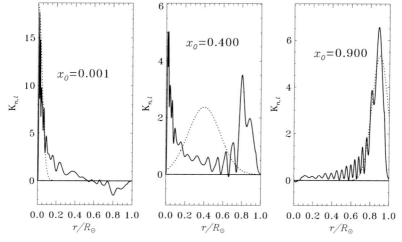

Figure 2. Averaging kernels (solid lines) at selected target radii for the inversion of $l = 1$ GOLF/*SOHO* data as obtained by using the SOLA method and including g modes. Dashed lines indicate the Gaussian target functions.

of stars other than the Sun appear to be much more complicated and less outstanding in terms of achievable results. The large stellar distances, the point-source character of the stars, and the low amplitude of the oscillations restrict the asteroseismic studies to the use of small sets of data characterized by modes with only low harmonic degrees ($l \leqslant 3$).

At present, rotational splittings are not yet available for main-sequence stars other than the Sun. However, it is possible to treat the Sun as a star by inverting a set of only low-degree modes. The variation of the Sun's angular velocity with the radius, shown in Fig. 1, has been determined by inversions of a set of 15 rotational splittings with $l = 1$, including 6 g modes and 9 p modes, detected by the GOLF instrument flying on *SOHO* (García *et al.* 2007, 2008). The results in the core and below the surface agree with those of García *et al.* (2011) obtained by inverting a full set of data with $l = 1 - 100$: the core appears to rotate faster than the surface. The ratio $\Omega_c/\Omega_{surf} \simeq 4 - 8$, where Ω_c and Ω_{surf} are respectively the angular velocity in the core and at the surface. It is possible to notice that the SOLA method allows one to find solutions better localized than the OLA method, but both fail to localize solutions between $0.2 - 0.8\, R_\odot$ as shown by the averaging kernels plotted in Fig. 2.

We have then considered the star KIC 4448777 located at the beginning of the ascending red-giant branch observed by the *Kepler* satellite (Di Mauro *et al.* 2013). The present asteroseismic inversion has been carried out by employing the rotational splittings of 15 mixed modes with $l = 1$. The gravity component of the mixed modes allows one to find solutions well localized in the core, while the acoustic component of the modes give the possibility to localize solutions in the upper layers. As in the case of the Sun, it is not possible to obtain localized kernels between the core and the upper layers.

We confirm the previous findings in other red giants (Deheuvels *et al.* 2012) that the core rotates faster than the surface, and in the considered star, at a rate of about 955 ± 25 nHz. The angular velocity, through the largest part of the convection zone, from the core towards the surface, decreases with increasing distance from the centre, reaching a value of about 84 ± 7 nHz at $0.9R$, so that $\Omega_c/\Omega_{surf} \simeq 10 - 12$. The spectroscopic value of $v \sin i$ is not known.

4. Conclusions

It is possible to conclude that helioseismic tools can be extended to other stars and the internal rotation of solar-type stars can be inferred by using inversion techniques to reconstruct the evolution of stellar rotation. Mixed modes, like the gravity modes, allow one to find solutions in the inner core.

The results seem to indicate that from the main sequence to early red-giant phases the rotation slows down both in the envelope and in the core, while the ratio Ω_c/Ω_{surf} is almost conserved.

Considering the fact that the internal rotation of the core, as predicted by the current theoretical models of red giants, is higher compared to our results, it is necessary to investigate more efficient mechanisms of angular momentum transport acting on the appropriate timescales during the different phases of stellar evolution.

Finally, we expect that measurements of rotational splittings, which will be soon available for subgiants and more evolved red-giant stars, will shed some light on the above picture and on the question of angular momentum transport.

References

Backus, G. E. & Gilbert, F. 1970, *Phil. Trans. R. Soc. London*, 266, 123
Baglin, A., Auvergne, M., Boisnard, L., *et al.* 2006, in: 36th COSPAR Scientific Assembly Plenary Meeting, 36, 3749
Beck, P., Montalbán, J., Kallinger, T., *et al.* 2012, *Nature*, 481, 55
Borucki, W. J., Koch, D., Basri, G., *et al.* 2010, *Science*, 327, 977
Cowling, T. G. & Newing, R. A. 1949, *ApJ*, 109, 149
Deheuvels, S., García, R. A., Chaplin, W. J., *et al.* 2012, *ApJ*, 756, 19
Di Mauro, M. P. 2003, *Lectures Notes in Physics*, 599, 31
Di Mauro, M. P. & Dziembowski, W. A. 1998, *MemSAIt*, 69, 559
Di Mauro, M. P., Cardini, D., Ventura, R., *et al.* 2013, in: 40th Liège International Astrophysical Colloquium *Ageing Low Mass Stars: From Red Giants to White Dwarfs*, EPJ Web of Conferences, 43, 03012
García, R. A., Turck-Chièze, S., Jiménez-Reyes, S. J., *et al.* 2007, *Science*, 316, 1591
García, R. A., Mathur, S., Ballot, J., Eff-Darwich, A., Jiménez-Reyes, S. J., & Korzennik, S. G. 2008, *Solar Phys.*, 251, 119
García, R. A., Salabert, D., Ballot, J., *et al.* 2011, *JPhCS*, 271, 2046
Mosser, B., Goupil, M.-J., Belkacem, K., *et al.* 2012, *A&A*, 548, A10
Paternò, L., Sofia, S., & Di Mauro, M. P. 1996, *A&A*, 314, 940
Pijpers, F. P. & Thompson, M. J. 1992, *A&A*, 262, L33
Scherrer, P. H., Bogart, R. S., Bush, R. I., *et al.* 1995, *Solar Phys.* 162, 129

Precision Asteroseismology
Proceedings IAU Symposium No. 301, 2013
J. A. Guzik, W. J. Chaplin, G. Handler & A. Pigulski, eds.

© International Astronomical Union 2014
doi:10.1017/S1743921313014579

Sunquakes and starquakes

Alexander G. Kosovichev[1,2,3]

[1] Hansen Experimental Physics Laboratory, Stanford University, Stanford, CA 94305, USA
email: `AKosovichev@solar.stanford.edu`
[2] Big Bear Solar Observatory, Big Bear City, CA 92314, USA
[3] Crimean Astrophysical Observatory, Nauchny, Crimea 98409, Ukraine

Abstract. In addition to well-known mechanisms of excitation of solar and stellar oscillations by turbulent convection and instabilities, the oscillations can be excited by an impulsive localized force caused by the energy release in solar and stellar flares. Such oscillations have been observed on the Sun ('sunquakes'), and created a lot of interesting discussions about physical mechanisms of the impulsive excitation and their relationship to the flare physics. The observation and theory have shown that most of a sunquake's energy is released in high-degree, high-frequency p modes. In addition, there have been reports on helioseismic observations of low-degree modes excited by strong solar flares. Much more powerful flares observed on other stars can cause 'starquakes' of substantially higher amplitude. Observations of such oscillations can provide new asteroseismic information and also constraints on mechanisms of stellar flares. I discuss the basic properties of sunquakes, and initial attempts to detect flare-excited oscillations in *Kepler* short-cadence data.

Keywords. Sun: flares, oscillations; stars: flares, oscillations

1. Introduction

Kepler observations have led to a discovery that stellar flares representing impulsive powerful energy releases occur not only in M-type dwarfs (UV Ceti-type variables) but also in a wide range of A-F type stars (Balona 2012; Maehara *et al.* 2012; Walkowicz *et al.* 2011). Previously, it was believed the F- and A-type stars do not have flaring activity. The discovery of super-flares on solar-type stars raised questions about the possibility of such flares on the Sun. These results triggered new debates about the physical mechanism of stellar flares and the relationship to the dynamo mechanism. Recent observations of solar flares from the Solar Dynamics Observatory (SDO) have found that the white-light flares are often accompanied by excitation of acoustic oscillations (p modes) in the solar interior, so-called 'sunquakes'. However, not all flares reveal the sunquakes. The relationship of the interior and atmospheric response to the energy release and optical emission is not understood, and is currently a subject of detailed studies in heliophysics. The flare-excited oscillations are mostly observed as local seismic waves, and while the theory predicts that the global acoustic waves are also excited, their amplitude is significantly lower than the amplitude of stochastically excited oscillations, therefore, the detection reports have been controversial. However, in the case of significantly more powerful stellar flares the impact on the star's surface is much greater. This can lead to excitation of the global low-degree oscillations to significantly higher amplitudes than on the Sun. Our preliminary study of the available short-cadence (SC) data provides indications of such 'starquakes'. However, a statistical study for a large sample of stars and longer observing intervals are needed.

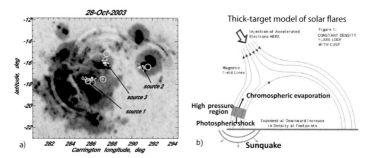

Figure 1. a) Observations of acoustic wavefronts (projected onto a sunspot image) during a large solar flare. Yellow and green circles show places of the hard X-ray and γ-ray emissions, and blue patches are Doppler-shift signals > 1 km s^{-1} from the localized flare impacts (Kosovichev 2006); b) illustration of excitation of 'sunquakes' in the 'thick-target' model of solar flares.

2. Solar flares and sunquakes

Recent observations of solar flares on the NASA space missions *SOHO* and *SDO* revealed that some flares result in excitation of acoustic oscillations in the solar interior, dubbed 'sunquakes'. The oscillations are observed in the Doppler shift and intensity variations as expanding circular waves (Fig. 1a). They are excited due to momentum and energy impulse in the solar photosphere, which is also observed in solar flares, but the mechanism of this impact is unknown. The expanding waves represent high-degree p modes; however, the stellar oscillation theory predicts that the global low-degree modes are also excited. Their detection has been reported from a statistical analysis of the correlation between the flare soft X-ray signals (observed by the *GOES* satellites) and the total solar irradiance measurements from the space observatory *SOHO* (Karoff & Kjeldsen 2008). However, there was no unambiguous detection of the whole-Sun oscillations caused by individual flare events. Indeed, the oscillation theory predicts that the amplitudes of the low-degree modes of sunquakes are significantly lower than the amplitudes of stochastically excited oscillations (Fig. 2a) (Kosovichev 2009).

The solar observations show that the sunquake events are mostly associated with compact or confined flares that do not produce coronal mass ejections and are characterized by the energy release in compact magnetic configurations in the lower atmosphere. Such flares usually produce white-light or continuum optical emission, and they do not necessarily have strong X-ray emission. The origin of such flare energy release is mysterious, and the subject of intensive investigation. These flares and sunquakes challenge the standard 'thick-target' flare model, which assumes that most of the flare energy is released in the form of high-energy particles due to magnetic reconnection processes in the coronal plasma. The particles travel along the magnetic field lines and heat the upper chromosphere to high temperature, creating a high-pressure region, which expands, producing a plasma eruption ('chromospheric evaporation') and a downward-traveling radiative shock (Fig. 1b). This shock can reach the photosphere and excite acoustic oscillations. However, the numerical simulations of the flare hydrodynamics show that the energy of the shock may be not sufficient to explain the observed oscillations, and also the photospheric impact is often observed at the beginning of the flare impulsive phase or even during the pre-heating phase, well before the maximum of the hard and soft X-ray emissions that are supposedly produced by the particle interaction with the higher chromosphere. Nevertheless, if a similar energy release mechanism works on other stars then the amplitudes of the flare-excited oscillations can be expected to be several orders of magnitude stronger than on the Sun.

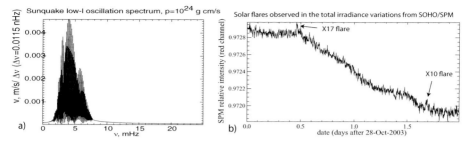

Figure 2. a) The theoretical spectrum of the flare excited oscillations has a maximum around the acoustic cut-off frequency ∼5 mHz (Kosovichev 2009); (b) The signal of two large solar flares observed in the total irradiance measurements (red channel) with the SPM instrument on the *SOHO* spacecraft.

3. Initial analysis of starquakes

Observations of stellar flares from *Kepler* provide a unique opportunity to investigate the relationship between the flare energy release and oscillations. The analysis of *Kepler* data by Balona (2012), Maehara *et al.* (2012) and Walkowicz *et al.* (2011) has led to the discovery of a large number of flaring stars of various spectral classes, from M- to A-class. While large flares on M-K stars were known from ground-based observations, the discovery of similar flares on F-A stars is surprising. However, most of the flare data were obtained with long cadence, and do not provide sufficient resolution to estimate the flare amplitude and duration of the various phases of the energy release, or to investigate the seismic response of the stars. For understanding the mechanisms of flares and the impact of these flares, it is important to obtain more short-cadence data for flaring stars.

The previous *Kepler* observations allowed us to select stars that showed multiple flares and with good quality oscillation data (when short-cadence runs are available). We visually inspected the light curves of the previously detected flaring stars, as well as the database of the ground observations.

The flare brightness signals observed by *Kepler* (as illustrated in Fig. 3a) are much stronger than those on the Sun (Fig. 2a). Thus, the photospheric impacts can be much stronger, and the flare oscillations can be observed in the *Kepler* data. In Fig. 3 we illustrate the potential detection of a 'starquake' event obtained from the SC *Kepler* observations of KIC 6106415. The *Kepler* light curve of this stellar flare is shown in panel (a). The oscillation power spectrum calculated for a 2-day period immediately after the flare (Fig. 3b) shows an enhancement of the p-mode amplitudes at ∼2.6 mHz, closer to the acoustic cut-off frequency for this star (∼3.5 mHz), compared to the oscillation spectrum taken during the star's quiet period (Fig. 3c), more than 5 days after the flare (the so-called 'numax' parameter at a frequency of ∼2.26 mHz, e.g., see Chaplin *et al.*, 2014). Of course, more statistical studies are needed to confirm this result.

Oscillations associated with stellar flares have previously been observed in optical and X-ray emissions with periods ranging from a few seconds to tens of minutes, and have usually been interpreted as transient oscillations of flaring loops in stellar atmospheres (Contadakis *et al.* 2012; Jakimiec & Tomczak 2012; Koljonen *et al.* 2011; Contadakis 2012; Qian *et al.* 2012). However, the atmospheric transient oscillations quickly decay, within $1-2$ hours after a flare impulse (Bryson *et al.* 2005), while the stellar oscillation modes will live for much longer, depending on the damping mechanism in stellar envelopes. Also, the maximum amplitude of the oscillations modes excited by flares is expected to be in the range of $3-20$ min, close to the cut-off period of acoustic oscillations. We cannot rule out that oscillations with longer periods, corresponding to g modes and mixed modes,

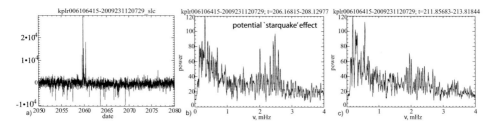

Figure 3. Illustration of a 'starquake' candidate obtained from the SC *Kepler* observations of KIC 6106415: (a) the *Kepler* light curve of a stellar flare. The oscillation power spectra: (b) calculated for a 2-day period immediately after the flare, and (c) taken during the star quiet period, more than 5 days after the flare.

are also excited. Nevertheless, we plan the initial search for the flare seismic responses in the frequency range around the acoustic cut-off frequency.

4. Discussion

The initial results from the *Kepler* mission have demonstrated its tremendous capability for studying stellar activity and oscillations. In particular, the previous observing programs have discovered that flares, during which stellar brightness dramatically increases for short periods of time, are common in both cool and hot stars. Previously, stellar flares were associated with active M-type stars, the UV Ceti variables, thought to be similar to solar flares, which represent a sudden release of magnetic energy accumulated in the coronal part of sunspot regions in the form of high-energy particles, which heat the lower atmosphere. However, the white-light emission in solar flares is rare, and there are only few cases when it has been unambiguously observed in the intergrated Sun-as-a-star observations. The stellar flares can be four orders-of-magnitude more powerful. This might be due to bigger sunspot regions generated by a more efficient dynamo process, because many of the flaring stars rotate faster than the Sun. However, there is an alternative point of view that the flares may be due to interactions with close companions, 'hot Jupiters'. The discovery of similar flares on hot A-type stars with a very shallow outer convection zone and without strong magnetic fields (Balona 2012) raises additional problems with the dynamo origin of the flare energy.

References

Balona, L. A. 2012, *MNRAS*, 423, 3420
Bryson, S., Kosovichev, A., & Levy, D. 2005, *Physica D Nonlinear Phenomena*, 201, 1
Chaplin, W. J., Basu, S., Huber, D., *et al.* 2014, *ApJS*, 210, 1
Contadakis, M. E. 2012, in: I. Papadakis & A. Anastasiadis (eds.), *10th Hellenic Astronomical Conference*, p. 27
Contadakis, M. E., Avgoloupis, S. J., & Seiradakis, J. H. 2012, *AN*, 333, 583
Jakimiec, J. & Tomczak, M. 2012, *Solar Phys.*, 278, 393
Karoff, C. & Kjeldsen, H. 2008, *ApJ*, 678, L73
Koljonen, K. I. I., Hannikainen, D. C., & McCollough, M. L. 2011, *MNRAS*, 416, L84
Kosovichev, A. G. 2006, *Solar Phys.*, 238, 1
Kosovichev, A. G. 2009, *AIP-CS*, 1170, 547
Maehara, H., Shibayama, T., Notsu, S., *et al.* 2012, *Nature*, 485, 478
Qian, S.-B., Zhang, J., Zhu, L.-Y., *et al.* 2012, *MNRAS*, 423, 3646
Walkowicz, L. M., Basri, G., Batalha, N., *et al.* 2011, *AJ*, 141, 50

Precision Asteroseismology
Proceedings IAU Symposium No. 301, 2013
J. A. Guzik, W. J. Chaplin, G. Handler & A. Pigulski, eds.

© International Astronomical Union 2014
doi:10.1017/S1743921313014580

Seismic studies of planet-harbouring stars

Sylvie Vauclair

Université de Toulouse, UPS-OMP, IRAP, France, and CNRS, IRAP,
14 avenue Edouard Belin, F-31400 Toulouse, France,
email: sylvie.vauclair@irap.omp.eu

Abstract. During the past decades, stellar oscillations and exoplanet searches were developed in parallel, and the observations were done with the same instruments: radial velocity method, essentially with ground-based instruments, and photometric methods (light curves) from space. The same observational data on one star could lead to planet discoveries at large time scales (days to years) and to the detection of stellar oscillations at small time scales (minutes), such as for the star μ Arae. Since the beginning, it seemed interesting to investigate the differences between stars with and without observed planets. Also, a precise determination of the stellar parameters is important to characterize the detected exoplanets. With the thousands of exoplanet candidates discovered by *Kepler*, automatic procedures and pipelines are needed with large data bases to characterize the central stars. However, precise asteroseismic studies of well-chosen stars are still important for a deeper insight.

Keywords. asteroseismology, exoplanets

1. Introduction

A review concerning the seismic studies of planet-harbouring stars should begin with the most well known of all these stars: our own Sun. This is the occasion for me to acknowledge all the work done by W.A. Dziembowski in that respect, and quote more specifically the only paper we cosigned, in which we studied the importance for helioseismology of the helium settling coupled with turbulence below the solar convection zone (Richard *et al.* 1996). However, in the following I will focus on exoplanet-harbouring stars.

There are many reasons why astrophysicists interested in exoplanets should bother about the asteroseismology of the central stars of planetary systems:

• The observations for stellar oscillations and exoplanet searches are done with the same instruments. In some cases, the same observations, analysed on different time scales, can lead to both planet detection and seismic studies. This was the case for the star μ Arae, observed with HARPS during eight nights in June 2004. These observations, aimed for asteroseismology, led to the discovery of the exoplanet μ Arae d (Santos *et al.* 2004b).

• "Some people's noise is other people's signal." Indeed, when searching for exoplanets, the signal-to-noise ratio is limited by the stellar oscillations, which appear as noise for the radial velocity variations induced by the planetary motions, whereas they represent in fact the stellar oscillation signal. Other limitations are related to granulation and stellar activity.

• Asteroseismology is important to obtain precise values of the parameters of exoplanet-hosting stars. Seismic studies, combined with spectroscopic observations, can lead to values of the stellar parameters which are much more precise than from spectroscopy alone.

• Asteroseismology can lead to the discovery of new planets. One case, already mentioned, is that of μ Arae. Studies of seismic period variations (the so-called "time delay method") can also lead to the discovery of planets like that detected around the extreme horizontal branch star V391 Peg (Silvotti *et al.* 2007). Note also the spectacular discovery of a compact planetary system which remained after the giant stage, around the hot B subdwarf KIC 05807616, by Charpinet *et al.* (2011).

• Finally, precise asteroseismology can lead to constraints on the internal structure of planet-hosting stars, and may help to understand star-planet interactions, such as angular momentum exchange, consequences of planetary matter accretion onto the star, tidal effects in the case of hot Jupiters, etc.

At the present time, the main goal of asteroseismology of exoplanet-hosting stars is to precisely derive their masses, radius, effective temperatures and ages, in order to obtain more precise results on the parameters of the detected planets themselves. Another objective consists in obtaining hints about the theories of planetary formation and migration. This can be obtained both from statistics of a large number of stars and from deep precise studies of some of these stars.

2.   The "old times"

2.1.   *The saga of μ Arae*

The exoplanet-hosting star μ Arae (HD 160691, HR 6585, GJ 691) is a G5 V star with a visual magnitude $V = 5.1$ mag, and an Hipparcos parallax $\pi = 65.5 \pm 0.8$ mas, which gives a distance to the Sun of 15.3 pc and a luminosity of $\log L/L_\odot = 0.28 \pm 0.012$. This star was observed for seismology in August 2004 with HARPS. At that time, two planets were known. The observations aimed for seismology lead to the discovery of a third planet, μ Ara d, with period 9.5 days (Santos *et al.* 2004b). Finally, evidence for a fourth planet was discovered by Pepe *et al.* (2007).

The HARPS seismic observations allowed identifying 43 oscillation modes of degrees $l = 0$ to $l = 3$ (Bouchy *et al.* 2005). The modelling was done with the TGEC (Toulouse-Geneva stellar evolution code). Atomic diffusion was included in all the models using the formalism derived by Paquette *et al.* (1986), as explained by Richard *et al.* (2004). The treatment of convection was done in the framework of the mixing-length theory, and the mixing-length parameter was adjusted as in the Sun. Adiabatic oscillation frequencies were computed using the adiabatic PULSE code (Brassard 1992).

From the analysis of the frequencies and comparison with models, the following values $T_{\mathrm{eff}} = 5770 \pm 50$ K and [Fe/H] $= 0.32 \pm 0.05$ dex were derived. Spectroscopic observations by various authors gave five different effective temperatures and metallicities (see references in Bazot *et al.* (2005)). The values obtained from seismology are much more precise than those obtained from spectroscopy alone.

2.2.   *The special case of ι Hor*

Among exoplanet-hosting stars, ι Hor is a special case for several reasons (see Laymand & Vauclair (2007) and Vauclair *et al.* (2008)). Three different groups have given different stellar parameters for this star: Gonzalez *et al.* (2001), Santos *et al.* (2004b) and Fischer & Valenti (2005). Meanwhile, Santos *et al.* (2004b) suggested a mass of $1.32\,M_\odot$ while Fischer & Valenti (2005) gave $1.17\,M_\odot$.

Some authors (Chereul *et al.* 1999, Grenon 2000, Montez *et al.* 2001) pointed out that this star has the same kinematical characteristics as the Hyades: its proper motion points towards the cluster convergence point. Two different reasons were possible for this behaviour: either the star formed together with the Hyades, in a region between the Sun and the centre of the Galaxy, which would explain its overmetallicity compared to that of the Sun, or it was dynamically cannabilized by chance (see Famaey *et al.* 2007).

Solar-type oscillations of ι Hor were detected with HARPS in November 2006. Up to 25 oscillation modes could be identified and compared with stellar models. The results led to the following conclusions for ι Hor (Vauclair *et al.* 2008): [Fe/H] is between 0.14 and 0.18; the helium abundance Y is small, 0.255 ± 0.015; the age of the star is 625 ± 5 Myr; the logarithm of the gravity is 4.40 ± 0.01 and its mass $1.25 \pm 0.01\,M_\odot$. The values obtained for the metallicity, helium abundance and age of this star are those characteristic of the Hyades cluster (Lebreton *et al.* 2001).

This star was particularly interesting to study for various reasons, and the results lead to important conclusions. Among these, the importance of the helium value has to be stressed. It is important to realize in all asteroseismic studies that the helium abundance does not always follow the metallicity. A high metallicity may very well be associated with a low helium, even if this seems in contradiction with normal chemical evolution. In this case, the results for the stellar parameters are very different from those obtained with a high helium abundance.

3. New ages

Asteroseismology was given a new boost with the launch of space telescopes partially devoted to the observations of solar-like oscillations in stars. The two recent most important space projects in this respect, *CoRoT* and *Kepler*, were conceived to detect both exoplanets and solar oscillations. Studying the oscillations of exoplanet-hosting stars is a logical by-product of the observations done with these two telescopes.

CoRoT (see Deheuvels, these proceedings) was launched on December 27, 2006. With a 27-cm mirror, it has two focal planes, one for the exoplanet field, one for the seismic field, and four CCD cameras. Its orbit is a polar one, at 900 kilometers. *Kepler* was launched on March 7, 2009. Its mirror diameter is 1.4 meters and its orbit is heliocentric. Both space missions have unfortunately finished their duties, but the amount of results is very large in both cases, although somewhat different in the way it is treated. They both led to a huge amount of data, which have to be treated in a statistical way, but *CoRoT* also led to deeper studies, at least for one very well observed exoplanet-hosting star HD 52265.

3.1. *Statistical studies with large data bases*

Thousands of exoplanet candidates were detected by *Kepler*, and about 700 planetary systems are now confirmed. This huge amount of data have to be treated with good and rapid methods to find the parameters of the central star with good enough precision, and derive the characteristics of the planets.

Large teams began to work several years ago on this subject, doing "hares and hounds" tests and comparing the modelling results obtained with various codes and methods. This is the case for the "asteroFLAG" team sponsored by the International Space Science Institute (ISSI) in Bern, Switzerland (Chaplin *et al.* 2008, Stello *et al.* 2009, Mathur *et al.* 2010, Benomar *et al.* 2012). They showed, for example, that the radius of the star may be derived with an accuracy of a few percent with knowledge of the large separation only. However, these studies generally assume a fixed stellar original abundance of helium, which remains a problem as discussed below.

Rapid derivations of the stellar parameters often use the scaling relations first proposed by Kjeldsen & Bedding (1995), which give the stellar mass and radius as a function of the large separation and the frequency at the maximum of the power spectrum. Improvements in the use of these relations are obtained by introducing the stellar gravity, which is known with much better precision from seismology than from spectroscopy.

Such a method has been used in a recent paper by Huber *et al.* (2013), in which the parameters of the central stars of planetary systems detected by *Kepler* are derived and compared with previous catalogs (Batalha *et al.* 2013). They find that the seismic parameters are similar to the spectroscopic ones for main-sequence stars, but not for evolved stars, for which the seismic values are systematically lower that the spectroscopic one, which leads to a larger stellar radius. They also can confirm previous statistics about planetary formation:

- Jupiter-like planets are more frequent around hotter stars.
- Sub-Neptune planets are found throughout the sample.
- There are more multiple systems around cool stars and more single planets around hot stars.
- No clear trend has been found for the orbital characteristics.

A crucial unknown parameter, very important for the determination of the stellar characteristics, is the helium abundance. Unfortunately, helium is not directly observable in the spectra of solar-type stars. Most studies assume that the helium abundance follows the metallicity in the same way as the general laws for the chemical evolution of galaxies. This is important for exoplanet-hosting stars, which are statistically overmetallic compared to stars without detected planets. However, as we have shown for the star ι Horologii, this may be wrong. It may happen that the star is overmetallic with a small helium abundance, as for the Hyades. In this case, the results obtained by using a high helium abundance may be wrong (Vauclair *et al.* 2008, Escobar 2013).

3.2. *Deep studies of one exoplanet-hosting star HD 52265*

Among all the main targets of the space telescope *CoRoT*, which were observed continuously during several months, one was specially chosen because it was known to harbour at least one planet. HD 52265 is a metal-rich main sequence star. Initially misclassified as a G0 III-IV star in the Bright Star Catalog, it has later been recognized as a G0 V dwarf.

It has a magnitude of $V = 6.301$ and a parallax of $\pi = 34.54 \pm 0.40$ mas, which leads to a distance of $d = 28.95 \pm 0.34$ pc and a luminosity of $\log L/L_\odot = 0.29 \pm 0.05$ (see references in Ballot *et al.* (2011) and in Escobar *et al.* (2012)).

The *CoRoT* observations were carried out during 117 consecutive days from 13 November 2008 to 3 March 2009, during the second long run in the galactic anti-center direction. The beautiful data which were obtained could lead to thirty one p-mode identifications, as well as precise determination of the atmospheric parameters and detailed element abundances. A large separation of $\Delta\nu = 98.4 \pm 0.1$ μHz and a small separation of $\delta\nu_{02} = 8.1 \pm 0.2$ μHz were derived.

A spectroscopic follow-up was done during the *CoRoT* observations with the NARVAL spectropolarimeter on TBL at the Pic du Midi Observatory (France). An upper limit of $\sim 1-2$ G was obtained for the magnetic field of this star (see Ballot *et al.* 2011).

Grids of stellar models were computed using the Toulouse-Geneva-stellar-Evolution Code (TGEC). Details about this code may be found in Hui-Bon-Hoa (2008), Théado *et al.* (2009, 2012). The unique capability of this code compared to other ones is that it can now include the effect of radiative accelerations in the computations of atomic diffusion. The radiative accelerations are computed with the SVP method (Single Value Parameter approximation, see Alecian & LeBlanc (2002), LeBlanc & Alecian (2004)). For each model, the oscillation frequencies were computed and compared with the observations using the various usual tests, including comparisons of frequency differences and detailed comparisons of echelle diagrams. Surface effects were included with the Kjeldsen *et al.* (2008) recipe.

Table 1. Results for the parameters of the exoplanet-hosting star HD 52265, observed with *CoRoT*, after Escobar *et al.* (2012).

$M/M_\odot = 1.24 \pm 0.02$	$[\mathrm{Fe/H}]_i = 0.27 \pm 0.04$
$R/R_\odot = 1.33 \pm 0.02$	$Y_i = 0.28 \pm 0.02$
$L/L_\odot = 2.23 \pm 0.03$	$[\mathrm{Fe/H}]_s = 0.20 \pm 0.04$
$\log g = 4.284 \pm 0.002$	$Y_s = 0.25 \pm 0.02$
Age (Gyr) $= 2.6 \pm 0.2$	T_{eff} (K) $= 6120 \pm 20$

Table 2. Results obtained for HD 52265 from AMP automatic analysis, using all observed frequencies (a) or only the most reliable frequencies (b), after Escobar *et al.* (2012).

	AMP(a)	AMP(b)		AMP(a)	AMP(b)
$M/M_\odot$	1.22	1.20	$[\mathrm{Fe/H}]$	0.23	0.215
$R/R_\odot$	1.321	1.310	Y_i	0.280	0.298
$L/L_\odot$	2.058	2.128	Age (Gyr)	3.00	2.38
$\log g$	4.282	4.282	T_{eff} (K)	6019	6097

The final results for this star, as given in Escobar *et al.* (2012), are quite precise (Table 1). They have been compared with the results obtained using two different automatic fits, the Asteroseismic Modeling Portal (AMP, Metcalfe *et al.* 2009) and the SEEK code (Quirion *et al.* 2010). The results are very good with AMP (Table 2). They are slightly different with SEEK, which gives a larger mass and radius and a smaller age: $M = 1.27 \pm 0.03 M_\odot$, $R = 1.34 \pm 0.02 R_\odot$ and age $= 2.37 \pm 0.29$ Gyr. These differences may be related to their convergence to a different helium value, as discussed in Escobar *et al.* (2012).

4. Conclusion

Asteroseismology on the one hand, and exoplanet detection on the other hand, are parts of a new era for astrophysics, especially for stellar physics. The instruments which are devoted to exoplanet searches also look for stellar oscillations. Coupling both studies is interesting and important to lead to precise values of the parameters of the central stars of planetary systems, for a better characterization of the planets.

At the present time, space projects lead to large amount of data, so that the large data bases have to be treated in a statistical way. Automatic procedures are important in that respect. However deep studies of individual stars are still necessary to test the physics and check the values obtained with statistical studies.

For solar type stars, the original helium abundance, which is not derived by spectroscopy, may lead to erroneous results if not evaluated correctly. This may be the largest difficulty at the present time for these kind of studies.

In the near future, we may expect that precise stellar studies will lead to a better understanding of star-planet interactions, abundance variations, angular momentum transfer, tidal effects, etc.

This is an open field in which many new results may be expected, with new projects such as the PLAnetary Transits and Oscillations of stars spacecraft (*PLATO*) coming soon.

References

Alecian, G. & LeBlanc, F. 2002, *MNRAS*, 332, 891

Ballot, J., Gizon, L., Samadi, R., *et al.* 2011, *ApJS*, 204, 24

Batalha, N. M.; Rowe, J. F., Bryson, S. T., *et al.* 2013, *A&A*, 530, A97

Bazot, M., Vauclair, S., Bouchy, F., & Santos, N. 2005, *A&A*, 440, 615

Benomar, O., Baudin, F., Chaplin, W. J., Elsworth, Y., & Appourchaux, T. 2012, *MNRAS*, 420, 2178

Bouchy, F., Bazot, M., Santos, N., Vauclair, S., & Sosnowska, D. 2005, *A&A*, 440, 609

Brassard, P. 1992, *ApJS*, 81, 747

Chaplin, W. J., Appourchaux, T., Arentoft, T., *et al.* 2008, *AN*, 329, 549

Charpinet, S., Fontaine, G., Brassard, P., *et al.* 2011, *Nature*, 480, 496

Chereul, E., Crézé, M., & Bienaymé, O. 1999, *A&AS*, 135,5

& Escobar, M. E., 2013, Ph.D. thesis, Université Paul Sabatier, Toulouse

Escobar, M. E., Théado, S., Vauclair, S., *et al.*, 2012, *A&A*, 543, A96

Famaey, B., Pont, F., Luri, X., *et al.* 2007, *A&A*, 461, 957

Fischer, D. A. & Valenti, J. 2005, *ApJ*, 622, 1102

Gizon, L., Ballot, J., Michel, E., *et al.* 2012, *PNAS*, 110, 13267

Gonzalez, G., Laws, C., Tyagi, S., & Reddy, B. E. 2001, *AJ*, 121, 432

Grenon, M. 2000, in: Matteucci & Giovanelli (eds.) *The evolution of the Milky Way*, p. 47

Huber, D., Chaplin, W. J., & Christensen-Dalsgaard, J. 2013, *ApJ*, 767, 127

Hui-Bon-Hoa, A. 2008, *Ap&SS*, 316, 55

Kjeldsen, H. & Bedding, T. R., 1995, *A&A*, 293, 87

Kjeldsen, H., Bedding, T. R., & Christensen-Dalsgaard, J. 2008, *ApJ*, 683, L175

Laymand, M. & Vauclair, S. 2007, *A&A*, 463, 657

LeBlanc, F. & Alecian, G. 2004, *MNRAS*, 352, 1329

Lebreton, Y., Fernandes, J., & Lejeune, T. 2001, *A&A*, 374, 540

Mathur, S., García, R. A., Régulo, C., *et al.* 2010, *A&A*, 511, 46

Montez, D., Lopez-Santiago, J., Galvez, M. C., *et al.* 2001, *MNRAS*, 328, 45

Paquette, C., Pelletier, C., Fontaine, G., & Michaud, G. 1986, *ApJS*, 61, 177

Pepe, F., Correia, A. C. M., Mayor, M., *et al.* 2007, *A&A*, 462, 769

Richard, O., Vauclair, S., Charbonnel, C., & Dziembowski, W. A. 1996, *A&A*, 312, 1000

Richard, O., Théado, S., & Vauclair, S. 2004, *Solar Phys.*, 220, 243

Santos, N. C., Israelian, G., & Mayor, M. 2004a, *A&A*, 415, 1153

Santos, N. C., Bouchy, F., Mayor, M., *et al.* 2004b, *A&A*, 426, L19

Silvotti, R., Schuh, S., Janulis, R., *et al.* 2007, *Nature*, 449, 189

Stello, D., Chaplin, W. J., Bruntt, H., *et al.* 2009, *ApJ*, 700, 1589

Théado, S., Vauclair, S., Alecian, G., LeBlanc, F., & Vauclair, S. 2009, *ApJ*, 704, 1262

Théado, S., Alecian, G., LeBlanc, F., & Vauclair, S. 2012, *A&A*, 546, A100

Vauclair, S., Laymand, M., Bouchy, F., Vauclair, G., Hui Bon Hoa, A., Charpinet, S., & Bazot, M. 2008, *A&A*, 482, L5

Precision Asteroseismology
Proceedings IAU Symposium No. 301, 2013
J. A. Guzik, W. J. Chaplin, G. Handler & A. Pigulski, eds.

© International Astronomical Union 2014
doi:10.1017/S1743921313014592

Concluding remarks

Jørgen Christensen-Dalsgaard[1] and Steven D. Kawaler[2]

[1] Stellar Astrophysics Centre, Department of Physics and Astronomy, Aarhus University, Ny Munkegade 120, 8000 Aarhus C, Denmark
email: jcd@phys.au.dk

[2] Department of Physics and Astronomy, Iowa State University
Ames, IA 50011 USA
email: sdk@iastate.edu

Abstract. We cannot presume to summarize all of the science we've discussed in the talks, posters, and informal discussions. Here, we discuss a few of the themes that emerged, concentrating on the theoretical basis that Wojtek Dziembowski and his colleagues have developed and explored over the past 40+ years. We connect those with observational results – especially those from recent ground-based surveys and space-based missions that have revolutionized the study of stellar variability.

1. Introduction

This five-day science conference was organized around the themes that Wojtek Dziembowski has explored in his exceptional career in astrophysics. In reality, though, we could only briefly touch upon the enormous strides in those fields that have been enabled by his work. In parallel with theoretical developments made by Wojtek and his friends in the theoretical and computational stellar astrophysics community, the observers have produced data at an ever-increasing rate. We now face a fire-hose of data coming from space-based observatories (*Kepler*, *CoRoT*, and hopefully soon *Gaia* and *TESS*) that can provide, long, continuous high-precision photometry. Ground-based large-scale variability surveys such as OGLE (described nicely by Grzegorz Pietrzyński on Monday) add to this deluge of data, as do data products from planet search projects such as WASP (as we heard from Barry Smalley).

Over 60 talks and over 70 posters discussed many aspects of stellar pulsation and stellar variability across the H-R diagram. Summarizing all of this work in a few pages is neither possible nor useful – the presenters have provided some excellent write-ups of their work in the pages of this volume, and we encourage you to take some time and enjoy reading through them. In this short summary, we highlight a few themes that emerged that illustrate the influence that Wojtek has had on the field. We close with a way to "measure" our links to Wojtek's specific contributions in a quantitative way.

2. Pulsation physics

It is fair to say that Wojtek's main scientific interest is in essence a simple one: the physics of stellar pulsations, inspired by the observed properties of pulsating stars. Application of stellar pulsations to the physics of stars and their evolution (what we now call "asteroseismology" — see Gough 1996), while also of interest, comes second. He admitted as much during his talk on the second day of the conference. However, the understanding obtained from investigating the properties of stellar pulsation is central to asteroseismology. A very important example is the development of the asymptotic theory of stellar pulsations to which Wojtek has played such a key role, and which is central to the analysis of observed frequencies. Although the early basis was established by the Belgian group (e.g. Ledoux 1962; Smeyers 1968) and by Vandakurov *et al.* (1967), Wojtek was probably the first to apply the results to the understanding of the properties of real stars. An important example is Wojtek's work on the pulsations of evolved stars in 1971 (Dziembowski 1971). † Although the paper uses a Cepheid model as the primary example, it essentially provides the foundation for the very rich investigations of red giants made possible by observations by *CoRoT* and *Kepler* nearly 40 years later.

Large-scale numerical computations of stellar internal hydrodynamics are becoming increasingly powerful, and perhaps increasingly realistic, as a tool to understand the properties of stellar pulsations. The potential for direct computation of the interaction between pulsations and convection, discussed by Friedrich Kupka on Wednesday, is promising significant progress on this otherwise intractable problem. Similarly, the computations presented by Irina Kitiashvili, also on Wednesday, point the way to a direct investigation of stochastically excited modes in solar-like pulsators, including the effects of magnetic fields and hence addressing the issues of stellar activity discussed in Travis Metcalfe's talk.

A very different area of research, but a similar level of computational effort, is represented by the huge grids of stellar models that are being created for the interpretation of stellar-oscillation data to obtain the underlying stellar parameters. This direction was exemplified by the studies of pre-main-sequence stars presented on Tuesday by Konstanze Zwintz. However, such computations do not guarantee insight or understanding and should be complemented with simpler treatments which might provide such insight from the computations.

This point was made, quite elegantly, by Richard Townsend in his discussion of pulsations in rotating stars on Tuesday, with reference to the following quote from a related paper by F. Soufi, Marie-Jo Goupil, and Wojtek (Soufi *et al.* 1998):

> "We feel perturbation theory calculations are still useful, because coding the formulae is in fact quite straightforward – far easier than deriving them. In contrats [sic], it is highly nontrivial to achieve a 10^{-3} precision in frequency calculations with a 2-D hydrocode, were one available. Undoubtedly, the use of such codes will ultimately be unavoidable, but then it will be very helpful to have a code based on the perturbational approach for comparisons at moderate equatorial velocities where both are valid."

Townsend also introduced the very useful concept of "narrative" to describe, for a given problem, the complete and coherent story that contains both the detailed results and the broader understanding. It is obvious that Wojtek has contributed greatly to this narrative in the study of pulsating stars.

The conference featured interesting new developments in several areas of pulsation physics. The Blazhko effect in RR Lyrae stars is one of the most long-standing puzzles in the physics of stellar pulsations. *Kepler* observations of a substantial number of RR Lyrae stars showing this effect, including RR Lyrae itself, as illustrated by Katrien Kolenberg's talk on Thursday, have provided data of unprecedented precision and extent for the study of this phenomenon. As Robert Szabó showed in his talk earlier on Thursday, an unexpected feature is the prevalence of, at least intermittent, period doubling in these stars. Together with nonlinear modelling this points to resonances, remarkably including the 9:2 resonance between the fundamental radial mode and a 'strange' mode, as being

---

† One of the present authors was reminded that some of the calculations for that paper were carried out on a Danish-built GIER computer, of similar type to the one on which he started his computational efforts.

involved. With further analysis and hydrodynamical modelling we may soon get closer to an understanding of the Blazhko effect, revealing also an unexpected richness in the physical behaviour of pulsating stars.

The basic processes of mode excitation are reasonably well-understood. Amongst cool stars with convective envelopes, the dominant process is stochastic excitation, by convection, of otherwise stable modes, This includes the remarkably rich behaviour found in the red giants, extending to what Wojtek and his collaborators (Soszyński *et al.* 2007) have named the OSARGs (*OGLE Small Amplitude Red Giants*). The recently established link between the OSARGs and the most luminous *Kepler* red giants (Mosser *et al.* 2013) is extremely interesting in this regard. Also, the instability of modes caused by the heat-engine mechanism (the κ mechanism) can increasingly be understood in terms of opacity perturbations; for hotter stars the revision of the opacities from iron-group elements was crucial (e.g., Dziembowski & Pamyatnykh 1993). However, there are still problems in reproducing the modes observed in specific stars, as discussed by Wojtek in his talk (see also Dziembowski & Pamyatnykh 2008). On Thursday, Gilles Fontaine discussed the interesting example of the GW Vir variables, at the beginning of the white-dwarf cooling sequence, where the details of the observed modes can be understood as the effect on the instability of competition between mass loss and settling.

For stars whose oscillations are excited by a heat-engine mechanism the processes limiting the amplitudes, and hence selecting those modes that reach observable amplitudes, remain poorly understood. Wojtek and his group have made major contributions to the study of this problem, emphasizing the potential importance of resonant interactions between unstable and stable modes, going back to Dziembowski (1982). Radek Smolec † gave a comprehensive overview of these issues on Thursday. He also discussed the cases of multi-mode pulsators detected in large-scale surveys, in particular OGLE, and the difficulties in understanding and interpreting the space-based observations of δ Scuti stars. An important issue in the latter case is to distinguish between actual modes of the star and combination frequencies resulting from nonlinear interactions. The conclusion appears to be that the problem of understanding mode selection, despite the very rich space-based data revealing extremely small amplitude pulsations, is still not solved.

While a deep understanding of the physics of stellar pulsations is of great interest in itself, it is important to emphasize the crucial need for comparison with asteroseismic observations or other diagnostic tools. A superficially simple example that came during the conference was the so-called p factor that relates the apparent and true velocity amplitude, illustrated by Nicolas Nardetto in his Tuesday presentation. The empirically determined p factor is essential for the use of the Baade-Wesselink method for distance determination. Also, the increasingly sophisticated understanding of the asymptotic properties of stellar oscillations has been crucial for interpreting of the observations of solar-like oscillations (for a recent example, see Goupil *et al.* 2013), and faces new challenges in the detailed interpretation of mixed modes in red-clump stars where composition discontinuities in the core give rise to very complex oscillation spectra. It should also be noted that a better understanding of the mode selection in, e.g., δ Scuti stars, as discussed above, could greatly help the asteroseismic use of observations of these stars for determining the properties of individual stars.

---

† Radek is Wojtek's 'grand-student' – his Ph.D. advisor was Paweł Moskalik, whose thesis advisor was Wojtek.

3. Precision in asteroseismology

The exquisite data from *CoRoT* and *Kepler* have been a major breakthrough for asteroseismology; photometry (though with less precision) of the huge number of stars observed with the OGLE project allows characterization of a very broad range of stellar variability, including rare types of pulsating stars. The challenge is to make full use of these outstanding data.

These challenges were addressed by Bill Chaplin on Monday, with particular emphasis on solar-like oscillations. The analysis of these data sets, with durations from months to years, requires correction of the raw photometry for discontinuous changes in level, and other irregularities resulting from instrumental and other non-stellar causes, e.g., with the changing orientation of the spacecraft from quarter to quarter for *Kepler*. With corrected data in hand, preliminary analysis in solar-like pulsators can be made in terms of the large frequency spacing and the frequency of maximum power, using apparently reliable scaling relations (Kjeldsen & Bedding 1995, 2011; Huber *et al.* 2011). However fully exploiting the data requires determination of the individual frequencies and, very importantly, a proper statistical characterization of the results. Reliable determination of the accuracy of the results is essential in the subsequent asteroseismic analysis.

Chris Engelbrecht, in his address on Tuesday morning, gave a comprehensive overview of the various analysis techniques that are available for frequency analyses of time-series data. An important part of the analysis, particularly for solar-like oscillations, is the determination of the non-oscillatory background, which in many cases is dominated by stellar effects (and hence in itself provides scientifically very interesting information). The individual steps in this analysis are relatively well understood, although the statistical properties of the observed frequencies, including possible correlations particularly for derived quantities such as frequency separations, are probably in general not adequately treated. This understanding has been applied to the analysis of individual stars, through considerable effort and time. A serious challenge will be to develop reliable automatic tools that can carry out the analysis at a similar level of reliability for thousands of stars, or indeed to determine how best to make scientific use of the results of such an analysis.

In the subsequent analysis of asteroseismic data the availability of reliable non-seismic characterization of the target stars is crucial (e.g., Uytterhoeven *et al.* 2010; Molenda-Żakowicz *et al.* 2013). This includes effective temperature, composition and, if possible, radius, obtained from photometry and spectroscopy. When available, interferometry provides vital constraints, and will be a powerful tool combined with determinations of distances which will be revolutionized by the *Gaia* mission. In the case of *Kepler* this is complicated by the fact that most stars observed are quite faint and hence require long observations to reach the spectral resolution needed for precise determination of stellar composition. The need for such data is becoming increasingly clear with the growing realization that the asteroseismic data, even at the precision offered by *Kepler*, in themselves are not sufficient fully to characterize the stars. This is particularly evident in a degeneracy in the results of fits to solar-like oscillation frequencies between the mass and initial helium abundance Y_0 of the star. Also, there seems to be a tendency that the fits in some cases prefer presumably unphysically low values of Y_0 (Mathur *et al.* 2012). Sufficiently precise non-seismic information may help break such degeneracies and remove the specter of unphysical results.

The asteroseismic inferences depend on making the best possible use of the data, ideally by devising diagnostics that provide information about global stellar properties or specific aspects of the stellar interior. For this purpose asymptotic analyses provide very valuable understanding of the oscillations and their dependence on stellar properties. As discussed by Sebastien Deheuvels in his talk on Monday afternoon, an important example for solar-like oscillations is the effect of acoustic glitches which have detectable signatures in the oscillation frequencies (e.g., Houdek & Gough 2007) and which are now being detected in data from *CoRoT* (Miglio *et al.* 2010; Mazumdar *et al.* 2012a) and *Kepler* (Mazumdar *et al.* 2012b). This may, for instance, provide an independent measure of the envelope helium abundance and hence help break the degeneracy between mass and initial helium abundance.

As we have struggled with for some time, δ Scuti stars present major problems for asteroseismic inferences with their very rich oscillation spectra and no clear systematics in the selection of the observed modes. As discussed by Katrien Uytterhoeven on Tuesday, additional observations that allow identification of at least some of the observed modes are very important. Not surprisingly, Wojtek's hand is present here, too, as the origin of these techniques, reflecting how the observational sensitivity to the modes depends on their degree, goes back to Dziembowski (1977) (see also Daszyńska-Daszkiewicz *et al.* 2005).

For other types of pulsating stars related procedures are available for asteroseismic inferences. A very interesting case are the subdwarf B stars, where fits to the observed frequencies provide very stringent constraints on the stars. In her talk on Thursday, Valerie Van Grootel (see also Van Grootel *et al.* 2013) discussed the case of acoustic modes in such a star and demonstrated a very high accuracy in the determination of the mass and radius of the star. The basis for the claim for accuracy, rather than only precision, was a careful analysis of the sensitivity of the results to uncertainties in the underlying stellar modelling. Such investigations, including potential numerical problems in the model and frequency computations, are indeed essential if we are to fulfill the full potential in precision asteroseismology. Another very interesting example is the analysis shown on Thursday by Noemi Giammichele of observations of a hydrogen-rich pulsating white dwarf. Based on five observed periods she was able to make a precise determination of several properties of the star, as characterized by, for example, mass, surface gravity and the thickness of the hydrogen and helium layers. In this case further investigations of the sensitivity to the model assumptions would probably be warranted.

4. Conclusion: the Dziembowski Number

Wojtek Dziembowski has been a steady influence on (now) generations of astrophysicists from Poland and many, many other nations throughout the world. His impact on stellar pulsation theory, and more generally on stellar astrophysics, has spanned over five decades. In that time, he has published many influential papers, and many astronomers have been honoured to collaborate with him and are proud to call themselves coauthors.

How can one quantify this reach of a single investigator's collaborations? Mathematicians have addressed this issue in at least one case – to recognize the extremely collaborative Paul Erdös (Goffman 1969), who published over 1500 papers in his career. The 'Erdös Number' was introduced to measure how closely one's coauthorship comes to a publication directly with Erdös: those who have published with Erdös directly have an Erdös number of 1; 511 scholars have an Erdös number of 1. Someone who has not published with Erdös but has published with someone who did coauthor a paper with Erdös earns an Erdös number of 2, and so on. Erdös himself has (had...) an Erdös number of 0. Currently, 9779 scholars have an Erdös number of 2. Interestingly, Wojtek has an Erdös number of 5, but his son, Stefan Dziembowski, has a 'better' Erdös number of 3 through his computer science collaboration with Ivan Damgård † (Cramer *et al.* 1999).

† Damgård (2) published with Pomerance (1) who published with Erdös (0).

So, what is your Dziembowski number? To date, Wojtek has published papers with 161 direct collaborators. If you are one of us, then you have a Dziembowski number of 1. If your Dziembowski number is greater than 1, then you should seek him out, and write a paper with him. But beyond improving your Dziembowski number, you will have the chance to experience the pure joy of collaborating directly with Wojtek, and seeing how much fun stellar pulsation physics can be in the process.

Acknowledgements

We take this opportunity to thank Wojtek for his huge contributions to our field, for his enthusiasm and inspiration and for his friendship over very many years. We are very grateful to Jagoda and Andrzej for organizing an extremely interesting and productive conference, and for giving us the pleasure and honour of providing the closing remarks. Funding for the Stellar Astrophysics Centre is provided by The Danish National Research Foundation (Grant DNRF106). The research is supported by the ASTERISK project (ASTERoseismic Investigations with SONG and *Kepler*) funded by the European Research Council (Grant agreement no.: 267864).

References

Cramer, R., Damgård, I., Dziembowski, S., Hirt, M., & Rabin, T. 1999, *Lecture Notes in Comput. Sci.*, 1592, 311

Daszyńska-Daszkiewicz, J., Dziembowski, W. A., Pamyatnykh, A. A., Breger, M., Zima, W., & Houdek, G. 2005, *A&A*, 438, 653

Dziembowski, W. 1971, *AcA*, 21, 289

Dziembowski, W. A. 1977, *AcA*, 27, 203

Dziembowski, W. 1982, *AcA*, 32, 147

Dziembowski, W. A. & Pamyatnykh, A. A. 1993, *MNRAS*, 262, 204

Dziembowski, W. A. & Pamyatnykh, A. A. 2008, *MNRAS*, 385, 2061

Goffman, C. 1969, *American Mathematical Monthly*, 76, 791

Gough, D. O. 1996, *Observatory*, 116, 313

Goupil, M. J., Mosser, B., Marques, J. P., et al. 2013, *A&A*, 549, A75

Houdek, G. & Gough, D. O. 2007, *MNRAS*, 375, 861

Huber, D., Bedding, T. R., Stello, D., et al. 2011, *ApJ*, 743, 143

Kjeldsen, H. & Bedding, T. R. 1995, *A&A*, 293, 87

Kjeldsen, H. & Bedding, T. R. 2011, *A&A*, 529, L8

Ledoux, P. 1962, *Bull. Acad. roy. Belg. Cl. Science, 5e série*, 48, 240

Mathur, S., Metcalfe, T. S., Woitaszek, M., et al. 2012, *ApJ*, 749, 152

Mazumdar, A., Michel, E., Antia, H. M., & Deheuvels, S. 2012a, *A&A*, 540, A31

Mazumdar, A., Monteiro, M. J. P. F. G., Ballot, J., et al. 2012b, *AN*, 333, 1040

Miglio, A., Montalbán, J., Carrier, F., et al. 2010, *A&A*, 520, L6

Molenda-Żakowicz, J., Sousa, S. G., Frasca, A., et al. 2013, *MNRAS*, 434, 1422

Mosser, B., Dziembowski, W. A., Belkacem, K., et al. 2013, *A&A*, 559, 137

Smeyers, P. 1968, *Ann. d'Astrophys.*, 31, 159

Soszyński, I., Dziembowski, W. A., Udalski, A., et al. 2007, *AcA*, 57, 201

Soufi, F., Goupil, M. J., & Dziembowski, W. A. 1998, *A&A*, 334, 911

Uytterhoeven, K., Briquet, M., Bruntt, H., et al., 2010, *AN*, 331, 993

Vandakurov, Yu. V. 1967, *Astron. Zh.*, 44, 786 (English translation: *Soviet Astron. AJ*, 11, 630)

Van Grootel, V., Charpinet, S., Brassard, P., Fontaine, G., & Green, E. M. 2013, *A&A*, 553, A97

PHOTOS FROM THE CONFERENCE

(taken by Alosha Pamyatnykh)

Focusing on pulsation-convection interaction (or an LOC announcement).
[All figures courtesy of Alexey A. Pamyatnykh]

Phil Goode explaining how much fun it is to work with Wojtek.

Paweł Moskalik demonstrating to Wojtek how easy seismology is.

Marie-Jo Goupil asking a difficult question.

Wojtek answering an even more difficult question.

Gérard Vauclair's comment attracts enormous attention.

Hideyuki Saio looking for a laser pointer speckle.

A happy Mike Jerzykiewicz.

Wojtek providing one of his many insightful remarks.

Steve Kawaler introducing the Dziembowski Number.

POSTERS

Precision Asteroseismology
Proceedings IAU Symposium No. 301, 2013
J. A. Guzik, W. J. Chaplin, G. Handler & A. Pigulski, eds.

© International Astronomical Union 2014
doi:10.1017/S1743921313014622

3D simulations of internal gravity waves in solar-like stars

Lucie Alvan, Allan Sacha Brun and Stéphane Mathis

Laboratoire AIM Paris-Saclay, CEA/DSM-CNRS-Université Paris Diderot, IRFU/SAp,
F-91191 Gif-sur-Yvette Cedex, France
email: lucie.alvan@cea.fr, allan-sacha.brun@cea.fr, stephane.mathis@cea.fr

Abstract. We perform numerical simulations of the whole Sun using the 3D anelastic spherical harmonic (ASH) code. In such models, the radiative and convective zones are non-linearly coupled and in the radiative interior a wave-like pattern is observed. For the first time, we are thus able to model in 3D the excitation and propagation of internal gravity waves (IGWs) in a solar-like star's radiative zone. We compare the properties of our waves to theoretical predictions and results of oscillation calculations. The obtained good agreement allows us to validate the consistency of our approach and to study the characteristics of IGWs. We find that a wave's spectrum is excited up to radial order $n = 58$. This spectrum evolves with depth and time; we show that the lifetime of the highest-frequency modes must be greater than 550 days. We also test the sensitivity of waves to rotation and are able to retrieve the rotation rate to within 5% error by measuring the frequency splitting.

Keywords. hydrodynamics, waves, stellar dynamics, methods: numerical

1. Spectrum of gravity waves

Because they can propagate deeply in radiation zones of stars, internal gravity waves (IGWs) are essential to probe stellar interiors (García *et al.* 2007). The spectrum presented in Fig. 1 (left) has been obtained with the ASH code (Brun *et al.* 2004). Our model nonlinearly couples the convective envelope to the stable radiative core of the Sun (Brun *et al.* 2011), assuming a realistic solar stratification from $r = 0$ up to $0.97\,R_\odot$, and gravity waves are excited by convective penetration.

The richness of this spectrum allows us to analyse quantitatively the properties of the waves and to compare them with linear or asymptotic predictions (e.g. Christensen-Dalsgaard 2003 and Kosovichev 2011). We retrieve for example the constant spacing in period between two modes of consecutive radial orders. The spectrum evolves depending on the depth from where it is extracted, showing the effect of radiative damping. And it evolves also with time, since g modes have a finite lifetime. We observe that high-frequency modes have a nearly constant amplitude over more than 550 days so their lifetime must be much longer.

2. Effect of rotation

One of the interests for our 3D simulations resides in our ability to distinguish between several values of the azimuthal order m. The model studied is rotating at the solar rotation rate ($\Omega_\odot = 2.6\,10^{-6}$ rad s^{-1}). We show in the right panel of Fig. 1 the rotational splitting affecting a g mode of degree $l = 2$. We expect this frequency shift to follow the asymptotic formula available in the frame rotating with the star, for $n \gg l$,

$$\omega_{lm} = \omega_{l0} + \frac{m}{l(l+1)}\Omega'_\odot. \qquad (2.1)$$

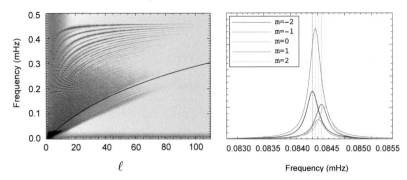

Figure 1. *Left*: Spectrum of IGWs for $r_0 = 0.33\,R_\odot$. The black line highlights radial order $n = 58$. *Right*: Rotationnal splitting for $l = 2$, $-2 < m < 2$ and $n = 8$. The peaks have been fitted with a Lorentzian to determine precisely the position of their maximum.

By repeating our measurements over several modes of different radial orders n, we show that the relative error between the real rotation rate $\Omega_\odot$ and $\Omega'_\odot$ is about 35% for $n = 5$, but is less than 5% for $n \geqslant 30$. This measure gives us a quantitative information about the validity of the asymptotic approximation. We also note the asymmetry observed between prograde and retrograde modes which is in agreement with the one predicted by Belkacem *et al.* (2009).

3. Conclusion

Thanks to these new results, a detailed analysis of IGWs in 3D non-linear dynamical simulations is possible. For the first time, we are able to model the behaviour of both propagative and standing waves in a realistic 3D cavity (Alvan *et al.* 2012, Brun *et al.* 2013). The complementarity between our simulations and asymptotic analysis leads to interesting predictions about the lifetime of the modes and the accuracy of the rotation-rate measurement. We are currently developing new models taking into account a velocity gradient deep in the radiative zone (instead of a flat profile) and we will be able to study both the effect of this differential rotation on waves and the angular momentum transport induced by IGWs.

Acknowledgement

We acknowledge financial support by ERC STARS2 207430 project, by CNES - GOLF and FP7-IRSES, and access to supercomputers through GENCI project 1623.

References

Alvan, L., Brun, A. S., & Mathis, S. 2012, *Proceedings SF2A-2012*, 289
Belkacem, K., Mathis, S., Goupil, M. J., & Samadi, R. 2009 *A&A*, 508, 345
Brun, A. S., Miesch, M. S., & Toomre, J. 2004, *ApJ*, 614, 1043
Brun, A. S., Miesch, M. S., & Toomre, J. 2011, *ApJ*, 742, 79
Brun, A. S., Alvan, L., Strugarek, A., Mathis, S., & García, R. A. 2013, *J. Phys.: Conf. Ser.*, 440, 012043
García R. A., Turck-Chièze, S., Jiménez-Reyes, S. J., *et al.* 2007, *Science*, 316, 1591
Christensen-Dalsgaard, J. 2003, *Lecture Notes on Stellar Oscillations*, p. 84
Kosovichev, A. G. 2011, *Lecture Notes in Physics*, 832, 3

Precision Asteroseismology
Proceedings IAU Symposium No. 301, 2013
J. A. Guzik, W. J. Chaplin, G. Handler & A. Pigulski, eds.

© International Astronomical Union 2014
doi:10.1017/S1743921313014634

Corotation resonances for gravity waves and their impact on angular momentum transport in stellar interiors

Lucie Alvan[1], Stéphane Mathis[1], and Thibaut Decressin[2]

[1] Laboratoire AIM Paris-Saclay, CEA/DSM-CNRS-Université Paris Diderot, IRFU/SAp,
F-91191 Gif-sur-Yvette Cedex, France
email: lucie.alvan@cea.fr, stephane.mathis@cea.fr

[2] Geneva Observatory, University of Geneva, chemin des Maillettes 51, 1290 Sauverny,
Switzerland
email: thibaut.decressin@unige.ch

Abstract. Gravity waves, which propagate in radiation zones, can extract or deposit angular momentum by radiative and viscous damping. Another process, poorly explored in stellar physics, concerns their direct interaction with the differential rotation and the related turbulence. In this work, we thus study their corotation resonances, also called critical layers, that occur where the Doppler-shifted frequency of the wave approaches zero. First, we study the adiabatic and non-adiabatic propagation of gravity waves near critical layers. Next, we derive the induced transport of angular momentum. Finally, we use the dynamical stellar evolution code STAREVOL to apply the results to the case of a solar-like star. The results depend on the value of the Richardson number at the critical layer. In the first stable case, the wave is damped. In the other unstable and turbulent case, the wave can be reflected and transmitted by the critical layer with a coefficient larger than one: the critical layer acts as a secondary source of excitation for gravity waves. These new results can have a strong impact on our understanding of angular momentum transport processes in stellar interiors along stellar evolution where strong gradients of angular velocity can develop.

Keywords. hydrodynamics, waves, turbulence, stars: rotation, stars: evolution

1. Damping or over-reflection

Critical layers (e.g. Booker & Bretherton 1967) occur in stellar radiation zones where the frequency of the wave is resonant with the mean-flow rotation rate. Mathematically, this means $\sigma(r) = \sigma_w + m\Delta\Omega(r) = 0$, where σ_w is the excitation frequency of the wave, m corresponds to a Fourier expansion along the longitudinal direction, and $\Delta\Omega(r) = \overline{\Omega}(r) - \Omega_{\mathrm{CZ}}$ is the difference between the angular velocity at the radius r and at the border with the convective zone, where the waves are excited. In the adiabatic case, the equation of propagation of IGWs in spherical coordinates is

$$\frac{\mathrm{d}^2 \Psi_{l,m}}{\mathrm{d}r^2} + \left[\frac{l(l+1)}{m^2} \frac{\mathrm{Ri}_c}{(r-r_c)^2} - k_{Hc}^2 \right] \Psi_{l,m} = 0, \qquad (1.1)$$

where $\Psi_{l,m}(r) = \bar{\rho}^{\frac{1}{2}} r^2 \hat{\xi}_{r;l,m}$, $\bar{\rho}$ is the average density and $\hat{\xi}_{r;l,m}$ the vertical displacement expanded in the spherical harmonics basis. The local Richardson number $\mathrm{Ri}_c = \left[N^2 / \left(r^2 (\mathrm{d}\bar{\Omega}/\mathrm{d}r)^2 \right) \right]_{r=r_c}$ depends on the Brunt-Väisälä frequency N. Its value determines the behaviour of the wave passing through a critical layer as shown in Fig. 1.

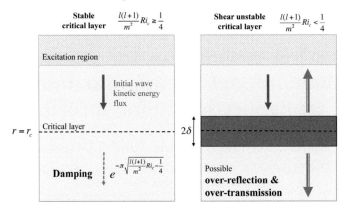

Figure 1. Two possible behaviours for a wave passing through a critical layer. 2δ is the thickness of the turbulent region.

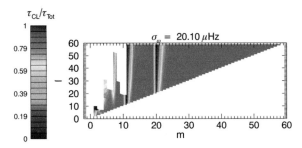

Figure 2. Comparison between τ_{CL} and $\tau_{tot} = \tau_{CL} + \tau_{rad}$. Red/dark tones correspond to regions where the action of the critical layers is dominant in comparison with the radiative damping.

2. Application to a solar-type star

We applied our theoretical prescriptions to a solar-type star calculated with the evolution code STAREVOL. It appears that only the first stable case can occur in the studied case. We show in Fig. 2 that the damping τ_{CL} produced by the critical layers competes with the radiative damping τ_{rad} described by Zahn *et al.* (1997). Both phenomena are complementary to explain the transport of angular momentum by IGWs (e.g. Charbonnel & Talon 2005, Alvan *et al.* 2013).

3. Conclusion

Further details about this study are available in Alvan *et al.* (2013). These results indicate that critical layers can lead to major modification of the transport of angular momentum in stellar interiors. A systematic exploration of different types of stars for different evolutionary stages will be undertaken and we expect to find stars where the unstable regime and possible tunneling or over-reflection/transmission take place.

References

Alvan, L., Mathis, S., & Decressin, T. 2013, *A&A*, 553, 86
Booker, J. R. & Bretherton, F. P. 1967, *J. Fluid Mech.*, 27, 513
Charbonnel, C. & Talon, S. 2005, *Science*, 309, 2189
Zahn, J.-P., Talon, S., & Matias, J. 1997, *A&A*, 322, 320

Precision Asteroseismology
Proceedings IAU Symposium No. 301, 2013   © International Astronomical Union 2014
J. A. Guzik, W. J. Chaplin, G. Handler & A. Pigulski, eds.   doi:10.1017/S1743921313014646

Precise and accurate interpolated stellar oscillation frequencies on the main sequence

Warrick H. Ball[1,2], Jesper Schou[2], Laurent Gizon[1,2], and João P. C. Marques[1]

[1] Institut für Astrophysik, Georg-August-Universität Göttingen,
Friedrich-Hund-Platz 1, 37077 Göttingen, Germany
email: wball@astro.physik.uni-goettingen.de

[2] Max-Planck-Institut für Sonnensystemforschung, Max-Planck-Str. 2,
37191 Katlenburg-Lindau, Germany

Abstract. High-quality data from space-based observatories present an opportunity to fit stellar models to observations of individually-identified oscillation frequencies, not just the large and small frequency separations. But such fits require the evaluation of a large number of accurate stellar models, which remains expensive. Here, we show that global-mode oscillation frequencies interpolated in a grid of stellar models are precise and accurate, at least in the neighbourhood of a solar model.

Keywords. stars: oscillations, methods: numerical

Asteroseismology from space presents an opportunity to fit stellar models using sets of individual mode frequencies for a large and growing number of solar-like oscillators. This wealth of data can tightly constrain stellar models but most fitting methods require either a large number of model evaluations for a single star or a fixed grid of low resolution to model many stars. Here, we show that interpolating in a grid of main-sequence models can provide precise and accurate stellar oscillation frequencies. The interpolation is much faster than the calculation of similarly accurate stellar models and overcomes deficiency in the grid resolution. Though stellar model frequencies have previously been interpolated linearly along evolutionary tracks (e.g. Kallinger *et al.* 2010), and thus in age, we are unaware of any previous attempts to interpolate in other stellar model parameters.

We computed a grid of models using the stellar evolution code CESTAM (Marques *et al.* 2013) on a regular grid with 61 ages t from 3 to 6 Gyr, eleven masses M from $0.975\,M_\odot$ to $1.025\,M_\odot$, eleven initial metallicities Z from 0.016 to 0.021, eleven initial hydrogen contents X from 0.69 to 0.74 and seven mixing-length parameters α from 1.6 to 1.9. These were chosen such that the central model is Sun-like and the range is large enough to characterize parameter uncertainties. The oscillation frequencies were calculated with ADIPLS (Christensen-Dalsgaard 2008). For any set of parameters within the grid's boundaries, we interpolate the oscillation frequencies of the corresponding model with cubic splines.

As a first test of the accuracy of the interpolation, we computed dense sequences (50 times the grid resolution) of models by taking the central parameter values and varying one of M, Z, X or α at a time. (Age is discussed in the next paragraph.) The fractional differences between the oscillation frequencies of these models (for modes $17 \leqslant n \leqslant 25$, $0 \leqslant l \leqslant 2$) for the metallicity Z are shown in Fig. 1. We note two points. First, there is scatter on the order of 10^{-6}. We attribute this to the accuracy of the stellar evolution code, which has a numerical tolerance of 10^{-6} in the mass co-ordinate. Second, there only appears to be one curve because the results for 27 different oscillations modes are plotted

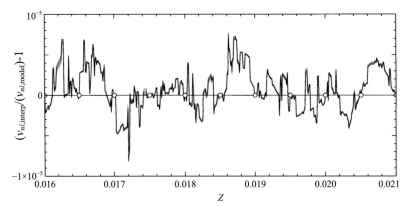

Figure 1. Fractional differences between modelled and interpolated frequencies along the metallicity Z, with the other parameters held fixed at $t = 4.5\,\mathrm{Gyr}$, $M = 1\,M_\odot$, $X = 0.715$ and $\alpha = 1.75$. There are 27 lines plotted, each corresponding to a different oscillation mode ($17 \leqslant n \leqslant 25$, $0 \leqslant l \leqslant 2$) but the lines lie mostly on top of each other and cannot be distinguished. The solid white circles show the interpolation points. The interpolation error is dominated by scatter that we attribute to numerical noise.

over each other. The error induced by interpolation thus behaves like a nearly-perfectly correlated fractional error in all the frequencies.

To model the interpolation error, we divided the grid into two parts. The first contained models spanning the grid with twice the step in each parameter, e.g., six metallicities instead of eleven. We then used these models to interpolate at the parameter values of the other models and computed the differences in the frequencies between the stellar models and the interpolated values. For each frequency, the fractional errors are approximately normally distributed with the same scatter of about 4×10^{-6} for all modes. We assume the errors are perfectly correlated and construct a covariance for the model errors. This is added to the observed covariance matrix when fitting stellar models to observed frequencies. Ideally, the interpolation error would be everywhere much smaller than the observed error. For our interpolation routine, the fractional errors correspond to absolute errors up to about $10\,\mathrm{nHz}$, which is several tens times smaller than frequencies derived from typical *Kepler* or *CoRoT* observations.

For the accuracy in the age t, we used every fourth model in the sequence of the central model to interpolate along that model's evolutionary track. The output is qualitatively similar to Fig. 1 but accurate everywhere to a fractional error smaller than 10^{-6}.

Thus, stellar oscillation frequencies can be precisely and accurately interpolated, provided that the stellar models are themselves accurately calculated. This may increase the computational cost of the grid but not of the interpolation itself. Finally, the interpolation should be tested for any given grid and, if non-negligible, appropriately characterized, noting that the model errors might be strongly correlated.

The authors acknowledge research funding by Deutsche Forschungsgemeinschaft (DFG) under grant SFB 963/1 "Astrophysical flow instabilities and turbulence" (Project A18).

References

Christensen-Dalsgaard, J. 2008, *Ap&SS*, 316, 113

Kallinger, T., Gruberbauer, M., Guenther, D. B., Fossati, L., & Weiss, W. W. 2010, *A&A*, 510, A106

Marques, J. P., Goupil, M.-J., Lebreton, Y., *et al.* 2013, *A&A*, 549, A74

Precision Asteroseismology
Proceedings IAU Symposium No. 301, 2013
J. A. Guzik, W. J. Chaplin, G. Handler & A. Pigulski, eds.

© International Astronomical Union 2014
doi:10.1017/S1743921313014658

Variability of M giant stars based on *Kepler* photometry: general characteristics

E. Bányai, L. L. Kiss and members of KASC WG12

Konkoly Observatory, Hungarian Academy of Sciences,
H-1121 Budapest, Konkoly Thege M. út 15-17, Hungary
email: ebanyai@konkoly.hu

Abstract. Our study based on the continuous, high-precision observations covering more than three years provided by the *Kepler* mission reveals a new systematic effect in the data, a possible transition region between solar-like and Mira-like oscillations, and gives an overview of M giant variability on a wide range of time-scales (hours to years).

Keywords. stars: variables: other, stars: AGB and post-AGB, techniques: photometric

1. Introduction

M giants are among the longest-period pulsating stars, hence their studies have traditionally been restricted to analyses of low-precision visual observations, or more recently, accurate ground-based CCD data.

We used thirteen quarters of *Kepler* long-cadence observations (one point per every 29.4 minutes) to analyse M giant variability, with a total time-span of over 1000 days. About two-thirds of our 317 stars have been selected from the ASAS-North survey of the *Kepler* field, with the rest supplemented from a randomly chosen M giant sample.

2. A new systematic effect – *Kepler*-year in the data

We found that the 23% of the total sample has small variations with sinusoidal modulation and period similar to the *Kepler*-year (372.5 days). Periods and phases indicated that those changes are likely to be caused by so far unrecognised systematics in the *Kepler* data. The phase of the *Kepler*-year variations varies among these stars. Further investigation revealed that the positions of the stars in the CCD-array display a clear correlation with the phase.

3. General characteristics

Based on the complexity in the time and the frequency domains, we sorted the stars into three groups. Stars in Group 1 have a wide range of periods between a few days and 100 days. Group 2 contains stars with very low-amplitude light curves that are mostly characterised by short-period oscillations, occasionally supplemented by slow changes that may be related to rotational modulation or instrumental drifts. Stars with light curves containing only a few periodic components (Miras and SRs) compose Group 3.

The power spectra and wavelets reveal very complex structures and rich behaviour. Peaks in the spectra are often transient in terms of time-dependent amplitudes. The overall picture is that of random variations presumably related to the stochasticity of the large convective envelope.

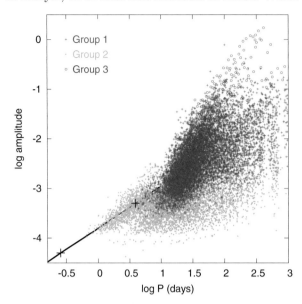

Figure 1. The period-amplitude relations for the whole sample using all the 50 frequencies and amplitudes. Different symbols distinguishes the three groups. Plus signs refer to two selected points in the top panel of Fig. 3 of Huber *et al.* (2011) related to solar-like oscillations. The black line is drawn through these points.

Determination of the significant period ratios provided the following results. The most populated clump is around $P_{\mathrm{short}}/P_{\mathrm{long}} \approx 0.7 - 0.8$, a ratio that is known to belong to the upper RGB stars. There is another distinct clump around period ratios of 0.5 that could be related to pulsation in the fundamental mode and first radial overtones (Takayama *et al.* 2013). The vertical concentrations at $\log P \approx 2.57$ and 2.87 result from the *Kepler*-year variability.

In Fig. 1 the bulk of the giants are spread in a triangular region. To the left of this upper envelope there is a distinct feature which shows strong correlation between the period and amplitude. To validate that the correlation is indeed in the extension of the ν_{max}-amplitude relation for the solar-like oscillations, we added two points, marked by the large plus signs, and a line drawn through these points. These points are taken from the top panel of Fig. 3 of Huber *et al.* (2011), where the oscillation amplitude vs. ν_{max} is shown for their entire *Kepler* sample. The excellent agreement between the line and the period-amplitude relation for Group 2, indicates that these stars are indeed the long-period extension of the solar-like oscillations.

Acknowledgements

This project has been supported by the Hungarian OTKA Grants K76816, K83790, K104607 and HUMAN MB08C 81013 grant of Mag Zrt., and the Lendület-2009 Young Researchers Program of the Hungarian Academy of Sciences.

References

Huber, D., Bedding, T. R., Stello, D., *et al.* 2011, *MNRAS*, 743, 143
Takayama, M., Saio, H., & Ita, Y. 2013, *MNRAS*, 431, 3189

Precision Asteroseismology
Proceedings IAU Symposium No. 301, 2013
J. A. Guzik, W. J. Chaplin, G. Handler & A. Pigulski, eds.

© International Astronomical Union 2014
doi:10.1017/S174392131301466X

Are RRab stars fully radial?

József M. Benkő and Róbert Szabó

Konkoly Observatory, MTA CSFK,
Konkoly Thege M. u. 15-17., H-1121, Budapest, Hungary
email: benko@konkoly.hu, rszabo@konkoly.hu

Abstract. Thanks to the space missions *CoRoT* and *Kepler* new oscillation frequencies have been discovered in the Fourier spectra of Blazhko RR Lyrae stars. The period doubling (PD) yields half-integer frequencies between the fundamental mode and its harmonics. In many cases the first and/or second radial overtone frequencies also appear temporally. Some stars show extra frequencies that were identified as potential non-radial modes. We show here that all these frequencies can be explained by pure radial pulsation as linear combinations of the frequencies of radial fundamental and overtone modes.

Keywords. stars: oscillations, stars: horizontal-branch, stars: variables: other

---

The first RRab star in which a non-radial mode was reported is the *CoRoT* target V1127 Aql. Chadid *et al.* (2010) explained 468 frequencies detected in this star by four independent frequencies, f_0, $f_{\rm m}$, $f' = 2f''$, $f_{\rm m1}$, and their combinations. The f_0 and $f_{\rm m}$ mean the main pulsation and modulation (Blazhko) frequency, respectively, f' (or f'') is the frequency of an independent, possibly non-radial mode, and $f_{\rm m1}$ is a secondary modulation frequency acting on 'additional modes' only (see Chadid *et al.* 2010 for the details). Later on, frequencies of possible non-radial modes have been found in the Fourier spectra of the Blazhko stars CoRoT 105288363 and V445 Lyr, V354 Lyr and V360 Lyr observed by *Kepler* (Benkő *et al.* 2010 = B10; Guggenberger *et al.* 2012 = G12). Many frequencies of these modes were found between the position of the radial first overtone and the PD frequencies and yielded period ratios P/P_0 around 0.7.

We homogeneously re-analyzed light curves of all the *CoRoT* and *Kepler* Blazhko RRab stars in which non-radial mode(s) were reported. We used the *CoRoT* 150-days-long data co-added to get 8-min sampling and *Kepler* 3-years-long long-cadence (30-min sampling) data covering Q1–Q12. The *CoRoT* white fluxes were cleaned and de-trended. In the case of *Kepler* targets, we used the raw pixel frames applying our own proper tailor-made apertures for each star and quarter separately (Benkő *et al.*, in preparation). The data were pre-whitened with the main pulsation frequencies and their harmonics, the modulation frequencies and as many modulation side peaks as possible. The resulting Fourier spectra and the frequency solutions from the literature are compared.

V1127 Aql (CoRoT 100689962). Our identification of f_0 and $f_{\rm m}$ are the same as in Chadid *et al.* (2010), but if we assume $f'' = f_2 - f_0$, then $f' = 2(f_2 - f_0)$, where $f_2 = 4.825397\,{\rm d}^{-1}$ is the frequency of the radial second overtone with the period ratio of 0.582. This identification eliminates the non-radial mode with its period ratio of 0.696. Poretti *et al.* (2010) have already noticed that V1127 Aql shows half-integer frequencies. If we accept the PD paradigm (Szabó *et al.* 2010), the frequency $f_{\rm m1}$ can also be interpreted as a linear combination: $1.5f_0 - f'$.

G12 found the following independent frequencies of *CoRoT 105288363*: f_0, $f_{\rm m}$, $f_{\rm s}$ (secondary Blazhko frequency), f_1, f_2, first and second radial overtone modes with the period ratios 0.745 and 0.590, respectively, and two non-radial modes, $f_{\rm N} = 2.442\,{\rm d}^{-1}$ $(P_{\rm N}/P_0 = 0.722)$ and $f_{\rm N2} = 2.2699\,{\rm d}^{-1}$. The multiple and time-dependent modulation of this star makes its frequency spectrum complicated. We removed more side peaks than G12, so we obtained a bit different peak structure. Now, the peak at f_{N2} seems to be insignificant while f_N can be identified as $2(f_2 - f_0)$.

In the case of $V445\,Lyr\,(KIC\,6186029)$ G12 reported f_0, f_m, f_s, $1.5f_0$ (PD), f_1, f_2 and non-radial mode $f_N = 2.7719$ d^{-1} ($P_N/P_0 = 0.703$). Many similarities between CoRoT 105288363 and V445 Lyr have been discussed by G12. In this study we find an additional one: the previously suggested non-radial mode f_N can also be identified as $2(f_2 - f_0)$.

According to B10, frequency content of $V354\,Lyr\,(KIC\,6183128)$ is the following: f_0, f_m, f_2 ($P_2/P_0 = 0.586$); two independent non-radial modes, $f' = 2.0810$ d^{-1} and $f''' = 2.6513$ d^{-1}, and $f'' = 2.4407$ d^{-1} which was identified as a possible radial first overtone mode with the period ratio of 0.729. The frequency spectrum of the Q1–Q12 data is a bit different from that for Q1–Q2 data (B10), because the amplitudes of the additional modes strongly depend on time (B10, Szabó et al., in preparation) and we removed more side peaks around harmonics eliminating more aliases. In consequence, the f'' frequency became insignificant. We explain $f' = (f_0 + f_1)/2$, where $f_1 = 2.3843$ d^{-1} and $f''' = 1.5f_0$ (PD). We detected two additional significant peaks at 2.999 d^{-1} and 2.300 d^{-1} which produced an equidistant triplet with the main PD frequency f'''.

Frequency solution for $V360\,Lyr\,(KIC\,9697825)$ from B10 is f_0, f_m, f_1 (first overtone mode with the period ratio $P_1/P_0 = 0.721$), and $f' = 2.6395$ d^{-1}, an independent non-radial mode. The star shows a consistent picture with the similar stars if we identify f' and its side peaks as a PD effect, and if f_1 is identified as $2(f_2 - f_0)$, where $f_2 = 3.046$ d^{-1}.

Summarizing: (i) Using linear combination frequencies of radial modes we obtained alternative solutions for all those Blazhko RRab stars in which non-radial modes were previously suggested. In other words, our mathematical description explains the frequency spectra solely by radial modes.

(ii) The amplitudes of the harmonics of combinations are many times higher than those of simple combination frequencies e.g. $A[2(f_2 - f_0)] \gg A(f_2 - f_0)$. This is unusual but a similar phenomenon, where the combination frequencies have higher amplitudes than their components, was reported by Balona et al. (2013) for a roAp star.

(iii) We searched for stars which show high-amplitude linear combination frequencies in their Fourier spectra and found at least two additional cases: CoRoT 103922434 and V366 Lyr (KIC 9578833).

Acknowledgements

This work was partially supported by the following grants: ESA PECS No 4000103541 /11/NL/KML, Hungarian OTKA Grant K-83790 and KTIA Urkut_10-1-2011-0019. RSz acknowledges the János Bolyai Research Scholarship of the Hungarian Academy of Sciences and the IAU travel grant.

References

Balona, L. A., Catanzaro, G., Crause, L., et al. 2013, MNRAS, 432, 2808
Benkő, J. M., Kolenberg, K., Szabó, R., et al. 2010, MNRAS, 409, 1585 (B10)
Chadid, M., Benkő, J. M., Szabó, R., et al. 2010, A&A, 510, A39
Guggenberger, E., Kolenberg, K., Nemec, J. M., et al. 2012, MNRAS, 424, 649 (G12)
Poretti, E. Paparó, M., Deleuil, M., et al. 2010, A&A, 520, A108
Szabó, R., Kolláth, Z., Molnár, L., et al. 2010, MNRAS, 409, 1244

Precision Asteroseismology
Proceedings IAU Symposium No. 301, 2013
J. A. Guzik, W. J. Chaplin, G. Handler & A. Pigulski, eds.

© International Astronomical Union 2014
doi:10.1017/S1743921313014671

Helio- and asteroseismology shedding light on "new physics"

Marek Biesiada

Department of Astrophysics and Cosmology, Institute of Physics,
University of Silesia, Uniwersytecka 4, 40-007 Katowice, Poland
email: biesiada@us.edu.pl

Abstract. Linkages between astronomy and physics have always been intimately close and mutually stimulating. Most often it was physics that served astronomy with its explanatory power. Today, however, we are increasingly witnessing the reverse: astrophysical considerations are being used to constrain new physics and moreover they are more efficient than laboratory experiments. This contribution reviews the ways helio- and white dwarf asteroseismology – branches in which Wojtek Dziembowski played a prominent role – are used for this purpose.

1. Introduction

There are two major problems in modern physics: Dark Matter (DM) and Dark Energy (DE), i.e. the accelerated expansion of the Universe. DM evidence is very strong (e.g., dynamical, gravitational lensing, cosmic microwave background radiation (CMBR) fluctuations, matter budget in the Universe). We expect DM composed of non-baryonic, neutral, stable and massive particles. Some hints come from particle physics, e.g. heavy particles, so called WIMPs, predicted by supersymmetry and axions, or light Nambu-Goldstone bosons invoked as a resolution of strong CP problem in QCD. Apart from solving the DM and DE problems other "exotic" possibilities may be contemplated, e.g. multidimensional worlds (Kaluza-Klein gravitons), primordial black holes, or varying fundamental constants. Exotic particles can serve as additional coolants (or heaters) in stellar interiors: in particular an additional coolant would heat up the star while it is ideal gas pressure supported (main sequence, horizontal branch), and cool it down while it is degenerate gas pressure supported (red giant branch, white dwarf, neutron star).

Helioseismology constrains the temperature profile inside the Sun, so it is sensitive to new channels of cooling/heating. Asteroseismology of pulsating white dwarfs (WD) is able to derive secular changes of period hence tracing the cooling rate. Comparison between evolutionary cooling and observations sheds light on possible new channels of cooling/heating.

2. Pulsating white dwarfs as a tool for astroparticle physics

G117-B15A belongs to the class of DAV WDs exhibiting non-radial pulsations in g-modes. DAV instability starts when the star is cool enough to develop a partial hydrogen ionisation zone sufficiently deep to excite pulsations. As the star cools further, the partial H-ionisation zone moves deeper, the thermal time-scale increases and so does the pulsation period. In 2005 after a total of 31 years of observations Kepler *et al.* (2005) obtained a measurement of the rate of period change with time for the largest amplitude periodicity at 215 s. Corrected for the proper motion, this rate can be compared with theoretical predictions setting limits to non-standard sources of energy or cooling.

Isern *et al.* (1992) raised for the first time the possibility of employing the measured rate of period change in G117-B15A to derive a constraint on the mass of axions. They considered the evolution of DAV WD models with and without axion emissivity, and compared the theoretical values of secular period rate for increasing masses of the axion with the observed rate of period change of G117-B15A at that time. The most recent estimate of axion mass from G117-B15A comes from Córsico *et al.* (2012) (see this paper for more details).

The interest in physical theories with extra spatial dimensions has recently experienced considerable revival in the context of DE. One can construct an effective theory of Kaluza-Klein (K-K) gravitons interacting with the standard model fields, calculate specific emissivity for gravi-bremsstrahlung of electrons and estimate the additional luminosity in K-K gravitons. Along these lines, using G117-B15A, Biesiada & Malec (2002) have obtained the bound on the energy scale for which extra dimensions might manifest themselves. This bound turned out to be one order of magnitude more stringent than the result obtained from LEP accelerator experiment. For the updated bound see Malec & Biesiada (2013).

There is also a debate in the literature over the issue of whether the quantities known as the constants of nature (such like G or fine structure constant) can vary with time. One of the reasons for this debate is connected with the string theory and associated ideas that the world we live in may have more than four dimensions. Because buoyancy is the restoring force for g-modes, the Brunt-Väisälä frequency is the most important quantity setting the scale in the pulsation spectrum. Using G117-B15A Biesiada & Malec (2004) obtained a bound on the rate of change of G. The most recent asteroseismological estimates are given by Córsico *et al.* (2013).

3. Helioseismology and primordial black holes

Primordial black holes (PBH) are expected to be formed in the early Universe - with a scale invariant power spectrum some overdensities will collapse into black holes. PBH production is enhanced during e.g. QCD phase transition when the pressure is suddenly reduced. The PBH mass range suggested by this mechanism ($10^{17} - 10^{26}$ g) cannot be probed by standard techniques like microlensing or gamma-rays. An interesting possibility to detect such a PBH was suggested by Kesden & Hanasoge (2011) who showed that passing through the Sun such PBH would excite distinctive oscillation patterns. It is very likely that PBHs can also leave unique transient imprints on echelle diagrams of pulsating stars.

References

Biesiada, M. & Malec, B. 2002, *Phys. Rev. D*, 65, 043008
Biesiada, M. & Malec, B. 2004, *MNRAS*, 350, 644
Córsico, A. H., Althaus, L. G., Miller Bertolami, M. M., *et al.* 2012, *MNRAS*, 424, 2792
Córsico, A. H., Althaus, L. G., Garcia-Berro, E., & Romero, A. D. 2013, *Journal of Cosmology and Astroparticle Physics*, 06, 032
Isern, J., Hernanz, M., & Garcia-Berro, E. 1992, *ApJ*, 392, L23
Kepler, S. O., Costa, J. E. S., Castanheira, B. G., *et al.* 2005, *ApJ*, 634, 1311
Kesden, M. & Hanasoge, S. 2011, *Phys. Rev. Lett.*, 107, 111101
Malec, B. & Biesiada, M. 2013, *ASP-CS*, 469, 21

Precision Asteroseismology
Proceedings IAU Symposium No. 301, 2013
J. A. Guzik, W. J. Chaplin, G. Handler & A. Pigulski, eds.

© International Astronomical Union 2014
doi:10.1017/S1743921313014683

Analysis of γ Doradus and δ Scuti stars observed by *Kepler*

Paul A. Bradley[1], Joyce A. Guzik[2], Lillian F. Miles[1], Jason Jackiewicz[3], Katrien Uytterhoeven[4], and Karen Kinemuchi[5]

[1] Los Alamos National Laboratory
XCP-6, MS F699, Los Alamos, NM 87545, USA
email: pbradley@lanl.gov, lfmiles@lanl.gov

[2] Los Alamos National Laboratory
XTD-NTA, MS T086, Los Alamos, NM 87545, USA
email: joy@lanl.gov

[3] Dept. of Astronomy, New Mexico State University
P.O. Box 30001, MSC 4500, Las Cruces, NM 88003, USA
email: jasonj@nmsu.edu

[4] Instituto de Astrofisica de Canarias (IAC)
E-38200, La Laguna, Tenerife, Spain
email: katrien@iac.es

[5] Apache Point Observatory,
P.O. Box 59, 2001 Apache Point Road, Sunspot, NM 88349, USA
email: kinemuchi@apo.nmsu.edu

Abstract. The *Kepler* spacecraft observed over 2000 faint stars that were part of our Guest Observer proposals. The stars were selected from the *Kepler* Input Catalog (KIC) to be in or near the γ Doradus or δ Scuti instability strips (8300 K $> T_{\rm eff} >$ 6200 K and 3.6 $< \log g <$ 4.7). The *Kepler* magnitude was $<$ 16 and the contamination factor was $< 10^{-2}$. The goal was to extend the search for "hybrid" δ Sct-γ Dor pulsators to fainter magnitudes. By inspecting the light curves and Fourier transforms, we find 42 δ Sct candidate stars, 299 γ Dor candidates, and 36 "hybrid" candidate stars showing both types of variations.

Keywords. techniques: photometric, stars: variables: δ Scuti, stars: variables: γ Doradus

1. Motivation, analysis and results

The *Kepler* spacecraft launched on 6 March, 2009 has revolutionized stellar pulsation studies with its ability to gather nearly continuous (duty cycle $>$ 90%) data with micro-magnitude precision. To better understand the statistics of δ Sct and γ Dor stars, we obtained Guest Observer (GO) data for multiple quarters. The first data set comes from Quarter 2 (Q2) and the last set we analyze in this paper is Q15. Almost all of the data are long cadence, with a few short-cadence data sets. Except for several of the 14 Cycle 1 Q2 – Q4 targets, these stars were chosen to have 8300 K $> T_{\rm eff} >$ 6200 K, 3.6 $< \log g <$ 4.7, and *Kepler* magnitude between 14.0 and 15.8. Recent *Kepler* observations show that the γ Dor and δ Sct stars have much overlap in the Hertzsprung-Russell and $\log g$ vs. $T_{\rm eff}$ diagrams (Grigahcène *et al.* 2010, Uytterhoeven *et al.* 2011).

In this paper, we take a "quick look" at the data to search for stars worthy of more detailed analysis. We use either MATLAB scripts written by J. Jackiewicz, or the "TOP-CAT" (Taylor 2011) program to extract time series of the raw and corrected fluxes in ASCII format. We then removed outlying data points, divided the light curve by the mean value and wrote output that could be read by our Fourier Transform (FT)

Table 1. Fraction of γ Dor, δ Sct, and hybrid stars from different studies.

Star type	Grigahcène *et al.*	Uytterhoeven *et al.*	This work
γ Dor	116 (55%)	100 (21%)	299 (79%)
δ Sct	67 (27%)	203 (43%)	42 (11%)
hybrid	51 (23%)	172 (36%)	36 (10%)

program (Tukey 1967). All of these steps were carried out in an automated manner via Python scripts. Stars with asteroseismic potential will be subjected to more rigorous analysis at a later date.

So far, we have analyzed data from 2251 stars for Quarters 2 through 15. 1021 of these stars show a signal consistent with random noise. There are 1230 that show variability in their light curves (> 20 ppm amplitude, with a range between 50 and 5000 ppm). Of these, 785 have longer period (> 3 d) variations and most of these stars probably have starspots rotating in and out of view. A number of stars show variations consistent with being a short period Cepheid or something similar. We found 67 eclipsing or ellipsoidal binary systems with periods ranging from several hours to about 20 days. The remaining 377 variable stars consist of 42 δ Sct candidate stars, 299 γ Dor candidates, and 36 "hybrid" candidate stars showing both δ Sct and γ Dor variations.

We compare our observational results to those of Grigahcène *et al.* (2010) and Uytterhoeven *et al.* (2011) in Table 1. Our data set shows mostly γ Dor stars, which is consistent with the findings of Grigahcène *et al.* (2010) but not of Uytterhoeven *et al.* (2011). One reason for the difference is that the *Kepler* Asteroseismic Science Consortium (KASC) sample studied by Grigahcène *et al.*, and the even larger KASC sample analyzed by Uytterhoeven *et al.* included nearly all of the brighter (*Kepler* mag < 14) stars in the *Kepler* field, as well as many previously known or suspected δ Sct stars; the KASC target stars also included more short-cadence observations able to identify high-frequency δ Sct stars. In contrast, our Guest Observer selection targeted stars in the *Kepler* input catalog with no prior observations, that were generally fainter and cooler, and so it is not surprising that a larger percentage of γ Dor variables were discovered. In future work, we plan to determine the relative fractions of γ Dor, δ Sct, and hybrid stars as a function of magnitude to see how the relative noise level affects the detection limits. We also plan to compare the H-R diagram location of the different types of variable stars relative to the boundaries of ground-based instability strips. The other readily apparent conclusions from these data are that *Kepler* can detect pulsations in 15th magnitude stars, and it can find δ Sct and hybrid stars even using only long-cadence data.

Acknowledgements

The authors acknowledge support from the NASA *Kepler* Guest Observer program. K.U. acknowledges support by the Spanish National Plan of R&D for 2010, AYA2010-17803. This project also benefitted from Project FP7-PEOPLE-IRSES:ASK No. 269194.

References

Grigahcène, A., Antoci, V., Balona, L., *et al.* 2010, *ApJ*, 713, L192
Taylor, M. 2011, TOPCAT is available via http://www.star.bris.ac.uk/~mbt/topcat, cited 27 Oct 2011.
Tukey, J. W., 1967, in: B. Harris (ed.), *Spectral Analysis of Time Series* (New York: Wiley)
Uytterhoeven, K., Moya, A., Grigahcène, A., *et al.* 2011, *A&A*, 534, A125

Precision Asteroseismology
Proceedings IAU Symposium No. 301, 2013
J. A. Guzik, W. J. Chaplin, G. Handler & A. Pigulski, eds.

© International Astronomical Union 2014
doi:10.1017/S1743921313014695

The IPoP method to measure Cepheid distances

J. Breitfelder[1,2], A. Mérand[1], P. Kervella[2], and A. Gallenne[3]

[1]European Southern Observatory, Alonso de Córdova 3107, Casilla 19001, Santiago 19, Chile
email: jbreitfe@eso.org

[2]LESIA, Observatoire de Paris, CNRS UMR 8109, UPMC, Université Paris Diderot, 5 place Jules Janssen, 92195 Meudon, France

[3]Universidad de Concepción, Departamento de Astronomía, Casilla 160-C, Concepción, Chile

Abstract. Cepheids are one of the most famous standard candles used to calibrate the Galactic distance scale. However, it is fundamental to develop and test independent tools to measure their distances, in order to reach a better calibration of their period-luminosity (P-L) relationship. We present here the first results obtained with the Integrated Parallax of Pulsation (IPoP) method, an extension of the classical Baade-Wesselink method that derives the distance by making a global modelisation of all the available data. With this method we aim to reach a 2% accuracy on distance measurements.

Cepheid masses are also an essential key for our comprehension of those objects. We briefly present an original approach to derive observational constraint on Cepheid masses. Unfortunately, it does not lead to promising results.

Keywords. gravitation, stars: fundamental parameters, stars: variables: Cepheids

1. General presentation of the IPoP method

Cepheids are a fundamental element of the extra-galactic distance ladder, but the calibration of their P-L relationship still has to be improved by measuring independent distances. We present here a strong computing tool (IPoP) that we developed to measure accurate distances of single Galactic Cepheids. This code is based on the classical Baade-Wesselink method but can integrate all the available observables in the modelisation (e.g. magnitudes and colours in all bands, interferometric angular diameters, radial velocities, effective temperatures and spectra), so we can get a good statistical accuracy. The redundancy of part of the data (e.g. photometry and interferometry to estimate the angular diameter) also results in a good robustness of the fitting process. The modelisation is also based on physical models (e.g. ATLAS9 atmospheric models) in order to control systematic errors.

Our method requires a large collection of data, that we mostly gather from online archives. We are also carrying out a large program of Cepheid observations in both hemispheres, with the VLTI and the CHARA interferometers.

2. Applications

2.1. Distance of δ Cep

The result that we obtained for δ Cep is presented in Fig. 1. In this study we used a p-factor of 1.27 (Mérand *et al.* 2005). We obtain a distance of $d = 314$ pc with the precision of 3%, consistent with the *HST* parallax value (Benedict *et al.* 2002): $\pi = 3.66 \pm 0.15$ milli-arcsec ($d = 273 \pm 66$ pc).

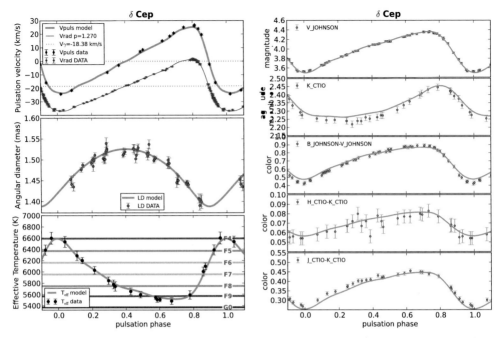

Figure 1. The IPoP method applied to δ Cep leads to a distance measurement with a 3% accuracy, $d = 314 \pm 10$ pc.

2.2. *Toward a new constraint on Cepheid masses*

The free fall acceleration, $a_{\rm ff}$, radius, R, and mass, M, are linked through: $M = a_{\rm ff} R^2/G$, where G is the gravitational constant. We can rewrite this equation by using our observables and a non-biased distance $d' = d/p$, where p is the p-factor, to get a minimum mass, $M_{\min}$:

$$M_{\min} = \frac{\theta^2 d'^2}{4G} \left| \frac{dv_{\rm rad}}{dt} \right|_{\rm ff,max} p^3, \qquad (2.1)$$

where θ is the angular diameter and $|dv_{\rm rad}/dt|_{\rm ff,max}$, the maximum acceleration in the phase of a contraction of a star.

Considering that other forces can oppose gravitation (e.g. radiation pressure or other pressure forces), this mass can only be considered as an inferior limit of the actual Cepheid mass. We implemented this equation in the IPoP code and found minimum masses much below the masses derived from models (Bono *et al.* 2011), so not really constraining. During the contraction of the star, the inward acceleration of the atmosphere amounts to about 25% of the free-fall acceleration, which means that the forces opposing gravity are not negligible. This original method was used by Lacour *et al.* (2009) to estimate the mass of a Mira star. Assuming a free fall acceleration, they fitted the diameter curve with a parabola to derive the acceleration, and then the corresponding gravitational mass.

References

Benedict, G. F., McArthur, B. E., Fredrick, L. W., *et al.* 2002, *AJ*, 124, 1695
Bono, G., Gieren, W. P., Marconi, M., & Fouqué, P., Caputo F. 2011, *ApJ*, 563, 319
Lacour, S., Thiébaut, E., & Perrin, G. 2009, *ApJ*, 707, 632
Mérand, A., Kervella, P., Coudé du Foresto, V., *et al.* 2005, *A&A*, 438, L9

Precision Asteroseismology
Proceedings IAU Symposium No. 301, 2013
J. A. Guzik, W. J. Chaplin, G. Handler & A. Pigulski, eds.

© International Astronomical Union 2014
doi:10.1017/S1743921313014701

Multicolour photometry of pulsating stars in the Galactic Bulge fields

Przemysław Bruś and Zbigniew Kołaczkowski

Instytut Astronomiczny Uniwersytetu Wrocławskiego
Kopernika 11, 51-622 Wrocław, Poland
email: brus@astro.uni.wroc.pl

Abstract. We present a study of photometric properties of very crowded stellar fields toward the Galactic Bulge. We performed a search for pulsating stars among thousands of variable stars from the OGLE-II survey supplementing the variability study with photometric measurements in four Johnson-Cousins $UBVI_C$ passbands. Using these data, we analysed the properties of objects located at different distances and, whenever possible, classified them.

Keywords. Galactic Bulge, pulsating stars, extinction

1. Archival data and follow-up observations

The main source of data used in this work is the entire OGLE-II DIA time-series I_C band photometric database of Galactic Bulge fields (Szymański 2005). All pulsating stars described here were discovered during our multiperiodicity search in the OGLE-II database. Most of them are not present in the catalog published by the OGLE Team (Woźniak et al. 2002) because of very small amplitudes (a few mmag).

The second part of our project is based on follow-up single-epoch observations of 25 selected fields carried out with the CTIO 1-m telescope in May and June 2007. Using the Y4kCam detector ($20' \times 20'$ field of view), we covered a relatively small part of the OGLE-II fields in four passbands: U (exposures: 1500 s), B (400 s), V (250 s), I_C (120 s). In each field we performed profile photometry by means of the DAOPHOT-II package (Stetson 1987). To perform the transformation of our photometry to the standard system, we carried out additional observations of a nearby standard field BWC (Paczyński et al. 1999) on a photometric night of June 3/4, 2007. Accurate astrometry carried out by means of the UCAC3 catalog (Zacharias *et al.* 2010) allowed us to perform reliable cross-identifications with the OGLE databases. Our standardized V and I_C measurements are in very good agreement with the OGLE-III photometry published by Szymański *et al.* (2011).

2. Analysis and results

In order to remove long-term and seasonal trends, we applied spline function fits as the first step of the time-series analysis. Next, we performed frequency analysis by means of the Fourier periodogram in the range between 0 and 40 d^{-1}. A given star was selected as a variable if the signal-to-noise ratio (S/N) of the dominant frequency exceeded a detection level equal to 5. For all stars selected through the S/N criterion, we applied a semi-automatic search for additional frequencies by means of consecutive prewhitening and repeated Fourier analysis of the residual light curve. In each iteration all parameters of the fit were updated. As a result, we found several hundreds of multiperiodic variables in each OGLE-II field. They are good candidates for pulsators, mainly those on the main-sequence. Moreover, we divided these stars into two groups: short-period pulsators with

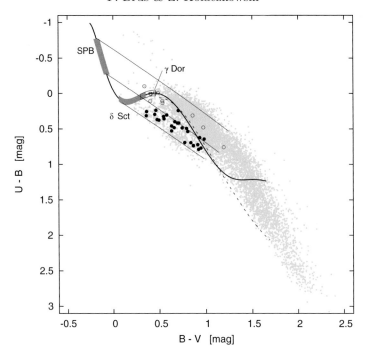

Figure 1. Observed $(U - B)$ vs. $(B - V)$ colour-colour diagram for stars in three selected fields toward the Galactic Bulge. All stars with $(U - B)$ errors smaller than 0.05 mag are shown as gray dots. The intrinsic relations for dwarfs and giants are plotted as solid and dashed curves, respectively. The schematic locations of SPB, δ Sct and γ Dor type pulsators are marked and labeled. Stars classified as δ Sct-type variables are shown as black filled dots, SPB or γ Dor variables as open circles. Three solid black lines indicate reddening line with a slope of 0.85.

dominant periods shorter than 0.3 d and long-period pulsators including the remaining multiperiodic variables. In the first group we expect β Cephei and δ Scuti-type stars; the second group consists of slowly pulsating B-type stars (SPB) and γ Doradus-type stars.

The preliminary multi-colour photometry of three selected fields allowed us to make use of colour-magnitude and colour-colour diagrams in our classification of variable stars. We found pulsating stars in a wide range of colour indices measured with an accuracy better than 0.05 mag (Fig. 1). The distribution of early A-type stars in the $(U - B)$ vs. $(B - V)$ diagram implies a significantly higher slope of the reddening line than the standard value. The whole sample of short-period variables turned out to contain only δ Scuti-type stars with $E(B - V)$ reddenings between 0.3 and 1 mag. The majority of the long-period pulsators cannot be unambiguously classified: they can be either highly reddened SPB or γ Doradus stars.

Acknowledgments. This work was supported by the National Science Center (NCN) grant No. 2011/03/B/ST9/02667.

References

Paczyński, B., Udalski, A., Szymański, M., *et al.* 1999, *AcA*, 49, 319
Stetson, P. B. 1987, *PASP*, 99, 191
Szymański, M. K. 2005, *AcA*, 55, 43
Szymański, M. K., Udalski A., Soszyński, I., *et al.* 2011, *AcA*, 61, 83
Woźniak, P. R., Udalski, A., Szymański, M., *et al.* 2002, *AcA*, 52, 129
Zacharias, N., Finch, C., Girard, T., *et al.* 2010, *AJ*, 139, 2184

Precision Asteroseismology
Proceedings IAU Symposium No. 301, 2013
J. A. Guzik, W. J. Chaplin, G. Handler & A. Pigulski, eds.

© International Astronomical Union 2014
doi:10.1017/S1743921313014713

Spectropolarimetric study of the classical Cepheid η Aql: pulsation and magnetic field

V. Butkovskaya[1], S. Plachinda[1], D. Baklanova[1], and V. Butkovskyi[2]

[1] Crimean Astrophysical Observatory of Taras Shevchenko National University of Kyiv,
98409, Nauchny, Crimea, Ukraine,
email: `varya@crao.crimea.ua`

[2] Taurida National V. I. Vernadsky University,
95007, Vernadskogo str. 4, Simferopol, Crimea, Ukraine

Abstract. We report the results of spectropolarimetric study of the classical Cepheid η Aql. We found that the longitudinal magnetic field of η Aql sinusoidally varies with the radial pulsation period, while the amplitude B, mean field B_0, and phases of maximum and minimum field change from year to year. We hypothesize that possible reasons of those variations are stellar axial rotation or dynamo mechanisms.

Keywords. stars: magnetic fields, stars: oscillations, stars: individual: η Aql

1. Introduction

Currently, the question of pulsation modulation of magnetic field in stars, both with convective and radiative envelopes, is still open.

RR Lyr. Babcock (1958) reported a detection of a magnetic field in RR Lyr. The longitudinal component of the field was found to be variable from -1580 to $+540$ G, but showed no correlation with the pulsation cycle of the star. Romanov *et al.* (1987, 1994) also registered significant magnetic field in RR Lyr, and found the field to be variable with an amplitude of up to 1.5 kG over the pulsation cycle. On the other hand, Preston (1967) and Chadid *et al.* (2004) detected no convincing evidence of a photospheric magnetic field in the star in the years 1963–1964 and 1999–2002, respectively.

η Aql. Photoelectric magnetometer observations of η Aql, performed by Borra *et al.* (1981, 1984), detected no magnetic field in this star. Plachinda (2000) was the first who detected magnetic field on η Aql and reported pulsation modulation of the longitudinal component from -100 to $+50$ G. Wade *et al.* (2002) detected no statistically significant longitudinal magnetic field in η Aql during 3 nights in 2001 and concluded that η Aql is a non-magnetic star, at least at a level of 10 G. Grunhut *et al.* (2010) registered clear Zeeman signatures in Stokes V parameter for η Aql and eight other supergiants.

γ Peg. Butkovskaya & Plachinda (2007) reported the modulation of the longitudinal magnetic field in β Cephei-type star γ Peg (B2 IV) with the amplitude of about 7 G over the 0.15-day pulsation period of the star.

In order to shed some light on the problem of pulsational modulation of stellar magnetic fields, we continued spectropolarimetric monitoring of η Aql during 60 nights between 2002 and 2012 using Coudé spectrograph at the 2.6-m Shajn telescope of the Crimean Astrophysical Observatory (Ukraine).

Table 1. Parameters of the variability of the longitudinal magnetic field of η Aql.

Year	Phase max	Phase min	Amplitude B, G	Mean B_0, G	F-test
2002	0.45	0.95	12.7 ± 2.2	5.5 ± 1.5	0.99
2004	0.78	0.28	13.9 ± 2.4	4.7 ± 1.5	0.99
2010	0.38	0.88	4.3 ± 1.1	-3.1 ± 1.1	0.98
2012	0.60	0.10	4.2 ± 2.3	-0.7 ± 1.5	0.69

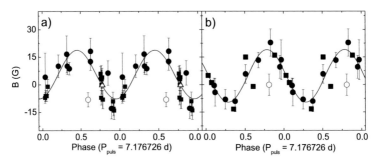

Figure 1. Longitudinal magnetic field of η Aql folded with the 7.176726-day pulsation period: a) our 2002 data (*closed and open circles*), data by Wade *et al.* (2002) (*black squares*), and Grunhut *et al.* (2010) (*open triangles*); b) our 2004 data (*closed and open circles*), data by Borra *et al.* (1981, 1984) (*black squares*). Fitted sinusoids are shown as continuous lines. Open circles represent our data that have not been taken into account in the fits.

2. Results

The technique of Zeeman splitting, used for the measurement of the longitudinal magnetic field, is described in detail by Butkovskaya & Plachinda (2007). We folded all values of the measured longitudinal magnetic field with the pulsational period according to the pulsation ephemeris: JD $= 2450100.861 + 7.176726E$ (Kiss & Vinkó 2000), where E is the number of pulsation cycles. As an example, the pulsation modulation of the longitudinal magnetic field of η Aql in 2002 and 2004 is illustrated in Fig. 1. We found that the magnetic field sinusoidally varies in phase with the radial pulsation of η Aql. This confirms the previous conclusions of Plachinda (2000). However, the amplitude B, mean field B_0, and phases of maximum and minimum field are changing from year to year (see Table 1 where the F-test indicates the statistical reliability of the detected variability for each year). The possible reason for those variations is stellar axial rotation or dynamo mechanisms.

References

Babcock, H. W. 1958, *ApJS*, 3, 141
Borra, E. F., Fletcher, J. M., & Poeckert, R. 1981, *ApJ*, 247, 569
Borra, E. F., Edwards, G., & Mayor, M. 1984, *ApJ*, 284, 211
Butkovskaya, V. & Plachinda, S. 2007, *A&A*, 469, 1069
Chadid, M., Wade, G. A., Shorlin, S. L. S., & Landstreet, J. D. 2004, *A&A*, 413, 1087
Grunhut, J. H., Wade, G. A., Hanes, D. A., & Alecian, E. 2010, *MNRAS*, 408, 2290
Kiss, L. L. & Vinkó, J. 2000, *MNRAS*, 314, 420
Plachinda, S. I. 2000, *A&A*, 360, 642
Preston, G. W. 1967, R. C. Cameron (ed.), *The Magnetic and Related Stars* (Baltimore: Mono Book Corporation), p. 26
Romanov, Yu. S., Udovichenko, S. N., & Frolov, M. S. 1987, *AZh Lett.*, 13, 69
Romanov, Yu. S., Udovichenko, S. N., & Frolov, M. S. 1994, *Bull. Spec. Astrophys. Obs.*, 38, 169
Wade, G. A., Chadid, M., Shorlin, S. L. S., Bagnulo, S., & Weiss, W. W. 2002, *A&A*, 392, L17

Precision Asteroseismology
Proceedings IAU Symposium No. 301, 2013
J. A. Guzik, W. J. Chaplin, G. Handler & A. Pigulski, eds.

© International Astronomical Union 2014
doi:10.1017/S1743921313014725

Stellar variability in the VVV survey: overview and first results

M. Catelan[1,2], **D. Minniti**[1,2], **P. W. Lucas**[3], **I. Dékány**[1,2],
R. K. Saito[1,2,4], **R. Angeloni**[1,2], **J. Alonso-García**[1,2], **M. Hempel**[1,2],
K. Hełminiak[1,2,5], **A. Jordán**[1,2], **R. Contreras Ramos**[1,2],
C. Navarrete[1,2], **J. C. Beamín**[1,2], **A. F. Rojas**[1,2], **F. Gran**[1,2],
C. E. Ferreira Lopes[1,2,6], **C. Contreras Peña**[3], **E. Kerins**[7],
L. Huckvale[7,8], **M. Rejkuba**[8], **R. Cohen**[9], **F. Mauro**[9], **J. Borissova**[10],
P. Amigo[1,2,10], **S. Eyheramendy**[11], **K. Pichara**[12], **N. Espinoza**[1,2],
C. Navarro[1,2,10], **G. Hajdu**[1,2], **D. N. Calderón Espinoza**[1,2],
G. A. Muro[1,2], **H. Andrews**[1,2,13], **V. Motta**[10], **R. Kurtev**[10],
J. P. Emerson[14], **C. Moni Bidin**[2,15], and **A.-N. Chené**[16]

[1] Pontificia Universidad Católica de Chile, Instituto de Astrofísica, Santiago, Chile
email: mcatelan@astro.puc.cl

[2] The Milky Way Millennium Nucleus, Santiago, Chile

[3] University of Hertfordshire, Hatfield, UK

[4] Universidade Federal de Sergipe, São Cristóvão, SE, Brazil

[5] Nicolaus Copernicus Astronomical Center, Toruń, Poland

[6] Universidade Federal do Rio Grande do Norte, Natal, Brazil

[7] The University of Manchester, Manchester, UK

[8] European Southern Observatory, Garching, Germany

[9] Universidad de Concepción, Concepción, Chile

[10] Universidad de Valparaíso, Valparaíso, Chile

[11] Pontificia Universidad Católica de Chile, Departamento de Estadística, Santiago, Chile

[12] Pontificia Universidad Católica de Chile, Facultad de Ingeniería, Santiago, Chile

[13] Leiden Observatory, Leiden, The Netherlands

[14] Queen Mary, University of London, London, UK

[15] Instituto de Astronomía, Universidad Católica del Norte, Antofagasta, Chile

[16] Gemini Observatory, Hawaii, USA

Abstract. The Vista Variables in the Vía Láctea (VVV) ESO Public Survey is an ongoing time-series, near-infrared (IR) survey of the Galactic bulge and an adjacent portion of the inner disk, covering 562 square degrees of the sky, using ESO's VISTA telescope. The survey has provided superb multi-color photometry in 5 broadband filters (Z, Y, J, H, and K_s), leading to the best map of the inner Milky Way ever obtained, particularly in the near-IR. The main part of the survey, which is focused on the variability in the K_s-band, is currently underway, with bulge fields observed between 34 and 73 times, and disk fields between 34 and 36 times. When the survey is complete, bulge (disk) fields will have been observed up to a total of 100 (60) times, providing unprecedented depth and time coverage in the near-IR. Here we provide a first overview of stellar variability in the VVV data.

Keywords. surveys, stars: novae, cataclysmic variables, stars: rotation, stars: spots, stars: variables: Cepheids, other

1. Overview

The VVV survey (Minniti *et al.* 2010, Catelan *et al.* 2011, Saito *et al.* 2012) has been monitoring the bulge and the southern disk in the K_s-band since 2010. It will provide, for the first time, a homogeneous database of long-baseline time-series photometry with up to 100 epochs for nearly 10^9 point sources. At present, when the extensive monitoring of the bulge fields has started, VVV has already provided a considerable number of epochs, suitable for analyses of stellar variability (Catelan *et al.* 2013). VVV provides a sparse time sampling, usually a single epoch for a few fields on a night (with an occasional second visit), distributed almost randomly over the seasonal visibility period of the area. Most of the currently available time-series data were taken in the third year of observations. An extensive overview of stellar variability in the current VVV Survey data, including detailed descriptions of the data, cadence, completeness, reduction and analysis techniques, and our efforts towards the automated classification of the VVV light curves, has recently been provided by Catelan *et al.* (2013), where one can also find examples of recent applications of these data, particularly in the context of Galactic structure. Sample light curves for many different variable star classes, including RR Lyrae, Cepheids (both classical and type II), long-period variables, eclipsing binaries, RS CVn systems, microlenses, novae, and transient events, are also provided.

2. Conclusions

VVV provides a treasure trove of scientific data that can be exploited in numerous scientific contexts. In terms of stellar variability, the project will provide up to several million (Catelan *et al.* 2013) calibrated K_s-band light curves for genuinely variable sources, including pulsating stars, eclipsing systems, rotating variables, cataclysmic stars, microlenses, planetary transits, and even transient events of unknown nature. At present, with the data-gathering phase of the VVV Survey having just crossed its half-way mark, we are really just taking the first steps in what will certainly be a long and exciting journey, during which it will be possible to address a myriad of time-domain astronomical applications. The latter include not only research on variable stars as such, but also their use as distance indicators and tracers of Galactic structure, origin, and evolution. Since VVV is a Public Survey, the data will quickly be made available to the entire astronomical community, opening the door to many additional applications and synergies with other ongoing and future projects that target the same fields as those covered by VVV.

Acknowledgements. This work is supported by the European Southern Observatory; the Basal Center for Astrophysics and Associated Technologies (PFB-06); the Chilean Ministry for the Economy, Development, and Tourism's Programa Iniciativa Científica Milenio through grant P07-021-F, awarded to The Milky Way Millennium Nucleus; and Fondecyt through grants #1110326 (M.C., I.D., J.A.-G.), 1120601 (J.B.), 1130140 (R.K.), 3130320 (R.C.R.), and 3130552 (J.A.-G.). C. Navarrete acknowledges grant CONICYT-PCHA/Magíster Nacional/2012-22121934.

References

Catelan, M., Minniti, D., Lucas, P. W., *et al.* 2011, *Carnegie Obs. Conf. Ser.*, 5, 145

Catelan, M., Minniti, D., Lucas, P. W., *et al.* 2013, in: K. Kinemuchi, H. A. Smith & N. De Lee (eds.), *40 Years of Variable Stars: A Celebration of Contributions by Horace A. Smith*, p. 139

Minniti, D., Lucas, P. W., Emerson, J. P., *et al.* 2010, *New Astron.*, 15, 433

Saito, R. K., Hempel, M., Minniti, D., *et al.* 2012, *A&A*, 537, A107

Precision Asteroseismology
Proceedings IAU Symposium No. 301, 2013
J. A. Guzik, W. J. Chaplin, G. Handler & A. Pigulski, eds.

© International Astronomical Union 2014
doi:10.1017/S1743921313014737

g-mode trapping and period spacings in hot B subdwarf stars

S. Charpinet[1,2], V. Van Grootel[3]†, P. Brassard[4], and G. Fontaine[4]

[1] Université de Toulouse, UPS-OMP, IRAP, Toulouse, France
email: stephane.charpinet@irap.omp.eu

[2] CNRS, IRAP, 14 avenue Edouard Belin, 31400 Toulouse, France

[3] Institut d'Astrophysique et de Géophysique de l'Université de Liège, Allée du 6 Août 17, B-4000 Liège, Belgium

[4] Université de Montréal, Pavillon Roger-Gaudry, Département de Physique, CP 6128, Succ. Centre-Ville, Montréal QC, H3C 3J7, Canada

Abstract. Hot B subdwarfs (sdB) are hot and compact helium core burning stars of nearly half a solar mass that can develop pulsational instabilities driving acoustic and/or gravity modes. These evolved stars are expected to be chemically stratified with an almost pure hydrogen envelope surrounding a helium mantle on top of a carbon/oxygen enriched core. However, the sdB stars pulsating in g-modes show regularities in their observed period distributions that, surprisingly (at first sight), are typical of the behavior of high order g-modes in chemically homogeneous (i.e., non-stratified) stars. This led to a claim that hot B subdwarfs could be much less chemically stratified than previously thought. Here, we reinvestigate trapping effects affecting g-modes in sdB stars. We show that standard stratified models of such stars can also produce nearly constant period spacings in the low frequency range similar to those found in g-mode spectra of sdB stars monitored with *Kepler*.

Keywords. stars: oscillations, stars: subdwarfs, stars: interiors

1. Introduction

The ultra high precision of white light photometry from space with *CoRoT* and *Kepler* reveals extremely rich g-mode period spectra. Among other properties, it was noticed that many modes seem to be nearly equally spaced in period, as one would expect for a homogeneous star showing g-mode oscillations in the asymptotic regime. This finding was in apparent contradiction with the strong g-mode trapping structures caused by the steep chemical gradients expected inside such evolved stratified stars (as illustrated in Charpinet *et al.* 2002), leading to the claim that sdB stars may be much less stratified than previously thought (Reed *et al.* 2011; Telting *et al.* 2012). However, extending the study up to the g-mode cutoff period, we show that the mode trapping efficiency in current stratified models strongly decreases with increasing radial order and that the "standard" models can in fact account quite well for the observed g-mode structure.

2. Period spacings and echelle diagram

Figure 1 shows the signature on the g-mode spectrum induced by the steep chemical gradients in sdB stars (dominated by the He to H transition at the base of the envelope). These produce distinct patterns of cyclic perturbations of the mean period spacings (left

† Chargé de recherches, Fonds de la Recherche Scientifique, FNRS, rue d'Egmont 5, B-1000 Bruxelles, Belgium.

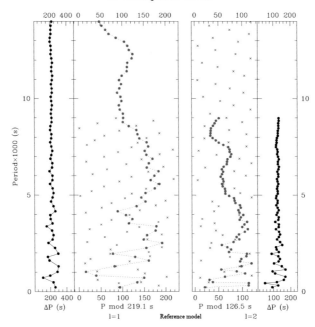

Figure 1. Theoretical period spacings, $\Delta P = P_{k+1} - P_k$ vs. P_k (left and right panels), and corresponding échelle diagrams (central panels) for the $l = 1$ and $l = 2$ g-modes in a representative sdB model. The filled circles are the $l = 1$ ($l = 2$) g-modes up to the cutoff period (near $k \sim 70$).

and right panels). However, these perturbations are not uniform along the period range of interest. From radial order $k = 1$ to $k \sim 20$ (periods less than $5\,000$ s for $l = 1$ modes), the pattern is comparable to the strong structures reported in Charpinet *et al.* (2002), but beyond this limit the fluctuations decrease in amplitude and the period spacings become nearly uniform. Consequently, the corresponding theoretical échelle diagrams (central panels in Fig. 1) show vertical ridges that develop at low frequencies. These are comparable to the ridges observed in, e.g., Fig. 6 of Telting *et al.* (2012). Moreover, the low radial order modes (with periods below $\sim 4\,000$ s) shown in Fig. 6 of Telting *et al.* (2012) appear to be scattered like in our reference stratified sdB model.

3. Conclusion

Sequences of modes with nearly equally spaced periods as found for several sdB pulsators monitored with *Kepler* cannot be interpreted as the signature of much smoother chemical transitions than expected from standard models of sdB stars. Strongly stratified models show nearly constant period spacings in the low frequency range of the observable g-mode spectrum and produce echelle diagrams very similar to current observations.

Acknowledgements

This work was supported in part by the Programme National de Physique Stellaire (PNPS, CNRS/INSU, France).

References

Charpinet, S., Fontaine, G., Brassard, P., & Dorman, B. 2002, *ApJS*, 139, 487
Reed, M. D., Baran, A., Quint, A. C., *et al.* 2011, *MNRAS*, 414, 2885
Telting, J. H., Østensen, R. H., Baran, A. S., *et al.* 2012, *A&A*, 544, A1

Precision Asteroseismology
Proceedings IAU Symposium No. 301, 2013
J. A. Guzik, W. J. Chaplin, G. Handler & A. Pigulski, eds.

© International Astronomical Union 2014
doi:10.1017/S1743921313014749

What asteroseismology can teach us about low-mass core helium burning models

Thomas N. Constantino, Simon W. Campbell and John C. Lattanzio

Monash Centre for Astrophysics, School of Mathematical Sciencies
Monash University, Victoria 3800, Australia
email: thomas.constantino@monash.edu

Abstract. Standard models of low-mass core helium burning stars typically give an asymptotic $l = 1$ g-mode period spacing well below that inferred from observed mixed modes. We find that most physical uncertainties, such as mixing beyond the fully convective core, are not significant enough to be responsible for such a discrepancy. The solution to the problem may lie in a deviation of structure away from its canonical form, such as a more massive H-exhausted core, which we briefly explore here.

Keywords. stars: horizontal-branch, stars: interiors, stars: late-type, stars: oscillations

1. Introduction

It has been posited for more than four decades that core helium burning (CHeB) stars develop a zone of slow mixing, or "semiconvection", beyond the fully convective core because of convective overshoot. The treatment of mixing in this region is a large source of uncertainty. Later evolution depends on the structure after this phase, and this is typically when the results of different stellar evolution codes begin to diverge.

Asteroseismology promises a unique insight into these stars and a chance to constrain our models of them. Bedding *et al.* (2011) have shown that it is possible to reliably distinguish CHeB stars from photometrically similar red giant branch (RGB) stars and Mosser *et al.* (2012) have developed a method to determine the gravity-dominated mixed mode period spacing $\Delta\Pi_1$ from the observed modes of mixed p and g character.

It has been found that our CHeB models from the Monash University stellar evolution code, irrespective of mixing scheme, have a range of asymptotic $\Delta\Pi_1$ below that inferred from observations of the *Kepler* red clump giants (Mosser *et al.* 2012). Similar results have been reported by Montalbán *et al.* (2013) using a different evolution code.

2. Possible sources of this discrepancy

The core mass of horizontal-branch stars has historically been constrained by matching stellar evolution isochrones to globular cluster colour-magnitude diagrams (e.g. Raffelt 1990). This method, however, depends on an interplay between RGB tip luminosity, initial helium abundance and a host of other factors (which are discussed in Catelan *et al.* 1996). If we assume that the red clump $\Delta\Pi_1$ determinations from Mosser *et al.* (2012) are representative, then we can more explicitly test our structure for zero-age horizontal-branch (ZAHB) models. ZAHB models should have a $\Delta\Pi_1$ that corresponds to the minimum observed since it then increases with age. This may in fact be the most direct probe of the mass of the H-exhausted core at the RGB tip. We have tested three different mixing schemes for the core: 1) using the Schwarzschild boundary with overshoot and 2) without overshoot, and 3) a new subroutine for semiconvection.

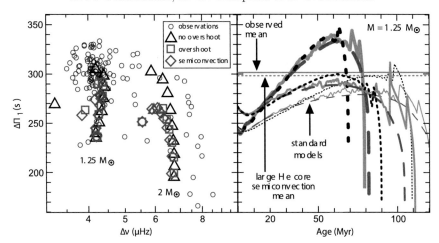

Figure 1. Left panel: $\Delta\Pi_1$ as a function of large frequency separation $\Delta\nu$ during the CHeB phase. *Kepler* observations are in circles (Mosser *et al.* 2012). Markers for $1.25\,M_\odot$ and $2\,M_\odot$ solar metallicity models are at 0.1 intervals in central helium mass fraction. No-overshooting, overshooting and semiconvection models are indicated by triangle, square and diamond markers respectively. Right panel: Evolution of $\Delta\Pi_1$ for standard red clump models (thin lines), models with a four-fold increase in neutrino emission (medium lines) and a case in which the H-exhausted core is artificially extended (thick lines). Models are $1.25\,M_\odot$ and solar-metallicity. The solid straight line is the observed $\Delta\Pi_1$ mean when $\Delta\nu < 5\,\mu$Hz. Short dashes, solid lines and long dashes correspond to mixing with no overshoot, overshoot and semiconvection respectively.

We find that none of these, however, can match the observations (Fig. 1). We also find that initial composition and microphysics such as reaction rates and equation of state are not significant factors.

Here we show that an experimental four-fold increase in neutrino emission on the RGB increases the core mass by $0.018\,M_\odot$ at the flash, and in turn initially gives a 20 s increase in $\Delta\Pi_1$ (Fig. 1). This difference then diminishes over time because of slower H-burning and the consequently slower He core growth. In order to match the high end of the observed $\Delta\Pi_1$ range we need to artificially grow the He core so that by the end of CHeB it is around $0.1\,M_\odot$ larger than in the normal case.

3. Conclusions

We have shown that uncertainties in mixing, reaction rates and composition do not appear to explain the discrepancy between the predicted g-mode period spacing from our models of red clump stars and that inferred from observations. We will conduct more detailed seismic analysis soon, with a view to determining whether or not there is indeed a shortcoming in our models and then uncovering the cause(s) of it. Other constraints such as cluster star counts will also be used.

References

Bedding, T. R., Mosser, B., Huber, D., *et al.* 2011, *Nature*, 471, 608

Catelan, M., de Freitas Pacheco, J. A., & Horvath, J. E. 1996, *ApJ*, 461, 231

Montalbán, J., Miglio, A., Noels, A., Dupret, M.-A., Scuflaire, R., & Ventura, P. 2013, *ApJ*, 766, 118

Mosser, B., Goupil, M. J., Belkacem, K., *et al.* 2012, *A&A*, 540, A143

Raffelt, G. G. 1990, *ApJ*, 365, 559

Precision Asteroseismology
Proceedings IAU Symposium No. 301, 2013
J. A. Guzik, W. J. Chaplin, G. Handler & A. Pigulski, eds.

© International Astronomical Union 2014
doi:10.1017/S1743921313014750

Asteroseismology with the new opacity bump at $\log T \approx 5.06$

H. Cugier

Astronomical Institute of the Wroclaw University,
51-622 Wroclaw, Poland,
email: cugier@astro.uni.wroc.pl

Abstract. Although the κ mechanism of pulsations is known for early-type stars, opacities and the equation of state are still uncertain. Stellar models calculated for the OP data implemented with the new Kurucz opacities at $\log T < 5.2$ were investigated for different chemical compositions of elements. The additional metallic opacity bump at $\log T \approx 5.06$ that occurs in the Kurucz data changes markedly the oscillation spectra of unstable modes. Basic properties of the new opacity bump and examples of seismic models are shown. B-type stars observed in the Galaxy, LMC and SMC were considered. The problem was studied using Dziembowski's computing codes for linear, non-adiabatic and non-radial oscillations.

Keywords. stars, stellar evolution, asteroseismology, radiative opacity

---

The opacity and equation of state are necessary ingredients for modelling internal structure, evolution and pulsation of stars. The well known OP and OPAL Rosseland-mean opacities agree reasonably well, but recent critical remarks concerning the atomic physics and accuracy of the OP (and OPAL) data are motivated by unsolved problems in helio- and astero-seismology, cf. e.g., Pradhan & Nahar (2009) and Dziembowski (2009). Recently, we show (cf. Cugier 2012) that the present-day Kurucz (2011) opacity project to include all spectral lines leads to the new opacity bump at $\log T \approx 5.06$, which has only a minor counterpart in the OP (and OPAL) data, cf. Seaton *et al.* (1994). This opacity bump is caused by metals and much more complete atom models than used by the OP and OPAL projects are essential for this effect, cf. Fig. 1a and b.

The H-R diagrams for stellar models $M > 30 \, M_\odot$ calculated with the opacity bump at $\log T \approx 5.06$ (K-OP models; case No. 2 in Fig. 1b) differ markedly from those corresponding to the OP (case No. 1 in Fig. 1b) and OPAL data, cf. Fig. 1c. Evolutionary tracks for $M < 30 \, M_\odot$ remain almost the same as in the OP and OPAL cases. The K-OP models mean that hybrid CK/OP opacity data were used, viz. the Castelli & Kurucz (2003) data at outer layers (up to $\log T = 5.2$) and the OP data at $\log T > 5.2$. No mass loss, no rotation and no core overshooting were assumed. For $M > 30 \, M_\odot$, the effect is caused by the gradient of radiation pressure near $\log T = 5.06$. The OP models for $Z = 0.0054$ are also shown in Fig. 1c.

The new opacity bump at $\log T \approx 5.06$ has important consequences for asteroseismology for all models studied in this paper $(M \geqslant 9 \, M_\odot)$. The proper behaviour of the Rosseland-mean opacities, eigenfunctions and thermal timescale of the driving zone are necessary for the κ mechanism to work, cf. e.g. Dziembowski (2009) and Pamyatnykh (1999). Figure 1d shows frequencies of unstable modes for stellar models calculated for $M = 10 \, M_\odot$ at the moment when the central hydrogen content was decreased to $X_c = 0.24$. The low-degree modes $(l = 0, 1$ and $2)$ are marked individually, while unstable modes with $l \geqslant 3$ are shown by small (black) points. The number of a model describes the opacity data used, cf. Fig. 1b. We found no unstable p modes for the K-OP models

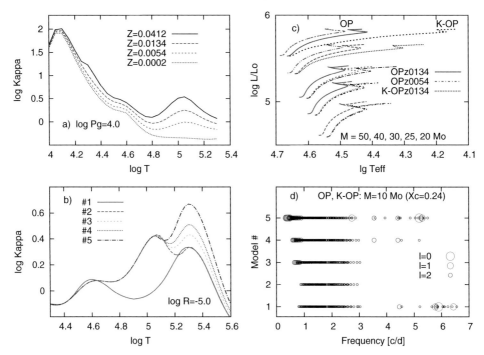

Figure 1. *a*) Examples of the Rosseland-mean opacity calculated by Castelli & Kurucz (2003) for different metallicity parameters $Z = 0.0412, 0.0134, 0.0054$ and 0.0002. (P_{g} means the gas pressure). *b*) The CK/OP opacities for $Z = 0.0134$ (case No. 2) are compared with the OP data (case No. 1) for $\log R = \log \rho - 3 \log T + 18 = -5$. The CK/OP data with increased OP bump at $\log T \approx 5.3$ are shown as cases $3-5$. *c*) The OP and K-OP models of stars are plotted in the H-R diagram for $Z = 0.0134$ and the OP models for $Z = 0.0054$. *d*) Seismic models of $M = 10\,M_\odot$ for different opacity data shown in panel *b*). 'Lo' means the solar luminosity and 'Mo' the solar mass.

with masses $10 \leqslant M \leqslant 20\,M_\odot$, including those calculated for $Z \geqslant 0.0264$, cf. model No. 2 in Fig. 1d. Using the Castelli & Kurucz (2003) opacity data at $\log T \leqslant 5.2$, an increase of the OP opacity bump at $\log T = 5.3$ by a factor of about 1.7 is necessary to obtain unstable *p*-modes for the Galactic early-B stars, cf. models No. $4-5$ in Fig. 1d. The same factor, about 1.7, is sufficient to drive pulsations for $Z = 0.0054$ (*p*- and *g*-modes) and for $Z = 0.003$ (*g*-modes). These values of Z are in a good agreement with those of derived spectroscopically for the Magellanic Clouds.

References

Castelli, F. & Kurucz, R. 2003, in: N. Piskunov, W. W. Weiss, & D. F. Gray (eds.), *Modelling of Stellar Atmospheres*, Proc. IAU Symposium No. 210 (Astronomical Society of the Pacific), Poster A20

Cugier, H. 2012, *A&A*, 547, 42

Dziembowski, W. 2009, *CoAst*, 158, 227

Kurucz, R. 2011, *Can. J. Phys.*, 89, 417

Pamyatnykh, A. A. 1999, *AcA*, 49, 119

Pradhan, A. K. & Nahar, S. N. 2009, *AIP-CP*, 1171, 52

Seaton, M. J., Yan, Y, Mihalas, D., & Pradhan, A. K. 1994, *MNRAS*, 266, 805

Precision Asteroseismology
Proceedings IAU Symposium No. 301, 2013
J. A. Guzik, W. J. Chaplin, G. Handler & A. Pigulski, eds.

© International Astronomical Union 2014
doi:10.1017/S1743921313014762

Spectroscopic and photometric study of two B-type pulsators in eclipsing systems

Dominik Drobek and Andrzej Pigulski

Instytut Astronomiczny, University of Wrocław,
ul. Kopernika 11, 51-622 Wrocław, Poland
email: drobek@astro.uni.wroc.pl

Abstract. Pulsating stars in eclipsing binary systems play an important role in asteroseismology. The combination of their spectroscopic and photometric orbital solutions can be used to determine, or at least to constrain, the masses and radii of components. To successfully perform any seismic modelling of a star, one has to identify at least some of the detected modes, which requires precise time-series photometric and spectroscopic observations. This work presents a progress report on the analysis of two β Cephei-type stars in eclipsing binaries: HD 101794 (V916 Cen) and HD 167003 (V4386 Sgr).

Keywords. stars: individual: HD 101794, HD 167003, stars: oscillations, binaries: eclipsing

1. Observations and data reduction

HD 101794 and HD 167003 have been observed at the South African Astronomical Observatory (SAAO) between May 2 and 19, 2009. The $UBVRI$ time-series photometry has been acquired using the UCT CCD detector at the 1.0-m telescope. Spectroscopic observations were carried out using the GIRAFFE échelle spectrograph at the 1.9-m telescope. We obtained spectra in the wavelength range between 4200 Å and 6900 Å with a resolution $R \approx 39\,000$. Stellar magnitudes were calculated using the program DAOPHOT II (Stetson 1987). Spectra were wavelength-calibrated and extracted using the IRAF software package (Tody 1993). The échelle orders were merged and normalised with a program of our own, and the radial velocities were calculated by cross-correlating the observed spectra with non-LTE models of Lanz & Hubeny (2007) using the method of Tonry & Davis (1979).

2. Data analysis and results

The spectrum of HD 101794 features very broad lines, the broadening being caused by rapid rotation. This is not surprising at all, since HD 101794 is a known Be star. Unfortunately, the resulting broadening reduces the accuracy of the radial velocities obtained with cross-correlation. For this reason the analysis of HD 101794 requires more consideration. The star is still under study, and the results will not be discussed here.

The radial velocity time-series data of HD 167003 were subjected to Fourier analysis. The orbital period was estimated at 10.88 d, which is close to the value of 10.79824 d obtained from the analysis of the All Sky Automated Survey phase 3 (ASAS-3) V-band photometry by Pigulski & Pojmański (2008). We adopted their period value in our subsequent analysis. Apart from the effects of orbital motion, at least four frequencies arising from stellar pulsations are present in the power spectrum. The frequency $f_1 = 7.351$ d^{-1} is only seen in the radial velocity data. Our $f_2 = 6.771$ d^{-1} and $f_3 = 7.023$ d^{-1} correspond to f_1 and f_3 found by Pigulski & Pojmański (2008). Our last frequency, $f_4 =$

8.451 d^{-1}, is not seen in their photometric data, while their f_4 was not detected in our radial velocity measurements. In our newly acquired SAAO multicolour photometry, we detected frequencies corresponding to f_1, f_2 and f_3 found by the aforementioned authors.

Once the frequencies of pulsation were determined, their contribution was removed from the original radial velocity data. We attempted to model the radial velocity changes arising from orbital motion, and arrived at the following set of parameters: orbital period $P_{orb} = 10.79824$ d (fixed); semi-amplitude $K = (31.8 \pm 0.6)$ km s^{-1}; eccentricity $e = 0.061 \pm 0.013$; argument of periastron $\omega = (299 \pm 8)°$; systemic velocity $\gamma = (-30 \pm 0.4)$ km s^{-1} and the time of periastron passage $T_0 = $ HJD 2454965.53 $\pm$ 0.23. The standard deviation from the fit amounts to 2.9 km s^{-1}.

3. Discussion

The results for HD 167003 are very encouraging. First of all, we have confirmed it is indeed a pulsating star in a binary system. In addition, we have confirmed that the orbital period amounts to 10.79824 d and is not twice as long, as initially suspected by Pigulski & Pojmański (2008). This star seems to be a single-lined spectroscopic binary, and our modelling suggests that the orbit is close to circular.

Pigulski & Pojmański (2008) detected only the primary eclipse in the ASAS-3 photometry of HD 167003. Our initial hypothesis was that the lack of the secondary eclipse is caused by a highly eccentric orbit. In light of the results of our modelling, we now know this cannot be the case. This suggests that the secondary eclipse is very shallow, and that the contribution of the secondary component to the total flux is small. While our present photometry is more accurate than the ASAS-3 photometry used by the previous investigators, we are also unable to find the secondary minimum in our observations. This could be because the orbital period of HD 167003 is quite long, and our phase coverage is incomplete. However, we managed to detect hints of a minute reflection effect.

Once the orbital inclination is known from the light curve modelling, we will be able to use the mass function to constrain the masses of components. From the fact that at least two of the photometrically detected modes are also seen in the radial velocity data, it seems probable that the attempts to identify mode degrees from amplitude ratios and phase differences will be successful. Therefore, this star seems to be a very good candidate for asteroseismic analysis.

Acknowledgements

The authors would like to thank G. Kopacki for acquiring photometric observations of our targets. D. Drobek would like to thank E. Niemczura for her help with the model spectra, and Z. Kołaczkowski and J. Molenda-Żakowicz for their assistance with the IRAF software package. This work has been supported by the National Science Centre grants no. 2011/01/N/ST9/00400 and no. 2011/03/B/ST9/02667.

References

Lanz, T. & Hubeny, I. 2007, *ApJS*, 169, 83
Pigulski, A. & Pojmański, G. 2008, *A&A*, 477, 917
Stetson, P. B. 1987, *PASP*, 99, 191
Tody, D. 1993, *ASP-CS*, 52, 173
Tonry, J. & Davis, M. 1979, *AJ*, 84, 1511

Precision Asteroseismology
Proceedings IAU Symposium No. 301, 2013
J. A. Guzik, W. J. Chaplin, G. Handler & A. Pigulski, eds.

© International Astronomical Union 2014
doi:10.1017/S1743921313014774

Strömgren photometry and medium-resolution spectroscopy of some δ Scuti and γ Doradus stars in the *Kepler* field

L. Fox-Machado

Instituto de Astronomía, Universidad Nacional Autónoma de México
email: lfox@astrosen.unam.mx

Abstract. We have obtained CCD photometry and medium-resolution spectroscopy of a number of δ Scuti and γ Doradus stars in the *Kepler* field-of-view as part of the ground-based observational efforts to support the *Kepler* space mission. In this work we present the preliminary results of these observations.

Keywords. stars: variables: δ Sct, stars: variables: γ Dor

1. Introduction

The *Kepler* space mission (Borucki *et al.* 2010) was successfully launched in March 2009 and since then it has been monitoring a huge number of stars in a region of 105 square degrees located between the constellations of Cygnus and Lyra. Although the main scientific goal of the mission is to discover Earth-sized planets, the high precision photometry provided by the *Kepler* satellite gives a unique opportunity to study the pulsational variability of thousands of stars across the H-R diagram in details by means of asteroseismic methods (Aerts *et al.* 2010). As is well known, asteroseismic studies require accurate and precise atmospheric parameters of the stars to produce reliable results. Since the precision of the physical parameters like effective temperature, gravity and metallicity available in the *Kepler Input Catalog* (KIC, Latham *et al.* 2005) is generally too low for asteroseismic modelling, to best exploit the *Kepler* data additional multi-colour and spectroscopic information is needed. In the framework of the *Kepler* Asteroseismic Science Consortium (KASC, *http://astro.phys.au.dk/KASC/*) several ground-based observational efforts have been undertaken to derive physical parameters of the *Kepler* stars with high precision (e.g. Uytterhoeven *et al.* 2011, Molenda-Żakowicz *et al.* 2011). This paper describes our observational efforts at the Observatorio Astronómico Nacional at San Pedro Mártir (OAN-SPM) in Baja California, Mexico to derive the physical parameters of several δ Scuti and γ Doradus stars in the *Kepler* field.

2. Observations, data reduction and conclusion

The CCD observations of 74 δ Scuti and γ Doradus stars in the *Kepler* field have been made with the 0.84-m f/15 Ritchey-Chrétien telescope at OAN-SPM, during six consecutive nights, from 2012 June 21 to June 26. The telescope hosted the filter-wheel 'Mexman' with the ESOPO (E2V) CCD camera, which has a 2048 × 4608 pixel array, with a pixel size of 15 × 15 μm^2. The typical field-of-view with this configuration is 8′ × 8′. The observations were taken with Strömgren *uvby* and Hβ filters to take advantage of the Strömgren-Crawford photometric system in deriving physical parameters of the

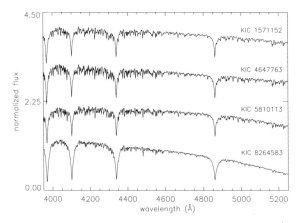

Figure 1. Reduced spectra of some *Kepler* targets.

stars. A set of standard stars from well observed open clusters (e.g. Peña *et al.* 2011) was also observed each night to transform instrumental observations onto the standard system and to correct for atmospheric extinction. The usual calibration procedures for CCD photometry have been carried out during our observing run. Sky flat fields, bias and dark exposures were taken every night. The data reduction of this CCD photometry has been carried out with the usual techniques and IRAF packages. The instrumental magnitudes and colours, once corrected for atmospheric extinction were transformed to the standard system. In this way, we have obtained the standard Strömgren indices not only of the 74 δ Scuti and γ Doradus target stars, but also of all the stars brighter than $V \sim 15$ mag located within each observed field.

The spectroscopic observations were conducted at the 2.12-m telescope of the OAN-SPM observatory during several short runs between 2010 and 2012. We used the same equipment as explained in Fox Machado *et al.* (2010). In particular, we used the Boller & Chivens spectrograph installed in the Cassegrain focus of the telescope. The 1200 lines/mm grating with a blaze angle of 13° was used. The grating angle was set to 19° to cover a wavelength range from 3950 Å to 5250 Å. A 2048×2048 E2V CCD camera was used for the observations. The typical resolution of the spectra is 2.2 Å and the dispersion 2.6 Å per pixel. The reduction procedure was performed with the standard routines of the IRAF package. Examples of the reduced spectra are shown in Fig. 1. The final results of these observations will be published elsewhere (Fox Machado *et al.*, in preparation).

The author acknowledges the financial support from the UNAM through grant PAPIIT IN104612 and from the IAU.

References

Aerts, C., Christensen-Dalsgaard, J., & Kurtz, D. W. 2010, *Asteroseismology* (Springer Science+Business Media B. V.)

Borucki, W. J., Koch, D., Basri, G., *et al.* 2010, *Science*, 327, 977

Fox Machado, L., Alvarez, M., Michel, R., *et al.* 2010, *New Astron.*, 15, 397

Latham, D. W., Brown, T. M., Monet, D. G., Everett, M., Esquerdo, G. A., & Hergenrother, C. W. 2005, *BAAS*, 37, 1340

Molenda-Żakowicz, J., Latham, D. W., Catanzaro, G., Frasca, A., & Quinn, S. N. 2011, *MNRAS*, 412, 1210

Peña, J. H., Fox Machado, L., García, H., *et al.* 2011, *Rev. Mexicana AyA*, 47, 309

Uytterhoeven, K., Moya, A., Grigahcène, A., *et al.* 2011, *A&A*, 534, A125

Precision Asteroseismology
Proceedings IAU Symposium No. 301, 2013
J. A. Guzik, W. J. Chaplin, G. Handler & A. Pigulski, eds.

© International Astronomical Union 2014
doi:10.1017/S1743921313014786

Characterizing the variability of Melotte 111 AV 1224: a new variable star in the Coma Berenices open cluster

L. Fox-Machado, R. Michel, M. Alvarez and J. H. Peña

Instituto de Astronomía, Universidad Nacional Autónoma de México
e-mails: lfox@astrosen.unam.mx, rmm@astrosen.unam.mx,
alvarez@astrosen.unam.mx, jhpena@astroscu.unam.mx

Abstract. A search for new pulsating stars in the Coma Berenices open cluster was carried out. As a result of this search, the cluster member Melotte 111 AV 1224 presented clear indications of photometric variability. In order to determine its physical parameters, Strömgren standard indices and low-resolution spectra were acquired. In this work, we present the preliminary results of these observations.

Keywords. stars: variables, open clusters and associations: individual: Melotte 111

1. Introduction

Coma Berenices (Melotte 111, RA = 12^h23^m, DEC = $+26°00'$, J2000.0) is the second closest open cluster to the Sun. The *Hipparcos* distance of Melotte 111 is $d = 89.0 \pm 2.1$ pc (van Leeuwen 1999), in agreement with older ground-based estimates (e.g. 85.4 ± 4.9 pc, Nicolet 1981). The metallicity of the cluster has been derived by several authors. For example, Cayrel de Strobel (1990) determined [Fe/H] $= -0.065 \pm 0.021$ dex, whereas Friel & Boesgaard (1992) found [Fe/H] $= 0.052 \pm 0.047$ dex. The age of the cluster is estimated between 400 and 500 Myr (Bounatiro & Arimoto 1993). As its physical parameters are well constrained, the variability studies in Melotte 111 are very important. We carried out a search for new pulsating stars in the direction of Melotte 111. As a result of this search, the star Melotte 111 AV 1224 was found to be a new variable star. This star was originally designated AV 1224 in the astrometric catalogue for the area of Coma Berenices (Abad & Vicente 1999). It is listed as a cluster member in the Simbad database. This work presents preliminary results aimed at characterizing the variability of this target.

2. Observations, data reduction and conclusion

The CCD observations of the Melotte 111 open cluster have been made with the 0.84-m f/15 Ritchey-Chrétien telescope at OAN-SPM observatory, during ten consecutive nights, between April 11 and 20, 2009. The telescope hosted the filter-wheel 'Mexman' with the Marconi (E2V) CCD camera, which has a 2048 × 2048 pixels array, with a pixel size of 15 × 15 μm^2. The typical field-of-view in this configuration amounts to $7' \times 7'$. The observations were obtained through a Johnson V filter. The usual calibration procedures for CCD photometry have been carried out during our observing run. Sky flat fields, bias and dark exposures were taken every night. The resulting light curve is not sinusoidal, but is strictly periodic; the frequency spectrum reveals two peaks, $2f_1 \sim 5.8$ d^{-1}, $A = 23.2$ mmag and $f_1 \sim 2.9$ d^{-1}, $A = 10.6$ mmag. The light curve of AV 1224 phased with its main period, $1/f_1 = 0.34$ d, is shown in Fig. 1a. We have also derived the following V

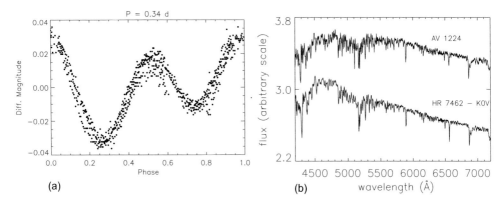

Figure 1. (a) Light curve of AV 1224 phased with the period of 0.34 d. (b) Spectrum of AV 1224 and HR 7462, a star of similar spectral type.

magnitude and indices in the Strömgren system for AV 1224: $V = 13.709$, $(b-y) = 0.526$, $m_1 = 0.276$, and $c_1 = 0.320$.

Spectroscopic observations of the star were conducted with the 2.12-m telescope of the OAN-SPM observatory in June 2011. We used the same equipment as explained by Baran *et al.* (2011). In particular, we used Boller & Chivens spectrograph installed in the Cassegrain focus of the telescope. The 400 lines/mm grating with a blaze angle of 4.18° was used. The grating angle was set to 7° to cover wavelength range from 4000 Å to 7500 Å. A 2048×2048 E2V CCD camera was used in the observations. The typical resolution of the recorded spectra is 8 Å and the dispersion amounts to 1.8 Å per pixel. The reduction procedure was performed with the standard routines of the IRAF package. Fig. 1(b) shows the reduced spectrum of AV 1224 and, for comparison, the spectrum of a standard star of spectral type K0 V taken on the same night. Considering the Strömgren indices and the stellar spectrum, the variability of AV 1224 due to pulsations can be ruled out. Its light curve resembles rather those observed in W Ursae Majoris-type variables (W UMa), also called EW stars. As it is known, the components of W UMa systems are in contact and are main-sequence stars of nearly the same spectral type, from around middle A to early K. Their orbital periods range from 0.2 to 1.4 days. An in-deep analysis of these observations will be given elsewhere.

The authors acknowledge the financial support from the UNAM through grant PAPIIT IN104612 and from the IAU.

References

Abad, C. & Vicente, B. 1999, *A&AS*, 136, 307
Baran, A. S., Fox Machado, L., Lykke, J., Nielsen, M., & Telting, J. H. 2011, *AcA*, 61, 325
Bounatiro, L. & Arimoto, N. 1993, *A&A*, 268, 829
Cayrel de Strobel, G. 1990, *MemSAIt*, 61, 613
Friel, E. D. & Boesgaard, A. M. 1992, *ApJ*, 387, 107
Nicolet, B. 1981, *A&A*, 104, 185
van Leeuwen, F. 1999, *A&A*, 344, L71

Precision Asteroseismology
Proceedings IAU Symposium No. 301, 2013
J. A. Guzik, W. J. Chaplin, G. Handler & A. Pigulski, eds.

© International Astronomical Union 2014
doi:10.1017/S1743921313014798

Asteroseismology from Dome A, Antarctica

J. N. Fu[1], W. K. Zong[1], Y. Yang[2,1], A. Moore[3], M. C. B. Ashley[4], X. Q. Cui[5], L. L. Feng[6], X. F. Gong[5], J. S. Lawrence[4,7], D. Luong-Van[4], Q. Liu[8], C. R. Pennypacker[9], Z. H. Shang[10], J. W. V. Storey[4], L. Z. Wang[8], L. F. Wang[6,2], H. G. Yang[11], X. Y. Yuan[6,12], D. G. York[13], X. Zhou[8], and Z. H. Zhu[1]

[1] Department of Astronomy, Beijing Normal University, Beijing 100875, China

[2] Department of Physics and Astronomy, Texas A&M University, College Station 77843, USA

[3] Caltech Optical Observatories, California Institute of Technology, Pasadena CA 91107, USA

[4] School of Physics, University of New South Wales, NSW 2052, Australia

[5] Nanjing Institute of Astronomical Optics and Technology, Nanjing 210042, China

[6] Purple Mountain Observatory, Chinese Academy of Sciences, Nanjing 210008, China

[7] Australian Astronomical Observatory, NSW 1710, Australia

[8] National Astronomical Obs. of China, Chinese Academy of Sciences, Beijing 100012, China

[9] Center for Astrophysics, Lawrence Berkeley National Laboratory, Berkeley, CA, USA

[10] Tianjin Normal University, Tianjin 300074, China

[11] Polar Research Institute of China, Pudong, Shanghai 200136, China

[12] Chinese Center for Antarctic Astronomy, Nanjing 210008, China

[13] Department of Astronomy and Astrophysics and Enrico Fermi Institute, University of Chicago, Chicago, IL 60637, USA

Abstract. Gattini and CSTAR have been installed at Dome A, Antarctica, which provide time-series photometric data for a large number of pulsating variable stars. We present the study for several variable stars with the data collected with the two facilities in 2009 to demonstrate the scientific potential of observations from Dome A for asteroseismology.

1. Introduction

Observations in polar sites open a new window for long, uninterrupted and consecutive time-series photometry/spectroscopy for pulsating variable stars, a powerful tool for asteroseismology. Dome A, located at longitude $77°06'57''$ E and latitude $80°25'08''$ S with a 4013-m elevation as the highest point on the Antarctic plateau, is widely predicted to be a very good astronomical site (Saunders *et al.* 2009). The first-generation facilities including Gattini (Moore *et al.* 2013) and CSTAR (Chinese Small Telescope ARray, Liu & Yuan 2009) were deployed on Dome A in 2008 January. We present some progress on asteroseismology, using the time-series multi-color photometric data collected with the two telescopes in the winter of 2009.

2. Bright variable stars observed with Gattini

Gattini is a 10.5-mm fish-eye lens equipped with an Apogee 2k×2k CCD camera. A filter rotator allows frames taken in Johnson BVR and a band of OH in turn. The field of view is $90°$. In 2009 winter, a total of 377 GB of data were taken. After data reduction and stellar photometry, light curves of 66 variable stars were obtained. Figure 1 shows the light curves of a W Vir type variable star κ Pav ($V = 4.4$ mag) in B, V, R in 2009.

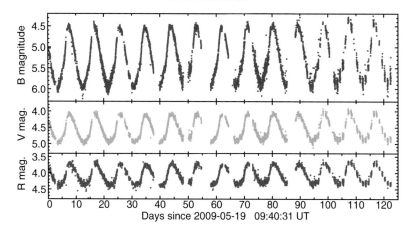

Figure 1. Light curves of κ Pav in B, V and R, data from 2009.

3. Variable stars observed with CSTAR

CSTAR has four telescopes with apertures of 15-cm, equipped with four 1k×1k CCD cameras. The filters of SDSS g', r' and i' are mounted to three of them while the fourth telescope has no filter. The pointing was fixed to the celestial southern pole with field of view of 4.5°×4.5°. Times-series CCD frames were taken in g' and r' in 2009 winter.

The light curves of HD 92277 ($V = 9.10$ mag, $B - V = 0.38$ mag) show that it is a new variable star. A total of 13 and 24 independent frequencies are detected in g' and r', respectively. Combining with the available parameters of the star, we classify it as a new δ Scuti star (Zong *et al.* 2013). In addition, light curves of three RR Lyrae stars in the field of view of CSTAR were obtained in g' and r' in 2009.

4. Summary

Variable stars have been studied with the time-series photometric data collected from Dome A, Antarctica. In an epoch of space missions (e.g. *CoRoT* and *Kepler*) which are providing high duty-cycle high precision time-series photometry for variables, observations for pulsating stars can be made through multiple colors simultaneously from Dome A, Antarctica, which provides important information for pulsation mode identification. This might bring a chance of a new break-through in asteroseismology.

Acknowledgement

This work is supported by the National Basic Research Program of China (973 Program 2013CB834900).

References

Liu, G. & Yuan, X. 2009, *Acta Astron. Sinica*, 50, 224
Moore, A., Yang, Y., Fu, J. N., *et al.* 2013, in: M. G. Burton, X. Cui, & N. F. H. Tothill (eds.), *Astrophysics from Antarctica*, Proc. IAU Symposium No. 288 (Cambridge: Cambridge University Press), p. 34
Saunders, W., Lawrence, J. S., Storey, J. W. V., *et al.* 2009, *PASP*, 121, 976
Zong, W. K., Fu, J. N., Niu, J. S., *et al.* 2013, *AJ*, submitted

Precision Asteroseismology
Proceedings IAU Symposium No. 301, 2013
J. A. Guzik, W. J. Chaplin, G. Handler & A. Pigulski, eds.

© International Astronomical Union 2014
doi:10.1017/S1743921313014804

An interferometric view on binarity and circumstellar envelopes of Cepheids

A. Gallenne[1], P. Kervella[2], A. Mérand[3], J. D. Monnier[4], J. Breitfleder[2,3], G. Pietrzyński[1,5] and W. Gieren[1]

[1] Universidad de Concepción, Departamento de Astronomía,
Casilla 160-C, Concepción, Chile
email: agallenne@astro-udec.cl

[2] LESIA, Observatoire de Paris, CNRS UMR 8109, UPMC,
Université Paris Diderot, 5 Place Jules Janssen, F-92195 Meudon, France

[3] European Southern Observatory, Alonso de Córdova 3107,
Casilla 19001, Santiago 19, Chile

[4] Astronomy Department, University of Michigan,
1034 Dennison Bldg, Ann Arbor, MI 48109-1090, USA

[5] Warsaw University Observatory, Al. Ujazdowskie 4, 00-478 Warsaw, Poland

Abstract. Optical interferometry is the only technique giving access to milli-arcsecond (mas) resolution at infrared wavelengths. For Cepheids, this is a powerful and unique tool to detect the orbiting companions and the circumstellar envelopes (CSE). CSEs are interesting because they might be used to trace the Cepheid evolution history, and more particularly they could impact the distance scale. Cepheids belonging to binary systems offer an unique opportunity to make progress in resolving the Cepheid mass discrepancy. The combination of spectroscopic and interferometric measurements will allow us to derive the orbital elements, distances, and dynamical masses. Here we focus on recent results using 2- to 6-telescopes beam combiners for the Cepheids X Sgr, T Mon and V1334 Cyg.

Keywords. stars: variables: Cepheids, stars: circumstellar matter, binaries

1. Introduction

Cepheid CSEs are interesting for several aspects. Firstly, they might be related to the past or ongoing stellar mass loss, and might be used to trace the Cepheid evolution history. Secondly, their presence may bias distance determination using Baade-Wesselink methods, and then bias the calibration of the IR period–luminosity (P–L) relation.

Galactic Cepheids in binary systems are also important to measure fundamental stellar parameters. The dynamical masses can be estimated, providing new constraints on the evolution and pulsation theory. This would give a new insight on Cepheid masses, and might settle the discrepancy between pulsational and evolutionary masses. Binary systems are also valuable for obtaining independent distances to Cepheids, needed to calibrate the P–L relation.

2. Circumstellar envelopes

The characterization of CSEs is essential as they give access to the present mass-loss rates of Cepheids. The CSEs were probably formed through past or ongoing mass loss, possibly generated by shock waves in the pulsating atmospheres of Cepheids. Our recent work using the mid-IR instrument VLTI/MIDI (2-beams combiner) enabled us to use the radiative transfer code DUSTY to model the CSEs of X Sgr and T Mon (Gallenne *et al.* 2013b). The fitted models gave mass-loss rates in the range 10^{-7}–10^{-8} $M_\odot$ yr^{-1}, consistent with the expected theoretical range. We also estimated a relative IR excess in the agreement with our previous work (Gallenne *et al.* 2011), and derived a mid-IR correlation between the relative excesses and the pulsation periods of Cepheids. This correlation shows that longer-period Cepheids have larger IR excesses.

In the near-IR the CSE flux emission might be negligible compared to the photospheric continuum, but this is not the case in the mid-IR, where the CSE emission dominates. This has an impact on the calibration of the P–L relation in the thermal domain (e.g. Monson *et al.* 2012), while the bias in the K band still needs to be more deeply studied.

3. Binary Cepheids

Most of Cepheid companions are located too close to a Cepheid ($\sim$ 1–40 mas) to be spatially resolved with a single-dish telescope. We are engaged in a long-term interferometric observing program, that aims at detecting and characterizing the companions of nearby Cepheids. Our main objectives are the determination of accurate masses and geometric distances from high-precision astrometry. The derived empirical masses will provide very valuable constraints in modeling the pulsation and evolution of intermediate-mass stars.

Gallenne *et al.* (2013a) presented the first results of this program for the Cepheid V1334 Cyg, using the CHARA/MIRC instrument (6-beams combiner). The companion was clearly detected in the closure phase signal. We combined our astrometric measurements with spectroscopic data to derive the complete set of orbital elements. We were also able to estimate the lower limit for the mass and distance.

Recently, the companion of AW Per and AX Cir was also detected using the IR combiner CHARA/MIRC and VLTI/PIONIER, respectively, and the results will be published soon.

4. Conclusion

Thanks to the high angular resolution provided by interferometry, we are able to study the close environment of Cepheids. It is important to keep analyzing the effect of the CSEs on the near- and mid-IR P–L relation, and probe the origin of their presence. Empirical masses of Cepheids are also particularly important to constrain theoretical models. Binary Cepheids are the best tools for such measurements. Interferometry coupled with radial velocity measurements seems to be a powerful way to estimate Cepheid masses.

Acknowledgments. AG acknowledges support from FONDECYT grant 3130361. JDM acknowledges funding from the NSF grants AST-0707927 and AST-0807577. WG and GP gratefully acknowledge financial support for this work from the BASAL Centro de Astrofísica y Tecnologías Afines (CATA) PFB-06/2007. Support from the Polish National Science Centre grant MAESTRO 2012/06/A/ST9/00269 and the Polish Ministry of Science grant Ideas Plus (awarded to GP) is also acknowledged.

References

Gallenne, A., Kervella, P., & Mérand, A. 2011, *A&A*, 538, A24
Gallenne, A., Monnier, J. D., Mérand, A., *et al.* 2013a, *A&A*, 552, A21
Gallenne, A., Mérand, A., Kervella P., *et al.* 2013b, *A&A*, 558, A140
Monson A. J., Freedman W. L., Madore, B. F., *et al.* 2012, *ApJ*, 759, 146

Precision Asteroseismology
Proceedings IAU Symposium No. 301, 2013
J. A. Guzik, W. J. Chaplin, G. Handler & A. Pigulski, eds.

© International Astronomical Union 2014
doi:10.1017/S1743921313014816

Searching for pulsations in *Kepler* eclipsing binary stars

Patrick Gaulme[1] and Joyce A. Guzik[2]

[1] Dept. of Astronomy, New Mexico State University
P.O. Box 30001, MSC 4500, Las Cruces, NM 88003 USA
email: gaulme@nmsu.edu

[2] Los Alamos National Laboratory, XTD-NTA, MS T086, Los Alamos, NM 87545 USA
email: joy@lanl.gov

Abstract. Eclipsing binaries can in principle provide additional constraints to facilitate astero-seismology of one or more pulsating components. We have identified 94 possible eclipsing binary systems in a sample of over 1800 stars observed in long cadence as part of the *Kepler* Guest Observer Program to search for γ Doradus and δ Scuti star candidates. We show the results of a procedure to fold the light curve to identify the potential binary period, subtract a fit to the binary light curve, and perform a Fourier analysis on the residuals to search for pulsation frequencies that may arise in one or both of the stellar components. From this sample, we have found a large variety of light curve types; about a dozen stars show frequencies consistent with δ Sct or γ Dor pulsations, or light curve features possibly produced by stellar activity (rotating spots). For several stars, the folded candidate 'binary' light curve resembles more closely that of an RR Lyr, Cepheid, or high-amplitude δ Sct star. We show highlights of our results and discuss the potential for asteroseismology of the most interesting objects.

Keywords. stars: binaries, stars: variables: δ Scuti, stars: variables: γ Doradus

---

Out of 1800 stars observed in long cadence as part of the *Kepler* Guest Observer Program to search for γ Dor and δ Sct star candidates, 94 present binary-like features. This means that these light curves are periodically modulated by systematic signals that resemble photometric dimmings produced by mutual eclipses of tight pairs of stars. Since these binary-like signals are large enough to be detected without specific tools, the signatures of δ Sct or γ Dor pulsations are not easily detectable, and we must subtract the eclipse modulations from the light curves to find the pulsations.

This cleaning process first requires precise measurement of orbital periods, which were unknown for all 94 candidates. For contact or semi-detached systems (usually with $P < 2$ days), we estimated the orbital periods by fitting, for each, the highest peak of the oversampled Fourier power spectrum. Then we are able to refine the estimates for detached systems (usually with $P > 2$ days) by fitting each eclipse with a function used to adjust for exoplanetary transits: the timing of each transit allows for an accurate measurement of the orbital period. Once the orbital periods are determined, the light curves are cleaned in two different ways depending on orbital periods. When a time series is long enough to contain more than about 20 orbits, we can consider that the photometric fluctuations of the signal coming from either stellar or instrumental origin may be averaged out by folding and rebinning the signal. We thus subtract from the light curve the mean folded light curve repeated on the whole set of orbits in the dataset. This method is best in principle, because of its simplicity and the absence of any assumption about the origin of the photometric fluctuations. For longer periods, the signal during eclipse is replaced by a second-order bridging polynomial (see Gaulme *et al.* 2013 for details).

413

Table 1. Properties of eight pulsators belonging to eclipsing binary systems. $T_{\rm eff}$, $\log g$, and metallicity [M/H] are from the *Kepler* Input Catalog. ELV = contact ellipsoidal variable; D = detached system; SD = semi-detached system.

KIC number	Binary type	Period (days)	Frequency (μHz)	$T_{\rm eff}$ (KIC)	$\log g$ (KIC)	[M/H] (KIC)
4570326	ELV	1.1	$100 - 230$			
4739791	D	0.9	$200 - 283$	7538	3.873	-0.089
5783368	SD (?)	3.7	$100 - 250$	7910	3.835	-0.300
5872506	ELV (?)	2.1	$100 - 283$	7571	3.864	-0.386
6048106	D	1.6	$5 - 38$	6777	4.166	-0.399
6220497	SD	1.3	$80 - 240$	7254	3.933	-0.198
6541245	ELV	1.6	$220 - 283$	6315	4.166	-1.516
11401845	D	2.2	$150 - 283$	7590	3.902	-0.276

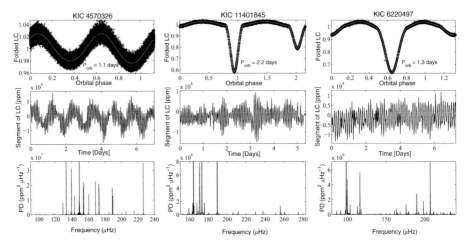

Figure 1. Folded light curve, light curve with binary signal subtracted, and power spectrum for KIC 4570326 (ellipsoidal variable), KIC 1140185 (detached system), and KIC 6220497 (semi-detached system). Each system shows frequencies in the δ Sct range.

Folding and cleaning the light curves allowed us to identify pulsations for eight binary systems (Table 1). Four candidates shown in our poster were eliminated here because they are likely δ Sct stars with a dominant large amplitude mode of period $0.1 - 0.2$ days. The remaining candidates are pulsating in the δ Sct frequency range, except for KIC 6048106 which is pulsating at γ Dor frequencies.

We have identified pulsators in about 10% of our eclipsing-binary candidates. The number of detections could certainly be increased by improving the filtering technique for eclipses. Indeed, for many contact or semi-detached systems, the eclipse depth actually fluctuates, preventing the folding approach from completely filtering out the eclipses. In addition, we must check that the pulsators truly belong to their associated eclipsing binaries. This can be done by checking the *Kepler* "target pixel files" and then obtaining radial velocity measurements to get masses and orbits of the eclipsing binary components. Modeling radial velocities coupled with light curves will lead to a complete characterization of these systems (masses, radii, orbital parameters).

Reference

Gaulme, P., McKeever, J., Rawls, M. L., Jackiewicz, J., Mosser, B., & Guzik, J. A. 2013, *ApJ*, 767, 82

Precision Asteroseismology
Proceedings IAU Symposium No. 301, 2013
J. A. Guzik, W. J. Chaplin, G. Handler & A. Pigulski, eds.

© International Astronomical Union 2014
doi:10.1017/S1743921313014828

Interaction of multidimensional convection and radial pulsation

Chris M. Geroux[1] and Robert G. Deupree[2]

[1]Institute for Computational Astrophysics and Department of Astronomy and Physics,
Saint Mary's University, Halifax, NS B3H 3C3 Canada.

[2]Now at Physics and Astronomy, University of Exeter, Stocker Road, Exeter, UK EX4 4QL
email: geroux@astro.ex.ac.uk

Abstract. We have previously calculated a number of 2D hydrodynamic simulations of convection and pulsation to full amplitude. These revealed a significantly better fit to the observed light curves near the red edge of the instability strip in the globular cluster M 3 than did previous 1D mixing length models. Here we compare those 2D results with our new 3D hydrodynamic simulations calculated with the same code. As expected, the horizontal spatial behaviour of convection in 2D and 3D is quite different, but the time dependence of the convective flux on pulsation phase is quite similar. The difference in pulsation growth rate is only about 0.1% per period, with the 3D models having more damping at each of the five effective temperatures considered. Full amplitude pulsation light curves in 2D and 3D are compared.

Keywords. convection, hydrodynamics, methods: numerical, stars: oscillations, stars: variables

1. Introduction

The importance of RR Lyrae and classical Cepheids as standard candles has led to a long history of trying to model the pulsation of these variables (e.g. Christy 1964, Bono & Stellingwerf 1994) with one-dimensional (1D) hydrodynamic simulations. It was speculated early on that convection in the ionization regions for models near the red edge would be important (Christy 1966, Cox *et al.* 1966).

We present an alternative to 1D time-dependent mixing length based approaches to convection in pulsation models (e.g. Stellingwerf 1982a,b; Kuhfuss 1986; Xiong 1989) with the aim to improve agreement with observations, specifically near the red edge of the RR Lyrae instability strip where the treatment of convection is especially important. Our approach (similar to Deupree 1977) is to follow the convective flow directly using the normal conservation laws of hydrodynamics in both 2D and 3D. Recently others have begun to perform 2D calculations of this nature (e.g. Mundprecht *et al.* 2013, Gastine & Dintrans 2011) but these have not yet been compared with observations. Our multidimensional full amplitude radial pulsation calculations have been made possible by using a moving grid system that adjusts the volume of the radial shells to keep the total mass in the shell constant (see Geroux & Deupree 2011, 2013 for details).

2. Key results

2D models compared with observations. We have compared our model light curves to light curves of RR Lyrae variables in M 3 as observed by Corwin & Carney (2001), finding reasonable agreement between our fundamental-mode model light curves from the near the fundamental blue edge ($T_{\rm eff} = 6700$ K) to near the fundamental red edge ($T_{\rm eff} = 6300$ K). Our single first-overtone model light curve also agrees reasonably well with observed light curves. A comparison of the modelled visual amplitude – effective temperature relation with that derived form data from Corwin & Carney shows good agreement for both our fundamental-mode models and our single first-overtone model.

3D convective flow patterns. The up-flow filling factor, f_{up} (fraction of the area with positive radial velocities) is phase dependent. When the model has expanded $f_{up} \approx 0.7$ and when contracted $f_{up} \approx 0.4$. The time average of the filling factor, $\langle f_{up} \rangle$, is about 0.6. The values of $\langle f_{up} \rangle$ and f_{up} during contraction and expansion are quite similar for both the 6300 K and 6700 K models with the value of $\langle f_{up} \rangle$ reasonably close to that found by other authors of about 2/3 (Magic *et al.* 2013, Stein & Nordlund 1998). Simplistic measurements of granule sizes produce $\log d_{gran} = 10.5$ and 10.3 for the 6300 K and 6700 K effective-temperature models respectively. This agrees with relations from Magic *et al.* (2013), derived from high resolution atmosphere models which gives $\log d_{gran} = 10.2$, and 10.18 for the 6300 K and 6700 K models respectively. This order of magnitude agreement suggests that our coarsely horizontally zoned calculations (20×20) are likely obtaining approximately the correct granule sizes and filling factors for the largest eddies. However, due to the coarse horizontal zoning we cannot hope to resolve smaller scale features.

Comparison of 2D and 3D models. We find good agreement both in terms of shape and amplitude between the 2D and 3D light curves of 6300 K and 6600 K models. The hotter model does, however, have some differences in shape during the descending light. The difference in pulsational growth rate between 2D and 3D models is only about 0.1% with 3D models having more damping.

Time dependence of 2D and 3D convection. The time dependence we found was very similar to that found by Deupree (1977) which led to the quenching of pulsation near the red edge, namely that the convective flux is a maximum during full contraction, and a minimum during full expansion. While the 2D and 3D calculations show the same time dependence, there are some differences. Most notably the 3D maximum convective fluxes are larger due to larger down flow velocities and slightly larger horizontal temperature variations. The total convective luminosity of the 3D calculations is actually slightly smaller. The reason this is true, while having larger maximum convective fluxes, is that the filling factors of the strong down-flows are smaller in 3D than 2D.

References

Bono, G. & Stellingwerf, R. F. 1994, *ApJS*, 93, 233
Christy, R. F. 1964, *Reviews of Modern Physics*, 36, 555
Christy, R. F. 1966, *ApJ*, 144, 108
Corwin, T. M. & Carney, B. W. 2001, *AJ*, 122, 3183
Cox, J. P., Cox, A. N., Olsen, K. H., King, D. S., & Eilers, D. D. 1966, *ApJ*, 144, 1038
Deupree, R. G. 1977, *ApJ*, 211, 509
Gastine, T. & Dintrans, B. 2011, *A&A*, 528, A6
Geroux, C. M. & Deupree, R. G. 2011, *ApJ*, 731, 18
Geroux, C. M. & Deupree, R. G. 2013, *ApJ*, 771, 113
Kuhfuss, R. 1986, *A&A*, 160, 116
Magic, Z., Collet, R., Asplund, M., *et al.* 2013, *A&A*, 557, A26
Mundprecht, E., Muthsam, H. J., & Kupka, F. 2013, *MNRAS*, 435, 3191
Stein, R. F. & Nordlund, Å. 1998, *ApJ*, 499, 914
Stellingwerf, R. F. 1982a, *ApJ*, 262, 330
Stellingwerf, R. F. 1982b, *ApJ*, 262, 339
Xiong, D. 1989, *A&A*, 209, 126

Precision Asteroseismology
Proceedings IAU Symposium No. 301, 2013
J. A. Guzik, W. J. Chaplin, G. Handler & A. Pigulski, eds.

© International Astronomical Union 2014
doi:10.1017/S174392131301483X

The diverse pulsational behaviour of β Cep stars: results from long-term monitoring

G. Handler

Nicolaus Copernicus Astronomical Center, Bartycka 18, 00-716 Warsaw, Poland
email: gerald@camk.edu.pl

Abstract. We studied seven β Cep stars photometrically over the past ten years. Some showed amplitude variations, some frequency changes, and others exhibited stable pulsations, with no consistent picture yet emerging. Additionally, 12 Lac appears to have a 6.7-yr binary companion.

Keywords. stars: early-type, stars: evolution, stars: individual (12 Lac, ν Eri, σ Sco, γ Peg), stars: oscillations (including pulsations), stars: rotation, binaries: general

1. Observations and results

This study is mostly based on measurements with the T6 automated telescope at Fairborn Observatory, Arizona (Strassmeier *et al.* 1997). Whereas the original idea was to collect data for asteroseismology, some interesting additional behaviour was observed...

1.1. *12 Lac*

For this star, we have 2381 h of data available, collected over nine seasons during 724 nights. It was noticed that the noise level in these measurements (as determined from the comparison stars) was higher than expected, especially around the known pulsation frequencies. Part of the explanation is likely binarity (Fig. 1).

However, even after correcting for the orbital light-time effect, the Fourier noise level in the data was still much higher than the measurement precision would suggest. We therefore observed 12 Lac with the *MOST* satellite (Walker *et al.* 2003), resulting in the detection of several more pulsation modes. It turned out that the high noise level in the ground-based data can readily be explained by the presence of these new pulsation modes, with seasonally variable amplitudes. In some subsets of data these amplitudes may be high enough to even facilitate photometric mode identification.

1.2. *ν Eri*

We have collected all literature data for the star and obtained 379 h of new observations during 91 nights last season. The amplitude spectrum of ν Eri is dominated by the fundamental radial mode and a rotationally split $l = 1$ triplet. The star also shows mild amplitude variability, but more striking are its frequency variations (Fig. 2).

1.3. *Other stars*

The pulsation spectrum of the hybrid β Cep/SPB star γ Peg (1497 h of data on 388 nights over six consecutive seasons) proved to be extremely stable over time, allowing to detect oscillations with amplitudes down to 0.35 mmag in the g-mode domain and down to 0.24 mmag at its p-mode frequencies.

σ Sco was monitored only last season (272 h of data during 81 nights). Its pulsation amplitude has dropped to less than 1/4 of what is was in 1972 (Jerzykiewicz & Sterken 1984). 15 CMa (803 hr/250 nights/6 seasons of new observations plus archival data)

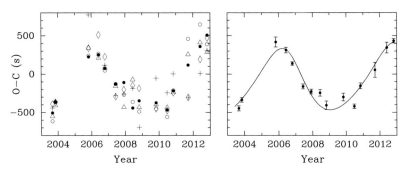

Figure 1. *Left:* (O−C) diagram of the four strongest pulsation modes and one combination frequency (different symbols) of 12 Lac. All vary in phase with the same amplitude. *Right:* orbital fit to the averaged (O−C) values, suggesting a $M > 1.4\,M_\odot$ companion in a 6.7-yr, $e \approx 0.3$ orbit.

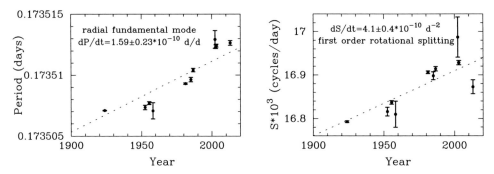

Figure 2. *Left:* the radial fundamental mode period of ν Eri over 90 years of measurement. The period increases, but about seven times faster than expected from stellar evolution calculations. *Right:* the first order rotational splitting ($S = (f_{m=+1} - f_{m=-1})/2)$) of the $l = 1, g_1$ mode of ν Eri, indicating a temporal *increase* of the rotation rate.

showed a fairly stable pulsation spectrum over the last 40 years, whereas HD 167743 (590 h/217 nights/4 seasons) exhibited amplitude variations of a few, but not all, its modes. Finally, at least the strongest modes of ALS 4680 (179 h/63 nights/3 seasons, plus *MOST* and ASAS data) underwent considerable seasonal amplitude changes.

Acknowledgements

Part of the data discussed here were acquired by Eloy Rodríguez, Refilwe Kgoadi, and Aleksander Schwarzenberg-Czerny. Some analyses benefitted from *MOST* observations. This research has been supported by the Polish NCN grant 2011/01/B/ST9/05448.

References

Jerzykiewicz, M. & Sterken, C., 1984, *MNRAS*, 211, 297
Strassmeier K. G., Boyd L. J., Epand D. H., & Granzer, T. 1997, *PASP*, 109, 697
Walker, G., Matthews, J., Kuschnig, R., *et al.* 2003, *PASP*, 115, 1023

Precision Asteroseismology
Proceedings IAU Symposium No. 301, 2013
J. A. Guzik, W. J. Chaplin, G. Handler & A. Pigulski, eds.

© International Astronomical Union 2014
doi:10.1017/S1743921313014841

KIC 6761539, a fast rotating γ Dor $-$ δ Sct hybrid star

W. Herzberg[1], D. Corre[1], K. Uytterhoeven[2,3], and M. Roth[1]

[1]Kiepenheuer-Institut für Sonnenphysik, 79104 Freiburg, Germany
email: herzberg@kis.uni-freiburg.de

[2]Instituto de Astrofísica de Canarias, 38205 La Laguna, Tenerife, Spain
[3]Departamento de Astrofísica, Universidad de La Laguna, 38200 La Laguna, Tenerife, Spain

Abstract. KIC 6761539 is one of many fast rotating γ Doradus – δ Scuti hybrid pulsators. A search for possible regularities in the frequency spectrum is performed and a first stellar model is presented.

Keywords. stars: individual KIC 6761539, stars: oscillations, stars: variables: δ Scuti

1. Introduction

The candidate γ Doradus - δ Scuti hybrid star KIC 6761539 (Uytterhoeven *et al.* 2011) is a rather fast rotator, with a value of 120 km s^{-1} for the projected rotational velocity ($v \sin i$). In this case study, we perform a search for all possible regularities in the frequency spectrum such as harmonics, combination frequencies and frequency spacings. Finally, we present a first preliminary model for the star.

For the analysis presented here, *Kepler* time series from Q0 to Q10 (May 2009 to September 2011) were used. The frequency extraction was carried out with the software tool 'Pysca' (Herzberg & Glogowski, these proceedings), which works via consecutive prewhitening of the time series. We decided to limit the analysis to frequencies with a signal-to-noise ratio higher than 9.5, yielding a set of 171 frequencies.

2. Harmonics and combination frequencies

Within our set of 171 frequencies, a search for harmonics was performed up to 5th order. This yields a total of three candidate harmonic frequencies:

$$F105 = 5 \cdot F6 \pm 3 \cdot 10^{-5} \text{ d}^{-1}$$
$$F44 = 5 \cdot F34 \pm 4 \cdot 10^{-4} \text{ d}^{-1}$$
$$F83 = 2 \cdot F55 \pm 1 \cdot 10^{-3} \text{ d}^{-1}$$

As the harmonics are either of high order ($n = 5$), or involve not the highest amplitude frequencies (F55), we suspect an agreement by chance in all three cases.

For the identification of combination frequencies, we included the fifteen highest amplitude frequencies as parent frequencies in our search. Eight candidate combination frequencies of second order were found:

$$F38 = F13 - F6 \pm 2 \cdot 10^{-4} \text{ d}^{-1} \qquad F128 = F8 - F5 \pm 3 \cdot 10^{-5} \text{ d}^{-1}$$
$$F49 = F14 - F5 \pm 1 \cdot 10^{-4} \text{ d}^{-1} \qquad F133 = F12 + F13 \pm 9 \cdot 10^{-4} \text{ d}^{-1}$$
$$F100 = F11 + F15 \pm 2 \cdot 10^{-3} \text{ d}^{-1} \qquad F137 = F10 - F12 \pm 3 \cdot 10^{-4} \text{ d}^{-1}$$
$$F112 = F14 - F8 \pm 6 \cdot 10^{-5} \text{ d}^{-1} \qquad F154 = F7 + F14 \pm 1 \cdot 10^{-3} \text{ d}^{-1}$$

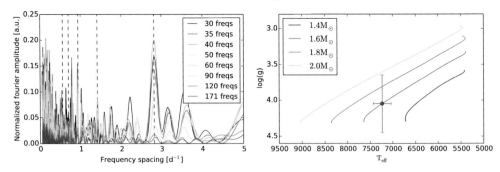

Figure 1. *Left*: DFT for different frequency sets. *Right*: Best fitting model for KIC 6761539. Four evolutionary tracks for different masses are shown as comparison.

It becomes evident that none of the parent frequencies is involved in a systematic way in these detections. Many higher order combinations were also detected, but since no systematic pattern appears, their physical significance remains unclear.

3. Preferred frequency spacing

We searched for a regular spacing in the frequencies by applying the discrete Fourier transform (DFT) to the frequency set, as presented in García-Hernández *et al.* (2013). We calculated the DFT for several groups of frequencies from the original set, starting with the 30 highest amplitude modes and slowly adding more and more lower amplitude modes (see Fig. 1, left panel). A preferred spacing of about 2.82 d^{-1} seems to be present, which is most pronounced between the first 40 frequencies. Submultiples up to 3rd order are visible.

4. First stellar model

A preliminary model for KIC 6761539 was calculated with the stellar evolution code MESA (Paxton *et al.* 2011). The best fitting model (see Fig. 1, right panel) was found with a least-squares minimization of the surface gravity $\log g$ and effective temperature T_{eff}, the values of which are taken from the *Kepler* Input Catalog (Brown *et al.* 2011). The model has a mass of 1.62 $M_{\odot}$ and an age of about 1.42 Gyr, corresponding to a hydrogen fraction of 0.48 in the core. The metallicity and chemical composition are taken to be solar, and the initial value of the rotation rate on the ZAMS is chosen equal to 30% of the critical rotation rate.

Acknowledgements

WH acknowledges financial support by the Deutsche Forschungsgemeinschaft (DFG) in the framework of project UY 52/1-1. KU acknowledges funding by the Spanish National Plan of R&D for 2010, project AYA2010-17803.

References

Brown, T. M., Latham, D. W., Everett, M. E., & Esquerdo, G. A. 2011, *AJ*, 142, 112
García-Hernández, A., Moya, A., Michel, E., *et al.* 2013, *A&A*, 559, A63
Paxton, B., Bildsten, L., Dotter, A., Herwig, F., Lesaffre, P., & Timmes, F. 2011, *ApJS*, 192, 3
Uytterhoeven, K., Moya, A., Grigahcène, A., *et al.* 2011, *A&A*, 534, A125

Precision Asteroseismology
Proceedings IAU Symposium No. 301, 2013
J. A. Guzik, W. J. Chaplin, G. Handler & A. Pigulski, eds.

© International Astronomical Union 2014
doi:10.1017/S1743921313014853

Pysca – Automated frequency extraction from photometric time series

Wiebke Herzberg and Kolja Glogowski

Kiepenheuer-Institut für Sonnenphysik, 79104 Freiburg, Germany
email: `herzberg@kis.uni-freiburg.de`, `glogowski@kis.uni-freiburg.de`

Abstract. Pysca, a Python software package for automated extraction of frequencies, amplitudes and phases from non-equally sampled photometric time series, is presented.

Keywords. methods: data analysis, stars: oscillations

1. Overview

Pysca is a software package that allows automated extraction of frequencies, amplitudes and phases from non-equally sampled photometric time series. Suited mainly for heat-driven pulsators, it can be regarded as an automated and more focused alternative to the feature rich and interactive Period04 (Lenz & Breger 2005). The use of Pysca is especially convenient when dealing with long time series of high signal-to-noise photometry, (e.g. as provided by the *Kepler* space telescope) where many frequencies are present in the data and the manual extraction can become very time consuming.

Pysca extracts frequencies by identifying the highest peaks in the Lomb-Scargle periodogram and fitting the time series with a superposition of harmonic functions of the corresponding frequencies. It is implemented as an iterative algorithm where the time series is progressively prewhitened up to a user defined termination condition. The signal-to-noise ratio is calculated for every frequency as a statistical measure of significance.

2. Algorithm

A schematic overview of the different steps involved in the frequency extraction process is shown in Fig. 1. The first step is to find the frequency of the highest peak in the Lomb-Scargle periodogram (LSP), which is calculated using the method described in Press & Rybicki (1989). The accuracy of the peak frequency is improved by applying a parabolic interpolation using 3 points. Next, the amplitudes A_i and phases ϕ_i of the new frequency and of all extracted frequencies ν_i from previous iterations are (re-)determined by fitting a model function of superimposed harmonic functions,

$$f(A_i, \phi_i; \nu_i) = \sum_i A_i \sin\left[2\pi(\nu_i t + \phi_i)\right],$$

to the original time series with fixed values of ν_i. The original time series is then prewhitened using $f(A_i, \phi_i; \nu_i)$ and a new LSP is computed where, due to the prewhitening, the peaks and sidelobes of all already handled modes are removed. The noise level at the peak positions is determined using the median of the data in neighborhood of ν_i from the "cleaned" periodogram and the signal-to-noise ratio is calculated using the A_i. After that the next iteration starts by finding the highest peak in the new LSP from which all previous peaks were removed. An example of the prewhitening performance is shown in Fig. 2.

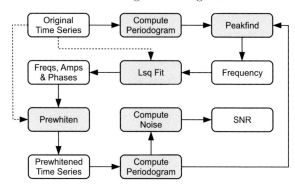

Figure 1. Steps of the frequency extraction procedure.

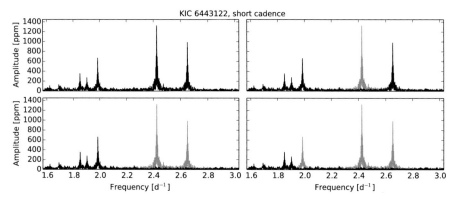

Figure 2. Peak extraction example for data obtained with the *Kepler* space telescope. The upper left panel shows a selected part of the original periodogram. The remaining three panels show the prewhitened periodograms (black) after peak extraction. The original peaks that were removed in this process are shown in gray.

3. Software package

Pysca is available via the Python Package Index (Herzberg & Glogowski 2013). The software package consists of a Python library and a ready-to-use command-line executable. The command-line program can be used to read in photometric time series from a file (supported are FITS files and plain text files) and to extract peak frequencies as well as the corresponding amplitudes and phases. The number of automatically extracted frequencies can be controlled by the various parameters. For developers it is also possible to use the library directly in order to create customized scripts for frequency extraction or incorporate Pysca into other Python programs.

Acknowledgements

WH acknowledges financial support by the Deutsche Forschungsgemeinschaft (DFG) in the framework of project UY 52/1-1.

References

Herzberg, W. & Glogowski, K. 2013, *PyPI*, http://pypi.python.org/pypi/pysca/
Lenz, P. & Breger, M. 2005, *CoAst*, 146, 53
Press, W. H. & Rybicki, G. B. 1989, *ApJ*, 338, 277

Precision Asteroseismology
Proceedings IAU Symposium No. 301, 2013
J. A. Guzik, W. J. Chaplin, G. Handler & A. Pigulski, eds.
© International Astronomical Union 2014
doi:10.1017/S1743921313014865

Rapidly varying A-type stars in the SuperWASP archive

Daniel L. Holdsworth and Barry Smalley

Astrophysics Group, Keele University, Staffordshire, ST5 5BG, United Kingdom
email: d.l.holdsworth@keele.ac.uk

Abstract. The searches for transiting exoplanets have produced a vast amount of time-resolved photometric data of many millions of stars. One of the leading ground-based surveys is the SuperWASP project. We present the initial results of a survey of over 1.5 million A-type stars in the search for high frequency pulsations using SuperWASP photometry. We are able to detect pulsations down to the 0.5 mmag level in the broad-band photometry. This has enabled the discovery of several rapidly oscillating Ap stars and over 200 δ Scuti stars with frequencies above 50 d^{-1}, and at least one pulsating sdB star. Such a large number of results allows us to statistically study the frequency overlap between roAp and δ Scuti stars and probe to higher frequency regimes with existing data.

Keywords. surveys, asteroseismology, techniques: photometric, stars: chemically peculiar, stars: variables: roAp, stars: variables: δ Scuti

1. SuperWASP

The Wide Angle Search for Planets (WASP) is a two site campaign in the search for transiting exoplanets (Pollacco *et al.* 2006). Each observatory consists of 8 telephoto lenses mounted in a 2×4 configuration. To date there are over 31 million objects in the WASP archive. Observations consist of two consecutive 30-s integrations followed by a 10 minute gap. The entire observable sky can be visited every 40 minutes. The short integrations, and non-uniform sampling, allow for a Nyquist frequency of up to 1440 d^{-1}.

Due to the observing strategy, observations of a single star can occur over many seasons, with the target sometimes appearing in more than one camera. This provides a long timebase of observations which can either be combined or split into individual data sets.

2. Methodology

We selected, using 2MASS colours, over 1.5 million A-type and earlier stars from the SuperWASP archive. We aimed to identify new pulsating systems with frequencies above 50 d^{-1}. Such a frequency range allows for the discovery of new δ Scuti stars, rapidly oscillating Ap (roAp) stars and compact pulsators. Thus, enabling a statistical study on potential frequency boundaries.

We calculated a periodogram for each WASP season for every individual object – over 9 million lightcurves. Periodograms for an object are cross-checked for peaks at the same frequency. An object is required to have two or more concurrent peaks to be thought of as genuine. Spectroscopic follow-up is sought for the most convincing candidates. For our northern targets we make use of the service mode on WHT/ISIS and UK SALT Consortium (UKSC) time on SALT/RSS for our southern targets.

Table 1. Current statistics from the WASP survey.

Pulsation Class	Candidates	Number with Spectra	Confirmed
roAp	21	21	10
δ Scuti (Am)	204	18*	29*(15)

*Some roAp candidates are found to be δ Scuti, hence the discrepancy.

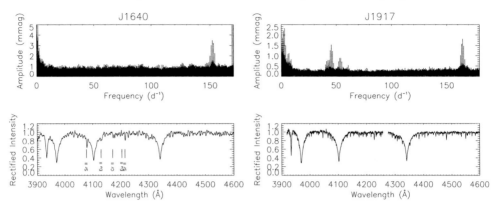

Figure 1. *Left:* An example of a WASP roAp periodogram and WHT/ISIS spectrum confirming the Ap nature of the star. *Right:* A hybrid target showing both roAp and δ Scuti pulsations. The SALT/RSS spectrum shows this is an Am star due to the weak Ca K line.

3. Results

3.1. *roAp stars*

To identify candidate roAp stars we require a target to show a single peak in the periodogram above 50 d^{-1}. This peak must appear in more than one season of WASP data to eliminate the possibility of spurious peaks due to noise. Using this criteria we identify 21 candidate stars. Classification spectra have been obtained for these targets, confirming at least 10 new roAp stars (e.g. Fig. 1, left).

3.2. *δ Scuti stars*

Candidate δ Scuti stars are identified as those objects with multiple frequencies higher than 50 d^{-1} and having their principle peak in multiple seasons. We have over 200 candidates up to a maximum frequency of 100 d^{-1}. Spectra have been secured for photometrically interesting targets such as J1917 (Fig. 1, right). 52% of the targets for which we currently have spectra are pulsating Am stars (also see Smalley *et al.* 2011).

4. Conclusions

The SuperWASP archive has provided a new approach in searching for stellar pulsations in the A-type stars. We have thus far identified 10 new roAp targets for further in depth study as well as highlighted over 200 potential new δ Scuti stars with periods shorter than ∼ 30 minutes. The archive holds many more examples of pulsating systems which are yet to be exploited.

References

Pollacco, D. L., Skillen, I., Collier Cameron, A., *et al.* 2006, *PASP*, 118, 140
Smalley, B., Kurtz, D. W., Smith, A. M. S., *et al.* 2011, *A&A*, 535, A3

Precision Asteroseismology
Proceedings IAU Symposium No. 301, 2013
J. A. Guzik, W. J. Chaplin, G. Handler & A. Pigulski, eds.

© International Astronomical Union 2014
doi:10.1017/S1743921313014877

Pulsation in extremely low-mass helium stars

C. S. Jeffery[1] and H. Saio[2]

[1] Armagh Observatory, College Hill, Armagh BT61 9DG, Northern Ireland

[2] Astronomical Institute, School of Science, Tohoku University, Sendai 980-8578, Japan

Abstract. We explore the stability of extremely low-mass stars ($M < 0.25\,M_\odot$) across a wide range of composition, effective temperature, and luminosity. We identify the instability boundaries associated with radial oscillations. These are a strong function of both composition and radial order ($0 \leqslant n \leqslant 13$). The classical blue edge shifts to higher effective temperature and luminosity with decreasing hydrogen abundance. Higher-order modes are more easily excited, and small islands of instability develop. Short-period oscillations have been discovered in the low-mass pre-white dwarf component of the eclipsing binary J0247–25. If its envelope is depleted in hydrogen, J0247–25B is unstable to intermediate-order p modes. Driving is by the classical κ mechanism operating in the second helium ionization zone. The observed periods, temperature and luminosity of J0247–25B require an envelope hydrogen abundance $0.2 \leqslant X \leqslant 0.3$.

Keywords. stars: early-type, stars: interiors, stars: oscillations, subdwarfs, white dwarfs

White dwarf masses extend from the Chandrasekhar limit to less than $0.17\,M_\odot$. Extremely low-mass white dwarfs (ELMs) require the prior removal of the hydrogen envelope from a low-mass giant, presumably following interaction in a close binary. Binary systems containing a pre-white dwarf can thus help explain the production of ELMs. Maxted *et al.* (2011) measured the unseen companion in the bright ecliping binary 1SWASP J024743.37–251549.2 (hereafter J0247–25) to have a mass of $0.23\,M_\odot$. They identify it as the core of a red giant stripped of its outer envelope before helium ignition. Maxted *et al.* (2013) show the light curve of J0247–25 to contain variations on timescales of $6-7$ and ≈ 40 minutes, in addition to two eclipses within its 0.668 d orbital period. They revise the companion mass downward to $0.19\,M_\odot$. The 40 minute variations are attributed to δ Scuti-type oscillations in the A-type primary. The multi-periodic $\approx 6-7$ minute oscillations are attributed to p-mode pulsations of order $n \approx 10$ in the pre-white dwarf (J0247–25B). The excitation mechanism for these oscillations is not known.

We (Jeffery & Saio 2013) have constructed linear non-adiabatic models for the stability of low-mass stellar envelopes against radial pulsation following methods described by Saio *et al.* (1983) and Jeffery & Saio (2006a,b) and using the unmodified OPAL tables adopted by Jeffery & Saio (2007). We considered a range of abundances with hydrogen-mass fraction $X = 0.90, 0.75, 0.50, 0.25, 0.01$ and 0.002 and a basic metal mass fraction $Z = 0.02$. We have assumed the iron and nickel abundances to retain their scaled solar values. A grid of model envelopes was computed for each chemical mixture and for: $M/M_\odot = 0.18(0.01)0.25$, $\log T_{\rm eff}/{\rm K} = 3.68(0.02)4.60$, and $\log(L/L_\odot)/(M/M_\odot) = -0.7(0.1)3.5$. The first 14 eigenfrequencies were located and stored, including real and imaginary components $\omega_{\rm r}$ and $\omega_{\rm i}$, the period Π and the number of nodes n in the eigensolution. Modes with $\omega_{\rm i} < 0$ were deemed unstable, *i.e.* pulsations could be excited. We identified instability regions according to (a) the number of unstable modes in each model and (b) the stability of specific modes. We examined the overall instability boundary for low-order modes up to $n = 13$. The results are invariant with the mass in the range $0.18-0.25\,M_\odot$. We also examined boundaries for individual modes as a function of $(T_{\rm eff}, L/M)$ and X. These were compared with the properties of J0247–25B.

For a conventional composition $X = 0.75, Z = 0.02$ we easily identify the classical instability strip where pulsations due to the second ionisation of helium are excited in the radial, first and second overtone modes ($n = 0, 1, 2$). The instability strip for $n = 0-3$ shifts to the blue and to higher L/M as the hydrogen abundance is reduced, and approaches but does not reach the latest location of J0247–25B at $X = 0.002$. The periods of the $n = 0-3$ modes at this location are $\approx 700-1000$ s, too long to account for the observed periods. For higher order modes, the instability finger for increasing order n extends to lower L/M. The blueward drift with decreasing X is replicated. A number of small instability islands develop blueward of the classical blue edge, for example at $X = 0.25$, $\log T_{\text{eff}} \approx 4.2-4.1$, $\log L/M \approx -0.5-1$, and $X = 0.002$, $\log T_{\text{eff}} \approx 4.15$, $\log L/M \approx 1$. The latest position of J0247–25B (Maxted *et al.* 2013) lies well within the instability island for $n = 8$ radial modes for a hydrogen-poor mixture $X = 0.25$ and with *exactly* the correct period if at least one of the observed modes is a classical radial mode. An $n = 9$ mode is also consistent with the observations. Low-degree non-radial modes of the same radial order may equally be excited.

To understand the excitation mechanism, we use a work integral

$$W(r) = 4\pi \int_0^r Pr^2 \Im \left(\frac{\delta P^*}{P} \frac{\delta\rho}{\rho} \right) dr,$$

where δP and $\delta\rho$ are the Lagrangian perturbations of the pressure and density, respectively, $\Im(\dots)$ means the imaginary part, and superscript '*' means complex-conjugate. For a pulsation mode, layers in which $dW/dr > 0$ (< 0) contribute to drive (damp) the pulsation. The pulsation is globally excited if $W(R) > 0$; *i.e.* if the work W is positive at the surface. By comparing two models with identical mass, luminosity and effective temperature, but different hydrogen abundances, we find that for J0247–25B the main driving comes from the second helium ionization zone at temperatures around $50\,000$ K, with some help from the first ionization zone, where the opacity derivative $\kappa_T = (\partial \ln \kappa / \partial \ln T)_\rho$ increases outward; *i.e.*, the κ (opacity) mechanism drives the pulsation. In the hydrogen-rich model, the mode is damped because of the high intrinsic opacity of hydrogen. The consequence of reducing this hydrogen-damping is initially to shift the blue-edge to the blue. Although the helium ionization zones are convective, no appreciable energy is carried by convection ($L_{\text{rad}}/L \approx 1$).

Intermediate-order p-mode pulsations have been detected in the very low-mass pre-white dwarf J0247–25B (Maxted *et al.* 2013). Hydrogen-rich stellar envelopes corresponding to the mass, luminosity and effective temperature of J0247–25B are stable against pulsation up to radial order $n = 13$. When the hydrogen abundance of the models is reduced, the blue edge of the classical (Cepheid) instability strip becomes bluer. For low-order modes it also becomes narrower. Reduced hydrogen damping allows intermediate-order modes to be excited in regions blueward of this edge. $n = 8-10$ pulsations in a $X \approx 0.25$ envelope are completely consistent with the observations of J0247–25B.

Bibliography

Jeffery, C. S. & Saio, H. 2006a, *MNRAS*, 371, 659
Jeffery, C. S. & Saio, H. 2006b, *MNRAS*, 372, L48
Jeffery, C. S. & Saio, H. 2007, *MNRAS*, 378, 379
Jeffery, C. S. & Saio, H. 2013, *MNRAS*, 435, 885
Maxted, P. F. L., Anderson, D. R., Burleigh, M. R., *et al.* 2011, *MNRAS* 418, 1156
Maxted, P. F. L., Serenelli, A. M., Miglio, A., *et al.* 2013, *Nature*, 498, 463
Saio, H., Winget, D. E., & Robinson, E. L. 1983, *ApJ*, 265, 982

Precision Asteroseismology
Proceedings IAU Symposium No. 301, 2013
J. A. Guzik, W. J. Chaplin, G. Handler & A. Pigulski, eds.

© International Astronomical Union 2014
doi:10.1017/S1743921313014889

Ground-based photometry for 42 *Kepler*-field RR Lyrae stars

Young-Beom Jeon[1], Chow-Choong Ngeow[2] and James M. Nemec[3]

[1] Korea Astronomy and Space Science Institute (KASI), Daejeon 305-348, Korea
email: ybjeon@kasi.re.kr

[2] Graduate Institution of Astronomy, National Central University, Taiwan
email: cngeow@gmail.com

[3] Department of Physics & Astronomy, Camosun College, Victoria, British Columbia, Canada
email: jmn@isr.bc.ca

Abstract. Follow-up $(U)BVRI$ photometric observations have been carried out for 42 RR Lyrae stars in the *Kepler* field. The new magnitude and color information will complement the available extensive high-precision *Kepler* photometry and recent spectroscopic results. The photometric observations were made with the following telescopes: 1-m and 41-cm telescopes of Lulin Observatory (Taiwan), 81-cm telescope of Tenagra Observatory (Arizona, USA), 1-m telescope at the Mt. Lemmon Optical Astronomy Observatory (LOAO, Arizona, USA), 1.8-m and 15-cm telescopes at the Bohyunsan Optical Astronomy Observatory (BOAO, Korea) and 61-cm telescope at the Sobaeksan Optical Astronomy Observatory (SOAO, Korea). The observations span from 2010 to 2013, with ∼200 to ∼600 data points per light curve. Preliminary results of the Korean observations were presented at the 5th KASC workshop in Hungary. In this work, we analyze all observations. These observations permit the construction of full light curves for these RR Lyrae stars and can be used to derive multi-filter Fourier parameters.

Keywords. stars: variable: RR Lyrae

We obtained ground-based $(U)BVRI$ photometric observations for 42 RR Lyrae stars in the *Kepler* field. Figure 1 shows a phased light curve of KIC 3864443 and the correlation between $\phi_{31}(Kp)$ and $\phi_{31}(V)$ for all Blazhko and non-Blazhko RRab stars. Table 1 presents the total amplitudes, A_{tot}, and ϕ_{31} parameters from Fourier analysis of RRab

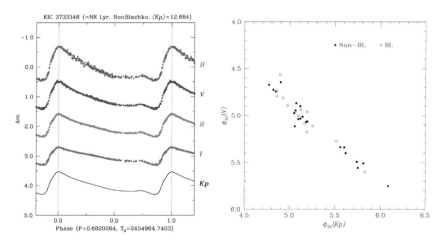

Figure 1. A sample phased light curve of KIC 3864443 (left) and the correlation between $\phi_{31}(Kp)$ and $\phi_{31}(V)$ for Blazhko and non-Blazhko RRab stars (right).

Table 1. Fourier parameters for RRab stars

KIC	GCVS	Period day	$A_{tot}(B)$ mag	$\phi_{31}(B)$ rad	$A_{tot}(V)$ mag	$\phi_{31}(V)$ rad	$A_{tot}(R)$ mag	$\phi_{31}(R)$ rad	$A_{tot}(I)$ mag	$\phi_{31}(I)$ rad
Blazhko:										
3864443	V2178 Cyg	0.48695	1.106	4.245	0.834	4.564	0.712	4.556	0.554	4.905
4484128	V808 Cyg	0.54786	1.398	4.856	1.115	5.055	0.922	5.254	0.707	5.621
5559631	V783 Cyg	0.6207	1.220	5.105	0.969	5.263	0.783	5.609	0.625	5.973
6183128	V354 Lyr	0.56169	1.028	4.942	0.821	5.168	0.639	5.496	0.490	6.082
6186029	V445 Lyr	0.51312	0.664	[6.514]	0.529	[6.964]	0.464	[7.106]	0.449	5.030
7257008	—	0.51178	1.252	4.632	0.997	4.769	0.739	5.149	0.594	5.057
7505345	V355 Lyr	0.4737	1.417	4.796	1.189	4.894	0.965	4.990	0.756	5.362
7671081	V450 Lyr	0.50461	1.122	4.519	0.873	4.919	0.736	4.884	0.614	5.381
9001926	V353 Lyr	0.5568	1.288	4.714	1.004	4.879	0.830	5.202	0.699	5.721
9578833	V366 Lyr	0.52703	1.344	4.935	1.067	5.090	0.869	5.385	0.662	5.576
9697825	V360 Lyr	0.55758	1.005	4.755	0.760	4.991	0.669	5.114	0.479	5.771
9973633	—	0.51075	1.506	5.240	1.315	5.107	0.909	5.206	0.732	5.526
10789273	V838 Cyg	0.48028	1.645	4.645	1.335	4.784	1.131	4.933	0.875	5.236
11125706	—	0.61322	0.706	5.302	0.540	5.614	0.452	5.970	0.352	[6.400]
12155928	V1104 Cyg	0.43639	1.603	4.657	1.295	4.689	1.062	4.851	0.891	5.242
Non-Blazhko:										
3733346	NR Lyr	0.68203	1.124	4.829	0.915	4.977	0.759	5.172	0.588	5.590
3866709	V715 Cyg	0.47071	1.525	4.702	1.226	4.643	1.015	4.929	0.841	5.269
5299596	V782 Cyg	0.52364	0.836	5.168	0.665	5.449	0.539	5.811	0.451	6.170
6070714	V784 Cyg	0.53409	1.055	5.477	0.830	5.839	0.616	6.191	0.482	[6.435]
6100702	—	0.48815	0.926	5.261	0.714	5.547	0.569	5.841	0.440	[6.357]
6763132	NQ Lyr	0.58779	1.185	4.811	0.954	4.958	0.770	5.105	0.606	5.628
6936115	FN Lyr	0.5274	1.482	4.622	1.226	4.758	0.996	4.857	0.813	5.204
7021124	—	0.62249	1.489	4.560	1.176	5.159	0.957	5.078	0.686	5.065
7030715	—	0.68361	0.944	5.162	0.732	5.437	0.605	5.639	0.473	6.202
7176080	V349 Lyr	0.50707	0.111	6.213	0.078	[8.054]	0.078	[6.859]	0.061	[7.009]
7742534	V368 Lyr	0.45649	1.666	4.551	1.327	4.705	1.104	4.932	0.976	5.174
7988343	V1510 Cyg	0.58114	1.496	4.850	1.247	4.963	0.970	5.122	0.759	5.427
8344381	V346 Lyr	0.57683	1.430	4.858	1.163	5.009	0.945	5.036	0.563	4.416
9508655	V350 Lyr	0.59424	1.452	4.809	1.189	4.955	0.974	5.130	0.758	5.523
9591503	V894 Cyg	0.57139	1.577	4.871	1.257	4.961	1.037	5.148	0.824	5.452
9658012	—	0.53321	1.402	4.811	1.125	5.012	0.943	5.103	0.719	5.489
9717032	—	0.55691	1.384	4.852	1.021	5.070	0.852	5.399	0.704	5.432
9947026	V2470 Cyg	0.54859	0.880	5.254	0.693	5.488	0.571	5.900	0.458	[6.344]
10136240	V1107 Cyg	0.56578	1.178	4.899	0.961	5.113	0.785	5.326	0.618	5.648
10136603	V1107 Cyg	0.43377	1.235	5.186	0.947	5.121	0.732	5.667	0.572	6.139
11802860	AW Dra	0.68722	1.321	5.240	1.045	5.431	0.848	5.639	0.652	6.045

Notes: [ ] denotes ϕ_{31} larger than 2π.

stars. RRc stars are expected to behave differently from the RRab stars and require special treatment, so they are not included. We used IRAF/digiphot/phot program to obtain the photometry. The periods in Table 1, as well as T_0 and Kp in Fig. 1, were taken from Nemec *et al.* (2011, 2013). The Fourier analysis was based on Géza Kovács' Fourier decomposition program.

From Table 1 we can calculate the mean differences between ϕ_{31} in $BVRI$ bands and $\phi_{31}(Kp)$ and their standard deviations. They are the following: $\Delta\phi_{31}(B) = \langle\phi_{31}(B) - \phi_{31}(Kp)\rangle = 0.430 \pm 0.107$ rad from 34 stars, $\Delta\phi_{31}(V) = 0.174 \pm 0.085$ rad from 34 stars, $\Delta\phi_{31}(R) = -0.018 \pm 0.053$ rad from 34 stars and $\Delta\phi_{31}(I) = -0.192 \pm 0.063$ rad from 35 stars. Earlier, $\Delta\phi_{31}(V) = 0.151$ rad was derived by Nemec *et al.* (2011) based on only three RR Lyrae stars. These results can help to derive metal abundances using Fourier parameters if $\phi_{31}(B, V, R, I)$ vs. $\phi_{31}(Kp)$ relations will be used to translate the former to the latter and then the [Fe/H] vs. $\phi_{31}(Kp)$ relation of Nemec *et al.* (2013) will be used to derive [Fe/H].

Transformation to standard photometry is currently in progress.

References

Nemec, J. M., Smolec, R., Benkő, J. M., *et al.* 2011, *MNRAS*, 417, 1022
Nemec, J. M., Cohen, J. G., Ripepi, V., *et al.* 2013, *ApJ*, 773, 181

Precision Asteroseismology
Proceedings IAU Symposium No. 301, 2013
J. A. Guzik, W. J. Chaplin, G. Handler & A. Pigulski, eds.

© International Astronomical Union 2014
doi:10.1017/S1743921313014890

The mean density and the surface gravity of the primary component of μ Eridani, an SPB variable in a single-lined spectroscopic and eclipsing system

Mikołaj Jerzykiewicz

Astronomical Institute, University of Wrocław, Kopernika 11, Pl-51-622 Wrocław, Poland
email: mjerz@astro.uni.wroc.pl

Abstract. Precision asteroseismology uses the observed effective temperature and luminosity or surface gravity in selecting evolutionary models for analysis. In the case of the primary component of μ Eri, an SPB variable, the surface gravity and luminosity can be derived from the parameters of the SB1 eclipsing system. We examine how the surface gravity and luminosity so derived help to select the evolutionary models suitable for asteroseismic analysis.

Keywords. stars: individual (μ Eri), stars: SB1, stars: eclipsing, stars: fundamental parameters

---

μ Eri (B5 IV, $V = 4.00$ mag) is a single-lined spectroscopic binary and a detached eclipsing variable with an invisible secondary component. The primary component is a slowly pulsating B (SPB) star. Recently, Jerzykiewicz *et al.* (2013) derived the parameters of the system using *MOST* (Walker *et al.* 2003) time-series photometry and ground-based radial-velocity observations. In addition, from archival *UBV* and *uvby* indices these authors derived the primary's effective temperature, $T_{\rm eff} = 15\,670 \pm 470$ K. For an assumed mass, M, the parameters of the system, viz., the relative radius of the primary component, the inclination of the orbit, the semi-amplitude of the primary's radial-velocity variation and the eccentricity, suffice to compute the mean density, $\langle\rho\rangle$, and surface gravity, g. Then, the luminosity, L, can be computed from $\langle\rho\rangle$ and $T_{\rm eff}$. For $M = 5$, 5.5 and 6 $M_\odot$, the results of this exercise are given in Table 1. As can be seen from the table, $\langle\rho\rangle$ is virtually independent of M, while $\log g$ varies slowly with M.

Figure 1 illustrates how $\log g$ and $\log L$ so derived can help to select the evolutionary models suitable for asteroseismic analysis. In the figure, the positions of the primary component of μ Eri (circles with error bars) are compared with the 5.5 and 6.0 $M_\odot$ evolutionary tracks, computed assuming $X = 0.7$ and $Z = 0.015$, the Asplund *et al.* (2009) solar mixture, the OPAL opacities (Iglesias & Rogers 1996), the equatorial rotation velocity of 140 km s^{-1} on the ZAMS, and two values of the convective-core overshooting parameter, $\alpha_{\rm ov} = 0.0$ (left panels) and 0.2 (right panels). The figure shows that in the case of $\alpha_{\rm ov} = 0.0$ (left panels) the post-main-sequence models are more suitable for asteroseismic analysis of μ Eri than the main-sequence models, while the reverse is true for

Table 1. Fundamental parameters of the primary component of μ Eri, computed for three assumed values of mass. M and L are in solar units, $\langle\rho\rangle$ and g, in cgs units.

M	$\langle\rho\rangle$	$\log g$	$\log L$
5.0	0.0345 ± 0.0032	3.597 ± 0.027	3.273 ± 0.058
5.5	0.0346 ± 0.0032	3.612 ± 0.027	3.299 ± 0.058
6.0	0.0348 ± 0.0032	3.625 ± 0.027	3.323 ± 0.058

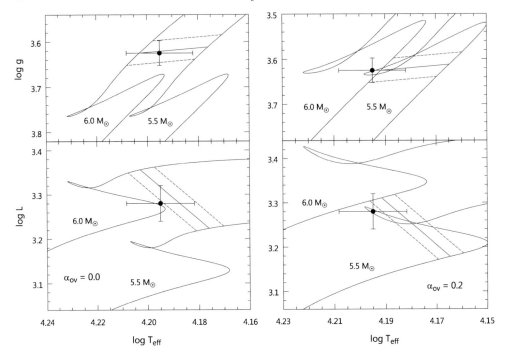

Figure 1. The positions of the primary component of μ Eri (circles with error bars) compared with the 5.5 and 6.0 $M_\odot$ evolutionary tracks, computed assuming $X = 0.7$ and $Z = 0.015$, the equatorial rotation velocity of 140 km s^{-1} on the ZAMS, and two values of the convective-core overshooting parameter, $\alpha_{ov} = 0.0$ (left panels) and 0.2 (right panels). Also shown are lines connecting points on the evolutionary tracks for which μ Eri's $\log g$ is equal to the evolutionary tracks' $\log g$ (upper panels) or μ Eri's $\log L$ is equal to the evolutionary tracks' $\log L$ (lower panels). The dashed lines run at $\pm\sigma$.

$\alpha_{ov} = 0.2$ (right panels). The former case would be an observational confirmation of the existence of pulsational instability in post-main-sequence massive-star models discovered recently by Daszyńska-Daszkiewicz *et al.* (2013).

Acknowledgements

We are indebted to Dr. Jadwiga Daszyńska-Daszkiewicz for computing the evolutionary tracks used in this paper and to Dr. Gerald Handler for pointing out an error in the original version of the paper. Support from MNiSW grant N N203 405139 is acknowledged. In this research, we have used the Aladin service, operated at CDS, Strasbourg, France, and the SAO/NASA Astrophysics Data System Abstract Service.

References

Asplund, M., Grevesse, N., Sauval, A. J., & Scott, P. 2009, *ARAA*, 47, 481
Daszyńska-Daszkiewicz, J., Ostrowski, J., & Pamyatnykh, A. A. 2013, *MNRAS*, 432, 3153
Iglesias, C. A. & Rogers, F. J. 1996, *ApJ*, 464, 943
Jerzykiewicz, M., Lehmann, H., Niemczura, E., *et al.* 2013, *MNRAS*, 432, 1032
Walker, G., Matthews, J., Kuschnig, R., *et al.* 2003, *PASP*, 115, 1023

Precision Asteroseismology
Proceedings IAU Symposium No. 301, 2013
J. A. Guzik, W. J. Chaplin, G. Handler & A. Pigulski, eds.

© International Astronomical Union 2014
doi:10.1017/S1743921313014907

Study of BL Her type pulsating variable stars using publicly available photometric databases

Monika Jurković[1] and László Szabados[2]

[1]Astronomical Observatory of Belgrade, Volgina 7, 11060 Belgrade, Serbia
email: mojur@aob.rs

[2]Konkoly Observatory of the Hungarian Academy of Sciences
H-1525 Budapest, P.O. Box 67., Hungary

Abstract. BL Her type pulsating variable stars are a subtype of Type II Cepheids, pulsating with periods in the range from 1 to 4 days. The General Catalog of Variable Stars lists 71 objects. For each star from this list, we searched for data in the publicly available photometric databases: AAVSO, ASAS, Catalina Sky Survey, *INTEGRAL* OMC, LINEAR, NSVS, SuperWASP. The analysis was done separately for each dataset. Here we present first results.

Keywords. stars: variable: Cepheids

1. Data and method

The list of BL Her variables was gathered from the General Catalogue of Variable Stars (GCVS). Time-series data for each star were collected by searching the publicly available databases: AAVSO, ASAS (Pojmański 1997, 2002, 2003; Pojmański & Maciejewski 2004, 2005; Pojmański *et al.* 2005), *INTEGRAL* OMC, Catalina Sky Survey, SuperWASP (Butters *et al.* 2010), etc. The data were analysed using Period04 (Lenz & Breger 2005).

2. First results

In the case of V553 Sco, V4410 Sgr, AT Tel, V714 Cyg, V742 Cyg and KO Lyr no data were found. In some cases there were not enough data points to perform the Fourier analysis. Here we present results on three stars we have studied. Our work has not been completed yet.

BD Cas – INTEGRAL data: instrumental errors in the light curve (Fig. 1).
NY Her - AAVSO data: SU UMa type dwarf nova (Kato *et al.* 2013), (Fig. 2).
UW For - ASAS data: an eclipsing binary (Fig. 3).

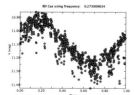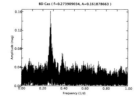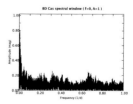

Figure 1. BD Cas from the *INTEGRAL* data. *Left:* phase diagram. *Middle:* Fourier spectrum, *Right:* spectral window.

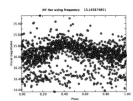 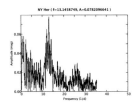 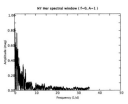

Figure 2. The same as in Fig. 1, but for the AAVSO data of NY Her.

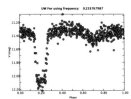 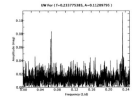 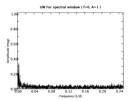

Figure 3. The same as in Fig. 1, but for the ASAS data of UW For.

3. Conclusion

As the presented three examples show, the GCVS classification for BL Her stars has to be taken into account with great care. With the completion of our study we hope to make a more reliable classification, and carry out further analysis of all objects.

Acknowledgements

The authors would like to thank for the IAU travel grant. The authors acknowledge financial support from the Ministry of Education, Science and Technical Development of the Republic of Serbia through project 176004, and the DOMUS 2013 program of the Hungarian Academy of Sciences. This reasearch is based on data from the OMC Archive at CAB (INTA-CSIC), pre-processed by ISDC, the Catalina Sky Survey, and AAVSO. This research has made use of VizieR catalogue access tool, CDS, Strasbourg, France.

References

Butters, O. W., West, R. G., Anderson, D. R., *et al.* 2010, *A&A*, 510, L10
Kato, T., Hambsch, F.-J., Maehara, H., *et al.* 2013, *PASJ*, 65, 23
Lenz, P. & Breger, M. 2005, *CoAst*, 146, 53
Pojmański, G. 1997, *AcA*, 47, 467
Pojmański, G. 2002, *AcA*, 52, 397
Pojmański, G. 2003, *AcA*, 53, 341
Pojmański, G. & Maciejewski, G. 2004, *AcA*, 54, 153
Pojmański, G. & Maciejewski, G. 2005, *AcA*, 55, 97
Pojmański, G., Pilecki, B., & Szczygieł, D. 2005, *AcA*, 55, 275

Precision Asteroseismology
Proceedings IAU Symposium No. 301, 2013
J. A. Guzik, W. J. Chaplin, G. Handler & A. Pigulski, eds.

© International Astronomical Union 2014
doi:10.1017/S1743921313014919

KIC 10486425: A *Kepler* eclipsing binary system with a pulsating component

Filiz Kahraman Aliçavuş[1,2] and Esin Soydugan[1,2]

[1] Department of Physics, Faculty of Arts and Sciences, Çanakkale Onsekiz Mart University
TR-17020 Çanakkale, Turkey
[2] Astrophysics Research Centre and Observatory, Çanakkale Onsekiz Mart University,
TR-17020 Çanakkale, Turkey
email: filizkahraman01@gmail.com

Abstract. We present frequency analysis of the *Kepler* light curve of KIC 10486425, an eclipsing binary system with a pulsating component. The parameters of the binary were obtained by modelling the light curve with the Wilson-Devinney program. The residuals from this modelling were subject to Fourier analysis which allowed us to detect 120 periodic terms characteristic for γ Dor-type pulsations. The dominant frequency of these changes amounts to 1.3189 d^{-1}.

Keywords. stars: binaries: eclipsing, stars: variables: γ Dor, stars: individual: KIC 10486425

1. Introduction

The *Kepler* satellite was launched in 2009 in order to discover Earth-like planets around solar-like stars by using transit method (Koch *et al.* 2010). The mission provided high-precision data for over 200 000 stars, contributing remarkably to many branches of astronomy, including asteroseismology. The detached eclipsing binary system KIC 10486425 observed by *Kepler* was discovered by the TrES project (Devor *et al.* 2009). In this study, we present light curves and frequency analysis of this star using high-precision *Kepler* data.

2. Light curve modelling and frequency analysis

The parameters of KIC 10486425 were derived by fitting the *Kepler* light curve by means of the Wilson-Devinney (WD) program (Wilson & Devinney 1971). For this purpose, we used *Kepler* Q1 and Q2 data. We adopted Mode 2 which corresponds to the detached configuration. In the analysis, some input parameters were assumed following the Catalog of Eclipsing Binary Stars (Prša *et al.* 2011). The other parameters were fixed. In particular, bolometric albedos were assumed to be equal to 0.5 (Rucinski 1969) and bolometric gravity-darkening coefficients were set to 0.32 (Lucy 1967). It was also assumed that both components have convective envelopes, that the effective temperature of the primary $T_1 = 7018$ K, and that the components of the system rotate synchronously. The other parameters ($e, i, T_2, \Omega_1, \Omega_2, q, L_2, L_3$) were adjusted. We derived two different solutions. In the first solution, a circular orbit was assumed. In the other one, eccentricity and the third-light contribution were chosen as free parameters, starting from 0.08 and 0.013 (Slawson *et al.* 2011), respectively. The parameters obtained from the light curve analysis are listed Table 1.

Having subtracted the theoretical light curve, we used *Kepler* Q1 to Q10 data and carried out frequency analysis of residuals with the PERIOD04 program (Lenz & Breger 2005). We obtained 120 frequencies above the significance limit (S/N $\geqslant$ 4) (Breger

F. Kahraman Aliçavuş & E. Soydugan

Table 1. Photometric solutions for KIC 10486425.

Parameter	Solution 1	Solution 2
T_0	54969.708432^*	54969.708432^*
P (day)	5.274816^*	5.274816^*
i $(^\circ)$	82.314 ± 0.030	82.549 ± 0.050
T_1 (K)	7018^*	7018^*
T_2 (K)	5210 ± 220	5727 ± 230
Ω_1	10.576 ± 0.006	12.129 ± 0.027
Ω_2	8.096 ± 0.053	8.055 ± 0.012
Φ	0.0000 ± 0.00002	0.0000 ± 0.00003
q	0.4039 ± 0.0310	0.5907 ± 0.0120
e	0.0^*	0.0350 ± 0.0008
L_3	0.0^*	0.0000 ± 0.0009
$L_1/(L_1 + L_2)$	0.90	0.70

*Fixed parameters.

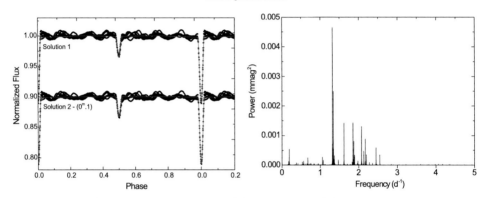

Figure 1. *Left:* Observed and theoretical (continuous line) light curves. *Right:* Power spectrum of the residuals from the WD fit.

et al. 1993). Some of these frequencies were harmonics of the system's orbital frequency or combination frequencies. The dominating frequency is equal to 1.3189 d^{-1}; its amplitude amounts to 4.152 mmag. The comparison of theoretical curves with the observational points, and the power spectrum of the residuals from the WD fit are shown in Fig. 1.

3. Summary and conclusion

We present the light curve and frequency analysis of KIC 10486425. As a result of this analysis, the parameters of the system were derived and 120 significant periodic terms were found in the residuals. Their frequencies indicate that the primary component of KIC 10486425 is a γ Dor-type star with the main pulsation period of about 0.76 d.

References

Breger, M., Stich, J., Garrido, R., *et al.* 1993, *A&A*, 271, 482
Devor, J., Charbonneau, D., O'Donovan, *et al.* 2009, *AJ*, 135, 850
Koch, D. G., Borucki, W. J., Basri, G., *et al.* 2010, *ApJ*, 713, L79
Lenz, P. & Breger, M. 2005, *CoAst*, 146, 53
Lucy, L. B. 1967, *ZfA*, 65, 89
Prša, A., Batalha, N., Slawson, R. W., *et al.* 2011, *AJ*, 141, 83
Rucinski, S. M. 1969, *AcA*, 19, 245
Slawson, R. W., Prša, A., Welsh, W. F., *et al.* 2011, *AJ*, 142, 160
Wilson, R. E. & Devinney, R. J. 1971, *ApJ*, 166, 605

Precision Asteroseismology
Proceedings IAU Symposium No. 301, 2013
J. A. Guzik, W. J. Chaplin, G. Handler & A. Pigulski, eds.

© International Astronomical Union 2014
doi:10.1017/S1743921313014920

LIMBO: A time-series Lucky Imaging survey of variability in Galactic globular clusters

N. Kains[1], D. M. Bramich[1], R. Figuera Jaimes[1,2], and J. Skottfelt[3]

[1]European Southern Observatory, Garching bei München, Germany

[2]School of Physics and Astronomy, University of St Andrews, United Kingdom

[3]Niel Bohr Institute, University of Copenhagen, Denmark

Abstract. We present a large observing project monitoring globular clusters (GC) over long time baselines, which will lead to a complete census of variable stars in those clusters down to several magnitudes below the horizontal branch (HB). The use of Lucky Imaging (LI) will allow us to obtain high-precision photometry for even faint objects, and long-term monitoring will also mean that observations are sensitive to detecting other slow transient phenomena, such as gravitational microlensing, the primary aim of this project.

Keywords. techniques: photometric, stars: variables, globular clusters: general

1. Introduction

The emergence of LI marks a significant advance in our ability to obtain high-resolution images from the ground, without the need for adaptive optics to correct for atmospheric turbulence. Carrying out complete variability censuses across entire clusters with conventional CCD observations is limited by our ability to obtain high-precision photometry in the crowded cores, where source blending often prevents the detection of low-amplitude variability, or of low-amplitude variability of stars much fainter than the HB.

In short, LI consists in taking a large number of very short (sub-second) exposures in order to "freeze" atmospheric turbulence. The best images are sorted according to a pre-determined criterion are then stacked to form a high signal-to-noise image with, in the best observing conditions, a resolution close to the diffraction limit. This then allows us to construct a high-resolution reference image, which we use to perform PSF difference imaging on all images and obtain very high-precision photometry of all sources.

In Fig. 1, we compare an image of the cluster NGC 6981 taken with a conventional CCD on a 2-m telescope (Bramich *et al.* 2011), with the same field observed with a LI camera on the Danish 1.54-m telescope (Skottfelt *et al.* 2013), in a similar seeing conditions. Using LI enabled us to detect two variables, V57 and V58, which were too close to a bright star on conventional CCD images to obtain high-precision photometry.

2. Observational set-up and cluster sample

Our sample consists of 30 clusters, observed using the 1.54-m Danish Telescope in La Silla, Chile. This programme is combined with microlensing monitoring of the Galactic Bulge by the MiNDSTEp consortium, making use of parts of the night when the Galactic Bulge is not observable. Our cluster sample was selected based on three main criteria:
- Each target should have a HB magnitude brighter than 18 in the V filter.
- No recent CCD time-series study of the target cluster has been conducted.
- The cluster has a dense core, i.e. a large density parameter ρ_0, which increases the probability of lensing by compact objects in the cluster core.

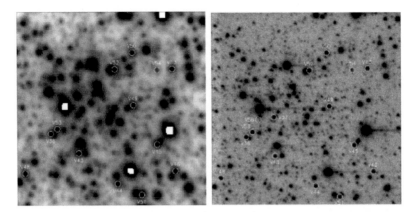

Figure 1. The core of globular cluster NGC 6981 observed with a conventional CCD mounted on the Hanle 2-meter telescope at the Indian Astrophysical Observatory, with a FWHM of $\sim 0.9''$ (Bramich *et al.* 2011, left), and a LI camera, mounted on the 1.54-m Danish Telescope in La Silla, with a FWHM of $\sim 0.4''$ (Skottfelt *et al.* 2013). LI photometry enabled us to detect variables V57 and V58, which were suffering from blending from a bright star on CCD images.

This set-up aims to maximise the potential of our survey, which will provide us with a complete census of variable stars in globular clusters, down to several magnitudes below the HB. This builds on our team's current work with conventional CCD camera (e.g. Figuera Jaimes *et al.* 2013, Kains *et al.* 2012, 2013, Arellano Ferro *et al.* 2013).

Furthermore, long-baseline high-precision photometry will allow us to detect changes and modulation to the variable signals due to effects like the Blazhko effect, as well as potential detection of rare objects, such as RR Lyrae in eclipsing binary systems. Finally, we also aim to detect transient phenomena such as gravitational microlensing by cluster members, which could enable us to detect objects such as stellar-mass or intermediate-mass black holes. The latter are predicted to reside in the cores of globular clusters, although there has not yet been any conclusive detection of these objects.

3. Conclusions

This survey will lead to a significant update of our knowledge of variable stars in our target clusters. Thanks to the large amount of observing time devoted to the project, this is also an excellent opportunity for systematic searches for gravitational microlensing in GC, including by stellar- and intermediate-mass black holes. In the coming years, the SONG (Stellar Observations Network Group) telescopes, which will also be equipped with LI cameras, may enable us to increase our sample of GC significantly, including clusters in the Northern Hemisphere.

References

Arellano Ferro, A., Bramich, D. M., Figuera Jaimes, R., *et al.* 2013, *MNRAS*, 434, 1220

Bramich, D. M., Figuera Jaimes, R., Giridhar, S., & Arellano Ferro, A. 2011, *MNRAS*, 413, 1275

Figuera Jaimes, J., Arellano Ferro, A., Bramich, D. M., Giridhar, S., & Kuppuswamy, K. 2013, *A&A*, 556, A20

Kains, N., Bramich, D. M., Figuera Jaimes, R., Arellano Ferro, A., Giridhar, S., & Kuppuswamy, K. 2012, *A&A*, 548, A92

Kains, N., Bramich, D. M., Arellano Ferro, A., *et al.* 2013, *A&A*, 555, A36

Skottfelt, J., Bramich, D. M., Figuera Jaimes, R., *et al.* 2013, *A&A*, 553, A111

Precision Asteroseismology
Proceedings IAU Symposium No. 301, 2013
J. A. Guzik, W. J. Chaplin, G. Handler & A. Pigulski, eds.

© International Astronomical Union 2014
doi:10.1017/S1743921313014932

Global Astrophysical Telescope System — telescope No. 2

Krzysztof Kamiński, Roman Baranowski, Monika Fagas, Wojciech Borczyk, Wojciech Dimitrov and Magdalena Polińska

Astronomical Observatory Institute, Faculty of Physics, A. Mickiewicz University,
Słoneczna 36, 60-286 Poznań, Poland
email: chrisk@amu.edu.pl

Abstract. We present the new, second spectroscopic telescope of Poznań Astronomical Observatory. The telescope allows automatic simultaneous spectroscopic and photometric observations and is scheduled to begin operation from Arizona in autumn 2013. Together with the telescope located in Borowiec, Poland, it will constitute a perfect instrument for nearly continuous spectroscopic observations of variable stars. With both instruments operational, the Global Astrophysical Telescope System will be established.

Keywords. instrumentation: spectrographs, telescopes

1. Introduction

Poznań Spectroscopic Telescope 2 (PST2) is the second spectroscopic telescope constructed at the Astronomical Observatory of Adam Mickiewicz University in Poznań (Poland). The first one (PST1) has been operational since 2007 in Borowiec observing station (Baranowski *et al.* 2009). PST2 will be transported from Poznań to Winer Observatory in Arizona (USA) in September 2013. Since both telescopes are equipped with similar echelle spectrographs, they will be perfectly suited for monitoring spectroscopic variations of pulsating stars in a nearly continuous way.

2. The telescope

PST2 is a slightly improved version of our first instrument. The new telescope is Planewave's model CDK700 of 0.7-m diameter with an alt-az robotic mount. The custom designed echelle spectrograph of $R = 40\,000$ resolution is fiber fed with an optional iodine cell and a simultaneous input light measurement. The main spectroscopic camera is an Andor iKon-L, whose low readout noise and negligible dark current enables us to record spectra with S/N ~ 5 for stars up to 11 mag, which is perfectly useful for radial velocity measurements of late spectral types. The telescope is equipped with a small guider scope for simultaneous photometry during spectroscopic exposures. Everything will be controlled remotely with our dedicated software. Fully autonomous operation mode is also being prepared.

3. Spectroscopy

The echelle spectrograph of PST2 is designed to maintain substantial thermal stability for both short and long periods of time. Any residual changes of temperature, as well as changes in atmospheric pressure, are recorded with a precision of 0.01°C and 0.01 hPa. This should allow for a post-observation correction of spectrograph instabilities. Our preliminary tests show that it should be possible to correct radial velocities to ~20 m s^{-1} r.m.s. using a single-point temperature and pressure measurement inside the spectrograph box. Further improvements of radial velocity measurements are possible with our optional iodine cell, but at the cost of reducing the telescope's effective limiting magnitude and contaminating the stellar spectrum with iodine lines.

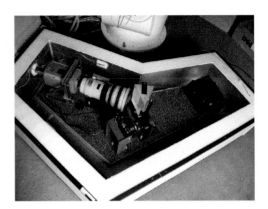

Figure 2. The PST2 echelle spectrograph inside its thermostatic box.

Figure 1. The PST2 telescope before installing additional equipment.

4. Photometry

The auxiliary guider scope is capable of making simultaneous photometric observations of even very bright spectroscopic targets. It is equipped with selectable two pairs of B and R Johnson filters. One pair of these filters is covered with a 0.1×0.1 mm metal wire mesh. Using these, we record interferometric patterns of all stars within the guider 0.5 degree field of view. By selecting a weaker first- or second-order image of a bright target star and a zero-order image of, usually much fainter, comparison star, we are able to collect simultaneously a comparable amount of light, and therefore obtain good quality relative photometry, even when the zero-order target star image is overexposed.

Additionally PST2 is also perfectly suited for photometry of fast moving NEAs. The telescope is capable of non-sidereal tracking with selectable, constant speed. The second Nasmyth focus of our telescope is equipped with an Andor iXon X$_3$ camera which is capable of delivering very short exposures with no dead time (by using the frame transfer technique) and no readout noise (by using electron-multiplying technology). All this combined allows us to follow fast moving asteroids for a substantial amount of time and make long series of very short exposures to combine them later, on both the target object and the comparison stars, without the cost of multiplying the camera's readout noise.

Reference

Baranowski, R., Smolec, R., Dimitrov, W., *et al.* 2009, *MNRAS*, 396, 2194

Precision Asteroseismology
Proceedings IAU Symposium No. 301, 2013
J. A. Guzik, W. J. Chaplin, G. Handler & A. Pigulski, eds.

© International Astronomical Union 2014
doi:10.1017/S1743921313014944

Chromospheric activity of 9 Aurigae

Anthony B. Kaye[1], Cristine Kristof[1], and Richard O. Gray[2]

[1] Department of Physics, Texas Tech University
Lubbock, Texas 79409-1051, United States of America
email:anthony.kaye@ttu.edu

[2] Department of Physics and Astronomy, Appalachian State University
Boone, North Carolina 28608, United States of America
email:grayro@appstate.edu

Abstract. The definition of γ Dor stars as a new class of variable stars by Kaye *et al.* (1999) was based on a number of criteria, including the failure to detect any emission in the Ca II H & K lines in these stars (Kaye & Strassmeier 1998; KS98). Over the last 30 years, efforts continue to look for the blue edge of chromospheric activity. As a part of this effort, we put γ Dor stars to the test to see if magnetic fields play a non-trivial role in their variability.

1. Introduction

The He I D_3 triplet has been known to be an indicator of chromospheric activity in main sequence stars as early as 1985 (Wolff *et al.* 1985); it is especially useful in stars where traditional signs of activity are absent (e.g., the Ca II H&K lines in F stars; KS98). Excellent studies on this topic can be found in Wolff *et al.* (1985, 1986), Wolff & Heasley (1987), Rachford (1997, 1998, 2000), and Rachford & Foight (2009; hereafter RF09).

The γ Dor stars were defined by Kaye *et al.* (1999) and their pulsation mechanism was identified by Guzik *et al.* (2000). To date, other than the null result of Kaye & Strassmeier (1998), there has been no systematic effort to investigate chromospheric activity in these stars. RF09 measured the D_3 triplet of 9 Aur twice, but very few (if any) γ Dor stars have enough D_3 measurements to carry out time series analysis to investigate variation in the D_3 line.

2. Observations and methodology

We obtained 196 high-resolution spectra of 9 Aur using the 0.9-meter Coudé Feed telescope and the Coudé spectrograph at the Kitt Peak National Observatory during 14 nights in December 1998. The IRAF-reduced spectra have a 2-pixel resolution of 0.21 Å, a 2-pixel signal-to-noise ratio (SNR) of ~500, and cover 315 Å between 5840 and 6155 Å.

In addition, one classification-resolution spectrum of 9 Aur was acquired with the Gray/Miller Cassegrain spectrograph on the 0.8-m telescope at the Dark Sky Observatory at Appalachian State University. This spectrum has a 2-pixel resolution of 1.8 Å, an SNR ~300, and spans a range of 800 Å centered on 4200 Å.

Measuring the equivalent width (W_λ) of the D_3 triplet is complicated since it is contaminated by both photospheric and telluric lines. Therefore, we followed this procedure:

(*a*) We used an ATLAS12 model (Kurucz 1993) with the standard known physical parameters for the Sun to adjust the gf values in the region of the D_3 triplet; the best model had the best fit with an observed spectrum of the Sun (Delbouille *et al.* 1990).

(*b*) We used the MD SIMPLEX method of Gray *et al.* (2001), ATLAS9/ATLAS12 models, and SPECTRUM (Gray & Corbally 1994) to model the 9 Aur photosphere and determine T_{eff}, $\log g$), and $[M/H]$; ξ_t was fixed at 2.0 km s^{-1}. We compared the model fluxes with published fluxes from the TD-1 satellite (Thompson $et\ al.$ 1978), from Strömgren uby photometry, and from fluxes from the spectrophotometry of Breger (1976) to ensure the model represented the star well.

(c) This model photosphere was convolved with the measured projected radial velocity of 9 Aur (17.0 km s^{-1}; Fekel 2003), a limb darkening coefficient of 0.6, and the two-pixel instrumental broadening. We then computed the W_λ of the D_3 triplet area between 5874.826 Å and 5876.781 Å. This allows us to subtract out all photospheric features in the area, leaving only the D_3 triplet and telluric lines.

(d) We then used a program that divides out the telluric lines with an "effective" air mass based upon the best telluric data of the observing run, smoothed with a 3-pixel low pass filter, rectified to a unit continuum, and measured each W_λ. Final D_3 W_λ's were computed by simply subtracting this result from the result obtained in step (c).

3. Conclusions

Our average W_λ is 14 mÅ and is consistent with the values reported by RF09. Using the 10 mÅ threshold criterion of Wolff $et\ al.$ (1986), we conclude that 9 Aur is moderately chromospherically active. We observe statistically significant variations in the D_3 line, and propose that some variation in our measurements of the D_3 line may be due to variations caused by the ongoing g-mode pulsations or to previously undiscovered acoustic modulations. Using the models generated by Warner, Kaye, & Guzik (2003), we are able to confirm the hypothesis of Kaye (1998) that claimed that at least some γ Dor stars have Rossby numbers of order unity (for 9 Aur, $Ro = 2.47$), so magnetic fields likely play a more significant role than originally thought.

References

Breger, M. 1976, $ApJS$, 32, 7

Delbouille, L., Roland, G., & Neven, L. 1990, $Atlas\ photometrique\ du\ spectra\ solaire\ de\ \lambda3000\ a\ \lambda10000$' (Liège: Université de Liège, Institut d'Astrophysique)

Fekel, F. C. 2003, $PASP$, 115, 807

Gray, R. O. & Corbally, C. J. 1994, AJ, 107, 742

Gray, R. O., Graham, P. W., & Hoyt, S. R. 2001, AJ, 121, 2159

Guzik, J. A., Kaye, A. B., Bradley, P. A., Cox, A. N., & Neuforge, C. 2000, ApJ, 542, L57

Kaye, A. B. 1998, Ph.D. thesis, Georgia State University

Kaye, A. B. & Strassmeier, K. G. 1998, $MNRAS$, 294, L35

Kaye, A. B., Handler, G., Krisciunas, K., Poretti, E., & Zerbi, F. M. 1999, $PASP$, 111, 840

Kurucz, R. L. 1993, CD-ROM 13, $ATLAS9\ Stellar\ Atmosphere\ Programs\ and\ 2\ km/s\ Grid$ (Cambridge: SAO)

Rachford, B. L. 1997, ApJ, 486, 994

Rachford, B. L. 1998, ApJ, 505, 255

Rachford, B. L. 2000, $MNRAS$, 315, 24

Rachford, B. L. & Foight, D. R. 2009, ApJ, 698, 786

Thompson, G. I., Nandy, K., Jamar, C., $et\ al.$ 1978, $Catalog\ of\ Stellar\ Ultraviolet\ Fluxes$ (NSSDC/ADC Cat. 2059B) (London: Sci. Res. Council)

Warner, P. B., Kaye, A. B., & Guzik, J. A. 2003, ApJ, 593, 1049

Wolff, S. C. & Heasley, J. N. 1987, $PASP$, 99, 957

Wolff, S. C., Heasley, J. N., & Varsik, J. 1985, $PASP$, 97, 707

Wolff, S. C., Boesgaard, A. M., & Simon, T. 1986, ApJ, 310, 360

Precision Asteroseismology
Proceedings IAU Symposium No. 301, 2013
J. A. Guzik, W. J. Chaplin, G. Handler & A. Pigulski, eds.

© International Astronomical Union 2014
doi:10.1017/S1743921313014956

Wide-field variability survey of the globular cluster NGC 4833

Grzegorz Kopacki

Instytut Astronomiczny, Uniwersytet Wrocławski,
Kopernika 11, 51-622 Wrocław, Poland
email: kopacki@astro.uni.wroc.pl

Abstract. We present preliminary results of the variability survey in the field of the globular cluster NGC 4833. We observed all 34 variable stars known in the cluster. In addition, we have found two new SX Phoenicis stars, one new RR Lyrae star, twelve new eclipsing systems mostly of the W Ursae Majoris type, nine new variable red giants, and ten new field-stars showing irregular variations. Properties of RR Lyrae stars indicate that NGC 4833 is an Oosterhoff's type II globular cluster.

Keywords. Galaxy: globular clusters: individual: NGC 4833, stars: Population II, stars: variables: RR Lyrae, stars: variables: SX Phoenicis

1. Introduction

Continuing our ongoing project aimed at the search and analysis of pulsating stars in globular clusters (see Kopacki 2013) we present preliminary results for NGC 4833. We used image subtraction method (ISM, Alard & Lupton 1998) which works well in crowded stellar fields like a cluster core and thus enables detection of many variable stars, such as RR Lyrae and SX Phoenicis stars.

NGC 4833 is the southern globular cluster of intermediate metallicity ([Fe/H] = −1.85). The most recent version of the Catalogue of Variable Stars in Globular Clusters (CVSGC, Clement *et al.* 2001) listed 34 objects in the field of this cluster including six SX Phoenicis stars and 20 RR Lyrae stars.

2. Observations and results

We used CCD observations obtained during a one-month observing run in Feb/Apr, 2008 using 40-inch telescope at Siding Spring Observatory, Australia. They consisted of 740 V-filter and 220 $I_{\rm C}$-filter CCD frames.

We confirmed all variable stars found recently in the cluster core by Darragh & Murphy (2012). In addition, we have detected two new SX Phoenicis stars, one new RR Lyrae star, twelve new eclipsing systems mostly of the W Ursae Majoris type, nine new variable red giants at the tip of the red giant branch, and ten new field-stars showing irregular variations. Equatorial coordinates of periodic variable stars we observed, together with derived periods, are given in Table 1. New variable stars are indicated with designations starting with letter 'n'.

The mean period of RRab stars in NGC 4833 is equal to $\langle P_{\rm ab} \rangle = 0.701$ d, and the relative percentage of RRc stars amounts to $N_{\rm c}/(N_{\rm ab} + N_{\rm c}) = 48$ %. With these values we find that NGC 4833 belongs to the Oosterhoff's II group of globular clusters.

Table 1. Equatorial coordinates and periods of periodic stars in NGC 4833.

Var	Type	α_{2000} [$^{\mathrm{h}\ \mathrm{m}\ \mathrm{s}}$]	δ_{2000} [$^{\circ}\ '\ ''$]	P [d]
v27	SXPhe	12 59 13.73	−70 52 09.5	0.0509813
v29	SXPhe	12 59 35.31	−70 52 40.9	0.059780
v30	SXPhe	12 59 42.56	−70 53 04.0	0.044256 0.045757
v31	SXPhe	12 59 47.98	−70 52 51.5	0.0533323 0.041726
v32	SXPhe	12 59 55.05	−70 52 24.2	0.0706795 0.072537
v33	SXPhe	12 59 57.64	−70 54 29.5	0.0719219 0.079772
n35	SXPhe	12 58 59.01	−70 52 05.1	0.037510 0.036708
n36	SXPhe	12 59 32.64	−70 53 15.9	0.040149
v01	RRLyr	12 58 36.53	−70 44 48.1	0.750082
v03	RRLyr	12 59 33.73	−70 52 13.2	0.74453
v04	RRLyr	12 59 33.76	−70 51 57.8	0.65577
v05	RRLyr	13 00 01.28	−70 53 17.3	0.629424
v06	RRLyr	12 59 56.18	−70 50 10.2	0.65400
v07	RRLyr	12 59 48.80	−70 52 21.1	0.66888
v12	RRLyr	12 59 37.91	−70 52 14.7	0.58980
v13	RRLyr	13 00 29.86	−70 52 55.9	0.36788
v14	RRLyr	12 59 31.04	−70 53 06.9	0.40842
v15	RRLyr	12 59 20.15	−70 53 25.0	0.66745
v17	RRLyr	12 59 43.98	−70 54 24.9	0.390263
v18	RRLyr	12 59 28.26	−70 54 25.5	0.42559
v19	RRLyr	12 59 06.02	−70 53 29.7	0.370658
v20	RRLyr	12 59 08.21	−70 52 23.5	0.2997 0.3012
v21	RRLyr	12 59 50.94	−70 50 36.5	0.39878
v22	RRLyr	12 59 45.10	−70 53 55.5	0.85095
v23	RRLyr	12 59 44.74	−70 51 27.0	0.406503
v24	RRLyr	12 59 36.68	−70 52 58.7	0.62612
v26	RRLyr	12 59 02.74	−70 52 51.9	0.31788
v28	RRLyr	12 59 21.22	−70 53 26.4	0.87401
n37	RRLyr	12 59 43.16	−70 53 36.5	0.30215
v25	Ecl	12 58 55.37	−70 51 45.9	0.72310
v34	Ecl	13 00 25.17	−70 49 16.4	0.36290
n38	Ecl	13 00 11.88	−70 51 26.4	0.257129
n39	Ecl	13 01 33.10	−70 49 22.6	0.28005
n40	Ecl	13 01 49.41	−70 51 42.1	0.282581
n41	Ecl	13 01 29.29	−70 44 58.0	0.29917
n42	Ecl	12 57 17.71	−70 47 51.0	0.3137
n43	Ecl	13 00 22.56	−70 55 25.8	0.3175
n44	Ecl	12 59 52.75	−70 44 32.7	0.3362
n45	Ecl	13 00 43.49	−70 47 26.9	0.362332
n46	Ecl	13 00 24.25	−70 56 01.4	0.38071
n47	Ecl	12 59 37.69	−70 51 28.4	0.48875
n48	Ecl	13 00 45.49	−70 47 38.2	0.49226
n49	Ecl	13 00 55.03	−70 54 15.9	1.1235
n59	Ukn	13 00 27.52	−70 54 01.6	2.889
n60	Ukn	12 59 14.62	−70 50 37.8	4.390
n61	Ukn	13 00 20.85	−70 49 56.4	6.300

Almost all observed SX Phoenicis stars show multiperiodic light changes (see Table 1) with one star, v31, exhibiting oscillation in two first radial modes. Moreover, we found in the RRc star v20 two closely-spaced frequencies.

Acknowledgement

This work was supported by the NCN grant 2011/03/B/ST9/02667.

References

Alard, C. & Lupton, R. 1998, *ApJ*, 503, 325

Clement, C., Muzzin, A., Dufton, Q., *et al.* 2001, *AJ*, 122, 2587

Darragh, A. & Murphy, B. 2012, *Journal of the Southeastern Association for Research in Astronomy*, 6, 72.

Kopacki, G. 2013, *AcA*, 63, 91

Precision Asteroseismology
Proceedings IAU Symposium No. 301, 2013
J. A. Guzik, W. J. Chaplin, G. Handler & A. Pigulski, eds.

© International Astronomical Union 2014
doi:10.1017/S1743921313014968

The rich frequency spectrum of the triple-mode variable AC And

Géza Kovács[1,2], Gáspár A. Bakos[3] and Joel D. Hartman[3]

[1]Dept. of Physics & Astrophysics, U. North Dakota, Grand Forks, ND

[2]Konkoly Obs., Budapest, Hungary

[3]Princeton University, Department of Astrophysical Sciences, Princeton

Abstract. Fourier analysis of the light curve of AC And from the HATNet database reveals the rich frequency structure of this object. Above 30 components are found down to the amplitude of 3 mmag. Several of these frequencies are not the linear combinations of the three basic components. We detect period increase in all three components that may lend support to the Population I classification of this variable.

Keywords. stars: variables, Cepheids, fundamental parameters

There are a handful of objects that seem to exhibit sustained pulsations in three radial modes (Wils *et al.* 2008). These are important objects with respect to the unique opportunity to derive their basic physical parameters (i.e., mass, luminosity and temperature) by using their periods only (assuming that the metallicity is known and that linear pulsation periods are close to the observed [nonlinear] ones – see Kovács & Buchler (1994) and Moskalik & Dziembowski (2005)).

By using the 9600 datapoint light curve gathered by HATNet†, we perform a Fourier frequency analysis on AC And, the prototype of radial three-mode pulsators. The analysis of the basic photometric data yields the three modes and their low-order linear combinations with high S/N. To reach the millimagnitude level, we need to minimize systematics. Because of the high amplitudes of the low-order components, we cannot use TFA (Kovács *et al.* 2005) in the frequency search mode, since it is based on the low S/N assumption. Therefore, we opted to use the method of Kovács & Bakos (2008). Here the most reliably determined signal components and the TFA template light curves are simultaneously fitted. After cleaning the data by the so-obtained TFA filter, we get clean, successively prewhitened spectra as shown on the left panel of Fig. 1. Successive prewhitening leads to the identification of many components, most of which are clearly the linear combinations of the 3 basic frequencies. However, both the frequency spectra (top right panel of Fig. 1) and the decomposed individual modes show the presence of non-fitting components even when considering linear combinations of the main frequencies up to order 10. This feature has also been noted in the discovery paper of Fitch & Szeidl (1976).

We also examined the rate of period change during the past 55 years. Knowledge of the period change is important, because if it is due to stellar evolution, it may help to select the right model if other parameters are ambiguous. From the analysis of Guman (1981, 1982) it seems that all three modes exhibit period increases much higher than expected

---

† The Hungarian-made Automated Telescope Network (Bakos *et al.* 2004) consists of 6 wide field-of-view, small aperture autonomous telescopes located in Hawaii (Mauna Kea) and Arizona (Fred Lawrence Whipple Observatory). The prime purpose of the project is to search for extrasolar planets via transit technique.

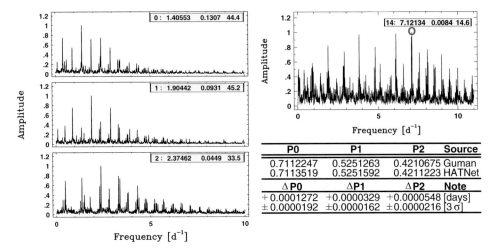

Figure 1. *Left:* Successively prewhitened frequency spectra of AC And; insets: prewhitening order; peak frequency [d^{-1}], amplitude [mag], S/N. *Top right:* Frequency spectrum of AC And after the 14th prewhitening. The circled peak corresponds to one of the handful of components that cannot be identified as the linear combination of the three main frequencies. *Bottom right:* Periods detected 55 years apart (upper two rows) and their differences together with the corresponding errors (lower two rows).

from an RR Lyrae star (Kovács & Buchler 1994). As seen in the table section of Fig. 1, the periods derived from the HATNet data confirm the period increase during the past 55 years. Note however that the speed of increase derived here is $\sim$4 times higher than given by Jurcsik *et al.* (2006) based on earlier data.

Except perhaps for the Strömgren photometry of Peña *et al.* (2005), AC And lacks good multicolor time series observations and deep spectroscopic work. As a result, the physical parameters of the star are still poorly known. Opposite to Peña *et al.* (2005), we think that the available photometric and pulsation data support that AC And is a relatively high ($\sim$3 $M_\odot$) mass star with $T_{\rm eff} \approx 5800 \pm 200$ K, $\log g \approx 2.5 \pm 0.5$ and [Fe/H] $\approx -0.5 \pm 0.5$.

Acknowledgement: G. K. thanks the Hungarian Scientific Research Foundation (OTKA) for support through grant K-81373.

References

Bakos, G. A., Noyes, R. W., Kovács, G., Stanek, K. Z., Sasselov, D. D., & Domsa, I. 2004, *PASP*, 116, 266
Fitch, W. S. & Szeidl, B. 1976, *ApJ*, 203, 616
Guman, I. 1981, *IBVS*, 2046
Guman, I. 1982, *Comm. Konkoly Obs.*, 78, 1
Jurcsik, J., Szeidl, B., Váradi, M., *et al.* 2006, *A&A*, 445, 617
Kovács, G. & Bakos, G. A. 2008, *CoAst*, 157, 82
Kovács, G. & Buchler, J. R. 1994, *A&A*, 281, 749
Kovács, G., Bakos, G. A., & Noyes, R. W. 2005, *MNRAS*, 356, 557
Moskalik, P. & Dziembowski, W. A. 2005, *A&A*, 434, 1077
Peña, J. H., Peniche, R., Hobart, M. A., de La Cruz, C., & Gallegos, A. A. 2005, *Rev. Mexicana AyA*, 41, 461
Wils, P., Rozakis, I., Kleidis, S., Hambsch, F.-J., & Bernhard, K. 2008, *A&A*, 478, 865

Precision Asteroseismology
Proceedings IAU Symposium No. 301, 2013
J. A. Guzik, W. J. Chaplin, G. Handler & A. Pigulski, eds.

Photometry using *Kepler* "superstamps" of open clusters NGC 6791 & NGC 6819

Charles A. Kuehn III, Jason Drury, Dennis Stello, and Timothy R. Bedding

Sydney Institute for Astronomy, School of Physics, The University of Sydney, NSW 2006,
Australia
email: kuehn@physics.usyd.edu.au

Abstract. The *Kepler* space telescope has proven to be a gold mine for the study of variable stars. Unfortunately, *Kepler* only returns a handful of pixels surrounding each star on the target list, which omits a large number of stars in the *Kepler* field. For the open clusters NGC 6791 and NGC 6819, *Kepler* also reads out larger superstamps which contain complete images of the central region of each cluster. These cluster images can potentially be used to study additional stars in the open clusters. We present preliminary results from using traditional photometric techniques to identify and analyze additional variable stars from these images.

The high photometric precision and the virtually uninterrupted observing cadence of the *Kepler* space telescope has made it a revolutionary tool for the study of stellar variability. In order to conserve bandwidth, the spacecraft downlinks a small postage stamp of pixels around each target star, thus photometric data is obtained for only a portion of the stars in the *Kepler* field.

Open clusters are ideal targets for asteroseismic studies since all stars in a cluster are thought to have the same age and composition, allowing us to better constrain the stellar models that are compared against the observed oscillation frequencies. Four open clusters spanning a range of ages and metallicities are located in the *Kepler* field of view (NGC 6791, NGC 6811, NGC 6819, and NGC 6866). Target stars were selected and observed in each of the open clusters but the majority of the cluster stars were not targeted. Results from these observations can be found in Corsaro *et al.* (2012) and references therein.

In addition to the postage stamps that were used for the work cited above, large 200×200 pixel (13.3 arcminutes on a side) superstamps of the clusters NGC 6791 and NGC 6819 were also obtained by *Kepler* in long cadence mode (Fig. 1). These superstamps cover the central regions of the two clusters, providing an opportunity to obtain photometric information on the non-target stars in these regions. The goal of this project is to use traditional photometric techniques to obtain light curves for all of the resolved stars in the superstamps.

The coarse pixel scale of *Kepler* (3.98 arcseconds/pixel) and the crowded nature of the clusters results in many of the stars being blended. The degree of blending prevents the use of aperture photometry on most of the cluster stars, however profile-fitting photometry is ideally suited for dealing with stars that are partially blended. Stars were identified in each image and photometry was obtained using Peter Stetson's Daophot/Allstar routines (Stetson 1994). The profiles used for the profile fitting were derived from the actual PSFs of bright, relatively uncrowded stars. Figure 2 shows the light curves for two stars using the data from quarter 1 of *Kepler* operation (12 May 2009 – 14 June 2009). While the variablity is easily evident in the light curves, there is still a large amount of noise and potential systematics that need to be removed. Much of the noise is probably from variations in the inter-pixel sensitivities. Correcting for this requires varying the PSFs

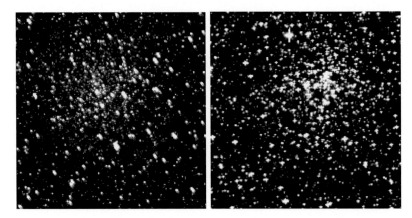

Figure 1. 200×200 pixel superstamp of NGC 6791 (left) and NGC 6819 (right). Superstamp images created using a routine by Ron Gilliland (private communication).

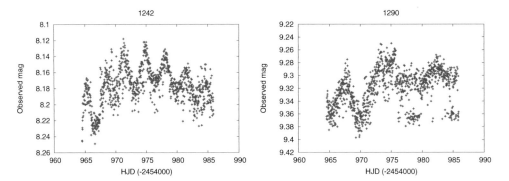

Figure 2. Sample quarter 1 light curves, obtained with profile-fitting photometry, of stars in NGC 6791. Magnitudes are given in the instrumental system and have not been converted to the standard system.

based on the location of the star on the CCD; we are currently working on implementing a fitting routine that can do this.

Image subtraction was also run on the superstamps in order to obtain photometry for stars that were too highly blended for profile fitting. Wojtek Pych's DIAPL2 package† was used to perform the image subtraction and obtain light curves. As with the profile-fitting, light curves obtained from the image subtraction method show a good deal of noise as well as systematic effects. We are currently in the process of refining the parameters of the image subtraction process to decrease the noise and systematic effects.

The preliminary results show that it is possible to obtain light curves from the *Kepler* superstamps of the open clusters NGC 6791 and NGC 6819, allowing the study of the variability of stars not included in the *Kepler* target list.

References

Corsaro, E., Stello, D., Huber, D., *et al.* 2012, *ApJ*, 757, 190
Stetson, P. B. 1994, *PASP*, 106, 250
Woźniak, P. R. 2000, *AcA*, 50, 421

† http://users.camk.edu.pl/pych/DIAPL/. DIAPL2 is an improved version of the DIA package (Woźniak 2000).

Precision Asteroseismology
Proceedings IAU Symposium No. 301, 2013
J. A. Guzik, W. J. Chaplin, G. Handler & A. Pigulski, eds.

© International Astronomical Union 2014
doi:10.1017/S1743921313014981

Modeling non-radial oscillations on components of close binaries

Olivera Latković and Attila Cséki

Astronomical Observatory, 11060 Belgrade, Serbia
email: olivia@aob.rs

Abstract. We developed an advanced binary system model that includes stellar oscillations on one or both stars, with the goal of mode identification by fitting of the photometric light curves. The oscillations are modeled as perturbations of the local surface temperature and the local gravitational potential. In the case of tidally distorted stars, it is assumed that the pulsation axis coincides with the direction connecting the centers of the components rather than with the rotation axis. The mode identification method, originally devised by B. Bíró, is similar to eclipse mapping in that it utilizes the amplitude, phase and frequency modulation of oscillations during the eclipse; but the identification is achieved by grid-fitting of the observed light curve rather than by image reconstruction. The proposed model and the mode identification method have so far been tested on synthetic data with encouraging results.

Keywords. asteroseismology, methods: data analysis, methods: numerical, techniques: photometric, binaries: close, binaries: eclipsing, stars: oscillations

---

The binary system modeling program, *Infinity*, was first presented on the IAU Symposium 282, "From Interacting Binaries to Exoplanets: Essential Modeling Tools" (Cséki & Latković 2012) as a work in progress. In the two years since the initial announcement, it has evolved into an elaborate tool for modeling a wide variety of binary stars: from well detached and eccentric, to close, interacting systems within the Roche formalism. The features that differentiate *Infinity* from other programs for modeling binary stars include:

- A novel approach to visibility detection (determining which elementary surfaces of the components are visible to the observer at a given time), using the "inverse painter's algorithm" that allows the inclusion of any number of additional components (like circumstellar and circumbinary disks, planets, or other stars and binaries in hierarchical systems) in the model;
- A novel approach to modeling the eclipse, using the "adaptive subdivision" of elementary surfaces that allows for arbitrary precision even when the eclipsing body is relatively small (like a planet);
- The capability of modeling conical and toroidal accretion disks;
- And finally, the feature most interesting in the context of asteroseismology: the capability of modeling non-radial oscillations on one or both components of the binary.

The oscillations are simulated as periodic perturbations in local effective temperature and/or local effective gravitational potential, where frequencies, amplitudes and phases, as well as the quantum numbers l and m of the associated spherical harmonic, are adjustable model parameters. The symmetry axis of the oscillations can also be specified in terms of its inclination to the rotational axis, which is presumed to be orthogonal to the orbital plane. It is therefore possible to use *Infinity* for modeling oblique pulsators. It is also possible to model oscillations on stars significantly distorted by tidal forces in close binary systems, with the pulsation axis aligned with the direction through the centres of the components.

The primary purpose of *Infinity* as a tool for modeling stellar oscillations is mode identification. Namely, the eclipse can be used as a spatial filter that samples the surface of the oscillating component and produces modulations in frequencies, amplitudes and phases of the oscillations. The "signature" of the modulation caused by the eclipse is unique for every mode, and can therefore be used to identify the mode. This is the principle that underpins the techniques of eclipse mapping (Bíró & Nuspl 2011) and direct light curve fitting (Latković & Bíró 2008).

In *Infinity*, the modes are identified by comparing the residual light curve with a grid of models that have different spherical harmonics representing each mode. The residual light curve is obtained by subtracting the orbital solution (the binary light curve) from the observations. Given the observed light curves in one or more passbands and the radial velocity curves, *Infinity* can be used to find the full spectrophotometric solution for a binary system by simultaneous fitting. After that, the residual light curve can be analyzed in order to measure the frequencies and amplitudes of oscillations. These are then used to create a grid of models in which all the binary and oscillation parameters other than the spherical harmonic quantum numbers, l and m, are fixed according to the results from previous steps. The identification is achieved by selecting the model which gives the best fit to the observations from the grid.

This procedure has been tested on simulated data with good results. The simulations were done for several representative system configurations (detached and close systems with total and partial eclipses) for all single modes with the degree $l \leqslant 4$, and for superpositions of up to three modes. In detached systems, where the pulsation axis coincides with the rotation axis, all the identifications were successful. In close systems, where the pulsation axis lies in the orbital plane as described above, the identifications were successful for single modes, but ambiguous for superpositions. This has to do with the fact that with an inclined pulsation axis, a single mode is detected as a multiplet with up to $2l + 1$ members, making for highly complex signatures in the light curves, as discussed in detail by Reed *et al.* (2005).

At the time of writing, we are making headway towards the first applications of mode identification on data from satellite observatories. The best candidates are short-period, but well-detached eclipsing binaries in which one of the components exhibits δ Sct or γ Dor oscillations.

References

Bíró, I. B. & Nuspl, J. 2011, *MNRAS*, 416, 1601

Cséki, A. & Latković, O. 2012, in: M. T. Richards & I. Hubeny (eds.), Proc. IAU Symposium 282: *From Interacting Binaries to Exoplanets: Essential Modeling Tools* (Cambridge: Cambridge University Press), p. 305

Latković, O. & Bíró, I. B. 2008, *CoAst*, 157, 330

Reed, M. D., Brondel, B. J., & Kawaler, S. D. 2005, *ApJ*, 634, 602

Precision Asteroseismology
Proceedings IAU Symposium No. 301, 2013
J. A. Guzik, W. J. Chaplin, G. Handler & A. Pigulski, eds.

© International Astronomical Union 2014
doi:10.1017/S1743921313014993

Nonlinear hydrodynamic simulations of radial pulsations in massive stars

Catherine Lovekin[1,2] and Joyce A. Guzik[2]

[1]Department of Physics, Mount Allison University
67 York St., Sackville, NB E4L 1E6 Canada
email: `clovekin@mta.ca`

[2]Los Alamos National Laboratory
XTD-NTA, MS T086, Los Alamos, NM 87545 USA
email: `joy@lanl.gov`

Abstract. We investigate the radial pulsation properties of massive main-sequence stars using both linear and non-linear calculations. Using 20, 40, 60 and 85 solar-mass models evolved by Meynet *et al.* (1994), we calculate nonlinear hydrodynamic envelope models including the effects of time-dependent convection. Many of these models are massive enough to lose a significant amount of mass as they evolve, which also reveals more helium-rich layers. This allows us to investigate the dependence of pulsation on mass, metallicity and surface helium abundance. We find that as a model loses mass, the periods become longer relative to the period predicted by the period-mean density relation (period $\times \sqrt{\bar{\rho}}$ is proportional to a constant, Q) for the initial model. Increased surface helium abundance causes a dramatic decrease in the period relative to that expected from Q, while changing the metallicity had little impact on the expected periods.

Keywords. stars: oscillations

1. Introduction

Many stars in the upper left quadrant of the H-R diagram are known to be pulsating (see, for example, Pamyatnykh 1999, Moffat 2010, Blomme *et al.* 2012, McNamara *et al.* 2012). In many cases, the evolution of these massive stars is not well understood, and pulsation characteristics can be used to constrain the evolutionary sequence. For this reason, we investigate radial pulsation in stars from 20 to 85 $M_\odot$, and look at how the pulsation periods and amplitudes are affected by changes in mass, metallicity and surface helium abundance. We use parameters taken from a grid of stellar models published by Meynet *et al.* (1994) to calculate linear, non-adiabatic (LNA) pulsations in envelope models. The zoning in these models was arranged such that the driving region near 2×10^5 K was sufficiently resolved, and the zones extended to depths at or below 2×10^6 K in order to capture the damping region. The LNA calculations are then used to initialize the linear hydrodynamic pulsation code DYNSTAR. The hydrodynamic code produces a radial velocity curve, from which growth rates and dominant pulsation periods are extracted. Details of the codes are discussed in more detail in Guzik & Lovekin (2012).

2. Results

We find that our instability regions are in agreement with the β Cep instability strip as found by Pamyatnykh (1999), and our hydrodynamic model amplitudes become very large for models at or beyond the Humphreys-Davidson limit (Humphreys & Davidson 1994).

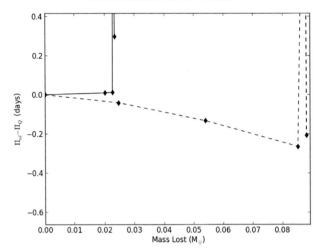

Figure 1. The difference between the nonlinear period and the Q prediction as a function of mass lost in $40\,M_\odot$ models with metallicity $Z = 0.004$ (solid) and $Z = 0.008$ (dashed). The sudden increase in the difference corresponds to the end of the main sequence in each model.

Our goal has been to investigate the changes in pulsation frequencies due to the effects of mass loss and changes in surface helium abundance as the models evolve. To do this, we use the Q parameter $(= \Pi\sqrt{\bar{\rho}})$, which is approximately constant for models with similar structure. The periods of our models range from < 0.5 d for the main-sequence $20\,M_\odot$ models, to approaching 80 d for the most massive evolved models. We find that, for our $20\,M_\odot$ models, Q can be used to predict the nonlinear pulsation frequencies quite well until the radius approximately doubles. We then applied the Q parameter to the $40\,M_\odot$ models, where the mass-loss rates are high enough to change the mass, but where the surface helium abundance stays nearly constant (Fig. 1). The spikes that appear around 0.02 and $0.09\,M_\odot$ lost are the end of the main sequence, where the change in the structure of the model is large enough that Q is no longer expected to be constant. In the most massive models, both mass loss and changes in surface helium abundance are significant. We find that an increase in the surface helium abundance decreases the periods relative to the predictions using Q. We also find that the magnitude of the decrease is much larger than the change produced by changes in metallicity or mass loss. We found that the metallicity had little effect on the periods.

We also investigated the amplitude of the pulsations in the hydrodynamic models, looking for trends with age, mass loss, surface helium abundance, and metallicity. We found that as the models lose mass, the amplitude of the pulsation increases, while increasing the surface helium abundance decreases the amplitude. The $20\,M_\odot$ models that lose almost no mass show a slight trend of decreasing amplitude as the model ages. We did not find a clear trend with metallicity.

References

Blomme, R., Briquet, M., Degroote, P., *et al.* 2012, *ASP-CS*, 465, 13

Guzik, J. A. & Lovekin, C. C. 2012, *Astron. Review*, 7, 13

Humphreys, R. M. & Davidson, K. 1994, *PASP*, 106, 1025

McNamara, B. J., Jackiewicz, J., & McKeever, J. 2012, *AJ*, 143, 101

Meynet, G., Maeder, A., Schaller, G., Schaerer, D., & Charbonnel, C. 1994, *A&AS*, 103, 97

Moffat, A. F. J. 2010, *Highlights of Astronomy*, 15, 366

Pamyatnykh, A. A. 1999, *AcA*, 49, 119

Precision Asteroseismology
Proceedings IAU Symposium No. 301, 2013
J. A. Guzik, W. J. Chaplin, G. Handler & A. Pigulski, eds.

© International Astronomical Union 2014
doi:10.1017/S1743921313015007

The Large Magellanic Cloud Cepheids: effects of helium content variations

Marcella Marconi[1], Roberta Carini[2], Enzo Brocato[2] and Gabriella Raimondo[3]

[1] INAF - Osservatorio Astronomico di Capodimonte, Via Moiariello 16, 80131 Napoli, Italy
email: marcella.marconi@oacn.inaf.it

[2] INAF - Osservatorio Astronomico di Roma, via Frascati 33, 00040 Monte Porzio Catone, Italy
email: carini@oa-roma.inaf.it, brocato@oa-roma.inaf.it

[3] INAF - Osservatorio Astronomico di Teramo, Mentore Maggini s.n.c., 64100 Teramo, Italy
email: raimondo@oa-teramo.inaf.it

Abstract. An extended set of evolutionary and pulsational models has been computed for two chemical compositions representative of classical Cepheids in the Large Magellanic Cloud. The comparison between the standard and He enhanced theoretical predictions is analysed and the implications for interpreting current observations and for defining the Cepheid based distance scale are discussed.

Keywords. stars: variables: Cepheids, stars: abundances, stars: distances

1. Nonlinear convective pulsation models for LMC Cepheids: the He abundance effect

The adopted pulsation models are nonlinear and include a nonlocal time-dependent treatment of convection (see e.g. Bono *et al.* 1999 and Marconi *et al.* 2005). A metallicity typical of LMC Cepheids, $Z = 0.008$ is assumed and two He contents, $Y = 0.25, 0.35$ are considered (Carini *et al.*, in preparation).

Figure 1 shows the predicted instability strip at $Z = 0.008$ as a function of the adopted He abundance. Solid and dashed lines are the fundamental strip edges, whereas dotted and long-dashed lines depict the first overtone ones for the labelled Y values. We notice that the instability strip gets hotter as the helium content increases at fixed Z, in agreement with previous results at solar metallicity (see e.g. Marconi *et al.* 2005). This effect reflects on the coefficients of the predicted optical Period-Luminosity (PL) relations. Indeed they are expected to get steeper as Y increases from $Y = 0.25$ to $Y = 0.35$.

2. Synthetic stellar populations including Cepheids

By combining evolutionary and pulsational models (Carini *et al.*, in preparation) we were able to produce synthetic stellar populations including Cepheids (the stars falling within the above shown instability strips) at both He contents and for various age assumptions. As an example we show in Fig. 2 the resulting populations for an age of 100 Myr. Synthetic stars falling within the predicted instability strip are represented by larger triangles. By applying the model based pulsation relation that connects the period to the mass, luminosity and effective temperature of synthetic stars falling within the instability strip, we are able to produce a period distribution for each assumption of the He content and age. Moreover, assuming a mixed population with a fraction of stars at $Y = 0.25$ and the remaining part at $Y = 0.35$, with a given age gap between the two subpopulations, we are able to predict the impact of the possible presence of

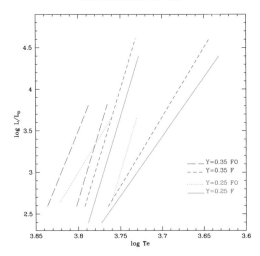

Figure 1. The predicted instability strip as a function of the adopted Y (see labels).

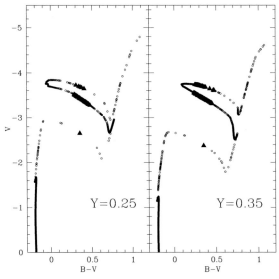

Figure 2. Synthetic stellar populations including Cepheids for an age of 100 Myr and the two labelled He contents.

multipopulations in investigated stellar systems on the observed period histogram (see Carini *et al.* for details). Multiple stellar populations could be in principle observed in very populous young and intermediate-age massive clusters and star forming regions, provided that spatial resolution and distance of the objects allows reliable photometric measurements of individual stars. Current model predictions suggest that the Cepheid-based distance scale could be affected by bias due to the unexpected presence of Cepheids with huge differences in their original He content and belonging to the same stellar system for which the distance is measured.

References

Bono, G., Marconi, M., & Stellingwerf, R. F. 1999, *ApJS*, 122, 167
Marconi, M., Musella, I., & Fiorentino, G. 2005, *ApJ*, 632, 590

Precision Asteroseismology
Proceedings IAU Symposium No. 301, 2013
J. A. Guzik, W. J. Chaplin, G. Handler & A. Pigulski, eds.

© International Astronomical Union 2014
doi:10.1017/S1743921313015019

Variable stars
in the young open cluster NGC 2244

Gabriela Michalska

Instytut Astronomiczny, Uniwersytet Wrocławski,
Kopernika 11, 51-622 Wrocław, Poland
email: michalska@astro.uni.wroc.pl

Abstract. We present results of a search for variable stars in the young open cluster NGC 2244. As a result we have found many eclipsing systems and pulsating stars, some of which are multiperiodic. Here we show only a few examples.

Keywords. galaxies: clusters: individual (NGC 2244), stars: pre–main-sequence

1. Introduction

NGC 2244 is a young open cluster associated with the Rosette Nebula and located in the Perseus Arm of the Galaxy. Its age is estimated with 2–6 Myr, the distance with $1.4 - 1.7$ kpc and the reddening with $E(B - V) = 0.47$ mag. The cluster is embedded in a H II region and is rich in OB stars. A photometric $UBVI$ and Hα study was performed by Park & Sung (2002). They found about 30 OB-type cluster members. They also discovered about 20 pre-main sequence (PMS) stars with Hα emission (four of them are massive Herbig BeAe stars) and six stars in NGC 2244 with X-ray emission. A large population of pre-main sequence stars was also found by Bonatto & Bica (2009). The cluster contains the double-lined eclipsing binary star V578 Mon (Hensberge *et al.* 2000) and several spectroscopic binaries. It contains also an Ap star (NGC 2244-334) having a strong magnetic field (Bagnulo *et al.* 2004).

2. Observations and reductions

The observations of NGC 2244 were carried out with the 1-m telescope at Cerro Tololo Inter-American Observatory (CTIO) in Chile. This telescope is equipped with a 4064 × 4064 CCD camera covering an area of about $20' \times 20'$ on the sky. Between December 24, 2009 and January 8, 2008 we collected about 2000 frames in the V filter and 170 frames in the I_C filter, 150 in the B filter and 70 in the U filter.

3. Analysis and results

Using profile and aperture photometry obtained with the DAOPHOT package, differential magnitudes of all detected stars were computed. The V-filter differential magnitudes were used in the search for variability. For each star, the light curve, the Fourier periodogram in the range between 0 and 60 d^{-1} and the phase diagram were inspected by eye. As a result we have found many variable stars. The Fourier periodogram of three PMS variables are shown in top panels of Fig. 1. One of them, V1, shows δ Scuti-type variations. The light curves of two other PMS variable stars, V4 and V5, and an eclipsing binary, V6, are shown in bottom panels of Fig. 1. The color-magnitude diagram and color-color diagram of the observed stars are shown in Fig. 2.

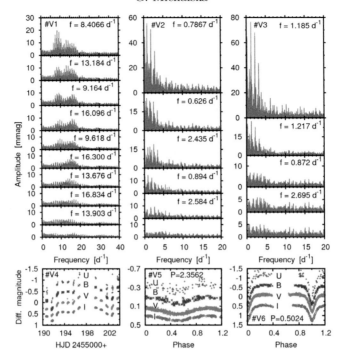

Figure 1. Fourier frequency spectra of V-filter data of three PMS pulsating stars (V1, V2 and V3) and light curves of two PMS variables, V4 and V5, and one eclipsing system, V6.

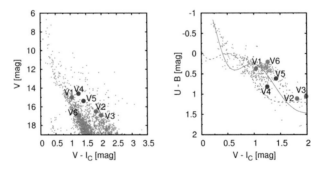

Figure 2. *Left*: The V vs. $(V - I_C)$ color-magnitude diagram for NGC 2244. *Right*: The $(U - B)$ vs. $(V - I_C)$ color-color diagram of NGC 2244. The dashed line shows the intrinsic color-color relation for main-sequence stars as given by Caldwell *et al.* (1993). The same relation for reddened stars with $E(B - V) = 0.47$ mag taken from Park & Sung (2002).

Acknowledgement

This work was supported by the NCN grant 2011/03/B/ST9/02667.

References

Bagnulo, S., Hensberge, H., Landstreet, J. D., Szeifert, T., & Wade, G. A. 2004, *A&A* 416, 1149
Bonatto, C. & Bica, E. 2009, *MNRAS*, 394, 2127
Caldwell, J. A. R., Cousins, A. W. J., Ahlers, C. C., van Wamelen, P., & Maritz, E. J. 1993, *South African Astronomical Observatory Circular*, 15, 1
Hensberge, H., Pavlovski, K. & Verschueren, W. 2000, *A&A*, 358, 553
Park, B.-G. & Sung, H. 2002, *AJ*, 123, 892

Precision Asteroseismology
Proceedings IAU Symposium No. 301, 2013
J. A. Guzik, W. J. Chaplin, G. Handler & A. Pigulski, eds.

© International Astronomical Union 2014
doi:10.1017/S1743921313015020

Asteroseismology of fast-rotating stars: the example of α Ophiuchi

Giovanni M. Mirouh[1,2], **Daniel R. Reese**[3], **Francisco Espinosa Lara**[1,2], **Jérôme Ballot**[1,2] **and Michel Rieutord**[1,2]

[1] Université de Toulouse, UPS-OMP, IRAP, Toulouse, France

[2] CNRS, IRAP, 14 avenue Edouard Belin, 31400 Toulouse, France

[3] Institut d'Astrophysique et Géophysique de l'Université de Liège, Allée du 6 Août 17, 4000 Liège, Belgium

Abstract. Many early-type stars have been measured with high angular velocities. In such stars, mode identification is difficult as the effects of fast and differential rotation are not well known. Using fundamental parameters measured by interferometry, the ESTER structure code and the TOP oscillation code, we investigate the oscillation spectrum of α Ophiuchi, for which observations by the *MOST* satellite found 57 oscillations frequencies. Results do not show a clear identification of the modes and highlight the difficulties of asteroseismology for such stars with a very complex oscillation spectrum.

Keywords. stars: oscillations, stars: rotation, stars: individual: α Oph, Rasalhague

1. Introduction

Intermediate- and high-mass stars are usually fast rotators. In some of these A-, B- and O-type stars, the κ mechanism excites eigenmodes. Because of the centrifugal flattening of the star and the strength of the Coriolis acceleration, the oscillation spectrum is much more complex than that of non-rotating stars. Interpretation of observed frequencies of these stars requires two-dimensional models and their oscillation spectrum computed in a non-perturbative way. Following the pioneering work of Deupree *et al.* (2012), we want to identify the observed modes of the fast-rotating star α Ophiuchi in order to better constrain its fundamental parameters. Compared to the previous work of Deupree, we improve the numerical resolution and get spectrally converged eigenfunctions that allow a reliable computation of visibilities and damping rates of the modes.

2. Models

We compute 2D models for the fast-rotating A-type star Rasalhague (α Ophiuchi) with the ESTER code (Rieutord & Espinosa Lara 2013), in which differential rotation is calculated self-consistently. Its surface equatorial rotation velocity of $240\,\mathrm{km\,s^{-1}}$ imposes a flatness of 0.168. Its equatorial and polar radii have been derived from interferometry (Zhao *et al.* 2009) and the stellar mass is constrained by an orbiting companion (Hinkley *et al.* 2011). Besides this, 57 oscillation frequencies have been measured by photometry (Monnier *et al.* 2010). To reproduce the fitted radii, luminosity, polar and equatorial temperatures, our model uses $M = 2.22\,M_\odot$, $\Omega = 0.62\Omega_{\mathrm{K}} = 1.65\,\mathrm{d^{-1}}$, mass fractions $Z = 0.02$ and $X = 0.7$ in the envelope, $X_{\mathrm{c}} = 0.26$ in the core (see Espinosa Lara & Rieutord 2013). The eigenvalue problem of adiabatic oscillations is solved with the TOP code (Reese *et al.* 2009) for modes with azimuthal orders $-4 \leqslant m \leqslant 4$, in the range of frequencies in which the p modes are thought to be observed (see Fig. 1).

Figure 1. From left to right: Rotation profile in α Oph (fast core, slow envelope), an $m = 0$ acoustic island-mode, an $m = -1$ g-mode.

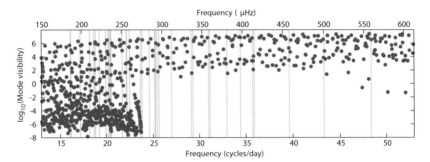

Figure 2. Visibilities for axisymmetric ($m = 0$) modes. The vertical lines indicate the frequencies measured with *MOST* (Monnier *et al.* 2010).

3. Comparisons between computed and observed modes

To select the modes that might be seen from Earth, we compute the mode visibilities, following Reese *et al.* (2013) and the thermal dissipation rates using the quasi-adiabatic approximation (Unno *et al.* 1989). In the domain that we investigated, our model yields only linearly stable modes. These are g modes and p modes modified by rotation. The g modes are the least-damped with large amplitude at the base of the envelope. Their visibility is much less than that of the p modes that exist in this interval, as their amplitude is evanescent at the surface. However, p modes are much more damped. As shown in Fig. 2, where we clearly distinguish the set of g modes on the low-frequency side, each observed frequency corresponds to several eigenmodes of the model. The damping rates do not seem to be able to lift this degeneracy of the matching. It may well be that our models are at the moment too simple, notably because of the chemical homogeneity of the envelope. Progress therefore calls for more realistic models that include the distribution of chemical elements resulting from time evolution.

References

Deupree, R. G., Castañeda, D., Peña, F., & Short, C. I. 2012, *ApJ*, 753, 20
Espinosa Lara, F. & Rieutord, M. 2013, *A&A*, 552, A35
Hinkley, S., Monnier, J. D., Oppenheimer, B. R., *et al.* 2011, *ApJ*, 726, 104
Monnier, J. D., Townsend, R. H. D., Che, X., *et al.* 2010, *ApJ*, 725, 1192
Reese, D. R., MacGregor, K. B., Jackson, S., *et al.* 2009, *A&A*, 506, A189
Reese, D. R., Prat, V., Barban, C., *et al.* 2013, *A&A*, 550, A77
Rieutord, M. & Espinosa Lara, F. 2013, in: *Lecture Notes in Physics*, vol. 865, p. 49
Unno, W., Osaki, Y., Ando, H., Saio, H., & Shibahashi, H. 1989, *Nonradial oscillations of stars* (Univ. of Tokyo Press: Tokyo)
Zhao, M., Monnier, J. D., Pedretti, E., *et al.* 2009, *ApJ*, 701, 209

Precision Asteroseismology
Proceedings IAU Symposium No. 301, 2013
J. A. Guzik, W. J. Chaplin, G. Handler & A. Pigulski, eds.

© International Astronomical Union 2014
doi:10.1017/S1743921313015032

LAMOST observations in the *Kepler* field

Joanna Molenda-Żakowicz[1], Peter De Cat[2], Jian-Ning Fu[3], Xiao-Hu Yang[3] and the LAMOST-*Kepler* collaboration

[1]Instytut Astronomiczny, Uniwersytet Wrocławski,
ul. Kopernika 11, 51-622 Wrocław, Poland
email: molenda@astro.uni.wroc.pl

[2]Royal Observatory of Belgium, Ringlaan 3, 1180 Brussels, Belgium
email: Peter.DeCat@oma.be

[3]Department of Astronomy, Beijing Normal University
19 Avenue Xinjiekouwai, Beijing 100875, China
email: jnfu@bnu.edu.cn, yaohoo@mail.bnu.edu.cn

Abstract. We present results of observations of 22 664 stars in the *Kepler* field of view acquired with the Large Sky Area Multi-Object Fiber Spectroscopic Telescope (LAMOST) in the years 2011–2012, and provide a database of the atmospheric parameters derived from those data.

Keywords. stars: fundamental parameters, spectrographs: LAMOST, space vehicles: *Kepler*

1. Introduction

Since its launch on 7 March 2009, the NASA space telescope *Kepler* has been providing quasi-uninterrupted photometric time series of micro-magnitude precision. Because those data have been acquired in one filter (Kp) of a wide bandpass (400–850 nm), they do not contain information about the atmospheric parameters of the stars or the interstellar extinction. The Kepler Input Catalogue (KIC, Latham *et al.* 2005) which provides the values of the effective temperature ($T_{\rm eff}$), surface gravity ($\log g$), and metallicity ([Fe/H]) of stars in the *Kepler* field-of-view (FoV) is useful only for the solar-type stars; for stars which are hotter, cooler, or chemically peculiar, the precision of the KIC drops significantly (Brown *et al.* 2011). Moreover, for many stars in the KIC the atmospheric parameters are lacking. Therefore, the ground-based follow-up observations are a key element for ensuring a complete and exhaustive exploitation of the *Kepler* data.

2. Observations with LAMOST

LAMOST (Large Sky Area Multi-Object Fiber Spectroscopic Telescope) is a new instrument with 4000 optical fibres attached to a 4-m telescope at the Xinglong station of National Astronomical Observatories of China (Luo *et al.* 2012). We applied for the observing time on that facility in order to observe 14 subfields covering practically the entire *Kepler* FoV. Our aim was to derive atmospheric parameters of the programme stars in an efficient and homogeneous way.

Our list of targets consisted of around 250 MK standard stars, 7 000 stars requested for observations by the Kepler Asteroseismic Science Consortium (KASC), 150 000 stars from the list of the planet-search group (Batalha *et al.* 2010), and 1 000 000 other targets in the *Kepler* FoV.

During the first two years of realization of our LAMOST-Kepler project (2011–2012), we observed four subfields (see Fig. 1) and acquired 22 664 useful spectra which correspond to 21 112 different targets. The full list of stars which have been observed will

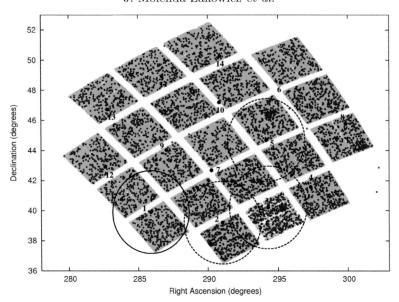

Figure 1. Distribution of our programme stars in the *Kepler* FoV. The centres of the 14 sub-fields selected for observations with LAMOST are indicated with dots labelled with numbers. The KASC targets are indicated with small black dots. The remaining targets which fill the *Kepler* FoV almost uniformly are indicated with grey dots. The sub-fields which have been observed in 2011 and 2012 are indicated with circles plotted with solid and dashed lines, respectively.

be made available by De Cat *et al.* (submitted). The spectra of 6420 stars which were observed in 2012 have been already analysed by means of the ROTFIT code (Frasca *et al.* 2003, 2006, Molenda-Żakowicz *et al.* 2013) and the IRAF software. The resulting values of T_{eff}, $\log g$, [Fe/H], and the radial velocity of these stars will be published elsewhere.

We note that the LAMOST spectra, which include, e.g., the lithium line at 6700 Å and the Ca II H&K lines at 3968.5 and 3933.7 Å, can be used also for other types of the scientific research like detection of stars with unusual abundance of Li or a search for stars which show chromospheric activity.

Acknowledgements

Guoshoujing Telescope (the Large Sky Area Multi-Object Fibre Spectroscopic Telescope LAMOST) is a National Major Scientific Project built by the Chinese Academy of Sciences. Funding for the project has been provided by the National Development and Reform Commission. LAMOST is operated and managed by the National Astronomical Observatories, Chinese Academy of Sciences. J.M-Ż acknowledges the MNiSW grant N N203 405139.

References

Batalha, N. M., Borucki, W. J., Koch, D. G., *et al.* 2010, *ApJ*, 713, L109
Brown, T. M., Latham, D. W., Everett, M. E., & Esquerdo, G. A. 2011, *AJ*, 142, 112
Frasca, A., Alcalá, J. M., Covino, E., *et al.* 2003, *A&A*, 405, 149.
Frasca, A., Guillout, P., Marilli, E., *et al.* 2006, *A&A*, 454, 301
Latham D. W., Brown, T. M., Monet, D. G., *et al.* 2005, *BAAS*, 37, 1340
Luo A.-L., Zhang, H.-T., Zhao, Y.-H., *et al.* 2012, *Research in Astronomy & Astroph.*, 12, 1243
Molenda-Żakowicz, J., Sousa, S. G., Frasca, A., *et al.* 2013, *MNRAS*, 434, 1422

Precision Asteroseismology
Proceedings IAU Symposium No. 301, 2013
J. A. Guzik, W. J. Chaplin, G. Handler & A. Pigulski, eds.
© International Astronomical Union 2014
doi:10.1017/S1743921313015044

Kepler RR Lyrae stars: beyond period doubling

L. Molnár[1,2], J. M. Benkő[1], R. Szabó[1], and Z. Kolláth[2,1]

[1] Konkoly Observatory, MTA CSFK
H-1121 Konkoly Thege Miklós út 15-17, Budapest, Hungary
email: molnar.laszlo@csfk.mta.hu

[2] Institute of Mathematics and Physics, Savaria Campus, University of West Hungary
H-9700 Szombathely, Károlyi Gáspár tér 4, Hungary

Abstract. We examined the complete short cadence sample of fundamental-mode *Kepler* RR Lyrae stars to further investigate the recently discovered dynamical effects such as period doubling and additional modes. Here we present the findings on four stars. V450 Lyr may be a non-classical double-mode RR Lyrae star pulsating in the fundamental mode and the second overtone. For the three remaining stars we observe the interaction of three different modes. Since the period ratios are close to resonant values, we observe quasi-repetitive patterns in the pulsation cycles in the stars. These findings support the mode-resonance explanations of the Blazhko effect.

Keywords. stars: variable: RR Lyrae, *Kepler*

1. Introduction

The discoveries of space-based photometric missions such as *Kepler*, *CoRoT* and *MOST* have reshaped our understanding of the pulsation of RR Lyrae stars. Not long ago, these stars were thought to be fairly simple, with three distinct classes: fundamental-mode, first-overtone and double-mode pulsators, with the only, though serious complicating factor being the Blazhko effect. It turned out, however, that the stars, especially the modulated ones show several dynamic phenomena: period doubling, additional modes and near-resonant states (Szabó, in this volume, and references therein).

2. Multi-mode RR Lyrae stars and their consequences

Due to their sharpness, light maxima of RR Lyrae stars are not well covered by the 29.4 min long-cadence sampling of *Kepler*. Therefore we analyzed only those RR Lyrae stars which have available short-cadence (SC) data. We searched for regularities and beating patterns in the light curves of all modulated RRab stars and compared them to the frequency content of the data. Beating can be separated from other proposed cycle-to-cycle irregularities (Gillet 2013) and temporal evolution can be followed with the visual method. We found four different types of behaviour occurring as an interaction between four different modes: the fundamental (FM), the 1st, 2nd and 9th overtones (O1, O2 and PD). The modes labeled O1 and O2 are not necessarily radial modes in all cases but have similar frequency ratios in all stars. The 9th overtone has been observed only through period doubling when it is in resonance with the fundamental mode.

• FM + PD + O1 (RR Lyr): the beating pattern of RR Lyr is dominated by period doubling but beating with the first overtone sometimes creates quasi-repetitive patterns lasting six pulsation cycles with differing maximum brightnesses (Molnár *et al.* 2012).

KIC 9697825 – Q8

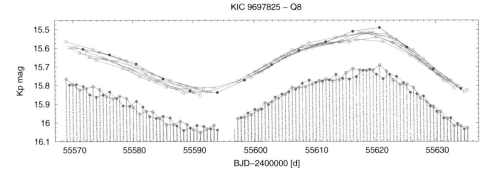

Figure 1. Dots: the bright part of the light curve of V360 Lyr (KIC 9697825). Every second maxima are connected with filled points and open circles, respectively. If we connect every eighth maxima instead (shifted vertically), the curves become much smoother, revealing the true structure of the beating pattern. One realization is highlighted with filled points.

- FM + O1 + PD (V360 Lyr): the first overtone has much higher amplitude than period doubling. The FM-O1 frequency ratio is close to 8:11, resulting in the quasi-repetition of an 8 FM-period long beating pattern (Fig. 1).
- FM + O2 (V460 Lyr): in this star only the FM and the 2nd overtone are visible, making it the third known FM/O2 double-mode star. The frequency ratio is close to 3:5, leading to a quasi-3-cycle beating pattern.
- FM + O2 + PD (KIC 7257008): the period ratio is the same as above but more complicated patterns are also observable. The frequency spectrum shows the signs of period doubling as well as peaks around the O1 mode but the latter are in fact $3f_0 - f_2$ combination peaks.

Although other stars show more complicated patterns in the cycle-to-cycle amplitude variations that are harder to describe, almost all show similar additional frequencies. These findings support the mode-interaction model of the Blazhko effect (Nowakowski & Dziembowski 2001, Buchler & Kolláth 2011). On the other hand, similar cycle-to-cycle irregularities described in the shock-interaction model (Gillet 2013) most likely arise simply from the combined effect of the beating of the modes and the uncertainties induced by the sparse sampling of the long cadence data instead of stochastic processes.

Acknowledgements

This work has been supported by by the Hungarian OTKA grant K83790, the MB08C 81013 Mobility-grant of the MAG Zrt., the KTIA URKUT_10-1-2011-0019 grant and the "Lendület-2009" Young Researchers' Programme. The work of L. Molnár was supported by the European Union and the State of Hungary, co-financed by the European Social Fund in the framework of TÁMOP 4.2.4. A/2-11-1-2012-0001 'National Excellence Program'. R.Sz. was supported by the János Bolyai Research Scholarship. L.M. and R.Sz. acknowledge the IAU grants for the conference.

References

Buchler, J. R. & Kolláth, Z. 2011, *ApJ*, 731, 24
Gillet, D. 2013, *A&A*, 554, A46
Molnár, L., Kolláth, Z., Szabó, R., *et al.* 2012, *ApJ*, 757, L13
Nowakowski, R. M. & Dziembowski, W. A. 2001, *AcA*, 51, 5

Precision Asteroseismology
Proceedings IAU Symposium No. 301, 2013
J. A. Guzik, W. J. Chaplin, G. Handler & A. Pigulski, eds.

© International Astronomical Union 2014
doi:10.1017/S1743921313015056

Revision of the [Fe/H] – ϕ_{31} – P relationship for RRc variables

Siobahn Morgan

Department of Earth Science, University of Northern Iowa,
Cedar Falls Iowa, USA
email: siobahn.morgan@uni.edu

Abstract. The relationship derived by Morgan *et al.* (2007) for type-c RR Lyrae variables (RRc) between values of [Fe/H] – ϕ_{31} – P has been revised and expanded. New relationships are based upon Fourier coefficients of 163 RRc variables in 19 Galactic globular clusters using the metallicity scales of Harris (2010), Zinn & West (1984) and Carretta *et al.* (2009). This larger database includes more low-metallicity clusters ([Fe/H] < −2.0), and the best fitting relations are found to depend upon values of $\log P$ rather than P. The new relations are applied to various populations of RRc including Milky Way field variables, LMC globular clusters variables, ω Cen RRc, and RRc in various OGLE III databases.

Keywords. stars: abundances, stars: variables: other

1. Introduction

The use of Fourier coefficients for deriving physical characteristics of variable stars is well established in the literature. Morgan *et al.* (2007, MWW hereafter), provided a relationship between the ϕ_{31} coefficient and two metallicity scales for 106 type-c RR Lyrae (RRc) variables in Galactic globular clusters. The present study expands this number to 163 variables, with the inclusion of data from seven additional globular clusters. All variables were examined for well defined light curves with low values of uncertainty in the ϕ_{31} coefficient, whenever available. In addition, the metallicity scales used in the present study have been revised to include values from Carretta *et al.* (2009), the recently revised values from Harris (2010), and Zinn & West (1984). The cluster metallicity values from Carretta *et al.* (2009), and Harris (2010) are very similar, as would be expected, though there is a slight deviation in values in metal rich clusters.

2. Method and application

Following the method of MWW, the values for the Fourier coefficient ϕ_{31} were fit to second order functions of P, ϕ_{31} and [Fe/H], as well as to $\log P$. It is noteworthy that the best fitting relations were consistently found for $\log P$-based formulae rather than those based upon P. In general, the values obtained for [Fe/H] using the relations found here and those from MWW were consistent in value, with differences typically less than 0.005 dex for the newer metallicity scales, while the average difference for the relationships based upon the Zinn & West (1984) values were closer to 0.02 dex. For simplicity only the relationship based upon the Carretta *et al.* (2009) values, which is

$$[\text{Fe/H}] = 4.86(\log P)^2 + 0.0183(\phi_{31})^2 - 0.820(\log P)\phi_{31} - 4.260, \qquad (2.1)$$

will be discussed here.

Following the procedure outlined in MMW, the [Fe/H] – ϕ_{31} – $\log(P)$ relation given above was applied to a variety of RRc variables to test its viability. The first population examined were field RRc variables. There was general agreement between the values from the literature and those from Equation 2.1, with the values from Equation 2.1 being slightly more metal poor compared to values obtained via ΔS values or spectroscopy. The metallicity of RRc variables in ω Cen (NGC 5139) were also calculated using equation 2.1. A total of 67 RRc stars with well defined light curves resulted in an average [Fe/H] $= -1.64\pm0.26$, with a range of metallicity values between -2.29 and -1.04. These results agree well with the range of known stellar populations for the cluster.

Equation 2.1 was also applied to RRc located in five LMC globular clusters. The results were varied, with metallicity values from the literature and those based upon the Fourier relation of RRab stars from Jurcsik & Kovács (1996) comparable to the results from Equation 2.1 for two of the clusters, NGC 1466, and NGC 1841. For the other three clusters (Reticulum, NGC 1786, NGC 2257), some of the variables included as RRc may actually be RRe variables (second overtone pulsators) and those should therefore be excluded. Their exclusion does bring the average metallicities based upon equation 2.1 closer to the other methods of [Fe/H] determination for two of the clusters, however the metallicity values based on Equation 2.1 for RRc in Reticulum remain significantly different from values in the literature.

Variables in the OGLE III database were also examined, though without any special discriminator other than that used in the creation of the database (see Soszyński et al. (2009) for methods of variable classification). Variables of an uncertain nature, or those that are foreground objects were removed, and the I magnitude based values for the Fourier coefficients were transformed to the V system using the method of Morgan et al. (1998), before Equation 2.1 was applied. For both the LMC and SMC variables, the average value for [Fe/H] from Equation 2.1 is significantly below that found in other studies of RR Lyrae variables in these galaxies, with $\langle[\text{Fe/H}]\rangle = -1.73\pm0.39$ for the LMC sample of 1974 variables, and $\langle[\text{Fe/H}]\rangle = -2.18 \pm 0.35$ for the 47 SMC variables. The average metallicity for the Milky Way Bulge sample of 4423 RRc stars was -1.22 ± 0.41.

3. Conclusions

The original metallicity relations based upon Fourier coefficients of MWW have been revised to utilize current metallicity scales of various authors as well as Fourier coefficients derived from recent high quality photometry programs. The relations are found to provide results similar to those of MWW. As more high-precision photometry of variables becomes available, it is likely that various relations between physical characteristics and Fourier coefficients will be utilized frequently. As with any method, these relations should be continually examined for consistency and accuracy.

References

Carretta, E., Bragaglia, A., Gratton, R., D'Orazi, V., & Lucatello, S. 2009, *A&A*, 508, 695
Harris, W. E. 2010, arXiv: 1012.3224
Jurcsik, J. & Kovács, G. 1996, *A&A*, 312, 111
Morgan, S. M., Simet, M., & Bargenquast, S. 1998, *AcA*, 48, 341
Morgan, S. M., Wahl, J. N., & Wieckhorst, R. M. 2007, *MNRAS*, 374, 1421
Soszyński, I., Udalski, A., Szymański, M. K., et al. 2009, *AcA*, 59, 1
Zinn, R. & West, M. J. 1984, *ApJS*, 55, 45

Precision Asteroseismology
Proceedings IAU Symposium No. 301, 2013
J. A. Guzik, W. J. Chaplin, G. Handler & A. Pigulski, eds.

© International Astronomical Union 2014
doi:10.1017/S1743921313015068

Long-term polarization observations of Mira variable stars suggest asymmetric structures

Hilding R. Neilson, Richard Ignace and Gary D. Henson

Dept. of Physics & Astronomy, East Tennessee State University, PO Box 70300, Johnson City,
TN 37614, USA
email: neilsonh@etsu.edu

Abstract. Mira and semi-regular variable stars have been studied for centuries but continue to be enigmatic. One unsolved mystery is the presence of polarization from these stars. In particular, we present 40 years of polarization measurements for the prototype o Ceti and V CVn and find very different phenomena for each star. The polarization fraction and position angle for Mira is found to be small and highly variable. On the other hand, the polarization fraction for V CVn is large and variable, from $2-7$ %, and its position angle is approximately constant, suggesting a long-term asymmetric structure. We suggest a number of potential scenarios to explain these observations.

Keywords. stars: circumstellar matter, stars: variables: other, techniques: polarimetric

1. Introduction

Mira and semi-regular variable stars have been observed almost continuously for centuries since the discovery by Fabricius in 1596 that the prototype was variable, although there is some evidence that variability was detected by earlier civilizations (e.g., Sahade & Wood 1978).

These continuous observations have been important for the understanding of stellar pulsation and shedding light on the theory of stellar evolution (Uttenthaler *et al.* 2011). Furthermore, the continuous observations have detected irregular pulsation in many evolved red giant stars. Mira variable stars are also important standard candles, that can be employed to measure distances to other galaxies (Whitelock *et al.* 2008).

Another probe to explore the structure and evolution of Mira variable stars is using linear polarization observations to probe asymmetric stellar properties (Serkowski & Shawl 2001).

2. Polarization data

We present polarization and position angle observations of the prototype Mira and the semi-regular variable star V CVn. These quantities are derived from observed Stokes Q and U parameters, where the polarization $p = (Q^2 + U^2)^{1/2}$, and the position angle $\psi = \tan^{-1}(U/Q)$. Using these parameters we explore asymmetric stellar structures, i.e., deviations from circular symmetry of the radiation field.

We plot in Fig. 1 the measured values of p, ψ, and V-band brightness variations for Mira (left) and V CVn (right) spanning a time scale of forty years. The latter data, denoted by dots are taken from the HPOL database (Wolff *et al.* 1996) while the triangles are from Magalhães *et al.* (1986a) and the squares from Poliakova (1981). The V-band observations are from the AAVSO database. For Mira, p varies from $0-1$% with large

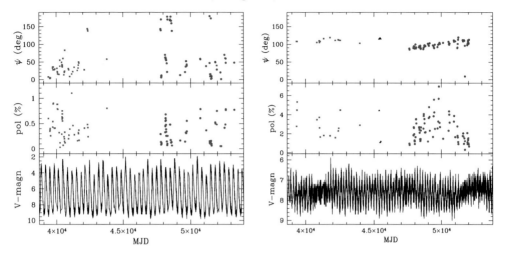

Figure 1. Measured position angle, ψ, in degrees; polarization percentage; and visual brightness for Mira (left) and V CVn (right) from the AAVSO database. Circles represent HPOL observations, triangles are those from Magalhães *et al.* (1986a) and the squares from Poliakova (1981).

variation of ψ, whereas, $p = 2 - 7\%$ for V CVn with a nearly constant value of ψ spanning 40 years.

3. Future work

We suggest that the polarization and position angle measurements for V CVn are consistent with the structure of a circumstellar disk (Neilson *et al.*, submitted), whereas the polarization for Mira is due to convective hot spots (Ignace *et al.*, in preparation).

We are also exploring polarization measurements of other Mira and semi-regular variable stars, such as R Leo. There are about 20 such stars in the HPOL database. However, none of the other stars in the HPOL database have detected polarization at similar levels as V CVn. Only the star L_2 Pup appears to have similar polarization (Magalhães *et al.* 1986b), suggesting that circumstellar disks about these stars is rare but is consistent with observations of disks about white dwarf stars (Farihi *et al.* 2005).

We continue to observe V CVn, Mira and a sample of variables in the northern hemisphere using the Lulin One-Meter Telescope in Taiwan. We are grateful for funding from NSF Grant AST-0807664 and to the many observers who contribute data to the AAVSO.

References

Farihi, J., Becklin, E. E., & Zuckerman, B. 2005, *ApJS*, 161, 394
Magalhães, A. M., Codina-Landaberry, S. J., Gneiding, C., & Coyne, G. V. 1986a, *A&A*, 154, 1
Magalhães, A. M., Coyne, G. V., & Benedetti, E. K. 1986b, *AJ*, 91, 919
Poliakova, T. A. 1981, *Leningradskii Universitet Vestnik Matematika Mekhanika Astronomiia*, 105
Sahade, J. & Wood, F. B. 1978, *Interacting Binary Stars* (Oxford: Pergamon Press)
Serkowski, K. & Shawl, S. J. 2001, *AJ*, 122, 2017
Uttenthaler, S., van Stiphout, K., Voet, K., *et al.* 2011, *A&A*, 531, A88
Whitelock, P. A., Feast, M. W., & van Leeuwen, F. 2008, *MNRAS*, 386, 313
Wolff, M. J., Nordsieck, K. H., & Nook, M. A. 1996, *AJ*, 111, 856

Precision Asteroseismology
Proceedings IAU Symposium No. 301, 2013
J. A. Guzik, W. J. Chaplin, G. Handler & A. Pigulski, eds.

© International Astronomical Union 2014
doi:10.1017/S174392131301507X

Making a Be star:
the role of rotation and pulsations

Coralie Neiner[1] and Stéphane Mathis[2,1]

[1]LESIA, UMR 8109 du CNRS, Observatoire de Paris, UPMC, Univ. Paris Diderot
5 place Jules Janssen, 92195 Meudon Cedex, France
email: coralie.neiner@obspm.fr

[2]Laboratoire AIM Paris-Saclay, CEA/DSM-CNRS-Université Paris Diderot
IRFU/SAp, Centre de Saclay, 91191 Gif-sur-Yvette Cedex, France
email: stephane.mathis@cea.fr

Abstract. The Be phenomenon, i.e. the ejection of matter from Be stars into a circumstellar disk, has been a long lasting mystery. In the last few years, the *CoRoT* satellite brought clear evidence that Be outbursts are directly correlated to pulsations and rapid rotation. In particular the stochastic excitation of gravito-inertial modes, such as those detected by *CoRoT* in the hot Be star HD 51452, is enhanced thanks to rapid rotation. These waves increase the transport of angular momentum and help to bring the already rapid stellar rotation to its critical value at the surface, allowing the star to eject material. Below we summarize the recent observational and theoretical findings and describe the new picture of the Be phenomenon which arose from these results.

Keywords. stars: emission-line, Be, stars: mass loss, stars: oscillations (including pulsations)

1. Stochastic excitation of pulsations in Be stars

Be stars are massive stars with a decretion disk that are known to pulsate thanks to the κ mechanism. The correlation between pulsations and the ejections of matter into the circumstellar disk was first proposed by Rivinius *et al.* (2001) and firmly established by Huat *et al.* (2009) thanks to *CoRoT* observations.

Sub-inertial gravito-inertial modes (below twice the rotation frequency) have recently been detected in the early Be star HD 51452 with *CoRoT* (Neiner *et al.* 2012). These modes cannot be excited by the κ mechanism usually invoked for those stars, because HD 51452 is too hot (B0 IVe) to be in the κ-driven g-mode instability strip. Since the observed modes have very low frequency and a short lifetime, we have proposed that they are excited stochastically in the convective core and at its interface with the surrounding radiative envelope.

In addition, low-frequency g modes have been observed with *CoRoT* in another early Be star, HD 49330, during an outburst (Huat *et al.* 2009). We propose that these modes are also stochastically excited, as suggested by their short lifetime. Indeed, these modes are only visible during the outburst, while the κ-driven p modes get destabilized during the outburst. However, in this case, the stochastic modes we observed are probably those excited just below the surface during the outburst rather than the ones excited in the convective core.

It was not expected that stochastically excited gravito-inertial modes could be observed in massive stars (Samadi *et al.* 2010). However, Be stars are very rapid rotators and stochastic excitation is enhanced in the presence of rapid rotation, through the Coriolis acceleration which modifies gravity waves. This has been demonstrated analytically by Mathis *et al.* (2013) and observed in numerical simulations by Rogers *et al.* (2013) (see also Browning *et al.* 2004). Indeed, in the convective zones, when rotation is rapid, gravity modes become less evanescent in the super-inertial regime and propagative inertial modes in the sub-inertial regime.

Such stochastic modes are thus probably present in all rapidly rotating massive stars. Therefore, in the case of rapid rotators, the identification of low-frequency modes should be considered carefully and not systematically attributed to the κ mechanism as has been done until recently.

2. Transport of angular momentum from the core to the surface

Lee (2013) showed that gravito-inertial modes excited by the κ mechanism transport angular momentum and could play a role in the Be phenomenon. However, in the sub-inertial regime, the transport of angular momentum was believed to become less efficient because of gravito-inertial waves equatorial trapping (Mathis *et al.* 2008, Mathis 2009). Our recent work shows that transport by trapped sub-inertial waves may be sustained in rapidly rotating stars thanks to the stronger stochastic excitation by turbulent convective flows. Moreover, sub-inertial gravito-inertial modes have very low frequencies and therefore they transport more angular momentum than modes with higher frequencies.

We thus propose that this mechanism allows to transport angular momentum from the convective core of Be stars, where sub-inertial gravito-inertial modes are excited, to their surface. The accumulation of angular momentum just below the surface of Be stars increases the surface velocity. The surface then reaches the critical velocity so that material gets ejected from the star.

3. Conclusions

Thanks to the discovery of stochastically excited gravito-inertial modes in the hot Be star HD 51452 and to the observation of the correlation between these pulsations and a Be outburst in HD 49330, both with the *CoRoT* satellite, we have shown that stochastic gravito-inertial waves play an important role in Be stars. We demonstrated analytically that it is rapid rotation that enhances those modes in Be stars. This is also confirmed in numerical simulations. Since sub-inertial gravito-inertial modes efficiently transport angular momentum, we propose that they could be the key to the Be phenomenon, i.e. to the ejection of material from the surface of Be stars into a circumstellar Keplerian disk.

References

Browning, M. K., Brun, A. S., & Toomre, J. 2004, *ApJ*, 601, 512
Huat, A.-L., Hubert, A.-M., Baudin, F., *et al.* 2009, *A&A*, 506, 95
Lee, U. 2013, *PASJ*, in press (arXiv: 1304.6471)
Mathis, S. 2009, *A&A*, 506, 811
Mathis, S., Talon, S., Pantillon, F.-P., & Zahn, J.-P. 2008, *Solar Phys.*, 251, 101
Mathis, S., Neiner C. & Tran Minh, N. 2013, *A&A*, in press
Neiner, C., Floquet, M., Samadi, R., *et al.* 2012, *A&A*, 546A, 47
Rivinius, T., Baade, D., Stefl, S., Townsend, R. H. D., Stahl, O., Wolf, B., & Kaufer, A. 2001, *A&A*, 369, 1058
Rogers, T. M., Lin, D. N. C., McElwaine, J. N., & Lau, H. H. B. 2013, *ApJ*, 772, 21
Samadi, R., Belkacem, K., Goupil, M. J., Dupret, M.-A., Brun, A. S., & Noels, A. 2010, *Ap&SS*, 328, 253

Precision Asteroseismology
Proceedings IAU Symposium No. 301, 2013
J. A. Guzik, W. J. Chaplin, G. Handler & A. Pigulski, eds.

© International Astronomical Union 2014
doi:10.1017/S1743921313015081

Spectroscopic survey of *Kepler* stars: high-resolution observations of A- and F-type stars

E. Niemczura[1], **B. Smalley**[2], **S. Murphy**[3], **G. Catanzaro**[4], **K. Uytterhoeven**[5,6], **D. Drobek**[1], **M. Briquet**[7], **P. De Cat**[8], **P. Marcos-Arenal**[9], **P. I. Pápics**[9], and **J. F. S. Gameiro**[10]

[1]Instytut Astronomiczny, Uniwersytet Wrocławski, Kopernika 11, 51-622 Wrocław, Poland email: eniem@astro.uni.wroc.pl; [2]Astrophysics Group, Keele University, Staffordshire, ST5 5BG, UK; [3]Jeremiah Horrocks Institute, University of Central Lancashire, Preston PR1 2HE, UK; [4]INAF-Osservatorio Astrofisico di Catania, Via S. Sofia 78, I-95123, Catania, Italy; [5]Instituto de Astrofisica de Canarias, E-38205 La Laguna, Tenerife, Spain; [6]Universidad de La Laguna, Departamento de Astrofisica, E-38206 La Laguna, Tenerife, Spain; [7]Institut d'Astrophysique et de Géophysique, Université de Liège, Allée du 6 Août 17, B-4000 Liège, Belgium; [8]Royal observatory of Belgium, Ringlaan 3, B-1180 Brussel, Belgium; [9]Instituut voor Sterrenkunde, KU Leuven, Celestijnenlaan 200D, 3001, Leuven, Belgium; [10]Centro de Astrofisica, Universidade do Porto, Rua das Estrelas

Abstract. Basic stellar parameters such as effective temperature, surface gravity, chemical composition, and projected rotational velocity, are important to classify stars and are crucial for successful asteroseismic modelling. However, the *Kepler* space data do not provide such information. Therefore, ground-based spectral and multi-colour observations of *Kepler* asteroseismic targets are necessary to complement the space data. For this purpose, in coordination with the KASC ground-based observational Working Groups, high-resolution spectroscopic data for more than 500 B, A, F and G-type stars were collected.

Keywords. stars: abundances, stars: chemically peculiar, stars: fundamental parameters

The NASA space mission *Kepler* was successfully launched on 7th March 2009. Since then, *Kepler* produced long photometric time series of an exceptional precision, which provide a unique opportunity to study the pulsational variability of thousands of stars across the H-R diagram in detail (Kjeldsen *et al.* 2010). We determined atmospheric parameters such as effective temperature T_{eff}, surface gravity $\log g$, microturbulence V_{turb}, rotational velocity $V \sin i$ and abundances of the chemical elements for more then 100 A-F stars from the *Kepler* field-of-view, for which we collected the high-resolution HERMES@Mercator spectra (Raskin *et al.* 2011). The pipeline-reduced spectra were normalised by the standard IRAF procedures. Spectrum classification was performed on the HERMES spectra smoothed to the appropriate low-resolution using the program SPECTRUM (Gray 1999). The classification process revealed over a dozen Am and other chemically peculiar stars in our sample. The atmospheric parameters were determined in two different ways.

The first method consists of spectral synthesis based on a least-squares optimisation algorithm (e.g. Niemczura *et al.* 2009). This method allows for the simultaneous determination of various parameters involved with stellar spectra. The synthetic spectrum depends on T_{eff}, $\log g$, V_{turb}, $V \sin i$ and abundances of chemical elements. The atmospheric models and synthetic spectra were computed with the LTE ATLAS9 and SYNTHE codes (Kurucz 1993). We derived T_{eff} using the sensitivity of Balmer line wings to temperature. Additionally, we adjust T_{eff}, $\log g$, and V_{turb} from the comparison of iron abundances determined from various Fe I and Fe II lines. The derived T_{eff} is considered accurate when there is no correlation between the iron abundances and excitation potentials of the atomic levels causing Fe I lines. The gravity is obtained by requiring the same abundances from both Fe I and Fe II lines.

In the second method, T_{eff} was determined from the spectral energy distribution (SED). These were constructed from literature photometry, e.g. 2MASS (Skrutskie *et al.* 2006), Tycho *B* and *V* magnitudes (Høg *et al.* 1997) supplemented with *TD-1* UV flux measurements (Carnochan 1979) where available. The interstellar reddenings were estimated from interstellar Na D lines (Munari & Zwitter 1997). These $E(B-V)$ values were used for the de-reddening of the SEDs. The stellar T_{eff} were determined by fitting Kurucz (1993) model flux to the de-reddened SEDs. The synthetic fluxes were convolved with photometric filter response functions. A Levenberg-Marquardt nonlinear least-squares procedure was used to find the solution that minimized the difference between the observed and model fluxes.

For most of the stars we have obtained consistent effective temperatures from all applied methods. The projected rotational velocities, $V \sin i$, have values ranging from 6 to $280 \, \text{km} \, \text{s}^{-1}$. The average value of $V \sin i$, $\sim 100 \, \text{km} \, \text{s}^{-1}$, is typical for A-type stars. The obtained microturbulence velocities for normal A-F stars equal to $2-4 \, \text{km} \, \text{s}^{-1}$, which is again typical for this kind of object. Our analysis allowed us to discover many interesting slowly rotating chemically peculiar stars:

• KIC 4768731 is one of a few rapidly oscillating Ap (roAp) stars known in the field-of-view of the *Kepler* satellite. It was classified as A5 Vp (SrCrEu) star. The roAp stars are the subgroup of CP2 stars (chemically peculiar stars with a global magnetic field). The discussion of global stellar parameters and pulsational properties of this object will be presented by Smalley *et al.* (in preparation).

• We analysed more than 10 metallic-line A stars (Am). All of them have typical abundance pattern characterized by underabundances of Ca and Sc, and overabundances of Sr and Y. These stars also have surface deficiencies in helium, which is why their pulsational driving is inefficient. All of the investigated Am stars lie in the δ Scuti instability strip (Murphy *et al.*, in preparation).

Acknowledgments. EN acknowledges the support from Fundação para a Ciência e a Tecnologia (FCT) through the grant 'Cooperação Cientica e Teccnologica FCT/Polonia 2011/2012 (Proc. 441.00 Polonia)' funded by FCT/MCTES, Portugal. This work was supported by MNiSW grant No. N N203 405139. DD would like to acknowledge the NSC grant No. 2011/01/N/ST9/00400. Calculations have been carried out using resources provided by WCSS, grant No. 214. KU acknowledges support by the Spanish National Plan of R&D for 2010, project AYA2010-17803. MB is F.R.S.-FNRS Postdoctoral Researcher, Belgium. PP acknowledges the FP7/20072013/ERC grant agreement No. 227224.

References

Carnochan, D. J. 1979, *Bull. Inf. CDS*, 17, 78

Gray, R. O. 1999, *Astrophysics Source Code Library*

Høg, E., Bässgen, G., Bastian, U., *et al.* 1997, *A&A*, 323, 57

Kjeldsen, H., Christensen-Dalsgaard, J., Handberg, R., *et al.* 2010, *AN*, 331, 966

Kurucz, R. L. 1993, *Kurucz CD-ROM 13*, SAO, Cambridge, USA.

Munari, U. & Zwitter, T. 1997, *A&A*, 318, 26

Niemczura, E., Morel, T., & Aerts, C. 2009, *A&A*, 506, 213

Raskin, G., van Winckel, H., & Hensberge, H. et al. 2011, *A&A*, 526 69

Skrutskie, M. F., Cutri, R. M., Stiening, R., *et al.* 2006, *AJ*, 131, 1163

Precision Asteroseismology
Proceedings IAU Symposium No. 301, 2013
J. A. Guzik, W. J. Chaplin, G. Handler & A. Pigulski, eds.

© International Astronomical Union 2014
doi:10.1017/S1743921313015093

New analysis of ZZ Ceti star PG 2303+243

E. Pakštienė[1], V. Laugalys[1], J. Qvam[2], and R. P. Boyle[3]

[1]Institute of Theoretical Physics and Astronomy, Vilnius University, Goštauto 12, Vilnius, Lithuania; email: erika.pakstiene@tfai.vu.lt
[2]Department of Physics, University of Oslo, P.O. Box 1048 Blindern, N-0316 Oslo, Norway
[3]Vatican Observatory Research Group, Steward Observatory, Tucson, Arizona 85721, U.S.A.

Abstract. The new photometric observations of PG 2303+243 were obtained in 2012 during a campaign carried out with three telescopes. The analysis of these observations is presented in this paper. We identified $l = 1$ and $l = 2$ pulsation modes. The pulsation periods were compared with theoretical ones for models of ZZ Ceti stars. This allowed us to estimate the physical parameters of PG 2303+243. The star seems to be cooler and has thicker hydrogen layer than it was thought before. We have derived $M_*/M_\odot = 0.66$, $T_{\rm eff} = 11014$ K and $\log(M_{\rm H}/M_*) = -4.246$ for this star.

Keywords. stars: variables, white dwarfs: ZZ Ceti: individual: PG 2303+243

1. Observations and analysis

PG 2303+243 was observed several times since its discovery as variable white dwarf (DAV) in 1987 (Vauclair *et al.* 1987). In our work we used the published photometric data obtained in 1990 (Vauclair *et al.* 1992), 2004 (Pakštienė *et al.* 2011) and 2005 (Pakštienė 2013). The new photometric observations of PG 2303+243 were obtained during a small campaign carried out between October 4 and 11, 2012, with three telescopes: 165-cm (Molėtai, Lithuania), 180-cm (VATT, Arizona) and 46-cm (Horten, Norway).

Despite unfavourable weather conditions in all observatories, the total duration of runs amounted to 55.88 h, which corresponds to a duty cycle of 32%. The detection limit of pulsations for a false alarm probability FAP = 1/1000 is 2.93 mma (milli-modulation amplitude) at 2000 μHz. This limit is lower than it was in observations made in 2005 (3.75 mma), but higher than in the 2004 campaign (1.79 mma). Fourier transform (FT) spectra of PG 2303+243 are clearly different for different seasons. Only some modes can be found at about the same frequency, but their amplitudes differ strongly. Observational data obtained during the 2004 campaign yielded the most detailed FT spectrum so far.

In total, we detected 39 periodic terms in the Fourier spectrum of the combined 2012 data. Frequencies of four pairs differed by less than 3 μHz, so we assumed that these are single peaks broadened by variable amplitudes of the modes. An additional 10 terms were removed from the list because they were identified as harmonics or combinations. We found that 4 frequencies might have harmonics. This left us with the list of 25 modes detected in the 2012 data that are likely independent. A more detailed description of these observations and their analysis will be published elsewhere.

2. Mode identification

We assume that only $l = 1$ modes reach high amplitudes and appear as dominant peaks in the FT spectrum. We found different dominant peaks for different seasons and finally picked up five dominant modes with the highest amplitudes from 1990, 2004, 2005 and 2012 observations. In order to find an appropriate model for PG 2303+243 we

Table 1. Comparison of the observed, Π_{obs}, and theoretical, Π_{th}, periods for PG 2303+243. Indicated years correspond to the season with the smallest $\Delta\Pi$. The asterisks (*) indicate periods found in data from more than one season.

Π_{obs} [s]	A [mma]	Π_{th} [s]	$\Delta\Pi$	l	k	year	Π_{obs} [s]	A [mma]	Π_{th} [s]	$\Delta\Pi$	l	k	year
261.8	2.3	261.9	−0.1	1	4	2004	778.5!	20.9	776.6	1.9	1	16	2005
270.1	3.1	271.5	−1.4	2	8	2012	816.2!	32.2	816.9	−0.7	1	17	2012*
335.5	2.5	336.5	−1.0	2	11	2005	820.3	30.5	821.6	−1.3	2	31	2012
390.7	2.8	390.3	0.5	2	13	2004*	845.7	5.6	845.9	−0.1	2	32	2012*
434.0	1.5	435.5	−1.5	2	15	1990	857.8	4.9	859.4	−1.6	1	18	2004
453.2	2.3	455.6	−2.3	2	16	1990*	873.2	3.9	873.0	0.2	2	33	2004
482.6	5.3	480.0	2.6	2	17	1990	925.3	7.9	927.2	−1.9	2	35	2012
577.9!	11.0	577.9	0.0	1	11	1990	940.6	7.5	942.7	−2.1	1	20	2005*
578.1	2.8	578.2	−0.1	2	21	2004	955.4	3.2	953.2	2.2	2	36	1990
606.4	4.4	604.2	2.2	2	22	2005	965.3	19.7	—	—	—	—	2004*
616.4!	31.3	616.0	0.4	1	12	2004*	998.3	6.7	1000.6	−2.3	2	38	2005
682.7	1.1	680.4	2.3	1	14	1990	1043.6	1.9	1046.3	−2.7	2	40	1990*
774.7	13.3	776.6	−2.0	1	16	2012	1227.3	1.8	1227.3	0.0	2	47	1990

used a set of models calculated by Romero *et al.* (2012). Our method for selection of the model is identical to the one described by Romero *et al.* (2012). We also used three quality functions, Φ, χ^2, and Ξ, which were based on differences between theoretical and observed periods. Minimum values of Φ, χ^2 and Ξ gave us the most appropriate seismic model for the star.

The most disappointing fact was that the best matching models do not predict pulsations with the period of 965.3 s ($A = 19.7$ mma), though earlier analyses (Romero *et al.* 2012, Pakštienė 2013) predicted it to be an $l = 1$ mode. Thus, in the final fit, we used only four periods: 577.93 s (11.0 mma), 616.4 s (31.4 mma), 778.5 s (20.9 mma) and 816.2 s (32.2 mma), flagged with exclamation marks in Table 1. These periods give the best match for the model having $M_*/M_\odot = 0.66$, $T_{\mathrm{eff}} = 11\,014$ K and $\log(M_{\mathrm{H}}/M_*) = -4.246$ with quality function $\Phi = 0.74$ s.

Then, we considered all periods observed in four seasons and compared them with periods taken from the selected DAV model of Romero *et al.* (2012). We have found that 25 observed periods agree within 2.7 s with the theoretical ones (Table 1). This identification results in a quality function $\Phi = 1.31$ s.

3. Conclusions

We have estimated that the mass of PG 2303+243 is exactly the same as estimated from spectroscopy by Bergeron *et al.* (2004) and is larger than was found by Romero *et al.* (2012). According to our analysis, the hydrogen layer of PG 2303+243 must be much thicker ($\log(M_{\mathrm{H}}/M_*) = -4.246$) than was found by Romero *et al.* (2012) and Bergeron *et al.* (2004). The effective temperature we estimated, when compared to values published by other authors, is the lowest one, but it fits well the parameters of cool DAV stars and is very close to the red edge of the DAV instability strip. This raises even larger interest to PG 2303+243.

References

Bergeron, P., Fontaine, G., Billères, M., & Boudreault, S., Green, E. M. 2004, *ApJ*, 600, 404
Pakštienė, E. 2013, *EPJ Web of Conferences*, 43, 05012
Pakštienė, E., Solheim, J. E., Handler, G., *et al.* 2011, *MNRAS*, 415, 1322
Romero, A. D., Corsico, A. H., Althaus, L. G., *et al.* 2012, *MNRAS*, 420, 1462
Vauclair, G., Dolez, N., & Chevreton, M. 1987, *A&A*, 175, L13
Vauclair, G., Belmonte, J. A., Pfeiffer, B., *et al.* 1992, *A&A*, 246, 547

Precision Asteroseismology
Proceedings IAU Symposium No. 301, 2013
J. A. Guzik, W. J. Chaplin, G. Handler & A. Pigulski, eds.

© International Astronomical Union 2014
doi:10.1017/S174392131301510X

Variable stars in open clusters

E. Paunzen[1], M. Zejda[1], Z. Mikulášek[1], J. Liška[1], J. Krtička[1],
J. Janík[1], M. Netopil[2], L. Fossati[3] and B. Baumann[2]

[1] Department of Theoretical Physics and Astrophysics, Masaryk University
Kotlářská 2, CZ 611 37, Brno, Czech Republic
email: epaunzen@physics.muni.cz

[2] Department of Astrophysics, Vienna University, Türkenschanzstr. 17, A-1180 Vienna, Austria

[3] Argelander-Institut für Astronomie der Universität Bonn
Auf dem Hügel 71, D-53121, Bonn, Germany

Abstract. We present our joint efforts to study variable stars in open clusters. This includes a new catalogue, a photometric survey for new variables, and the database WEBDA. Our tools will shed more light on stellar variability in open clusters.

Keywords. open clusters and associations: general, stars: variables: other, techniques: photometric, astronomical data bases: miscellaneous, surveys, catalogs

1. Introduction

The study of an individual star provides only limited, frequently inaccurate and uncertain information about it and possibly about the interstellar medium between this object and us. On the other hand, open clusters provide an ideal opportunity to simultaneously study a group of stars located in a relatively small space, at the same distance from the Sun, and with the same age and initial chemical composition. The detection of any variable star in such stellar aggregates and its use to gather further information make research of open clusters very effective. For example, variable stars in open clusters allow one to obtain crucial information on both variable stars and open clusters in general. This improves our knowledge about both variable stars and open clusters and yields new data for the study of the dynamics, evolution, and structure of the whole Milky Way.

2. A new catalogue of variable stars in open cluster fields

In Zejda *et al.* (2012), we presented a new catalogue of variable stars in open cluster fields. For the compilation of the catalogue, the most complete database of variable stars, managed by the American Association of Variable Star Observers, AAVSO, as the International Variable Star Index (VSX, http://www.aavso.org/vsx) was used. The list of stars included in the VSX and the version 3.2. Catalogue of Open Clusters DAML02 (http://www.astro.iag.usp.br/~wilton/) were matched. We divided the open clusters into two categories according to their sizes, where the limiting diameter was chosen as 60′. We restricted our sample to clusters with diameters of less than five degrees, with the exception of the Hyades (Melotte 25). For both samples of open clusters, we generated a list of all suspected variables and variable stars located within the fields of open clusters. We checked the cluster fields and vicinities up to two times the given cluster radius. In the first group of 461 open clusters smaller than 60′, we found 8 938 variable stars. In the second group of 74 open clusters, we located 9 127 objects.

As a first heuristic approximation to a detailed membership analysis, we present the dependence of areal density of variable stars on the distance to the published cluster centers. In the area of open clusters larger than 60′ (mostly nearby clusters), the variables are strongly dominated by background objects.

3. Variable star survey in Galactic open clusters

We have started a comprehensive study in order to photometrically monitor all accessible members of several well chosen open clusters. The telescopes used are typically in the one meter class or smaller. The target clusters will cover the known metallicity and age range for such aggregates and the time base of the observations will span from several minutes to several months, which will allow us to study several kinds of variability in much more details. In general, we expect to find pulsation of red giants, variability of young stars due to their accretion disks, planet transits, β Cephei or δ Scuti variability, slowly pulsating B-type stars, stellar winds of the most massive objects, variability due to spots of chemically peculiar (CP) and PMS objects, emission and accretion episodes of Be and shell stars, as well as eclipses of binary systems.

Because we intend to observe a complete sample of members for a given mass and luminosity range, a unique global picture of variability for several open clusters and their member stars will arise. Numerous important astrophysical processes (for example diffusion, mass loss, and rotation) can be analyzed in correlation with the local as well as global Galactic environment. In particular, e.g. the evolution of individual variable groups will be studied in dependence on the age and Galactic location. As a spin-off, we also observe possible variable field stars in the area of the target clusters.

4. Using WEBDA for variable star research

WEBDA (http://webda.physics.muni.cz) is a site devoted to observational data of stellar clusters in the Milky Way and the Small Magellanic Cloud. It is intended to provide a reliable presentation of the available data and knowledge about these objects. The success of WEBDA is documented by its worldwide usage and the related acknowledgements in the literature: more than 750 refereed publications since the year 2000 acknowledge its use. The database content includes astrometric data in the form of coordinates, rectangular positions, and proper motions, photometric data in the major systems in which star clusters have been observed, but also spectroscopic data like spectral classification, radial velocities, and rotational velocities. It also contains miscellaneous types of supplementary data like membership probabilities, orbital elements of spectroscopic binaries, and periods for different kinds of variable stars as well as an extensive bibliography. To date, about four million individual measurements are included in the database.

Currently, there are about 4300 periods and frequencies for about 3300 objects in 90 open clusters listed in WEBDA.

Acknowledgements

This project was supported by the SoMoPro II Programme (3SGA5916), co-financed by the European Union and the South Moravian Region, grant GA ČR P209/12/0217, 7AMB12AT003, #LG12001 (Czech Ministry of Education, Youth and Sports), FWF P22691-N16, WTZ BG 03/2013, and CZ-10/2012.

Reference

Zejda, M., Paunzen, E., Baumann, B., Mikulášek, Z., & Liška, J. 2012, *A&A*, 548, A97

Precision Asteroseismology
Proceedings IAU Symposium No. 301, 2013
J. A. Guzik, W. J. Chaplin, G. Handler & A. Pigulski, eds.

© International Astronomical Union 2014
doi:10.1017/S1743921313015111

On the interchange of alternating-amplitude pulsation cycles

E. Plachy[1], L. Molnár[1,2], Z. Kolláth[2,1], J. M. Benkő[1] and K. Kolenberg[3]

[1]Konkoly Observatory, Konkoly Thege Miklós út 15-17, H-1121, Budapest, Hungary
email: eplachy@konkoly.hu

[2]Institute of Mathematics and Physics, Savaria Campus, University of West Hungary
H-9700 Szombathely, Károlyi Gáspár tér 4, Hungary

[3]Harvard-Smithsonian Center for Astrophysics, 60 Garden Street, Cambridge, MA 02138, USA

Abstract. We characterized the time intervals between the interchanges of the alternating high- and low-amplitude extrema of three RV Tauri and three RR Lyrae stars.

1. Introduction

The brightness variations of RV Tauri stars are characterized by alternating primary and secondary minima. The alternation is not strictly regular; the order of the deep and shallow minima shows occasional reversals (Wallerstein 2002). According to the most likely explanation the alternating light variation is caused by the 2:1 resonance between the fundamental and the first-overtone mode (Fokin 1994). In this near-resonance state the system may display low-dimensional chaotic behaviour, as proposed by Buchler & Kovács (1987) and Buchler *et al.* (1996). The chaos hypothesis is also supported by the observed interchanges of deep and shallow minima.

Similar interchanges were recently discovered in the order of the period-doubled pulsation cycles in RR Lyrae stars observed by the *Kepler* space telescope (Szabó *et al.* 2010). In these cases the period doubling of the fundamental mode is caused by a 9:2 resonance with the ninth overtone. At the time of writing, the four-year long photometric data of *Kepler* have become available to study the nature of the interchanges of the alternating maxima in these RR Lyrae stars and to compare to the similar phenomenon in RV Tauri variables.

2. Data and method

We analysed three RV Tauri stars (TT Oph, UZ Oph, and U Mon) and three RR Lyrae stars (RR Lyr, V808 Cyg, V355 Lyr) that clearly show the interchanges. RV Tauri light curves are based on several data sources: visual and photometric measurements of amateur astronomers in the AAVSO, AFOEV and VSOLJ databases, data from the ASAS (Pojmański 1997), NSVS (Woźniak *et al.* 2004) and Catalina Sky surveys (Drake *et al.* 2009), as well as the observations of the *Hipparcos* satellite (ESA 1997) and the Optical Monitoring Camera aboard the Integral satellite (Mas-Hesse *et al.* 2003). In the case of RR Lyrae stars we used all the available *Kepler* photometric data between the Q1–Q15 runs.

To determine the extrema of the cycles we applied least-squares parabola and cubic spline fitting. We followed the method used by Molnár *et al.* (2012) by connecting the even- and odd-numbered cycles respectively. This representation significantly helps to recognize the epoch of the interchanges.

3. Results and conclusions

All three RV Tauri stars show interchanges on short and long timescales. The characteristic interval between two interchanges are $14-21$ cycles for TT Oph, and $3-9$ cycles for U Mon. The longer timescale denote $112-119$ cycles for TT Oph and $39-51$ for U Mon. Interestingly, the light curve of TT Oph remains in the same phase during the first 249 cycles. U Mon also shows a section of 136 cycles without interchanges. The total number of interchanges is low for UZ Oph, but a short timescale is suspected around $5-25$ cycles and a long one around $45-65$ cycles.

The timescales of the interchanges of RR Lyrae stars are not divided into two groups; instead they spread to a broad interval between 3 and 39, 86 or 60 cycles for V808 Cyg, V355 Lyr and RR Lyr respectively. Most of the intervals are around $6-30$ cycles for all RR Lyrae stars. We note that determination of the timescales of the interchanges was not unambiguous for V355 Lyr and RR Lyr, where the alternation seemingly disappears occasionally, making the lengths between the interchanges uncertain up to a few cycles.

No periodicities were recognized in the occurrences of the interchanges for any of the analysed stars. RV Tauri stars can stay for long periods of time in one phase without interchanges. Similar behaviour was not observed in the RR Lyrae stars. The Blazhko periods are $\sim$170 cycles long for V808 Cyg, $\sim$66 cycles for V355 Lyr and $\sim$70 cycles for RR Lyr: the time intervals between the interchanges do not show any correlation with the Blazhko periods.

Acknowledgements

This work has been supported by the Hungarian OTKA grant K83790, the MB08C 81013 Mobility-grant of the MAG Zrt., the "Lendület-2009" Young Researchers' Programme of the Hungarian Academy of Sciences and the KTIA URKUT_10-1-2011-0019 grant. The *Kepler* Team and the *Kepler* Guest Observer Office are recognized for helping to make the mission and these data possible. We acknowledge with thanks the variable star observations from the AAVSO International Database, the AFOEV database, operated at CDS, France, and the VSOLJ database contributed by observers worldwide. KK is grateful for the support of her Marie Curie Fellowship (255267- SAS-RRL, FP7).

References

Buchler, J. R. & Kovács, G. 1987, *ApJ*, 320, 57
Buchler, J. R., Kolláth, Z., Serre, T., & Mattei, J. 1996, *ApJ*, 462, 489
Drake, A. J., Djorgovski, S. G., Mahabal, A., *et al.* 2009, *ApJ*, 696, 870
ESA, 1997, *The Hipparcos and Tycho Catalogues*, ESA SP-1200 (Noordwijk)
Fokin, A. B. 1994, *A&A*, 292, 133
Mas-Hesse, J. M., Giménez, A., Culhane, J. L., *et al.* 2003, *A&A*, 411, L261
Molnár, L., Kolláth, Z., Szabó, R., *et al.* 2012, *ApJ*, 757, L13
Pojmański, G. 1997, *AcA*, 47, 467
Szabó, R., Kolláth, Z., Molnár, L., *et al.* 2010, *MNRAS*, 409, 1244
Wallerstein, G. 2002, *PASP*, 144, 689
Woźniak, P. R., Vestrand, W. T., Akerlof, C. W., *et al.* 2004, *AJ*, 127, 2436

Precision Asteroseismology
Proceedings IAU Symposium No. 301, 2013   © International Astronomical Union 2014
J. A. Guzik, W. J. Chaplin, G. Handler & A. Pigulski, eds.   doi:10.1017/S1743921313015123

Global Astrophysical Telescope System – GATS

M. Polińska, K. Kamiński, W. Dimitrov, M. Fagas, W. Borczyk, T. Kwiatkowski, R. Baranowski, P. Bartczak, and A. Schwarzenberg–Czerny

Astronomical Observatory Institute, Faculty of Physics, Adam Mickiewicz University,
Słoneczna Street 36, 60-286 Poznań, Poland
email: polinska@amu.edu.pl

Abstract. The Global Astronomical Telescope System is a project managed by the Astronomical Observatory Institute of Adam Mickiewicz University in Poznań (Poland) and it is primarily intended for stellar medium/high resolution spectroscopy. The system will be operating as a global network of robotic telescopes. The GATS consists of two telescopes: PST 1 in Poland (near Poznań) and PST 2 in the USA (Arizona). The GATS project is also intended to cooperate with the BRITE satellites and supplement their photometry with spectroscopic observations.

Keywords. instrumentation: spectrographs, techniques: radial velocities, spectroscopic, stars: binaries, oscillations (including pulsations).

The first telescope, PST 1 (Poznań Spectroscopic Telescopes), was designed as a spectroscope producing medium/high resolution spectra. The telescope has been in operation at Borowiec Astrogeodynamic Observatory near Poznań (Poland) since summer 2007, and its description can be found in Baranowski *et al.* (2009). The other telescope, PST 2, is scheduled to start regular observations in the end of 2013 at a site in Arizona, USA (for more information see Kamiński *et al.*, these proceedings).

The PST 1 consists of a twin 0.5-m Newtonian telescopes, fibre-fed echelle spectrograph with a resolution $R \sim 35\,000$, and a low-noise back-illuminated 2k × 2k Andor DZ 436 CCD camera. The telescope is embedded on an equatorial, robotic mount and is capable of performing remotely-controlled observations. The PST 1 may be used in several different modes. In the basic mode only a single mirror is used for spectroscopy. In the second mode both mirrors are used to produce two simultaneous echelle spectra of the same object. Such parallel spectra can be used to increase signal-to-noise (S/N) ratio or help to remove cosmic rays. The third mode allows for simultaneous photometric and spectroscopic observations.

Results of radial velocity measurements done so far with the PST 1 show a stability at a level of $100\ \mathrm{m\,s^{-1}}$. This corresponds to about 1/25 of the pixel size of our CCD (13.5 μm). Thanks to its light-efficient design and low-noise camera, we are able to achieve S/N ~ 100 spectrum for a $V = 7$ mag star with a 60-min exposure time. This means that we can precisely measure radial velocities for late-type stars down to magnitude of about 11.5 mag. The spectral range covers 57 echelle orders with wavelength from 4280 Å to 7500 Å.

The PST 2 is to be placed at least 120 degrees apart from the PST 1 in longitude. Both telescopes will form the GATS system which will allow for a high duty-cycle, nearly continuous monitoring of selected objects (see Fig. 1).

The GATS will be used to do research in the following topics:
- asteroseismology of hybrid pulsating stars,

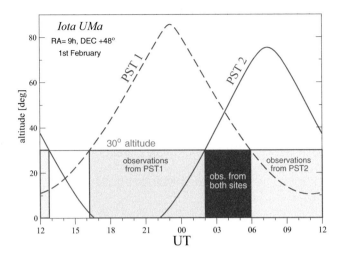

Figure 1. Changes of altitude of ι UMa as observed by the PST 1 (dashed line) and PST 2 (solid line), computed for 1st of February. The dark rectangle indicates the time when the object can be observed by both telescopes.

- stellar rotation and dynamical evolution of binary eclipsing stars,
- stellar activity cycles.

For more information, please visit the website of the GATS project: http://www.astro.amu.edu.pl/GATS.

Reference

Baranowski, R., Smolec, R., Dimitrov, W., *et al.* 2009, *MNRAS*, 396, 2194

Precision Asteroseismology
Proceedings IAU Symposium No. 301, 2013
J. A. Guzik, W. J. Chaplin, G. Handler & A. Pigulski, eds.

© International Astronomical Union 2014
doi:10.1017/S1743921313015135

Mode identification from spectroscopy of gravity-mode pulsators

K. R. Pollard[1], E. Brunsden[1], P. L. Cottrell[1], M. Davie[1], A. Greenwood[1], D. J. Wright[2], and P. De Cat[3]

[1] Dept of Physics & Astronomy, University of Canterbury, Christchurch, New Zealand
email: karen.pollard@canterbury.ac.nz

[2] Department of Astrophysics, University of New South Wales, Sydney, Australia

[3] Royal Observatory of Belgium, Ringlaan 3, 1180 Brussel, Belgium

Abstract. The gravity modes present in γ Doradus stars probe the deep stellar interiors and are thus of particular interest in asteroseismology. For the MUSICIAN programme at the University of Canterbury, we obtain extensive high-resolution echelle spectra of γ Dor stars from the Mt John University Observatory in New Zealand. We analyze these to obtain the pulsational frequencies and identify these with the multiple pulsational modes excited in the star. A summary of recent results from our spectroscopic mode-identification programme is given.

Keywords. line: profiles, techniques: spectroscopic, stars: oscillations (including pulsations)

1. Introduction: why are γ Dor stars interesting?

The γ Doradus stars are of particular interest in asteroseismology for a number of reasons. First, the γ Dor stars are defined as A-F type main sequence (or close to main sequence) stars that display gravity-modes (g-modes) of high radial order. These g-modes probe the deep interior of the stars and thus provide us with a probe of the structure and evolutionary changes that occur in regions as deep as the core. Second, some "hybrid" γ Dor/δ Scuti stars also undergo simultaneous pressure-mode pulsations which probe the outer convective envelopes of these stars. Finally, it is possible that solar-like oscillations may also occur in such stars, which would supply information about time and distance scales for convective processes. Although a star showing all three types of variability (g-modes, p-modes and solar-like variability) is yet to be discovered, it is intriguing that this region of the H-R diagram would allow such a star to exist.

2. Results from our extensive high-precision time series spectroscopy

For our MUSICIAN programme at the University of Canterbury, we have obtained extensive, high-resolution echelle spectra of the g-mode γ Dor pulsators and related stars, using the 1.0-m McLellan telescope and the high-resolution fibre-fed HERCULES spectrograph at the Mt John University Observatory. The observatory is at latitude $43°59.2'$ S and longitude $170°27.9'$ E and is thus at a useful longitude to obtain multi-site data in association with sites in Chile, Australia and South Africa. This is an ideal set-up for undertaking long time-series echelle spectra of bright stars ($V \lesssim 9$).

A description of our observational, reduction and analysis techniques is given in Brunsden *et al.* (these proceedings). Our intent is to obtain reliable mode identifications that can constrain the theoretical models of γ Dor stars. In particular we are looking at the way that g-modes differ from p-mode pulsations and the effects of stellar rotation on the mode identification. Rotational effects appear to be important since the rotational

Table 1. A summary of our spectroscopic mode identifications of γ Dor stars.

Identifier	$v \sin i$ (km s^{-1})	Frequency (d^{-1})	Photometric	Spectroscopic	Reference
HD 12901	64	1.3959(2)	1	1,1	1, 3, 5
		1.1862(2)	1	1,1	1, 3, 5
		1.6812(2)	1	1,1	3, 5
		1.2156(2)	1	1,1	1, 3, 5
		1.5596(2)	1	1,1	3, 5
HD 27290	57	1.3641(2)		1,1	2, 5
(γ Dor)		1.3209(2)		3,3 or 1,1	2, 5
		1.4712(3)		1,1 or 2,0	2, 5
		1.8783(2)		1,1	5
HD 40745	44	0.7523(5)		2,-1	8
		1.0863(7)		3,-3 or 2,-2	8
HD 55892	51	0.055972(4)		1,-1	6, 9
(QW Pup)		0.064846(4)		4,-1	6, 9
		5.219398(2)		4, 2	6, 9
HD 65526	59	2.616		1,-1	7
		1.840		1,-1	7
HD 112429	116	0.0515(3)		1,-1	6
HD 135825	40	1.3150(3)		1,1	4, 5
		0.2902(4)		2,-2	4, 5
		1.4045(5)		4,0	4, 5
		1.8829(5)		1,1	4, 5
HD 139095	64	2.353		2$\pm$1,1$\pm$1	9
HD 189631	38	1.6774(2)		1,1	5, 7
		1.4172(2)		1,1	5, 7
		0.0714(2)		2,-2	5, 7
		1.8228(2)		1,1 or 1,1	5, 7

References: [1] Aerts *et al.* (2004); [2] Balona *et al.* (1996); [3] Brunsden (2012a); [4] Brunsden (2012b); [5] Brunsden (2013); [6] Davie (2013); [7] Greenwood (2012); [8] Maisonneuve *et al.* (2011); [9] Wright (2008).

and pulsational frequencies are of the same magnitude. We note that identification of modes as either p-mode or g-mode based purely on the observed frequency, rather than the co-rotating frequency, is uncertain unless the rotational frequency is known.

Table 1 summarises the significant number of γ Doradus stars spectroscopic pulsation mode identifications that we have obtained. This is credited to extensive, high-resolution datasets and sophisticated analysis techniques. We can contribute these, and further results, to constrain stellar structure and seismic models of γ Doradus stars. We hope to further refine our analysis methods, and thus our inputs to the stellar structure models, through a more thorough investigation and understanding of the effects of rotation on stellar pulsations.

References

Aerts, C., Cuypers, J., De Cat, P., *et al.* 2004, *A&A* 415, 1079

Balona L. A., Böhm T., Foing, B. H., *et al.* 1996, *MNRAS* 281, 1315

Brunsden E., Pollard K. R., Cottrell P. L., Wright D. J., & De Cat P. 2012a, *MNRAS* 427, 2512

Brunsden E., Pollard K. R., Cottrell P. L., Wright D. J., De Cat P., & Kilmartin P. M. 2012b, *MNRAS* 422, 3535

Brunsden E. 2013, Ph.D. thesis, University of Canterbury

Davie, M. W. 2013, M.Sc. thesis, University of Canterbury

Greenwood, A. 2012, Astr391 project, University of Canterbury

Maisonneuve, F., Pollard, K. R., Cottrell, P. L., *et al.* 2011, *MNRAS* 415, 2977

Wright, D. J. 2008, Ph.D. thesis, University of Canterbury

Precision Asteroseismology
Proceedings IAU Symposium No. 301, 2013
J. A. Guzik, W. J. Chaplin, G. Handler & A. Pigulski, eds.

© International Astronomical Union 2014
doi:10.1017/S1743921313015147

An observational asteroseismic study of the pulsating B-type stars in the open cluster NGC 884

S. Saesen[1,2], M. Briquet[2,3,†], C. Aerts[2,4], A. Miglio[5], and F. Carrier[2]

[1] Observatoire de Genève, Université de Genève, Switzerland
email: sophie.saesen@unige.ch

[2] Instituut voor Sterrenkunde, Katholieke Universiteit Leuven, Belgium
[3] Institut d'Astrophysique et de Géophysique de l'Université de Liège, Belgium
[4] Department of Astrophysics, Radboud University Nijmegen, The Netherlands
[5] School of Physics and Astronomy, University of Birmingham, UK

Abstract. Recent progress in the seismic interpretation of field β Cep stars has resulted in improvements of the physical description in the stellar structure and evolution model computations of massive stars. Further asteroseismic constraints can be obtained from studying ensembles of stars in a young open cluster, which all have similar age, distance and chemical composition. We present an observational asteroseismic study based on the discovery of numerous multi-periodic and mono-periodic B-type stars in the open cluster NGC 884 (χ Persei). Our study illustrates the current status of ensemble asteroseismology of this young open cluster.

Keywords. open cluster and associations: individual (NGC 884), stars: early-type, stars: oscillations, techniques: photometric

1. Data

We exploited the differential time-resolved multi-colour photometry of a selected field of NGC 884 presented in Saesen *et al.* (2010). The photometry was gathered at 12 sites using 15 different instruments and resulted in almost 77 500 CCD images and 92 hours of photo-electric data. The data span three observation seasons (2005 – 2007) and the resulting precision of the light curves of the brightest stars is 5.7 mmag in V, 6.9 mmag in B, 5.0 mmag in I and 5.3 mmag in U. We identified 75 periodic B-type stars through an automated frequency analysis of the V-filter data and focus on these stars in the current study. We derived effective temperature and luminosity values from absolute photometry in the seven Geneva filters and we combined them with literature values to get the position of the B-type stars in the H-R diagram.

2. Derivation of the pulsational properties of the B-type stars

We performed a detailed frequency analysis with PERIOD04 (Lenz & Breger 2005). We used a weighted and non-weighted frequency analysis for the V light curve, and only a weighted analysis for B, I and U. We considered a peak significant if it has a signal-to-noise ratio above 4. A classical prewhitening scheme was adopted to determine all significant frequencies present in a certain star. As a result, we keep 65 periodic B-type

† Postdoctoral Researcher of the Fund for Scientific Research, Fonds de la Recherche Scientifique – FNRS, Belgium

stars of which 36 are mono-periodic, 16 are bi-periodic, 10 are tri-periodic, 2 are quadru-periodic and one star has 9 independent frequencies. The results of the multi-frequency fits in the different colours are presented in Table 4 of Saesen *et al.* (2013).

Since none of the detected modes shows any phase differences in the different filters, we restricted for the mode identification to the photometric amplitude ratios to determine the degree l, following the method of Dupret *et al.* (2003). For each pulsator, we selected stellar equilibrium models that fitted the observed position in the H-R diagram within a 2σ-error box. We then selected for each model and each degree l between 0 and 4 the theoretical frequencies that fit the observed ones, taking into account rotational splitting. We confronted the theoretical amplitude ratios with those observed and made a mode identification by eliminating the degrees for which the theoretical amplitude ratios do not match the observed ones. We securely identified the degree for 12 of the 114 detected frequencies.

3. Relations between pulsation and basic stellar parameters

For the following analyses, we exclude three stars: one ellipsoidal binary and two non-member δ Sct stars. All other 62 B-type stars can be considered members of NGC 884 based on photometric diagrams and membership studies in the literature. We did not find a correlation between the projected rotational velocity and the amplitude or frequency of the modes. The amplitudes of the oscillations, however, decrease as the frequency values increase. A comparison between the observed and theoretical frequency-radius relation allowed us to identify eight β Cep stars in our sample, namely Oo 2246, Oo 2299, Oo 2444, Oo 2488, Oo 2520, Oo 2572, Oo 2601 and Oo 2694.

4. Age estimate of the cluster

The seismic properties of the selected β Cep stars were confronted with those predicted for a dense full grid of stellar models (Briquet *et al.* 2011). We searched for a consistent cluster solution by requesting common cluster parameters, i.e., equal age and initial chemical composition, without specifying their values. We eliminated models based on the observed position of the stars in the H-R diagram and the observed frequency spectrum, allowing for rotational splitting in the first-order approximation. Imposing the identified degrees and measured frequencies of the radial, dipole and quadrupole modes of five β Cep stars led to a seismic cluster age estimate of $\log(\text{age/yr}) = 7.12 - 7.28$. This is fully compatible with the age estimate obtained from modeling of an eclipsing binary in the cluster ($\log(\text{age/yr}) > 7.10$, Southworth *et al.* 2004) and illustrates the valuable alternative that cluster asteroseismology can offer for age determinations.

5. Further details

For more information, we refer the interested reader to Saesen *et al.* (2010, 2013).

References

Briquet, M., Aerts, C., Baglin, A. *et al.* 2011, *A&A*, 527, A112
Dupret, M.-A., De Ridder, J., De Cat, P. *et al.* 2003, *A&A*, 398, 677
Lenz, P. & Breger, M. 2005, *CoAst*, 146, 53
Saesen, S., Carrier, F., Pigulski, A., *et al.* 2010, *A&A*, 515, A16
Saesen, S., Briquet, M., Aerts, C., *et al.* 2013, *AJ*, 146, 102
Southworth, J., Zucker, S., Maxted, P. F. L., & Smalley, B. 2004, *MNRAS*, 355, 986

Precision Asteroseismology
Proceedings IAU Symposium No. 301, 2013
J. A. Guzik, W. J. Chaplin, G. Handler & A. Pigulski, eds.

© International Astronomical Union 2014
doi:10.1017/S1743921313015159

On the information content
of stellar spectra

Jesper Schou

Max Planck Institute for Solar System Research
Max-Planck-Str. 2, 37191 Katlenburg-Lindau, Germany
email: schou@mps.mpg.de

Abstract. With the increasing quality of asteroseismic observations it is important to minimize the random and systematic errors in mode parameter estimates. To this end it is important to understand how the oscillations relate to the directly observed quantities, such as intensities and spectra, and to derived quantities, such as Doppler velocity. Here I list some of the effects we need to take into account and show an example of the impact of some of them.

Keywords. stars: oscillations

1. Calculation of mode sensitivities

To do precision asteroseismology, we need precise and accurate observations, preferably for a large number of modes. We thus need to understand how the oscillations of a star relate to the observed quantities, such as spectra and broad-band intensity measurements.

Often the intensity perturbation caused by a mode is calculated by multiplying a spherical harmonic with the limb darkening and integrating over the disk. For a velocity observation, a velocity projection factor is also included. This neglects many physical, instrumental and data analysis effects and can cause systematic errors, loss of S/N and failure to make the best use of the observations. It is thus planned to systematically attack these problems, noting that significant work has been done in other contexts (e.g. Zima 2006). This is becoming increasingly important in view of the increased quantity and quality of data soon to be available from the SONG project (Grundahl *et al.* 2009).

For intensity observations some effects to consider include accurate limb darkening, that the intensity perturbation is not necessarily given by the limb darkening times the spherical harmonic (due to such effects as the observing height change with incidence angle), that there are nonadiabatic effects, that the mode phases depend on height (Baldner & Schou 2012), causing center-to-limb phase shifts, the interaction of the oscillations with convection causing spatially variable perturbations and so forth. For spectrally-resolved observations (e.g. Doppler velocity) the situation is further complicated by the nonradial motion near the surface, change of height of formation as a function of spectral line, position in the spectral line, line broadening, convective blueshift etc.

To illustrate the importance of some of these effects, Fig. 1 shows the relative mode visibilities for different values of the degree l, and angular order m, as a function of inclination i. To calculate these results, a snapshot of a solar MHD calculation (courtesy of Bob Stein, using the Stagger code (Beeck *et al.* 2012) and the EOS and opacities from Nordlund (1982)) was used to synthesize the Fe I 617.3 nm line as a function of viewing angle. These profiles were then shifted according to a solid-body rotation law and averaged over the disk to create a reference line profile. The velocity pattern from several low-degree spherical harmonics was then added to the velocity profile, integrated and cross-correlated with the reference profile to determine the mode sensitivity.

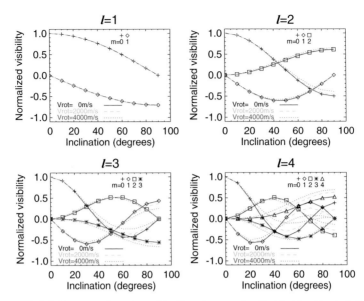

Figure 1. Visibilities as a function of inclination angle, divided by the total visibility (summed in quadrature over m). The results are labeled by the equatorial velocity.

Figure 1 shows that even a modest solar-like rotation causes a significant change in the mode visibilities, especially at $l = 3$ and $l = 4$, which could cause a significant mis-estimation of the inclination. For unresolved modes it may also lead to mis-estimation of the rotational splittings. The zero-velocity cases follow the results of Gizon & Solanki (2003), as expected. On the other hand many potentially significant effects were not taken into account. The oscillations were assumed to be radial at the surface, constant as a function of height and unaffected by the granulation. Also, no account was taken of differential rotation or the possible variation of the phase or amplitude with height described by Baldner & Schou (2012).

2. Conclusion

It is clear that there are many effects we need to take into account if we wish to accurately model stellar spectra and extract the maximum information. Figure 1 only shows one example and the plan is to systematically investigate all the relevant effects.

The author would like to thank Bob Stein, Björn Löptien, Aaron Birch, Robert Cameron, Charles Baldner, Regner Trampedach and Warrick Ball for insights and help with various aspects of the analysis.

References

Baldner, C. S. & Schou, J. 2012, *ApJ*, 760, L1
Beeck, B., Collet, R., Steffen, M., *et al.* 2012, *A&A*, 539, A121
Gizon, L. & Solanki, S. K. 2003, *ApJ*, 589, 1009
Grundahl, F., Christensen-Dalsgaard, J., Arentoft, T., *et al.* 2009, *CoAst*, 158, 345
Nordlund, A. 1982, *A&A*, 107, 1
Zima, W. 2006, *A&A*, 455, 227

Precision Asteroseismology
Proceedings IAU Symposium No. 301, 2013
J. A. Guzik, W. J. Chaplin, G. Handler & A. Pigulski, eds.

© International Astronomical Union 2014
doi:10.1017/S1743921313015160

δ Scuti-type pulsation in the hot component of the Algol-type binary system BG Peg

T. Şenyüz[1] and E. Soydugan[1,2]

[1] Astrophysics Research Centre and Observatory, Çanakkale Onsekiz Mart University,
Çanakkale, Turkey
email: tuncsenyuz@gmail.com
[2] Department of Physics, Faculty of Arts and Sciences,
Çanakkale Onsekiz Mart University, Çanakkale, Turkey
email: esoydugan@comu.edu.tr

Abstract. In this study, 23 Algol-type binary systems, which were selected as candidate binaries with pulsating components, were observed at the Çanakkale Onsekiz Mart University Observatory. One of these systems was BG Peg. Its hotter component shows δ Scuti-type light variations. Physical parameters of BG Peg were derived from modelling the V light curve using the Wilson-Devinney code. The frequency analysis shows that the pulsational component of the BG Peg system pulsates in two modes with periods of 0.039 and 0.047 d. Mode identification indicates that both modes are most likely non-radial $l = 2$ modes.

Keywords. techniques: photometric, stars: binaries: eclipsing, stars: individual (BG Peg)

1. Introduction

Algol-type eclipsing binary system BG Peg ($V = 10.50$ mag, $P_{orb} = 1.952443$ d) was listed in the catalogue of Algol-type binary stars (Svechnikov & Kuznetsova 1990). It was also listed by Soydugan *et al.* (2006) as a system with the component in the δ Sct instability strip. Soydugan *et al.* (2009) discovered δ Scuti-type light variations in the hotter component of the system. The δ Scuti-type variability in BG Peg was also found independently by Dvorak (2009). In this paper, photometric study of BG Peg made using only V-filter data are presented.

2. Photometric observations

BG Peg was photometrically observed in Johnson's B and V filters over 13 nights during 2008 observing season at the Çanakkale Onsekiz Mart University Observatory using the 40-cm Schmidt-Cassegrain telescope equipped with a SBIG STL-1001E CCD camera. TYC 1698.1052 and TYC 1698.1142 were selected as comparison and check stars, respectively. The standard error of this photometry was about 0.010 mag in both filters.

3. Light curve and frequency analysis of BG Peg

The light curve of BG Peg was analyzed using the Wilson-Devinney (WD) program in a semi-detached configuration. From the analysis of the light curve, we derived the following parameters: phase shift, orbital inclination, dimensionless potential of the primary component, mass ratio and fractional luminosity of the primary component. The other parameters were kept fixed. The temperature of the primary was adopted following Popper (1980), namely 8770 K for A2 spectral type. Our WD solution is the following: $i = 83.2°$, $T_{eff,2} = 5155$ K, $\Omega_1 = 3.628$, $\Omega_2 = 2.313$, $q = 0.233$, $r_{1,mean} = 0.296$, and

T. Şenyüz & E. Soydugan

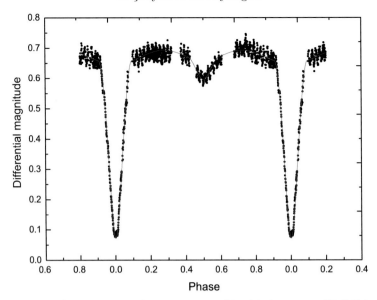

Figure 1. Synthetic (continuous line) and observed (dots) light curve of BG Peg in V filter.

$r_{2,\text{mean}} = 0.265$. The comparison of the theoretical and observed light curves is presented in Fig. 1.

A frequency analysis of the V-filter data was made using PERIOD 04 program (Lenz & Breger 2005). Pulsational frequencies of the primary component were found to be equal to $25.54\ \mathrm{d}^{-1}$ and $21.05\ \mathrm{d}^{-1}$. The pulsation amplitudes in V filter we determined are equal to 0.0153 and 0.0072 mag for the two modes, respectively.

4. Conclusion

Stars, which were selected from the catalogue of eclipsing binaries with candidate δ Scuti-type components prepared by Soydugan *et al.* (2006), have been observed since 2007. We discovered δ Scuti-type pulsations in the hotter components of BG Peg, DY Aqr and IO UMa. Two pulsational frequencies were detected in the hotter component of BG Peg. Using the FAMIAS program, we obtained spherical harmonic degrees to be $l = 2$ for both modes observed in this star.

Acknowledgement

We wish to thank the Turkish Scientific and Technical Research Council (Grant no. 107T634) for supporting this study.

References

Dvorak, S. 2009, *CoAst*, 160, 64
Lenz, P. & Breger, M. 2005, *CoAst*, 146, 53
Popper, D. M. 1980, *ARAA*, 18, 115
Soydugan, E., Soydugan, F., & Demircan, O., İbanoğlu, C. 2006, *MNRAS*, 370, 2013
Soydugan, E., Soydugan, F., Şenyüz, T., *et al.* 2009, *IBVS*, 5902
Svechnikov, M. A. & Kuznetsova, E. F. 1990, *Catalogue of Approximate Photometric and Absolute Elements of Eclipsing Variable Stars* (Izd-vo Ural'skogo universiteta)
Wilson, R. E. & Devinney, R. J. 1971, *ApJ*, 166, 605
Zima, W. 2008, *CoAst*, 155, 17

Precision Asteroseismology
Proceedings IAU Symposium No. 301, 2013
J. A. Guzik, W. J. Chaplin, G. Handler & A. Pigulski, eds.

© International Astronomical Union 2014
doi:10.1017/S1743921313015172

The BlaSGalF database

Marek Skarka

Department of Theoretical Physics and Astrophysics, Faculty of Science, Masaryk University,
Kotlářská 2, 611 37 Brno, Czech Republic
email: maska@physics.muni.cz

Abstract. BLASGALF is the name of an online, regularly updated list of known RR Lyraes exhibiting the Blazhko effect. It is an acronym of Blazhko Stars of Galactic Field. The list currently contains about 270 stars. It gives basic information about positions, brightnesses, pulsation and modulation periods based on data from catalogs and more than sixty papers. Using this database we found a lack of Blazhko stars with appropriate period characteristics. We introduce the present form of the list and discuss future plans for this database.

Keywords. RR Lyr, Blazhko effect, databases

1. Introduction

The idea of the database was induced by the actual need of an overall list of RR Lyraes that show the Blazhko effect†. Although there were few lists available (e.g. Smith 1995, Kolenberg & Uluş 2008, Le Borgne *et al.* 2012), none of them was comprehensive.

Our work started with the publication of the first, hard copy version of the list containing 242 Blazhko variables (Skarka 2013). To keep the list up to date, we subsequently launched the online version at http://physics.muni.cz/~blasgalf/, which is regularly updated.

At the beginning of August 2013 BLASGALF contained 270 stars. This number fluctuates as new stars are added or false Blazhko stars are removed. BLASGALF is thus an excellent tool that provides an actual overview of Blazhko variables.

2. Application of the database

BLASGALF could assist with the provision of the statistics on the numbers and incidences of Blazhko stars. There are
- 218 RRab type stars (205 single modulated, 4 variables with changing Blazhko period, 9 stars with multiple modulation period),
- 52 first overtone Blazhko RR Lyraes (50 single modulated, 2 multi modulated),

in the database. Fundamental mode RR Lyraes constitute 81% of known field Blazhko stars. About 5% of them show compound Blazhko effect with more than one modulation period. If we consider the incidence rate of Blazhko stars (270) among the stars in the GCVS catalogue (altogether 6143 stars of RRab and RRc type), this percentage is only about 4%, which is a very small percentage – recent estimates are about 47% (Jurcsik *et al.* 2009). We stress that our database does not contain Blazhko stars from large sky surveys like Catalina (http://www.lpl.arizona.edu/css/), which would certainly affect such statistics.

In addition to providing a quick overview of the stars, BLASGALF offers the possibility to compile basic statistics for the Blazhko variables. Figure 1 shows the distribution of

---

† Amplitude and/or phase modulation of the light curve, which was firstly noted in RW Dra by S. N. Blazhko (1907).

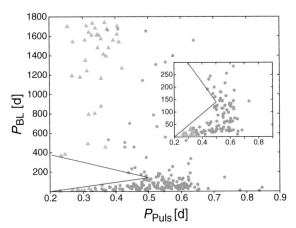

Figure 1. Distribution of Blazhko periods of stars according to their basic pulsation periods. RRc stars are plotted with triangles, while circles correspond to RRab stars.

Blazhko RR Lyraes in a P_{BL} vs. P_{Puls} diagram. It is clear that the stars are spread out more or less randomly when the Blazhko period exceeds 300 d. Below this limit the distribution has a very interesting shape. There is a lack of stars in the area delimited by the red lines, which intersect at a point near $P_{Puls} = 0.5$ d and $P_{BL} = 140$ d. The slope of the lower line is approximately 467, while the slope of the upper line is -800. There is no explanation for this interesting feature.

3. Future plans and conclusions

We would like to make our website more comprehensive and user friendly. The most important, soon available, improvements would be links to the VSX (Watson *et al.* 2006) and GEOS databases (Le Borgne *et al.* 2007) for each star. Other intended changes relate to the organization of the website – sorting according to various characteristics, listing with regard to particular constellation, a new section with stars only suspected of the Blazhko effect until now, etc. Other ideas are to extend the current information with light curves, amplitudes of the Blazhko effect and Fourier parameters. In addition, basic information about the Blazhko effect might be provided on the website.

BLASGALF offers the possibility of a quick overview of Blazhko stars. One application of this list was demonstrated on the Blazhko period distribution.

References

Blazhko, S. N. 1907, *AN*, 175, 325

Jurcsik, J., Sódor, Á., Szeidl, B., *et al.* 2009, *MNRAS*, 400, 1006

Kolenberg, K. & Uluş, N. D. 2008, http://www.univie.ac.at/tops/blazhko/Blazhkolist.html

Le Borgne, J.-F., Paschke, A., Vandenbroere, J., *et al.* 2007, *A&A*, 476, 307

Le Borgne, J.-F., Klotz, A., Poretti, E., *et al.* 2012, *AJ*, 144, 39

Samus, N. N., Durlevich, O. V., Kazarovets E. V., *et al.* 2012, *GCVS, VizieR On-line Data Catalog: B/gcvs*

Skarka, M. 2013, *A&A*, 549, A101

Smith, H. A. 1995, *RR Lyrae stars*, Cambridge Astrophysics Series, Vol. 27 (Cambridge, UK: Cambridge University Press)

Watson, C. L. 2006, *Society for Astronomical Sciences Annual Symposium*, 25, 47

Precision Asteroseismology
Proceedings IAU Symposium No. 301, 2013
J. A. Guzik, W. J. Chaplin, G. Handler & A. Pigulski, eds.

© International Astronomical Union 2014
doi:10.1017/S1743921313015184

Review of the ultrafast time resolution photopolarimeters based on SPADs

Aga Słowikowska[1], Gottfried Kanbach[2], Krzysztof Goździewski[3], Krzysztof Krzeszowski[1] and Arne Rau[2]

[1] Kepler Institute of Astronomy, University of Zielona Góra, Zielona Góra, Poland

[2] Max Planck Institute for Extraterrestrial Physics, Garching, Germany

[3] Toruń Centre for Astronomy, Nicolaus Copernicus University, Toruń, Poland

Abstract. We review photopolarimeters that are based on the Single Photon Avalanche Diodes (SPADs) and were designed, built, developed, and extensively used for high time resolution studies of astrophysical sources. Examples of such detectors are OPTIMA, GASP, AquEYE, and IquEYE which can measure the time of arrival of single optical photons with an accuracy of down to 50 picoseconds. We describe the most exciting results obtained with the SPADs detectors starting from the best existing optical polarimetric measurements of the Crab pulsar, the discovery of the first optical magnetar and its quasi-periodic oscillations, as well as a verification of exoplanets around eclipsing cataclysmic variables. Additionally, we discuss possible applications of such detectors for asteroseismology.

Keywords. photometers, polarization, pulsars, magnetars, cataclysmic variables, exoplanets

1. Single Photon Avalanche Diodes (SPAD) instruments

OPTIMA – Optical Pulsar Timing Analyzer (Straubmeier 2001, Kanbach *et al.* 2003 and 2008, Stefanescu 2011) is a very high time resolution, single-photon sensitive optical photometer and polarimeter. It uses optical fibres to gather light from fixed apertures in the focal plane to SPADs, while the field surrounding the apertures is imaged using a standard CCD detector. Single photons are recorded with absolute UTC time-scale tagging accuracy of ~ 5 ns. The quantum efficiency of the SPADs reaches a maximum of 60% at 750 nm and lies above 20% in the range 450–950 nm. The system was designed from scratch as a guest instrument, easily adapted to different telescopes. It can be reconfigured for photometric or polarimetric use within one observing run.

GASP – Galway Astronomical Stokes Polarimeter is a fast, full Stokes, astronomical imaging polarimeter. Its construction allows to measure all four elements of the Stokes vector (I, Q, U, V) simultaneously over a broad wavelength range (400–800 nm). Measuring linear and circular polarization with a time resolution of the order of microseconds makes GASP a unique astronomical instrument (Kyne *et al.* 2010).

IquEYE and AquEYE – ultra-high-speed photometers capable of tagging the arrival time of each photon with a resolution and accuracy of 50 picoseconds, for hours of continuous acquisition, and with a dynamic range of more than 6 orders of magnitude (Barbieri *et al.* 2009, 2012, Naletto *et al.* 2009).

All these detectors are portable and were used successfully at several observatories.

2. Example results

Crab pulsar and its nebula. The linear polarization of the Crab pulsar and its close environment was derived with a time resolution as short as 11 μs, which corresponds to a phase interval of 1/3000 of the pulsar rotation (33 ms). High sampling allows to derive polarization details never achieved before (Słowikowska *et al.* 2009). These results were recently confirmed by Moran *et al.* (2013). On the other hand, AquEYE allowed the first sub-microsecond optical timing of the Crab (Germanà *et al.* 2012).

First optical magnetar. We observed extremely bright and rapid optical flaring in the Galactic transient SWIFT J195509.6+261406 (Stefanescu *et al.* 2008). Its optical light curves are phenomenologically similar to high-energy light curves of soft γ-ray repeaters and anomalous X-ray pulsars, which are thought to be neutron stars with extremely high magnetic fields – magnetars.

HU Aqr – a Jovian companion? The eclipsing binary HU Aqr is a polar (mWD + M4 V) with an orbital period of 125 minutes. This binary exhibits periodic variations of the observed-minus-calculated (O−C) curve. We performed a detailed study of archival and new OPTIMA observations, in terms of a new Light Travel Time (LTT) ephemeris model, formulated with respect to the Jacobi coordinates with the origin in the mass center of the binary. We found that the observations are best explained by one periodic signal, which can be interpreted by the presence of a $\sim$7 Jupiter-mass planet, in a $\sim$10 yr quasi-circular orbit (Goździewski *et al.* 2012, Słowikowska *et al.* 2013).

3. Conclusions

High time resolution instruments based on SPADs allow to time tag single optical photons with accuracy even down to 50 picoseconds. They have been successfully used to study rapidly changing sources such as optical pulsars, cataclysmic variables, including polars and intermediate polars (Nasiroglu *et al.* 2012), as well as very fast flares from optical magnetars. SPAD instruments can be successfully used also in case of pulsating stars to study different oscillation modes, especially at high frequencies. Such oscillations penetrate to different depths inside the stars and provide information about their otherwise unobservable interiors.

Acknowledgements

AS and KK acknowledge support from NCN, grant UMO-2011/03/D/ST9/00656.

References

Barbieri, C., Naletto, G., Capraro, I., *et al.* 2009, *Proc. SPIE*, 7355, 15
Barbieri, C., Naletto, G., Zampieri, L., *et al.* 2012, in: R. E. M. Griffin, R. J. Hanisch, & R. Seaman (eds.), *New Horizons in Time-Domain Astronomy*, Proc. IAU Symposium No. 285 (Cambridge: Cambridge University Press), p. 280
Germanà, C., Zampieri, L., Barbieri, C., *et al.* 2012, *A&A*, 548, A47
Goździewski, K., Nasiroglu, I., Słowikowska, A., *et al.* 2012, *MNRAS*, 425, 930
Kanbach, G., Kellner, S., Schrey, F. Z., *et al.* 2003, *Proc. SPIE*, 4841, 82
Kanbach, G., Stefanescu, A., Duscha, S., *et al.* 2008, *ASSL*, 351, 153
Kyne, G., Sheehan, B., Collins, P., Redfern, M., & Shearer, A. 2010, *EPJ Web of Conferences*, 5, 5003
Moran, P., Shearer, A., Mignani, R. P., *et al.* 2013, *MNRAS*, 433, 2564
Naletto G., Barbieri, C., Occhipinti, T., *et al.* 2009, *A&A*, 508, 531
Nasiroglu, I., Słowikowska, A., Kanbach, G., & Haberl, F. 2012, *MNRAS*, 420, 3350
Słowikowska, A., Kanbach, G., Kramer, M., & Stefanescu, A. 2009, *MNRAS*, 397, 103
Słowikowska, A., Goździewski, K., Nasiroglu, I., *et al.* 2013, *ASP-CS*, 469, 363
Stefanescu, A. 2011, Ph.D. Thesis, Technische Universität München
Stefanescu, A., Kanbach, G., Słowikowska, A., *et al.* 2008, *Nature*, 455, 503
Straubmeier, C. 2001, Ph.D. Thesis, Technische Universität München

Precision Asteroseismology
Proceedings IAU Symposium No. 301, 2013
J. A. Guzik, W. J. Chaplin, G. Handler & A. Pigulski, eds.

© International Astronomical Union 2014
doi:10.1017/S1743921313015196

Order and chaos in hydrodynamic BL Her models

Radosław Smolec and Paweł Moskalik

Nicolaus Copernicus Astronomical Centre,
ul. Bartycka 18, 00-716 Warszawa, Poland
email: smolec@camk.edu.pl

Abstract. Many dynamical systems of different complexity, e.g. 1D logistic map, the Lorentz equations, or real phenomena, like turbulent convection, show chaotic behaviour. Despite huge differences, the dynamical scenarios for these systems are strikingly similar: chaotic bands are born through the series of period doubling bifurcations and merge through interior crises. Within chaotic bands periodic windows are born through the tangent bifurcations, preceded by the intermittent behaviour. We demonstrate such behaviour in models of pulsating stars.

Keywords. convection, hydrodynamics, methods: numerical, stars: oscillations, Cepheids

During our study of hydrodynamic BL Her-type models showing periodic and quasi-periodic modulation of pulsation akin to the Blazhko Effect (Smolec & Moskalik 2012) we also found the domains of chaotic behaviour. Here we report the initial analysis of these models, focusing on the universal behaviour they display.

All of our models were computed with the Warsaw nonlinear convective pulsation codes (Smolec & Moskalik 2008). The models have the same mass ($0.55\,M_\odot$) and chemical composition ($X/Z = 0.76/0.0001$) and were computed along a horizontal strip ($136\,L_\odot$) in the H-R diagram with a maximum step in effective temperature of $1\,$K. Convective parameters are the same as in Smolec & Moskalik (2012). We note that the eddy-viscous dissipation was strongly decreased, resulting in too-large pulsation amplitudes as compared with observations. Yet, the models show a wealth of dynamical behaviours characteristic of deterministic chaos. Except for the period-doubling effect (Smolec *et al.* 2012), such behaviour was not detected in any BL Her star so far, but these models may help to understand the pulsation of more luminous, longer-period, irregular variables.

In Fig. 1 (top) we show the bifurcation diagram for our models: the histogram of possible values of maximum radii (gray shaded) plotted as a function of the control parameter for which we choose the effective temperature, $T_{\rm eff}$. This diagram is compared with the bifurcation diagram for the simplest chaotic system, namely the iteration of the logistic map, $x_{n+1} = kx_n(1 - x_n)$ (Fig. 1, bottom). Despite the huge differences between the two systems, striking similarities are apparent.

The chaotic bands are born through the period-doubling cascades (within dashed frames in Fig. 1). In case of our models the single-periodic (period-one) cycle (one point on the bifurcation diagram), found at both the cool and the hot edge of the computational domain, undergoes a series of period-doubling bifurcations en route to chaos. Period-two, period-four and period-eight cycles are all detected. In case of the logistic map the same scenario is computed as k is increased beyond $k = 3.0$.

Within the chaotic bands, periodic windows of order emerge. There are several such windows in case of our models, e.g. period-5, period-6, period-7 or period-9 windows. We stress that, contrary to recent claims (Plachy *et al.* 2013), stable periodic cycles are not necessarily caused by the resonances among pulsation modes, but may be an intrinsic

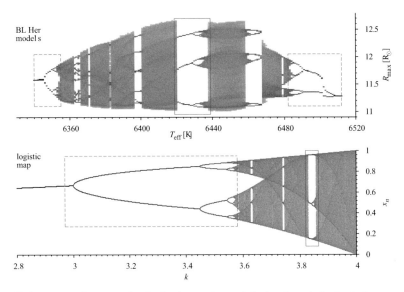

Figure 1. Bifurcation diagrams for hydrodynamic models (top) and the logistic map (bottom).

property of the chaotic systems. We focus our attention on parameter ranges indicated with solid frames in Fig. 1. Similarity between our models and the logistic map is apparent again. The well understood properties of simple logistic mapping allow us to explain the dynamics of the much more complex hydrodynamic models. As $k(T_{\mathrm{eff}})$ is increased the stable period-3 cycle is born through the tangent bifurcation. The bifurcation is preceded with *intermittent* behaviour (Pomeau & Mannevile 1980) – evolution of the system is characterized by long phases of almost periodic behaviour interrupted with shorter bursts of chaos. As $k(T_{\mathrm{eff}})$ is increased further, the period-3 cycle undergoes a series of period-doubling bifurcations to form three chaotic bands. These bands finally merge through *interior crises* (Grebogi et al. 1982) – a bifurcation in which the volume occupied by chaotic attractor suddenly changes. The crises occurs when three chaotic bands collide with the unstable period-3 cycle born along with stable period-3 cycle in the tangent bifurcation. Detailed analysis of the presented models is in preparation.

Acknowledgements

We acknowledge the IAU grants for the conference. This research is supported by the Polish Ministry of Science and Higher Education through Iuventus+ grant (IP2012 036572) awarded to RS.

References

Grebogi, C., Ott, E., & Yorke, J. A. 1982, *Phys. Rev. Lett.*, 48, 1507
Plachy, E., Kolláth, Z., & Molnár, L. 2013, *MNRAS*, 433, 3590
Pomeau, Y. & Mannevile, P. 1980, *Comm. Math. Phys.*, 74, 189
Smolec, R. & Moskalik, P. 2008, *AcA*, 58, 193
Smolec, R. & Moskalik, P. 2012, *MNRAS*, 426, 108
Smolec, R., Soszyński, I., Moskalik, P., et al. 2012, *MNRAS*, 419, 2407

Precision Asteroseismology
Proceedings IAU Symposium No. 301, 2013
J. A. Guzik, W. J. Chaplin, G. Handler & A. Pigulski, eds.

© International Astronomical Union 2014
doi:10.1017/S1743921313015202

Extensive spectroscopic and photometric study of HD 25558, a long orbital-period binary with two SPB components

Á. Sódor[1,2], P. De Cat[1], D. J. Wright[1,3], C. Neiner[4], M. Briquet[5],
R. J. Dukes[6], F. C. Fekel[7], G. W. Henry[7], M. H. Williamson[7],
M. W. Muterspaugh[7], E. Brunsden[8], K. R. Pollard[8], P. L. Cottrell[8],
F. Maisonneuve[8], P. M. Kilmartin[8], J. M. Matthews[9], T. Kallinger[10],
P. G. Beck[11], E. Kambe[12], C. A. Engelbrecht[13], R. J. Czanik[14],
S. Yang[15], O. Hashimoto[16], S. Honda[16,17], J.-N. Fu[18],
B. Castanheira[19], H. Lehmann[20], N. Behara[21], H. Van Winckel[11],
S. Scaringi[11], J. Menu[11], A. Lobel[1], P. Lampens[1], and P. Mathias[22]

[1] Royal Observatory of Belgium, Brussel, Belgium
[2] Konkoly Observatory, Hungarian Academy of Sciences, Budapest, Hungary
[3] School of Physics, University of New South Wales, Sydney, Australia
[4] LESIA, CNRS, Observatoire de Paris, UMPC, Université Paris Diderot, Meudon, France
[5] Institut d'Astrophysique et de Géophysique, Université de Liège, Liège, Belgium
[6] Department of Physics and Astronomy, The College of Charleston, Charleston, SC, USA
[7] Center of Excellence in Information Systems, Tennessee State University, Nashville, TN, USA
[8] Department of Physics and Astronomy, University of Canterbury, Christchurch, New Zealand
[9] Dept. of Physics and Astronomy, University of British Columbia, Vancouver, BC, Canada
[10] Institute for Astronomy, University of Vienna, Vienna, Austria
[11] Instituut voor Sterrenkunde, K. U. Leuven, Leuven, Belgium
[12] Okayama Astrophysical Observatory, National Astronomical Observatory, Okayama, Japan
[13] Department of Physics, University of Johannesburg, Johannesburg, South Africa
[14] Dept. of Phys., Potchefstroom Campus, North-West University, Potchefstroom, South Africa
[15] Department of Physics and Astronomy, University of Victoria, Victoria, BC, Canada
[16] Gunma Astronomical Observatory, Takayama-mura, Agatsuma, Gunma, Japan
[17] Kwasan Observatory, Kyoto University, Kyoto, Japan
[18] Department of Astronomy, Beijing Normal University, Beijing, China
[19] Department of Astronomy, The University of Texas at Austin, Austin, TX, USA
[20] Thüringer Landessternwarte Tautenburg, Tautenburg, Germany
[21] Université Libre de Bruxelles, Bruxelles, Belgium
[22] Université de Toulouse, UPS-OMP, IRAP, CNRS, Toulouse, France

Abstract. We carried out an extensive photometric and spectroscopic investigation of the SPB binary, HD 25558 (see Fig. 1 for the time and geographic distribution of the observations). The ∼2000 spectra obtained at 13 observatories during 5 observing seasons, the ground-based multi-colour light curves and the photometric data from the *MOST* satellite revealed that this object is a double-lined spectroscopic binary with a very long orbital period of about 9 years. We determined the physical parameters of the components, and have found that both lie within the SPB instability strip. Accordingly, both components show line-profile variations consistent with stellar pulsations. Altogether, 11 independent frequencies and one harmonic frequency were identified in the data. The observational data do not allow the inference of a reliable orbital solution, thus, disentangling cannot be performed on the spectra. Since the lines of the two components are never completely separated, the analysis is very complicated. Nevertheless, pixel-by-pixel variability analysis of the cross-correlated line profiles was successful, and we were able to attribute all the frequencies to the primary or secondary component. Spectroscopic and photometric mode-identification was also performed for several of these frequencies of both binary components. The spectroscopic mode-identification results suggest that the inclination and rotation of the two components are rather different. While the primary is a slow rotator with ∼6 d rotation period, seen at ∼60° inclination, the secondary rotates fast with ∼1.2 d rotation period, and is seen at ∼20° inclination. Our spectropolarimetric measurements revealed that the secondary component has a magnetic field with at least a few hundred Gauss strength, while no magnetic field was detected in the primary.

The detailed analysis and results of this study will be published elsewhere.

Keywords. line: profiles, stars: binaries: general, stars: oscillations (including pulsations), stars: rotation, stars: variables: other

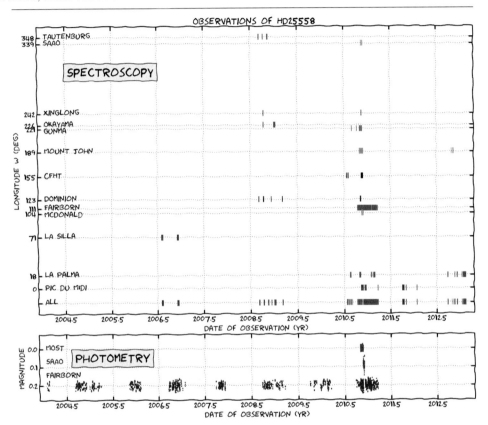

Figure 1. Time and geographic distribution of the spectroscopic and photometric observations of HD 25558. The main campaign was organised in the 2010/11 season.

Acknowledgements. Á.S. acknowledges support of the Belgian Federal Science Policy (project M0/33/029) and of the Eötvös Scholarship from the Hungarian Scholarship Board Office. M.B. is F.R.S.-FNRS Postdoctoral Researcher, Belgium. T.K. acknowledges financial support from the Austrian Science Fund (FWF P23608). The observational data we used in this study were obtained with the following telescopes and instruments: 2-m AST (T13) and 75-cm APT (T5) at Fairborn Observatory, McLellan tel.+HERCULES at Mt John Observatory, Otto Struve tel. at McDonald Observatory, 1.9-m tel.+GIRAFFE and 50-cm tel. at SAAO, Mercator tel.+HERMES at La Palma, 2.2-m tel. at Xinglong Observatory, 1.5-m tel.+GAOES at GAO, Alfred Jensch tel. at Karl Schwarzschild Observatory, 1.9-m tel.+HIDES at OAO, 1.2-m tel.+McKellar at DAO, Bernard Lyot tel.+NARVAL at Obs. Pic du Midi, CFHT+ESPaDOnS at Hawaii, Euler tel.+CORALIE at La Silla, the Canadian *MOST* satellite.

Precision Asteroseismology
Proceedings IAU Symposium No. 301, 2013
J. A. Guzik, W. J. Chaplin, G. Handler & A. Pigulski, eds.

© International Astronomical Union 2014
doi:10.1017/S1743921313015214

Analysis of strange-mode instability with time-dependent convection in hot massive stars

Takafumi Sonoi and Hiromoto Shibahashi

Department of Astronomy, University of Tokyo, Tokyo, 113-0033, Japan
email: sonoi@astron.s.u-tokyo.ac.jp, shibahashi@astron.s.u-tokyo.ac.jp

Abstract. We carry out nonadiabatic analysis of strange-modes in hot massive stars with time-dependent convection (TDC) for the first time. Although convective luminosity in envelopes of hot massive stars is not as dominative as in stars near the red edge of the classical Cepheid instability strip in the Hertzsprung-Russell (H-R) diagram, we have found that the strange-mode instability can be affected by the treatment of convection. However, existence of the instability around and over the Humphreys-Davidson (H-D) limit is independent of the treatment. This implies that the strange-mode instability could be responsible for the lack of observed stars over the H-D limit regardless of uncertainties on convection theories.

Keywords. stars: Hertzsprung-Russell diagram, stars: oscillations (including pulsations)

"Strange" modes with extremely high growth rates appearing in very luminous stars with $L/M \gtrsim 10^4 L_\odot/M_\odot$ have significantly different characteristics from ordinary p- and g-modes (Wood 1976, Shibahashi & Osaki 1981). Stability of strange-modes in hot massive stars has been analyzed with frozen-in convection (FC) by Glatzel & Mehren (1996), Godart *et al.* (2011), Saio *et al.* (2013), and others. In envelopes of hot massive stars, convective luminosity is not as dominant as in stars near the red edge of the classical Cepheid instability strip. But the strange-modes are excited at the convection zones, and we cannot definitely conclude that effects of convection are negligible.

We construct stellar models with $X = 0.70$, $Z = 0.02$ by using MESA (Paxton *et al.* 2011), and analyze their radial modes with the nonadiabatic code developed by Sonoi & Shibahashi (2012). We use the time-dependent convection (TDC) formulation derived by Unno (1967) and developed later by Gabriel *et al.* (1974) and by Gabriel (1996). This TDC theory has already been independently implemented by Grigahcène *et al.* (2005). Figure 1 shows the results for $60\,M_\odot$ with two types of FC; zero Lagrangian and Eulerian perturbations of convective luminosity, $\delta L_{\rm C} = 0$ and $L'_{\rm C} = 0$. The ascending sequences such as A1 and A2 correspond to ordinary modes, while the descending ones such as D1, D2 and D3 are strange-modes. On the whole, instability is suppressed in the $L'_{\rm C} = 0$ case compared with the $\delta L_{\rm C} = 0$ case. The pulsations with $L'_{\rm C} = 0$ are damped due to the substantial gradient of convective luminosity near the boundaries of convective layers, while there is no convective damping with $\delta L_{\rm C} = 0$.

Figure 2 shows the result with TDC. The degree of the instability is in between the two types of FC shown in Fig. 1. While unstable modes of D1 and D2 are excited around the convective layer caused by the Fe opacity bump, where the ratio of the convective to the total luminosity $L_{\rm C}/L_{\rm r}$ is $\sim 10^{-1}$, the sequence of D3 is pulsationally unstable invariably in all the three treatments, as seen in Figs. 1 and 2. These unstable modes of D3 are excited around the layer of He opacity bump, where $L_{\rm C}/L_{\rm r}$ is negligibly small, and hence we conclude that those modes are definitely unstable regardless of uncertainties of convection theories. We note that the instability of D3 appears around and above

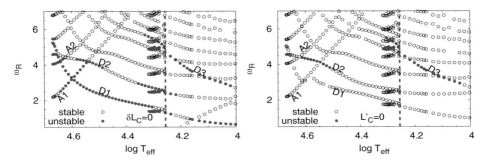

Figure 1. Modal diagrams of radial modes for 60 $M_\odot$ with FC of zero Lagrangian (*left panel*) and Eulerian (*right panel*) perturbations of convective luminosity. The open and filled circles denote stable and unstable modes, respectively, and the vertical dashed line in each panel shows the cross point of the evolutionary track and the H-D limit.

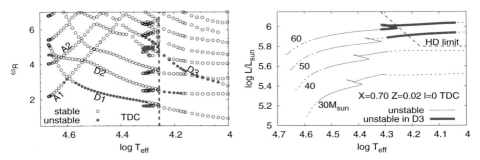

Figure 2. Results with TDC. Left: Same type of modal diagram as Fig. 1. Right: Instability ranges on the evolutionary tracks derived with TDC on the H-R diagram. The thin dashed lines are evolutionary tracks, and the solid parts correspond to evolutionary stages having at least one unstable mode. Instability of D3 is particularly indicated with the thick solid lines.

the H-D (Humphreys & Davidson 1979) limit. The resultant phenomena by the strange-mode instability have been proposed to be pulsationally driven mass loss by nonlinear analyses (e.g., Grott *et al.* 2005) and by observation (Aerts *et al.* 2010). More detailed investigations are worth doing.

References

Aerts C., Lefever, K., Baglin, A., *et al.* 2010, *A&A*, 513, L11
Gabriel, M. 1996, *Bull. of the Astron. Soc. of India*, 24, 233
Gabriel, M., Scuflaire, R., Noels, A., & Boury, A. 1974, *Bull. Acad. Royale de Belgique*, 60, 866
Glatzel, W. & Mehren, S. 1996, *MNRAS*, 282, 1470
Godart, M., Dupret, M.-A., Noels, A., *et al.* 2011, in: C. Neiner, G. Wade, G. Meynet, & G. Peters (eds.), *Active OB stars: structure, evolution, mass loss, and critical limits*, Proc. IAU Symposium No. 272 (San Francisco: ASP), p. 503
Grigahcène, A., Dupret, M.-A., Gabriel, M., Garrido, R., & Scuflaire, R. 2005, *A&A*, 434, 1055
Grott, M., Chernigovski, S., & Glatzel, W. 2005, *MNRAS*, 360, 1532
Humphreys, R. M. & Davidson, K. 1979, *ApJ*, 232, 409
Paxton, B., Bildsten, L., Dotter, A., Herwig, F., Lesaffre, P., & Timmes, F. 2011, *ApJS*, 192, 3
Saio, H., Georgy, C., & Meynet, G. 2013, *MNRAS*, 433, 1246
Shibahashi, H. & Osaki, Y. 1981, *PASJ*, 33, 427
Sonoi, T. & Shibahashi, H. 2012, *PASJ*, 64, 2
Unno, W. 1967, *PASJ*, 19, 140
Wood, P. R. 1976, *MNRAS*, 174, 531

Precision Asteroseismology
Proceedings IAU Symposium No. 301, 2013
J. A. Guzik, W. J. Chaplin, G. Handler & A. Pigulski, eds.

© International Astronomical Union 2014
doi:10.1017/S1743921313015226

Variable stars in the field of the young open cluster Roslund 2

P. Sowicka[1], G. Handler[2], R. Taubner[3], M. Brunner[3], V.-M. Passegger[3], F. Bauer[3], and E. Paunzen[4]

[1] Astronomical Observatory, Jagiellonian University, Orla 171, 30-244 Kraków, Poland,
email: paula@byk.oa.uj.edu.pl

[2] Nicolaus Copernicus Astronomical Center, Bartycka 18, 00-716 Warsaw, Poland

[3] Institut für Astronomie, Universität Wien, Türkenschanzstrasse 17, 1180 Wien, Austria

[4] Department of Theoretical Physics and Astrophysics, Masaryk University, Kotlarska 2, 611 37 Brno, Czech Republic

Abstract. The study of variable stars in open clusters via asteroseismology is a powerful tool for the study of stellar evolution and stars in general. That is because stars in clusters can be assumed to originate from the same interstellar cloud, so they share similar properties such as age and overall metallicity. We performed a search for variable stars in the field of the young open star cluster Roslund 2, with photoelectric and CCD photometry acquired at two different telescopes. Within the resulting light curves we have found 12 variable stars. Our measurements confirm three previously known variables.

Keywords. open clusters and associations: individual (Roslund 2), stars: variables

1. Introduction

The cluster Roslund 2 is a young (∼8 Myr, Kharchenko *et al.* 2005) open cluster, found in a survey of high luminosity stars in the northern Milky Way (Roslund 1960). It is the biggest cluster found in this survey, with a diameter of about 45′. No dedicated search for variable stars in this cluster is available in the literature.

Studying pulsating stars in open clusters has many advantages. Most importantly we can assume the same origin for all stars, i.e. we have tight constraints on their ages, metallicity and distance, which helps in asteroseismic investigations. The B-type pulsators, which naturally occur in open clusters, are good targets for asteroseismology. So it is important to find more such stars in open clusters and that was the purpose of our measurements.

2. Observations, data processing and searching for variable stars

The observations were performed using the 0.8-m "vienna little telescope" (vlt) at the Institute of Astronomy of the University of Vienna and the 0.75-m Automatic Photoelectric Telescope (APT) T6 at Fairborn Observatory in Arizona. We used 56 hours of CCD photometry in the *V* filter (vlt) and 42 hours of data in the Strömgren *uvy* filters (APT).

The reduction of the vlt data was made using standard procedures. For photometry we used two programs: DAOPHOT/ALLSTAR and DIAPL2 (Difference Image Analysis Package) (W. Pych, private communication). After visual examination of our light curves we found three variables. We cleaned the light curves by rejecting points by 3σ level and

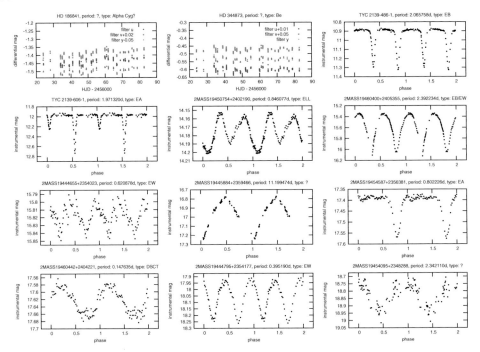

Figure 1. Resulting light curves of the variable stars

we found five additional variable stars. Afterwards we used program PERLARGE to compute Fourier amplitude spectra for all stars, and we found two more variables.

The APT data were reduced following standard photoelectric photometry schemes. The targets were: HD 186841, HD 344870, HD 344873 and HD 344878. Because of a nearby contaminating star, the measurements of HD 344878 were useless. HD 186841 was confirmed to be variable. HD 344870 was most constant and was used as a comparison star. HD 344873 was discovered to be variable.

For most of found variables we determined primary values of periods, based on the nice shape of the phased light curve. Figure 1 shows the resulting light curves, with the values of periods (if determined) and first classifications of the variable star types.

3. Summary

We performed a search for variable stars in the field of the young open star cluster Roslund 2. Three of the variables we observed were previously known, and we determined periods for ten stars. Unfortunately, none of these variables is likely to be of β Cephei type.

Acknowledgements. We are grateful to Wojtek Pych for permission to use and continued support of his DIAPL2 software. This research has been supported by the Polish NCN grant 2011/01/B/ST9/05448. PS acknowledges the Summer Student Program at the Nicolaus Copernicus Astronomical Center in Warsaw and IAU Grant.

References

Roslund, C. 1960, *PASP*, 72, 205

Kharchenko, N. V., Piskunov, A. E., Röser, S., Schilbach, E., & Scholz, R.-D. 2005, *A&A*, 438, 1163

Precision Asteroseismology
Proceedings IAU Symposium No. 301, 2013
J. A. Guzik, W. J. Chaplin, G. Handler & A. Pigulski, eds.

© International Astronomical Union 2014
doi:10.1017/S1743921313015238

Analysis of rosette modes of oscillations in rotating stars

Masao Takata[1] and Hideyuki Saio[2]

[1]Department of Astronomy, School of Science, The University of Tokyo,
7-3-1 Hongo, Bunkyo-ku, Tokyo 113-0033, Japan
email: takata@astron.s.u-tokyo.ac.jp

[2]Astronomical Institute, Graduate School of Science, Tohoku University,
6-3 Aramaki, Aoba-ku, Sendai 980-8578, Japan
email: saio@astr.tohoku.ac.jp

Abstract. We analyse the rosette modes of oscillations in rotating stars, which are characterised by rosette patterns of the kinetic-energy distribution on the meridional plane. Following our previous argument that these modes are generated by the rotation-induced interaction among eigenmodes with almost the same frequency, we discuss the structure of the rosette patterns base on the JWKB analysis. We also demonstrate that there exist nonaxisymmetric rosette modes.

Keywords. stars: interiors, stars: oscillations, stars: rotation

1. Introduction

The rosette modes constitute a new class of eigenmodes of oscillations in rotating stars. They have been discovered by Ballot *et al.* (2012), who numerically computed adiabatic eigenmodes of a polytropic model with index 3 that rotates uniformly at a few tens of percent of the break-up rate. Although the frequency of the modes is found in the range of that of gravity modes, it is not too low to be in the inertial domain, where the amplitude of eigenmodes is confined in the equatorial region. The most remarkable property of the modes is the fact that their kinetic-energy density distributes along rosette patterns on the meridional plane. Takata & Saio (2013) have identified the physical origin of the rosette modes as close degeneracy in the frequency among eigenmodes that have successive values of the spherical degree with the same parity (in the absence of rotation). Those modes strongly interact with each other due to the Coriolis force to form a family of rosette modes. This process can be precisely described by quasi-degenerate perturbation theory. As further analysis along the same lines, this paper concentrates on the JWKB analysis and nonaxisymmetric rosette modes.

2. Analysis and discussions

An asymptotic analysis can be developed to explain the structure of the rosette patterns in a simple way. After deriving expressions of eigenfrequencies and eigenfunctions of gravity modes for a large spherical degree, we substitute those expressions into the formula of the quasi-degenerate perturbation theory. Assuming that eigenmodes are completely degenerate in the absence of rotation, and that the number of degenerate modes is infinite, we obtain simple expressions of the rosette patterns, which are depicted in Fig. 1. It is found that the rosette pattern of each mode is composed of eight curves (in general), which are specified by two parameters. The first parameter, K, indicates a common non-negative integer to all rosette modes that come from a given family of

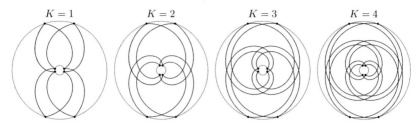

Figure 1. Schematic asymptotic structure of rosette modes on the meridional plane. The parameter q is chosen to be $q = 0.2$, whereas the parameter K is given above each panel. Refer to the main text for the meanings of the parameters. The dashed circles represent the inner and outer turning points.

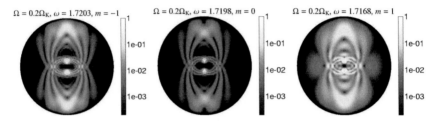

Figure 2. Distribution of the kinetic-energy density (multiplied by the square of the radius) of rosette modes on the meridional plane. The rotation rate, Ω (in units of the break-up rate, Ω_K), the dimensionless frequency, ω, and the azimuthal order, m, are given above each panel.

close degeneracy. The larger is K, the lower is the frequency of the corresponding family. The difference in the polar angle between the two end points of each constituent curve is equal to $K\pi/2$. The second parameter, q, is a fractional number between 0 and 1. The tilt angle of each rosette pattern from the rotation axis is equal to $q\pi/2$.

In the analysis of nonaxisymmetric rosette modes based on the quasi-degenerate perturbation theory, it turns out to be essential to assume that the Ledoux constant, $C_{n,l}$, is a small quantity on the same order as the rotation rate, Ω. This assumption has been checked not only numerically, but also asymptotically for large values of the spherical degree. Because of this assumption, the leading-order perturbation to the frequency (in the corotating frame), $-mC_{n,l}\Omega$, where m is the azimuthal order, is regarded as a second-order quantity. The rest of the analysis is essentially the same as in the axisymmetric cases. Examples of the thus-calculated nonaxisymmetric rosette modes are shown in Fig. 2. It is observed that the retrograde mode ($m = -1$) shows a clearer rosette pattern than the prograde mode ($m = 1$).

3. Conclusion and prospects

The rosette modes have revealed a new aspect of the effect of rotaion on oscillations. Even more analysis is awaited to discuss topics such as the possibility of their excitation and their influence on angular-momentum transport in stars.

References

Ballot, J., Lignières, F., Prat, V., Reese, D. R., & Rieutord, M. 2012, *ASP-CS*, 462, 389

Takata, M. & Saio, H. 2013, *PASJ*, 65, 68

Precision Asteroseismology
Proceedings IAU Symposium No. 301, 2013
J. A. Guzik, W. J. Chaplin, G. Handler & A. Pigulski, eds.

© International Astronomical Union 2014
doi:10.1017/S174392131301524X

The fast rotating δ Scuti pulsator V376 Per: Frequency analysis and mode identification

Nathalie Themessl[1], Veronique Fritz[1,2], Michel Breger[1,3], Sabine Karrer[1] and Barbara G. Castanheira[3]

[1]Institute for Astrophysics, University of Vienna,
Türkenschanzstraße 12, A-1180 Vienna, Austria
email: nathalie.themessl@univie.ac.at

[2]Kapteyn Astronomical Institute, University of Groningen, The Netherlands

[3]Department of Astronomy, University of Texas at Austin, USA

Abstract. Our simultaneous analysis of ground-based photometric and high-resolution spectroscopic data of the δ Scuti star V376 Per revealed eight individual frequencies from 82 nights of two-color photometry and six frequencies from the line-profile variations using 769 stellar spectra. Additionally, we identified the corresponding pulsation modes and derived reliable estimates of the line profile and pulsation mode parameters.

Keywords. stars: variables: δ Scuti, stars: interiors, line: profiles, methods: data analysis

1. Introduction

The δ Scuti star V376 Per (= HR 1170) was first detected to be variable by Breger (1969). The latest photometric and spectroscopic campaign (2008 – 2011) provided more than 3800 hours of multi-color data obtained by the twin automatic photoelectric 75-cm telescopes Wolfgang & Amadeus at Fairborn Observatory in Arizona and 769 stellar spectra obtained with the 2.1-m Otto Struve telescope at McDonald Observatory in Texas, USA. While the multi-color measurements were corrected according to the three-star-technique (Breger 1993), we reduced the spectra with standard `IRAF` routines in order to remove effects caused by the instrument itself, the atmosphere and Earth's motion.

2. Frequency analysis and mode identification

In the first step, we derived the individual frequencies from the photometric lightcurves using `Period04` (Lenz & Breger 2004) and we extracted the spectroscopic frequencies in the pixels across the unblended Fe II line at 4508 Å with `FAMIAS` (Zima 2008). Figure 1 illustrates both frequency spectra with the one-day aliases clearly visible in the spectral windows. The 12 most significant frequencies with an amplitude signal-to-noise ratio (SNR) $\geqslant 4$ and the results of the subsequent mode identification are listed in Table 1. Due to the high rotational velocity of V376 Per we used the Fourier parameter fit method, embedded in `FAMIAS`, to determine the pulsation modes by means of the following stellar parameters: $M = 1.71\,M_\odot$, $R = 2.39\,R_\odot$ (Kim & Lee 1990), $T_{\rm eff} = 7100$ K, solar metallicity and $\log g = 3.54$ (from Strömgren photometry).

3. Discussion and conclusion

The complementary photometric and spectroscopic studies provided useful constraints on the different modes of each detected frequency. In summary, the analysis provided

Table 1. The most significant frequencies in the photometric $(f_1 - f_8)$ and the spectroscopic $(\nu_1 - \nu_6)$ dataset including the best solution for the mode identification.

Frequency [d^{-1}]	Amplitude in v [mag]	Amplitude in y [mag]	Amplitude [km s^{-1}]	SNR	Mode identification
$f_1 = 10.06$	0.0499	0.0346		189	$l = 0$
$f_2 = 11.91$	0.0240	0.0166		91	$l = 1$
$f_3 = 18.60$	0.0076	0.0053		21	$l = 1$
$f_4 = 9.69$	0.0049	0.0033		18	$l = 0$
$f_5 = 21.97$	0.0031	0.0022		13	—
$f_6 = 10.88$	0.0041	0.0029		16	$l = 0, 1$
$f_7 = 19.71$	0.0022	0.0017		7	$l = 0, 1$
$f_8 = 15.63$	0.0027	0.0019		8	$l = 1$
$\nu_1 = 17.92$			1.176	13	$l = 10, m = 8$
$\nu_2 = 11.91$			0.710	11	$l = 3\text{--}4, m = 2$
$\nu_3 = 14.10$			0.622	9	$l = 6, m = 1, 2, 3, 4$
$\nu_4 = 16.97$			0.594	7	$l = 9, m = 3, 5, 7, 9$
$\nu_5 = 10.06$			0.447	11	$l = 0, m = 0$
$\nu_6 = 23.67$			0.428	5	$l = 10\text{--}11, m = 4, 5, 6$

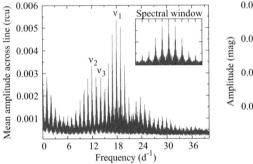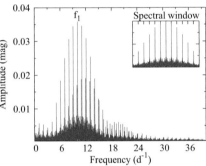

Figure 1. *Left*: The one-dimensional Fourier spectrum of the original spectroscopic dataset revealing the first three frequencies with the highest amplitude (in relative continuum units). *Right*: The general frequency spectrum of the combined photometric datasets (y and v) with the highest peak at 10.06 d^{-1}. The one-day alias peaks are clearly visible in both spectral windows.

low l-values for the photometrically detected frequencies and $l > 5$ for those detected in the spectroscopic dataset. We note that the line-profile analysis is capable of detecting (and also favoring) high-degree modes with an uncertainty of ± 1 in l. For this reason, we attained a slightly higher degree for $\nu_2 = 11.91$ d^{-1} by applying the spectroscopic method compared to the more reliable photometric determination of the same frequency. Moreover, the projected equatorial rotational velocity ($v \sin i$) could be constrained to 100 ± 2 km s^{-1} suggesting a favored inclination between 35 and 75 degrees. Additionally, the computation of different evolution models revealed f_1 as the radial first overtone, which in turn specifies the evolutionary state of V396 Per. The evolutionary track of the model also suggests that the star no longer resides on the main sequence.

References

Breger, M. 1969, *AJ*, 74, 166
Breger, M. 1993, *Proc. of the IAU Colloquium*, 136, 106
Kim, S. & Lee, S. 1990, *Journal of the Korean Astron. Society*, 23, 1
Lenz, P. & Breger, M. 2004, *CoAst*, 146, 53
Zima, W. 2008, *CoAst*, 155, 17

Precision Asteroseismology
Proceedings IAU Symposium No. 301, 2013
J. A. Guzik, W. J. Chaplin, G. Handler & A. Pigulski, eds.

© International Astronomical Union 2014
doi:10.1017/S1743921313015251

Oscillation parameters of red giants observed in the *CoRoT* exofield

N. Themessl[1], T. Kallinger[1], J. Montalbán[2], and R. A. García[3]

[1]Institute for Astrophysics, University of Vienna,
Türkenschanzstraße 12, A-1180 Vienna, Austria
email: nathalie.themessl@univie.ac.at

[2]Institut d'Astrophysique et de Géophysique, Université de Liège, Belgium

[3]Laboratoire AIM, CEA/DSM-CNRS, Université Paris 7 Diderot, IRFU/SAp, France

Abstract. The high-precision data obtained by the *CoRoT* (Convection, Rotation and planetary Transits) space mission allows firm detections of solar-like oscillations in a great variety of different stars. We derived reliable estimates of the frequency of maximum oscillation power $\nu_{\max}$ and the large frequency separation $\Delta\nu$ of more than 300 cool red giants, that were observed by *CoRoT*. A detailed study of their seismic parameters provides an exclusive view of the population of the observed giants.

Keywords. stars: late-type, stars: oscillations, stars: fundamental parameters

1. Introduction

After the depletion of hydrogen in their cores, stars reach the giant branch. In these cool G and K stars solar-like oscillations are expected with long periods in the range of hours to several days. They exhibit radial modes of high radial order and nonradial mixed p/g modes, which reveal important information about their innermost structure. Nowadays, many different methods allow the analysis of these very complex spectra including the extraction of individual frequencies (Mosser *et al.* 2012). However, long continuous observations over several years are necessary to properly resolve the oscillations and to enable unambiguous detections. Just recently, space missions such as *CoRoT* (Baglin *et al.* 2006) and *Kepler* have provided the necessary data accuracy, which allows to probe and understand the oscillation spectra of these evolved stars.

2. Data and observation

In this project, we use more than 450 red-giant targets, that were observed by the *CoRoT* space telescope. The available data consist of reduced chromatic and monochromatic lightcurves that were obtained during the first (LRc01) and the second long run (LRa01) of the satellite.

3. Global oscillation parameters

The power spectrum of solar-like oscillations shows a white noise component, a frequency-dependent background component modified by several super-Lorentzian functions (with the exponent equal to 4 instead of 2) and a power excess generated through pulsation, whose shape can be approximated by a Gaussian (Fig. 1). These characteristic features defined our global model, which was used to derive the oscillation parameters.

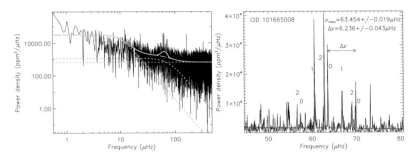

Figure 1. *Left*: The power density spectrum including a global fit to the spectrum with (solid yellow) and without (large dashed blue) a Gaussian, and the three different background components (dashed green). *Right*: The residual power density spectrum of the same star with the red line characterizing the best model fit, which is used to derive the p-mode spacings.

After the visual inspection of each individual power spectrum about 100 red-giant targets were excluded from the analysis due to the lack of a distinctive power excess hump. Following Kallinger *et al.* (2010, 2012), we fitted the global model to the observed power density spectrum by using a Bayesian nested sampling algorithm (*MultiNest*; Feroz *et al.* 2009). Besides characterizing the granulation background signal, the global fit gives the frequency of maximum oscillation power $\nu_{\max}$ (i.e. the centre of the oscillation power excess). The oscillation power excess itself consists of a sequence of near regularly spaced radial and nonradial modes. Most importantly, these regularities can be characterized by the so-called large and small frequency separation $\Delta\nu$ and $\delta\nu$, respectively, which represent the frequency difference between consecutive modes with the same spherical degree and consecutive modes but with a difference in spherical degree of two. In order to derive $\Delta\nu$ and $\delta\nu$ we fitted a sequence of Lorentzian profiles to the central part of the power excess, again using *MultiNest*. The advantage of our approach is that *MultiNest* scans the parameter space for a real global maximum in the likelihood "mountains" and delivers a very robust solution preventing us from over-fitting the data.

4. Discussion

The firm detection of solar-like oscillations in many different stars would not only provide a better understanding of the characteristics of stochastically excited oscillations, but would also improve current state-of-the-art stellar structure and evolution models. We analyzed a large number of red-giant stars and derived the frequency of maximum oscillation power, the large frequency separation and the mode degree. Since both seismic parameters scale with the stellar mass, radius, and effective temperature (Kjeldsen & Bedding 1995), they can be further used to determine the fundamental parameters of a star. As these parameters change with stellar evolution, they provide crucial insights into the population of the observed giants (Miglio *et al.* 2012).

References

Baglin, A., Auvergne, M., Barge, P., *et al.* 2006, *ESA SP-1306*, 33
Feroz, F., Hobson, M. P., & Bridges, M. 2009, *MNRAS*, 398, 1601
Kallinger, T., Hekker, S., Mosser, B., *et al.* 2012, *A&A*, 541, A51
Kallinger, T., Mosser, B., Hekker, S., *et al.* 2010, *A&A*, 522, A1
Kjeldsen, H. & Bedding, T. R. 1995, *A&A*, 293, 87
Miglio, A., Brogaard, K., Stello, D., *et al.* 2012, *MNRAS*, 419, 2077
Mosser, B., Elsworth, Y., Hekker, S., *et al.* 2012, *A&A*, 537, A30

Precision Asteroseismology
Proceedings IAU Symposium No. 301, 2013
J. A. Guzik, W. J. Chaplin, G. Handler & A. Pigulski, eds.

© International Astronomical Union 2014
doi:10.1017/S1743921313015263

Pulsations in the late-type B supergiant star HD 202850†

Sanja Tomić[1], Michaela Kraus[2] and Mary Oksala[2]

[1]Department of Astronomy, Faculty of Mathematics, University of Belgrade
Studentski trg 16, 11000 Belgrade, Serbia
email: sanja@sunstel.asu.cas.cz

[2]Astronomický ústav, Akademie věd České republiky
Fričova 298, 25165 Ondřejov, Czech Republic

Abstract. HD 202850 is a late B-type supergiant. It is known that photospheric lines of such stars vary. Due to macroturbulence the lines are much wider than expected. Macroturbulence has been linked to stellar pulsations. It has been reported that there are several B supergiants that undergo pulsations. In our previous work, we detected a pulsational period of 1.59 hours in this object from data taken with the Ondřejov 2-m telescope. We continued to investigate this object and we took several time series with the DAO 1.2-m telescope. Our new data suggest that there may be some additional pulsational periods in this star. We present our new results in this poster.

Keywords. stars: supergiants, oscillations, late-type, techniques: spectroscopic, radial velocities

1. Introduction

B supergiants are very important for stellar and galactic evolution, as they enrich their environments with chemically processed material via their line-driven winds. They show strong line profile variability. Their lines are wider than expected from their parameters. The excessive width is due to macroturbulence. Both line profile variability and macroturbulence are indications of stellar pulsations. However, so far only very few such supergiants were investigated to determine their pulsation periods.

2. HD 202850 (= σ Cyg)

HD 202850 is a late B-type supergiant star. Its stellar parameters (Markova & Puls 2008) are given in Table 1. It has been classified as B9 Iab, and is located in the OB association Cyg OB 4 at a distance of ≈ 1 kpc. It falls out of any previously calculated instability domains (Saio 2011). In our previous work, we described the 1.59 h pulsation period we detected (see Kraus *et al.* 2012).

3. New preliminary results

In 2012 we took a new set of 294 spectra distributed over 5 nights with the DAO 1.2-m telescope. Exposures were five minutes long, with a signal-to-noise ratio between 150 and 250. The moment analysis showed variability in all three moments, and the first and the third moment seem to vary in phase. Due to high noise, the FFT analysis did not show any pronounced peaks, therefore the period(s) were estimated by fitting a combination

† Based on observations acquired at the Dominion Astrophysical Observatory, Herzberg Institute of Astrophysics, National Research Council of Canada

Table 1. Parameters of HD 202850

T_{eff}	$\log L/L_\odot$	$\log g$	R_*	M	$v\sin i$	v_{macro}
[K]			$[R_\odot]$	$[M_\odot]$	$[\text{km s}^{-1}]$	$[\text{km s}^{-1}]$
11 000	4.59	1.87	54	8^{+4}_{-3}	33 ± 2	33 ± 2

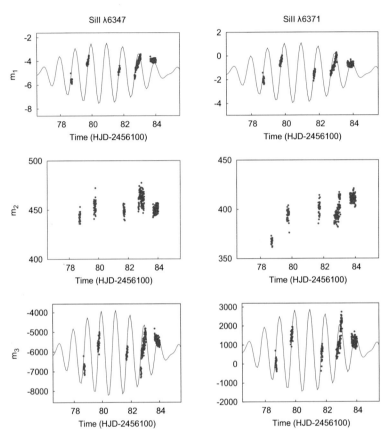

Figure 1. First three small moments. A combination of possible two periods is fit through the first and the third moment.

of sine curves (Fig. 1). We found two new possible periods (a 22.2 h period and a 25.2 h period).

Acknowledgements

S.T. acknowledges financial support from an IAU grant. M.K. and M.E.O. acknowledge financial support from GAČR under grant number P209/11/1198. The Astronomical Institute Ondřejov is supported by the project RVO:67985815.

References

Kraus, M., Tomić, S., Oksala, M. E., & Smole, M. 2012, *A&A*, 542, L32

Markova, N. & Puls, J. 2008, *A&A*, 478, 823

Saio, H. 2011, in: C. Neiner, G. Wade, G. Meynet, & G. Peters (eds.), *Active OB Stars: Structure, Evolution, Mass Loss, and Critical Limits*, Proc. IAU Symposium No. 272 (Cambridge: Cambridge University Press) p. 468

Precision Asteroseismology
Proceedings IAU Symposium No. 301, 2013
J. A. Guzik, W. J. Chaplin, G. Handler & A. Pigulski, eds.

© International Astronomical Union 2014
doi:10.1017/S1743921313015275

GYRE: A new open-source stellar oscillation code

Rich Townsend[1]†, Seth Teitler[1] and Bill Paxton[2]

[1]Department of Astronomy, University of Wisconsin-Madison, Madison, WI 53706, USA
[2]Kavli Institute for Theoretical Physics, University of California
Santa Barbara, CA 93106, USA

Abstract. We introduce GYRE, a new open-source stellar oscillation code which solves the adiabatic/non-adiabatic pulsation equations using a novel Magnus Multiple Shooting (MMS) numerical scheme. The code has a global error scaling of up to 6th order in the grid spacing, and can therefore achieve high accuracy with few grid points. It is moreover robust and efficiently makes use of multiple processor cores and/or nodes. We present an example calculation using GYRE, and discuss recent work to integrate GYRE into the asteroseismic optimization module of the MESA stellar evolution code.

Keywords. methods: numerical, stars: evolution, stars: interiors, stars: oscillations

---

Interpreting the wealth of new observations provided by *MOST*, *CoRoT* and *Kepler* requires the theorist's analog to the telescope: a stellar oscillation code which calculates the eigenfrequency spectrum of an arbitrary input stellar model. Comparing a calculated spectrum against a measured one provides a concrete metric for evaluating a model, and therefore constitutes the bread and butter of quantitative asteroseismology.

There's no shortage of oscillation codes available to the community; the nine codes reviewed in Moya *et al.* (2008) are likely only a fraction of those being used on a day-to-day basis. However, automated asteroseismic optimization tools such as AMP (Metcalfe *et al.* 2009) and MESA (Paxton *et al.* 2013) are placing ever-increasing demands on these codes. A code will typically be executed hundreds or thousands of times during an optimization run, and must therefore make efficient use of available computational resources such as multi-processor hardware. The code must be robust, running and producing sensible output without manual intervention. The code must have an accuracy that matches or exceeds the frequency precision now achievable by satellite missions. Finally, it is preferable that the code address the various physical processes that inevitably complicate calculations, such as non-adiabaticity, rotation, and magnetic fields.

Currently, there are no publicly available oscillation codes which address all of these requirements. This motivated us to develop another code, 'GYRE', which is built on a novel Magnus Multiple Shooting (MMS) scheme for solving both adiabatic and non-adiabatic pulsation problems. GYRE and the MMS scheme are described in detail in a forthcoming paper (Townsend & Teitler 2013). The code is written in standard-conforming Fortran 2008 with a modular architecture that allows straightforward extension to handle more complicated problems. To leverage multiple processor cores and/or cluster nodes, the code is parallelized using a combination of OpenMP and MPI.

As an illustration of GYRE in action, Fig. 1 presents results from a simulation exploring how the dipole-mode oscillation frequencies of a $1.5\,M_\odot$ MESA stellar model change as the star evolves through the so-called red bump phase. (This phase occurs when H-burning shell reaches the composition discontinuity left by the convective envelope after

† email: townsend@astro.wisc.edu

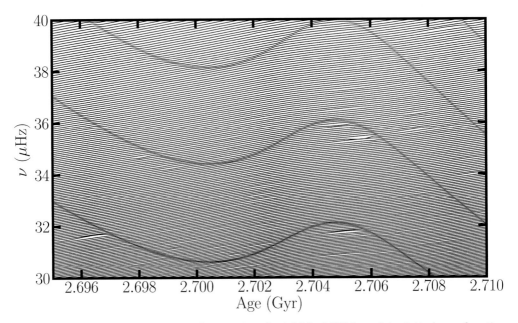

Figure 1. Dipole-mode linear eigenfrequencies of a 1.5 $M_\odot$ MESA model, plotted as a function of stellar age during evolution through the RGB bump phase. Coupling between envelope p modes and core g modes is revealed in the avoided crossings; the reversal in the time evolution of the crossing frequencies, between 2.700 and 2.705 Gyr, arises from the temporary contraction of the stellar envelope during the bump phase.

first dredge-up, causing a temporary reversal in the star's luminosity growth.) The figure reveals that the effects of bump passage can clearly be seen in the avoided crossings which arise from coupling between core g modes and envelope p modes: as the envelope contracts during the bump phase, the frequencies of the avoided crossings correspondingly increase.

Since revision 5232 MESA includes GYRE as one of the oscillation codes underpinning its asteroseismic optimization module (the other, currently, is ADIPLS). Communication between MESA and GYRE is accomplished through a simple application programming interface: MESA passes a model to GYRE, which then returns a list of modes having eigenfrequencies in a given range.

GYRE is open for use and distribution under the GNU General Public License; our hope is that a community of practice will arise around the code, bringing together users and developers to shape the code's future evolution in ways that best serve the field and its participants. Source code, documentation and other materials can be found at http://www.astro.wisc.edu/~townsend/gyre/.

Acknowledgements

We acknowledge support from NSF awards AST-0908688 and AST-0904607 and NASA award NNX12AC72G.

References

Metcalfe, T. S., Creevey, O. L., & Christensen-Dalsgaard, J. 2009, *ApJ*, 699, 373
Moya, A., Christensen-Dalsgaard, J., Charpinet, S., *et al.* 2008, *Ap&SS*, 316, 231
Paxton, B., Cantiello, M., Arras, P., *et al.* 2013, *ApJS*, 208, 4
Townsend, R. H. D. & Teitler, S. A. 2013, *MNRAS*, 435, 3406

Precision Asteroseismology
Proceedings IAU Symposium No. 301, 2013
J. A. Guzik, W. J. Chaplin, G. Handler & A. Pigulski, eds.

© International Astronomical Union 2014
doi:10.1017/S1743921313015287

A search for pulsations in two Algol-type systems V1241 Tau and GQ Dra

Burak Ulaş[1], Ceren Ulusoy[2], Kosmas Gazeas[3], Naci Erkan[4], and Alexios Liakos[5]

[1] İzmir Turk College Planetarium, 8019/21 sok., No: 22, İzmir, Turkey
email: bulash@gmail.com

[2] College of Graduate Studies, University of South Africa, PO Box 392, UNISA 0003, Pretoria, South Africa
email: cerenuastro@gmail.com

[3] Department of Astrophysics, Astronomy and Mechanics, National and Kapodistrian University of Athens, GR-157 84, Zografos, Athens, Greece
email: kgaze@phys.uoa.gr

[4] Department of Physics, Faculty of Arts and Sciences, Çanakkale Onsekiz Mart University, Terzioglu Campus, TR-17100, Çanakkale, Turkey
email: nacierkan@comu.edu.tr

[5] Institute for Astronomy & Astrophysics, Space Applications & Remote Sensing, National Observatory of Athens, I.Metaxa & Vas. Pavlou St., GR-15236, Palaia Penteli, Greece
email: alliakos@phys.uoa.gr

Abstract. We present new photometric observations of two eclipsing binary systems, V1241 Tau and GQ Dra. We use the following methodology: initially, the Wilson-Devinney code is applied to the light curves in order to determine the photometric elements of the systems. Then, the residuals are analysed using Fourier techniques. The results are the following. One frequency can be possibly attributed to a real light variation of V1241 Tau, while there is no evidence of pulsations in the light curve of GQ Dra.

Keywords. stars: binaries: eclipsing, stars: oscillations (including pulsations)

1. Introduction

V1241 Tau (= WX Eri) was found to be an eclipsing binary by Henrietta Leavitt (Pickering 1908). Sarma & Abhyankar (1979) claimed that the primary component shows periodic variations in its light. However, the periods of this variation were simply 1/5 and 1/6 of the orbital period which is likely a result of the rectification process they applied. Arentoft *et al.* (2004) found no trace of pulsations. The light variation of GQ Dra was discovered by *Hipparcos* (ESA 1997).

2. Observations and solution of the light curves

BVRI light curves of V1241 Tau were obtained with the 0.4-m telescope of the University of Athens Observatory in November 2012. GQ Dra was observed with the 1.22-m telescope of the Onsekiz Mart University Observatory on 7 nights in March and April 2013. Light curves of both systems were analysed using PHOEBE (Prša & Zwitter 2005) software. The results are the following. V1241 Tau has a semi-detached configuration with the inclination of about 81.5° and the mass ratio of 0.44. The hotter and cooler components have the effective temperatures equal to 7500 K and 4906 K, respectively. We find that 91 percent of the light in *V* filter comes from the primary component.

507

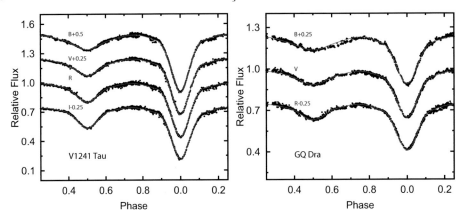

Figure 1. Observed (points) and theoretical (lines) light curves of the two systems. Some curves are shifted in flux axis for the sake of clarity.

Our solution for the semi-detached binary GQ Dra is the first light curve solution in the literature. Mass ratio of the system is found to be 0.25 and the orbital inclination was calculated to be 75.3°. The effective temperature of the secondary was found to be equal to 5050 K while the primary's $T_{\rm eff}$ was fixed at 8750 K. The comparison between observations and model light curve is shown in Fig. 1.

3. Search for pulsations

We searched the residuals from the PHEOBE fits for periodic variability using Fourier techniques. In the case of V1241 Tau, we found a significant peak at frequency of $f = 2.13$ d^{-1} and amplitude of 0.0097 mag which possibly can be attributed to real changes in one of the components. However, it is worth to emphasize that the variability can be also due to instrumental effects. The same method was applied to the residuals from the solution of the light curve of GQ Dra. No remarkable periodic variability was found.

4. Conclusions

The main result regarding the pulsational variability in the systems is that there is no convincing evidence of pulsations in either of them. The residual light curve of the system V1241 Tau shows periodic variation with a small amplitude, but its origin is probably an instrumental effect.

Acknowledgements

CU sincerely thanks the South African National Research Foundation (NRF) for the award of NRF Multi-Wavelength Astronomy Research Programme (MWGR), Grant No: 86563 to Prof. L.L. Leeuw at UNISA, Reference: MWA1203150687.

References

Arentoft, T., Lampens, P., Van Cauteren, P., Duerbeck, H. W., García-Melendo, E., & Sterken, C. 2004, *A&A*, 418, 249
ESA, 1997, *The Hipparcos & Tycho Catalogues*, ESA SP-1200
Pickering, E. C. 1908, *AN*, 178, 157
Prša, A. & Zwitter, T. 2005, *ApJ*, 628, 426
Sarma, M. B. K. & Abhyankar, K. D. 1979, *Ap&SS* 65, 443

Precision Asteroseismology
Proceedings IAU Symposium No. 301, 2013
J. A. Guzik, W. J. Chaplin, G. Handler & A. Pigulski, eds.

© International Astronomical Union 2014
doi:10.1017/S1743921313015299

Seismic investigation of the γ Dor star KIC 6462033: The first results of *Kepler* and ground-based follow up observations

C. Ulusoy[1], B. Ulaş[2], M. Damasso[3,4], A. Carbognani[3], D. Cenadelli[3], I. Stateva[5], I. Kh. Iliev[5], and D. Dimitrov[5]

[1] College of Graduate Studies, University of South Africa, PO Box 392, UNISA 0003, Pretoria, South Africa
email: `cerenuastro@gmail.com`

[2] Department of Astronomy and Space Sciences, Faculty of Sciences, University of Ege, Bornova, 35100, İzmir, Turkey

[3] Astronomical Observatory of the Autonomous Region of the Aosta Valley (OAVdA) Loc. Lignan 39, 11020 Nus (Aosta), Italy

[4] Department of Physics and Astronomy, University of Padova, vicolo dell'Osservatorio 3, I-35122 Padova, Italy

[5] Institute of Astronomy with NAO, Bulgarian Academy of Sciences, blvd. Tsarigradsko chaussee 72, Sofia 1784, Bulgaria

Abstract. Preliminary results on the analysis of the *Kepler* light curve and photometric ground-based time series of γ Dor star KIC 6462033 (TYC 3144-646-1, $V = 10.83$, $P = 0.69686$ d) are presented in order to determine pulsation frequencies.

Keywords. stars: individual: KIC 6462033, stars: oscillations, stars: variables: γ Dor

1. Introduction

γ Dor variables are multiperiodic nonradial pulsators that oscillate in high-order g-modes with periods of the order of a day (Balona *et al.* 2011). Their position in the H-R diagram partially overlaps with the cool part of δ Sct instability strip. This means that stars showing both types of pulsations may exist (Uytterhoeven *et al.* 2011, Balona & Dziembowski 2011). It should therefore be noted that searches for such objects are highly important for understanding the oscillation mechanisms of simultaneously excited p- and g-modes in a star.

2. Ground-based and *Kepler* photometry

Ground-based observations of KIC 6462033 were carried out at the Astronomical Observatory of the Autonomous Region of the Aosta Valley (OAVdA) between the Julian dates JD 2 455 740.4 − 2 455 776.5. All data were obtained with the FLI PL3041-1-BB CCD camera attached to the 81-cm telescope in the $UBVI$ photometric passbands. The observed light variation of the star in the B filter is shown in Fig. 1.

In order to derive the frequency content of the variability of the star, we analyzed *Kepler* short cadence (SC) data which consist of 38016 points taken in *Kepler* quarter Q3.3. The data were prepared for analysis using `KEPCOTREND` package (Fraquelli & Thompson 2012, Christiansen *et al.* 2012, Ulusoy *et al.* 2013).

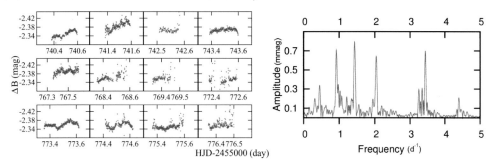

Figure 1. *Left:* B-filter light curves of KIC 6462033. *Right:* Fourier spectrum of SC *Kepler* data for KIC 6462033.

3. Fourier analysis

We used `PERIOD04` (Lenz & Breger 2005) software for the analysis of both SC and ground-based time series. Following Breger *et al.* (2011), the signal to noise ratio (S/N) equal to 3.5 was adopted as a detection threshold. From the analysis of the *Kepler* data, we find that the light curve is dominated by four independent modes with frequencies $f_1 = 0.9242$, $f_2 = 1.4363$, $f_3 = 2.0409$ and $f_4 = 3.4257\,\mathrm{d}^{-1}$. Due to lack of data with sufficient quality, we were able to detect only the first two frequencies from the ground-based time series. The ground-based B-filter light curve and frequency spectrum of the *Kepler* data are shown in Fig. 1.

4. Conclusions

The star KIC 6462033 was recently classified as a γ Dor-type star by Uytterhoeven *et al.* (2011). We have performed frequency analysis of the obtained ground-based data as well as the *Kepler* data for KIC 6462033. This is a first step of our scheduled study. We confirm that KIC 6462033 pulsates in the frequency range of γ Dor type pulsators. In order to try seismic modeling, mode identification will be an essential requirement at least for a few frequencies. We expect that new photometric and spectroscopic observations will provide more data that can be used for this purpose.

Acknowledgements

CU sincerely thanks the South African National Research Foundation (NRF) for the award of NRF Multi-Wavelength Astronomy Research Programme (MWGR), Grant No: 86563 to Prof. L.L. Leeuw at UNISA, Reference: MWA1203150687.

References

Balona, L. A. & Dziembowski, W. A. 2011, *MNRAS*, 417, 591
Balona, L. A., Guzik, J. A., Uytterhoeven, K., *et al.* 2011, *MNRAS*, 415, 3531
Breger, M., Balona, L., Lenz, P., *et al.* 2011, *MNRAS*, 414, 1721
Christiansen, J. L., Barclay, T., Jenkins, J. M., *et al.* 2012, *Kepler Data Release 14 Notes (KSCI-19054-001)*
Fraquelli, D. & Thompson, S. E. 2012, *Kepler Archive Manual (KDMC-10008-004)*
Lenz, P. & Breger, M. 2005, *CoAst*, 146, 53
Ulusoy, C., Ulaş, B., Gülmez, T., *et al.* 2013, *MNRAS*, 433, 394
Uytterhoeven, K., Moya, A., Grigahcène, A., *et al.* 2011, *A&A*, 534, A125

Precision Asteroseismology
Proceedings IAU Symposium No. 301, 2013
J. A. Guzik, W. J. Chaplin, G. Handler & A. Pigulski, eds.
© International Astronomical Union 2014
doi:10.1017/S1743921313015305

Antarctica photometric survey using PDM13

Cyrus Zalian and Merieme Chadid

Université de Nice Sophia-Antipolis, Observatoire de la Côte d'Azur

Abstract. PDM13 is a new graphic interface program dedicated to frequency domain analysis based on the Phase Dispersion Minimization technique (PDM, Stellingwerf 2012). In this paper, we will present the different algorithms running in PDM13, including the Auto-Segmentation, the Gauss-Newton and the PDM algorithms. More details on this triptych are available in our recent paper (Zalian *et al.*, submitted). Their aim is to offer a simple and powerful way to extract frequency. Amongst the numerous improvements offered by the program, we will particularly focus on the reduction of aliases and the ability to look directly for multiple-period phenomena and the Blazhko effect. After that, we will show the first results from PDM13 using the Antarctica photometric survey.

Keywords. Frequency analysis, phase dispersion minimization, graphical interface, Antarctica photometric survey

1. Introduction

Frequency analysis of light curves is one of the many tools provided by asteroseismology to understand the underlying mechanisms of pulsating stars. Several extraction algorithms exist based on Fourier transform (Lenz & Breger 2005), minimisation of standard deviation (Stellingwerf 1978), entropy (Ulrych & Bishop 1975). Among these methods, phase dispersion minimization is a well-known method, particularly efficient in poor-time coverage data and large gaps. Unfortunately, like many other powerful algorithms, the lack of intuitive, user-friendly interface prevent their use within a large community of researchers. PDM13 tries to overcome this problem by including a graphical interface as well as complementary algorithms, resulting in a simple and powerful program.

2. Interface and algorithms

2.1. *PDM algorithm*

For a given frequency, the PDM algorithm folds the data to obtain a phase plot and divides the obtained diagram into 10 or 100 bins depending on the number of data. For each bin, a mean value is calculated and an interpolated curve is computed. Finally the program calculates the corresponding variance and divides it by the initial variance according to equation (2.1). Thus, good guesses are characterized by low values of Θ:

$$\Theta = \frac{s^2}{\theta^2},\qquad(2.1)$$

where s^2 is the variance for a trial period and θ^2 the overall variance.

Different significance algorithms are also available (Schwarzenberg-Czerny 1997, Linnell Nemec & Nemec 1985) and can be included using the option menu.

2.2. *Gauss-Newton algorithm*

Unfortunately, PDM does not provide the amplitude of an extracted frequency as a Fourier transform analysis would do. Hence, we have added an amplitude fitting algorithm based on the Gauss-Newton algorithm (Björck 1996).

2.3. *Auto-segmentation*

Another feature added to program is the auto-segmentation algorithm. Previous versions of PDM included the possibility to divide the data into segments which were studied independently so that aliases caused by gaps become less prominent. Still, the segmentation method was trivial and not optimised. Based on the Lee and Heghinian method implemented (Lee & Heghinian 1977), we are able to detect gaps accurately.

2.4. *Graphical interface and process*

After each PDM run, the residuals are calculated following Stobie's method (Stobie 1970) enabling another run to determine multiple periods. The process is shortened and harmonic amplitudes are obtained directly with the Gauss-Newton fit. Moreover, exploiting the co-cyclarity of the modulated signal, PDM13 is able to look directly for Blazhko modulation.

3. S Arae: Case study using PDM13

S Arae was observed during 700 h using PAIX – Photometer AntarctIca eXtinction – attached to a 40-cm Ritchey-Chrétien optical telescope located at Dome Charlie in Antarctica, designed and built by PaixTeam at Université Nice Sophia-Antipolis (Chadid *et al.* 2010). It is a RR Lyrae Blazhko star with a main frequency of $f_1 = 2.2129$ d^{-1} and modulation frequency of $f_m = 0.02$ d^{-1}.

We have been able, using our program, to extract up to 10th-order harmonics with Blazhko multiplet structure. Auto-segmentation combined with the Gauss-Newton algorithm simplifies the analysis giving us quick and accurate results.

References

Björck, Å. 1996, *Numerical Methods for Least Squares Problems*, (Philadelphia, SIAM 1996)
Chadid, M., Vernin, J., Mekarnia, D., Chapellier, E., Trinquet, H., & Bono, G. 2010, *A&A*, 516, L15
Lee, A. F. S. & Heghinian, S. M. 1977, *Technometrics*, 19, 503
Lenz, P. & Breger, M. 2005, *CoAst*, 146, 53
Linnell Nemec, A. F. & Nemec, J. M. 1985, *AJ*, 90, 2317
Schwarzenberg-Czerny, A. 1997, *ApJ*, 489, 941
Stellingwerf, R. F. 1978, *ApJ*, 224, 953
Stobie, R. S. 1970, *Observatory*, 90, 20
Ulrych, T. J. & Bishop, T. N. 1975, *Reviews of Geophysics and Space Physics*, 13, 183

Author Index

Printed in the United States
by Baker & Taylor Publisher Services